BASIC MECHANICAL ENGINEERING

BASIC MECHANICAL ENGINEERING

(Common to all branches of B.E. First Year)

[According to the New Syllabus of RGPV Bhopal (MP)]

By

Dr. Mohan Sen

BE (Hons.), ME (Hons.), Ph.D., FIE

Deputy Director

Directorate of Technical Education

Bhopal, Madhya Pradesh

UNIVERSITY SCIENCE PRESS

(An Imprint of Laxmi Publications Pvt. Ltd.)

BANGALORE • CHENNAI • COCHIN • GUWAHATI • HYDERABAD
JALANDHAR • KOLKATA • LUCKNOW • MUMBAI • RANCHI • NEW DELHI
INDIA • USA • GHANA • KENYA

BASIC MECHANICAL ENGINEERING

Printed and bound in India
Typeset at ABRO Enterprises
First Edition : 2006; Second Edition : 2015
ISBN 978-81-318-0748-4

PUBLISHED IN INDIA BY

UNIVERSITY SCIENCE PRESS
(An Imprint of Laxmi Publications Pvt.Ltd.)
An ISO 9001:2008 Certified Company
113, GOLDEN HOUSE, DARYAGANJ,
NEW DELHI - 110002, INDIA
Telephone : 91-11-4353 2500, 4353 2501
Fax : 91-11-2325 2572, 4353 2528
www.laxmipublications.com info@laxmipublications.com

Branches

✆	Bangalore	080-26 75 69 30	
✆	Chennai	044-24 34 47 26,	24 35 95 07
✆	Cochin	0484-237 70 04,	405 13 03
✆	Guwahati	0361-254 36 69,	251 38 81
✆	Hyderabad	040-27 55 53 83,	27 55 53 93
✆	Jalandhar	0181-222 12 72	
✆	Kolkata	033-22 27 43 84	
✆	Lucknow	0522-220 99 16	
✆	Mumbai	022-24 91 54 15,	24 92 78 69
✆	Ranchi	0651-220 44 64	

C—
Printed at : Repro Knowledgecast Limited, Thane

This book is dedicated to the
Loving Memory of My Father

Contents

Arun Nahar
Director

Directorate of Technical Education, M. P.
Satpura Bhawan, 4th Floor, Bhopal - 462004
www.mptechedu.org
Telegram : Taksanchalak
Telephone : Office : 0755-2576751
Resi. : 0755-2576768
Fax : 0755-2552219
Email : arunnahar19@gmail.com
dtemp@mp.nic.in

No. Sr.8.A/2013/286

Bhopal Dated : 14-8-2013

Foreword

I appreciate Dr. Mohan Sen's efforts in writing a useful book on "Basic Mechanical Engineering" for the students of BE First Year of all branches of Engineering. This book covers the entire syllabus of Basic Mechanical Engineering (Subject code: BE-203) adopted by Rajiv Gandhi Proudyogiki Vishwavidyalaya (RGPV) Bhopal (MP). An earnest attempt has been made by the author to present the subject matter on diverse topics in a form which is comprehensive but simple and easy to understand. With a strong emphasis on basic concepts and principles, all efforts have been made to make the text simple with the help of line diagrams wherever they were needed.

A large number of solved examples have been included in the book for the clear and easy understanding of the subject matter. The objective type questions and review questions are also included to enable the students to consolidate their knowledge of the subject.

I hope that the student community will receive the book with great enthusiasm. I congratulate the author for his efforts in bringing out this book.

(Arun Nahar)

Preface to the Second Edition

This book is intended to serve as a textbook covering the syllabus of "Basic Mechanical Engineering" (Course code: BE-203) adopted by Rajiv Gandhi Proudyogiki Vishwavidyalaya (RGPV) Bhopal (MP). The second edition of this book attempts to cover the diverse topics of the syllabus in the best possible manner, without making the contents voluminous. Throughout the textbook, attempt has been made to present the subject matter in a simple, lucid and precise manner. Care has been taken in making the text student friendly with greater emphasis on the basic concepts of the subjects.

I wish to express my profound thanks to all those who helped me in making this book a reality. Suggestions for the improvements of the book from the readers will be acknowledged and deeply appreciated.

—Author

Preface to the First Edition

The Syllabus of the subject "BASIC Mechanical Engineering (BE 204)" as prescribed by the RGPV (Rajiv Gandhi Proudyogiki Vishwavidyalaya) Bhopal for the BE First Year Course (all branches) consists of three units from the Workshop Technology and two units from the Heat Engines. The diversity in different units of the syllabus makes it difficult for a novice to learn this important subject. Despite the fact that several books on this subject of "Basic Mechanical Engineering" are available in the market, but the need of a textbook covering the fundamentals of the subject matter in a simple language and supported by simplified line diagrams has prompted the author to write this book.

In preparing this book, the content is confined to basic facts of the subject useful to students of all the branches. All efforts have been made to make this book as complete as possible. But as the subject has wide ramifications in several areas of workshop technology and heat engines, this small volume cannot claim to be a very comprehensive treatment of the subject at large. All suggestions with a view to making the book more useful will be gratefully acknowledged and highly appreciated.

I take this opportunity to thank Mr. Saurabh Gupta, Managing Director, M/s Laxmi Publications (P) Ltd., New Delhi for getting this book published with a nice and clear print.

This book would not have been possible without the cooperation of my wife Pushplata as she has provided me the much necessary support during the preparation of the manuscript of the book.

—Author

Course Contents

Category of Course	Course Title	Course Code	Credits – 6C			Theory Papers (ES)
			L	T	P	
Interdisciplinary DID-2	Basic Mechanical Engineering	**BE204**	3	1	2	Max. Marks-100 Min. Marks-35 Duration-3hrs.

Branch : Common Semester

Course : BE204 Basic Mechanical Engineering

Unit I

Materials: Classification of engineering material, Composition of cast iron and carbon steels on Iron-Carbon diagram and their mechanical properties. Alloy steels and their application. Stress-strain diagram, Hooks law and modulus of elasticity. Tensile, shear, hardness and fatigue testing of materials.

Unit II

Measurement: Temperature, pressure, velocity, flow strain, force and torque measurements, concept of measurement error and uncertainty analysis, Vernier caliper, micrometer, dial gauges, slip gauges, sinebar and combination set. Introduction to lath, drilling, milling and shaping machines.

Unit III

Fluids: Fluid properties pressure, density and viscosity. Pressure variation with depth, static and kinetic energy, **Bernauli's** equation for incompressible fluids, Viscous and turbulent flow, working principle of fluid coupling, pumps, compressors turbines, positive displacement machines and pneumatic machines. Hydraulic power and pumped storage plants for peak load management as compared to base load plants.

Unit IV

Thermodynamics: First and second law of thermodynamic; steam properties, steam processes at constant pressure, volume, enthalpy and entropy, classification and working of boilers, efficiency and performance analysis, natural and artificial draught, calculation of chimney height. Refrigeration, vapour absorption and compression cycles, coefficient of perform (COP), refrigerant properties and eco friendly refrigerants.

Unit V

Reciprocating Machines: Steam engines, hypothetical and actual indicator diagram, Carnot cycle and ideal efficiency, Otto and diesel cycles, working of two stroke and four stroke petrol and diesel IC engines.

1 Engineering Materials

1.1 INTRODUCTION

Engineers from all disciplines should have some basic and applied knowledge of engineering materials so that they can do their work more efficiently while using them. Since ancient times, human civilization right from the beginning of stone age to iron and steel age and now to silicon age has seen the impact of these materials on the quality of human life. In all engineering disciplines, the desired performance of a component or structure depends on the right choice of a material. Mechanical engineers want light weight high strength and durable material for producing mechanical components such as connecting rods, cranks, etc. Aerospace engineers demand light weight, high strength and high temperature resistant materials like glass fiber reinforced materials. Electrical engineers demand good quality ceramic and insulating materials so that their switches can react quickly and insulators can withstand high temperatures efficiently. Civil engineers will naturally like to select those materials from which they can design and construct safe, reliable and corrosion free structures like bridges, dams, and flyovers etc. As usual, chemical engineers prefer to use corrosion resistant and eco-friendly materials for the pipeline and other storing chemicals in industries.

1.2 CLASSIFICATION OF ENGINEERING MATERIALS

The engineering materials can be classified in different ways. As per the recent trends, the engineering materials have been classified in five categories. These are:

(1) Metals and alloys

(2) Ceramics

(3) Polymers

(4) Composites

(5) Semiconductors.

Figure 1.1 shows the example of the application of each category of materials.

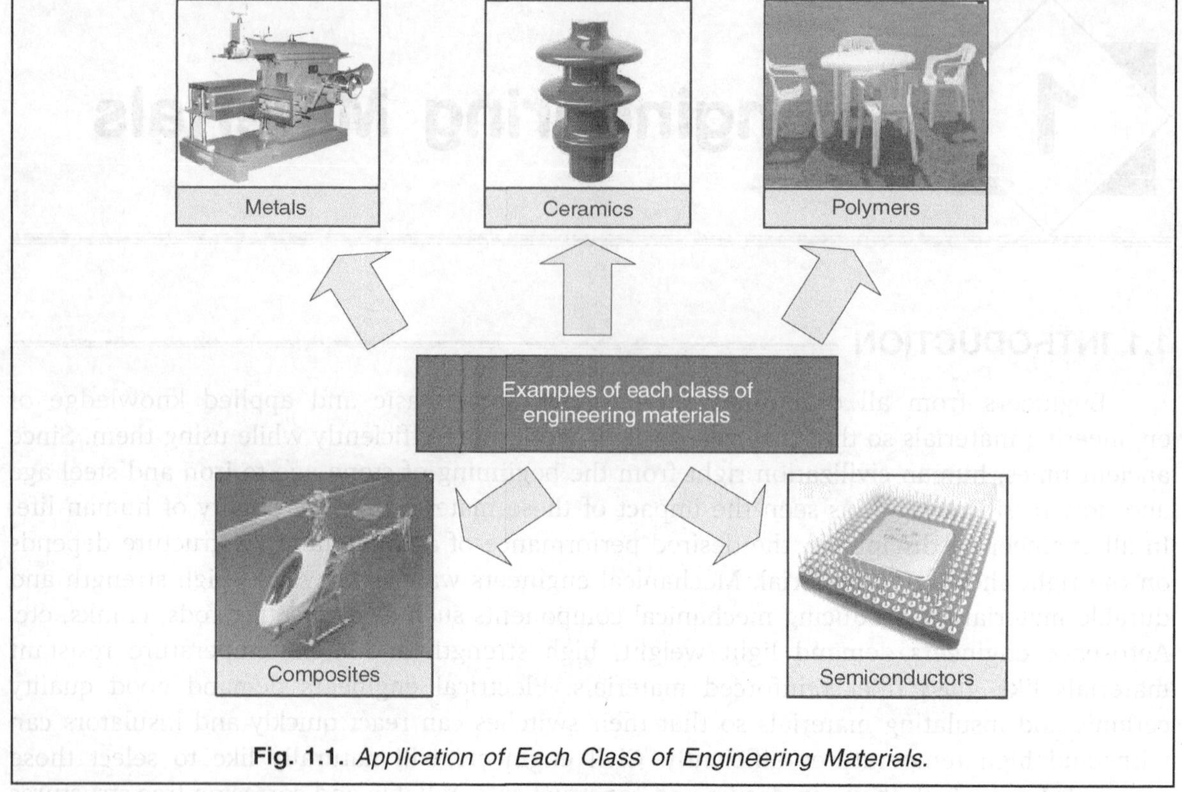

Fig. 1.1. *Application of Each Class of Engineering Materials.*

1.2.1 Metals and Alloys

Metals are inorganic substances which are composed of one or more metallic elements and some non-metal elements. The examples of metals and alloys are iron, steel, copper, aluminium, nickel and titanium. Non-metals like carbon, nitrogen, hydrogen and oxygen may be present in some form in metals and alloys. The general features of these materials are:

(*i*) good ductility and formability

(*ii*) good electrical and thermal conductivity

(*iii*) better resistance against shock

(*iv*) high strength

(*v*) high stiffness.

Metals are very useful for the structural or load bearing applications. In practice, pure metals are not very useful. Mostly, in all real-life applications, combination of one metal with other metals is used. This combination of two different metals is called alloys. By combining one metal with the other, one can achieve desirable properties in the material.

Metals and alloys can be further classified into two groups:

- Ferrous Metals and alloys
- Non-ferrous metals and alloys.

Ferrous Metals and Alloys	*Nonferrous Metals and Alloys*
These contain a large percentage of iron. Examples: Steels and Cast Irons. Pure iron has limited commercial value but when it is combined with carbon, iron becomes commercially very useful than any other metal. Alloys of iron and carbon form steels and cast irons.	These materials do not contain iron or possess very small percentage of iron. Examples: Aluminium, Copper, Zinc, Titanium and Cobalt.

1.2.2 Ceramics

Ceramics are non-metallic inorganic solids which consist of semi-metals and non-metals chemically bonded together. Ceramics have several attractive features. Advantages of ceramics materials for engineering applications include light weight, high strength, good abrasion resistance, reduced friction and insulating properties. The new generation of ceramics have the ability to retain their strength at high temperature and possess better resistance against wear and corrosion. These materials also have poor electrical and thermal conductivities but also possess some desirable optical, electrical and thermal properties.

Examples of traditional ceramics include clay, glass, brick, porcelain and alumina whereas those of advanced ceramics include alumina, silicon nitrate, and silicon carbide etc.

Applications:

- Window glass
- Ceramic tiles in space shuttle
- Spark plug
- Transducers for stereo record player.

Limitations: Ceramics have poor ductility, poor formability and shock resistance. Due to these reasons, Ceramics are not preferred for structural and load bearing applications.

1.2.3 Polymers

Polymers are produced by creating large molecular chains or networks from organic molecules obtained from petroleum or agriculture products by a process named polymerization. Polymers generally contain carbon and one or more elements such as nitrogen, oxygen, hydrogen and chlorine. Polymers have low electrical and thermal conductivity. Some polymers being good insulators are used for electrical insulating applications. The other features of polymers include:

- their low strength
- unsuitable for use at high temperature
- light weight and
- their excellent resistance to corrosion.

Polymers can be classified in three categories as shown in Fig. 1.2.

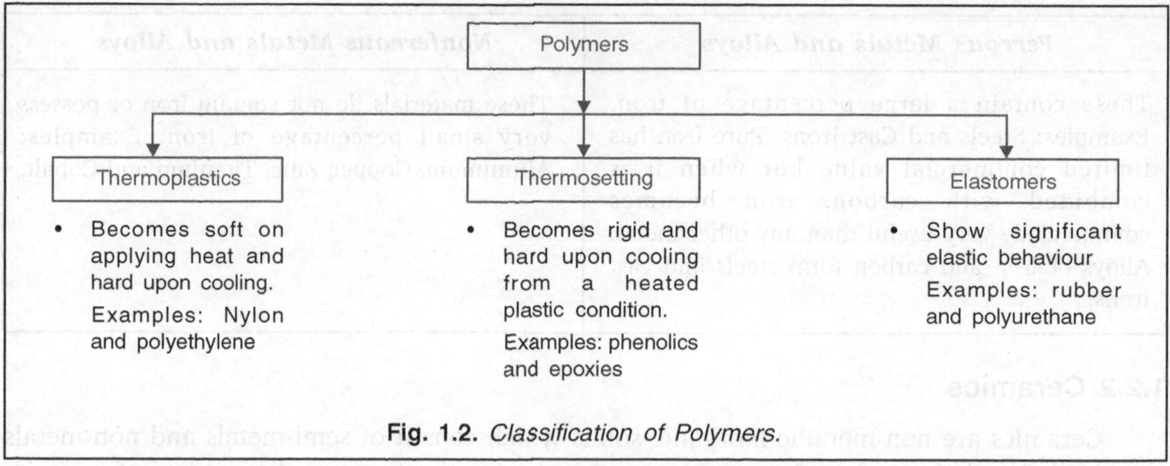

Fig. 1.2. *Classification of Polymers.*

Examples of polymers include rubber, plastics (nylon, polyethylene and polyvinyl chloride), acrylic coatings and phenolics etc.

Applications:

- In the manufacturing of DVDs (digital video disks).
- Polyethylene is used for food packaging.
- Epoxy encapsulation of integrated circuits.
- Acrylic coatings as coating materials for tennis courts.

1.2.4 Composites

A composite material is formed by artificial combination of two or more materials so as to attain properties that can not be achieved by a single material. Composites are also found in nature for example wood composites can also be produced synthetically. Usually, structure of a composite consists of particles or fibers of one phase mixed in a second phase called matrix. In general composites can be classified on the basis of their matrix materials as mentioned below:

- Polymer matrix
- Cement matrix
- Carbon matrix
- Metal matrix
- Ceramic matrix.

The outstanding characteristics of these materials include their:

(*i*) High strength

(*ii*) Good toughness

(*iii*) High stiffness

(*iv*) Light weight

(*v*) Excellent resistance to corrosion

Examples of composites are: Glass fiber reinforced plastics (GFRP), Reinforced cement concrete (RCC), cemented carbide and plywood. In GFRP, a light weight high strength structural composite is obtained by embedding continuous carbon fibers in one or more orientation in a polyester or epoxy resin. Another composite material which is very important to civil engineers is RCC. This is obtained by combining cement (the matrix, *i.e.*, binder, obtained by a reaction called hydration between cement and water), sand, gravel and optionally other ingredients known as admixtures. The examples of admixtures are short fibers and silica fumes. The other important composites which find extensive use in metal cutting industry are cemented carbides as these are hard and strong and have the capability to sustain high temperature

Applications:

The higher strength-to-weight ratios of composite materials make them hot favourite for the automobile and aviation industry. Composites are used in producing aircraft components, car bodies, tennis rackets, fishing rods and boat hulls.

1.2.5 Semiconductors

Semiconductor is usually defined as a material that has an electrical resistivity between that of a typical metal and typical insulator. The commonly accepted value of electrical resistivity for semiconductor lies in the range of 10^{-3} to 10^{9} Ωcm.

Semiconductor can also be defined as a material whose energy gap for electronic excitations lies between zero to 4 eV (electron volt).

The best known semiconductor material is Si (silicon) and GaAs (gallium arsenic). Nature is also the source of few semiconductor materials such as zinc blends (ZnS), Cuprite (Cu_2O) and galena (PbS) etc.

Semiconductors have negative temperature of coefficient of resistance. The electrical conductivity of semiconductors can be varied widely as a function of *e.g.*, temperature, optical excitation and impurity content.

Applications:

A semiconductor like Si is the main source of producing computer chips. These materials are used in various electronic devices such as transistors, various optoelectronic devices for generation and detection of electromagnetic radiation including data transmission through fiber optic networks. GaAs finds extensive use in light emitting devices and radiation detectors as GaAs possesses superior electronic mobility.

How to select the right material?

As many engineering materials are available in the market, an engineer should try to find out the answer to the following questions before selecting a material for a structure or a component as desired or needed by society.

(*a*) Ease of using the selected materials.

(*b*) Whether selected materials possess the required mechanical, electrical, magnetic and optical properties or not?

(*c*) What is the likely cost of using the selected materials?

(*d*) What effect the environmental conditions such as wind, rain, corrosive medium etc., will have on the performance of the selected materials in real life conditions?

(*e*) The ability of the selected materials to sustain the static or dynamic loading.

1.3 PROPERTIES OF MATERIALS

Engineering materials possess different types of properties and behave differently under the working conditions. While selecting a material, the basic objective of an engineer is to match material properties to the real life requirements of a product. An engineer must know at least three things about the materials.

(*a*) Which properties to consider?

(*b*) How to find out the required properties? and

(*c*) Limitations of the selected materials.

The properties of materials can be classified as shown in Fig. 1.3.

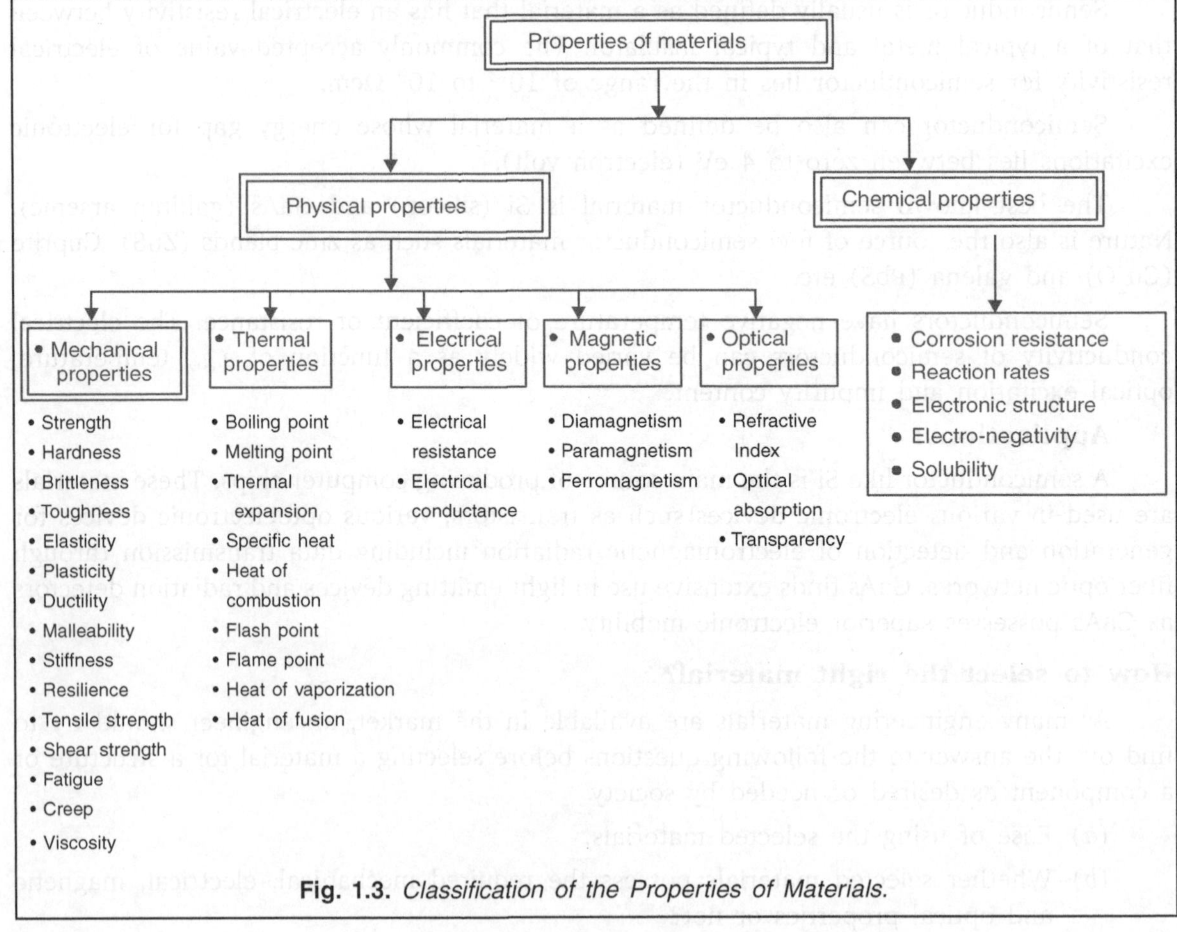

Fig. 1.3. *Classification of the Properties of Materials.*

It is easy to distinguish one particular material from other by comparing their physical properties but the idea about how a selected material will react to applied load or actual operating conditions can only be had by studying mechanical properties of the selected material.

1.3.1 Mechanical Properties of Materials

The important mechanical properties of the materials are:

Strength

Strength is the measure of the ability of a material to withstand external forces. It may be defined as the resistance offered by the material against deformation due to application of the external forces. The forces which generally act upon a structure or machine component may be tensile, compressive and shear in nature. Therefore the strength is also termed as tensile strength, compressive strength and shear strength.

Hardness

Hardness is a measure of the resistance to penetration of the surface of a material by a hard or sharp object.

Brittleness

Brittleness is defined as the tendency of a material to fracture without noticeable plastic deformation. Materials with less than 5% elongation may be considered as brittle materials.

Toughness

Toughness is defined as the measure of energy required to break a material. It is also defined as the ability to resist fracture under shock loading when loading has crossed the elastic limit.

Ductility

Ductility may be defined as the ability of material to withstand plastic deformation without rupture. It is also defined as the ability of material to be drawn into wire. Ductility is expressed in terms of percent elongation and percent reduction of area. Materials with more than 15% elongation may be treated as ductile materials.

Malleability

Malleability is defined as the ability of material to withstand the deformation in all directions without cracking. Malleability is different than ductility in case of wire drawing. These two properties are same as regard the forming of material into sheets or strips but not for wire drawing. For example lead is malleable and suitable to be rolled into sheets but is not fit for drawing into the wire.

Elasticity

Elasticity is defined as the ability of material to return to its original form after the removal of the external forces responsible for creating the deformation. Elastic limit is the measure of elasticity.

Plasticity

Plasticity of a material can be defined as its ability to be deformed permanently without fracture. Permanent deformation occurs only when the material is stressed beyond the yield point.

Stiffness

It is defined as the property of material which enables it to resist deflection or elastic deformation under the applied load. Modulus of rigidity is the measure of stiffness.

Resilience

Resilience is defined as the ability of the material to store energy and to resist shock and impact. Resilience is measured by the amount of energy absorbed per unit volume in deforming a material up to elastic limit.

Fatigue

In static loading, the applied load does not vary with time whereas in cyclic loading the applied load changes with time. The failure caused by cyclic loading is termed as fatigue. Fatigue failures usually occur at sharp corners or notches or because of inclusions inside a material.

Creep

Creep is one of the important mechanical properties of the material. It can be defined as the gradual deformation developed in a material over a specific period of time due to the application of a constant stress or constant strain. This gradual deformation may result in a machine component failing much below the tensile strength of the material.

1.4 STRESS, STRAIN AND STRESS-STRAIN DIAGRAM

The following definitions and units are commonly used in describing mechanical properties of materials.

Stress—stress can be defined as the internal resistance offered by the material under the application of force (load).

Consider a cylindrical (steel) specimen of initial length L_0 and initial cross sectional area A_0 with force F applied as shown in Fig. 1.4.

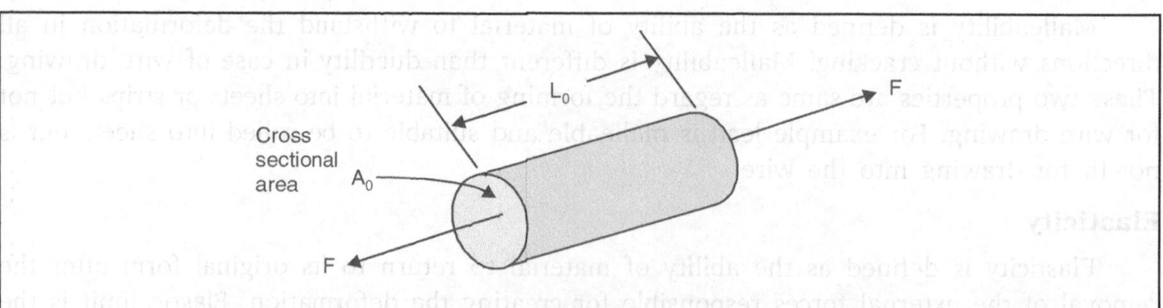

Fig. 1.4. *A Schematic Diagram of a Round Steel Bar Depicting the Tensile Deformation.*

Stress σ is defined as the force F per unit area A_o.

$$\sigma = \frac{F}{A_o} \qquad ...(1.1)$$

This definition of stress as given by equation (1.1) represents the engineering stress however in this equation if A_0 is substituted by A_a then the same equation defines the true stress, where A_a represents actual cross-sectional area of the specimen when force F is applied.

$$\sigma_{\text{actual}} = \frac{F}{A_a} \qquad ...(1.2)$$

In testing of the materials, it is important to distinguish between engineering stress and actual stress as specimen cross-sectional are decreases rapidly when applied force is of tensile type and increases when applied force is compressive type.

Unit of stress: (in SI) Newton per square metre (**N/m²**)

$1 N/m^2 = 1 Pa$

10^3 Pa = 1 kiloPascal = 1000 Pa

10^6 Pa = 1 megapascal = 1 MPa

(in MKS) **Kgf/cm²**

There are two types of basic stresses (*i*) Normal stresses and (*ii*) shearing stresses

(*i*) Normal Stresses

These stresses, also known as '*direct stresses*', are of two types:

(*a*) Tensile stresses and

(*b*) Compressive stresses.

These stresses act normal to the stressed surfaces under consideration. If the forces are directed away from the workpiece then the workpiece tends to increase in length under the action of applied forces and stresses developed in the workpiece are called as tensile stresses. If the applied force acts inward *i.e.*, it tends to shorten the workpiece in length (or increases in diameter) then the stresses experienced by the workpiece are known as compressive stresses.

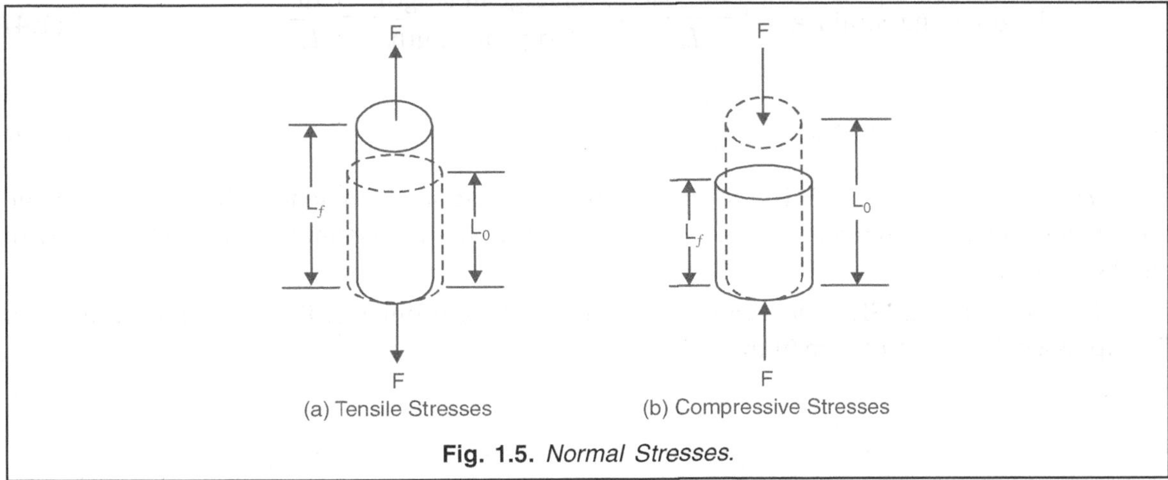

Fig. 1.5. *Normal Stresses.*

Where L_0 = original length of the specimen before application of the load

L_f = the final length of the specimen after application of respective loadings.

Shearing Stresses

When applied forces act tangentially on the workpiece, then the stresses experienced by the workpiece are called as shearing stresses.

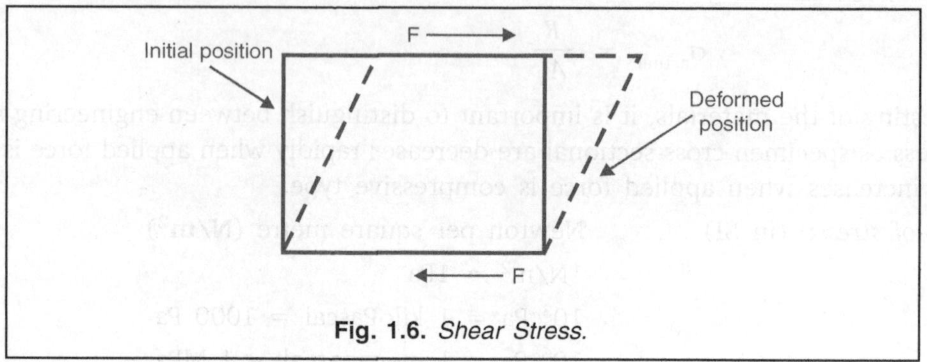

Fig. 1.6. *Shear Stress.*

The shear stress is usually designated by τ and is given by

$$\text{shear stress } \tau = \frac{F}{A} = \frac{\text{Applied force}}{\text{Shear area}} \quad \text{...(1.3)}$$

Strain

Strain is defined as the change in length to the original length of the specimen.

Upon applying a tensile force, the specimen becomes longer and when compressive force is applied it becomes shorter. For example as shown in Fig. 1.5 (*a*) the original length of the workpiece is L_0 and L_f becomes the final length of the specimen after the applied force F has been removed. For a specimen of original length L_0 the engineering strain and true strain are defined by equations (1.4) and (1.5) respectively.

$$\text{Engineering strain } \varepsilon = \frac{L_f - L_0}{L_0} = \frac{\text{Change in length}}{\text{Original length}} = \frac{\Delta L}{L_0} \quad \text{...(1.4)}$$

$$\text{True strain } \varepsilon_{\text{true}} = \frac{\Delta L}{L} \quad \text{...(1.5)}$$

Where ($L = L_0 + \Delta L$) is the specimen length at a given load. The reason for the distinction between engineering and true strain is that the specimen length changes with applied stress.

Unit of strain (in SI): The strain has no units (dimensionless). But sometimes the strain is expressed in terms of mm/mm.

1.5 TESTING OF MATERIALS

The mechanical properties of the materials can be determined by conducting the few standard tests. These standard tests are necessary so that a customer can buy the required materials directly from market by just looking at the required values of some properties without bothering about the actual details of the tests. These standard tests are:

(*i*) Tensile test
(*ii*) Hardness test
(*iii*) Impact test
(*iv*) Fatigue test
(*v*) Creep test
(*vi*) Compression test
(*vii*) Shear test.

1.5.1 Tensile Test

Tensile test measures the resistance of a material to a static or gradually applied force. Tensile properties include the resistance of materials to pulling or stretching forces. The following mechanical properties of materials can be obtained from a tensile test:

(*i*) Modulus of elasticity
(*ii*) Yield strength
(*iii*) Ultimate tensile strength
(*iv*) Percentage elongation
(*v*) Percentage reduction in area.

Universal Testing Machine popularly known as UTM (as shown in Fig. 1.7) is mostly used to perform tensile, compression, and bending tests on different materials.

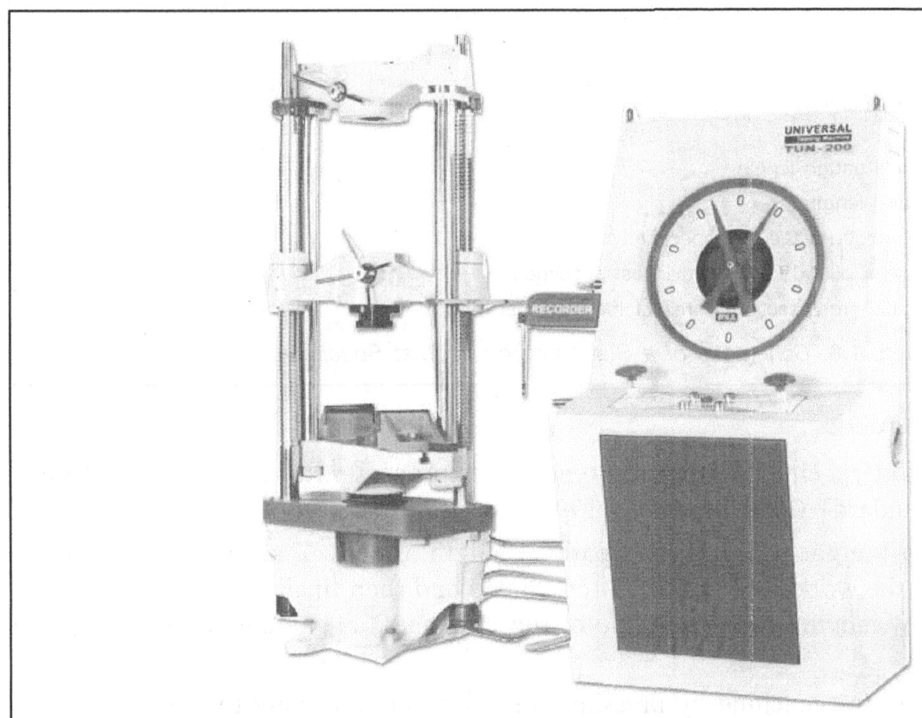

Fig. 1.7. *Universal Testing Machine (Courtesy: Fine Spavy Associate Private Ltd).*

BIS standards and requirements as per IS: 432–1982 which is used for structural work are given below:

Sr. No.	Type of steel	Bar diameter	Yield stress (N/mm²)	Ultimate stress (N/mm²)	Minimum elongation on gauge length
1.	Mild steel	≤ 20 mm	250	410	23%
	Grade-I	>20 mm and ≤ 50 mm	240	410	23%
2.	Mild steel	≤ 20 mm	225	370	23%
	Grade-II	>20 mm and ≤ 50 mm	215	370	23%

- The gauge length for tension test specimen shall be $5.65\sqrt{A_0}$ where A_0 is the cross-sectional area.
- For round bar, gauge length is 5 times the nominal diameter of bar.

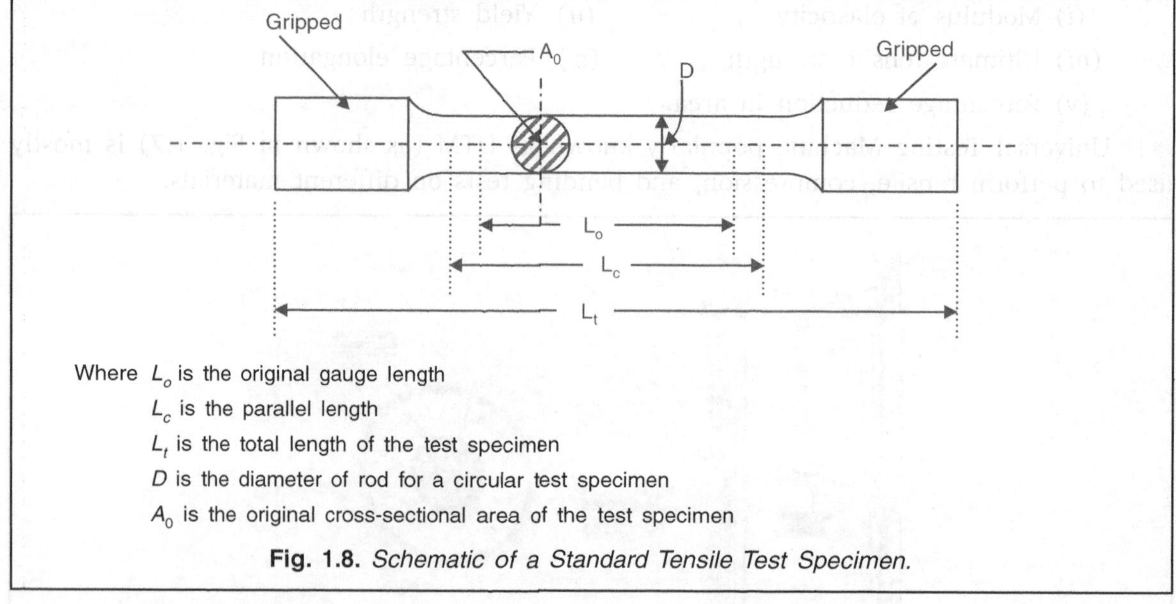

Where L_o is the original gauge length
L_c is the parallel length
L_t is the total length of the test specimen
D is the diameter of rod for a circular test specimen
A_0 is the original cross-sectional area of the test specimen

Fig. 1.8. *Schematic of a Standard Tensile Test Specimen.*

Tensile test procedure:

1. Prepare the test specimen as per the tensile test procedure as given in BIS (Bureau of Indian Standard) Code IS: 1608: 1995.
2. After marking the gauge length and parallel length as shown in Fig. 1.8, first clamp upper end of the workpiece with UTM crosshead and then finally clamp its lower end very tightly by moving the lower side of the crosshead as per the parallel length of the specimen.
3. Now switch on the machine by pressing green button and apply the load gradually by turning the load valve.

4. Note down the elongation corresponding to each increment in load (Say for 1 ton, 2 ton, 3 ton and so on).
5. Continue loading the specimen by gradually increasing the load till the workpiece breaks. Note down the reading corresponding to this failure.
6. Next release the pressure by turning the load valve in opposite direction.
7. Now takeout the broken pieces from the cross-head jaws by gently hammering the workpiece and tightening lever.
8. Next join the broken pieces of test specimen and measure the final length.
9. Draw the stress-strain diagram after making necessary calculations.

Observations

Observations before test:

1. Diameter of bar:

   (*a*) d_1 = ... mm, (*b*) d_2 = ... mm, (*c*) d_3 = ... mm.
2. Average diameter = $d = (d_1 + d_2 + d_3)/3$ = ... mm.
3. Gauge length = $L_0 = 5d$ = ... = ... mm.
4. Least count of extensometer = L.C. = ... mm.
5. A_0 is the original cross sectional area = ... mm^2.

Observation during test:

1. Range of loading = ... kN
2. Maximum load = ... kN
3. Breaking load = ... kN

Observation table for tension test

Sr. No.	*Load in KN (P)*	*Extensometer reading in divisions (EMD)*			*Extension (L.C. × EMD) (mm)*	*Stress (σ = P/A) (N/mm²)*	*Strain (Extension) /gauge length) (ε)*
		Left reading	*Right reading*	*Average*			
1							
2							
3							
4							
5							
6							

7							
8							
9							
10							
11							
12							
13							
14							
15							

Observation after test:

L_f = Final length of the specimen obtained after joining the broken pieces.

From the load and elongation readings, stress and strain can be calculated for each set of readings. Next a stress-strain curve can be plotted by plotting calculated stress values in N/m^2 or MPa along the *y*-axis and strain along the *x*-axis.

To study the material's response to test conditions it is necessary to remove size effects. The size effect comes in play due to difference in loading with different sizes of the specimen and due to variation in gauge length because of different gauge lengths of workpiece. If load is divided by original cross sectional area and the elongation is divided by the original gauge length, the size effects are eliminated. The results of a single test apply to all sizes and shapes of specimens for a given material if the load value is converted to stress and elongation in gauge length to strain. Engineering stress and strain can be calculated using equations (1.1, 1.2, 1.4 and 1.5). Using these values a stress-strain curve can be plotted. The nature of the curve depends on the type of material used for testing purposes. The stress-strain curve for low carbon steel (popularly called mild steel) is shown in Fig. 1.9 and Fig. 1.10 illustrates the stress-strain diagram with the changes as seen in the geometry of the specimen during testing. Fig. 1.11 shows the stress strain diagram for other categories of materials like polymer, cast iron and ceramic.

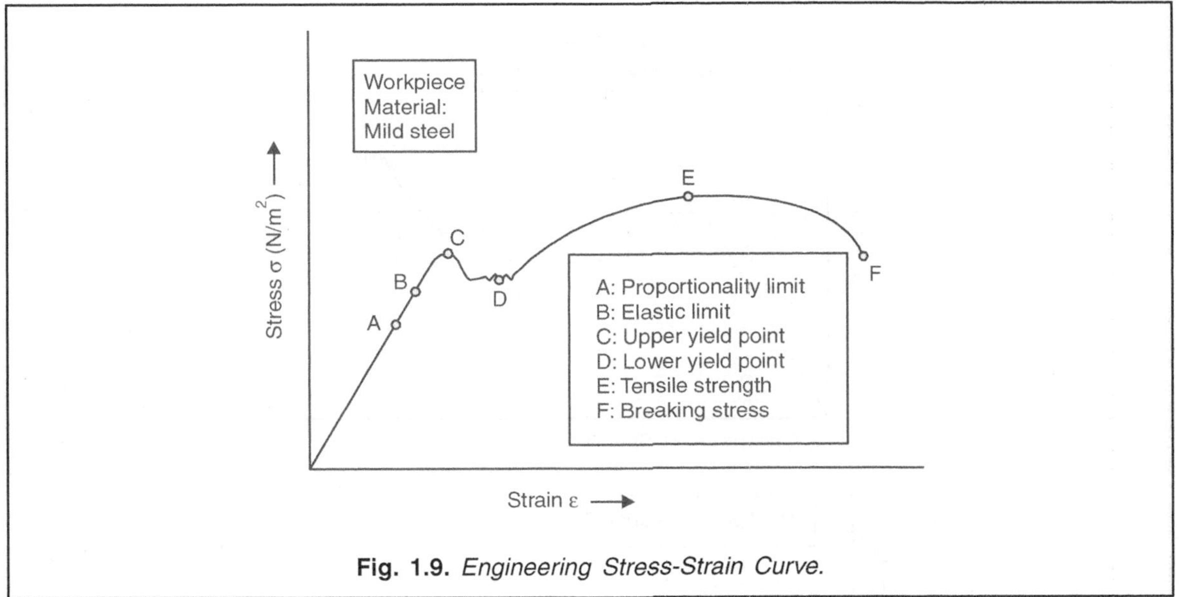

Fig. 1.9. *Engineering Stress-Strain Curve.*

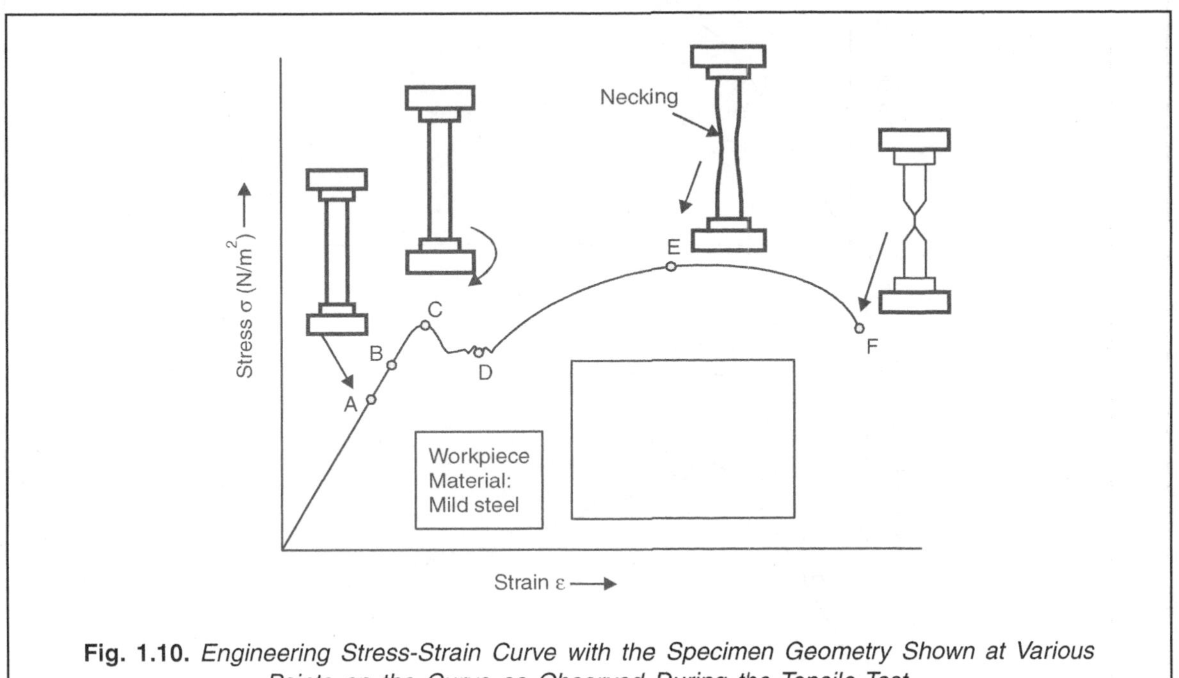

Fig. 1.10. *Engineering Stress-Strain Curve with the Specimen Geometry Shown at Various Points on the Curve as Observed During the Tensile Test.*

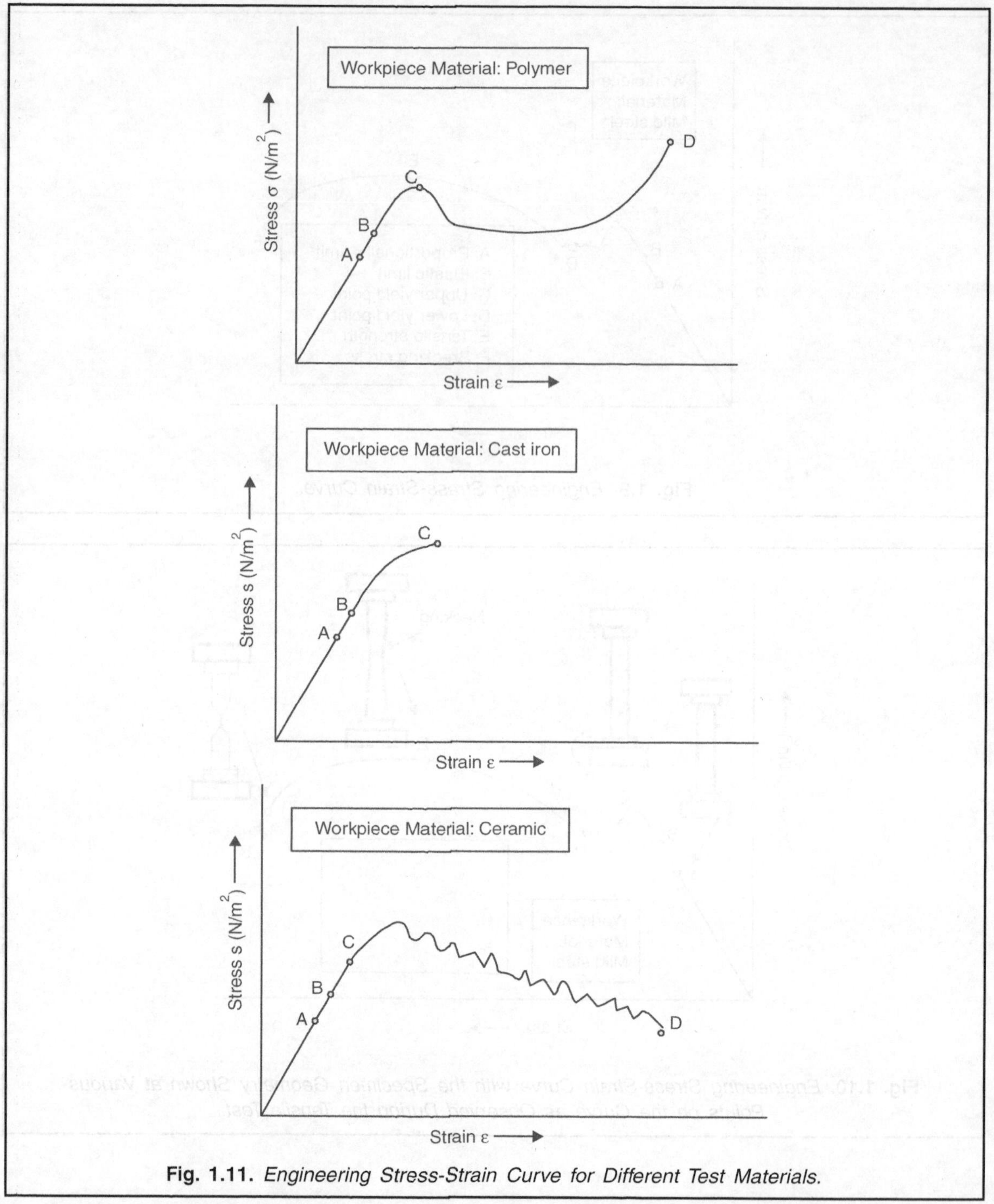

Fig. 1.11. *Engineering Stress-Strain Curve for Different Test Materials.*

Initially the left side of the stress-strain curve is a straight line. This region is known as elastic region. In this region the material will return to its original length if the applied force is removed from the specimen. The relationship between stress and strain in the elastic region is defined by Hooke's Law.

Hooke's Law: Stress is directly proportional to strain under elastic deformation.

$$\text{Stress} \propto \text{Strain}$$

$$\sigma \propto \varepsilon$$

$$\sigma = E\varepsilon$$

$$\therefore \quad E = \frac{\sigma}{\varepsilon} \quad \text{...(1.6)}$$

Where E is the modulus of elasticity,

Modulus of elasticity: Modulus of elasticity is a measure of stiffness. It indicates the ability of a material to resist deflection under the influence of a force. The high value of modulus of elasticity is an indication of the fact that material is relatively stiff and does not deflect easily. For example aluminium and its alloys have low value of modulus of elasticity whereas the steels possess high value of modulus of elasticity.

As shown in Fig 1.9 the point A represents the *'proportionality limit'*. Beyond point 'A' the stress no longer remains linearly proportional to the strain. The maximum stress for which truly elastic behavior exists is called *elastic limit*. The curve area beyond the elastic limit is known as plastic deformation or the plastic region.

It is difficult to find out when elastic limit ends on the curve and plastic deformation starts. This can be determined by drawing a line parallel to elastic portion of the curve offset by a distance of 0.002 m/m. The intersection of this line with the curve is called *yield point*. This is also referred to as 0.2% yield strength or 0.2% offset yield strength. Mostly engineers prefer to use yield stress as the maximum strength of the material for structural design purposes. The curve shown in Fig 1.9 is for low carbon steel and important feature of this curve is the presence of upper and the lower yield points at the curve. The highest value of the stress (at point C) just before extensive strain takes place is called *upper yield point*. The point D is referred to as *lower yield point* at which sudden increase in strain takes place without any appreciable increase in the force.

Most materials do not exhibit a well-defined yield point on a stress-strain curve. For such materials the elastic to plastic transition region is defined by the offset yield strength, which is also known as proof stress. Such type of stress strain curve is shown in Fig.1.12.

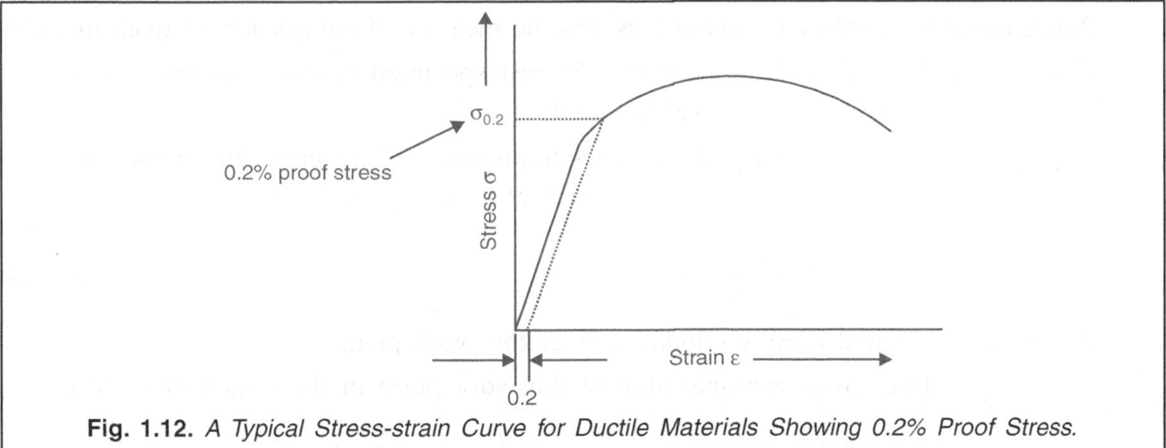

Fig. 1.12. *A Typical Stress-strain Curve for Ductile Materials Showing 0.2% Proof Stress.*

Tensile Strength

The highest value of the stress that a material can sustain is called tensile strength. Point E on the curve (Fig. 1.9) denotes the tensile strength. After this point the cross-sectional area of specimen starts decreasing very fast. This phenomenon is known as 'necking' as shown in Fig. 1.10.

Tensile strength can be determined by drawing a horizontal line from the highest point on the stress-strain curve on to the stress axis. The intersection of this line with the y-axis gives the tensile strength. Engineers use ultimate tensile strength in some of the structural design work. The stress corresponding to breaking point is called the breaking strength of the material.

Percentage Elongation

It is the measure of ductility. The percentage elongation can be determined by using equation (1.7). For this the original length and final length of the test specimen need to be determined. For calculating the final length, place both parts of the fractured work piece together on a table as shown in Fig. 1.13 (B) and then with the help of a scale or a caliper one can measure the final length of the work piece.

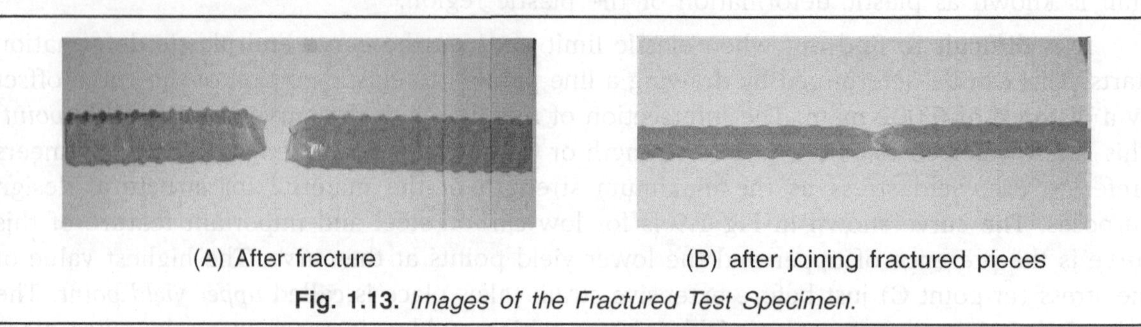

(A) After fracture (B) after joining fractured pieces

Fig. 1.13. *Images of the Fractured Test Specimen.*

$$\text{Percentage Elongation} = \frac{\text{Final Length} - \text{Initial Length}}{\text{Initial Length}} \times 100 \qquad ...(1.7)$$

$$= \frac{L_f - L_i}{L_i} \times 100$$

Percentage reduction in area: It is also the measure of the ductility of work material.

This can be determined by measuring the test specimen diameter at the start of the tensile test and final diameter in necking region.

$$\% \text{ Reduction in area} = \frac{\text{Original cross sectional area} - \text{Final cross sectional area}}{\text{Original corss sectional area}}$$

$$= \frac{A_0 - A_f}{A_0} \qquad ...(1.8)$$

Where A_0 = original cross sectional area of the work piece

A_f = final cross sectional area of the work piece in the region of necking

Poisson Ratio

The absolute value of the ratio between the lateral strain (shrinkage) and the longitudinal strain (elongation) is called the **Poisson's ratio** of the material. It is denoted by η (eta). This is measured only in the elastic region of the stress-strain curve.

$$\text{Poisson's ratio } \eta = \frac{-\dfrac{D_f - D_0}{D_0}}{\dfrac{L_f - L_0}{L_0}} \qquad ...(1.9)$$

Its maximum value is 0.5 which is also the value of Poisson's ratio for perfectly incompressible material. For majority of materials, its value lies in the range of 0.25 to 0.50. For steel its value is 0.303 and for rubber 0.49.

Utility of Tensile Test

- Tensile test is used to determine the value of safe or allowable stresses for a given material. Higher is the yield stress, better is the material.
- It is used to identify whether the specimen material is ductile or brittle. If specimen material is ductile than it will have reasonable percentage elongation and have distinct plastic deformation zone as shown in Fig. 1.10 whereas brittle material will show little plastic behaviour as shown in Fig. 1.11 and have a low percentage elongation.
- It is also used to check the effectiveness of heat treatment processes on the work material.

1.5.2 Hardness Test

Hardness is a measure of the resistance to penetration of the surface of a material by a hard or sharp object. Hardness represents resistance to scratching or indentation by a hard object on a metal surface. Hardness cannot be defined precisely. Hardness is a qualitative measure of strength of the material. The hardness of a metal is measured by forcing an indenter into its surface. Usually the shape of an indenter material is a ball, pyramid or cone. These are made of a material much harder than the material being tested. For example commonly used indenter materials are hardened steel, tungsten carbide and diamond.

From industrial point of view, the two important hardness test used in practice are (*a*) Brinell test and (*b*) Rockwell test. The different types of indenters used in hardness tests are shown in Fig. 1.14. Hardness numbers are used as a qualitative basis for the comparison of materials, for preparing the specifications for manufacturing and heat treatment, quality control and correlation with other material's properties.

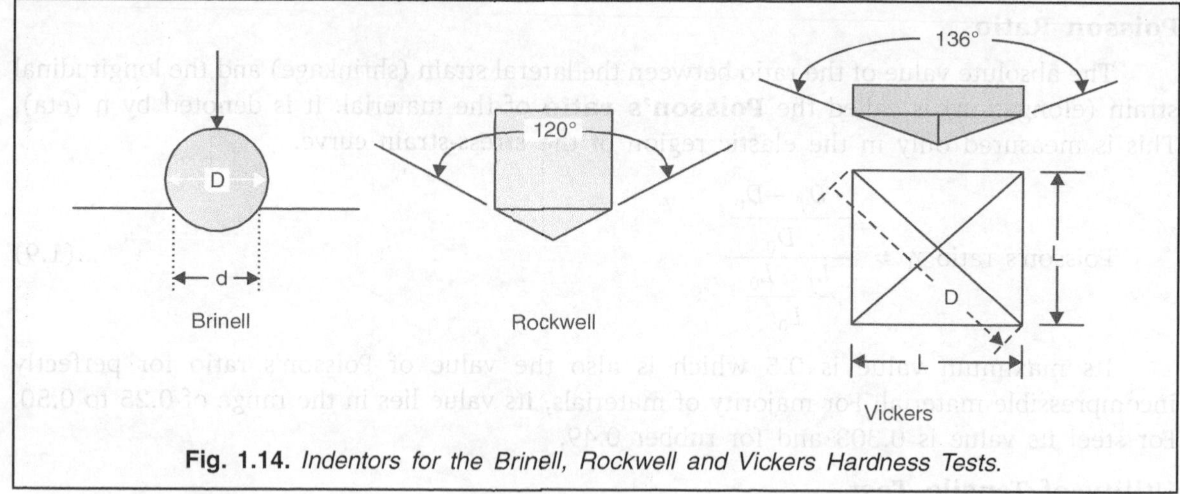

Fig. 1.14. *Indentors for the Brinell, Rockwell and Vickers Hardness Tests.*

Test Procedure

In the hardness tests, a known load is applied slowly by pressing the indenter at 90° onto the metal surface being tested. After indentation has been made, the indenter is withdrawn from the workpiece surface. As shown in Fig. 1.15, an empirical number is then noted from the dial guage provided on a hardness tester. This recorded number from the dial guage or a digital readout is based on the cross-sectional area or depth of impression created by an indenter.

(*a*) Brinell Hardness Test

Brinell hardness test is based on penetrating the workpiece surface by a hardened steel ball (indenter) at a predetermined load. To determine the hardness of a steel specimen, a 10 mm diameter steel ball is pressed hard by a load of 3000 kg. For soft material like aluminium and copper a load of 500 kg is used. After removal of the load, the surface area of the indentation is measured. Brinell hardness is obtained by dividing the applied load by the surface area of the indentation. The diameter of the impression typically 2 to 5 mm is measured and the Brinell hardness number (called as BHN) is calculated from the following equation:

$$HBN = \frac{F}{\frac{\pi D}{2}\left[D - \left(\sqrt{(D^2 - d^2)}\right)\right]} \quad ...(1.10)$$

Where F is applied load in kgs

D is diameter of the indenter in mm

d is diameter of the impression in mm

Though Brinell hardness has the same unit as that of the stress but is generally expressed as a number without assigning any unit. A Brinell hardness number can be obtained in just a few minutes with virtually no penetration of the work surface and even without breaking the workpiece, yet it provides a close approximation of the tensile strength of the material. Fig. 1.15 shows a schematic of a Brinell hardness testing machine.

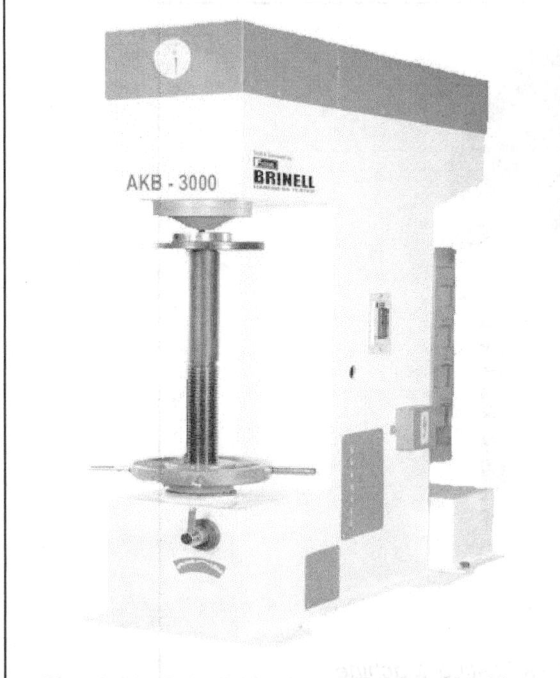

Fig. 1.15. *Brinell Hardness Testing Machine. (Courtesy: Fine Spavy Associate Private Ltd).*

Fig. 1.16. *Rockwell Hardness Testing Machine (Courtesy: Fine Spavy Associate Private Ltd).*

(*b*) Rockwell Hardness Test

In this hardness test, a small diameter steel ball is used for soft materials and a diamond cone is used for harder materials. In Rockwell hardness test, the depth of the penetration is a parameter to assess the hardness of a material whereas in Brinell hardness test surface area was taken as a parameter to find out the hardness. The depth of penetration of the indenter is automatically measured by the testing machine as shown in Fig.1.16 and convert it to a Rockwell hardness number *i.e.,* Hardness Rockwell (HR). In this test, the depth of penetration is not measured. Instead of that, the hardness values can be noted directly from a dial provided on the machine. The Rockwell hardness number can not be directly related to strength of metals and alloys. Rockwell test is more popular than Brinell test. There are two reasons behind this. Firstly it takes little time in test and secondly a very small indentation is made on the surface. This second feature enables the finished part to undergo this test.

Different variations of this test are used. A Rockwell C (HRC) test is used for hard steels whereas a Rockwell test (HRF) is preferred for soft materials like aluminium. Rockwell tests provide a hardness number that has no units.

Knoop hardness test

The Knoop hardness test is a microhardness test forming very small indentations that need a high quality microscope to obtain the measurement. Less than 2N load is applied in this hardness test.

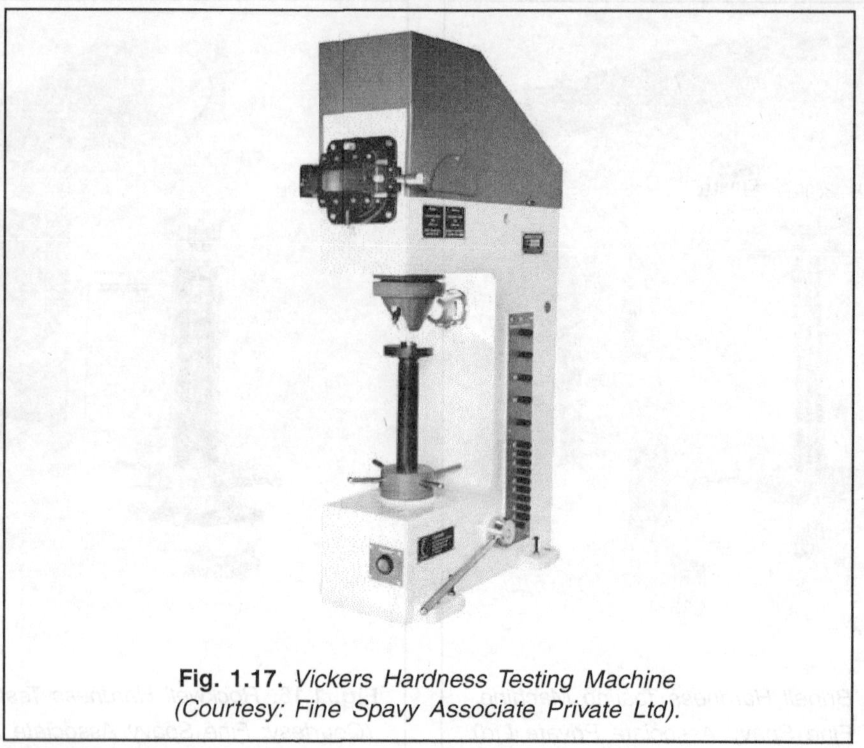

Fig. 1.17. *Vickers Hardness Testing Machine (Courtesy: Fine Spavy Associate Private Ltd).*

Vickers hardness test

The Vickers hardness test is also known as Vickers diamond pyramid test. Fig.1.17 shows a schematic of a Vickers hardness testing machine. This test can be conducted either as a macro and microhardness test. A square based diamond pyramid is used in this test. The included angle between opposite faces of the pyramid is 136°. The vicker hardness number (VHN) is obtained by dividing the applied force by the surface area of indentation. It is obtained by equation given below

$$VHN = \frac{2F}{L^2}\frac{\sin\alpha}{2} \qquad ...(1.11)$$

Where F is the indentation force which range from 0.1 to 1 kg but may be used up to 120 kgs.

L is the diagonal length of the indentation

α is the included angle of 136°.

The main benefit of vicker hardness test is that the result is independent of the load. Since a high quality polished surface is required for the test specimen therefore this test is generally preferred for the research work.

1.5.3 Compression Test

In this test the forces are pushed towards each other (*i.e.*, act inwards). Compression tests are also performed on materials but the important thing is that the work piece should not be too long as compared to the thickness otherwise buckling may occur. The nature of failure due to compressive forces depend upon structure of the material. A typical stress-strain plot for the compressive test is shown in Fig. 1.18.

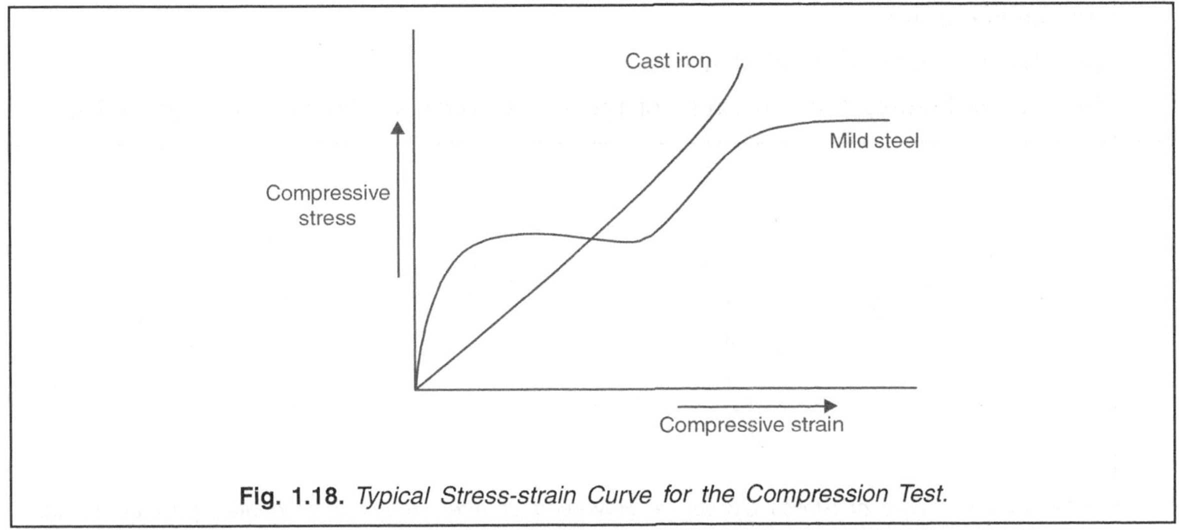

Fig. 1.18. *Typical Stress-strain Curve for the Compression Test.*

1.5.4 Fatigue Test

Fatigue test measures the number of cycles that need to be applied for a particular load until failure occurs.

There are two types of loading which act on engineering structures and machine components. These are (*a*) Static loading and (*b*) cyclic or dynamic loading. In static loading the applied load does not change with time whereas in cyclic loading, the applied load changes with time. The metal parts working under the cyclic loading will fail at a stress considerable lower than that a component can sustain under the influence of static loading. Such types of failures that occur under repeated and cyclic stresses are called fatigue failures. The common examples of machine parts failing due to fatigue are shafts, connecting rods, gears, and turbine blades etc. It has been estimated that fatigue failures happen to be the cause of nearly 80 to 85% failure of all the machine parts. A fatigue failure occurs without any warning.

Engineers should have clear idea about the fatigue failure and should design the components in such a way so that fatigue failures can be reduced.

A fatigue failure usually originates at a point of stress concentration such as sharp corners or notch, or a metallurgical inclusion or flaw. A fatigue failure can be recognized from the appearance of the fractured surface, which shows two distinct regions. One is a smooth region and the other is a rough region. Smooth region results due to rubbing action as the crack propagates through the section whereas rough region is after-result of material's inability to sustain load.

The main reason behinds the fatigue failure can be:

(*i*) maximum tensile stress of high value.

(*ii*) wide fluctuation in applied stress.

(*iii*) large number of cycles of applied stress.

(*iv*) stress concentration.

(*v*) corrosion.

(*vi*) temperature.

(*vii*) flaws in metallurgical structure.

The fatigue failures occur because of the stress cycles as shown in the Fig. 1.19.

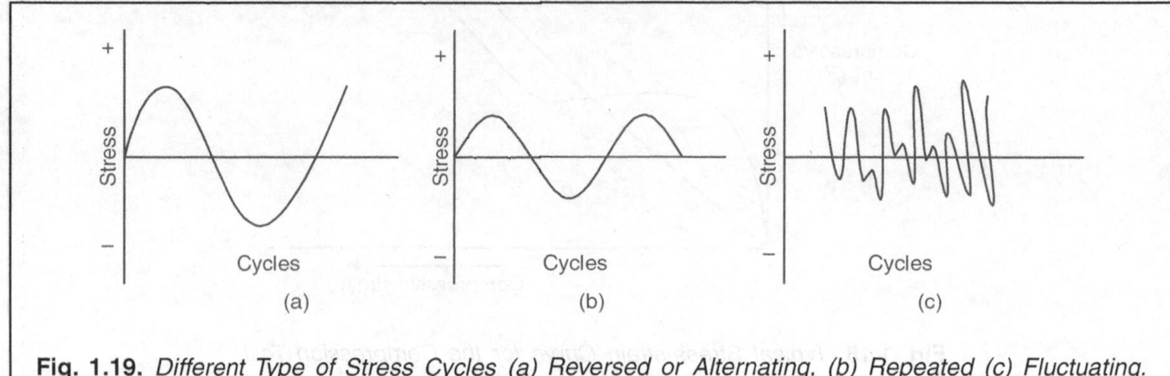

Fig. 1.19. *Different Type of Stress Cycles (a) Reversed or Alternating, (b) Repeated (c) Fluctuating.*

Many testing machines are available to determine the fatigue life of a material. The few notable fatigue testing machines in use are:

(*i*) Rotating beam fatigue testing machine

(*a*) Constant bending moment fatigue testing machine

(*b*) Cantilever type fatigue testing machine

(*ii*) Reciprocating bending fatigue testing machine

(*iii*) Multi-axial stress fatigue testing machine

(*iv*) Repeated torsion machine

(*v*) Computer controlled closed cycle fatigue testing machine

The most popular among the above listed machines is rotating beam fatigue testing machines (Fig. 1.20).The rotating beam test is commonly preferred to carry out the fatigue test. In this test the rotating work piece is subjected to alternating compressive and tensile stresses of equal magnitude. The surface of this work piece is specially prepared, polished and tapered towards the center. Fatigue test consumes lot of time and needs a large number of specimens to generate the data for the S-N curve. In the S-N curve, S (stress) is plotted along the *y*-axis and N (the number of cycles) is plotted along *x*-axis.

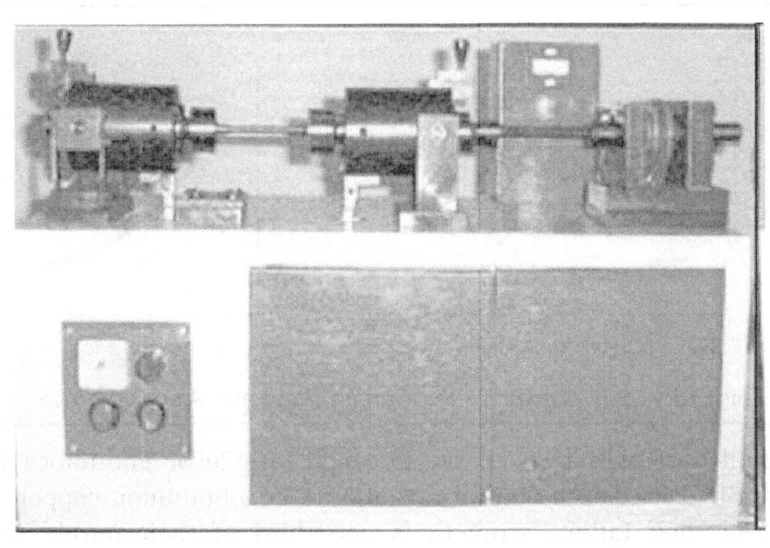

Fig. 1.20. *Rotating Beam Fatigue Testing Machine (Courtesy: Adinath Impex).*

The test procedure:

1. Prepare the test specimen as per the instruction manual of the machine. If the test is to be done for commercial purpose then follow the test procedure as given in BIS code.
2. Measure the surface roughness of the specimen.
3. Mount the specimen in the rotating beam testing machine.
4. Record the weight and measure the distance needed to calculate the bending moment.
5. Set cycle counter at zero.
6. Start the machine and wait till the specimen fails.
7. Record the number of cycles from the counter.
8. Repeat step 1 to 7 for other specimens by changing weights.
9. Plot the S-N curve from the experimental data recorded for the test material.
10. Save the images of the fractured test pieces on scanning electronic microscope or by using optical metallurgical microscope.

Data from this test are plotted in the form of S-N curve. The S-N curve for the test material (mild steel) is shown in Fig. 1.21. This curve helps us to find out the expected fatigue life at a particular stress level and the fatigue limit for the workpiece material. The fatigue life tells us how long a component survives when a stress is repeatedly applied to the component material. Fatigue limit is the stress below which the material will not fail.

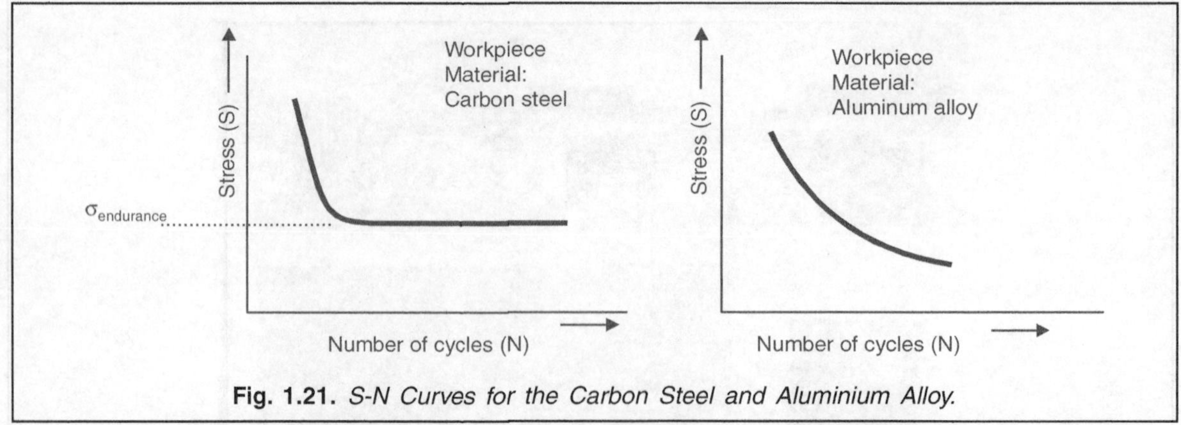

Fig. 1.21. *S-N Curves for the Carbon Steel and Aluminium Alloy.*

The horizontal part of the S-N curve is called fatigue or endurance limit and it lies between 10^6 and 10^{10} cycles. The non-ferrous alloys like aluminium, copper etc. do not have a fatigue limit and their fatigue strength is one third of their tensile strength. For the nonferrous materials, fatigue life is a critical design consideration. Applied stress on these materials should be low enough so that fatigue failure does not occur in the lifetime of a component.

1.5.5 Creep

Creep can be defined as the gradual deformation developed in a material over a specified period of time due to the application of a constant stress or constant strain. This gradual or slow deformation may result in a machine part or component failing much below the tensile strength of the material. Creep has a considerable influence on the performance of the parts particularly at high temperature. High temperature has significant effect on mechanical properties of materials. As we know the strength of metal decreases at high temperature. High temperature is associated with numerous applications such as:

(*i*) Steam turbines, boilers and cracking units may exceed temperature of 500°C.

(*ii*) The temperature of turbine buckets is generally above 750°C in case of gas turbine.

(*iii*) Oil refineries.

(*iv*) Chemical plants.

(*v*) Missiles, Spacecraft, rocket parts are also subjected to the very high temperature.

In keeping view of the cost involved in running and manufacturing of the above listed applications, creep resistant material should be used in their design.

Creep limit it is defined as the maximum strength that will result in creep at a rate lower than some specified rate. Creep tests are carried out in a similar fashion as the tensile tests and using a similar test piece. A creep test usually takes several hours or weeks as compared to tensile test that hardly needs few minutes. It is a general practice to maintain the load constant instead of constant stress while conducting an engineering creep test. An ideal creep curve is shown in Fig 1.22. The creep rate is given by the slope of this curve ($d\varepsilon/dt$). Following an initial rapid elongation of the work piece, the creep rate decreases with time, then reaches essentially a steady state in which the creep rate increases rapidly with time until failure takes place.

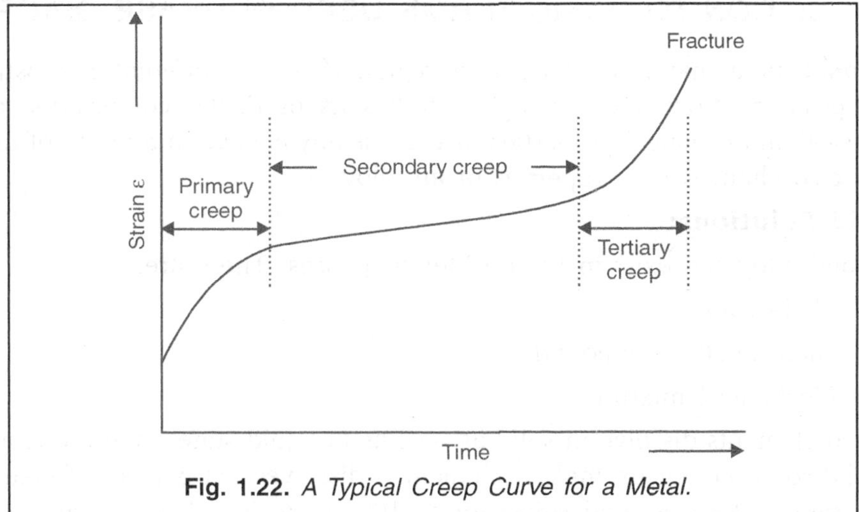

Fig. 1.22. *A Typical Creep Curve for a Metal.*

1.5.6 Torsion

Torsion is defined as the twisting action of one side of a body with respect to an adjacent side. A torsion test is used to determine the shearing properties of hollow and solid circular sections. This test enables the engineers to find out the following properties:

(*i*) Proportional limit

(*ii*) Shearing yield strength

(*iii*) Shearing resilience

(*iv*) Stiffness

(*v*) Ductility

(*vi*) Toughness.

Torsion testing machines are available for an extensive range of torque measurements in order to test the bar; tube or wire etc. The strain is measured by a torsion indicator called troptometer. Torsion tests are very important as it enables the designer to know the efficiency of mechanical parts such as axles, shafts, and twist drills and also for materials lacking toughness under twisting stresses. A typical torsion test curve is shown in Fig. 1.23.

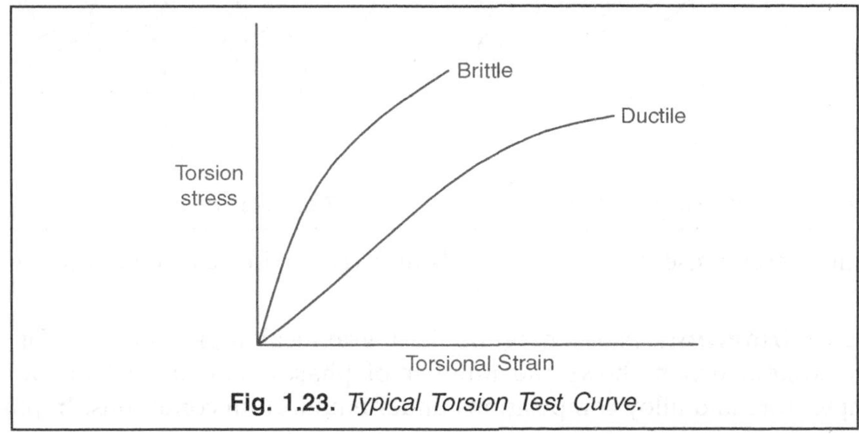

Fig. 1.23. *Typical Torsion Test Curve.*

1.6 INTRODUCTION TO BASIC TERMS USED IN PHASE DIAGRAM

(*a*) **Phase:** In a material, phase is a region that has different microstructure and compositions from another region. It has its particular composition and distinct physical and chemical properties. If there is any change in a phase of an alloy than this also changes the properties of an alloy.

(*b*) **Solid Solutions:**

A solid alloy can exist in three different phases. These are:

(*i*) Solid solutions

(*ii*) Intermetallic compound

(*iii*) Mechanical mixture

When two elements dissolve in solid as well as in liquid state, then the resulting phase is called a solid solution. For example if element A dissolves ten percent of element B, then element B is said to have a solid solubility of 10% in A. The base metal A is known as 'Solvent' and the alloying element 'B' as solute. During formation of solid solution, the atoms of solute occupy certain places in the crystal lattice structure of the solvent.

The solid solutions are of two types:

(*i*) Substitutional solid solution

A substitutional solid solution is formed when some of the atoms of solvent are replaced by the solute atoms at their lattice points as shown in Fig. 1.24.

(*ii*) Interstitial solid solution

In Interstitial solid solution, the solute atom does not displace a solvent atom but occupy the empty spaces or interstices of the crystal lattice structure of the solvent atoms as shown in Fig. 1.25. Interstitial solid solutions have limited solid solubility. Example: interstitial solid solution of carbon in iron.

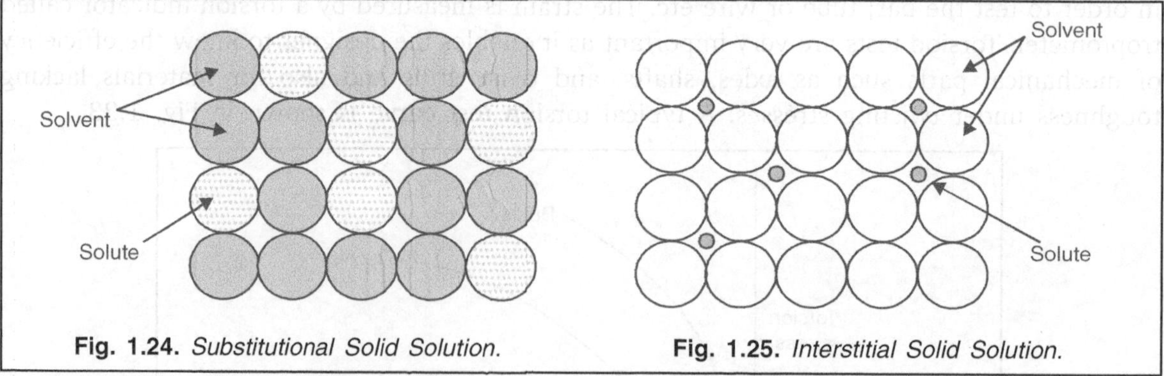

Fig. 1.24. *Substitutional Solid Solution.* **Fig. 1.25.** *Interstitial Solid Solution.*

Important phases present in carbon steels are: (*i*) Ferrite (*ii*) Cementite, (*iii*) Austenite, (*iv*) Pearlite.

(*c*) **Phase Diagram:** Phase diagram is a graphical representation for a particular alloy system which shows the number of phases that are present at a particular temperature and alloy compositions under equilibrium conditions. In phase diagram, temperature is plotted along y-axis and composition of alloy (by weight %) is plotted along *x*-axis.

(*d*) **Allotropy of Metals:** Most of the metals change their arrangement of atoms and the crystal lattice structure due to changes in operating conditions such as pressure and temperature. This type of change of phase in solid state is called allotropy or polymorphism. Iron crystal has three allotropic forms namely alpha (α), gamma (γ) and delta (δ) irons. These allotropic forms are achieved when iron from the molten state is cooled down to room temperature.

1.6.1 Cooling Curves

Cooling curve is used to get the idea about the temperatures at which changes in phase of pure metals and alloys occur. A cooling curve is a plot between the temperatures recorded during cooling of an alloy from its melting point to room temperatures and the time taken for this cooling.

A cooling curve provides information about the solid state phase transformation in metals and alloys. Figure 1.26 shows a cooling curve of a pure iron. The melting point of pure iron is 1539°C. When the iron starts freezing from this temperature, a high temperature solid of BCC structure is obtained. It is also called **δ-iron**. During further cooling of iron, when temperature reaches at about 1394°C iron changes its phase from δ-iron BCC structure to **γ-iron** FCC structure. With further cooling, a second phase transformation takes place at about 910°C where γ-iron FCC structure changes into **α-iron** BCC structure. When temperature drops down to 768°C the paramagnetic properties of iron changes into ferromagnetic with no change in its crystal structure.

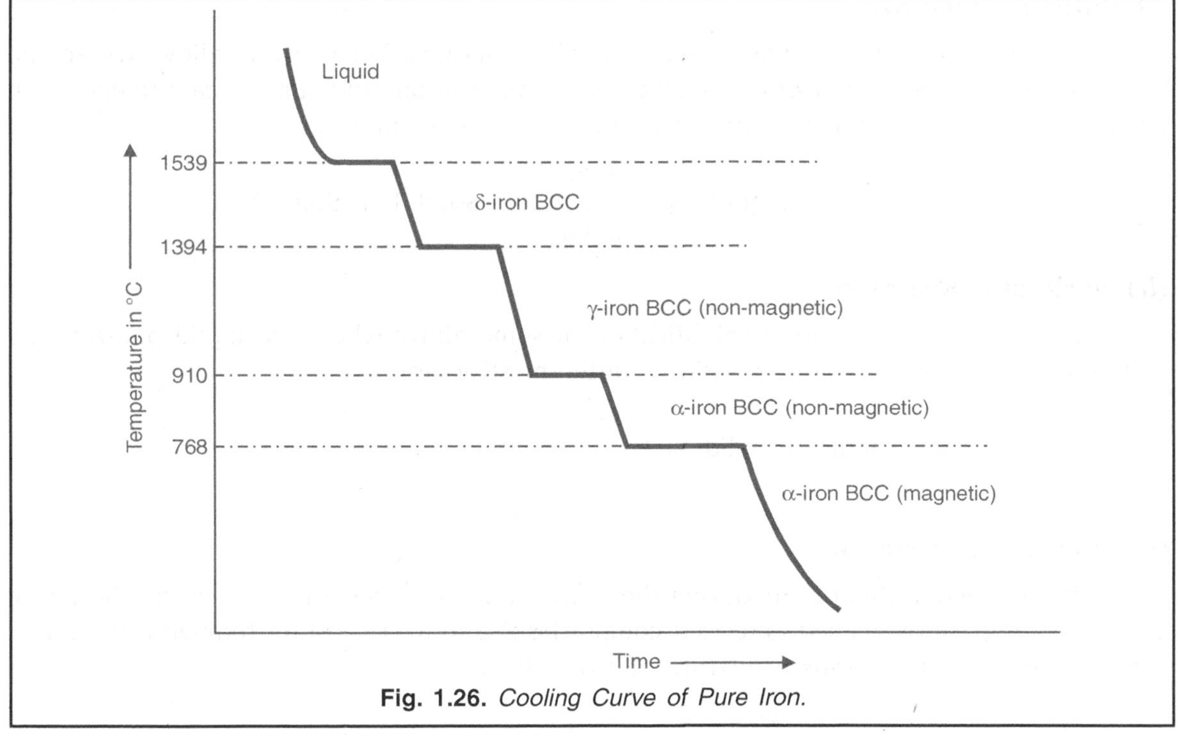

Fig. 1.26. *Cooling Curve of Pure Iron.*

1.6.2 Phases of Steel

(*i*) **Ferrite.** It is an interstitial solid solution of carbon in α-iron. It has a BCC crystal structure and is very soft and highly ductile and highly magnetic.

(*ii*) **Cementite.** It is a chemical compound of iron and carbon called iron carbide (Fe_3C). Cementite consists of 6.67% carbon and it is very hard and brittle.

(*iii*) **Austenite.** It is an interstitial solid solution of carbon in γ-iron. Austenite is formed when steel is heated above critical temperature (AC_3). Austenite is soft and non-magnetic and when cooled below 723°C, it changes into ferrite and pearlite.

(*iv*) **Pearlite.** Pearlite is a eutectoid mixture composed of alternating layers of ferrite and cementite. The strength of iron and steel is due to Pearlite. It consists 87% of ferrite and 13% of cementite by weight. It is obtained by transformation of austenite (having 0.8% carbon) during slow cooling at constant temperature AC_1. The rate of cooling affects the size of pearlite.

(*v*) **Ledeburite.** Ledeburite is a eutectic mixture consists of alternate layers of austenite and cementite. It contains 4.3% carbon and is obtained at 1130°C.

(*vi*) **Martensite.** Martensite is the supersaturated solid solution of carbon in α-iron. It is magnetic and made up of a needle like fibrous structure. It is formed when plain carbon steel from austenitic state is rapidly cooled down to room temperature by quenching it in water.

1.6.3 Different Reaction in Phase Diagrams

(*i*) Eutectic Reaction

Eutectic reaction is one which occurs during solidification of some alloys where the liquid phase at a fixed composition (called eutectic composition) and at a constant temperature (called eutectic temperature) forms a mixture of two solid phases.

$$\text{Liquid} \underset{\text{Heating}}{\overset{\text{Cooling}}{\rightleftarrows}} \text{Solid 1} + \text{Solid 2}$$

(*ii*) Peritectic Reaction

This reaction occurs during solidification of some alloys where the liquid phase reacts with a solid phase to give a new solid of different structure.

$$\text{Liquid} + \text{Solid} \underset{\text{Heating}}{\overset{\text{Cooling}}{\rightleftarrows}} \text{New Solid}$$

(*iii*) Eutectoid Reaction

It is the reaction that occurs during the solidification of some alloys where a solid phase at a fixed composition (called eutectoid composition) and at a constant temperature (called eutectoid temperature) forms a mixture of two solids.

$$\text{Solid} \underset{\text{Heating}}{\overset{\text{Cooling}}{\rightleftharpoons}} \text{Solid 1} + \text{Solid 2}$$

1.7 WROUGHT IRON

It is manufactured from pig iron by the pudding process, which removes impurities such as carbon, sulphur and phosphorus leaving nearly pure iron. A typical composition of wrought iron is

Iron:	99.5–99.8%
Carbon:	0.01–0.03%
Silicon:	0.0–0.1%
Phosphorus:	0.04–0.2%
Sulphur:	0.02–0.04%
Manganese:	0.0–0.25%

Important properties of wrought iron are:

- Its good corrosion resistance.
- Its good ductility.
- Its low cost.

Applications: It is used in cases where visual warning of failure is required. For example **link chains** are made of wrought iron where overloading may cause a considerable increase in length of the links before the actual failure occurs.

1.8 IRON-CARBON EQUILIBRIUM DIAGRAM

It is also known as iron-iron carbide phase or simply iron-carbon diagram. Iron-carbon diagram shows the different phases in iron carbon alloys that are present at a particular temperature for different percentages of carbon. In the iron-carbon phase diagram the percentage of carbon varies from 0 to 6.67%. The iron-carbon phase diagram, does not show the percentage of carbon from 0 to 100 because except graphite, no other important phase is present after 6.67% of carbon. Iron carbon diagram is plotted with carbon percentage along x-axis and temperature along y-axis. It is very important tool for an engineer to get the detailed information on phases, microstructures, temperatures and percentage of carbon in the alloys of iron and carbon namely steels and cast irons.

As we have already defined different allotropic forms of iron, let us now discuss the iron-carbon diagram as shown in Fig. 1.27 with reference to the definitions of α-iron, γ-iron, δ-iron and cementite (*i.e.*, Fe_3C). Let us also define here steel and cast iron so that one can easily understand this iron-carbon diagram.

Steel is an alloy of iron and carbon in which maximum percentage of carbon can go up to 2% whereas cast irons are also the alloys of iron and carbon in which the percentage of carbon is more than 2%.

- Line AB represents pure iron and point E which is at 1539°C, represents the melting point of pure iron.
- The curve line EHJK is a liquidus line. The alloy of iron and carbon is in liquid form above this line EHJK.
- Line EFGIJL is a solidus line below which alloy of iron and carbon will precipitate in solid form.
- In between these two curve lines *i.e.*, liquidus and solidus, the alloy is in partly solid and is in partly liquid form.
- Along line JK, the cementite crystal starts precipitating from the liquid alloy. Line JL is a solidus line below which solid of cementite and ledeburite takes shape.

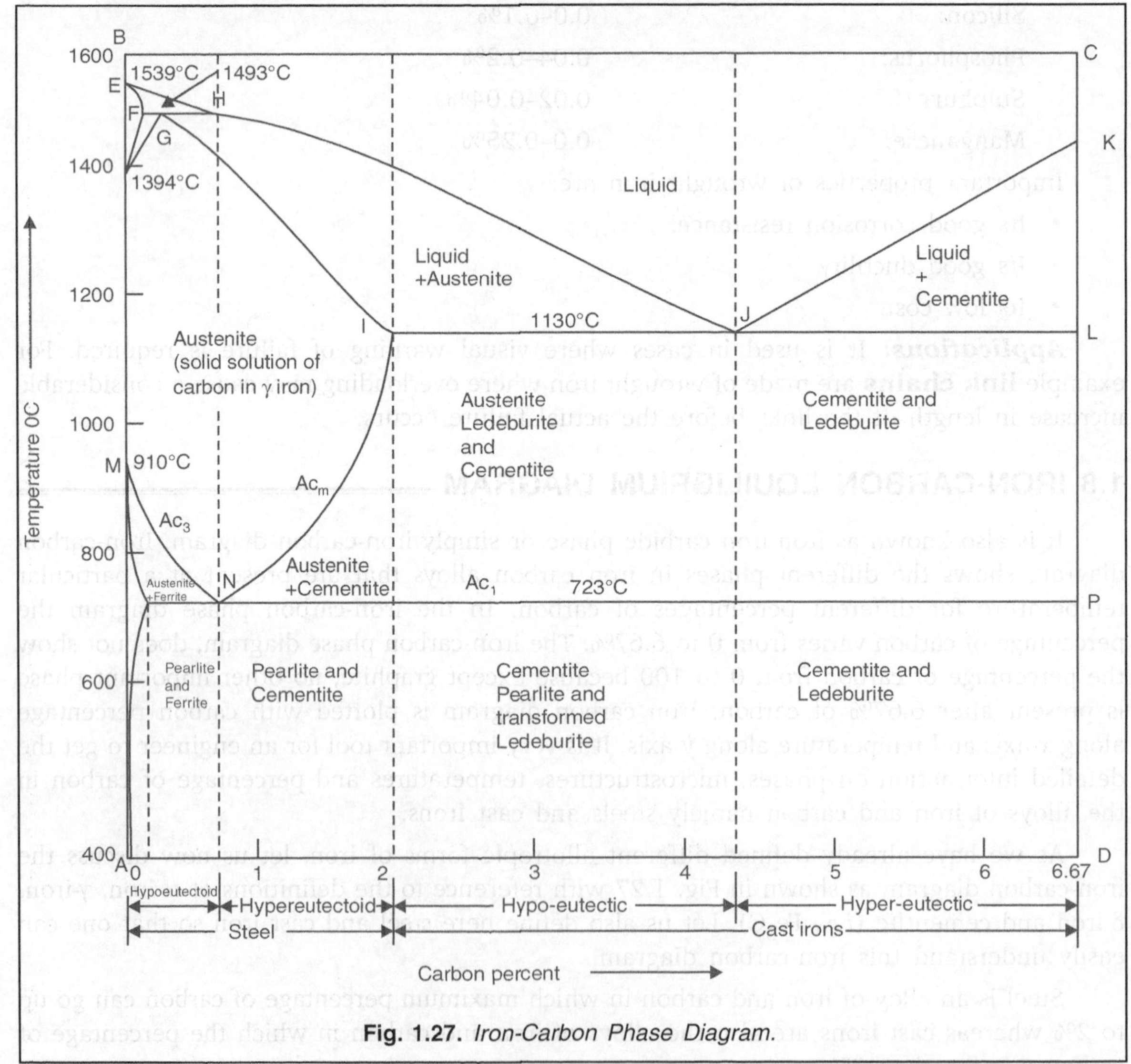

Fig. 1.27. *Iron-Carbon Phase Diagram.*

- As we know that below solidus line FGIJL, all iron-carbon alloy will solidify. As defined earlier, steel with 2% carbon will solidify along line FGI and cast irons with carbon percentage more than 2 will solidify along line IJL.
- At 1130°C eutectic reaction takes place along line IJL. In iron-carbon diagram point J represents a eutectic point at which the percentage of carbon is 4.3. At this point, austenite and cementite precipitate from liquid alloy and form ledeburite which is an eutectic alloy. This allotropy is not observable because austenite is unstable at room temperature.
- Curve line *MNI* represents an eutectoid reaction which results in solid-solid transformation. N is the eutectoid point. At this point the percentage of carbon in alloy of iron and carbon is 0.83. It is the lowest point in the diagram at which austenite disappear when alloy is cooled down slowly. The resultant crystal structure obtained at this point is known as 'Pearlite'. It is a mechanical mixture of ferrite and cementite. It consists of laminar layers of ferrite and cementite.
- Along curve line *MN*, austenite decomposes resulting in to the separation of ferrite from cementite. The line *MN* is designated by Ac_3. Similarly along curve line *NI*, cementite precipitates from austenite. This line is called upper critical temperature line and is designated by letters 'Ac_m'.
- The line *ONP* is designated by Ac_1 for heating and Ar_1 for cooling purposes. This line is known as lower critical temperature line.
- These designations like Ac_3, Ac_m, and Ac_1 have lot of importance for engineers interested in improving the properties of steel by using various heat treatment processes for different compositions.
- Point '*O*' shows the maximum solubility of carbon in α-iron.
- Iron with carbon content up to 0.83% is called hypoeutectoid steel and carbon content more than 0.83% but less than 2% is called hypereutectoid steel.
- Alloys of Iron and carbon in which carbon content varies from 2 to 4.3% is called hypoeutectic cast irons and those with carbon varying from 4.3 to 6.67% are called hypereutectic cast irons.

1.9 STEEL

Steel is an alloy of iron and carbon with some other alloying elements. It contains carbon up to 2% in iron-carbide form. Steel never contains graphite or free carbon.

The steel is produced by oxidizing the carbon and other impurities in the molten pig iron until the carbon content of the iron is reduced to the required level. Fig. 1.28 shows the steel making procedure and the subsequent structural shapes produced from it.

The impurities present in steel affect its properties significantly. The major impurities in steel are (*i*) phosphorus (*ii*) Sulfur.

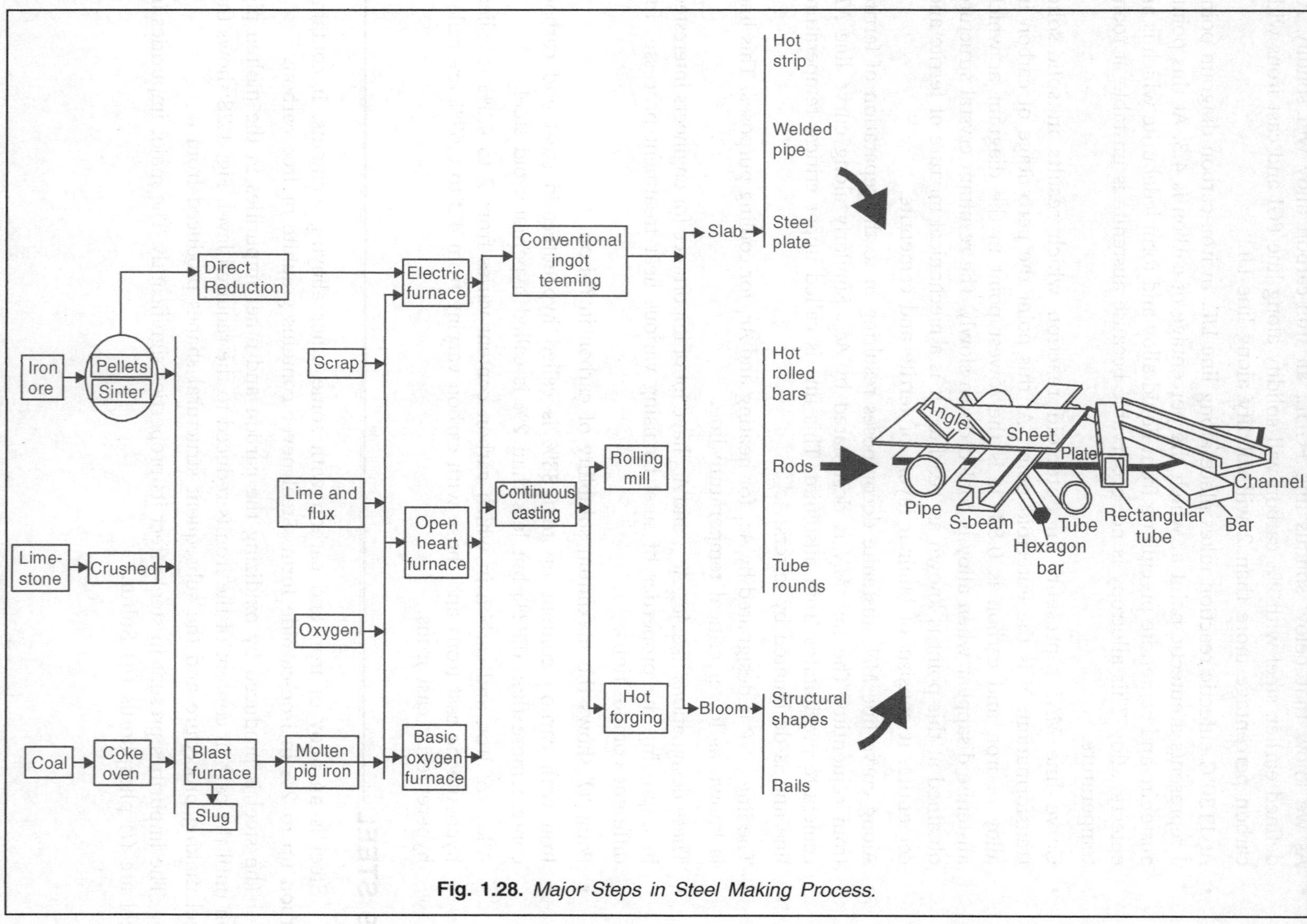

Fig. 1.28. *Major Steps in Steel Making Process.*

Phosphorus

Phosphorus is not desirable in steel. Phosphorus has a negative effect on hardness of steel and lowers its melting point due to formation of iron sulphide. Phosphorus content in steel is strictly controlled, with a maximum of 0.05% is allowed. Its presence in high carbon steel increases the brittleness whereas 0.15% of phosphorus in low carbon steels slightly improves the strength and corrosion resistance.

Sulfur

Sulfur is an impurity which forms iron sulfide. The presence of sulfur in steel makes it brittle, lowers its melting point, and causes problem during welding. This makes the steel unsuitable for mechanical working (hot or cold working). The sulfur content in steel is generally kept less than 0.05 percent.

1.9.1 Classification of Steel

There are different basis to classify the steel. These are as follows:

(*i*) **On the basis of carbon content**

On the basis of carbon contents, steel can be categorized in three groups:

(*a*) Low carbon steel

(*b*) Medium carbon steel

(*c*) High carbon steel

(*ii*) **On the basis of total amount of alloying elements**

On the basis of the total composition of alloying elements, steel is of two types:

(*a*) Low alloy steels: In these steels total alloy elements are less than 8%.

(*b*) High alloy steel: In high alloy steel, the total alloying elements are more than 8%.

(*iii*) **On the basis of principal alloying element**

On the basis of principal alloying element, steel can be classified after its name *e.g.*,

(*a*) Nickel steel

(*b*) Chromium steel

(*c*) Manganese Steel

(*d*) Cobalt Steel

(*e*) Molybdenum Steel

(*iv*) **On the basis of application**

On the basis of application mode of using steel, it can be classified as

(*a*) Structural steel

(*b*) Tool steel

(*c*) Stainless steel

(*v*) **On the basis of deoxidization**

Depending upon the degree of deoxidization of steel during its production, steels can be of three types.

(*a*) Killed steel (*b*) Semi-killed steel (*c*) Rimmed steel

(*a*) **Killed steels.** Killed steels are produced when the steel in an ingot mould remains covered with a deoxidizing material (*e.g.*, silicon) until the deoxidization is complete. These steels possess high strength and are easily forged.

(*b*) **Semi-killed steels.** If deoxidization remains incomplete, then semi-killed steel is produced. In semi-killed steel, gas evolution takes place during operations.

(*c*) **Rimmed steels.** Under this category of steel gas evolution takes place from the outer rim of the ingot during initial solidification period from liquid state.

(*vi*) **On the basis of mechanical working**

On the basis of mechanical working of steel during production, steels can be of two types:

(*a*) Hot rolled steel (*b*) Cold rolled steel

A general classification of steel is shown in Fig. 1.29.

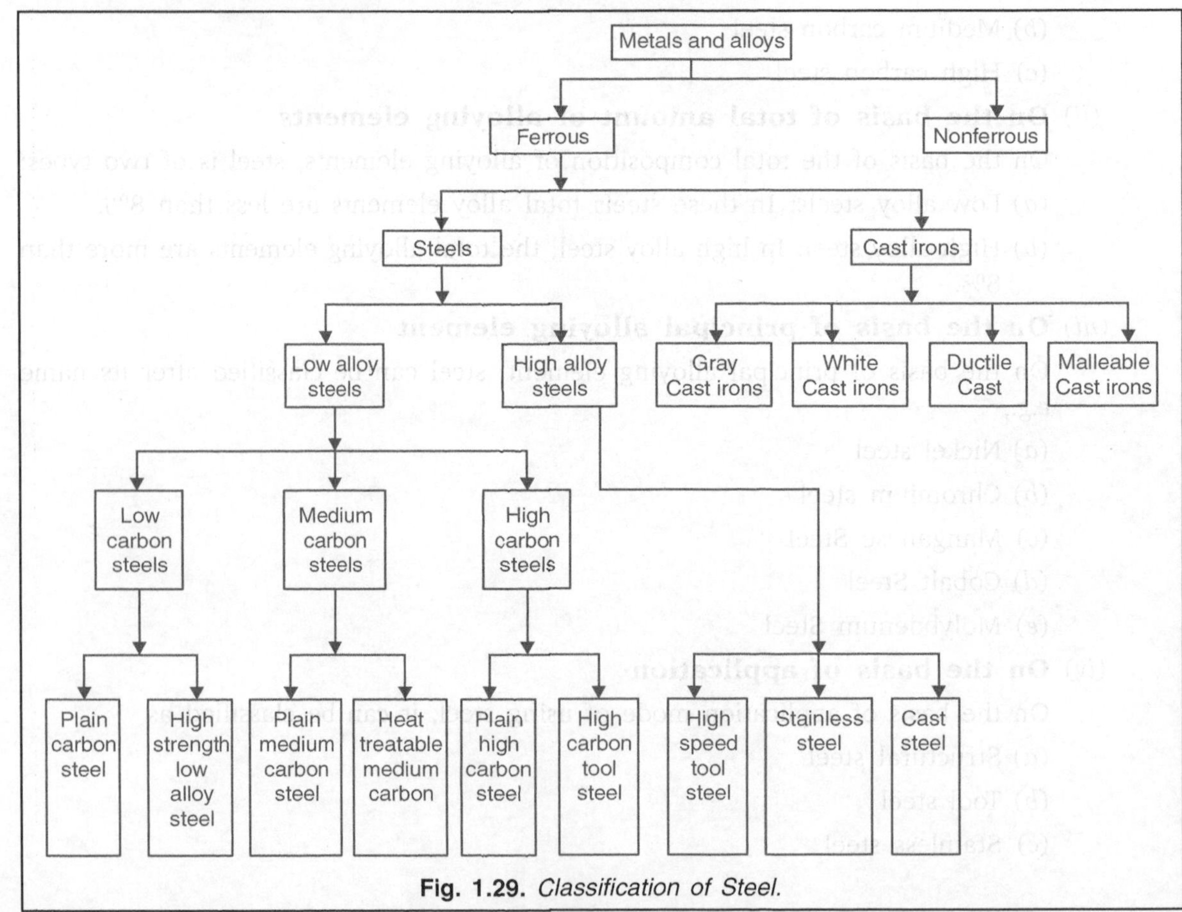

Fig. 1.29. *Classification of Steel.*

1.9.2 Plain Carbon Steel

It is an alloy of iron and carbon with varying percentage of certain alloying elements like phosphorus, manganese and sulphur. It contains maximum of 1.6% carbon along with these impurities. The maximum content of these impurities should not cross the limit as mentioned below.

Manganese	*Silicon*	*Copper*	*Phosphorus*	*Sulphur*
1.65%	0.6	0.6	0.4	0.035

As mentioned earlier plain carbon steel is of three types:

(*a*) Low Carbon Steel

(*b*) Medium Carbon Steel

(*c*) High Carbon Steel

(*a*) Low Carbon Steel

It is commercially known as 'Mild steel'. Its chemical composition is as follows:

Carbon: 0.15 to 0.3%

Manganese: 0.25 to 1.0%

Phosphorus: less than 0.05%

Sulphur: less than 0.05%

Properties:

- Good toughness
- Better strength
- Good machinability
- Good ductility
- Good weldability

Applications:

- In construction applications (skyscrapers, bridges, and buildings)
- Structural shapes (angle, channel, I-section)
- Pressure vessels (boiler plates)
- Automobiles (motor vehicle frames)
- Home utensils
- Casting, stamping.

(*b*) Medium Carbon Steel

Medium carbon steels contain carbon between 0.3 to 0.6%.

Properties:

- High tensile strength
- Improved toughness

- Good bending and torsion strength
- Better hardness
- Good machinability

Applications:

- Loco wheels
- Agricultural tools and other implements
- Automobile parts
- Springs,
- Gears, Shafts,
- Drop forgings

(c) High Carbon Steel

The carbon percentage in these steels varies from 0.8 and 1.6.

Properties:

- High yield strength
- Large torque capacity
- Higher hardness
- Reduced impact strength

Applications:

- Railway rails,
- Drills, band saws, lathe tools, chisels, punches, knives, milling cutters etc.
- Cylinder linings,
- High tension wire,
- Ball races and ball for bearings, wood cutting tools.

Limitations of plain carbon steel are:

- Plain carbon steel experiences drop in ductility and impact resistance if strengthened beyond 650 MPa.
- Poor corrosion resistance.
- Poor oxidation resistance.

Due to poor physical and mechanical properties, pure metals are not used for engineering purposes. To achieve desired properties, alloying elements are added to the pure metals. For example steel is an alloy of iron and carbon in which carbon is added as alloying element in iron. The presence of carbon in steel increases strength, hardness and other properties of the alloy.

1.9.3 Alloy Steel

Plain carbon steel with different quantities of alloying elements such as manganese, silicon, suplhur, nickel, chromium, tungsten and vanadium etc., is termed alloy steel. The answer to the limitations of plain carbon steel lies in the development of alloy steels in which alloying elements are added to achieve improvement in its mechanical properties. In order to obtain high tensile strength and hardness for large sized components at higher temperature, alloy steels are preferred. Alloy steels cost more as compared to plain carbon steels.

The alloy steels are generally classified into two classes:

(*a*) Low alloy steels: In low alloy steels the total percentage of alloying elements is less than 8%.

(*b*) High alloy steels: In high alloy steels, the total percentage of alloying elements is more than 8%.

1.9.4 The Effect of Major Alloying Elements

The effect of the major alloying elements is given in Table 1.1.

Table 1.1: Principal Effects of Major Alloying Elements in Steels

Alloying Elements	*Primary Function and its General Effect*	*Percentage*
Manganese (Mn)	Used for the deoxidization and desulphurization of liquid steel, improves strength, hardness and magnetic properties.	0.25–13
Silicon(Si)	Acts as a cheap deoxidizer and graphitizer, provides hardness and high magnetic permeability, increases resilience and strength.	0.2–5.0
Chromium (Cr)	Chromium addition results in the formation of chromium carbide which is very hard. It causes changes in the grain structure of steel so that toughness of steel increases. When it is added in small amounts (0.5–1.5%) it increases strength, impact resistance and toughness. When it is added in large amounts (12–30%), the high temperature properties and corrosion resistance of steel are greatly improved.	0.5–1.5 12–30
Nickel(Ni)	Addition of nickel increases strength of steel without loss of ductility. Nickel along with chromium improves the toughness of steel. The presence of nickel in steel improves fatigue and abrasion resistance.	2.0–20
Tungsten (W)	Tungsten is a strong carbide former which forms extremely hard and very stable carbide. It improves hardness at high temperature.	3.0–10
Molybdenum (Mo)	Molybdenum is a strong carbide forming element. It improves high temperature creep resistance.	0.2–5.0
Vanadium (V)	Vanadium is a powerful deoxidizer, a strong carbide former, inhibits grain growth and is very expensive. It presence in steel increases strength while retaining ductility.	0.15–1.0

Copper (Cu)	Improves Corrosion resistance	0.1–0.4
Aluminium (Al)	Promotes nitriding properties	0.95–1.3
Boron (B)	Increases hardenability	0.001–0.003
Titanium (Ti)	Titanium is a strong carbide former element in steel and refines the grains structure during cooling. It increases yield strength and improves weldability of steel.	0.02–0.05
Lead (Pb)	Improves machinability.	0.15–0.35
Sulphur (S)	Sulphur is a harmful element in steel and its permissible percentage is 0.05% in most of the commercial steels. Its presence is detrimental to hot forming properties *i.e.*, cracks may occur during hot rolling and forging processes.	0.05
Cobalt (Co)	Presence of cobalt in steel improves its creep resistance and strength at high temperature. Its higher contents decreases impact resistance of steel	14 (max.)

(*a*) Low Alloy Steel

Its microstructure is similar to plain carbon steels and requires the same heat treatment. The low alloy steels may contain one or more than one alloying elements. The alloying elements usually vary from 3 to 4%. These alloying elements increase the strength, toughness and hardenability.

Applications: shafts, axles, gears, and springs.

(*b*) High Alloy Steels

The percentage of alloying elements can go up to 50% in high alloy steels. Microstructure and heat treatment processes required for high alloy steels are quite different as compared to plain carbon steel. High speed steel and stainless steel are the few examples of high alloy steel.

1.9.5 High Speed Steel

High Speed Steel (HSS) was first introduced in 1900 by Taylor & White. The acronym high speed steel is due to the fact that these steels can be used at twice the cutting speed of high carbon steel for the same tool life. HSS is much superior as compared to high carbon tool steel as it retains cutting edge hardness up to a temperature of 600°C. The ability of cutting tool material to hold their cutting edge at high temperature is known as hot hardness. To improve the machining properties of high speed steel, certain alloying elements are added in these steels such as chromium, tungsten, molybdenum, cobalt, and vanadium. The addition of tungsten improves the hot hardness, vanadium increases the wear resistance and hot hardness, chromium increases the hardenability whereas cobalt improves the hot hardness. These steels have excellent heat resistance, wear resistance and high compressive strength.

HSS usually possess hardness of about 64 HRC. (AISI) American Iron and Steel Institute has established a system for classifying the high speed steel as:

(*i*) T-type HSS

(*ii*) M-type HSS

T-type high speed steels have tungsten as one of their major alloying elements and M-type has molybdenum as their major alloying element. A number preceding T and M does not carry any special purpose but it enables the user to select HSS of his choice from available types *e.g.,* M1, M2, M7, M10, T1, T15 etc.

One of the most commonly preferred type of HSS is 18-4-1 type HSS designated T1 as per AISI system, contains 18% tungsten, 4% chromium, 1% vanadium. The microstructure of HSS reveals hard, brittle, refractory carbide particles held in a lower melting point matrix. The HSS is also classified as the tungsten type and the molybdenum type. The **tungsten type** HSS contains 12 to 20% W (tungsten) along with chromium, vanadium and cobalt as other alloying elements. Whereas **molybdenum type** HSS contains 4 to 10% Mo with chromium, vanadium, tungsten and cobalt as other alloying elements. One typical composition of tungsten type HSS and molybdenum type HSS is given below.

Applications: single point cutting tools, drills, reamers, milling cutters, hobs, saws, and broaches etc.

1.9.6 Stainless Steels

The most important engineering material of everyone's life is stainless steel. It is preferred due to its excellent corrosion resistance in different working environments. The corrosion resistance of stainless steels is due to the presence of high contents of chromium. A minimum of 12% chromium is required to make the steel "stainless". The corrosion resistance is due to formation of a strongly adherent chromium oxide on the surface of the metal.

1.9.7 Cast Steel

Cast steel is not a special type of steel, but it is a term often used when the molten material is poured directly in the castings.

1.10 CAST IRONS

Cast iron is an alloy of iron and carbon in which carbon content varies from 2 to 6.67%. The term 'cast' in the Cast irons represents that these materials are to be cast (it means by using any casting processe, for details please refer any chapter on foundry in a textbook on manufacturing processes) into the desired shape rather than worked in solid state. Different properties of cast irons can be obtained by varying the amount, type, size and distribution of the various forms of carbon. The carbon in cast irons is generally present in two forms.

(*i*) Iron carbide (Fe_3C) which is called as combined iron.

(*ii*) Graphite which is called free carbon.

Besides carbon, another important element in cast irons is silicon. The silicon acts as a graphite stabilizer. Silicon presence reduces carbon content in the cast irons. Usually Carbon is found in carbide form in cast irons when silicon is not present in cast irons. Addition of silicon to cast irons increases the temperature difference between stable and metastable eutectic reactions during cooling of the cast irons. This helps in nucleation and growth of stable iron graphite eutectic reaction.

Cast irons are used extensively as casting materials due to their easy melting, good fluidity and high machinability. Cast irons have low impact resistance and poor ductility.

1.10.1 Classification of Cast Irons

Cast Irons are generally classified on the basis of the appearance of their fractured surfaces, microstructure or their properties. There are five types of cast irons. These are:

(A) Gray cast iron
(B) White cast iron
(C) Malleable cast iron
(D) Ductile cast iron
(E) Compacted graphite iron

(A) Gray cast irons

It is the most commonly used cast iron. The types of cast irons those have gray fracture appearances are known as gray cast irons. The gray fracture appearance in cast iron is due to the presence of high volume fraction of graphite flakes. The graphite flakes have a gray appearance. Gray cast irons have high silicon content as silicon promotes formation of graphite during solidification.

Gray cast iron does not possess ductility because of the presence of graphite flakes. However gray iron is least expensive of all engineering materials. It flows easily when in liquid state and thus allows to be cast easily into intricate shapes. It important properties are:

- High resistance to wear
- Good damping capacity
- Hard and brittle
- Weak in tension
- Good machinability

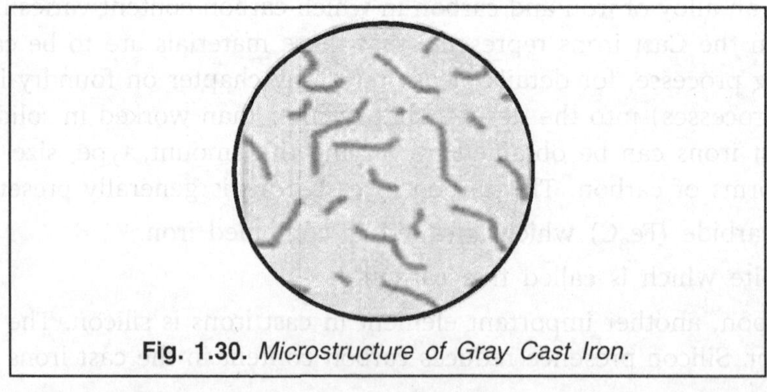

Fig. 1.30. *Microstructure of Gray Cast Iron.*

Gray cast iron when subjected to a blow or (an impact load) shatters easily. That is why hammers should never be made of gray cast irons. Figure 1.30 shows the microstructure of gray cast iron.

Applications:

- It is used in making cylinders of IC engines.
- Machine tool beds.
- Structural members loaded in compression.

(B) White cast iron

Cast irons which on fracture have a white or bright crystalline fractured surface are known as white cast irons. In white cast irons, the entire carbon content is present in the form of cementite and pearlite and no free carbon is present.

White cast iron is produced by rapidly cooling the molten cast iron. The fast solidification of molten cast iron transforms a large percentage of carbon into iron carbide instead of graphite. White appearance of the fractured surfaces results duo to presence of an iron carbide network. It is also known as chilled cast irons.

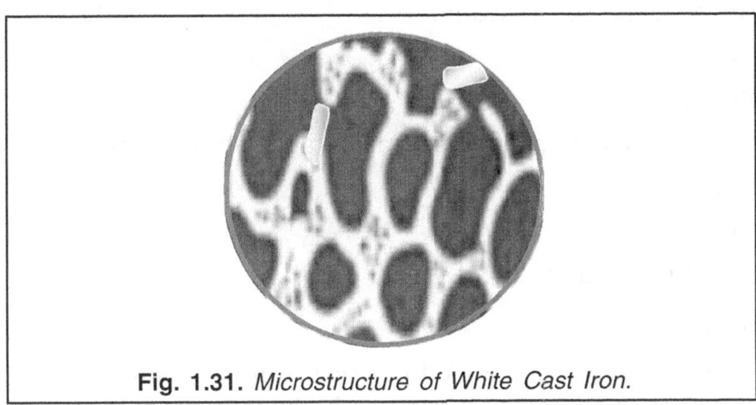

Fig. 1.31. *Microstructure of White Cast Iron.*

White cast irons can also be produced by adjusting the composition of gray cast irons in such a way that carbon and silicon contents minimized. The lower contents of silicon reduce the chances of forming graphite and thus the resulting structure has carbon in the form of iron carbide. Figure 1.31 shows the microstructure of white cast iron.

These irons are extremely hard and possess high wear resistance.

Applications:

- Brake shoes
- Rollers of rolling mills
- Plow shaves
- Wear plates.

(C) Malleable cast iron

Malleable cast iron is produced by considerable heating of white cast iron at high temperature. Usually white cast iron is heated to 800–950°C for a very long period of 30 hours or even more. During this time interval, irregularly shaped nodules of graphite form in the absence of hard and brittle iron carbide and this makes the iron 'malleable'. Figure 1.32 shows the microstructure of malleable cast iron. Its microstructure consists of free carbon nodules and ferrite.

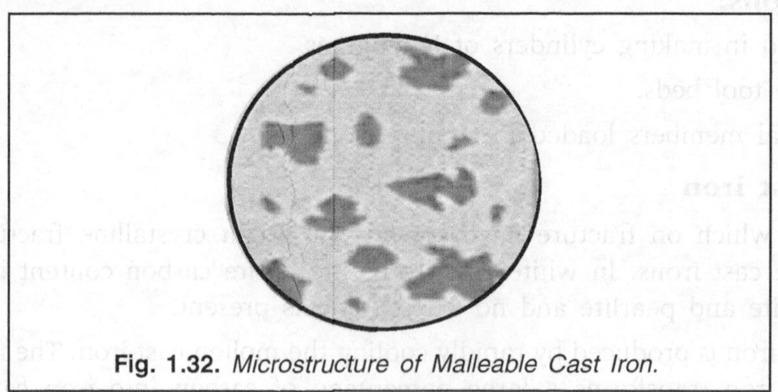

Fig. 1.32. *Microstructure of Malleable Cast Iron.*

Malleable cast iron has superior properties in comparison with gray cast iron. Malleable cast irons possess good properties such as castability, machinability, good corrosion resistance, better resistance to bending and twisting.

Applications:

- Automobile parts
- Agricultural tools and implements (tractor and plough parts)
- Gear housing
- Small tools
- Hardware and pipefitting.

(D) Ductile cast iron

Ductile cast iron is also known as nodular cast iron or spheroidal graphite cast iron. In ductile cast iron, graphite is present in the form of spheres or nodules. These nodules are not irregularly shaped as in malleable iron and are formed during solidification as shown in Fig. 1.33.

Ductile iron is produced by adding a small amount of magnesium or cerium to the molten gray cast iron as these desulphurizes iron and upon solidification, nodular cast iron is formed.

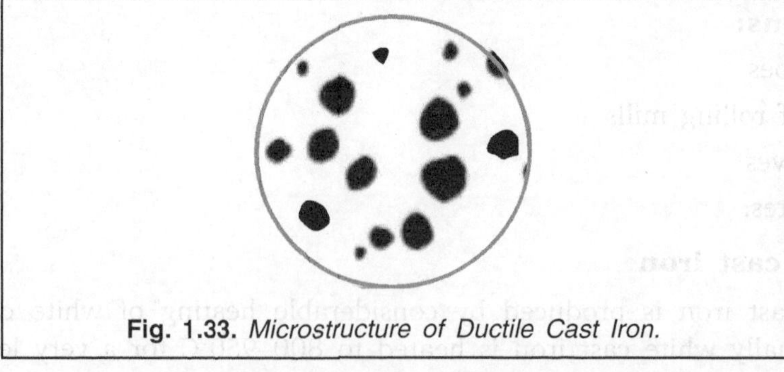

Fig. 1.33. *Microstructure of Ductile Cast Iron.*

Excellent engineering properties of ductile cast irons are attributed to the presence of spherical nodules of graphite in its internal structure. Ductile cast irons possess good fluidity, castability, good wear resistance, excellent machinability, high strength, toughness and ductility.

Applications:

- Door locks
- Crankshafts
- Piston and cylinder heads
- Pulleys
- Huge press rolls

(E) Compacted graphite iron

Compacted graphite iron has a graphite shape between those of flake graphite as found in gray cast irons and nodular graphite in ductile cast iron. Compacted graphite iron has properties in between those of gray and ductile cast irons. Figure 1.34 shows the microstructure of compacted graphite cast iron.

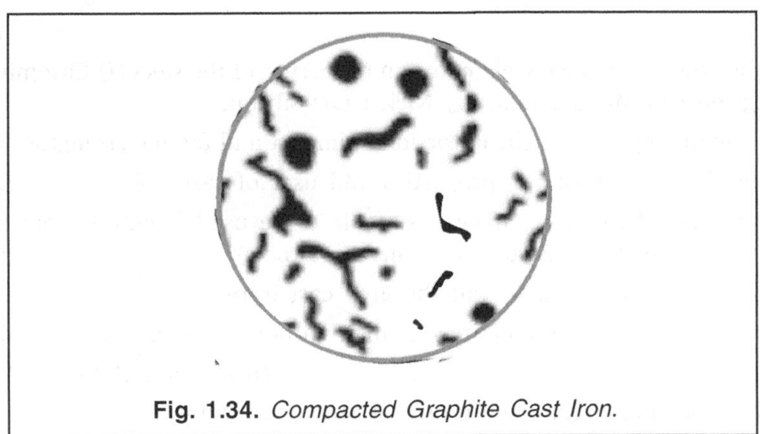

Fig. 1.34. *Compacted Graphite Cast Iron.*

The typical compositions of all the five types of cast irons are given in Table 1.2.

Table 1.2: Typical Cast Iron Compositions (% of Major Elements)

Element	*Gray Cast Iron*	*White Cast Iron*	*Ductile Cast Iron*	*Malleable Cast Iron*	*Compacted Graphite Iron*
Carbon	2.5–4.0	1.8–3.5	3.0–4.0	2.0–2.9	2.4–4.0
Silicon	1.0–3.0	0.5–1.9	1.8–2.8	1.10–1.60	1.0–3.0
Manganese	0.2–1.0	0.25–0.80	0.1–1.0	0.15–1.20	0.2–1.0
Sulphur	0.02–0.2	0.06–0.20	0.01–0.03	0.02–0.2	0.01–0.10
Phosphorus	0.05–1.0	0.06–0.20	0.01–0.10	0.02–0.20	0.01–0.03

Review Questions

1. How do engineering materials differ from other materials? *(RGPV DEC. 2005)*
2. Explain in brief the various properties of engineering materials. *(RGPV DEC. 2004)*
3. Write in details about the classification of materials. *(RGPV Dec. 2004, June 2007, and June 2008)*
4. Write about the various mechanical properties of metals. *(RGPV June 2005, Jan./Feb 2006, June 2008)*
5. Distinguish between true stress and engineering stress and between true strain and engineering strain. *(RGPV June 2006)*
6. Draw stress-strain diagram for low carbon steel and mark characteristics points. *(RGPV June 2006)*
7. What are alloy steels? Why is alloying done? *(RGPV Jan./Feb. 2006)*
8. Write Carbon percentage in cast iron and in low medium and high carbon steels. *(RGPV DEC.2005)*
9. What is the effect of following elements on properties of the steel (*i*) Chromium (*ii*) Manganese (*iii*) Tungsten (*iv*) Molybdenum (*v*) Nickel (*vi*) silicon.
10. Differentiate among tensile, shear, torsional, impact and fatigue strengths. *(RGPV DEC.2005)*
11. Give in brief the composition, properties and uses of cast iron. *(RGPV DEC. 2004)*
12. What properties of cast iron make it suitable to be used largely for machine frames, beds, base plates etc. in the construction of machine tools? *(RGPV-June 2001)*
13. Distinguish between malleable and the grey cast irons.
14. Give the composition and properties of high carbon steel and high speed steel. *(RGPV-June 2002, RGPV Nov./Dec. 2007)*
15. Define the following engineering terms (*i*) Stress (*ii*) strain (*iii*) shear stress (*iv*) Modulus of elasticity.
16. Differentiate between elastic and plastic range in case of stress-strain diagram for the low carbon steel.
17. Discuss the composition, properties and uses of cast iron. *(RGPV-Dec.2001) (RGPV-June 2007, RGPV Nov./Dec. 2007)*
18. Explain the following properties observed in stress-strain diagram: (*i*) proportionality limit (*ii*) upper and lower yield point (*iii*) ultimate strength. *(RGPV Jan./Feb. 2006)*
19. What are the different types of steel? *(RGPV June 2006)*
20. How is the tensile strength determined from the engineering stress-strain diagram?
21. Write short notes on mechanical properties of materials. *(RGPV-Dec.2001)*
22. Define fatigue limit or endurance limit. Explain a metal fatigue failure.
23. Explain S-N curve? List any four factors, which affect the fatigue strength of a metal?
24. What makes the stainless steel “stainless”?
25. Give the composition, mechanical properties and uses of cast iron, mild steel, high carbon steel and high speed steel. *(RGPV June 2005)*

OBJECTIVE TYPE QUESTIONS

1. In a tensile test on cast iron, the order of percentage reduction of area may be

(a) 15% (b) 80%

(c) 25% (d) 0%

2. For a mild steel in a tensile test, the breaking stress as compared to ultimate tensile stress is

(a) same (b) more

(c) less (d) any value

3. For a mild steel, the value of the modulus of elasticity is of the order of

(a) 380 GPa (b) 210 GPa

(c) 2100 GPA (d) 0.1 GPa

4. Poisson's ratio is defined as the ratio of

(a) longitudinal stress and lateral stress

(b) lateral stress and lateral strain

(c) lateral strain and longitudinal strain

(d) none of the above

5. The extension of the material takes place more quickly as compared to the increase in load at a point called

(a) yielding point of the material (b) elastic point of the material

(c) breaking point of the material (d) None of the above

6. The property of a material due to which it breaks with little permanent distortion is called

(a) brittleness (b) ductility

(c) malleability (d) plasticity

7. Match list A with list B and select the correct answer using the codes given below the lists:

List A	***List B***
A. Tensile strength B. Impact Strength C. Bending Strength D. Fatigue strength	1. Rotating bending machine 2. Three point loading machine 3. Universal Testing Machine 4. Izod Testing machine

(a) A4-B3-C2-D1 (b) A3-B2-C1-D4

(c) A2-B1-C4-D3 (d) A3-B4-C2-D1

8. Match list A with list B and select the correct answer using the codes given below:

List A (Mechanical properties)	***List B (Meaning of properties)***
A. Ductility B. Hardness C. Malleability D. Toughness	1. Resistance to indentation 2. Ability to absorb energy during plastic deformation 3. Percentage of elongation 4. Ability to be rolled into flat products

(*a*) A1-B4-C3-D2 (*b*) A3-B4-C2-D1

(*c*) A2-B3-C4-D1 (*d*) A3-B1-C4-D2

9. Match List A with List B and select the correct answer using the codes given below the list:

List A	*List B*
A. Ultimate strength	1. Internal structure
B. Natural strain	2. Change of length per unit instantaneous length
C. Conventional strain	3. Change of length per unit gauge length
D. Stress	4. Load per unit area

(*a*) A1-B2-C3-D4 (*b*) A4-B3-C2-D1

(*c*) A1-B3-C2-D4 (*d*) A4-B2-C3-D1

10. An idealized stress-strain curve for a plastic material is given by

(*a*)

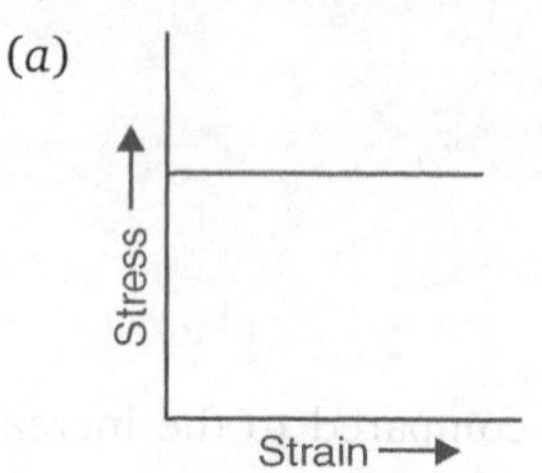

(*b*)

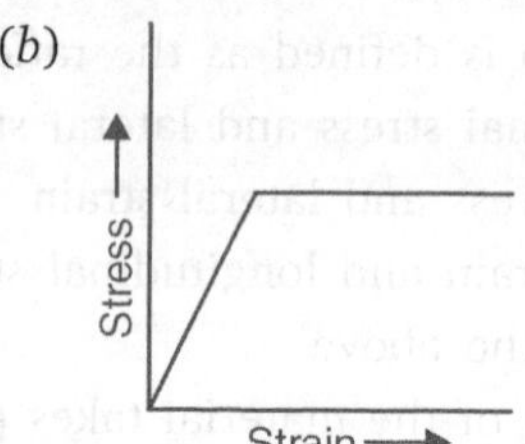

(*c*)

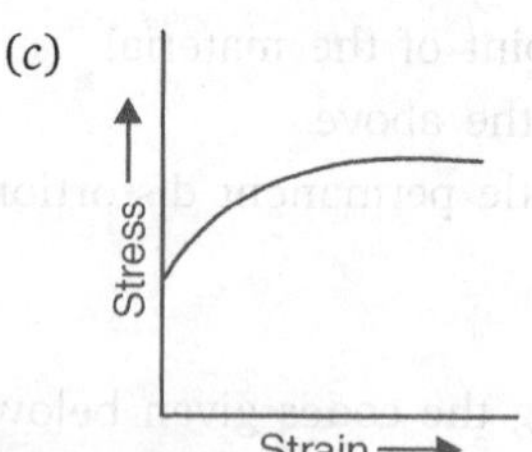

(*d*)

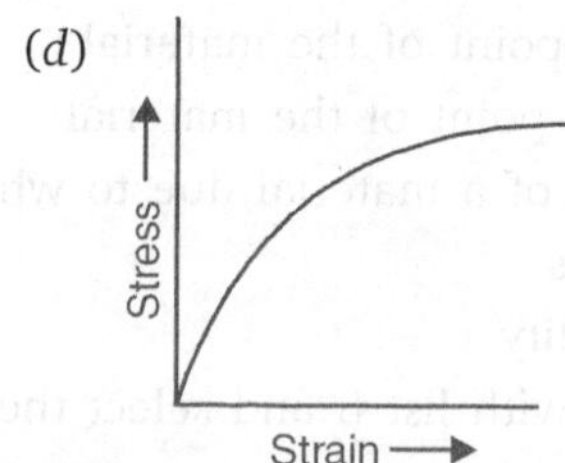

11. In a ductile material the point beyond which the linearity between stress and strain ceases is called

(*a*) Yield point (*b*) Proportional limit

(*c*) Elastic point (*d*) Ultimate limit

12. The property of material that determines its behaviour during repeated/cyclic loading is called

(*a*) Brittleness (*b*) Fatigue

(*c*) Toughness (*d*) Hardness

13. According to BIS (Bureau of Indian Standards) specifications, a plain carbon steel designated as C30 means that that the carbon content is

(*a*) 0.03% (*b*) 0.25 to 0.35%

(*c*) 0.3 to 0.5% (*d*) 0.5 to 0.6%

14. The alloy mainly used for corrosion resistance in stainless steel is

(*a*) Chromium (*b*) cobalt

(*c*) manganese (*d*) silicon

15. The most elastic material is

(*a*) rubber (*b*) glass

(*c*) plastic (*d*) steel

16. The property of a material to absorb large amount of energy before fracture is known as

(*a*) Toughness (*b*) plasticity

(*c*) ductility (*d*) shock proof

17. Hooke's law states that stress and strain

(*a*) are curvilinearly related (*b*) have unpredictable relationship

(*c*) are inversely proportional (*d*) are directly proportional

18. Poisson ratio is the ratio between

(*a*) $\frac{\text{lateral strain}}{\text{linear strain}}$ (*b*) $\frac{\text{linear strain}}{\text{lateral strain}}$

(*c*) $\frac{\text{linear stress}}{\text{linear strain}}$ (*d*) $\frac{\text{linear stress}}{\text{lateral stress}}$

19. Select the proper sequence

1. Proportional limit 2. Elastic limit 3. Yielding 4. Failure

(*a*) 2, 3, 1, 4 (*b*) 2, 1, 3, 4

(*c*) 1, 3, 2, 4 (*d*) 1, 2, 3, 4

20. Hardness as a function of carbon content is shown in

(*a*)

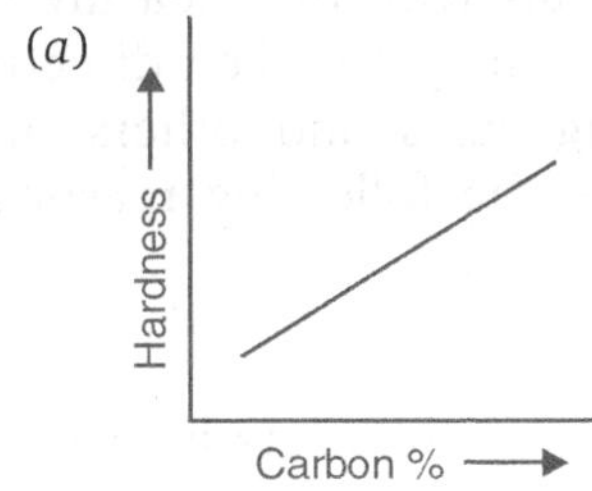

(*b*)

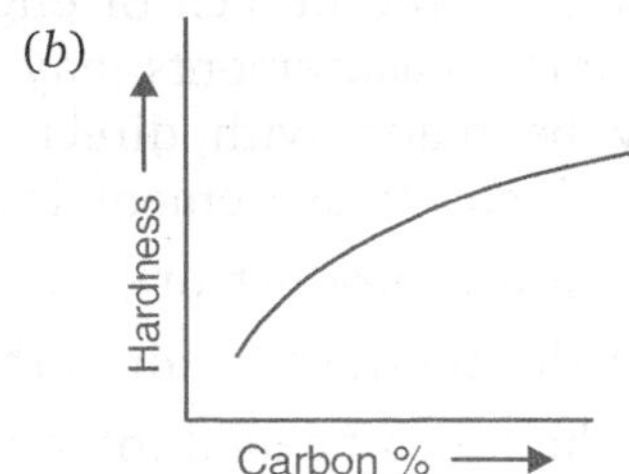

(*c*)

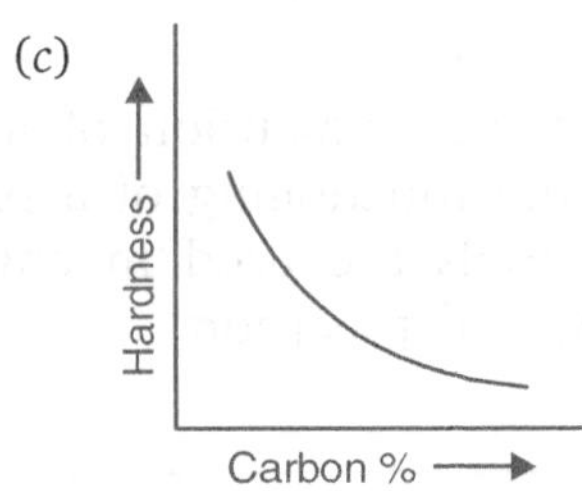

(*d*)

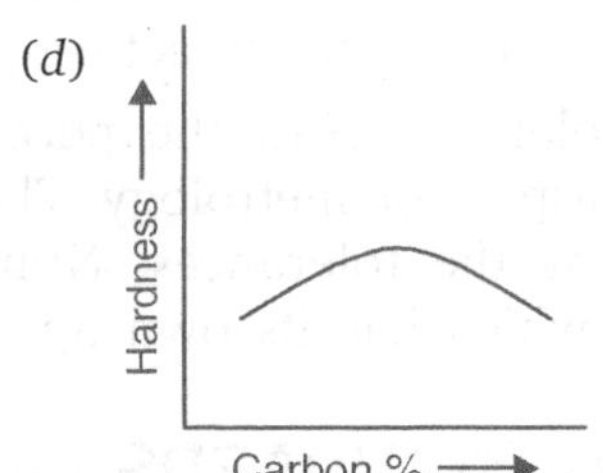

ANSWERS

1. (*a*)	**2.** (*d*)	**3.** (*a*)	**4.** (*c*)	**5.** (*a*)
6. (*c*)	**7.** (*d*)	**8.** (*d*)	**9.** (*a*)	**10.** (*a*)
11. (*a*)	**12.** (*b*)	**13.** (*b*)	**14.** (*a*)	**15.** (*a*)
16. (*a*)	**17.** (*d*)	**18.** (*a*)	**19.** (*d*)	**20.** (*b*).

Measurement and Instrumentation

2.1 INTRODUCTION

Each measurement can be written or expressed as a dimension. Dimensional measurements are the foundation of every industry whether manufacturing or service, business and commerce. Dimensional measurements are needed not only in the field of design of the machine parts or products, but the measurements play a very significant role in the success of a building design, and assembling work. Different types of measuring tools and gauges (also called gages) are used for the measurement and inspection for determining the accuracy of parts or products manufactured. Measuring instruments with digital readout devices allow to take direct reading of the values. In these instruments, the scale is replaced by a read-out device, which digitally displays the value of the quantity measured.

Metrology is one of the branch of engineering which deals with the measurement and measuring tools. Some measurements may be computed others may be taken directly. Actual measurement may be made with direct reading measuring rules and others measuring instruments. Dimensional measurement is needed because of the following reasons:

- To manufacture a product or any other object.
- To describe the product or any other object.
- To provide the much needed interchangeability of parts to the users by exercising the control over the dimensions of the parts to be produced. Controlled measurements are necessary in the manufacturing of interchangeable parts.

Dimensional tolerance (*i.e.,* the permissible variation in the dimension of a part) is another important aspect of metrology. The cost as well as the functioning of a product is greatly influenced by the tolerances. Numerous measuring tools are used in engineering metrology, each of which has its own application, sensitivity, and precision.

2.2 MEASURING STANDARDS

Earlier the dimension of a part used to be measured by comparing its dimension with a known standard. This concept of measurement has not changed ever since and still continuing. The efforts were made at the international level to adapt an uniform standard of measurement. Standards are material representation of the fundamental units as the legal and scientific bases to which all measurements must be referred.

Two common standards of length are:

1. International standard metre (SI units)
2. Imperial standard yard (British standard units).

All over the world the metre has now been adopted as a standard. The word ***'metre'*** is derived from the Greek word ***'metron'*** meaning ***measure***. Originally one metre was defined as one-ten millionth of the distance between the north pole and the equator. Today the metre is now a Systeme International d' Unites or SI unit of length, and is the international standard. In 1960, the metre was officially defined as 1650763.73 wavelengths in a vacuum of the orange light given off by electrically excited Krypton 86 (a rare gas) at 15°C and 760 mm pressure of Hg.

Thus,

$$1 \text{ metre} = 1650763.73 \text{ wavelength of Kr-86}$$

For practical viewpoint, the international prototype metre was defined as the distance at 0°C between the centre positions of two lines graduated on the polished surface of a bar of pure platinum-iridium alloy (10% iridium) of 1020 mm total length and 16 mm^2 cross-section area.

Table 2.1 shows the types of measurement and the instruments required for measuring it.

Table 2.1: Types of Measurements and Instruments Used

S. No.	*Types of Measurements*	*Instruments*
1.	Linear	Steel rule Vernier caliper Micrometer
2.	Angular	Sine bar Bevel protractor
3.	Comparative Length	Dial gauge / Dial indicator Electronic gauge Gauge blocks or Slip gauges
4.	Straightness	Autocollimator Laser Beams
5.	Flatness	Interferometry
6.	Roundness	Dial indicator Circular tracing
7.	Profile	Dial indicator Radius gauge Optical comparator Coordinate measuring machine
8.	Go-Not Go	Plug gauge Ring gauge Snap gauge
9.	Microscopes	Tool maker's microscope Optical microscope Scanning Electronic Microscope (SEM)

2.3 CHARACTERISTICS OF A MEASURING INSTRUMENT

The function of a measuring instrument is to sense or detect a parameter encountered in a scientific process or research, such as temperature, pressure, resistance, current, voltage, flow and motion etc. The measuring instrument should have the capability of detecting any changes that occur in the measured parameter.

The measuring instruments should possess the following characteristics:

Range and span

The range of an instrument defines the minimum and maximum value of a quantity that an instrument can measure. Span of an instrument is equal to algebraic difference between its upper and lower range. For example a micrometer measuring from 0 to 50 mm is said to have a range of 0–50 mm and span 50 mm.

Accuracy

Accuracy is defined as the closeness with which the reading approaches an acceptable standard value or the true value. It is numerically equal to the degree of error in the final result. In any experiment, accuracy of the measured quantity is influenced by the limits of the intrinsic error, limits of variation in indication, accuracy of the observer and the environment.

Correction

The static correction is defined as the difference between the true value and the measured value of a quantity. Correction in the instrument reading is of the same magnitude as the error but opposite in sign.

Precision

Precision is the degree to which an instrument gives repeated measurements of the same standard. Accuracy of observation is the accuracy attainable in reading the scale of an instrument. This accuracy depends on the quality of scale, the width of the pointer, the space between the pointer and the scale, the illumination of the scale and the skill of the observer.

Readability

This factor is concerned only with the pointer and scale on which the reading is to be seen. The readability is proportional to the total length of the scale, and is affected by the spacing of graduations, thickness of the pointer, graduation marks and parallax effect. Longer the scale length, the greater is the number of significant figures that may be read from it.

Threshold

If the instrument input is gradually increased from zero, there will be some minimum value below which no output change can be observed. This minimum value represents the threshold of an instrument.

Linearity

Linearity is defined as the ability of the instrument to reproduce the input characteristics symmetrically. It can be expressed by the equation:

$$Y = mX + C \quad \text{...(2.1)}$$

Where Y is the output, X is the input, m is the slope and C is the intercept.

The working range of many instruments provide the linear relationship between the output and input. It is normally desirable that the output reading of a measuring instrument is linearly proportional to the quantity being measured. The 'x' marked on (Fig. 2.1) shows a plot of the typical output readings of an instrument when a sequence of input quantities are applied to it. Normal procedure is to draw a good fit straight line through the 'x' as shown in (Fig. 2.1). The non-linearity is then defined as the maximum deviation of any of the output readings marked 'x' from this straight line.

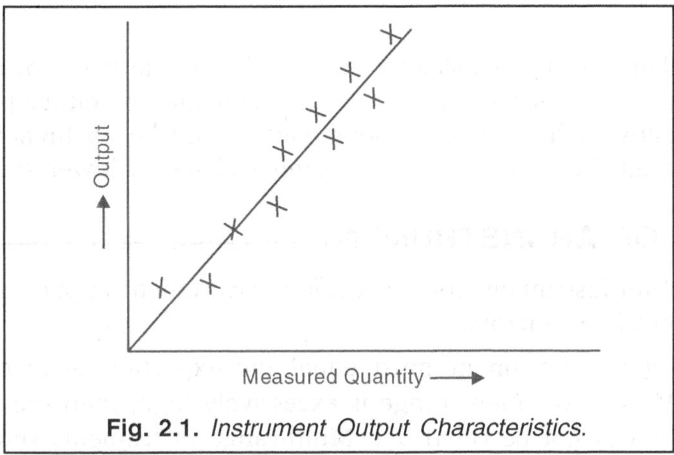

Fig. 2.1. *Instrument Output Characteristics.*

Static Sensitivity (Scale Factor)

It is defined as the ratio of the magnitude of output signal to the magnitude of the quantity being measured *i.e.,* input signal.

$$\text{Static Sensitivity} = \frac{\text{Change of output signal}}{\text{Change in input signal}}$$

Hysteresis

It is defined as the magnitude of error caused in the output for a given value of input, when this value is approached first in increasing order and then in decreasing order. This is observed mostly in mechanical devices. Its value depends upon the backlash of gears, coulomb friction, elapsed time of previous cycle, and elasticity of materials.

Backlash

It can be defined as the maximum distance (linear or angular) through which any part of a mechanical system may be moved in one direction without moving the other parts.

Drift

It is defined as the undesirable variation in the output over a period of time that is not related with changes in input variables or operating environments.

Dead zone (Dead Band)

It is defined as the maximum change through which an input signal can be varied without getting any response from the measuring instruments.

Resolution

Resolution is the ability of a measuring system to discriminate between nearly equal values.

Calibration

It can be defined as the comparison of specific values of the input and output of a measuring instrument with a corresponding standard reference.

Error

The difference between the measured value and the true value of a quantity is called static error or absolute error of measurement. The error may be either positive or negative. If the error is on positive side, the instrument reading will be on higher side whereas if it is on negative side then the corresponding reading will be on lower side.

2.4 SELECTION OF AN INSTRUMENT

The selection of an instrument for a specific measurement application depends upon a large number of interrelated factors.

- ☞ The range of an instrument must equal the expected range of a variable to be measured. If the instrument range is excessively high, then the possibility of a loss in readability can not be ruled out. Multi-range instruments should be selected for measuring the variables having wide-range.
- ☞ The sensitivity and readability of an instrument requires careful consideration. The influence of small variation in each physical quantity on the final result of the test determines the minimum sensitivity of each instrument in relation to the required accuracy of the result.
- ☞ An instrument used for an important measurement should be easy to calibrate and calibration errors must be stable during the test period.
- ☞ The expected errors of an instrument due to ambient temperature, pressure and humidity changes during the test must be either negligible or well defined.
- ☞ One of the most important factor that has a decisive influence on the selection of an instrument is the convenience and ease of reading the instrument from a particular location.

2.5 CLASSIFICATION OF THE MEASURING INSTRUMENTS

(A) On the basis of accuracy

(*i*) *Precision instruments.* Precision instruments are those instruments which have the ability to measure within an accuracy of 0.01 mm or more.

(*ii*) *Non-precision instruments.* The instruments in this category can measure to a visible line graduated on an instrument.

(B) On the basis of taking readings

(i) *Direct reading type.* The instruments of this category read dimensions directly from the workpiece. A direct measuring instrument is used to measure the actual dimensions of the parts. Vernier scales, Micrometer are examples of direct reading type instruments.

(ii) *Indirect reading type.* The instruments of this category are used to transfer the measurement from a workpiece to the direct measuring instrument such as steel rule. Calipers and dividers without any graduated scales are few examples of indirect reading type instruments.

(C) On the basis of types of measurement

Generally the instruments are categorized into four classes on the basis of the types of measurement as shown in (Fig. 2.2). These four classes are:

(i) Instruments for the linear measurement

(ii) Instruments for the angular measurement

(iii) Instruments for the taper measurement

(iv) Instruments for the surface measurement

(D) Further the instruments can also be classified as:

(i) *Line measuring instruments:* The ends of the work being measured are aligned with graduation of scale from which the dimension is read directly. Steel rules are the line measuring tools.

(ii) *End measuring instruments:* These instruments take the measurement between two ends of the workpiece as it happens in the case of a micrometer and gauge blocks, etc.

2.6 ERRORS IN MEASUREMENT

No physical quantity can be measured with perfect certainty. Errors accompany any measurement, however well it is conducted. There are always errors in any measurement. It means that if we measure some quantity and then repeat the measurement, we will almost certainly measure a different value for the second time. Only measurement related with counting are absolutely accurate (*i.e.,* integers) and all others have associated 'error' or uncertainty.

Error analysis is the evaluation of uncertainty in measurement. Experimental error is the difference between a measurement and the true value or between two measured values.

During experimentation it is important to know the difference between the precision and accuracy. Precision tells you about the quality of measurement. It does not guarantee that the measurement is correct. Whereas 'accuracy' tells you how close your measured value is from its ideal value.

One can get a fair idea about the precision and accuracy terms by considering an air-rifle shooting example. Four shooters having different skill levels compete for a slot in shooting team. At the end of their qualifying round Fig. 2.3 shows the result.

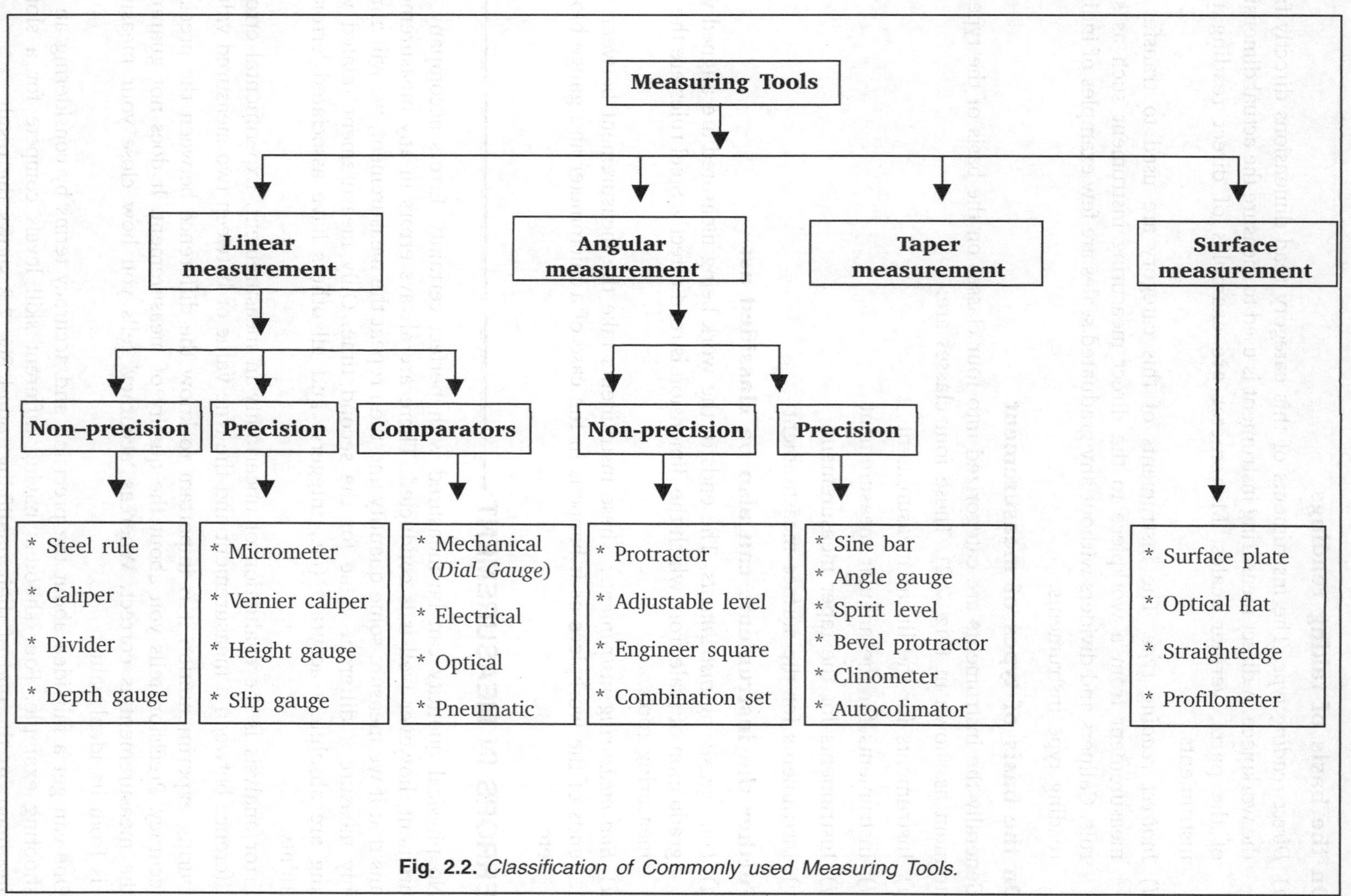

Fig. 2.2. *Classification of Commonly used Measuring Tools.*

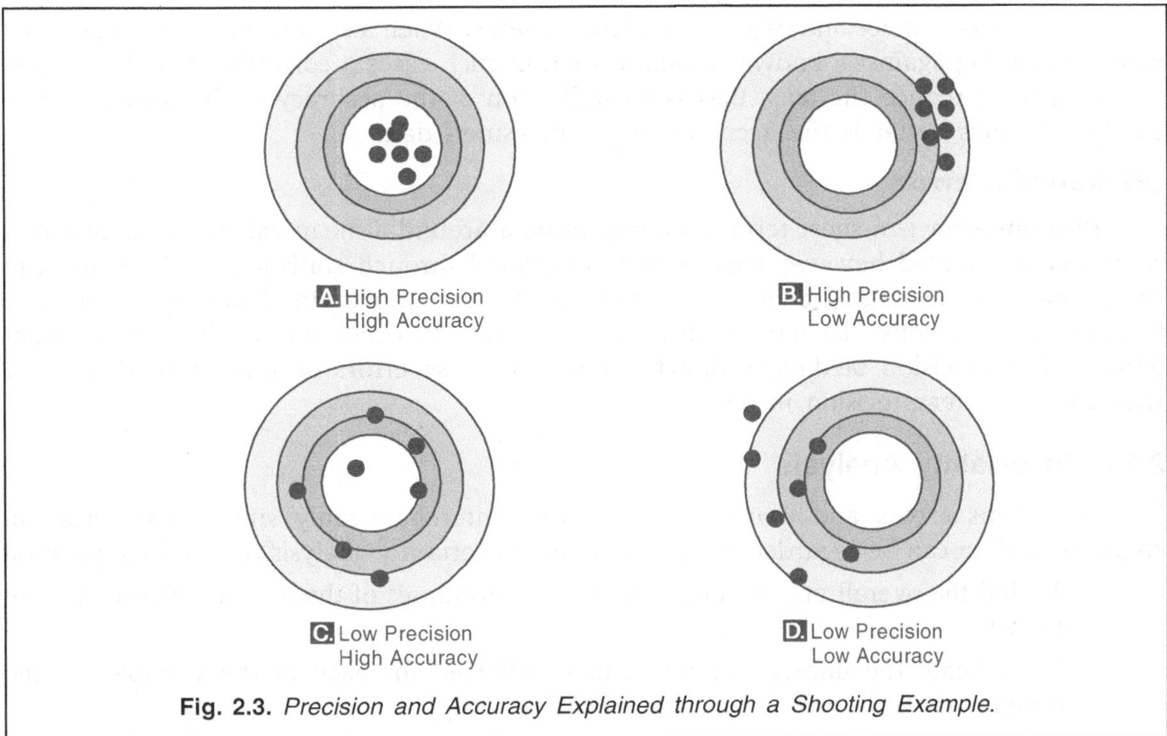

Fig. 2.3. *Precision and Accuracy Explained through a Shooting Example.*

From the Fig. 2.3 it is clear that player A is highly skilled and competent enough to get the vacant slot in the team. His all the shots have hit the innermost circle. This result indicates high accuracy as well as high precision. Player B is precise but not accurate. Similarly player C is less precise and more accurate whereas Player D is poor in both the precision and accuracy. The analogy of this example holds good with those performing experiments in the laboratories. One has to work hard and apply himself for coming close to the results of player A.

Percentage error

It is a measure of the accuracy of a measurement. It is the ratio of difference between the measured value and true value to true value.

$$\% \text{ error} = \frac{|\text{Measured value} - \text{True value}|}{\text{Truevalue}} \qquad ...(2.2)$$

2.6.1 Types of Errors

Measurement errors may be classified as either systematic or random, depending on the conditions and procedures of the measurement. An instrument can give random error in one situation and systematic in another.

(A) Systematic errors (Bias)

Systematic error is a permanent deflection in the same direction from the true value. These errors results due to faulty or improperly calibrated instruments and are difficult to detect. Systematic errors can not be analyzed statistically however these errors can be reduced by proper choice and calibration of instruments. When a systematic error is identified while calibrating against a known standard, then in such cases a correction may be applied to eliminate or reduce the bias. Bias is an indication of the accuracy of the measurement. Smaller the bias better is the accuracy of the measured data.

(B) Random error

Random error is a short term scattering of data around a mean value. Random errors cannot be eliminated however these can be evaluated through statistical analysis and can be minimized by averaging over a large number of observations. Random errors occur due to unspecified reasons like natural disturbances that may occur during the measurement process. The precision of data is directly linked to these errors as smaller random error signifies the better precision in data.

2.6.2 Uncertainty Analysis

The terms errors and uncertainty are used interchangeably since most errors in engineering data can be regarded as uncertainties. Uncertainty analysis serves three purpose

1. To find the overall uncertainty, when the uncertainties of the individual variables are known.
2. To estimate the uncertainty that can be allowed for each of the variables in the design, when the overall error is specified.
3. To select alternative methods for testing.

It is very difficult to measure any physical quantity exactly. One can measure its value with a certain range of uncertainty. This is mathematically expressed as

Measured value of quantity $X = X' \pm \Delta X$ ...(2.3)

Where X' is the best estimate of the measured variable and ΔX is called as uncertainty or error or margin of error.

This statement clearly means that the experimenter's best estimate for the quantity concerned is X and experimenter is confident that the value of this quantity X lies between $X' - \Delta X$ and $X' + \Delta X$.

2.6.3 Fractional Uncertainty

It is given by the ratio of the uncertainty divided by the average value.

2.6.4 Causes of Uncertainty

The major causes of uncertainty are:

- Problem associated in defining the variable to be measured.
- Operating characteristics of the instrument.
- Interaction of the user with the instrument.
- Interaction between the instrument and the system under measurement.
- Ambient operating conditions.
- Measurement techniques.

2.6.5 Propagation of Uncertainties

The measurement of variables such as temperature, pressure, density etc have uncertainties involved with their measurement. The uncertainties in the individual variables used to propagate through a data reduction equation in to the experimental results. Just to know the effect of uncertainties on result, one has to use the uncertainty analysis.

2.6.6 General Uncertainty Analysis

Consider the result y of an experiment is a function of x number of independent variables.

$$y = f(x_1, x_2, x_3, \ldots, x_n) \qquad \ldots(2.4)$$

These independent measured variables $x_1, x_2, x_3, \ldots, x_n$ have uncertainties $u_1, u_2, u_3 \ldots u_n$.

A first order Taylor expansion gives

$$y(x_1 + u_1, x_2 + u_2, x_3 + u_3, \ldots, x_n + u_n)$$

$$\approx y(x_1, x_2, x_3, \ldots x_n) + \frac{\partial y}{\partial x} u_1 + \frac{\partial y}{\partial x_2} u_2 + \ldots + \frac{\partial y}{\partial x_n} u_n \qquad \ldots(2.5)$$

Equation (2.4) is the data reduction equation used to find out result y from the measured values of variables x_1.

If the individual uncertainties are statistically independent, the uncertainty of y is expressed by equation (2.6)

$$u_y = \sqrt{\left(\frac{\partial y}{\partial x_1} u_1\right)^2 + \left(\frac{\partial y}{\partial x_2} u_2\right)^2 + \left(\frac{\partial y}{\partial x_3} u_3\right)^2 + \ldots + \left(\frac{\partial y}{\partial x_n} u_n\right)^2} \qquad \ldots(2.6)$$

Determining uncertainty in repeated measurements

The mean or average and the standard deviation are generally used to get the idea about the variation in repeated measurements.

Mean

For a set of n measured values for some quantity x, the mean of x is designated by symbol $\overline{x}$ and is calculated by the equation (2.7)

$$\overline{x} = \frac{1}{n}\sum_{i=1}^{n} x_i = \frac{x_1 + x_2 + x_3 + \ldots + x_n}{n} \qquad \ldots(2.7)$$

$$\overline{x} = \frac{\text{sum of measured values}}{\text{number of measured values}}$$

Standard Deviation

It describes the deviation or spread of the measured data about the mean. The standard deviation is measured by the symbol σ and is calculated using equation (2.8)

$$\sigma = \sqrt{\frac{1}{n-1}\sum_{i=1}^{n}(x_i - \bar{x})^2} \qquad ...(2.8)$$

The smaller the standard deviation less spread is the data about the mean.

2.7 SURFACE PLATE

Uses : It is used to verify the flatness of the work surface (Fig. 2.4). It is used universally to provide a reference surface. The surface plate is an auxiliary accessory. As an accessory it is mostly used in conjunction with other measuring and inspection tools and test equipments. Surface plate is used mainly for the layout work at the shop floor. Since layout lines marked on the workpiece guide the removal of material from the workpiece. For this purpose the work is placed on the surface plate and desired layout lines are marked on the workpiece using a Vernier height gauge.

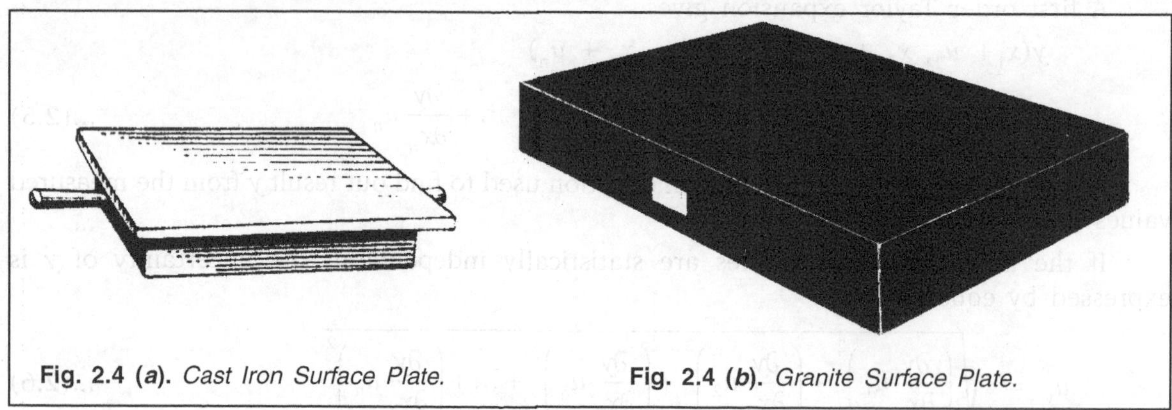

Fig. 2.4 (*a*). *Cast Iron Surface Plate.* **Fig. 2.4 (*b*).** *Granite Surface Plate.*

Construction of surface plate

A surface plate is simple, flat plate that has been accurately machined. Surface plates are usually made of cast iron or semi-steel or granite. Ribs are provided underneath of the plate to prevent it from warping. The cast iron or semi-steel surface plates suffer from various limitations. These are:

- Heavy work piece or layout accessories should be placed gently on the cast iron surface plate.
- Cast iron surface plate is prone to rust when exposed to atmospheric conditions.
- The use of rust inhibitor usually collects dirt or grit that will abrade the surface of the plate.

To overcome the problems associated with the use of cast iron surface plate, the ***GRANITE*** is preferred as the material for making the surface plate (Fig. 2.4 (*b*)). The granite surface plate has distinct advantages over the cast iron surface plate. These are:

- High degree of surface flatness
- High stability. Granite plate maintains surface flatness as there is no warpage.
- Capability to keep the flatness intact even when chipped whereas the cast iron plate tends to develop tiny crack.
- Simple maintenance, and rust proof surfaces.

Surface plates are available in numerous sizes. The size of a surface plate may vary from few square centimetres in area to one to two metres.

Care

- Surface plate should be handled properly.
- It should be kept as dry as possible and lightly oiled.
- Surface plate should be covered when not in use. This will protect the working surface from damage, which could impair accuracy.
- Surface plate should never be used as an anvil otherwise hammering will result in permanent damage to the flat working surface.
- Surface plate should be kept clean if layouts and measurements are to be carried out with maximum reliability and accuracy.
- Heavy work pieces or layout accessories should be placed gently on the cast iron surface plate.

2.8 STRAIGHTEDGE

Straightedge, as shown in Fig. 2.5 is used for measuring the straightness and flatness of the work surface. A straightedge consists of a length of steel or any other suitable material usually of narrow and deep section. It is available in various lengths varying from few centimetres to metres. Straightedge of 300 mm length or more should have ribbed cast iron pattern. The surface to be checked should be tested with the straightedge in different directions. The usual shape of the straightedge is rectangular but for precision work one edge may be formed into a knife-edge. The narrow edge of the straightedge serves as the working surface whereas the wide edge serves as its base when not in use and this also provides the rigidity to the tool.

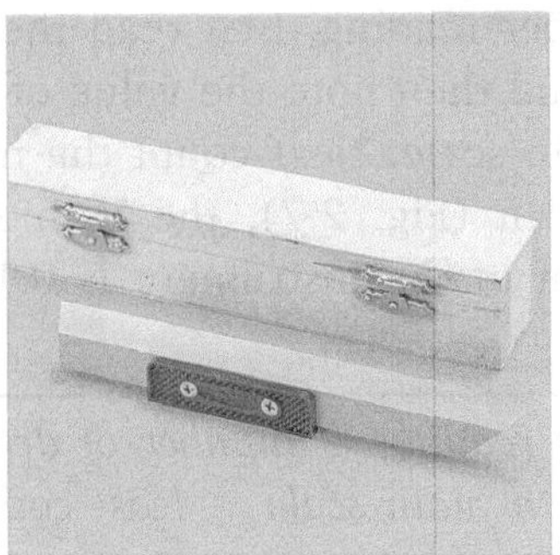

Fig. 2.5. *Straightedge.*

The straightedge and surface plate are the most commonly used tools for verifying the flatness of the work surface. The accuracy of the straightedge is assessed on the basis of the accuracy of surface plate. The working edge of the straight edge is placed on the surface to be tested for flatness or straightness and kept horizontally at the eye level watching some light source between the edge and the work surface under test. Any small discrepancy can be detected by the appearance of light between the two surfaces. Light observed at places would indicate the deviations from the straightness.

2.9 VERNIER CALIPER

Vernier caliper is one of the most important measuring instrument on the shop floor. It has got its name after "Pierre Vernier" the famous French scientist and the great mathematician. According to him, the eye cannot notice the exact distance between the two lines but can easily identify when the two lines coincide so as to form a single straight line. His 'Vernier Principle' is developed on the basis of this fact. The Vernier principle is used in calipers, micrometers, and protractors etc.

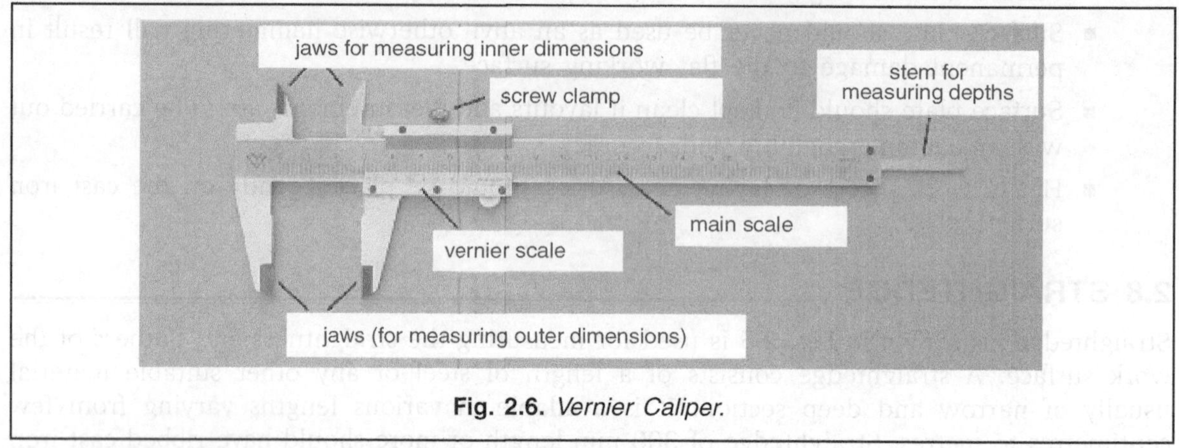

Fig. 2.6. *Vernier Caliper.*

Vernier caliper (Fig. 2.6) consists of one graduated beam popularly called the main scale and a sliding jaw with a Vernier (called Vernier scale). The Vernier scale is placed close to the main scale so that the graduation on both scales can be observed together. The spacing of the Vernier scale graduations is shorter than the spacing of the main scale graduations.

How to read the Vernier caliper?

To find out the Vernier caliper reading, first read the main scale corresponding to the left of the zero of Vernier scale and then note the value of whole mm to the left of the zero of the Vernier scale from the main scale. Next count the number of divisions of the Vernier scale coinciding the main scale. In (Fig. 2.7), the Vernier scale is divided into 10 equal divisions and thus the least count of the instrument is 0.1 mm.

Thus,

Final reading = main scale reading + number of divisions on vernier coinciding with main scale × least count

For example as shown in (Fig. 2.7) the main scale reading to the left of the zero of the Vernier scale is 15 mm. Next at the 8th division the Vernier scale coincides with the main scale. Since least count of the instrument is equal to 0.1 mm therefore the Vernier scale reading is equal to $8 \times 0.1 = 0.8$ mm.

Thus the final reading is

$$= 15 + 8 \times 0.1$$
$$= 15.8 \text{ mm}$$

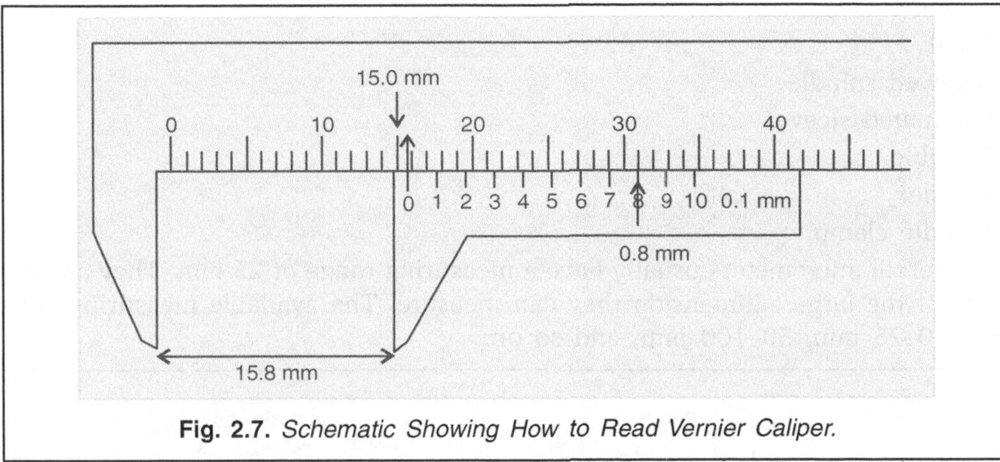

Fig. 2.7. *Schematic Showing How to Read Vernier Caliper.*

Applications

- A simple Vernier caliper may be used for outside and inside linear measurement.
- A "Cross Horn" Vernier caliper may be used to make outside, inside and depth measurements.
- It serves as a layout tool when used on a height gauge. Vernier height gauge is used to measure the height of the parts to an accuracy of 0.01 mm.
- Vernier when used on different gauges can measure the depth of holes, recesses, steps and slots.
- It is used on different protractors for the purpose of angular measurement.

2.10 MICROMETER

Micrometer is one of the most commonly used precision measuring tools in manufacturing industries. A micrometer (Fig. 2.8) is used to measure length, width, thickness, and diameter of different parts. Correct use of micrometer is essential to everyone engaged in making or inspecting the machine parts. Commonly used micrometers are of three types:

- Outside (External) micrometer
- Inside (Internal) micrometer
- Depth micrometer

Other types of micrometers are:

- Blade micrometer
- Disc micrometer
- Tube micrometer
- Screw thread micrometer
- Taper micrometer

2.10.1 Outside or External Micrometer

The outside or external micrometer is mainly used to measure external dimensions. The main parts of the external micrometer as shown in (Fig. 2.8) are:

1. Frame
2. Anvil
3. Screwed spindle
4. Graduated sleeve
5. Thimble
6. Ratchet
7. Spindle clamp.

The external micrometers usually have a measuring range of 25 mm. They are identified by the size of the largest dimension they can measure. The available measuring ranges are 0–50 mm, 50–75 mm, 50–100 mm, and so on.

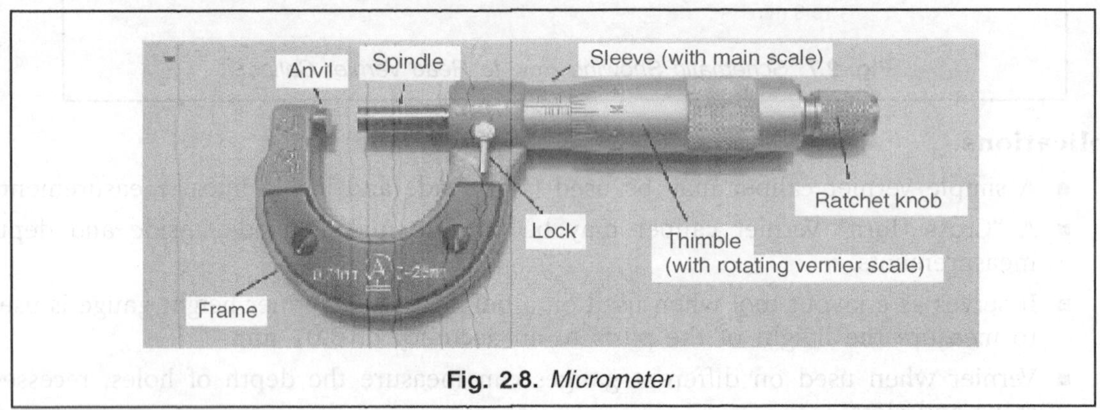

Fig. 2.8. *Micrometer.*

The micrometer spindle has 50 threads per 25 mm. Therefore 2 revolutions of the spindle are required to move the spindle by 1 mm. The graduation on the barrel or sleeve is in two parts. The set below the datum line reads in 1 mm, and the set above the line reads ½ mm. The thimble is graduated with 50 equal divisions with every fifth division numbered. If one revolution of the thimble is 0.5 mm, then each division on the thimble is equal to 0.5 mm divided by 50 or 0.5/50 = 0.01 mm. It is the value of one division on the thimble, which is the least that can be correctly read with the help of a micrometer and is called the least count. When the micrometer is closed *i.e.,* its anvil face touches the spindle face, the 0 of the main scale coincide '0' on the circumference. If it does not, then zero error exists there and the micrometer needs adjustment.

2.10.2 Using the Micrometer

The micrometer has threaded spindle and a graduated thimble and sleeve. A micrometer should be gripped by the frame leaving the thumb and forefinger free to operate the thimble. As the thimble is rotated by the thumb and fingers, the spindle moves and approaches towards the anvil. For measuring a dimension, the workpiece is held between the anvil face and the spindle face. Apply enough pressure on spindle and the anvil to yield a reliable result. Applying excessive pressure or too low pressure will result in inaccurate measurement. The micrometers should be held in both hands whenever possible. The cylindrical workpieces should be checked at least twice with measurements made 90° apart. This is to check for an out-of-round condition. All important measurement should be made at room temperature (preferably at 24–25°C). The anvils on the micrometers can be equipped with conical or ball contacts. They are used to measure inside recesses, threaded rod diameters, and wall thickness of tubes and curved sheets.

2.10.2.1 *How to Read a Micrometer*

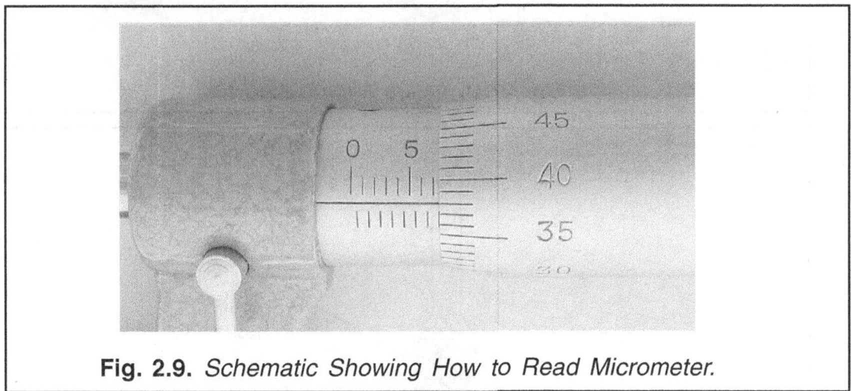

Fig. 2.9. *Schematic Showing How to Read Micrometer.*

1. First note the number of divisions in mm on the sleeve (main scale) up to 0.5 mm.
2. Next note the number of divisions on the thimble coinciding with the index line (or central line) on the sleeve.
3. Now multiply the number of divisions noted in step 2 on the thimble by the least count.

Thus the final reading will be

Final reading = sleeve reading + No. of divisions on thimble coinciding with central line on sleeve × Least count

For example as shown in (Fig. 2.9) the number of divisions on the sleeve is 7 whereas 38 division on the thimble is coinciding with the central line on the sleeve. Thus the final reading is

$$\text{Final reading} = 7 + 38 \times 0.01$$
$$= 7.38 \text{ mm.}$$

2.10.3 *Inside (Internal) Micrometer*

The inside micrometer (Fig. 2.10) also operates on the principle of external micrometer. The internal micrometer is used for the internal measurements such as measuring the bore diameter of a cylinder to an accuracy of 0.01 mm. A handle is provided to hold the instrument into regions where holding the instrument directly would be difficult. The measuring range of the inside micrometer can be changed by attaching the extension rods. Take at least two readings 90° apart to obtain the size of a hole or bore. The readings should be identical. Inside micrometer do not have a spindle lock. Therefore to prevent the spindle from turning, the adjusting nut should be maintained slightly tighter than normal.

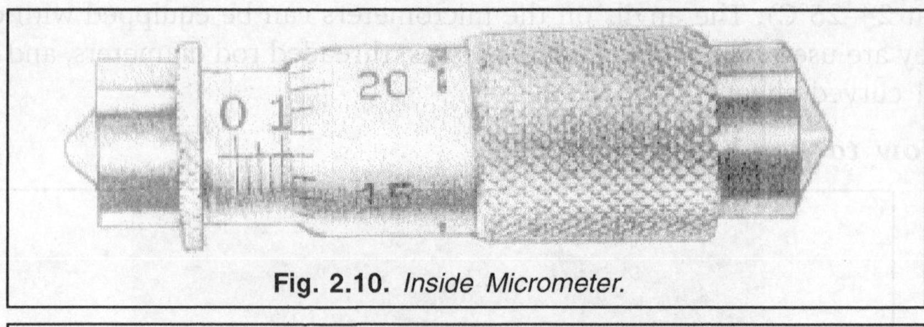

Fig. 2.10. *Inside Micrometer.*

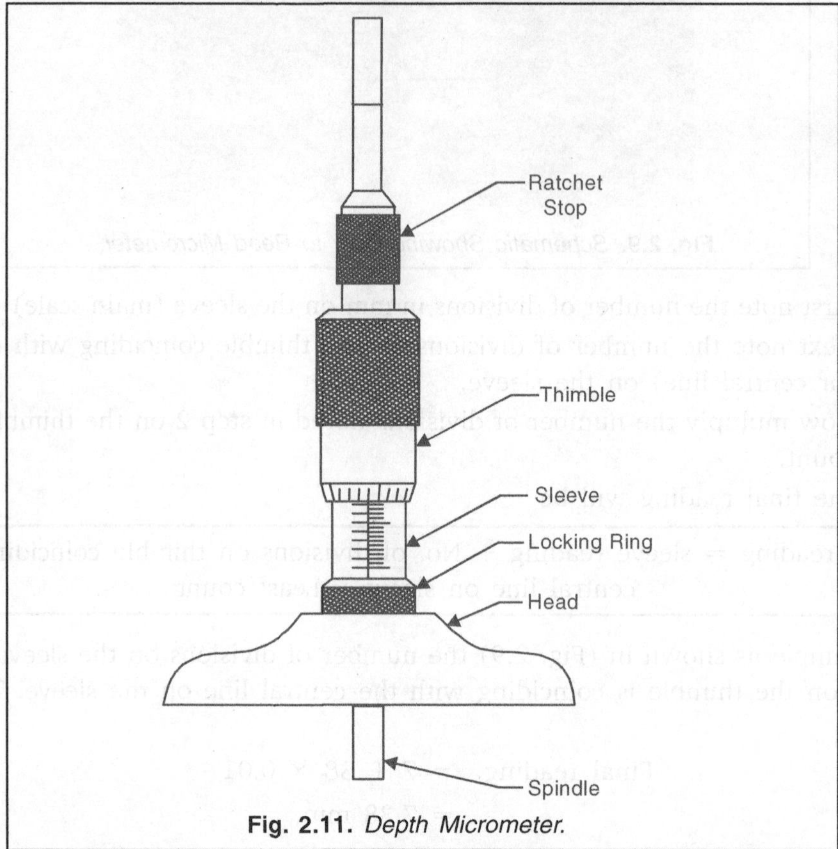

Fig. 2.11. *Depth Micrometer.*

2.10.4 Depth Micrometer

A depth micrometer (Fig. 2.11) is used to precisely measure depths of holes, grooves, shoulders and recesses. The depth micrometers are provided a set of inter-changeable rods to accommodate different depth measurements. The principle of measurement is similar to that of an outside micrometer. Depth micrometers are available in measuring ranges of 50 to 75 mm, 75 to 150 mm, 150 to 600 mm, 650 to 1200 mm.

2.11 DIAL GAUGE

A dial gauge or dial indicator is the most commonly used mechanical type comparator. Comparators are used to transfer or compare a measurement with a known standard with a high degree of accuracy. Dial gauge is used to make a comparative (relative) measurement between the work-piece and gauge block from the same reference plane.

The two main functions of a dial indicator are:

- To measure a length. This is the distance between a standard dimension to which the dial indicator is set and the length of the part being measured.
- To determine the errors in geometrical form (Ovality, out of roundness, lobed form, taper etc.) or positional errors of surface (in parallelism, squareness, alignment etc.).

The dial gauge consists of a small clock and a plunger (stylus probe). Very slight upward pressure when applied on the plunger (stylus probe) moves it upward. This upward movement (or linear movement) of the plunger is converted into the rotary movement of an indicator on the dial gauge. The dial is divided into 100 divisions. One complete revolution of the indicator corresponds to 1 mm linear travel of plunger. Thus each division on the dial indicates a movement of 0.01 mm. The indicator is set to zero at a certain reference surface, and the instrument or the surface to be measured is brought into contact with the plunger. The movement of the indicator is read directly on the circular dial.

The dial gauge is a sensitive instrument. The mechanism of a dial indicator is shown in the (Fig. 2.12). The internal mechanism consists of small gear-trains or levers. The movement

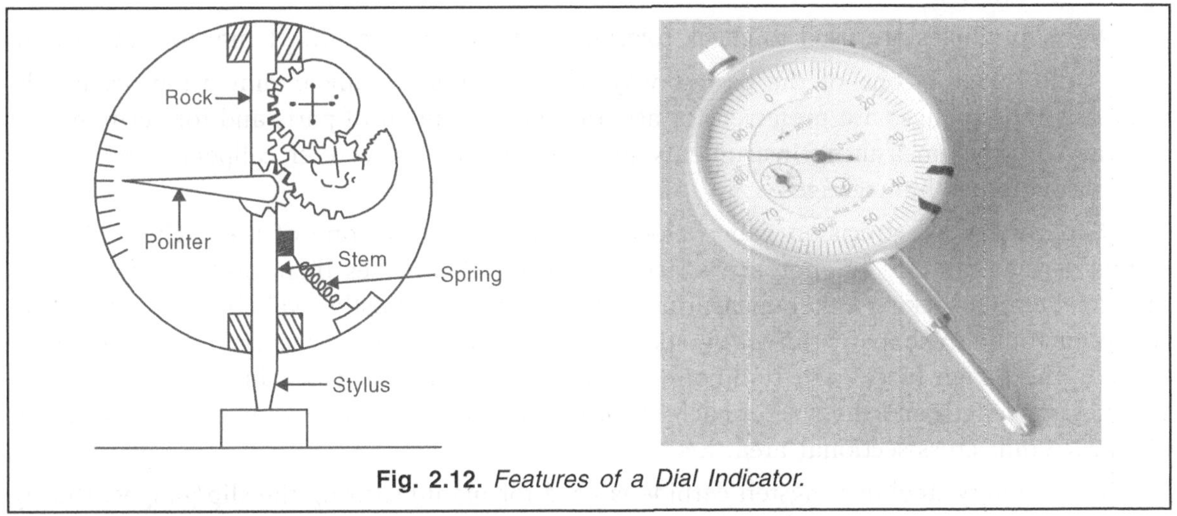

Fig. 2.12. *Features of a Dial Indicator.*

of the plunger is transmitted by means of a toothed rack through a compound gear train to an indicator, which rotates across the face of a stationary graduated dial. The desired measuring pressure is provided by small spring provided in the mechanism.

Applications of dial gauge

1. *It may be used to check the out of roundness of a part on a lathe machine.* The dial gauge is clamped in a holder which in turn is mounted in a lathe tool holder. Any roundness variation on the workpiece is detected by slowly rotating the workpiece and bringing this in contact with the dial gauge contact point (plunger). Any roundness variation causes the contact point of the indicator to move upwards and downwards.
2. It may be used to test the holes for true size and to check other surface irregularities such as concentricity (out of roundness). Such type of dial gauge is called dial bore

gauge. It is used for checking engine block cylinders for size, taper, bell mouth and ovality etc.

3. The dial gauge is preferred extensively for inside, outside and depth measurement. Such type of dial gauge is called dial caliper gauge.
4. Dial indicating snap gauges are used for determining the acceptance of a workpiece as per the specified tolerances.

The dial gauges (indicators) of several designs are available for use as portable or benchtop units. The dial gauges are required for the different measurement ranges. Sensitive dial indicators may have a range of 0–1 mm or 0–5 mm. Dial gauges with least count of 0.01 mm may have the range of 0–5 mm and 0–10 mm. The dial gauges possess the accuracy of the order of 0.01mm. The dial gauges with digital readout are also available.

2.12 SLIP GAUGES

- ☞ Also known as **Gauge Blocks.**
- ☞ The slip gauges or the gauge blocks are also called **Johansson gauges** after their originator.
- ☞ Slip gauges are used to check tolerances in the range of 0.001 mm to 0.0005 mm.

A slip gauge is one of the easiest way of measuring the linear dimension accurately. Generally these gauges are preferred for accurate measurement of parts and for verifying the accuracy of other measuring instruments such as micrometer, vernier calipers, comparators and various other limit gauges.

Slip gauges are precision gauge blocks. Gauge blocks are one of the highest physical standards that can closely approach exact dimension. This makes them useful as measuring instrument for checking other measuring tools. These are mostly available in rectangular, square or round in shape with two parallel diametrically opposite measuring sides accurately lapped. The gauge blocks are commonly available in sets. Each block in a set has a given thickness or length and the size of each block is marked on it. These gauges mostly come in 32mm × 9mm cross-sectional area.

High quality steel or tungsten carbide is used for manufacturing the slip gauges. During their manufacture, they are put through many heating and cooling cycles that stabilize their dimension. For the smooth functioning of the slip gauge, its surface should be extremely smooth and flat. The slip gauges are almost always used in combination with what is called gauge block stack. The secret of slip gauge use lies in the ability to place two or more gauge blocks together in such a way that the air between them is displaced. The space between the wrung slip gauge is called wring interval. Once this task is accomplished atmospheric pressure will hold the stack together. Different slip gauges are wrung together in order to establish a definite dimension. Extreme care is needed in handling the slip gauges.

Slip gauges are usually of three types (Table 2.2).

Table 2.2: Types of Slip Gauges

Types of Gauge Blocks (Slip Gauges)	*Accuracy*
Master blocks	0.00005 mm
Inspection blocks	0.0001 mm
Working blocks	0.0002 mm

Master and inspection blocks are used in air-conditioned laboratories. Some sets include two additional wear blocks. The wear blocks are required in those cases where direct contact is made. The usual thickness of wear blocks is "0.050 and 0.100". Generally the slip gauges are manufactured into five grades of accuracy as shown in Table 2.3.

Table 2.3: Classification of the Slip Gauges as Per Accuracy

Accuracy Grade	*Purpose*
Grade I slip gauges	On shop floor in producing components, tools and gauges.
Grade II slip gauges	Rough work
Grade 0 slip gauges	Checking the accuracy of other gauges
Grade 00 slip gauges	Checking the accuracy of other gauges
Calibration grade slip gauges	Calibration of the comparators, dial gauges and various other instruments

A common gauge block set contains 81 to 88 blocks ranging from 0.5 to 100 mm in thickness. The measuring range of the sets varies from "25 to 84". A M88 Grade I slip gauge set contains 88 blocks. The detail of this set is shown in Table 2.4.

Table 2.4: Details of a M88 Grade I Slip Gauge Set

Size of Range	*Increments*	*Number of Blocks*
0.5, 1.00, 1.0005	—	3
1.001–1.009	0.001	9
1.01–1.49	0.010	49
1.50–9.50	0.500	17
10.0–100	10.00	10

Applications:

1. Slip gauges are used in metrology laboratory, tool room and machine shop for the calibration of precision measuring instruments.
2. Slip gauges are used in setting sine bars for establishing angles.
3. These gauges are used to set other measuring instruments such as snap gauges.
4. These are used as auxiliary measuring system on milling machine.

2.13 COMBINATION SET

Combination set is used in fitting and machine shop. It is a combination of a square head, center-head, bevel protractor, spirit level and graduated rule. A schematic diagram of a combination set is shown in Fig. 2.13.

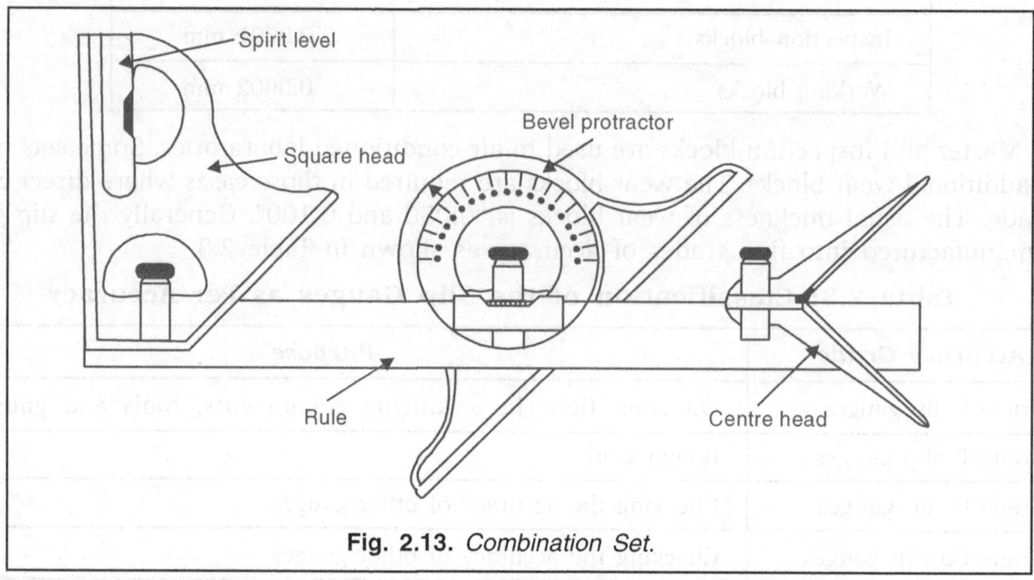

Fig. 2.13. *Combination Set.*

The square head slides on the graduated rule and can be located at any position. This feature makes the tool useful for layout as the square head can be set according to the rule graduations. The square head has two edges. One edge is at right angle (90°) to the rule whereas the other edge makes an angle 45° with the rule. This feature makes it possible to measure and mark the angle from 45° to 90° with respect to rule.

Center head may be used to extend a line around a corner. It has two arms set at right angles to each other and is positioned on the graduated rule in such a manner that this angle is divided in two by the edge of the rule. Center head is used to determine the center of a round workpiece.

The angular measurement can be accomplished by the bevel protractor which is attached to the rule. The bevel protractor can be moved along the rule and locked in any position. The bevel protractor has a flat base permitting it to rest squarely on the work surface. A spirit level is also positioned in a combination set. Its purpose is to check the level of the work with reference to scale edge.

2.14 SINE BAR

The purpose of a sine bar is to measure the angles. The bevel protractor measures the angles directly whereas the sine bar measures the angles indirectly. The sine bar (Fig. 2.14) consists of a bar of stabilized, highly finished chrome steel, resting on two cylindrical rollers, the centres of which are 100 mm, 200 mm, 250 mm or 300 mm apart depending upon the requirements. The sine bar may come in different styles, shapes or lengths, but its purpose and principles are same. The sine bar has the capability to reduce all angles in terms of a right angle, the simplest form of trigonometrical solution.

The Sine bar is set using slip or gauge blocks in such a way that the two plugs of the sine bar rest on slip gauges wrung to such heights that the bottom of the sine bar rests perfectly over the workpiece surface whose inclination is to be found.

$$\text{Sin } \theta = (h_1 - h_2)/L \qquad ...(2.9)$$

where θ is the angle of inclination of the workpiece surface to be determined?

h_1 and h_2 are the heights built up by slip gauges

L is the length of the sine bar.

2.14.1 Theory of Sine Bar

The sine of an angle is defined as the ratio of the side opposite the angle whose sine is to be found over the hypotenuse of the right triangle. As shown in (Fig. 2.14), the sine of angle X is side x over side z, the hypotenuse; the sine of the angle Y is y/z. Angle Z is the right angle, which makes triangle XYZ a right angle. Sine bar will always be the hypotenuse in any solution as long it is represented by side z. The length of the hypotenuse, or side z, is always known since it can be only 100 mm if a 100 mm sine bar is used, or 250 mm if 250 mm sine bar is used. The question is how to find out the angle between the hypotenuse and the side y, or the angle X. Since we know side x or one can find it by using slip gauges equal to it.

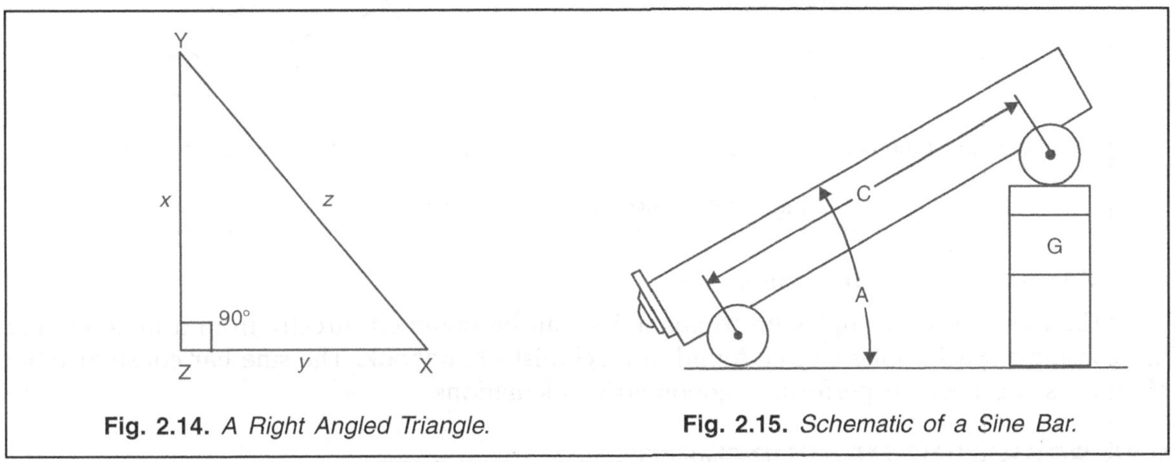

Fig. 2.14. *A Right Angled Triangle.*

Fig. 2.15. *Schematic of a Sine Bar.*

2.14.2 How to Use Sine Bar?

Sine bar as shown in Fig. 2.15 is used in different set-ups and inspection problems. Its important applications are as follows:

- In checking the flatness of surfaces
- In checking of bevel gears
- In checking of a taper plug gauge
- In checking a taper key.

Fig. 2.16 illustrates the few applications of a sine bar. The measurement of angles by sine bar involves placing the part on an inclined bar or plate and adjusting the angles by placing slip gauges on a surface plate. After the part is placed on the sine bar, a dial indicator is used to scan the top surface of the part. Slip gauges are added or removed as necessary until the top surface is parallel to the surface plate. The angle on the part is then calculated using trigonometric relationships.

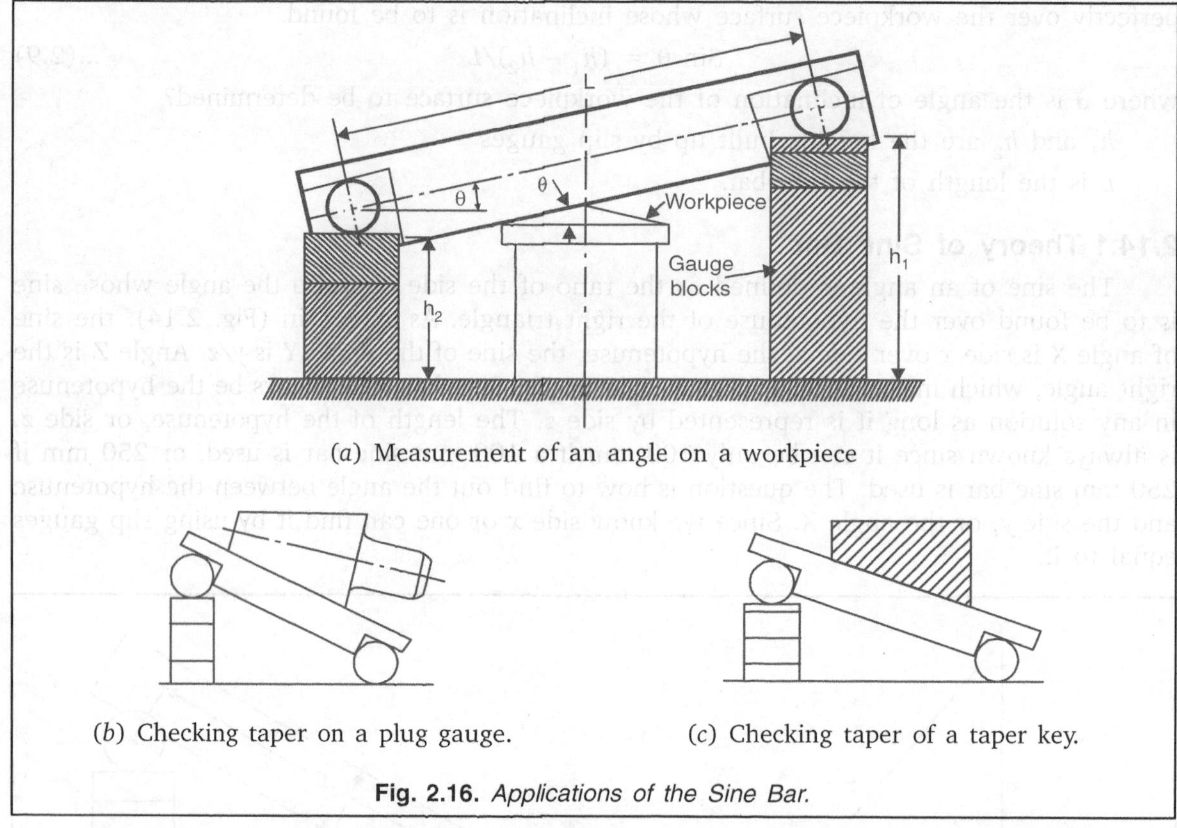

(*a*) Measurement of an angle on a workpiece

(*b*) Checking taper on a plug gauge.

(*c*) Checking taper of a taper key.

Fig. 2.16. *Applications of the Sine Bar.*

Sine bar constant tables

The elevations for angles up to about 55° can be obtained directly from a table of sine bar constants. Such tables can be found in machinist's handbook. The sine bar constant table eliminates the need to perform trigonometric calculations.

2.15 PRESSURE MEASUREMENT

2.15.1 Introduction

Pressure is defined as the force exerted by a fluid per unit area, measured over a given point or over a surface. The basic unit of pressure is Pascal and denoted by 'Pa'. When a fluid applies a force of 1N over an area of 1 m^2, then the pressure is equivalent to 1 Pascal.

i.e., $$1\text{Pa} = 1\ \frac{N}{m^2}$$

Since Pascal is a very small unit, so for the purpose of calculation and other derivation its higher value is generally expressed in *kPa* (kilopascal) or *MPa* (megapascal).

$$1\ kPa = 1\ kilopascal = 1000\ Pa = 1000\ \frac{N}{m^2}$$

$$1\ bar = 100\ kPa = 10^5\ kPa = 10^5\ Pa = 10^5\ \frac{N}{m^2}$$

The other important pressure units are:

$$1\frac{kgf}{cm^2} = 9.807\frac{N}{cm^2} = 9.807 \times 10^4\ \frac{N}{m^2} = 9.81 \times 10^4\ Pa$$

$$1\frac{kgf}{cm^2} = 14.223\ psi$$

As we know that the atmosphere around us has air at some pressure therefore zero pressure point can only be defined with reference to some datum. Atmospheric pressure provides most suitable reference for pressure measurement. The atmospheric pressure is due to the weight of the earth's atmosphere. The standard value of atmospheric pressure has been taken as $P_{atm} = 101.3\ kPa$. The pressure of atmosphere does not remain constant at a given location due to changes in climatic conditions.

There are three categories of pressure measurements namely:

(*a*) Absolute pressure

(*b*) Gauge pressure

(*c*) Differential pressure

(*a*) Absolute pressure

Absolute pressure is the difference between the pressure at a particular point and the absolute zero of pressure. Absolute zero of pressure will exist only in a completely evacuated space (perfect vacuum).

(*b*) Gauge pressure

Mostly pressure measuring devices are calibrated to read zero in the atmosphere and so these devices indicate the difference between the absolute pressure and local atmospheric pressure. This difference is called the gauge pressure. Simply speaking pressure above the atmospheric value is called gauge pressure. Pressure measured below the atmospheric pressure is called vacuum pressure. Vacuum gauges are used to measure vacuum pressures that indicate the difference between the atmospheric pressure and absolute zero pressure. Absolute, gauge and vacuum pressures are all positive quantities. Relationship between pressures stated for different references is shown in (Fig. 2.17).

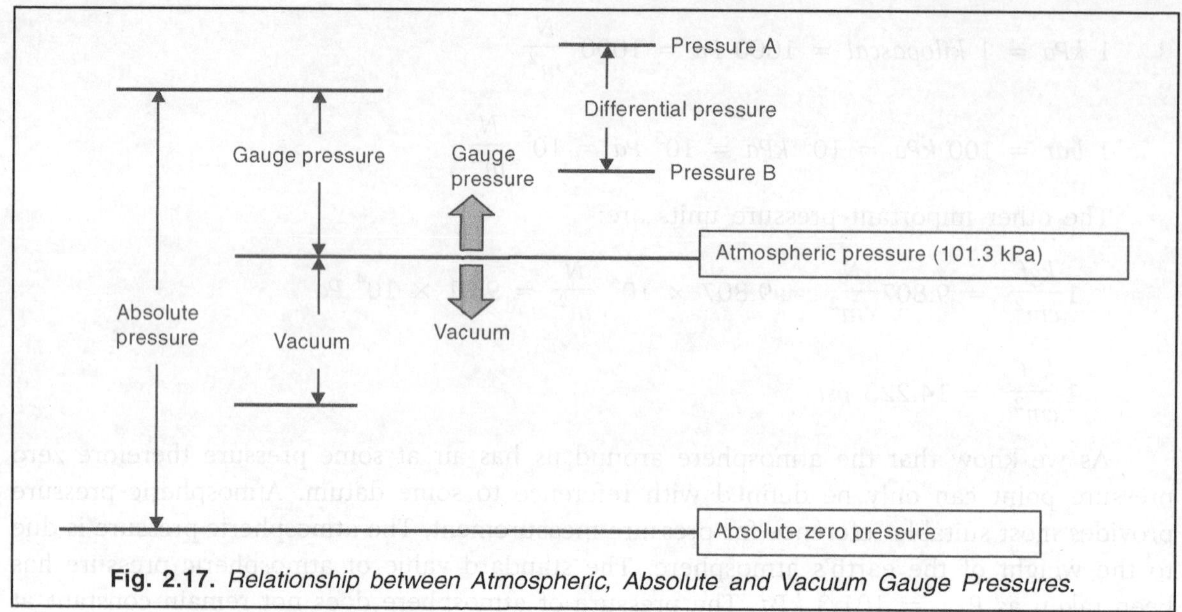

Fig. 2.17. *Relationship between Atmospheric, Absolute and Vacuum Gauge Pressures.*

$$P_{gauge} = P_{abs} - P_{atm} \quad \text{...(2.10)}$$

$$P_{vaccum} = P_{atm} - P_{gauge} \quad \text{...(2.11)}$$

(c) Differential pressure

When the pressure measuring device measures the difference between the two unknown pressures, neither of which atmospheric pressure, then measurement is known as the differential pressure.

2.15.2 Pressure Measuring Instruments

Pressure measuring instruments are classified in two categories.

(*i*) Gravitational type

(*ii*) Elastic type

2.15.2.1 *Gravitational Type Pressure Measuring Instruments*

In this category of pressure measuring instruments, the pressure exerted by the fluid is balanced by a liquid column or its weight. The important instruments in this category are: manometers, McLeod gauge etc.

(*a*) Manometer

Manometer is commonly used to measure pressure and vacuum. It is used to measure small and moderate pressure differences. A manometers is mainly consists of a glass or plastic U-tube containing one or more fluids such as mercury, water, alcohol or oil. Fig. 2.18 shows a glass U-tube manometer partially filled with liquid and both ends initially open to the atmosphere.

When a gauge pressure P is to be measured, it is applied to the top of one of the columns and top of the other column remains open. When the liquid in the tube is mercury, then the pressure is expressed in mm of mercury (mm of Hg). The rise of liquid gives the pressure head at that point. Pressure P_1 is given by equation (2.12)

$$= pgh \ N/m^2 \qquad ...(2.12)$$

The difference in pressure $P_1 - P_2$ is given by the basic equation of fluid statics.

$$P_1 - P_2 = gh\ (p_2 - p_1) \qquad ...(2.13)$$

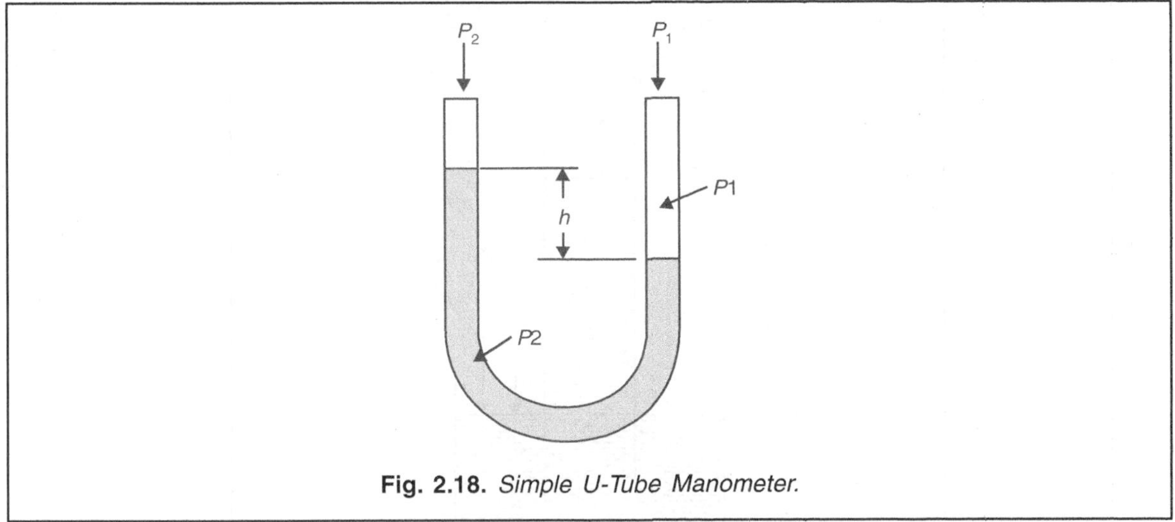

Fig. 2.18. *Simple U-Tube Manometer.*

When the difference in pressure is small, the displacement of liquid column h is amplified by inclining one or both limbs of the manometer or by using a combination of liquids.

(*b*) Barometer

Barometer is used to measure the atmospheric pressure. That is why the atmospheric pressure is also known as barometric pressure. A simple barometer can be constructed from a glass tube that is closed at one end and open at the other. The length of the tube should be greater than 762 mm. This tube is first filled with mercury and then inverted into a mercury container that is open to atmosphere as shown in (Fig 2.19). When the tube is inverted in a container, the mercury in the tube will drop by a certain amount, thus creating a vacuum at the top end of the tube so the pressure at A is zero where as the pressure at point B is equal to atmospheric pressure. The height of the column as measured in (Fig. 2.19) and expressed in mm of mercury (Hg) will be proportional to atmospheric pressure.

Considering the balance of mercury column of height h due to the forces acting over it, we can write:

$$P_{atm} = \rho gh \qquad ...(2.14)$$

As the atmospheric pressure varies, the height of the mercury column varies. So the height of measuring column can be calibrated to measure the atmospheric pressure.

Now let us find out the height of a mercury column for one atmospheric pressure.

$$P_{atm} = 1atm = 101.3 \text{ kPa}$$

From above equation (2.14) we get

$$h = \frac{P_{atm}}{\rho g} = \frac{101.3 \times 1000}{13.6 \times 9.81} = 0.760\ m = 760\ mm$$

Based on this calculation, one atmospheric pressure is defined as the pressure equivalent of a column of mercury that is exactly 760 mm in height at 0°C.

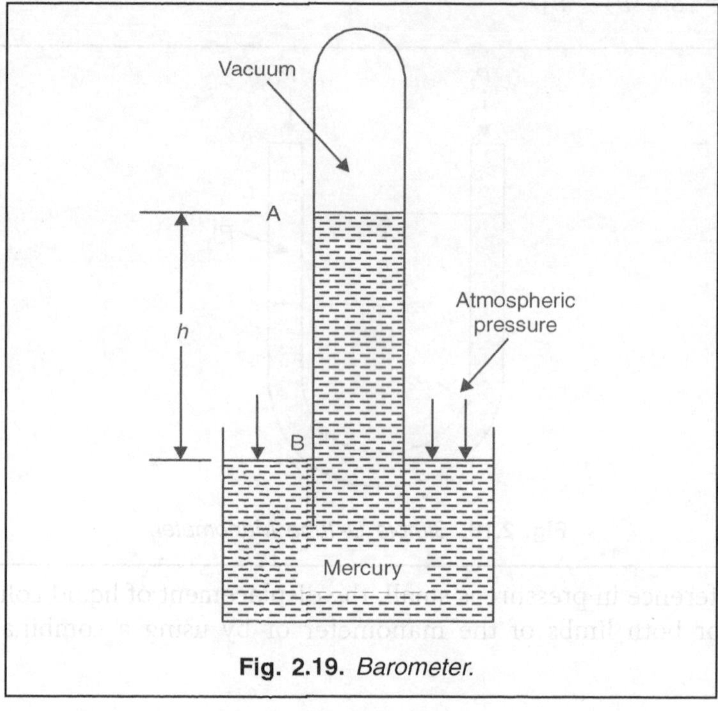

Fig. 2.19. *Barometer.*

(c) McLeod Gauge

McLeod gauge is used to measure low pressure ranging from 10^3 microns to 10^{-3} microns. It is mainly used for vacuum measurement. Fig. 2.20 shows a schematic diagram of McLeod gauge. As shown in (Fig. 2.20), It consists of a capillary C of cross sectional area of 0.02 cm^2 which is connected to a bulb B. The unknown pressure source is connected at the point A and the mercury level is lowered to the cut-off position. Under these conditions, a known volume of gas v_1 and unknown pressure p_1 is admitted into the measuring capillary, the bulb and into the tube down to the cut off points. Mercury is then forced out of the reservoir R to fill in the space occupied by the bulb B and other tubes. This is done by the plunger motion which continues such that the mercury level in the reference capillary touches the zero reading. The earlier trapped gas is now gets compressed and its new volume v_2 in the capillary can be read directly from the scale. As shown in (Fig. 2.20) height 'h' represents the measure of the compressed gas volume trapped into the measuring capillary. The difference in the height of the two columns gives the pressure p_2. By using the Boyle's law $p_1 v_1 = p_2 v_2$ one can easily find out the unknown pressure p_1.

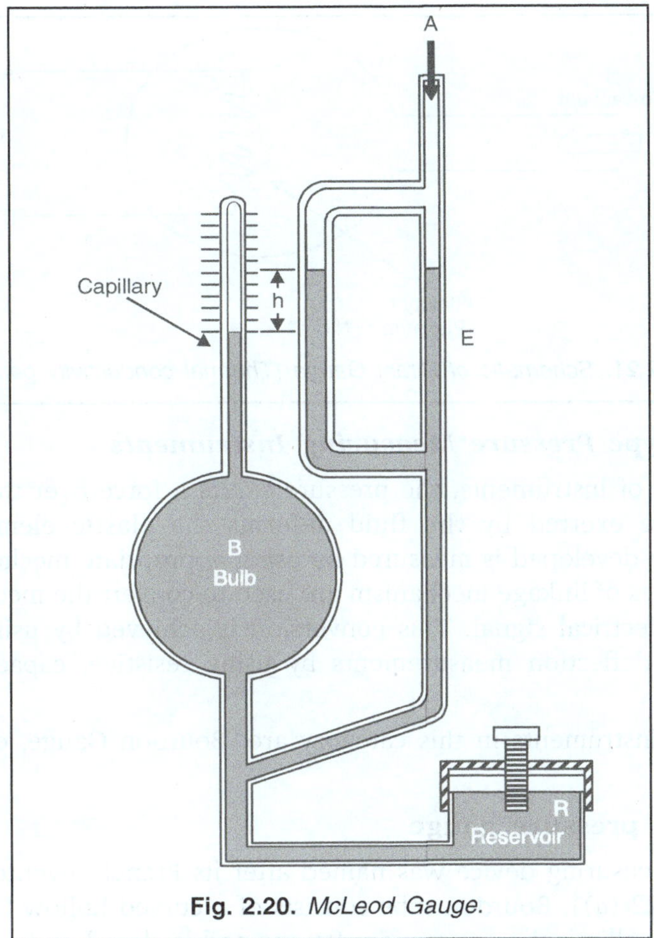

Fig. 2.20. *McLeod Gauge.*

The pressure measured by McLeod gauge is based on linear dimensions so this gauge is used as a reference to calibrate all other vacuum gauges in industries. McLeod gauge is not suitable for measuring the pressure of gases containing vapours as vapours gets condensed during compression.

(*d*) Thermal conductivity gauge or pirani gauge

Pirani gauge is one of the gauges based on the property of variation of thermal conductivity with pressure. It is used to measure pressure in a vacuum chamber and its pressure range lies from 1 to 10^{-1} Pa. At low pressures, kinetic theory of gases predicts a linear relationship between pressure and thermal conductivity. The Pirani gauge as shown in (Fig. 2.21) consists of a platinum or nickel filament through which a small current is passed, causing it to heat up. By making the filament one arm of a Wheatstone bridge circuit, the change in wire resistance can be determined which in turn is calibrated to measure the pressure.

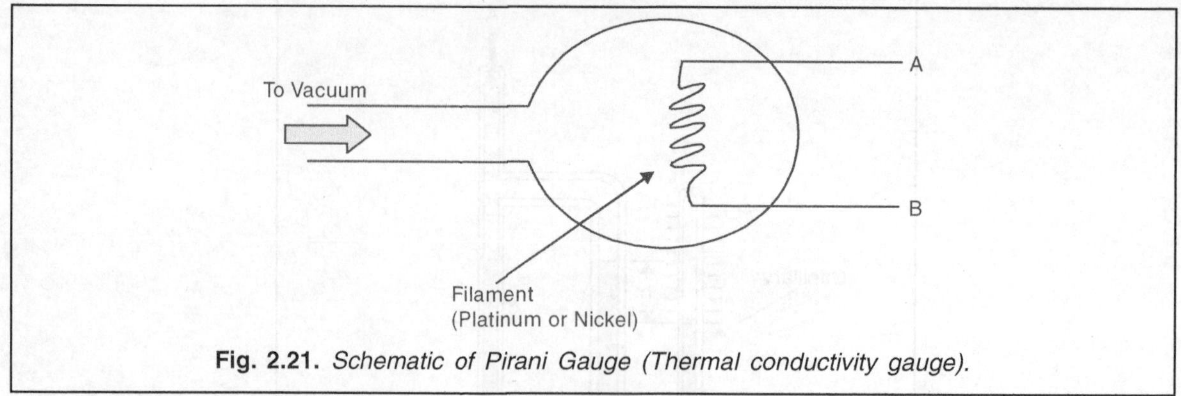

Fig. 2.21. *Schematic of Pirani Gauge (Thermal conductivity gauge).*

2.15.2.2 *Elastic Type Pressure Measuring Instruments*

In this category of instruments, the pressure exerts a force over the area of an elastic device. The pressure exerted by the fluid deforms the elastic element. The resultant deformation or strain developed is measured by using appropriate mechanical and electrical sensors. Different types of linkage mechanism are used to convert the mechanical phenomena into an equivalent electrical signal. This conversion is achieved by using the transduction principles based on deflection measurements by using resistive, capacitive and inductive principles.

The important instruments in this category are: Bourdon Gauge, diaphragm, bellows, capsule etc.

(*a*) Bourdon tube pressure gauge

This pressure measuring device was named after its French inventor Eugene Bourdon. As shown in (Fig. 2.22 (*a*)), Bourdon tube consists of a curved hollow C-shaped metal tube having rectangular or elliptical cross section. Its one end is closed and connected to the dial indicator needle whereas its other end is connected to any container whose pressure is to be determined like a steam boiler. The Bourdon tube does not deflect when it is not connected to any pressure vessel. The needle on the dial indicator is calibrated to read zero gauge pressure when the tube end is open to atmosphere and in undeflected position. Bourdon tube has the property of changing its shape with internal pressure variations. When the pressure inside the tube increases, this stretches the tube and moves the needle in proportion to the pressure applied.

A wide range of materials such as brass, phosphor bronze, beryllium copper, Monel, and various stainless steel alloys are used for making the Bourdon tube.

Bourdon tube is used for measuring pressures in the range of 10 kPa to 100 MPa with an accuracy of about ± 1% of full scale.

The helical and spiral shaped Bourdon tubes find wide use in pressure measurement as shown in (Figs. 2.22 (*b*) and (*c*)). In case of helical shaped Bourdon tube, whenever the pressure inside the tube increases, the close end of the tube rotates. This rotation of tube is used by helical shaped Bourdon tube to move the slider of a potentiometer and so the pressure is measured in terms of electrical output. Helical shaped Bourdon tube can measure pressure up to 50 MPa.

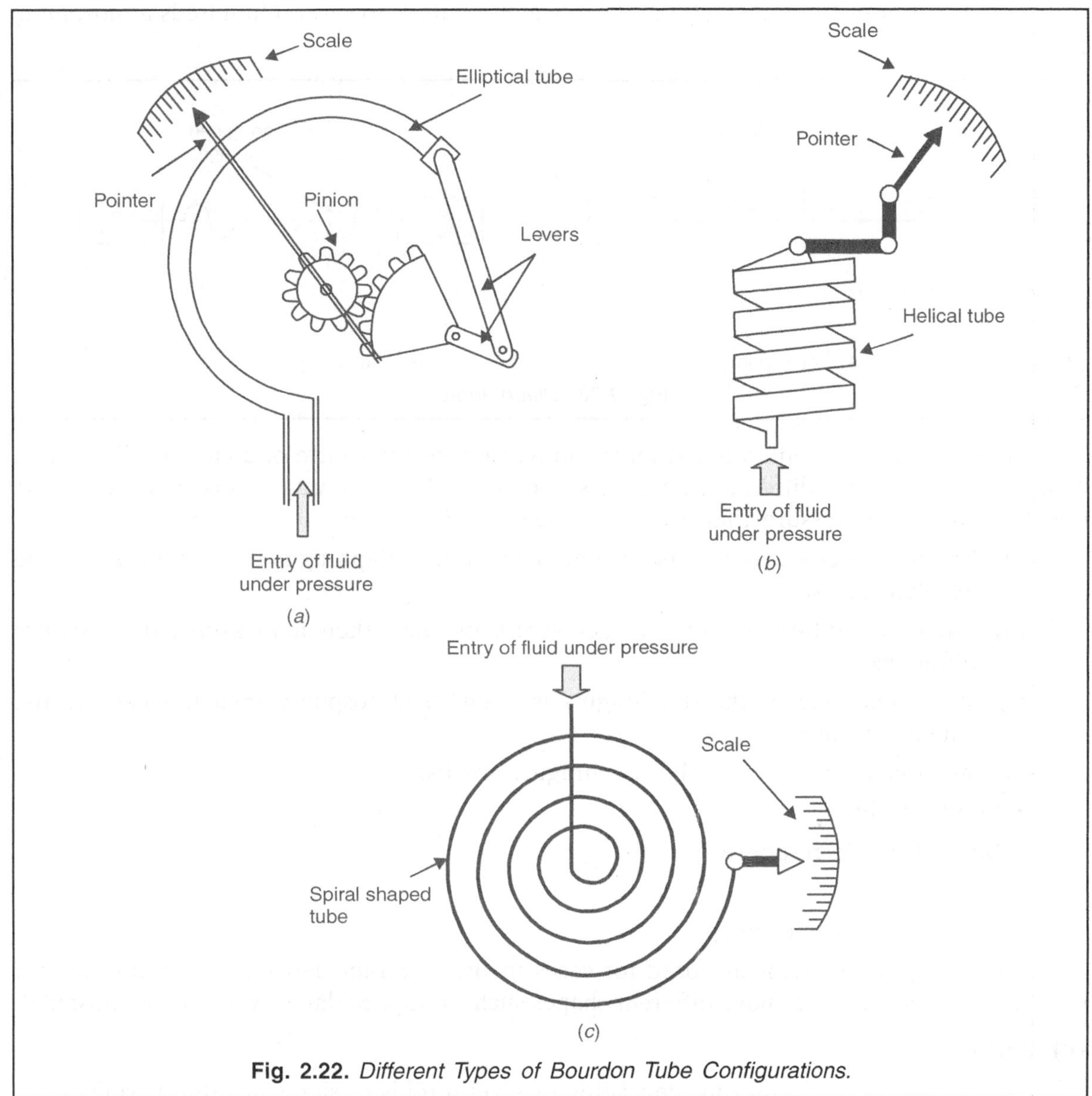

Fig. 2.22. *Different Types of Bourdon Tube Configurations.*

(*b*) Diaphragm

It is the most common type of pressure sensing elastic element that is used to convert pressure to deflection. A diaphragm is a thin circular plate which is stretched and clamped firmly around its edges. Commonly used materials for making diaphragms include brass, stainless steel, phosphor bronze, beryllium copper, titanium and haste alloy. The main considerations in the selection of the material for a diaphragm are the chemical nature of fluid which is likely to come in contact with the diaphragm material and effect of shock and vibration. The structure of the diaphragm may be flat or corrugated as shown in (Fig. 2.23). Diaphragms are used because of their high accuracy and good dynamic response. The typical pressure range varies from few millimeters of water columns to several hundreds atmospheric pressure.

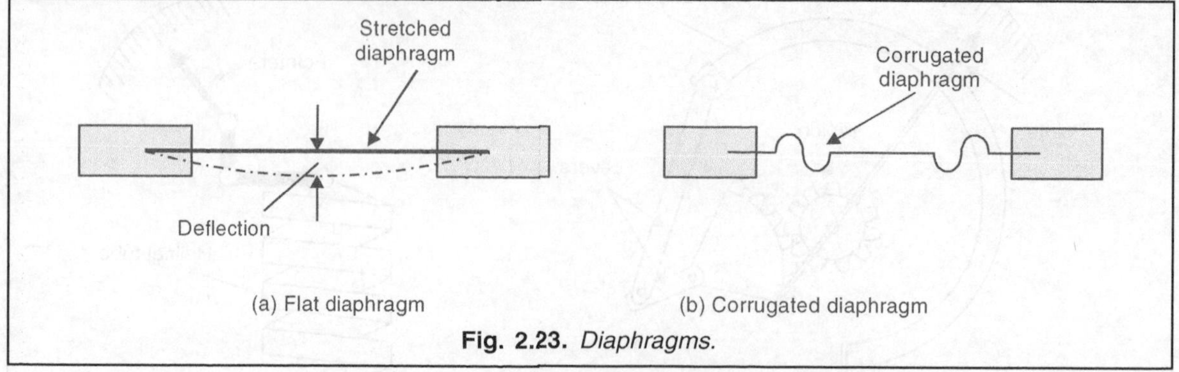

Fig. 2.23. *Diaphragms.*

Pressure measurement is based on the movement of the centre of a circular diaphragm as a result of pressure difference across its two sides. The diaphragm gets deflected as its one side has more pressure than the other side. In diaphragms:

(*a*) When perfect vacuum acts on one side of the diaphragm then it measures the absolute pressure.

(*b*) When two different pressures act across its sides then it measures the pressure difference.

(*c*) When one side of the diaphragms is open to atmosphere then it measures the gauge pressure.

The main characteristics of the diaphragms are its:

- Good stability
- Good dynamic response
- High reliability
- Low hysteresis and creep.

Corrugated diaphragms are used for cases involving larger deflection. The profile of a corrugated diaphragm may have different shapes such as trapezoidal, saw tooth, or sinusoidal.

(c) Bellows

This is a thin wall-metal tube consisting of several folds of side walls that permits axial expansion and contraction as shown in (Fig. 2.24). Bellows are sealed from one end. The sealed end moves axially when pressure is applied to the other end. Most bellows are made from seamless tubes, and their folds (or convolutions) either are hydraulically formed or mechanically rolled. Most commonly used material for making the bellows are brass, stainless steel, phosphor bronze, and beryllium copper. The selection of material is based on certain considerations such as pressure range, hysteresis and creep, and corrosion etc. Similar to diaphragm gauges, bellows are the low pressure measuring devices. These are used to measure pressure in the range of few millimeters of water column to about 1.0 MPa.

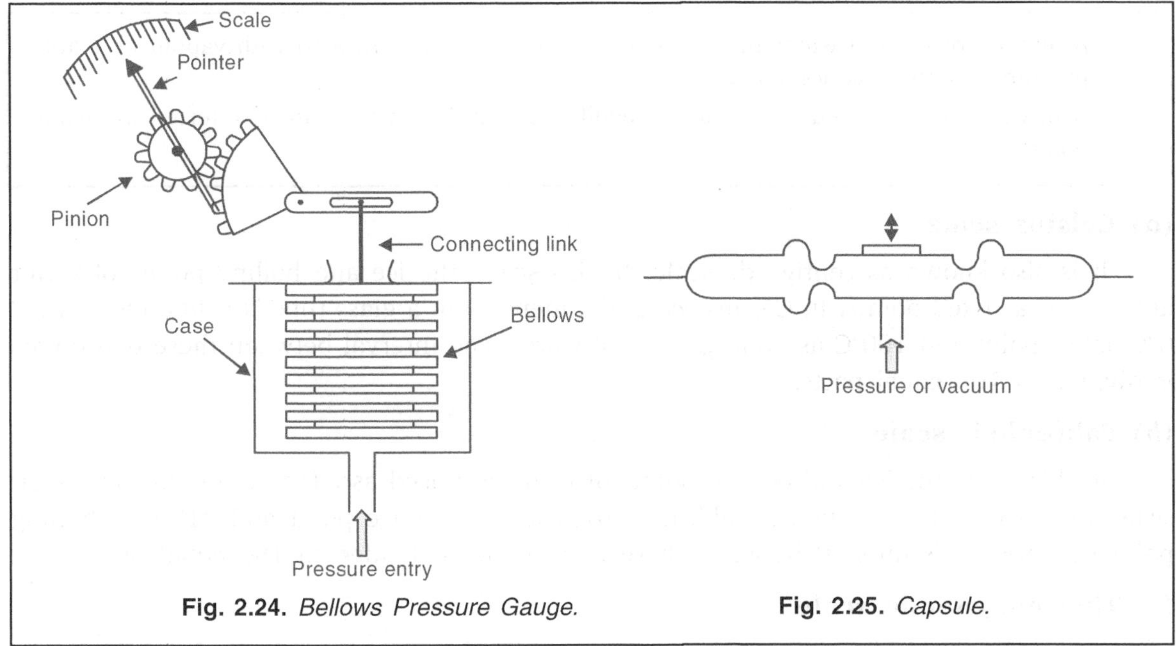

Fig. 2.24. *Bellows Pressure Gauge.*

Fig. 2.25. *Capsule.*

(*d*) Capsule

A capsule as shown in (Fig. 2.25) also referred to as an 'aneroid' is formed by joining the peripheries of two diaphragms through soldering. Pressure to be measured is admitted through one diaphragm and other diaphragm is linked to a mechanical member. The displacement of this member is proportional to the difference of the inner and outer pressures. These configurations are also used in aircrafts.

2.16 TEMPERATURE MEASUREMENT

Temperature may be defined as the degree of hotness or coldness of a body measured on a definite scale. When a body is heated or cooled down one can observe different effects and one of these effects can be put into use for the measurement of temperature. The major effects include:

- Change in the physical state of a body
- Change in chemical state of a body
- Variations in electrical properties
- Generation of an electromotive force (or current flow) at the junction of two dissimilar metals.
- Changes in the intensity of the total radiation emitted.

The different methods of temperature measurements are discussed below:

2.16.1 Temperature Scales

Temperature scales provide us a common basis for measuring temperatures. All temperature scales are based on some easily reproducible states such as freezing and boiling point of water. Freezing point is also known as ice point and boiling point as steam point.

> A mixture of ice and water that is in equilibrium with air saturated with vapour at 1 atm pressure is termed as ice point.
>
> A mixture of water and its vapour in equilibrium at 1 atm pressure is referred as steam point.

(a) Celsius scale

It is also known as centigrade scale. In this scale, the ice and boiling points of water are treated as fixed points. In Celsius scale, the mercury in a glass tube is calibrated to read 0°C as ice point and 100°C as boiling point of water. The interval between these two points is divided in 100 equal parts.

(b) Fahrenheit scale

In this scale, the ice and boiling points of water are used as fixed points. In Fahrenheit scale, mercury in a glass tube is calibrated to read 32°F as ice point and 212°F as boiling point of water. This interval between these two points is divided in 180 equal parts.

(c) Thermodynamic scale

It is really desirable to have a temperature scale that is independent of the properties of any substance or matter. Such type of a temperature scale is known as thermodynamic temperature scale.

(*i*) **Kelvin scale**: In SI units, the thermodynamic scale is known as Kelvin scale. The temperature unit on this scale is the 'Kelvin' which is denoted by 'K'. The lowest temperature on the Kelvin scale is 0 K. No molecular activity takes place at absolute zero Kelvin and therefore body does not possess any heat content. The temperature at this condition is the lowest temperature possible and is referred to as absolute zero. Kelvin scale has a value of 273.16°K and a value of 373.16°K at 100°C.

(*ii*) **Rankine scale:** The thermodynamic scale in FPS or English system is known as Rankine cycle. The temperature unit on this scale is the Rankine which is denoted by 'R'. In this scale absolute zero is marked at 460°F.

The following expressions are used to convert one scale reading into another.

$$^\circ C = \frac{5}{9}(^\circ F - 32)$$

$$K = (^\circ C + 273.16) \qquad \text{...(2.15)}$$

$$R = (^\circ F + 460)$$

(Fig. 2.26) shows the comparison of various temperature scales.

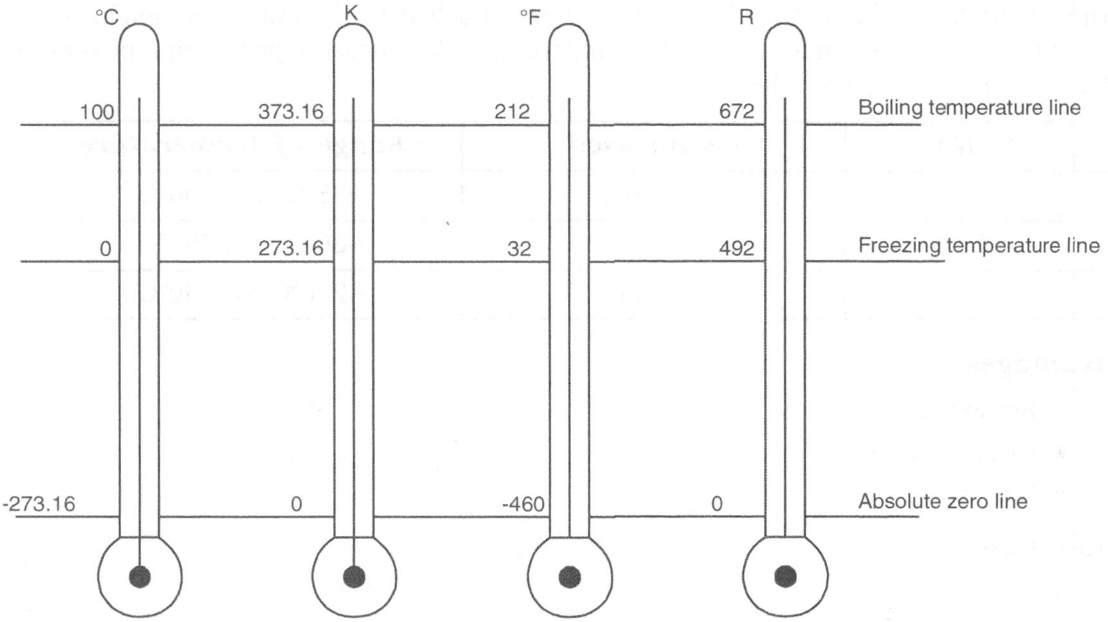

Fig. 2.26. *Comparison of Various Temperature Scales.*

2.16.2 Mechanical Temperature Sensors

Mechanically operated elements sensitive to temperatures are used extensively to measure the temperature. A large number of devices consist of a sensor bulb filled with a liquid sensitive to temperature or an enclosed volume sensitive to pressure or volume changes. These are known as filled system thermometers.

Advantages

- Simple design
- Low cost
- Self contained operation
- Rugged construction.

Limitations

- Limited temperature coverage
- Large size
- Not compatible with electrical system of measurement
- Poor dynamic response.

Filled system thermometers are of two types those responding to volume changes *i.e.,* liquid in glass thermometer and those responding to pressure changes such as vapour pressure thermometer.

(*a*) Liquid in glass thermometer

The liquid in glass thermometer as shown in (Fig. 2.27) measures the temperature in terms of the level of liquid in the glass capillary attached to the bulb. The height to which the liquid expands is a measure of the temperature. The measurement range possible with different liquids is given below.

S. No.	*Liquid used*	*Range of temperature*
1	Mercury	–35°C to +600°C
2	Alcohol	–80°C to +70°C
3	Pentane	–200°C to +30°C

Advantages

- Reasonable accuracy
- Direct reading
- Low cost.

Limitations

- Fragile
- Slow response to temperature changes.

Example: Standard Clinical Thermometer

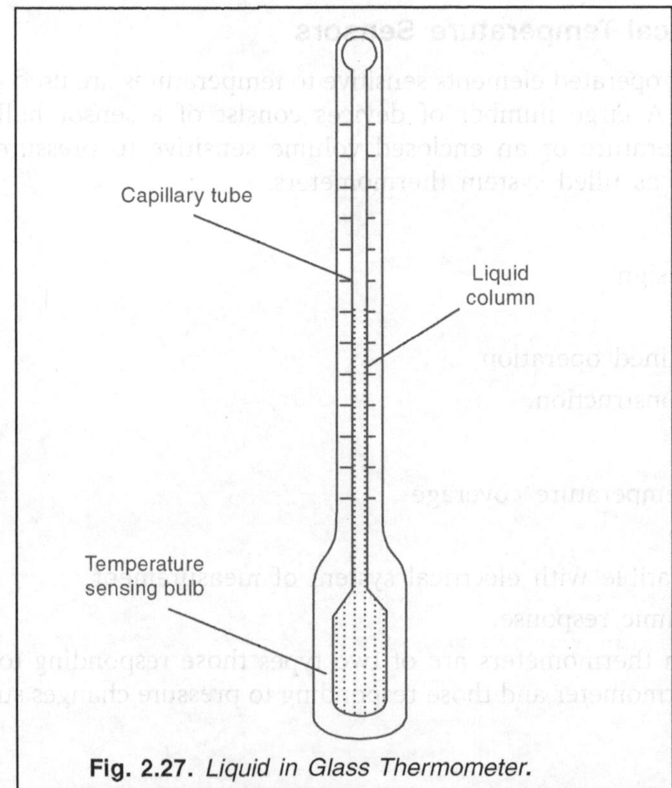

Fig. 2.27. *Liquid in Glass Thermometer.*

(*b*) Liquid filled system thermometers

The fragility of liquid in glass thermometer has been overcome by using metal containers filled with expanding liquids or gases. A standard liquid filled system thermometer as shown in Fig. 2.28 consists of the following main parts:

(*i*) A temperature sensor in the form of an immersible bulb with expanding fluid.

(*ii*) An elastic measuring element like spiral spring, diaphragm, bourdon tube etc.

(*iii*) Capillary tube

(*iv*) Indicating mechanism.

When this thermometer is exposed to the medium, whose temperature needs to be determined, the liquid contained in the bulb expands and this change in volume of the liquid which moves through the capillary and drives the elastic measuring element. Mercury and toluene are used as filling fluids.

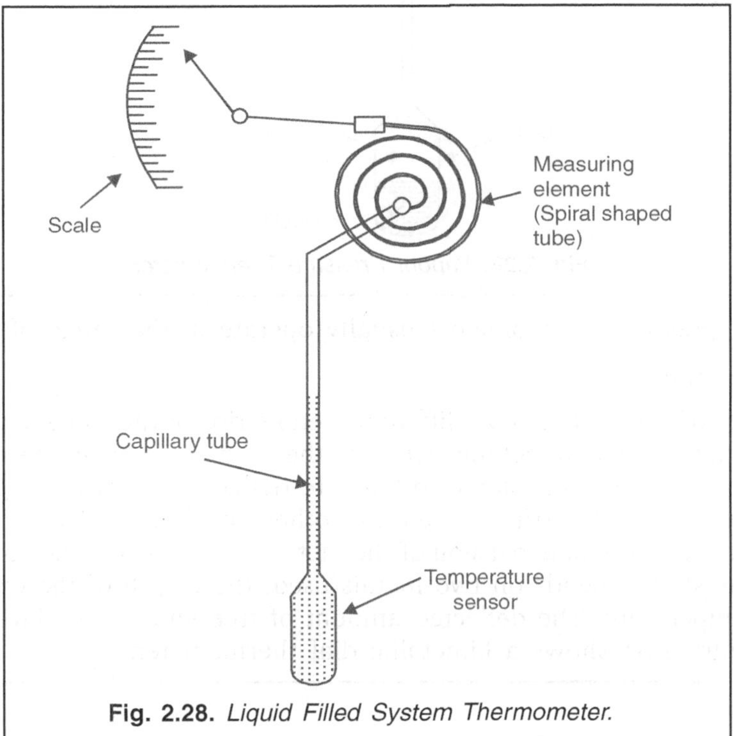

Fig. 2.28. *Liquid Filled System Thermometer.*

(c) Vapour pressure thermometers

The design and construction of vapour pressure thermometer is same as that of liquid filled system thermometers except that the bulb is partially filled with a low boiling liquid and rest is filled with its vapour. The change in vapour pressure in the bulb over the liquid column is a function of temperature. Increase in temperature at the bulb results in further vaporization. This increase in vapour pressure can be measured by using a suitable elastic measuring elements coupled to an indicator as shown in (Fig. 2.29). The liquids used in this category of thermometers are: ethylene, ether, hexane and toluene etc. One of the disadvantages with this thermometer is that its indication is not linear with the temperature.

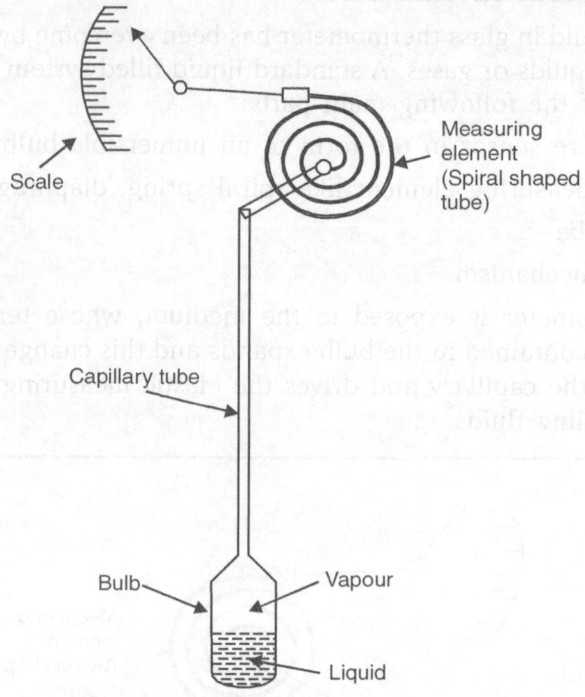

Fig. 2.29. *Vapour Pressure Thermometer.*

The vapour pressure thermometers usually operate in the range of –10°C to 200°C.

(*d*) Bimetallic strips

A bimetallic strip consists of two different metal strips of the same length firmly bonded together by welding. For a bimetallic strip in the form of a cantilever beam, when the temperature increases the composite strip tends to deflect as shown in Fig. 2.30. Cantilever beam means one end of the strip is fixed and other end is free. This happens due to the different coefficient of thermal expansion of the metals used in strip. The amount of deflection in free end of the strip depends on two metals used, the length of the composite strip and the change in temperature. The deflected amount of free end of cantilever is a measure of temperature. Figure 2.31 shows a bimetallic dial thermometer.

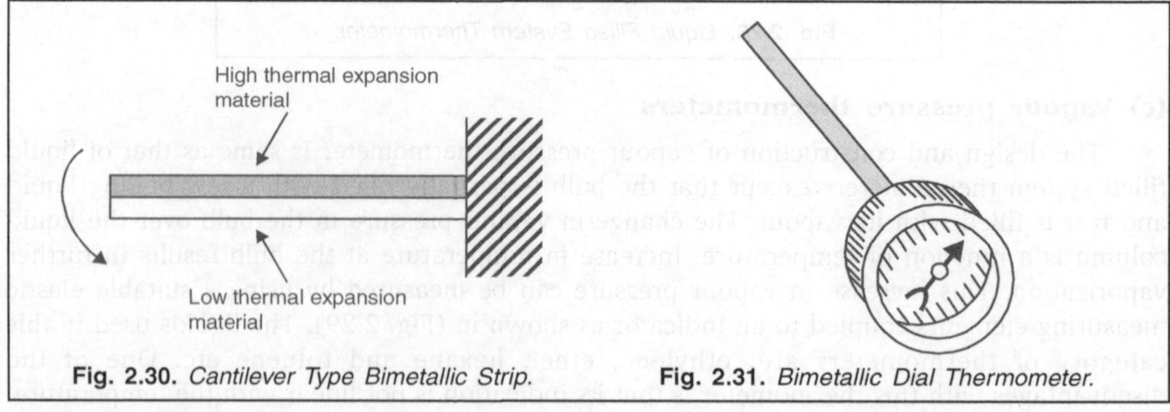

Fig. 2.30. *Cantilever Type Bimetallic Strip.* **Fig. 2.31.** *Bimetallic Dial Thermometer.*

The mechanical temperature sensing devices are of little use in high precision instrumentation. Electrical based temperature measuring sensors are a better choice due to their high accuracy, quick response and good sensitivity.

The two most important electrical type temperatures measuring devices are resistance thermometers and thermocouples.

(*e*) Resistance thermometers

It has been observed with most of the metals that their resistance increases linearly with the temperature but at a low sensitivity. This change in resistance with temperature is used for measuring temperature and devices based on utilizing this phenomenon are known as resistance thermometers. The equation (2.16) given below represents the relationship between the temperature and resistance of a metal.

$$R_t = R_0 (1 + \alpha t) \quad ...(2.16)$$

Where R_t is the resistance at any temperature t°C

R_0 is the resistance at 0°C

α is a constant (also termed as temperature coefficient of resistance)

A resistance thermometer comprises a very fine resistance wire wound on to an insulating ceramic former as shown in Fig. 2.32. The upper limit of temperature measurement rests on the type of the former on which the resistance wire is wound. Ceramic formers can sustain temperature up to about 850°C. The selected wire must have a known characteristic relating resistance with temperature and should not change with time. The resistance element is placed inside a sheath to protect it against possible damages.

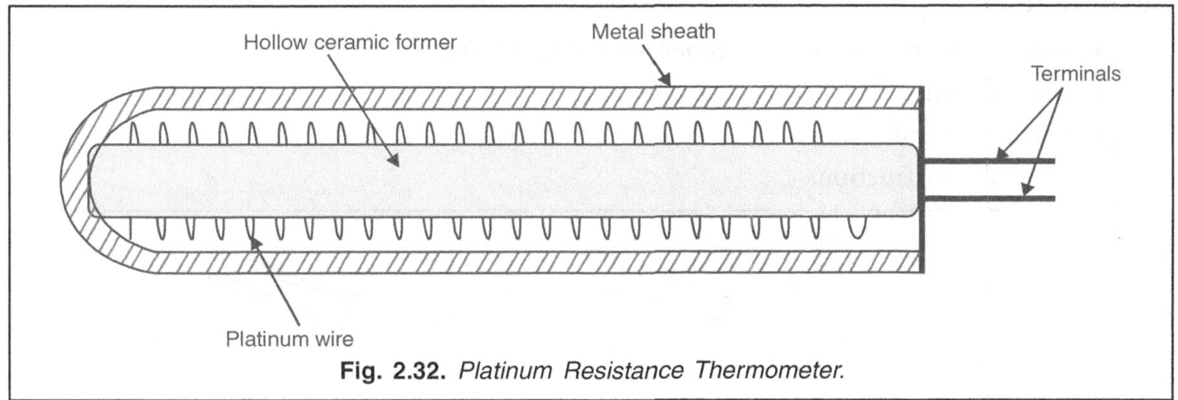

Fig. 2.32. *Platinum Resistance Thermometer.*

The resistance element is usually made of pure metals which include metals like Ni, Cu and Pt. Among these materials, Cu has very low resistance. Due to this resistance element made of thin wire of copper requires a long length. Ni based resistance thermometers can be used with good reliability and repeatability in the range of 100 to 450 K. However Ni based resistance thermometers have less stability and these are more prone to environmental attacks. Platinum is considered as best option for resistance thermometers as platinum based resistance thermometers have following advantages;

- High linearity
- Good repeatability

- Long term stability
- Accuracy attainable is of the order of ± 0.01°C up to 500°C
- Working temperature range: –200 to 850°C

Mostly platinum is preferred widely as material for resistance element in industrial resistance thermometers which are also known as resistance temperature detectors (RTD).

(f) Thermistors

As we know that the resistivity of metals increases with an increase in temperature. It means that the temperature coefficient is positive whereas in some semiconductors the resistance decreases with increase in temperature. It means that semiconductor materials have negative temperature coefficient. Temperature sensors based on resistance changes in semiconductor materials are called as thermistors. The word 'Thermistor' is actually a contraction of the words 'thermal resistor'. Thermistors are made of sintered ceramics, usually from the mixtures of metal oxides such as chromium, cobalt, iron, manganese and nickel. Thermistors are made in the form of beads, discs, and rods. The thermistors are preferred where high sensor output and lost cost are the priorities.

Most thermistors have negative temperature coefficient but thermistors having positive temperature coefficient are also available in the market but used occasionally due to their high sensitivity. Positive temperature coefficient thermistors have temperature coefficient positive only over a limited temperature range; beyond this range their temperature coefficient is negative.

- Working temperature range: 170 to 470 K

Advantages

- Well suited for precision temperature measurements
- High sensitivity
- Small physical size
- Rugged construction.

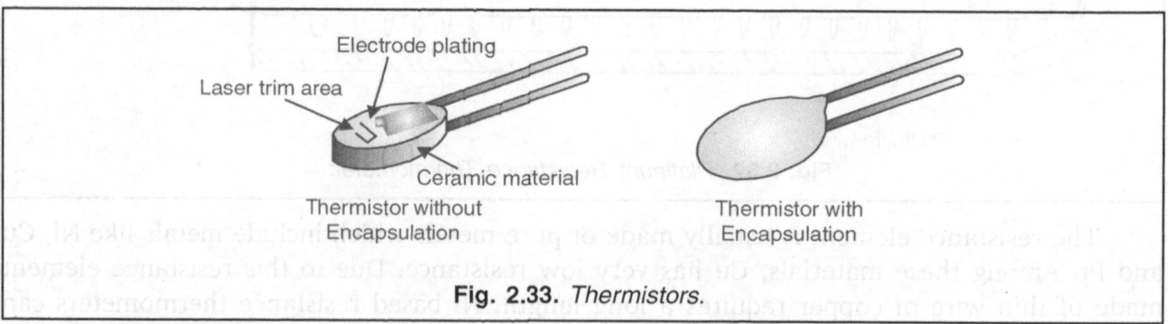

Fig. 2.33. *Thermistors.*

(g) Thermocouple

It is the most commonly used electrical device for temperature measurement. A thermocouple circuit is made by joining two wires of different materials A and B at their ends so as to form a closed loop. A schematic diagram of a thermoelectric circuit is shown in (Fig. 2.34). The temperature measurement is based on the principle that current flows in a closed circuit made up of two dissimilar metals if the junctions of the two metals are kept at different temperature. This principle is known as 'Seebeck effect'. Due to Seebeck effect, a net emf is generated in the circuit which depends on the difference in temperature of the two junctions. In practice the temperature measurement is carried out with two junctions. One of them, the reference junction is kept at a constant temperature usually at 0°C and the other is exposed to medium whose temperature is to be measured. Experimentally it has been found that the magnitude of the current is directly related to the two materials of thermocouple wires and the temperature difference (T_1–T_2).

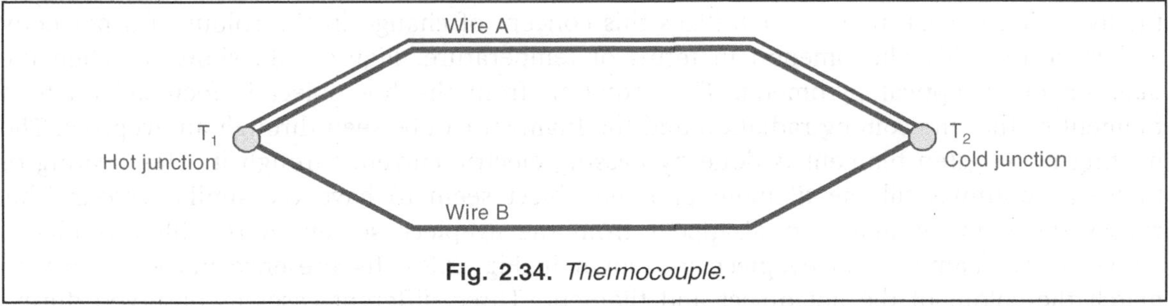

Fig. 2.34. *Thermocouple.*

Practically a suitable device such as millivoltmeter can be used in thermocouple circuit to measure emf or flow of current. A thermocouple can be calibrated by measuring the thermal emf at different known temperatures.

Working range: –272°C to 2000°C

The material used for thermocouples should have:

(*i*) high sensitivity and linearity

(*ii*) time-invariant behaviour

(*iii*) physical strength and stability at high temperature

(*iv*) low cost.

Certain combinations of materials also give thermal emf proportional to the temperature differences between their junctions. The commonly used material combinations are copper-constantan, chromel-constantan, iron-constantan and chromel-alumel, platinum-platinum-rhodium etc.

In order to avoid measurement error through thermocouple one should taken into consideration the effect of heating and cooling. For this one should know the Peltier effect and Joule and Thomson effect.

Peltier effect: When electric current flows through a thermocouple circuit as shown in (Fig. 2.34), then joule heating will occur. It has been observed that current flowing through the junction causes temperature increase at one junction and decreases temperature at the other end.

Thompson effect: A current passing through a conducting material when subjected to a temperature gradient causes either an evolution or absorption of heat which depends on temperature gradient and the current flow direction.

Advantages

- Due to its small mass, it comes to thermal equilibrium with system quickly whose temperature is to be measured.
- Best suited for measurement of exhaust gas temperatures.

(*h*) Optical pyrometer

It is also known as disappearing filament pyrometer or heated filament pyrometer. As we know that when a particular metal is heated then it first becomes red then orange and finally white. Optical pyrometer utilizes this concept of change in the colour of a hot body and interprets this phenomenon in terms of temperature. Figure 2.35 shows a schematic diagram of an optical pyrometer. The radiation from the hot object is focussed on to a filament so that the coming radiation and the filament can be seen through an eyepiece. The heating of tungsten filament is done by passing electric current through it. This heating of filament continues till the filament and the object seem to have the similar colour. This causes the filament image to disappear from the eyepiece screen. A red filter is placed between the filament and eyepiece as shown in Fig. 2.35. Its presence makes it easy to match the colour of the hot object and filament. Three different cases as observed during experimentation are also shown in Fig. 2.35.

Working temperature range: 600°C to 3000°C

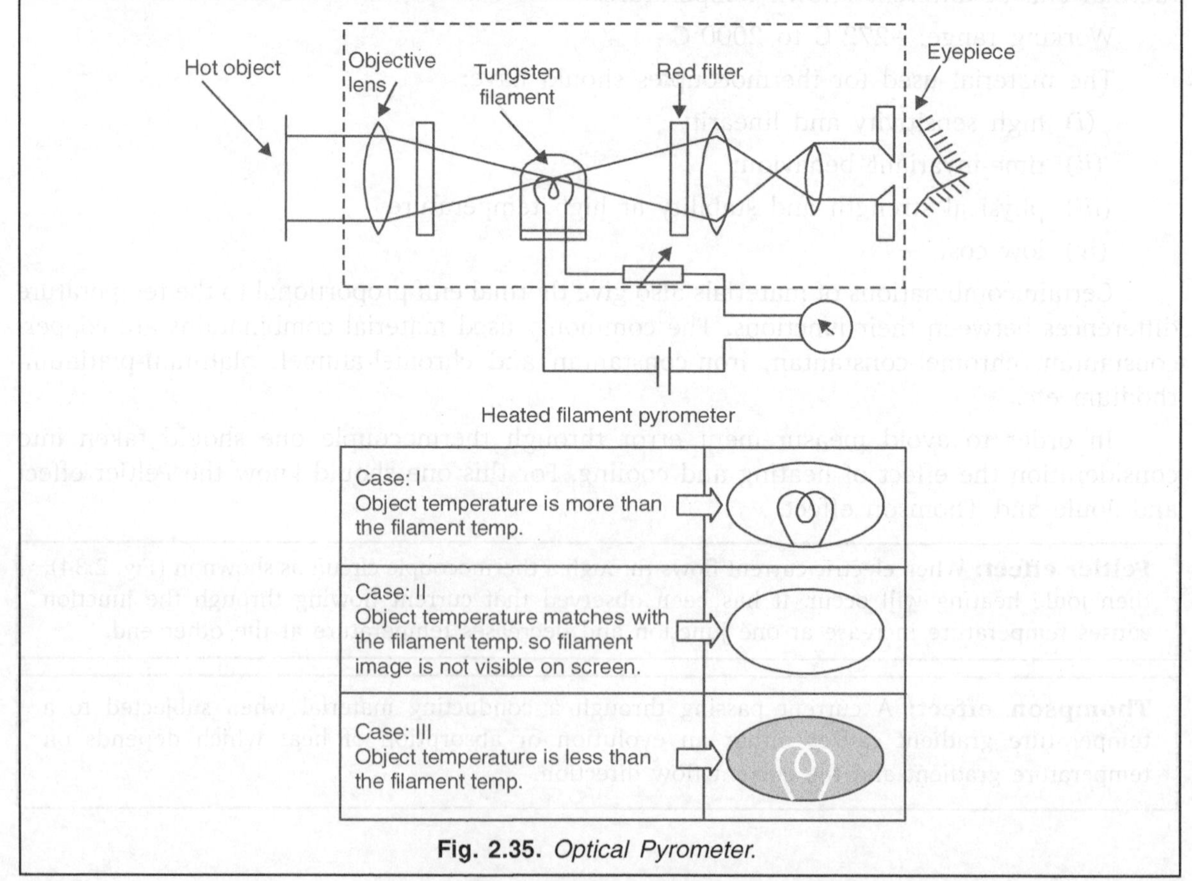

Fig. 2.35. *Optical Pyrometer.*

Advantages

- Non-contact type measuring device
- Higher accuracy
- Suitable for distant objects.

2.17 STRAIN MEASUREMENT

As we know strain is defined as the ratio of change in length to the original length. The strain measurements are usually carried out on the free surface of a body. The strain gauge is a device which is used to measure dimensional changes on the surface of a test specimen.

Stain gauges can be classified in three broad categories depending upon the principle of operation and their constructional features. These are:

- Electrical strain gauges
- Mechanical strain gauges
- Optical strain gauges

Out of these three types, electrical resistance type stain gauges are used widely as these possess some inherent advantages in the process of measurement.

In mechanical type strain gauges, by using gears and levers, the change in length under observation is magnified mechanically.

Optical strain gauges are same as the mechanical type strain gauges except multiple mirrors or prisms are used as reflectors to achieve the desired magnification.

2.17.1 Electrical Strain Gauges

The principle of an electrical strain gauge is based upon measuring the changes in resistance, capacitance or inductance of the gauge material. These changes are proportional to the strain transferred from the test sample to the gauge element. In case of a resistance type electrical strain gauge, this fractional change in resistance $\Delta R/R$ of the wire material is proportional to the strain ε.

$$\frac{\Delta R}{R} \propto \varepsilon$$

$$\therefore \quad \frac{\Delta R}{R} = G\varepsilon \qquad \text{...(2.17)}$$

Where G is the constant of proportionality and is referred to as the 'Gauge factor'.

A resistance type electrical strain gauge consists of metal foil or a strip of semiconductor material or a metal wire subjected to the same strain as that of the specimen under test, achieved through proper bonding of the gauge to the test sample. The resistance of the wire changes as a function of strain, increasing with tension and decreasing with compression. This change in resistance can be measured by using a Wheatstone bridge. The resistance type electrical strain gauges work on this principle. (Fig. 2.36) shows a typical classification of electrical strain gauges.

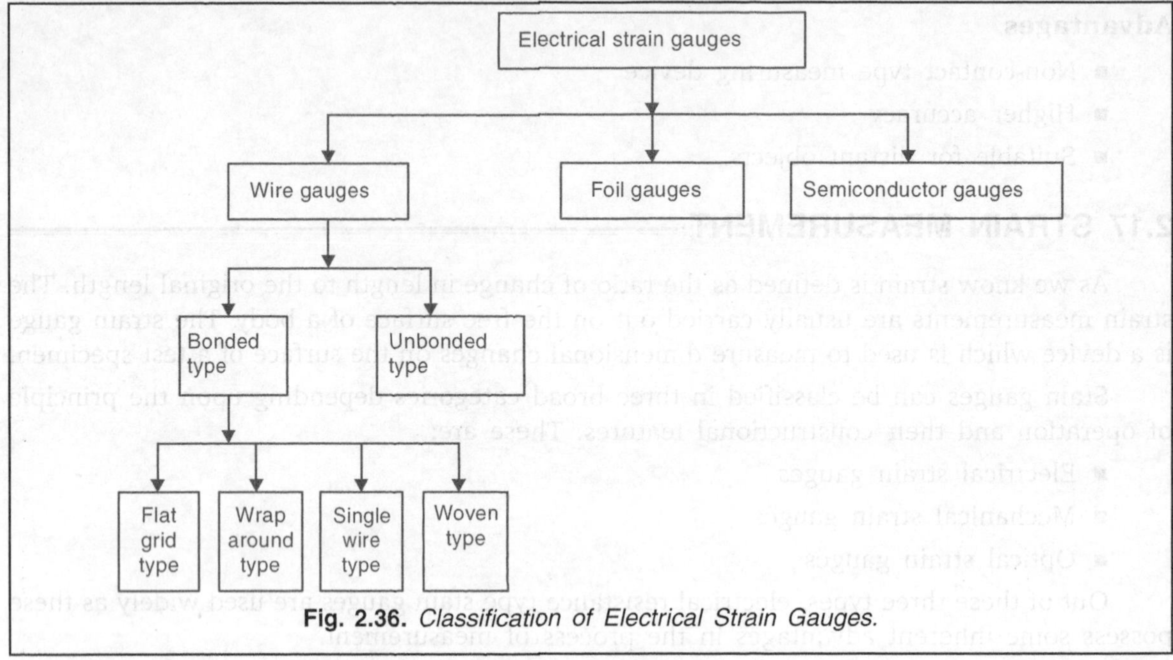

Fig. 2.36. *Classification of Electrical Strain Gauges.*

In the bonded type, the strain gauge is bonded directly to the surface of the specimen being tested with a thin layer of adhesive cement which in turn transmits the strain from the specimen to the gauge wires as shown in (Fig. 2.37). In case of an unbonded type strain gauge, strain is transferred to the resistance wire directly without any backing.

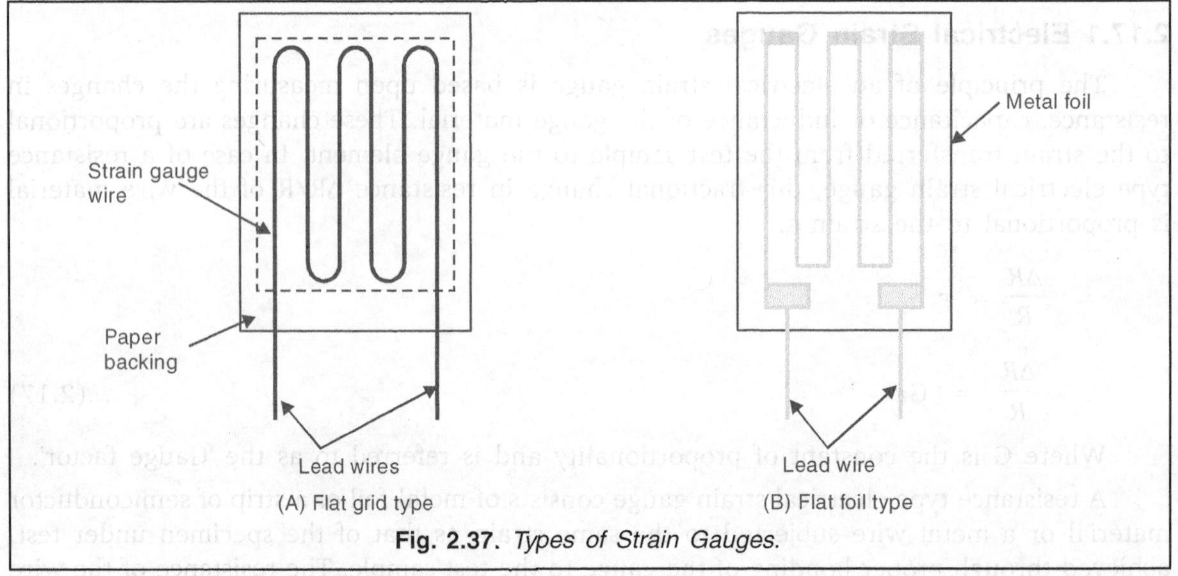

Fig. 2.37. *Types of Strain Gauges.*

The main problem associated with the strain gauge is that the resistance of the gauge changes when the temperature changes. Therefore some techniques need to be used to compensate for such changes by eliminating the effect of temperature.

2.18 FORCE MEASUREMENTS

Force

Force is a vector quantity which changes or tends to change the relative motion or shape of a body to which it is applied. Therefore the measurement of force involves the determination of its magnitude and direction. The force applied to a body or structure is also known as load.

Newton's second law of motion defines the force as mentioned below:

> The net external force exerted on a body is equal to the product of its mass and its acceleration.
>
> $F = ma$

There are two reasons behind the creation of mechanical forces (*i*) variation in the stored potential energy of a body due to change in its physical dimension, (*ii*) rate of change of momentum of a mass with reference to time.

Forces which deals with potential energy are static in nature and those deals with rate of change of momentum are dynamic in nature.

There are two methods of measurement of forces.

(A) Direct methods: these involve a direct comparison against a known gravitational force on a standard mass. This category of measurement of forces includes scales and balances. These devices should able to sustain the full effect of the force either in tension or compression.

(B) Indirect methods: these involve the measurement of elastic deformation of a body under the influence of applied force. In this category, the measurement of force can be done by

(*i*) Translating the force to a fluid pressure and then measuring the resultant pressure. Examples: hydraulic and pneumatic load cell.

(*ii*) By applying the forces on a known mass and then measuring its acceleration.

(*iii*) Applying the force on an elastic member and then measuring its subsequent deformation. Examples: proving ring.

2.18.1 Mechanical Weighing Systems

Under this category of force measurement, force is measured by making a comparison between the force due to gravity acting on a standard mass and the force due to gravity acting on the unknown mass. Examples: Balance, multi lever system, pendulum force measuring mechanism. Direction of force being parallel to the gravitational force implies that only its magnitude needs to be determined.

(*a*) Equal arm beam balance

In an equal arm balance as shown in (Fig. 2.38) consists of a beam pivoted on a knife edge fulcrum at the centre. The standard masses are added to a pan on one side of a pivoted beam until the magnitude of the gravity force on it balances the magnitude of the gravitational force on the unknown mass acting at the other end of the beam.

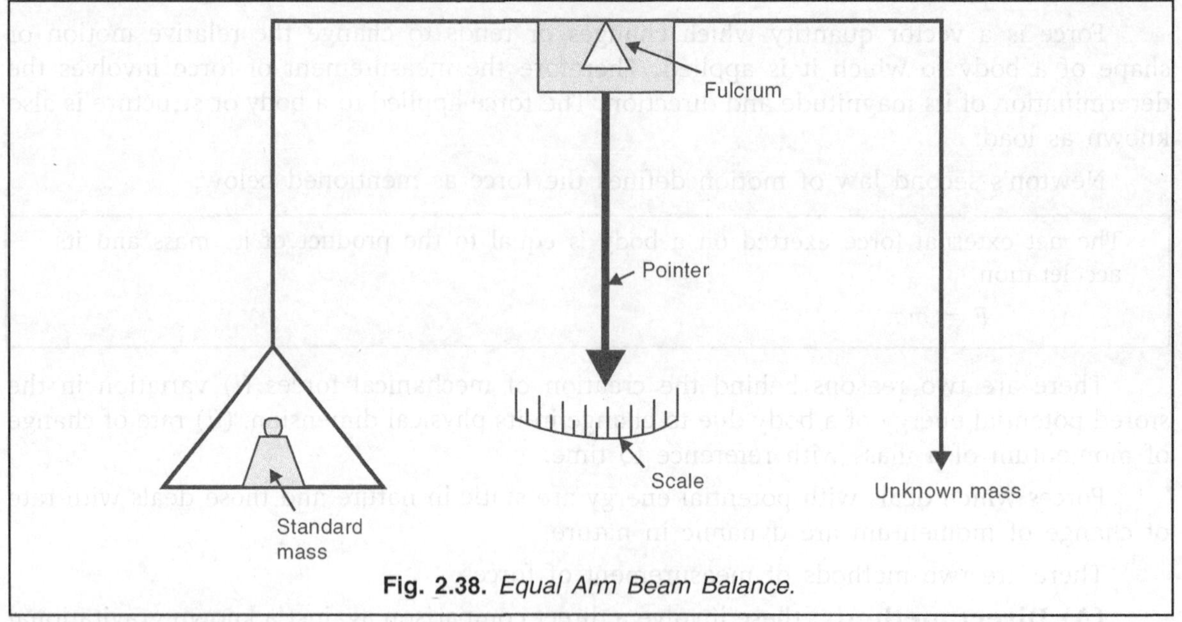

Fig. 2.38. *Equal Arm Beam Balance.*

(*b*) Weigh beam

The operating principle of weigh beam is same as that of equal arm beam balance (Fig. 2.39). It is having more rugged construction in comparison with beam balance. One can see this type of balance at a railway parcel room. Here standard masses are added to balance the unknown mass. The proper adjustment is achieved by moving known masses on a notched graduated bar until the pointer is brought to the null setting.

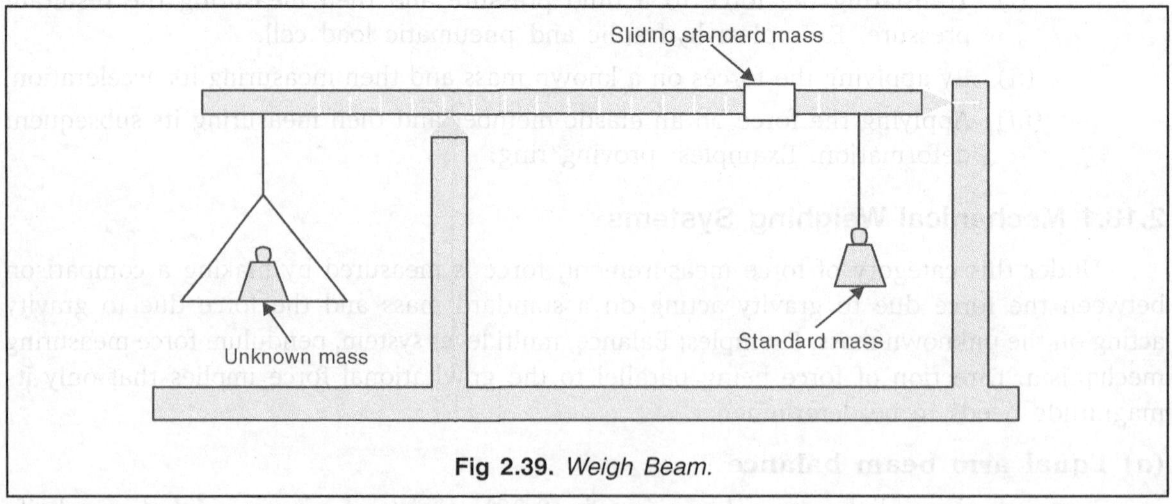

Fig 2.39. *Weigh Beam.*

2.18.2 Measurement of Forces with Elastic Elements

Elastic elements are commonly used for measuring the static as well as dynamic forces due to their long working range, continuous monitoring, simple functioning and robustness.

By measuring the displacement or deflection of these elastic members one can estimate the magnitude of the applied force. Few of these are discussed below:

(a) Springs

A simple coil spring as shown in (Fig. 2.40) is an example of a force displacement type transducer. If δ is the deflection in the spring from its equilibrium position then the force F can be determined by using equations (2.18) and (2.19).

$$F = k\delta \qquad \text{...(2.18)}$$

where

$$\delta = \qquad \text{...(2.19)}$$

Here δ is axial deflection of the spring

F is the applied force

G is the modulus of rigidity

D is the mean coil diameter

d is wire diameter

N number of active coils

As we know

$$\frac{8FD^3NGd^4}{\delta Gd^4\ 8D^3N}$$

Fig. 2.40. *Coil Spring.*

$$K = \qquad \text{...(2.20)}$$

Where K is the stiffness of the spring and is defined as the force required to produce unit deflection.

Measurement of forces by using springs provides reasonably satisfactory results for steady state measurements.

(b) Axially loaded members

This method uses a bar or a rod (Fig. 2.41 (*a*)) which is axially loaded either in tension or in compression. The displacement y at the free end when a force F is applied is given by using the equation (2.21) mentioned below.

$$y = \frac{FL}{AE} \qquad \text{...(2.21)}$$

Where A is the area of the bar

L is the length of the bar

E is the Young's modulus of elasticity

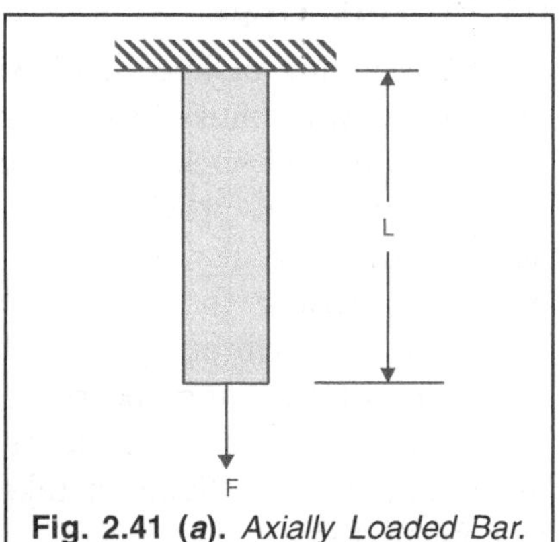

Fig. 2.41 (*a*). *Axially Loaded Bar.*

As we know

$$F = K\,y$$

where, K is given by $\dfrac{AE}{L}$ ...(2.22)

For a simply supported beam (Fig. 2.41 (*b*)) $y = \dfrac{1}{48}\dfrac{FL^3}{EI}$ and $K = \dfrac{48EI}{L^3}$

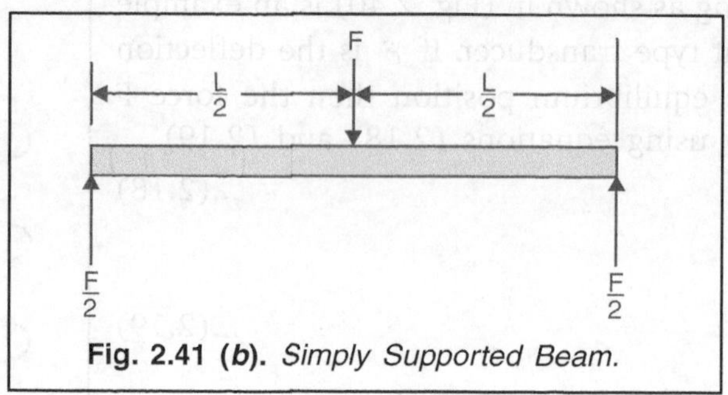

Fig. 2.41 (*b*). *Simply Supported Beam.*

For a cantilever beam $y = \dfrac{FL^3}{3EI}$ and $K = \dfrac{3EI}{L^3}$

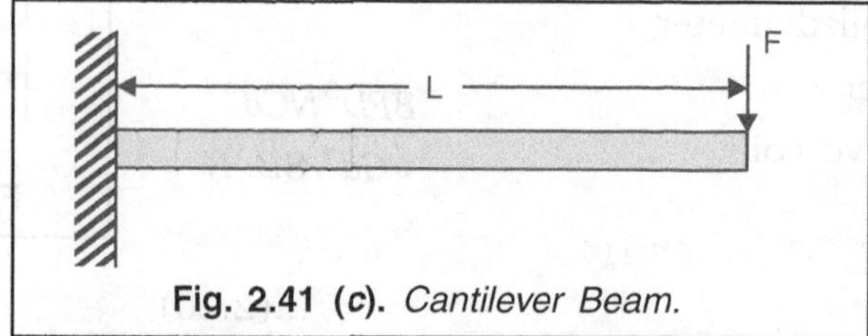

Fig. 2.41 (*c*). *Cantilever Beam.*

Cantilevers are best suited for the measurement of bending moments about two mutually perpendicular axes of the beam. If the point of application of force is known then bending moment can be directly translated into forces.

The strains developed due to applied force are measured by the strain gauges.

(c) Proving rings

It is a ring of known physical dimensions and mechanical properties. Proving ring is used to calibrate those instruments and machine where accurate measurement of large static loads is required such as tensile testing machines. (Fig. 2.42) shows a schematic diagram of a compression proving ring. In this the load applied is of compressive nature which deforms the ring. For the measurement of ring deflection, a highly precision micrometer is used. To ensure the desired precision in measurement, one edge of the micrometer is mounted on a vibrating reed device which is used to obtain a vibratory motion. At the periphery of the ring, bosses are provided to clamp the ring rigidly to avoid rotation. Whenever an external tensile or compressive force is applied over to these external bosses, the change in ring diameter takes place. This change in diameter of the ring is proportional to applied force.

As discussed earlier the deflection in the ring is measured by a micrometer screw and a vibrating reed provided inside the proving ring as shown in Fig. 2.42. The difference in the micrometer readings before and after the application of the force is the measure of the compression in the ring. If applied load is tensile in nature then the difference in the micrometer readings before and after the application of load becomes the measure of the amount of elongation in the proving ring. A linear variable differential transformer popularly called LVDT is used to monitor changes in the deflection of the proving ring by means of changes in the voltage.

Octagonal and extended octagonal type proving rings are preferred for the force measurement as these are designed in such a way that their top surface do not rotate when load is applied.

Working range: up to 1500 kN

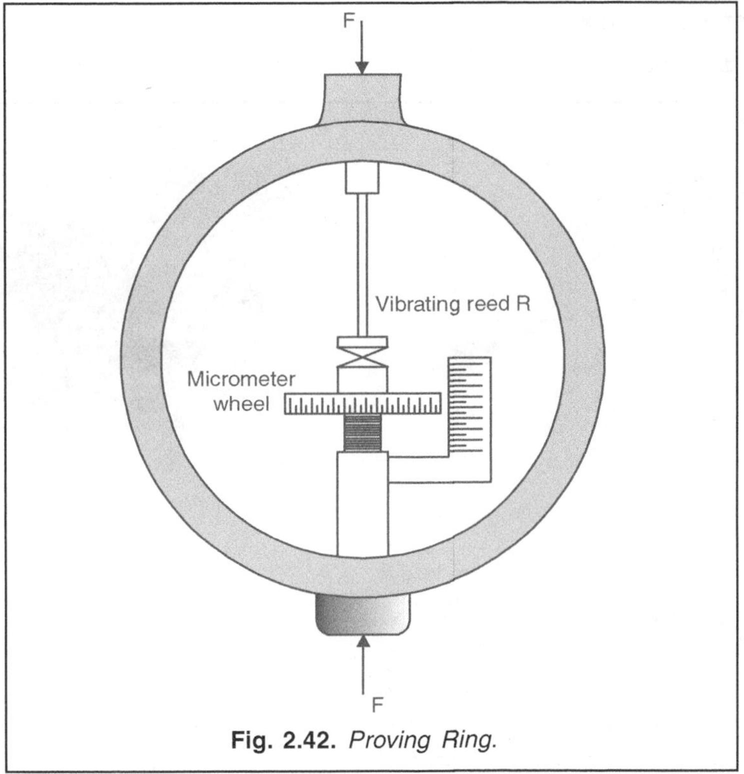

Fig. 2.42. *Proving Ring.*

(*d*) Load cells

A load cell is one of most popular electromechanical type transducer which is used to measure the forces (both types: static and dynamic), pressure and other loads. It converts force into a measurable electrical output. The load cells are generally classified on the basis of their output signals. These outputs can be hydraulic, pneumatic or electrical type. So the major categories of the load cells are:

(*i*) Hydraulic load cell,

(*ii*) Pneumatic load cell and

(*iii*) Electric load cells.

Load cell derives its output from the deformation of an elastic member having high tensile strength.

(i) Hydraulic Load Cell

In hydraulic load cell as shown in Fig. 2.43 consists of a diaphragm which transmits the force to a liquid when some force is applied over it. The liquid medium contained in a confined space has a preload pressure of about 196 kPa. Liquid pressure increases on applying the force over the diaphragm whose value can be obtained by using equation $p = \frac{F}{A}$, where F is the applied force and A is the effective cross sectional area of the diaphragm. This pressure is recorded by a precision pressure gauge calibrated directly in terms of some force units.

Working range: 25 MN

Accuracy: ± 0.1%

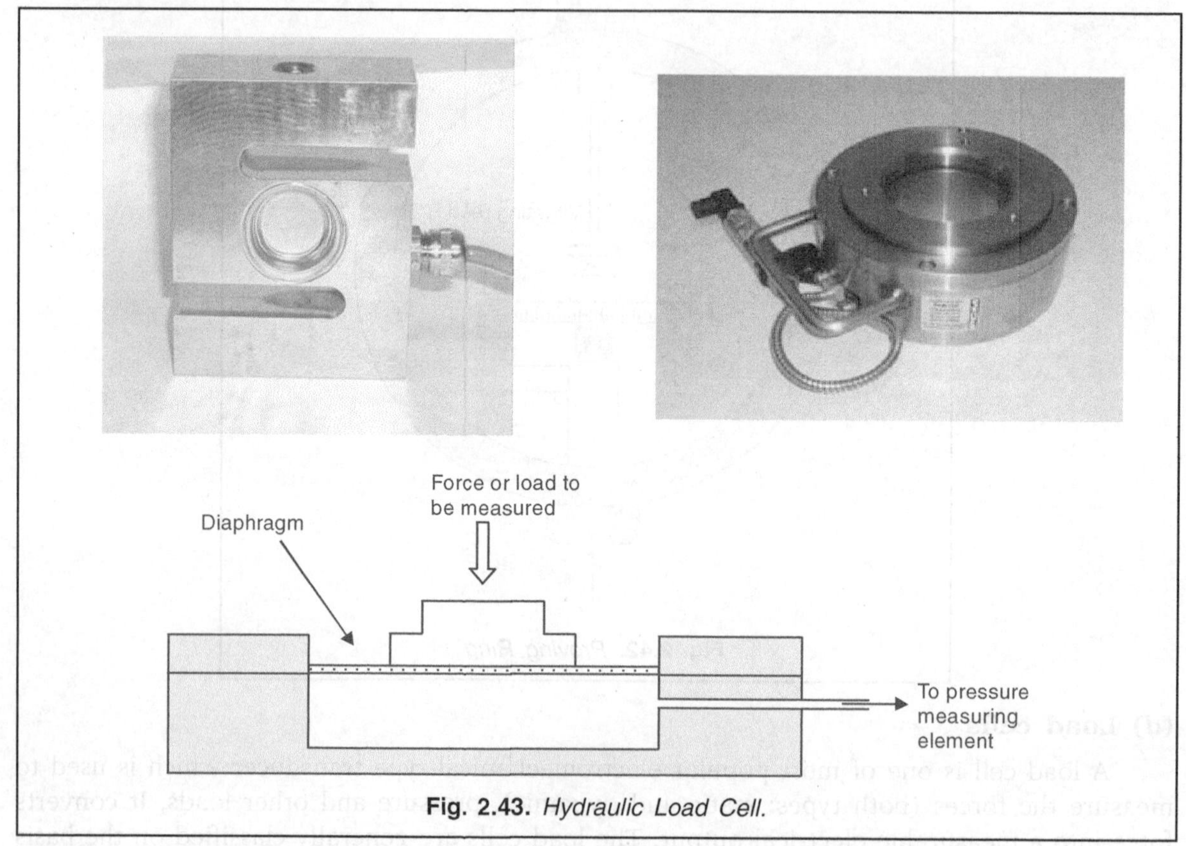

Fig. 2.43. *Hydraulic Load Cell.*

(ii) Pneumatic Load Cell

Pneumatic load cell as shown in (Fig. 2.44) operates on the force-balance principle and uses a nozzle-flapper type transducer. When force is applied over the flapper of the transducer comes near to the nozzle and this causes deflection in the diaphragm. In this load cell, the downward force is balanced by an upward force of air pressure against the effective area of the diaphragm. When diaphragm is deflected, the nozzle opening is closed down which in turn increases the system back pressure. This increased pressure when acts on the diaphragm, produces an effective upward force which pushes the diaphragm to obtains its preload position. For any constant applied force the system attains equilibrium at a specific nozzle opening and a corresponding pressure is indicated by the height of a mercury column in a manometer.

Working range: 2.5×10^3 kgf

Accuracy: $\pm$ 0.5%

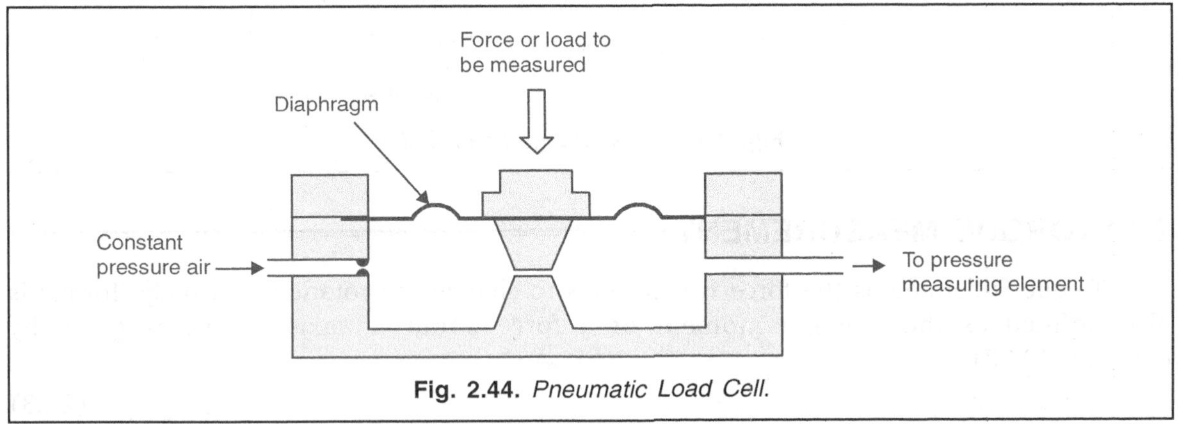

Fig. 2.44. *Pneumatic Load Cell.*

(*iii*) Electrical Type Load Cell

Electric type load cells find extensive used due to their linear characteristics and small size in comparison with hydraulic and pneumatic load cells. Electric load cell uses either inductive material, magnetostrictive transducer or a resistive material. The most commonly used among these is resistive load cells popularly called as stain gauge load cell.

(*iv*) Strain Gauge Load Cell

This type of load cell converts the load acting on them into electrical signals. The strain gauges used are resistive pick up type transducers. The strain gauges are bonded over a beam or structural member that deforms when force is applied. When force is applied strain changes the electrical resistance of the gauges in proportion to the load. The resistive type load cell design as shown in Fig. 2.45 uses four strain gauges to obtain maximum sensitivity and temperature compensation. These four strain gauges are adhered to a cylindrical tube as shown in Fig. 2.45 two of the gauges are usually in tension and two in compression. The positive aspect of the strain gauge load cell is that the effect of temperature can be compensated.

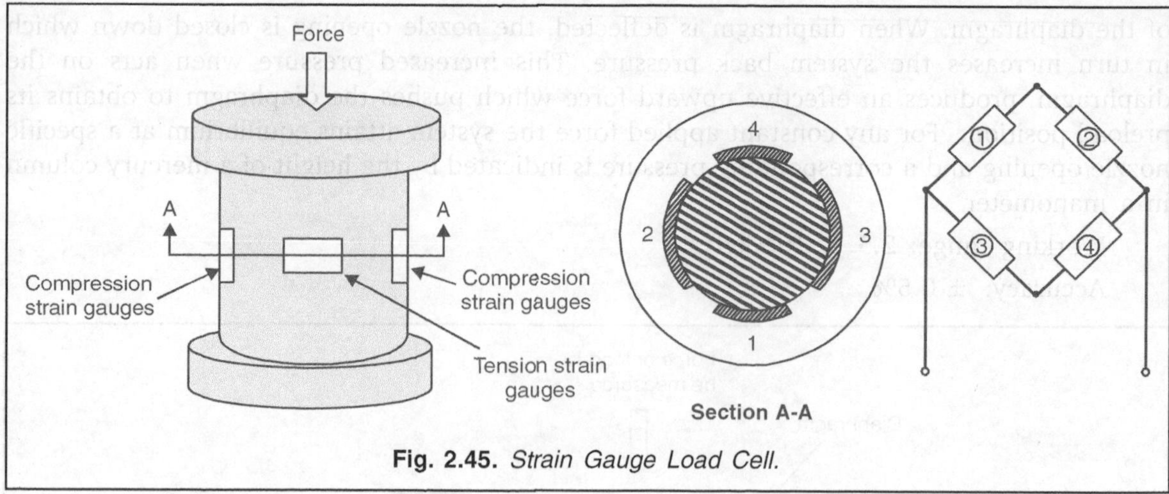

Fig. 2.45. *Strain Gauge Load Cell.*

2.19 TORQUE MEASUREMENT

Torque is defined as the force which tends to change the rotation of a body. Torque is also defined as the twisting moment of a force about an axis. Toque is given by equation (2.23).

$$T = F \times r \qquad ...(2.23)$$

Where T is the torque

F is the force, and

r is perpendicular distance from the axis of rotation of the line of action of the force.

The measurement of torque is directly related to the measurement of force. Therefore torque standards as such are not necessary, since force and perpendicular distance are sufficient to define the torque.

Measurement of applied torque is very important for all rotating bodies so as to ensure the design of the rotating element is safe under shear stresses.

There are three techniques of measuring the torque. These are:

(*a*) Prony brake torque measurement method.

(*b*) By measuring the strains produced in a rotating body due to an applied torque.

(*c*) By measuring the reaction forces in shaft bearings.

Dynamometer:

Dynamometer refers to a device that absorbs and measures the power output of a prime mover.

Prime mover:

Prime movers are the mechanical power producing machines which receive input energy from natural sources such as air, moving water, coal and petroleum etc which gets converted into the machanical energy. Examples of prime movers are internal combustion engines such as petrol engines, diesel engines, steam engines, steam turbines, gas turbines, wind mills etc.

2.19.1 Prony Brake Torque Measurement

'Prony brake' is the oldest device that is used for measurement of torque and dissipation of power from rotary machines. Prony brake is an example of an absorbing dynamometer.

Figure 2.46 shows a schematic diagram of a 'Prony brake'. In this wooden blocks are mounted on a flexible band or rope which is connected to the arm. Some arrangement is provided to tighten the rope to increase the friction resistance between the blocks and the rotating flywheel of the machine. The torque exerted on the Prony brake is given by

$T = F \times r$

Where F is the force which can be measured by conventional force measuring instrument as mechanical balances or by load cell.

Prony brake is cheap but difficult to adjust and maintain a specific load.

The power dissipated in the brake is calculated from equation (2.24).

$$\textbf{Power developed} = \frac{2\pi NT}{60 \times 1000} kW \qquad \text{...(2.24)}$$

Where T is the torque in kW and N is the speed of rotation in RPM

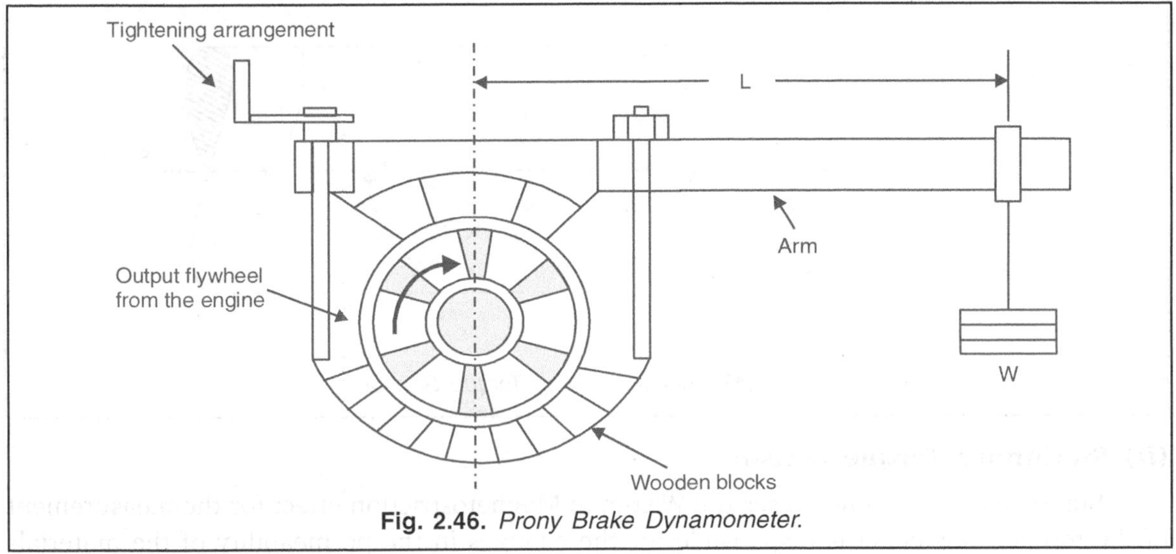

Fig. 2.46. *Prony Brake Dynamometer.*

2.19.2 Torque Sensors

Two types of torque sensors are available for torque measurements. These are:

(*i*) Rotating torque sensor

(*ii*) Stationary torque sensor.

(*i*) Rotating Torque Sensor

Rotating toque sensor is also known as in-line rotating torque sensor. It is used for the torque measurement of machines and other equipments having rotary motion. It consists of a metal shaft with bonded strain gauges electronically connected in the form of a Wheatstone bridge. The stresses acting on a rotating shaft subject to torsion are shown in Fig. 2.47. The strain gauges are kept on the shaft at precisely 45° to the shaft axis to sense tensile and compressive deformation due to torsion. The strain gauges 1 and 3 must be diametrically opposite, similarly strain gauges 2 and 4 must also be placed diametrically opposite. One can use the strain rosettes to ensure the accurate placement of the strain gauges. As we know from strength of material that along one direction, at 45° to the shaft axis, pure tensile stress exists whereas 45° in the other direction, pure compressive stress exists.

The strain elements are located on the shaft to sense compressive and tensile deformation due to torsion. The Wheatstone bridge output is in proportion to torsion and hence to the applied torque on the shaft. Bridge power and output voltage are connected to sensor through slip rings and brushes. The five rotating rings form the slip rings used in the system. These five rings support five pairs of brushes conducting excitation to and signals from the torque sensor. In this torque sensor, the contact resistance, contact friction, and heating effect greatly influence the measurement of torque. When rotation speed exceeds 30m/s then non-contact type rotary sensor is preferred for measuring the torque. This rotating torque sensor should be installed between flexible couplings so as to protect the transducer from any kind of the damage.

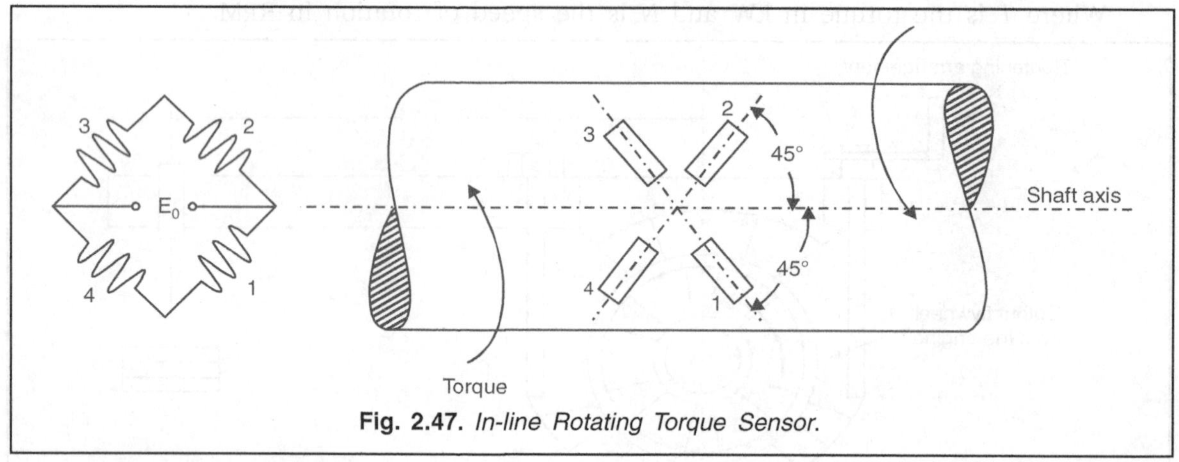

Fig. 2.47. *In-line Rotating Torque Sensor.*

(ii) Stationary Torque Sensor

Stationary torque sensor uses the Weidman Magnetostriction effect for the measurement of the torque. This effect is responsible for the changes in the permeability of the materials subjected to the tensile or compressive stresses. Permeability is a magnetic property related to the ability of the material to concentrate magnetic flux. Permeability tends to increase under tensile stresses and decreases under compressive stresses.

This sensor consists of two primary and two secondary winding mounted near the rotating shaft as shown in Fig. 2.48. The secondary coils S_1 and S_2 are coupled by magnetic lines of flux through the shaft to the primary windings. Whenever there the shaft is not loaded, the permeability of the shaft material is uniform and equal but opposite voltages are induced in secondary coils S_1 and S_2. When shaft starts transmitting the torque, permeability and the number of the magnetic lines of flux increase in the direction of tensile force and decrease in the direction of compressive force. When shaft is not transmitting the torque *i.e.,* there is no loading on the shaft then the Wheatstone bridge is in balanced condition and the output voltage is zero. The algebraic sum of voltage induced in the secondary coils S_1 and S_2 is proportional to the torque.

In stationary torque sensor, the primary and secondary windings are designed in such a way that produces a circuit analogous to the Wheatstone bridge as shown in (Fig. 2.48).

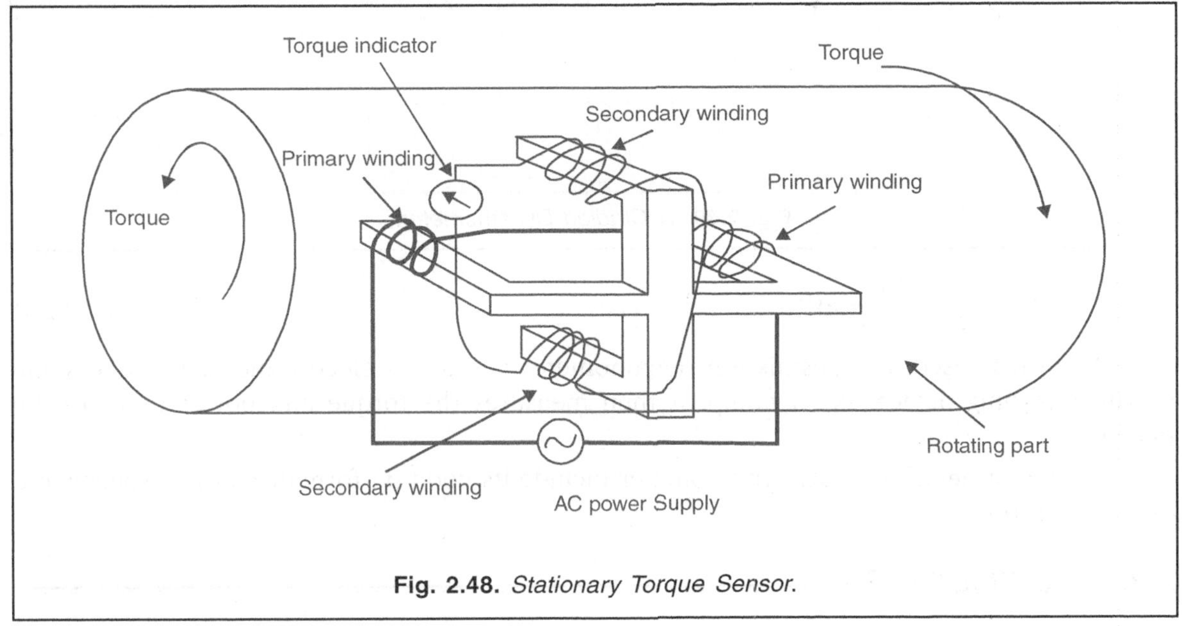

Fig. 2.48. *Stationary Torque Sensor.*

2.19.3 DC Cradled Dynamometer

The DC cradled dynamometer is perhaps the most widely used device for power and torque measurements on internal combustion engines, small steam turbines, pumps and other mechanical rotary machines. The basic arrangement of this device is shown in (Fig. 2.49). A DC dynamometer is basically a DC motor with an option to use it as a DC generator. In this DC dynamometer the input mechanical energy gets converted in to electrical energy which can either be dissipated through a resistance grid or recovered for use. A DC generator is mounted on bearings with a moment arm extending from the body of the motor to a force measurement device which is usually a pendulum scale. When the device is connected to power producing machine it acts as a DC generator whose output may be varied by dissipating the power through resistance grids. The torque impressed on the dynamometer is measured with the moment arm and the output power can be determined by using equation (2.25).

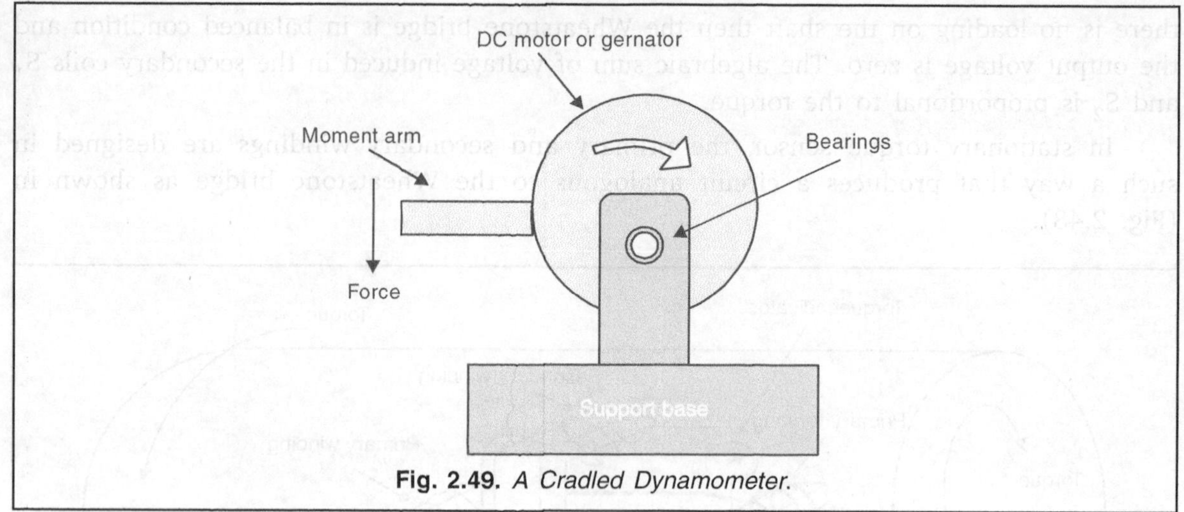

Fig. 2.49. *A Cradled Dynamometer.*

$$P = \frac{2\pi NT}{60 \times 1000} kW \quad ...(2.25)$$

When it is used as transmission dynamometer *i.e.*, as an electric motor to drive some power absorbing device like a pump, it then measures the torque and power input to the machine.

The major benefits of DC dynamometer include its good performance at low speeds and ease of control.

2.20 FLOWMETERS

All flowmeters can be classified under the following types depending on the principal of operation as well as other characteristics

1. Constant area—variable head flowmeters
2. Constant head—variable area flowmeters
3. Electromagnetic flowmeters
4. Mechanical flowmeters
5. Anemometer
6. Ultrasonic flowmeters
7. Vortex flowmeters.

2.20.1 Orifice Flowmeter

The orifice meter is the simplest and cheapest type of differential flowmeter for medium and large pipe sizes. The orifice plate when fitted in a pipeline causes an increase in the flow velocity and corresponding decrease in pressure. The orifice plate connected in the pipe line is basically a thin plate of metal with a circular opening. There are three types of orifice shapes in use as shown in (Fig. 2.50). These are (*a*) concentric, (*b*) eccentric and (*c*) segmented type. The most commonly preferred shape is concentric type. The type of orifice plate used depends upon the characteristics of the fluids whose flow rate etc., is to be measured. The other two types of orifice plates are used for fluids containing abrasive particles.

There are various techniques for the selection of the pressure measuring points on the pipeline called as pressure tapping. The usual practice is to locate tapping at distance D and D/2 in the upstream and downstream respectively where D is the diameter of the pipe.

(Figure 2.51) shows a schematic of a orifice plate flowmeter fitted across a pipe line.

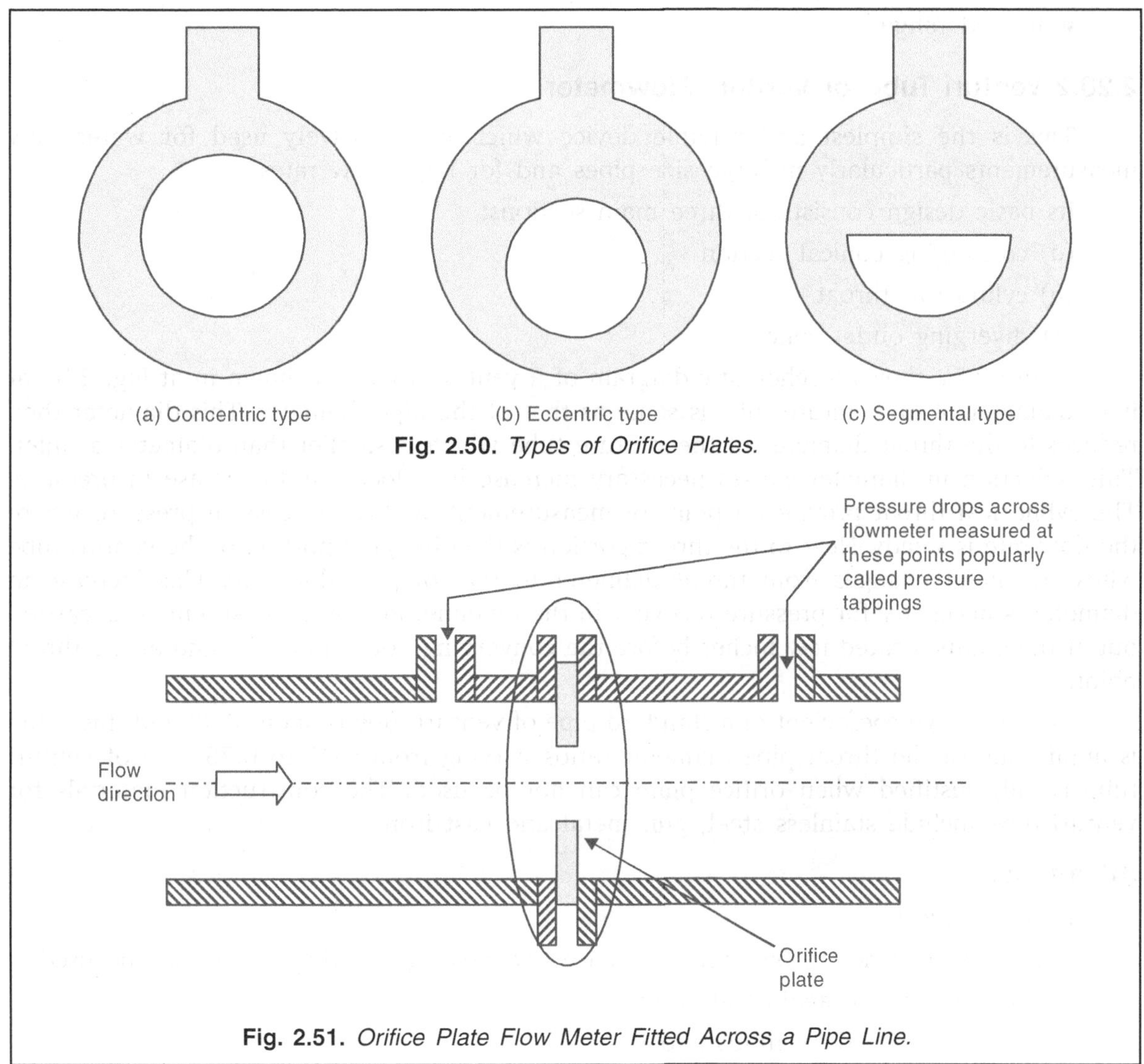

Fig. 2.50. *Types of Orifice Plates.*

Fig. 2.51. *Orifice Plate Flow Meter Fitted Across a Pipe Line.*

The construction materials for orifice plate include mild steel, stainless steel, gun metal and phosphor bronze.

Advantages

- simple construction
- no moving parts

- high reliability
- used with wide range of pipes (1.25 to 150 cm).

Limitations

- poor accuracy
- high pressure losses
- not suitable for high velocities
- limited range.

2.20.2 Venturi Tube or Venturi Flowmeter

This is the simplest and reliable device which is extensively used for water flow measurements particularly in large size pipes and for large flow rates.

Its basic design consists of three main sections;

(*a*) converging conical section

(*b*) cylindrical throat

(*c*) diverging outlet cone.

Figure 2.52 shows a schematic diagram of a venture tube. As shown in at Fig. 2.52 at inlet diameter of the venture tube is same as that of the pipe diameter. This diameter then reduces to the throat diameter of the venture tube which is smaller than diameter at inlet. This reduction in diameter causes necessary increase in velocity and decrease in pressure. The cylindrical throat provides a point of measurement of this decrease in pressure where the flow rate is steady. Next to the throat portion is the diverging portion of the venturi tube whose diameter changes from throat diameter to that of pipe diameter. This increase in diameter is necessary for pressure recovery in the pipeline. Pressure measurement is carried out at the points located few inches before the converging portion of tube and at the throat point.

The discharge coefficient of a standard pipe of venturi tube is about 0.99 and this value is applicable for the throat pipe diameter ratios varying from 0.25 to 0.75. Use of venturi tube is only justified when orifice plate can not be used. The construction materials for venturi tube include stainless steel, gun metal and cast iron.

Advantages

- smooth design
- section need not to be circular, square and rectangular shapes can also be used
- better resistance against abrasion
- well suited for dirty fluids and gases.

Limitations

- large size
- high cost

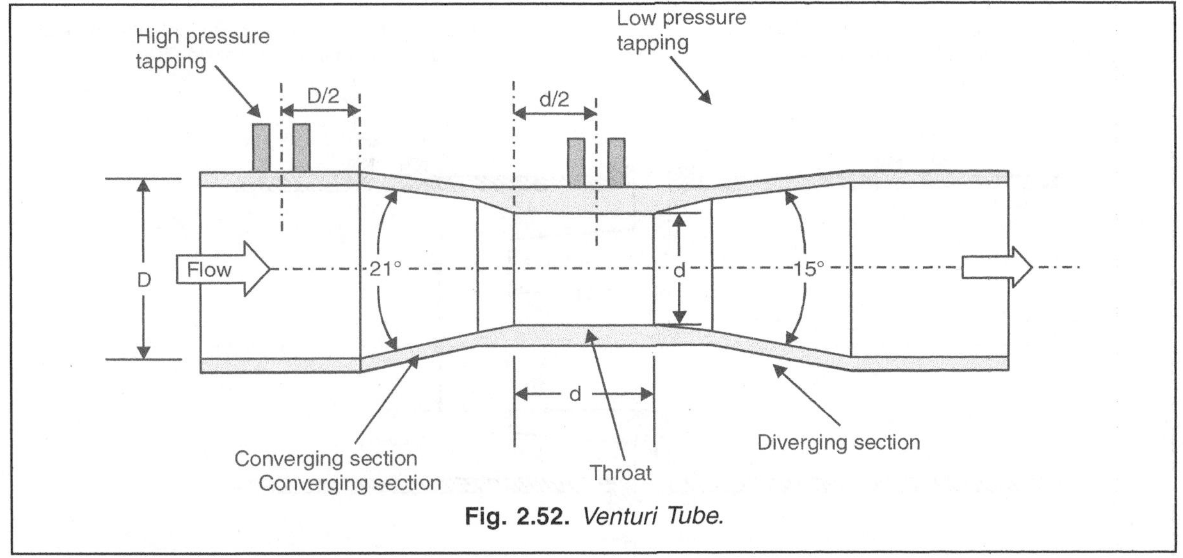

Fig. 2.52. *Venturi Tube.*

2.20.3 Flow Nozzle

The flow nozzle as shown in (Fig. 2.53) is a primary flow metering device which combines the best features of the orifice plate and venturi tube. It has a discharge coefficient close to one. Flow nozzles have some similarity with venturi tube as it has curved form of approach. Due to this, it features gradual change in cross sectional area and thus have coefficient of discharge similar to venturi tube. The curvature on this flow nozzle makes it more suitable for dirty fluids or fluids carrying abrasive particles etc. Its construction is simpler than a venturi tube. Flow nozzle is commonly used for high velocity, high temperature applications such as metering of steam.

Advantages

- no sharp edges to erode
- cheaper than venturi tube
- suitable for dirty and abrasive fluids.

Limitations

- high cost
- difficult installation
- not available for above 120 cm diameter
- poor pressure recovery

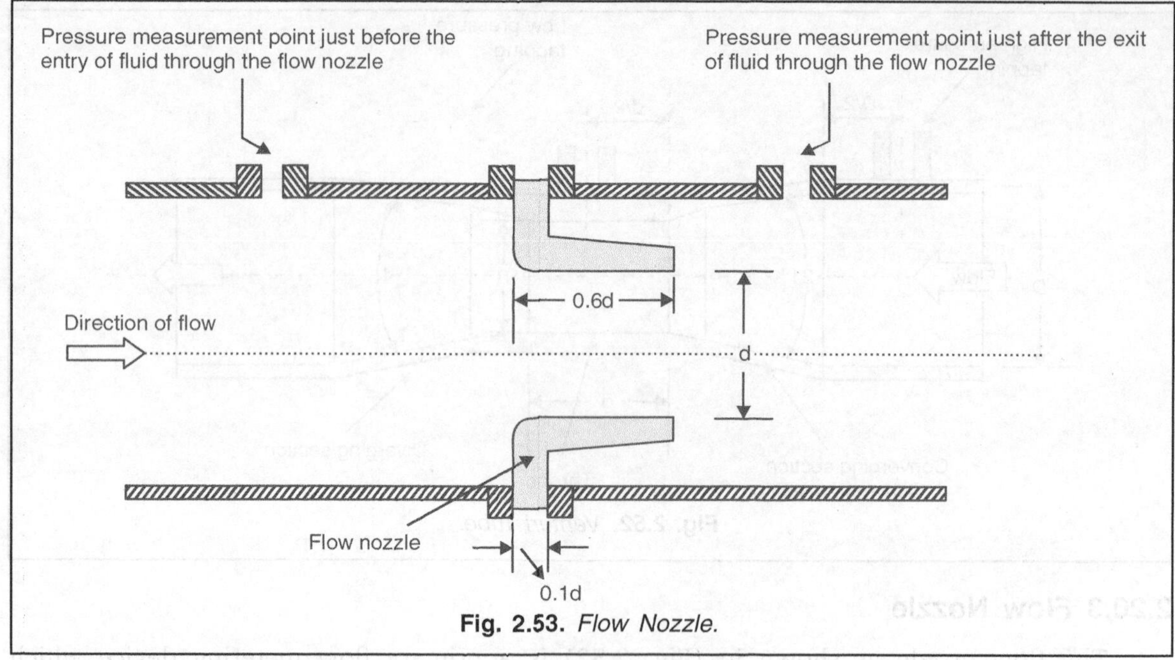

Fig. 2.53. *Flow Nozzle.*

2.20.4 Pitot Tube

It is used to measure the local fluid velocity at a point in the channel or duct through the fluid is flowing. Pitot tube as shown in Fig. 2.54 is mainly used for the measurement of air or gas velocities. For example air velocity measurements on board an airplane. It consists of a cylindrical probe inserted into the fluid stream. Pitot tube converts the velocity head into an impact pressure. The difference between the static pressure and the impact pressure is a measure of the flow rate.

The basic design of Pitot tube involves two concentrically placed tubes bent at 90° as shown in Fig. 2.54. Since flow does not take place through the Pitot tube so flow comes to rest at the Pitot tube tip. Corresponding to this position the new pressure is developed due to the obstruction in fluid flow. This new pressure developed at the Pitot tube is called impact pressure and is designated by p_i. Few centimeters before the placement of Pitot tube, static pressure of the fluid flow is measured which is designated by p_s. By using the equation (2.26), flow velocity can be determined.

$$v_f = \sqrt{\frac{2g(p_i - p_s)}{\rho}} \quad \text{...(2.26)}$$

This equation is deduced by applying the Bernoulli's Theorem. At the static pressure point $p_1 = p_s$, the flow velocity is $v_1 = v_f$ and at impact pressure point $p_2 = p_i$, the flow velocity $v_2 = 0$.

Advantages

- Suitable for high flow velocity fluids
- Cheap
- Easy to install.

Limitations

- Measures velocity at only one position in the cross section of the tube
- Poor accuracy
- Not suitable for dirty fluids
- Sensitive for the fluid flow disturbance if any before the tube tip.

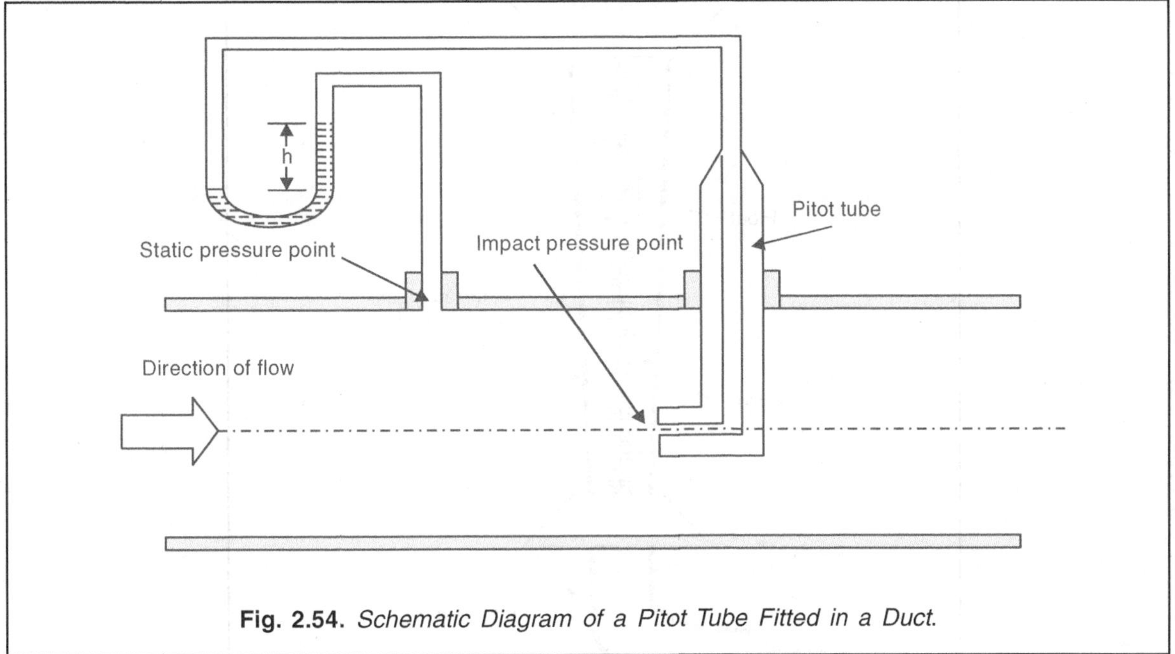

Fig. 2.54. *Schematic Diagram of a Pitot Tube Fitted in a Duct.*

2.20.5 Rotameter

A rotameter as shown in Fig. 2.55 is also referred to as variable area flow meters. A rotameter consists of a vertical tube with a tapered cone and inside which is a float which moves on corresponding to each flow rate through the tube. The conical tube is made of glass or stainless steel and the floats are made of brass, special plastics or stainless steel.

As the flow increases the drag on the float increases and causes it to rise in the tube. The rise of a float to a section of a tube where the cross-sectional area of tube is more causes the reduction in the drag due to annular flow around the float. The float will thus attain a particular height corresponding to the condition where the upwards drag of the flow around the float precisely balances the gravitational force on the float.

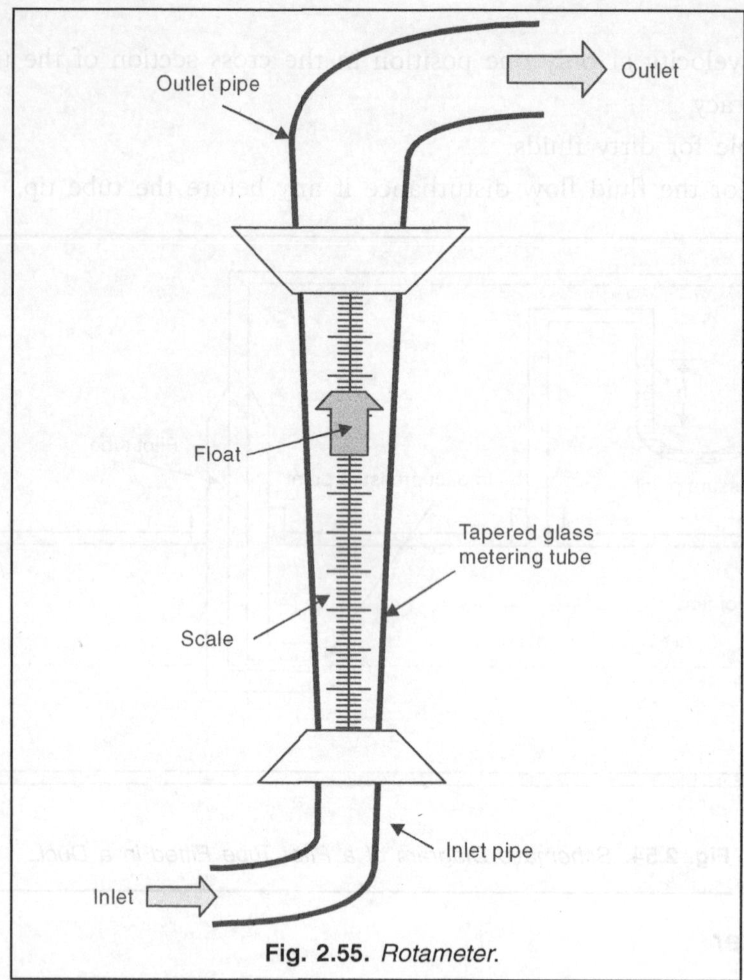

Fig. 2.55. *Rotameter.*

The fundamental equation for an incompressible flow through a rotameter is given by equation (2.27).

$$Q = C_d A \sqrt{2g \frac{V_f}{A_f} \left(\frac{\rho_2 - \rho_1}{\rho_1} \right)} \quad ...(2.27)$$

Where

Q = the flow rate

C_d = the discharge coefficient

A = the gap area between the float and the tube

V_f = the volume of float

A_f = surface area of the float

ρ_2 = the density of float material

ρ_1 = the density of liquid

The lift on the float is against gravity so during measurement the tube should be mounted vertically. Rotameter is used to measure gas and liquid flows in the range of 0.1×10^{-8} to 0.1 m^3/s for a wide range of liquids.

Applications

- It finds extensive applications in the laboratory, testing and production lines.
- Liquid or gas flowmeter where high accuracy in measurement is not essential.
- Where visual indicator of flow is required.

Advantages

- Inexpensive device.
- Simple operation.
- Ease of reading and installation.
- Visual inspection of flow rate.

Disadvantages

- Low accuracy in comparison with venture and orifice meters.
- Sensitive to viscosity and density of the fluid.
- Affected by oscillations during pulsating flows.
- Limited capacity.

2.20.6 Mechanical Flowmeters

Mechanical flowmeters signifies the quantity of fluid that flows past a given point in a specified period of time. The average flow rate can be estimated by dividing the total flow quantity by the time taken to flow. The two most commonly used flowmeters are:

(*a*) Positive displacement meter

(*b*) Turbine flowmeter

(*a*) Positive displacement meter

These are basically designed to perform as flow quantity meters rather than flow rate meter. These are used to measure the volume flow rates. The important flowmeters belonging to this category are: sliding vane flowmeter, rotary vane flowmeter, lobed-impeller flowmeter, reciprocating piston meter etc. All these meters are self powered and drive mechanical counters. In these flowmeters, rotation of the mechanical elements carries the fluid in to the self contained compartments. These close compartments are formed by means of sliding vanes or by specially shaped rotors which keep contact as they rotate and allow only a certain volume to flow through the flowmeter from its inlet to exit passage

Applications: for measuring household water consumption, LPG flow rate etc.

(*i*) Sliding vane positive displacement flowmeter

The sliding vane positive displacement meter is shown in Fig. 2.56. These flowmeters are the most accurate. In this type of meters, the vanes moves radially out to form compartment and goes back to their seats to release the fluid and then move back to the inlet side. Volume passed is proportional to the rotation and the flow rate is proportional to rotational speed.

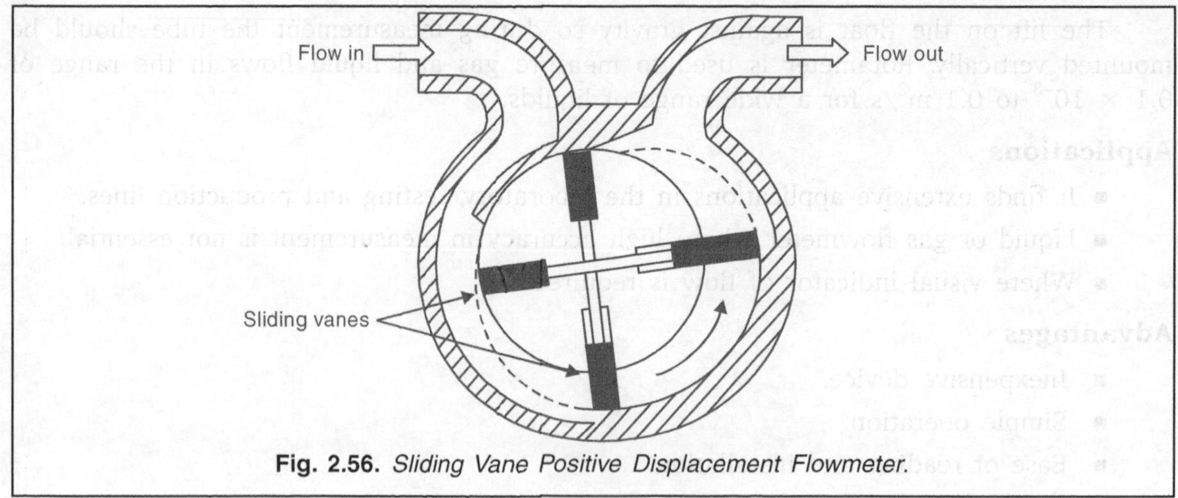

Fig. 2.56. *Sliding Vane Positive Displacement Flowmeter.*

(*ii*) Impeller type flow meter

Impeller flow meter as shown in (Fig. 2.57) finds extensive use in the measurement of solvents and dry liquids. It consists of two identical oval shaped rotors that mesh together similar to a gear drive. In this flowmeter the rotational speed is proportional to the volume of the fluid flow. The swept volume between each rotor and the chamber wall is passed four times per revolution by this flowmeter.

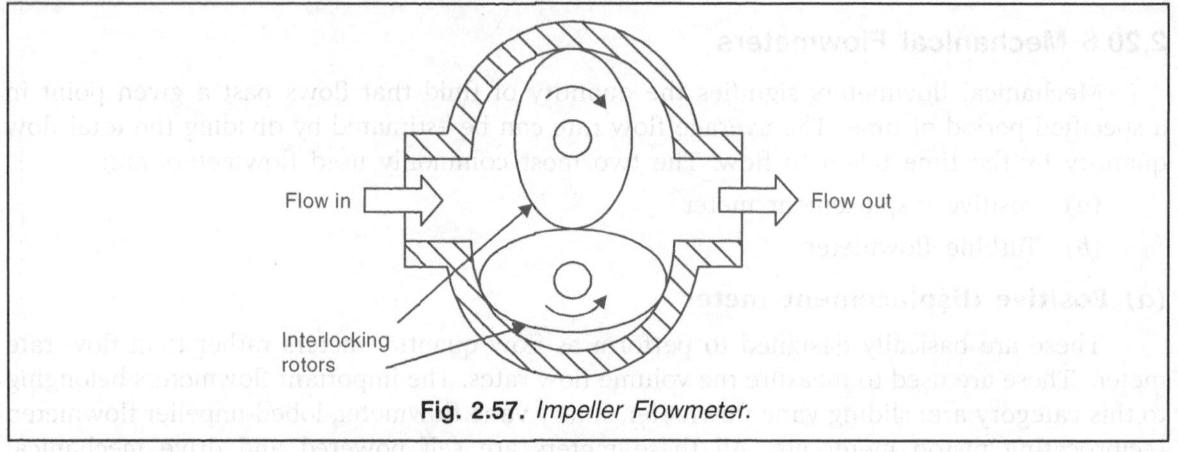

Fig. 2.57. *Impeller Flowmeter.*

(*b*) Turbine flowmeter

It is the non-frictional positive displacement type flowmeter. Turbine flowmeter as shown in (Fig. 2.58) drives its name from the use of its operating principle. This flowmeter uses a turbine wheel which is axially mounted.

This turbine wheel is placed in the path of fluid flow. The flowing fluid spins the turbine wheel when the fluid stream impinges on the turbine blades. The angular speed of the turbine wheel is proportional to the flow velocity as a function of blade angle. When a steady angular speed is achieved, the angular speed can be taken as proportional to the volumetric flow rates of fluid. (Figure 2.58) shows a schematic diagram of a turbine flowmeter.

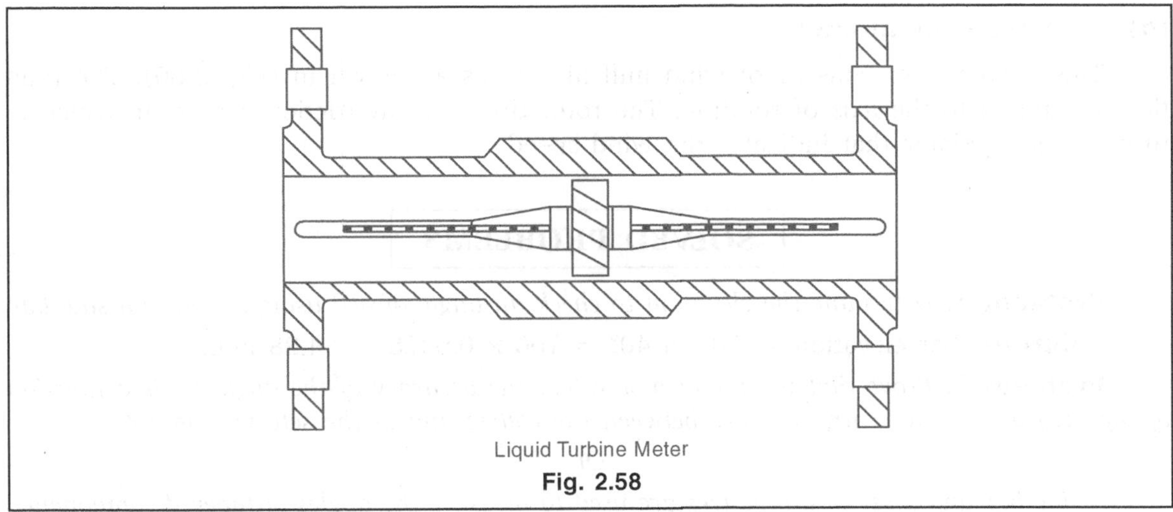

Liquid Turbine Meter

Fig. 2.58

2.21 ANEMOMETER

Anemometers are used to measure velocities of fluid streams. It is used to measure fluid flow velocity in wind tunnel, water flow in a closed channel, and the wind speed.

There are two widely used anemometers are:

(*a*) Cup type anemometer

(*b*) Vane type anemometer.

(*a*) Cup type anemometer

Cup anemometer consists of a vertical spindle to which three or four cups are attached as shown in (Fig. 2.59). The spindle carrying cups is mounted on bearings and rotates freely about the vertical axis. When flow stream like air passes through the anemometer, drag forces act on the cup to cause the rotation of vertical spindle. For example if air is blowing from left direction, drag on the cup A will be more than at cup B. The resultant torque rotate the assembly in anticlockwise direction. The rotation of the spindle can be read from the dial gauge provided with the unit. The frequency of rotation gives a measure of the average air velocity in the region.

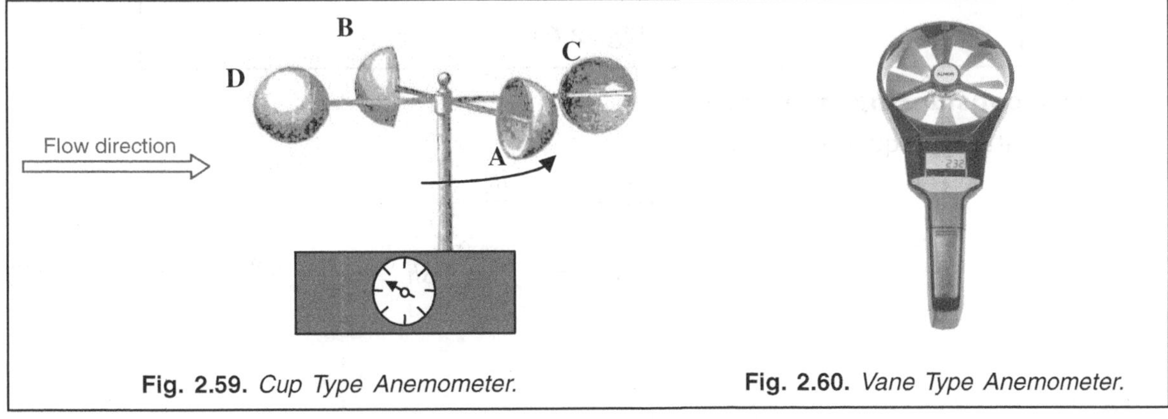

Fig. 2.59. *Cup Type Anemometer.* **Fig. 2.60.** *Vane Type Anemometer.*

(*b*) Vane type anemometer

This anemometer consists of wind mill like vanes as shown in (Fig. 2.60). The fluid flow is parallel to the axis of rotation. The rotor drives a low friction gear train which in turn drives a pointer that indicates the wind speed.

SOLVED PROBLEMS

Problem 1. *Determine the elevation to check an angle of 40° using a 100 mm sine bar.*

Solution. Bar elevation $= 100 \sin 40° = 100 \times 0.6428 = 64.28$ mm.

Problem 2. *Prove that in case of a sine bar, the accuracy of the angle set is a function of the accuracy of the centre distance between the rollers and of the setting height ?*

Or

A sine bar and a set of slip gauges are used to measure the angles of tapered components. Neglecting errors due to the use of slip gauges prove that a given sine bar achieves greater accuracy while measuring smaller angles then while measuring large angles.

Solution. As we can see from the (Fig. 2.61)

$$\sin\theta = h/L \qquad \text{... } (i)$$

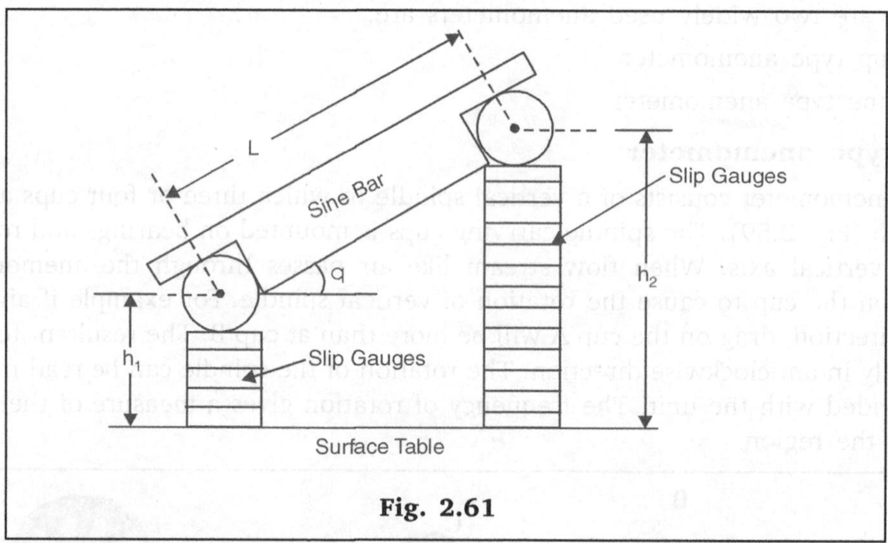

Fig. 2.61

First considering h to vary

Differentiating equation (*i*) we get

$$\cos\theta = \frac{dh}{L}$$

$$\Delta\theta_1 = \frac{\sec\theta}{L}\,\Delta h \qquad \text{... } (ii)$$

Secondly considering L to vary

Differentiating equation (*i*), we get

$$\cos\theta\, d\theta_2 = \frac{h}{L^2}\, dL$$

or

$$d\theta_2 = \frac{1}{\cos\theta}\frac{h}{L}\frac{dL}{L}$$

$$d\theta_2 = \frac{\sin\theta}{\cos\theta}\frac{dL}{L}$$

$$d\theta_2 = \frac{\tan\theta}{L}\Delta l \qquad \text{... (iii)}$$

$$\text{Total error in } \theta = \Delta\theta = \frac{\sec\theta}{L}\Delta h - \frac{\tan\theta}{L}\Delta l \text{ radians} \qquad \text{... (iv)}$$

It is clear from equation (*iv*) that

- the accuracy of angular setting is high for the higher value of L whereas other things remaining constant.
- the accuracy of angular setting is low for the higher value of θ. Since at θ→90°, the value of sec θ and tan θ is ∞ (infinity) and it is unlikely that dimensions L and h are totally error free.

Problem 3. *A 100 mm Sine bar is used to find the included angle of a plug gauge as shown in (Fig. 2.62). The height of the slip gauges below the lower cylinder is 25mm.*

(*a*) *If the cylinder is 30° what is the height of slip gauges below the upper cylinder ?*

(*b*) *What would the height of the slip gauges (Fig. 2.63) below the upper cylinder be if the length of Sine bar was 100.005 mm and the upper cylinder 0.005 mm larger ?*

Solution. (*a*) Let d_1 and d_2 to be cylinder diameters on which the sine bar rests and x be the height below the upper cylinder.

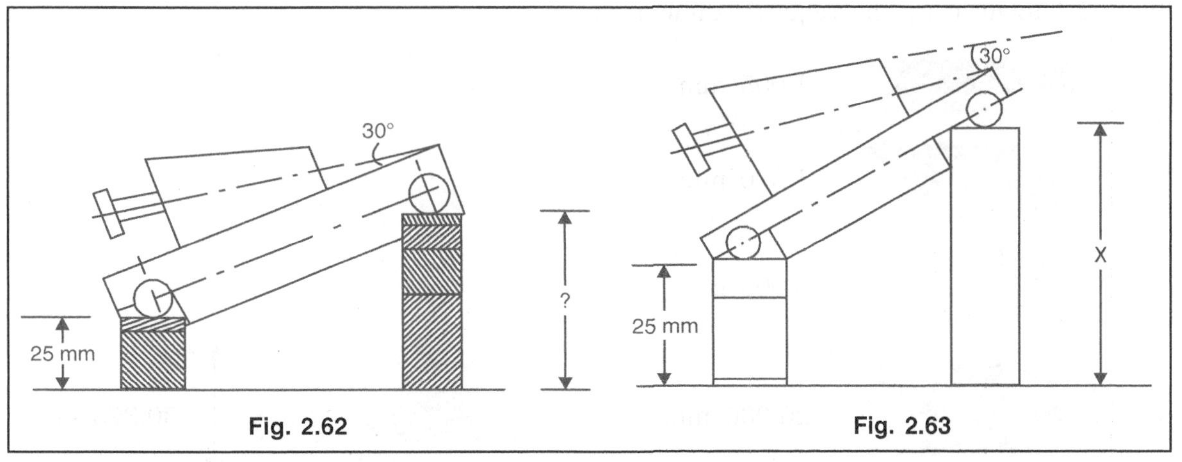

$$h_1 - h_2 = L\sin\theta_2 \qquad \text{... (i)}$$

$$\text{Height of upper cylinder } h_1 = x + \frac{d_1}{2}$$

$$\text{Height of lower cylinder } h_2 = 25 + \frac{d_2}{2}$$

On considering equal cylinder diameters

$$d_1 = d_2$$

From equation (*i*), we get,

$$x + \frac{d_1}{2} - 25 - \frac{d_2}{2} = 100 \sin 30^\circ$$

$$\therefore \quad x = 25 + 100 \sin 30^\circ = 75 \text{ mm} \qquad \textbf{Ans.}$$

(*b*) If $L = 100.005$ mm

and $d_1 = d_2 + 0.005$

$$\text{Height of upper cylinder } h_1 = x + \frac{d_1}{2}$$

$$\text{Height of lower cylinder } h_2 = 25 + \frac{d_2}{2}$$

From equation (*i*), $h_1 - h_2 = L \sin \theta$

$$\therefore \quad x = 25 + \frac{d_2}{2} + L \sin \theta - \left[\frac{d_2 + 0.005}{2}\right]$$

$$x = 25 + 100.005 \sin 30^\circ - \frac{0.005}{2} = 75.0025 - 0.0025 = 75 \text{ mm.}$$

Problem 4. *How will you combine the slip gauges to measure 30.375 mm.*

Solution. One can select the slip gauge dimension from a set of 88 pieces as given in Table 2.4 to measure the required dimension.

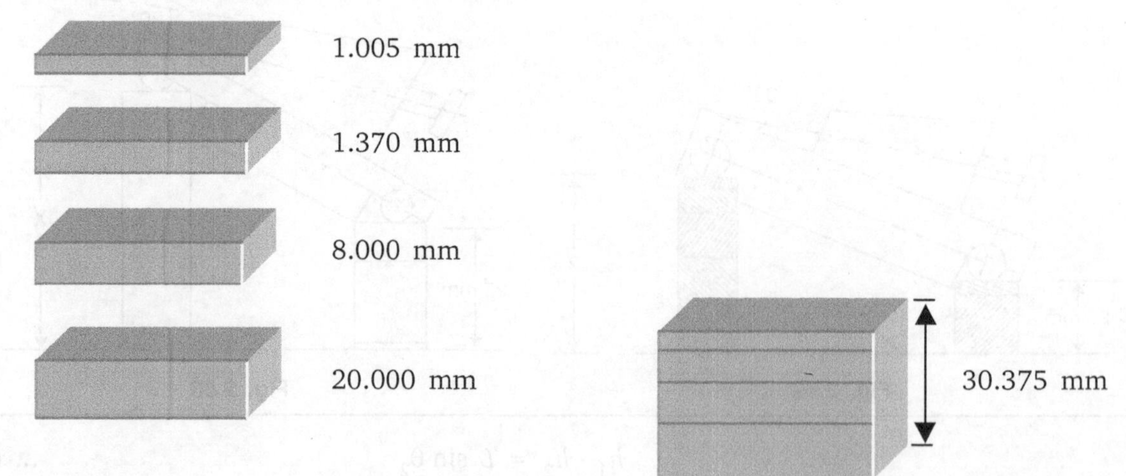

REVIEW QUESTIONS

1. Classify the different measuring tools used in modern workshop.

2. What is the function of the surface plate? List the advantages of granite surface plate over the cast iron surface plate.

3. Explain the construction and working of a Combination Set. *(RGPV, June 2002)*

4. Describe the construction and use of a Combination Set. *(RGPV, Dec. 2001)*

5. How the angular measurement is carried out with the help of a Sine bar? Explain.

6. List down the four applications of a dial gauge.

7. Describe the principle of a Vernier caliper.

8. Distinguish between the direct and indirect reading linear measurements. Name the instruments used in each category.

9. What is a sine bar? Why does it have this name?

10. Draw a neat sketch of a Combination Set. Show different parts and state the different functions that it can perform.

11. Give a detailed classification of micrometer.

12. How will you find out the least count of a micrometer? Sketch the diagram of a micrometer and label its different parts.

13. Explain the following:

(*a*) Inside Micrometer

(*b*) Outside Micrometer

(*c*) Depth Micrometer.

14. What are the Slip gauges? Where can you use them?

15. What is the Bevel protractor? How do you use it for precision angular measurements? Explain.

16. Give the constructional details of a Sine bar. List its important applications.

17. Explain the importance of surface plate and straight edge in the engineering metrology lab.

18. Discuss the use of slip gauges in the industry. How can you combine them to measure 40.175 mm?

19. List the different grades of slip gauges or gauge blocks used in industry.

20. Name the suitable instrument for checking

(*a*) Flatness

(*b*) Roundness

(*c*) Concentricity.

21. Draw a neat sketch to illustrate the use of 'sine bar' for measurment of taper angle of taper plug guage and explain it briefly. *(RGPV, Dec. 2003)*

22. Write short notes on the following:

(*a*) Dial guage

(*b*) combination set *(RGPV, June, 2004)*

23. How are elastic elements used for force or torque measurements?

24. Explain with the help of a sketch the construction and working of a proving ring.

25. Explain different methods of measuring the flow rates.

26. Explain the construction and application of Prony brake dynamometer.

27. Explain the following properties of any measuring instrument:

(*a*) Hysteresis,

(*b*) Sensitivity,

(*c*) Response time, and

(*d*) Precision and accuracy. *(RGPV April 2009)*

28. Explain the construction, working and uses of a micrometer. *(RGTU June 2008)*

29. Write short notes on the following

(*a*) Errors of measurement

(*b*) Uncertainty analysis. *(RGTU June 2008)*

30. What are thermistors? How do they differ from resistance thermometers?

OBJECTIVE TYPE QUESTIONS

Select the most appropriate answer:

1. Sine bar is specified by

(A) the size of roller diameters (B) weight of the sine bar

(C) the centre distance between the two rollers

(D) its total length

2. The surface finish of all the working surfaces and the cylindrical surfaces of the rollers of the sine bar is of the order of

(A) 0.2μ (B) 0.5μ

(C) 1.0μ (D) 10μ

3. Straightedge is used to measure

(A) ovality (B) flatness

(C) parallelism (D) perpendicularity

4. The slip gauges are classified in .... categories according to accuracy.

(A) five (B) four

(C) three (D) two

5. The standard length of a sine bar is measured

(A) from edge to edge of the bar.

(B) between inner circumference of two rollers.

(C) between outer circumference of two rollers.

(D) between the centres of two rollers.

6. The vital characteristic of a measuring instrument in general is ....

(A) repeatability (B) accuracy

(C) sensitivity (D) precision.

7. The number of Slip gauges in a set are

(A) 88 (B) 103

(C) 45 (D) all of the above.

8. Surface plate is usually made of granite because

(A) it possesses high degree of surface flatness.

(B) there is no tendency to warp.

(C) it is free from built-in residual stresses.

(D) its surface is rust proof.

(E) all of the above.

9. Accuracy of setting a sine bar

(A) decrease with steep angle. (B) is not good for small angles.

(C) is very good for large angles. (D) none of the above.

10. The combination set is used to

(A) scribe lines. (B) to check angular surfaces.

(C) draw ellipses. (D) none of the above

11. The advantage of vernier caliper over the micrometer is that it

(A) is easy and quick to use. (B) is more accurate.

(C) can be used to make inside as well as outside measurement.

(D) is more rigid.

12. For the precision measurement, two gauge blocks (slip gauges) are joined by

(A) adhesion (B) welding

(C) riveting (D) wringing.

13. Bevel protractor is used for

(A) angular measurement. (B) linear measurement.

(C) length measurement. (D) angular and linear measurement.

14. An angle measuring device is

(A) Bevel protractor (B) sine bar

(C) Combination square (D) all of the above.

15. Millimetre (mm) scale in case of micrometer is marked on

(A) barrel (B) anvil

(C) thimble (D) spindle.

16. circular scale of the micrometer is marked on

(A) anvil (B) thimble

(C) ratchet (D) barrel.

17. Accuracy of an instrument is defined as

(A) closeness with which the reading approaches an accepted standard value or the true value

(B) the ratio of output signal to input signal

(C) closeness with which individual measurement are distributed about their mean value
(D) the smallest change in the measured value.

18. Repeatability of an instrument is defined as
(A) closeness with which the reading approaches an accepted standard value or the true value
(B) the ratio of output signal to input signal
(C) closeness with which individual measurement are distributed about their mean value
(D) the measure of the deviation of test results from the mean value.

19. Calibration is defined as
(A) closeness with which the reading approaches an accepted standard value or the true value
(B) closeness with which individual measurement are distributed about their mean value
(C) the comparison of specific values of input and output of an instrument with a corresponding reference standard.
(D) the smallest change in the measured value.

20. Precision is defined as
(A) closeness with which the reading approaches an accepted standard value or the true value
(B) the ratio of output signal to input signal
(C) closeness with which individual measurement are distributed about their mean value
(D) the smallest change in the measured value.

21. the smallest change in measured value to which the instrument will respond is called
(A) sensitivity (B) resolution
(C) accuracy (D) hysteresis.

22. Load cell is a
(A) force measuring sensor (B) current measuring sensor
(C) humidity measuring sensor (D) temperature measuring sensor.

23. Local velocity is measured by
(A) venturi tube (B) pitot tube
(C) orifice plate (D) none of the above.

24. Dynamometers are
(A) force measuring devices (B) energy measuring device
(C) torque measuring devices (D) Power measuring devices.

25. The minimum value of input which is necessary to cause a detectable change from zero output is called
(A) least count (B) threshold
(C) precision (D) resolution.

26. Which of the following thermocouple generates highest emf for a given temperature
(A) chromel-alumel (B) platinum-rhodium platinum
(C) iron-constantan (D) copper-constantan.

27. The function of the flow nozzle is to measure

(A) flow rate (B) temperature

(C) pressure (D) voltage.

28. Proving rings are used to measure

(A) acceleration (B) Velocity

(C) Displacement (D) Force.

29. Match List A with List B and select the correct answer using the codes given below the list:

List A	**List B**
A. Rotameter	1. Vena contracta
B. Venturimeter	2. End contraction
C. Orifice meter	3. Tapering tube
D. Flow nozzle	4. Convergent-divergent
	5. Bell mouth entry

(A) A1-B2-C3-D4 (B) A3-B4-C1-D5

(C) A5-B4-C2-D1 (D) A3-B5-C1-D2

30. The ratio of output signal to a change in input variable is called

(A) accuracy (B) drift

(C) threshold (D) sensitivity.

31. Which one of the following is measured by a rotameter?

(A) velocity of fluids (B) discharge of fluids

(C) Viscosity of fluids (D) Rotational speeds of solid shafts.

32. The differential manometer connected to a Pitot static tube used for measuring fluid velocity gives:

(A) static pressure (B) Total pressure

(C) Dynamic pressure

(D) Difference between total pressure and dynamic pressure.

33. Error of a measuring instrument is

(A) closeness with which the reading approaches an accepted standard value or the true value

(B) the ratio of output signal to input signal

(C) closeness with which individual measurement are distributed about their mean value

(D) the difference between the measured value and actual value.

34. Systematic errors are

(A) unknown errors (B) random errors

(C) regularly and repetitive in nature (D) unpredictable errors.

35. In general, the most important characteristic of a measuring instrument is

(A) readability (B) precision

(C) adaptability (D) repeatability

(E) accuracy.

36. ....are usually found in the measuring system and it is very difficult to eliminate them.

(A) random errors (B) calibration errors

(C) manual errors (D) systematic errors.

ANSWERS

1. (C)	**2.** (A)	**3.** (B)	**4.** (A)	**5.** (D)	**6.** (A)
7. (D)	**8.** (E)	**9.** (A)	**10.** (B)	**11.** (C)	**12.** (D)
13. (A)	**14.** (D)	**15.** (A)	**16.** (B)	**17.** (A)	**18.** (D)
19. (C)	**20.** (C)	**21.** (B)	**22.** (A)	**23.** (B)	**24.** (D)
25. (B)	**26.** (C)	**27.** (A)	**28.** (D)	**29.** (B)	**30.** (D)
31. (B)	**32.** (C)	**33.** (D)	**34.** (C)	**35.** (D)	**36.** (A).

■■■

3 Machine Tools

3.1 INTRODUCTION

A machine tool is a machine for making components by removing material from the workpiece by using a cutting tool. Machine tools are capable of producing themselves therefore the machine tools are often referred to as mother machine. In the manufacturing processes, machine tools are the most versatile and almost any product can be produced with them. The machine tools should able to fulfill the following requirements:

Requirements of a machine tool

- The machine tool should be able to produce consistently the components of specified shape and size, dimensional accuracy and surface finish. These features depend upon the rigidity of the machine elements.
- Machine tool should be able to provide the flexibility of machining the material at various cutting speeds and at different material removal rates.
- The machine tools should be ergonomically designed so as to enable the operator to setup and control the machine without any problems.

3.2 TYPES OF MACHINE TOOLS

The different ways of classifying the machine tools are:

3.2.1 Based on the Purpose

The machine tools can be classified in four categories on the basis of their purpose.

(A) General purpose machine tools

(B) Production machine tools

(C) Special purpose or single-purpose machine tools

(D) Limited purpose machine tools.

(A) General purpose machine tool

The machine tools of this category can perform nearly all metal cutting operations on a wide range of parts. Commonly used general-purpose machine tools are:

1. Lathe machine
2. Drilling machine

3. Boring machine
4. Shaper and Planer
5. Grinding machine
6. Milling machine.

(B) Production machine tools

These machine tools are used for producing the components at a faster rate and to reduce the machining cost. These are designed to machine a given part by employing multiple tools so that the tools can work simultaneously and perform operations automatically and successively. Following machines fall under this category:

1. Multiple tool lathes
2. Multiple head milling machines
3. Capstan and Turret lathes
4. Semi automatic lathe
5. Automatic screw machine
6. Production milling machine.

(C) Special purpose or single-purpose machine tools

These machines are used for producing similar parts in large quantities. This reduces manufacturing cost when repeated jobs are produced. The following machines come under this category:

1. Piston turning lathe
2. Camshaft grinder
3. Gear generator.

(D) Limited purpose machine tools

Under this category, machines are capable of completing few operations on different types of workpieces. Example of this category is automatic cutting off machine.

3.2.2 Based on the Degree of Accuracy

On the basis of degree of accuracy, the machine tools can be classified into following categories:

(A) Normal accuracy machine tools
(B) Above normal accuracy machine tools
(C) Precision machine tools
(D) High precision machine tools
(E) Super high precision machine tools.

3.2.3 Based on the Weight

According to the weight, the machine tools are classified as

(A) Light weight machine tools – up to 1 ton
(B) Medium weight machine tools – up to 10 tons
(C) Heavy weight machine tools – above 10 tons.

3.2.4 Based on the Type of the Surface Generated

On the basis of the surface generated the machine tools are of the following types:

Machine tools for the cylindrical work	Centre lathe, Turret and Capstan lathe, Boring machine, Cylindrical grinder, Centreless grinder
Machine tools for the surface work	*Based on reciprocating motion-* Planing, Shaping, Slotter, Broaching machine
	Based on rotary motion- Lathe, Milling, Grinder
Machine tools for the holes making	Lathe, Drilling, Honing machine
Machine tools for the threads	Lathe, Screw cutting machine, Thread milling
Machine tools for the Gear cutting	Milling machine, Hobbing machine

3.2.5 Based on the Type of Tools Used

The machine tools can also be classified as:

(A) Single point cutting tools

(B) Multi-point cutting tools

(C) Abrasive wheels.

3.3 OBJECTIVE OF THE MACHINE TOOLS

The basic objectives of the machine tools are:

(*i*) To hold the cutting tools.

(*ii*) To hold and support the workpiece.

(*iii*) To provide the ways and means of producing the relative motion between tools and workpiece so as to enable the cutting tool to produce parts of required shape and size as per the requirement.

3.4 TYPES OF CUTTING MOTION IN MACHINE TOOLS

The cutting motions are of two types:

1. Primary motion.
2. Auxiliary motion.

Primary motion. It includes the cutting motion and the feed motion. The primary motion is the main motion provided by a machine tool or manually to cause relative motion between the tool and workpiece so that the face of the tool approaches the workpiece material. The primary motion absorbs most of the power required to perform a machining operation. *Primary motion has the purpose of removing the metal (chip) from the workpiece.*

The feed motion is a motion that may be provided to a tool or workpiece by a machine tool which, when added to the primary motion, leads to a repeated or continuous chip removal and the creation of a machined surface of desired geometry. This motion may proceed in steps or continuously, in either case it usually absorbs a small proportion of the total power required to perform a machining operation. The speed of the cutting motion is determined by the optimum cutting speed, while the speed of the feed movement depend on the required degree of surface finish.

Auxiliary motion. It does not take part in direct removal of material from a workpiece. The auxiliary motion aids the primary motion to fulfill the assigned function.

The auxiliary motion includes the following:

- Clamping and unclamping of the workpiece.
- Control motion during the cutting process.
- Setup motion as per the shape and size of the workpiece.
- Adjusts the machine tool according to the required speeds and feeds.
- Motion to fix and release particular machine member.

Auxiliary motion can be performed either manually or automatically.

3.5 TYPES OF CUTTING AND FEED MOTIONS

The cutting motion, which plays the lead role in a machine tool, is of two types.

(*i*) Rotating.

(*ii*) Reciprocating.

The cutting motion may be imparted to either the workpiece or to the cutting tool.

Machine	*Cutting Motion*
Lathe	Rotation of workpiece
Milling	Rotation of cutting tool
Drilling	Rotation of cutting tool
Grinding	Rotation of cutting tool
Broaching	Reciprocating motion of cutting tool
Slotting	Reciprocating motion of cutting tool
Planer	Reciprocating motion of the workpiece
Shaper	Reciprocating motion of cutting tool

In certain machine tools however the principal movement results from simultaneous rotation of both the workpiece and the cutting tools, as in drilling small diameter holes on a multiple spindle automatic lathe.

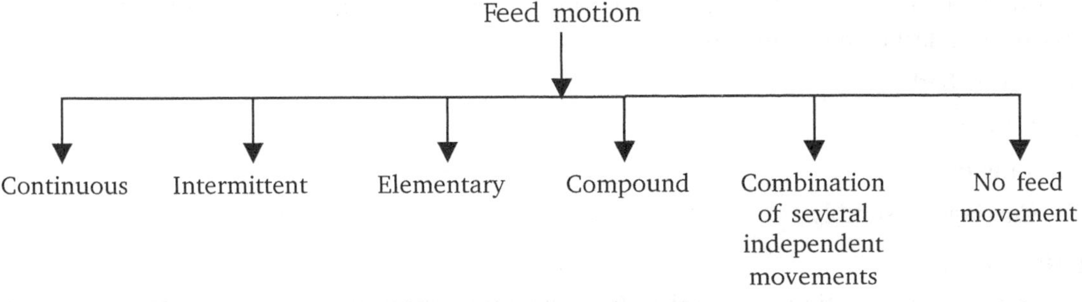

Machine	*Feed Motion*
Lathe	Continuous
Milling	Continuous
Drilling	Continuous
Planer	Intermittent
Broaching	No feed

3.6 LATHE MACHINE

One of the most important machine tools in metal industry is the lathe machine. *The main function of a lathe is to remove material from a workpiece so as to give it the required shape and size.*

3.6.1 Main Parts of a Lathe Machine

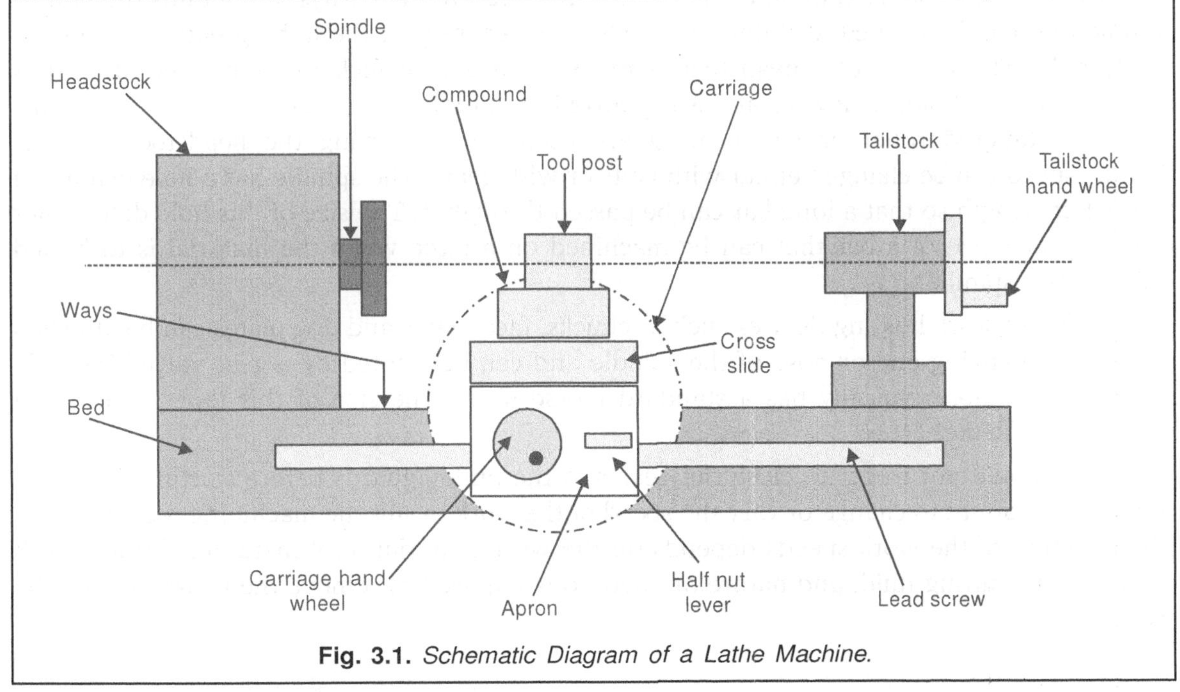

Fig. 3.1. *Schematic Diagram of a Lathe Machine.*

Figure (3.1) shows the schematic diagram of a lathe and its most important parts. The main parts of a lathe machine are:

(*a*) Lathe bed

(*b*) Headstock

(*c*) Tailstock

(*d*) Carriage assembly.

(*a*) Lathe bed

It forms the foundation of a lathe. It is the main supporting structure of the lathe on which the headstock and tailstock are mounted. Its rigidity and alignment affects the accuracy of the machined parts. Therefore the construction of a lathe bed is carried out in such a way so as to enable it to resist the stresses created by heavy machining cuts. The material used for the construction of a lathe bed, should possess high compressive strength, good wear resistance and good damping characteristics. Cast iron in combination with nickel and chromium is generally preferred for producing the lathe bed.

On top of a lathe bed there are two sets of guideways. Guideways consist of two inverted vees and two flat bearing surfaces. One guideway is outer ways to provide bearing and sliding surface for the carriage and the other is inner ways for the tailstock. In lathe, the ways are surface hardened and machined precisely. Proper care should be taken to ensure that the ways are not damaged. Any inaccuracy in them usually means that the accuracy of the entire lathe is destroyed. Generally, the guideways of the lathe are accurately machined by grinding or by milling and hand scrapping.

(*b*) Headstock

The headstock is fixed on left side of the lathe bed on inner ways. It contains the hollow spindle that can be rotated at different speeds. The accuracy of the lathe greatly depends on the spindle. The spindle of a headstock is made of carbon or nickel chrome steel. It carries the various work holding devices and is supported by spindle bearings on each end. Generally preloaded tapered roller or ball bearing are used for supporting the headstock spindle. Spindle speed can be changed either with belts or with gears. The spindle has a hole extending through its length so that a long bar can be passed through it. The size of this hole determines the maximum size of a bar that can be machined on a lathe when the material is to be fed through the spindle.

Different work holding devices such as chucks, face plates and dog plates can be attached to the over hanging end or nose of the spindle and can be rotated by a gear valve. The hole in the spindle nose typically has a standard morse taper. The size of this taper varies with the size of a lathe.

One should not forget to close the gear and the pulley guards before starting the lathe. It is very important to change or vary the speed of the work to suit the machining requirements. The selection of the work speeds depends on the work material, tool material, surface finish requirement, cutting fluid, and nature of operations required to remove the material from the workpiece.

Workpiece can be rotated at different speed on the lathe by using:

(A) V-belt drive or cone pulley on the headstock spindle with or without the provision of back gear.

(B) a gear drive.

(*c*) Tailstock

Tailstock is located on the inner ways at the right hand side of the lathe bed. Tailstock is used to support one end of a workpiece for machining or to hold cutting tools such as drills, reamers and taps. The tailstock assembly usually consists of three parts namely:

(*i*) A lower casing unit.

(*ii*) An upper casing unit.

(*iii*) Tailstock quill (sliding spindle).

A lower casing unit of tailstock fits on the inner ways of the lathe bed. It can slide longitudinally and clamped in any position along the bed. An upper casing unit fits on the lower unit. The tailstock quill is a hollow steel cylinder, which can be moved longitudinally in and out of the upper casing by operating it by a hand wheel and locked in position with a spindle clamp lever. The open end of the quill has a morse taper into which the dead centre or other tools fit. A work holding device, when used on tailstock is called dead centre by the user and when used on headstock is called live centre.

(*d*) Carriage assembly

It provides the means for mounting, moving and controlling the cutting tools. The carriage assembly consists of the following parts:

(*i*) Saddle

(*ii*) Cross slide

(*iii*) Compound rest

(*iv*) Tool post

(*v*) Apron.

The saddle is a flat H–shaped casting that fits on the outer ways on the bed. It carries the cross slide and the compound rest. The cross slide is mounted on the carriage. The cross slide is moved perpendicular to the axis of the lathe by manually turning the cross feed screw handle or by engaging the cross feed lever for automatic power feed.

The cross slide provides a means for moving the lathe tool in the facing or cut off direction. The compound rest is mounted on the cross slide and can be swiveled to any angle horizontally with the lathe axis in order to produce bevels and tapers. **The compound rest** Fig. (3.2) can only be moved manually by turning the compound rest feed screw handle. In majority of lathes, the tool post is mounted on the compound rest. The compound rest has a circular base graduated in degrees. It is used for obtaining the angular cuts and small tapers. **The tool post** is used to hold the tool. **The apron** is the part of the carriage facing the operator. It contains the gears, clutches and levers for operating the carriage and the cross slide by hand and power feed.

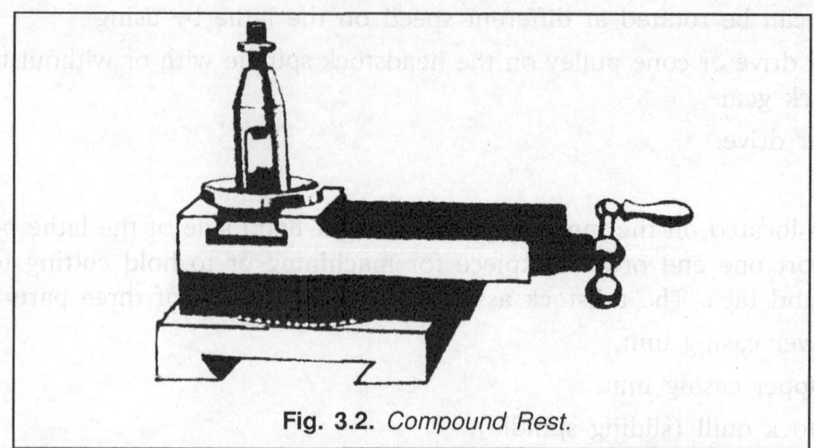

Fig. 3.2. *Compound Rest.*

For cutting internal and external threads a lead screw is used. During cutting threads, split nut is required to be engaged with the lead screw. Modern lathes have quick-change gearbox driven by the spindle and can easily connect feed rod and lead screw for cutting internal threads.

3.6.2 Size and Specification of Lathes

The size of a lathe as shown in Fig. 3.3 is usually specified by:

(*i*) Height of the centres measured from the lathe bed.

(*ii*) The swing of the workpiece diameter over bed: This is the maximum diameter of work that can be rotated on a lathe.

(*iii*) The maximum distance between the centres: This is the maximum length of a workpiece that can be mounted between the lathe centres.

(*iv*) The length of the bed: This indicates the approximate floor space occupied by a lathe.

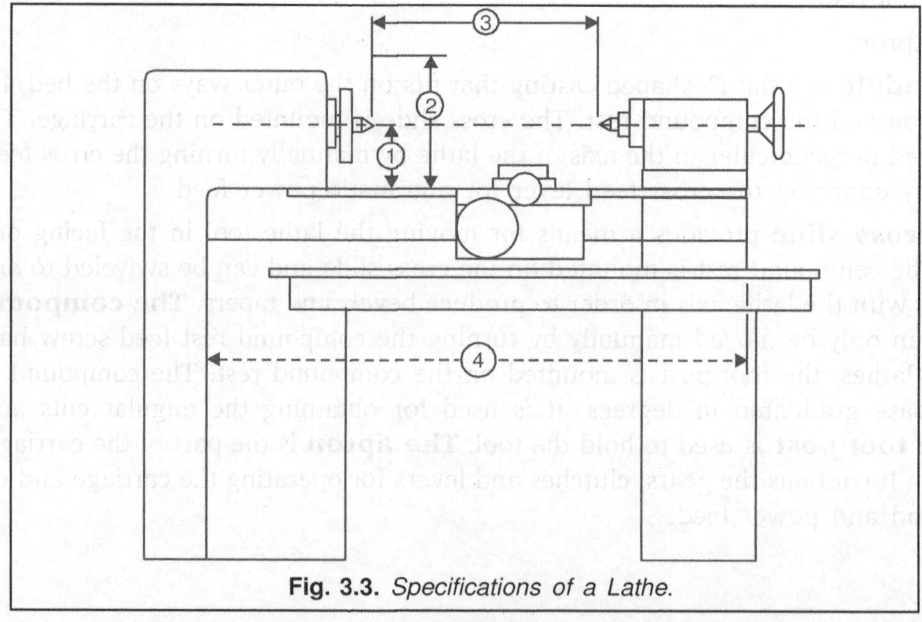

Fig. 3.3. *Specifications of a Lathe.*

A lathe size indicated by **356 mm × 762 mm × 1800 mm** means maximum diameter of workpiece is 356 mm, 762 mm is the length between the lathe centres and 1800 mm is the length of the lathe bed.

3.6.3 Technical Specifications of a Lathe Machine

Every engineer should know how to write or prepare the technical specifications for a common purpose lathe machine. The technical specifications of a medium duty lathe machine (courtesy "Premier Lathe Manufacturing Company Rajkot") are given below:

Capacity

Center height	:	215 mm
Swing over bed	:	405 mm
Swing over cross slide	:	250 mm
Swing in gap	:	685 mm

Bed

Bed Length	:	1370 mm
Bed type	:	2 V and 2 flat
Bed width	:	280 mm
Length of gap in front of face plate	:	150 mm

Headstock

Number of spindle speeds	:	6
Range of spindle speeds	:	30 to 550 RPM
Spindle bore	:	50 mm
Spindle threads	:	6 TPI
Spindle Morse Taper	:	MT-4
Electric Motor	:	2 HP

Standard Accessories: Chuck back plate, Tool post spanner, Dead centre -2 Nos., Motor pulley etc.

Optional Accessories: Face plate, Steady rest, Follow rest, Carrier plate, Rear tool post with extended cross slide, taper turning attachment, machine lamp etc.

3.6.4 Classification of Lathes

A suitable classification of lathe machines is difficult due to many variations available in their size, design, methods of drive and applications. The commonly used classification of the lathe is as follows:

(*i*) Speed Lathe

This is simplest of all types of lathes. It consists of a bed, a headstock, a tailstock, and an adjustable slide for supporting the tool. The tool is mounted on an adjustable slide and is fed into the work by hand control. The lathe is driven at high speed with spindle speeds of about 4000 RPM. The reason of using high speed being use of small depth of cut and hand tools.

The speed lathe is principally used for turning wood, patterns, and centring cylindrical parts prior to further work on the engine lathe. The "Speed lathe" has been so named because of the high speed of the head stock spindle. The work can be held between centres on a chuck or attached to a faceplate on the headstock.

(ii) Engine Lathe or Centre Lathe

An engine lathe is a general-purpose lathe machine. It is used to perform a variety of operations. The engine lathe derives its name from the early lathes, which were powered by steam engines.

It differs from a speed lathe in that it has additional features for controlling the spindle speed and for supporting and controlling the feed of the fixed cutting tool. Engine lathes are provided with a tailstock to facilitate the holding of the work between centres and permit the use of tool like drills, reamers, tapes etc. Centre lathes are built to handle work upto 1000 mm in diameter and 1000–4000 mm in length. Several variations in the design of the headstock are available through which the power can be supplied to the machine.

1. Light or medium duty lathes receive their power through a short belt from a motor or from a small cone-pulley counter shaft driven by the motor.
2. Heavy-duty lathes receive their power by a geared head drive. The different spindle speeds are obtained by changing the gear trains through the position of levers on the headstock. A constant speed motor mounted on a lathe usually drives such lathes, but in some cases variable speed motors are also used.

Fig. 3.4 shows a schematic diagram of a centre lathe machine.

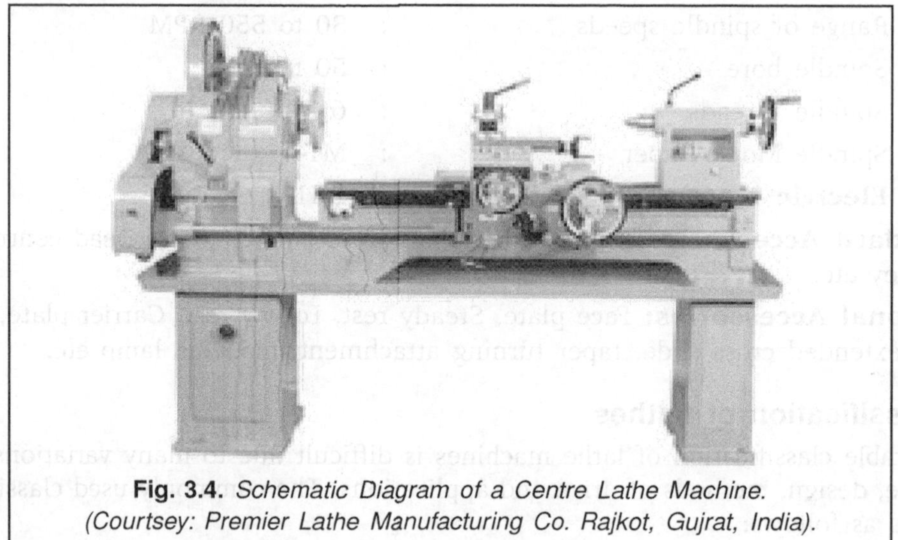

Fig. 3.4. *Schematic Diagram of a Centre Lathe Machine.*
(Courtsey: Premier Lathe Manufacturing Co. Rajkot, Gujrat, India).

(iii) Bench Lathe

The name bench lathe is given to a small engine lathe that is mounted on a workbench. In design it has the same features as speed or engine lathes and differs only in size and mounting. It is adapted to small work, having a maximum swing capacity of 255 mm (10″) at the faceplate. Bench lathe is used for small precision work like instrument parts.

(*iv*) Special Purpose Lathe

Several types of special purpose lathes are made to machine specific types of work. These include wheel lathes, hollow spindle lathes and gap frame lathes.

1. **Wheel lathe:** Permits the turning of journals and wheel threads of railroad car wheel and axle assemblies.
2. **Hollow spindle lathe:** Permits the loading of a tubular or shaft like workpieces through hollow spindle in the rear of a headstock. Workpiece may then be extended as required through the headstock into the work area, gripped by jaw type holder.
3. **Gap-bed lathe:** It is used for machining large work with irregular protrusions. On this lathe, a section of bed, adjacent to the headstock can be removed to machine a large size workpiece.

(*v*) Duplicating Lathe

Duplicating lathes include copying, tracer, profiling, numerical control and continuous path turning lathes. All these lathes are versatile production machines and are adaptable to the turning of external and internal contours. Tracer and numerical control lathes are the most commonly used types of duplicating lathes. Each employs a different actuating system to guide a single point cutting tool that generates the desired profile as the workpiece is rotated.

(*vi*) Tool Room Lathes

A tool room lathe has features similar to an engine lathes. It is built more accurately and has wide range of spindle speeds varying from very low to high speed of about 2500 rpm. This lathe is ideal for producing small tools, dies, and gauges and in machining works where accuracy is needed. It is costlier than engine lathe of the same size.

(*vii*) Automatic Lathe

A machine that moves the work and tools at the proper rates and sequences through a cycle without the attention of the operator to perform an operation on one piece is called an automatic lathe.

Strictly speaking, the machine is a semi automatic if an operator is required to load and unload the machine and start each cycle. Automatic lathes have the basic units of simple lathes-bed, headstock, tool slides and sometimes a tailstock. In addition, an automatic lathe drives the tools through all the steps of a cycle without an operator once the machine has been setup. The workpiece is rotated between centres. The tools are carried on blocks on the front and rear slides. The front slides moves along the bed, and the tools in this case make straight cuts along the workpiece, retract at the end of cut and are withdrawn to the starting position. Often an operator can look several machines in a group.

- In these lathes, all operating sequences, including changing of speeds and feeds, tool movements, indexing of tool turret and bar feed are carried out automatically.
- All automatically lathes perform basically similar functions, but they appear in a variety of forms. Many have no tailstock. Some automatic lathes have two or three

work spindles so that two or three sets of tools can perform the simultaneous cuts on two or three workpiece at a time.

- Automatic lathes, particularly vertical machines, without tailstock are commonly called chucking machines. The distinguishing feature of this lathe is that all the tools cut the workpiece at the same time.

(*viii*) Turret Lathe

A turret lathe is a manual lathe having a hexagonal tool holding turret in place of the tailstock of an engine lathe. These lathes are designed for short and continuous production runs. It has an indexable square tool post on the cross slide for mounting the turning tools. Tools for different operations such as drilling, counter sinking, reaming, tapping etc. mounted on different stations of the turret, can be successively brought into working positions by indexing the turret. A less skilled operator can run this machine.

(*ix*) Capstan Lathe

Capstan lathes employed for machining small parts are similar to turret lathes, except that the turret slide (known as the capstan slide) moves on an auxiliary slide. The stationary auxiliary slide can be conveniently clamped on the bed to facilitate the work length. It is ideal for fast production of small parts.

(*x*) CNC Lathe

In computer numerically controlled lathe, the path of the cutting tools to produce the desired shape of workpiece along with other auxiliary functions like speed and feed, tailstock positioning etc. are controlled by the computer. It is very useful for the production of small sized parts required in the precision machinery.

3.6.5 Work Holding Devices for Lathes

Five methods are commonly used for supporting workpiece in lathes. The workpiece can be:

- Held between centres.
- Held in a chuck.
- Held in a collet.
- Mounted on a faceplate.
- Mounted on a carriage.

(A) Lathe Centres

The most common method of holding the work in a lathe is to use the two centres popularly called by users as *live centre* and *dead centre*. Out of the two lathe centres one is used in the spindle hole and other in the tailstock spindle. These two centres take up the thrust due to metal cutting and the entire load of the workpiece on small bearing surface. Therefore these are made of very hard materials to resist deflection and wear. Dead centre is made of hardened steel with a morse taper on one end so that it can fit into the spindle hole. The other end is ground to a 60° taper. Some times the tip of this taper is made of tungsten carbide to provide better wear resistance. Dead centre shown in Fig. 3.5 is mounted in the tailstock spindle. Live centres of the type shown in Fig. (3.6) are designed so that the end that fits into the headstock spindle is mounted on ball or roller bearings.

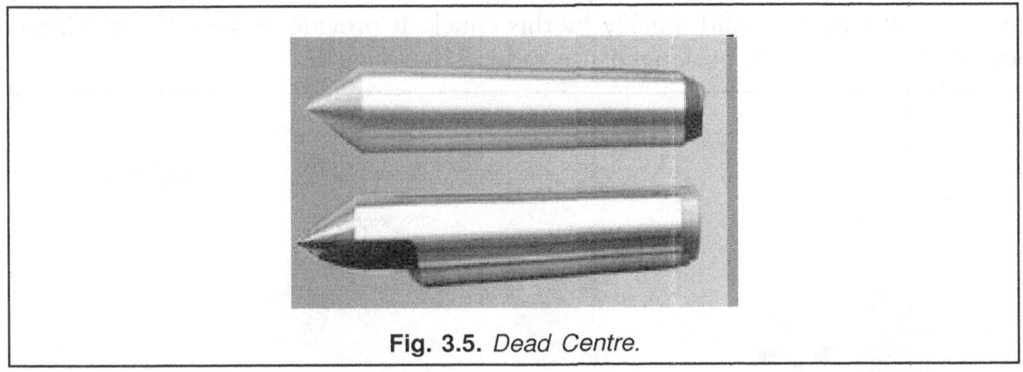

Fig. 3.5. *Dead Centre.*

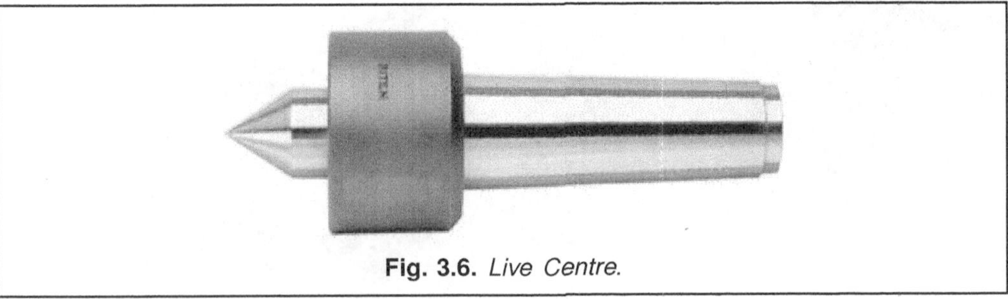

Fig. 3.6. *Live Centre.*

It is free to rotate and no lubrication is required. Because work and centre at the headstock end rotate together, no lubricant is needed in the centre hole at this end. Adequate lubrication must be provided at the centre in the tailstock end which does not rotate.

(B) Lathe Chucks

Lathe chuck is an important device for holding and rotating a workpiece on a lathe. Workpiece of short length and large diameter or of irregular shape those cannot be conveniently mounted between centres, can be held quickly and rigidly in a chuck. Lathe chucks permit more operations to be performed in comparison to when work piece is held between centres. A chuck is attached to the lathe spindle by means of bolts with the back plate screwed on to the spindle nose.

The most commonly used chuck types are:

(*i*) Three jaw universal chuck

(*ii*) Four jaw independent chuck

(*iii*) Combination chuck

(*iv*) Magnetic chuck

(*v*) Collet chuck.

(i) Three jaw universal chuck

Three jaw universal chuck as shown in Fig. (3.7), also known as self-centreing chuck is used to hold the workpiece of similar shape or that has a round or hexagonal cross-section. The job is centred automatically and quickly by this chuck. It provides automatic centring with an accuracy of about 0.001. Its prolonged use reduces its accuracy.

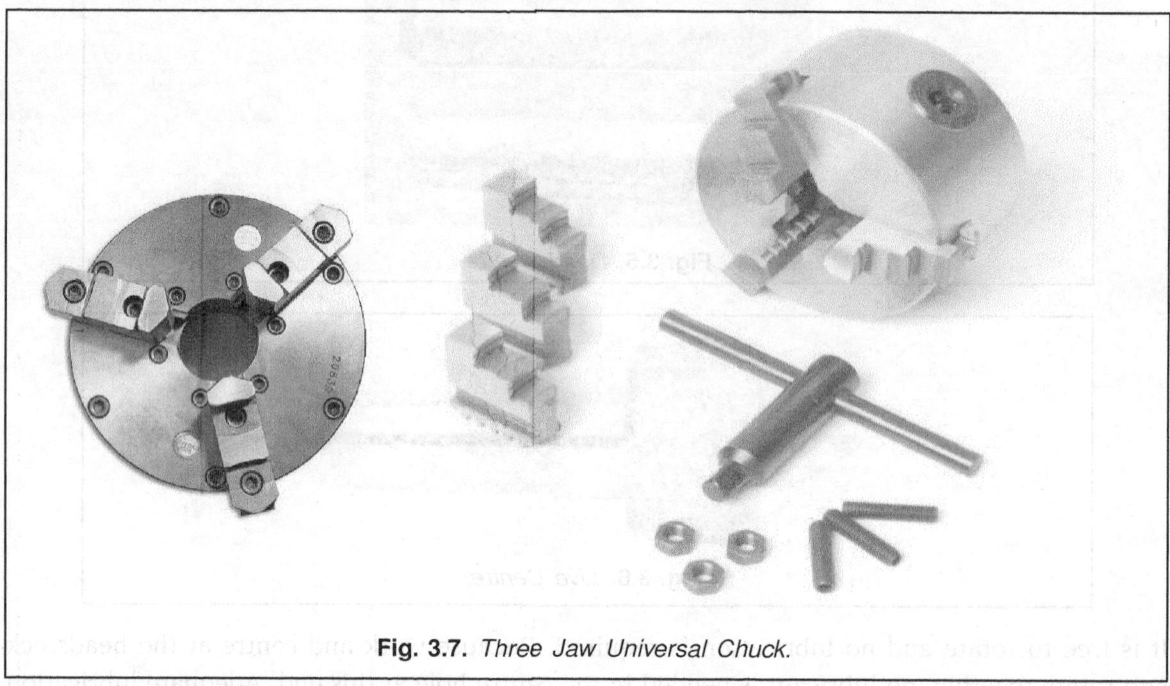

Fig. 3.7. *Three Jaw Universal Chuck.*

In this chuck, all the jaws slide simultaneously by an equal amount within the slots provided on the body by rotating any one of the three pinions, which meshes with the teeth cut on the under side of the scroll disc.

(ii) Four jaw independent chuck

In this type of chuck each jaw can be moved inward and outward independently by means of a key. Thus it can be used to support different types of workpieces. Four jaw chucks are heavier and more rugged than the three jaw type as undue pressure on one jaw does not destroy the accuracy of this chuck. It is generally used for the heavy work.

This type of chuck as shown in Fig. (3.8) is preferred for holding non-symmetrical workpieces, rough castings, and parts of irregular shape.

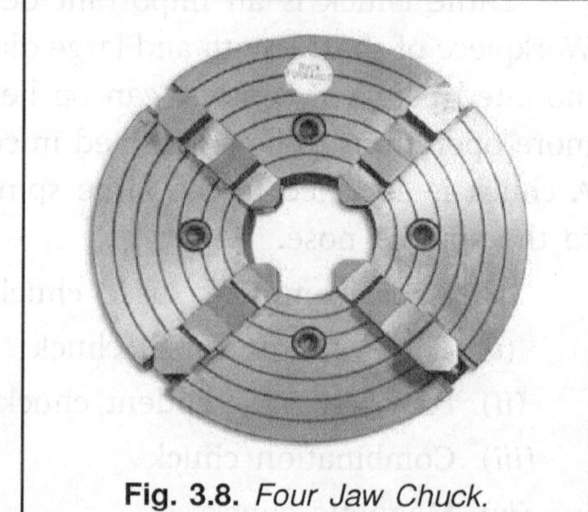

Fig. 3.8. *Four Jaw Chuck.*

(iii) Combination chuck

As the name implies a combination chuck provides the advantages of both the three jaw and four jaw chucks. It may be used both as a self-centering and as an independent chuck.

(*iv*) Magnetic chuck

Magnetic chuck is preferred for holding workpieces provided that the cutting and other forces are not high. The holding force is distributed over the entire contact surface of the work. Demagnetization of the workpiece is essential after holding it on a magnetic chuck. The chuck is used for holding thin workpieces of magnetic material, as these cannot be held in an ordinary chuck. The holding power of the chuck is obtained by the magnetic flux radiating either from the electro-magnets or from the permanent magnets placed within the chuck.

(*v*) Collet chuck

It is used for holding bars of small sizes. Different sizes of collets are used for holding different sizes of the bar stock. These are commonly used in capstan and turret lathes.

(C) Mandrels

Workpieces, those need machining on both ends are often mounted on mandrels for turning between centres. A mandrel is a device for holding and rotating a hollow workpiece that has been previously drilled or bored Fig. (3.9).

Four common types of mandrel are:

(*i*) Plain mandrel

(*ii*) Solid mandrel

(*iii*) Gang mandrel

(*iv*) Cone mandrel.

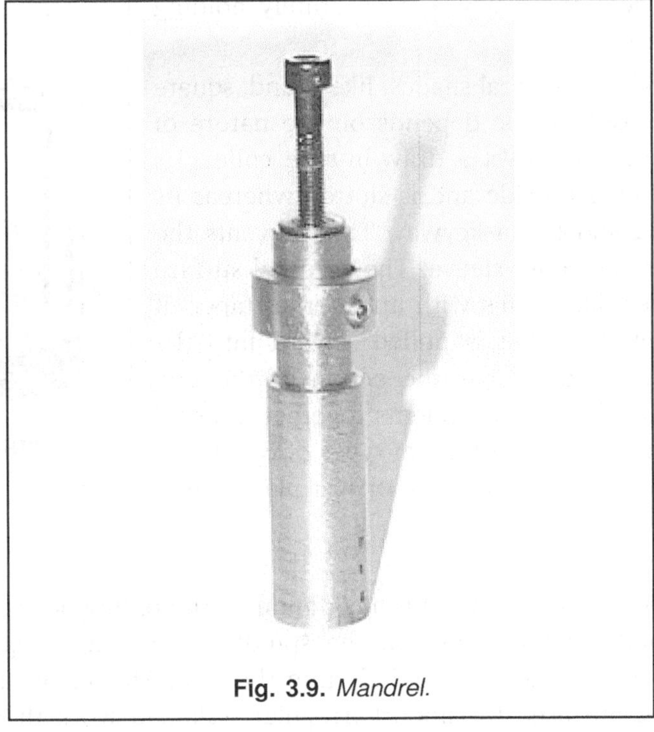

Fig. 3.9. *Mandrel.*

(i) Plain mandrel

This type of mandrel is most commonly used in workshops and finds wide applications where a large number of identical pieces having standard sized holes are required to be mounted on it.

(ii) Solid mandrel

Solid mandrels usually range from 100 to 300 mm in length and are accurately ground with a taper of 1 to 2 mm per 100 mm length. The mandrel should be mounted between centres as the cutting force tends to tighten the work on the mandrel taper. Solid mandrels permit the work to be machined on both ends as well as on the cylindrical surface.

(iii) Gang mandrel

Gang mandrel is used when only the cylindrical surface of a workpiece is to be machined. This has a fixed collar at one end and a movable collar at the threaded end that may be adjusted to its position by a nut.

(iv) Cone mandrel

This mandrel has the advantage that it can be used to centre workpieces having a range of hole sizes.

(D) Collet

A collet is used for holding small semi-finished or finished parts so that additional operations may be performed. Collets are precisely machined. Their design features make collet a practical device for quickly and accurately holding symmetrical workpieces.

Collets are available in several shapes, like round, square and hexagonal. The inside shape depends on the nature of the workpiece. Fig. (3.10) shows a draw-in-type collet. Its one end is tapered on the outside and is slotted whereas its outer end is threaded and has a keyway. This prevents the collet from turning in the collet sleeve. The external surface of the collet is a taper that mates with an internal taper of a collet sleeve. When the collet is pulled inward into the spindle by means of the draw bar the action of the two mating tapers squeezes the collet segments together, causing them to grip the workpiece. Four different types of collet are: spring collet, draw in collet, collet chuck or just plain collet.

Fig. 3.10. *Collet.*

(E) Faceplate

A faceplate as shown in Fig. (3.11) is a common work-holding device on a lathe machine. A faceplate is a round plate that fits on the spindle nose. The faceplate has a series of elongated T-slots. These slots are cut radially into the face. The T-slots accommodate T-head bolt heads. Faceplates are used to support irregularly shaped work that cannot be gripped

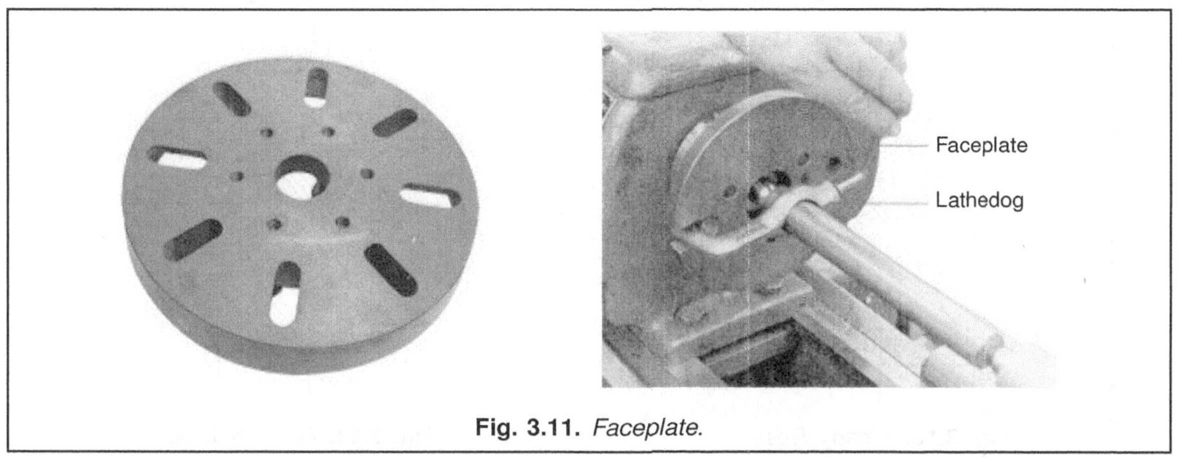

Fig. 3.11. *Faceplate.*

easily in chucks or collets. For the irregular parts it should first be positioned on an angle plate. The angle plate then can be strapped to the faceplate to permit machining of an irregular shaped part at a particular axis. Because of their shape, other workpieces that require facing, drilling, boring, reaming or threading may be screwed to the faceplate. Care must be taken in faceplate work to use balancing weights. These offset any centrifugal forces that may be created by the unequal distribution of the mass. Any slight out of balance produces an unsafe condition. Unless the workpiece and the work holding devices are secured and balanced, the spindle speed may cause the parts to fly out and cause personal injury and damage to the machine.

(F) Driver Plate

A great deal of work is machined on the lathe between centres. The centred work-piece is driven by the lathe dog and driver plate (or dog plate). The driver plate is a cylindrical plate that attaches to the spindle nose. The driver plate usually has two through, elongated radial slots. The tail of the lathe dog fits into a slot. The body of the lathe dog is clamped over the workpiece. As the spindle and driver plate turn, the motion through the lathe dog to the workpiece is imparted.

(G) Steady Rest and Follower Rest

Long workpieces must be supported at regular intervals and continuously while a cut is being taken. The steady rest as shown in Fig. (3.12) is an accessory that performs this function. It helps prevent springing of the workpiece under the cutting action of a tool. The steady rest is clamped to the lathe ways and has three movable fingers those can be adjusted to contact the work and align it. A steady rest can also be used in place of the tailstock for supporting the long workpieces (for drilling, boring and tapping).

The follower rest performs a similar function except that it follows the cutting action. A follower rest as shown in Fig. (3.13) has two adjustable jaws that support the workpiece. The rest is bolted to the back end of the carriage and moves with it. The work-supporting shoes of the follower rest support the revolving workpiece at two places. The shoes prevent the forces of the cutting tool against the workpiece from springing it.

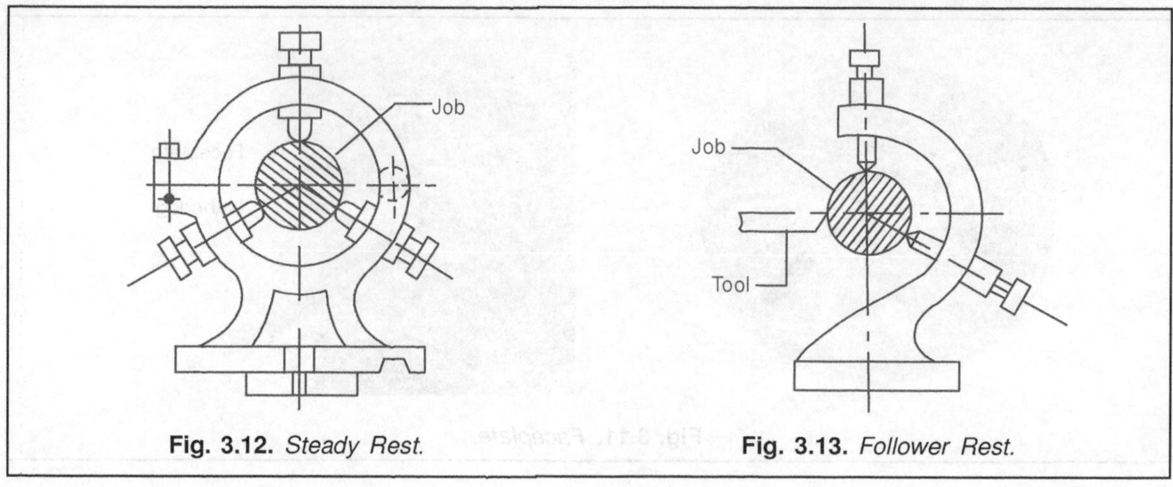

Fig. 3.12. *Steady Rest.*

Fig. 3.13. *Follower Rest.*

(H) Lathe Dog or Carrier

A carrier or lathe dog Fig. (3.14) is a device attached to a workpiece to be turned between centres.

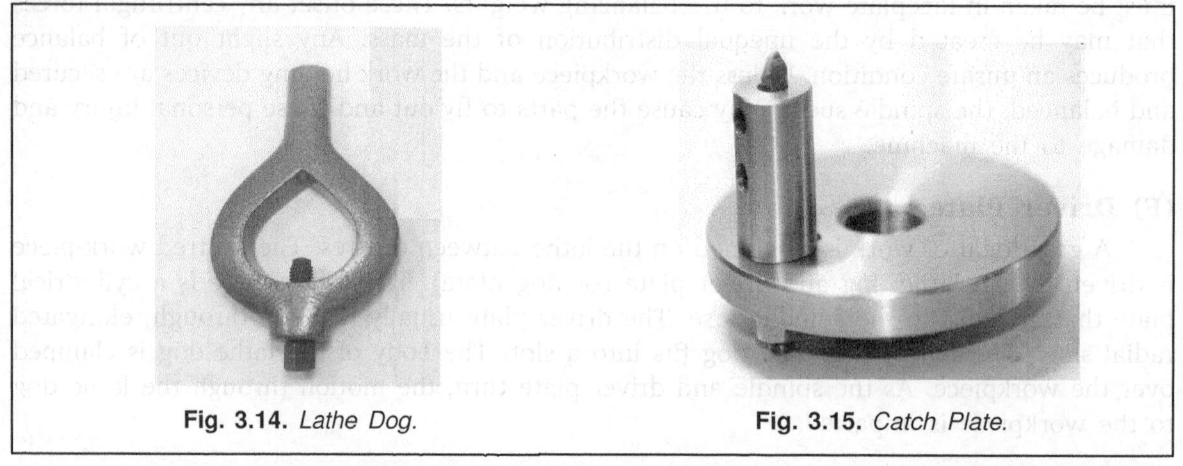

Fig. 3.14. *Lathe Dog.*

Fig. 3.15. *Catch Plate.*

The lathe dog as shown in Fig (3.14) and catch plate as shown in Fig (3.15) provide a mechanical connection between the workpiece and the spindle. The lathe dog is clamped to the work. The tail of the dog enters a slot in the catch plate, which is attached to the lathe spindle in the same manner as a lathe chuck.

3.6.6 Lathe Operations

Following operations can be performed on a lathe machine Fig. (3.16):

Turning	Threading	Drilling	Grinding	Undercutting
Facing	Boring	Milling	Polishing	Counter boring
Knurling	Parting	Reaming	Grooving	Spinning

(i) Turning

Turning is the process of the removal of material from cylindrical and conical surfaces externally. To produce a circular section or cylindrical section, the workpiece is made to rotate about its axis by the lathe. The workpiece is rotated into a longitudinally fed, single point cutting tool.

In Taper turning, the tool is fed at an angle to the axis of rotation. Taper turning produces an external conical surface.

When the workpiece having different diameters is turned, the surface forming the step from one diameter to the other is called the shoulder, and machining this part of the workpiece is called shoulder turning.

☞ The workpieces are given cylindrical shape by simple turning, plane surfaces by facing or transverse turning, taper by taper turning, profiles by profile turning or threads by thread turning.

(ii) Facing

Facing is the process of machining a flat surface across the face of a workpiece. In facing, the tool is fed at 90° to the axis of rotation.

(iii) Chamfering

Chamfering is the operation of beveling the extreme end of a workpiece. This is done to remove the burrs, to protect the end of the workpiece being damaged and to have a better look.

(iv) Parting

Parting is the operation by which one section of workpiece is severed from the remainder by means of a cutoff tool as shown in Fig. 3.16. Because parting tools are thin and have considerable over-hang, this process is less accurate and more difficult. The tool should be set exactly at the height of the axis of rotation, be kept sharp, have proper clearance angles and be fed into the workpiece with uniform feed rate.

(v) Knurling

Knurling is the process of embossing a diamond shaped pattern on a workpiece surface. The purpose of knurling is to provide an effective gripping surface on a workpiece so as to prevent it from slipping when used by hand. Knurling not only provides a good gripping surface but is also enhances the appearance of a part.

(vi) Polishing

It is one of the final operations required on a workpiece. Polishing produces a fine – grained and smooth surface finish and removes tool marks. A polished surface is produced by holding and moving an abrasive cloth across the work surface as it rotates at high speed from 1500 to 2000 m/min. Limited amount of material is removed in the process.

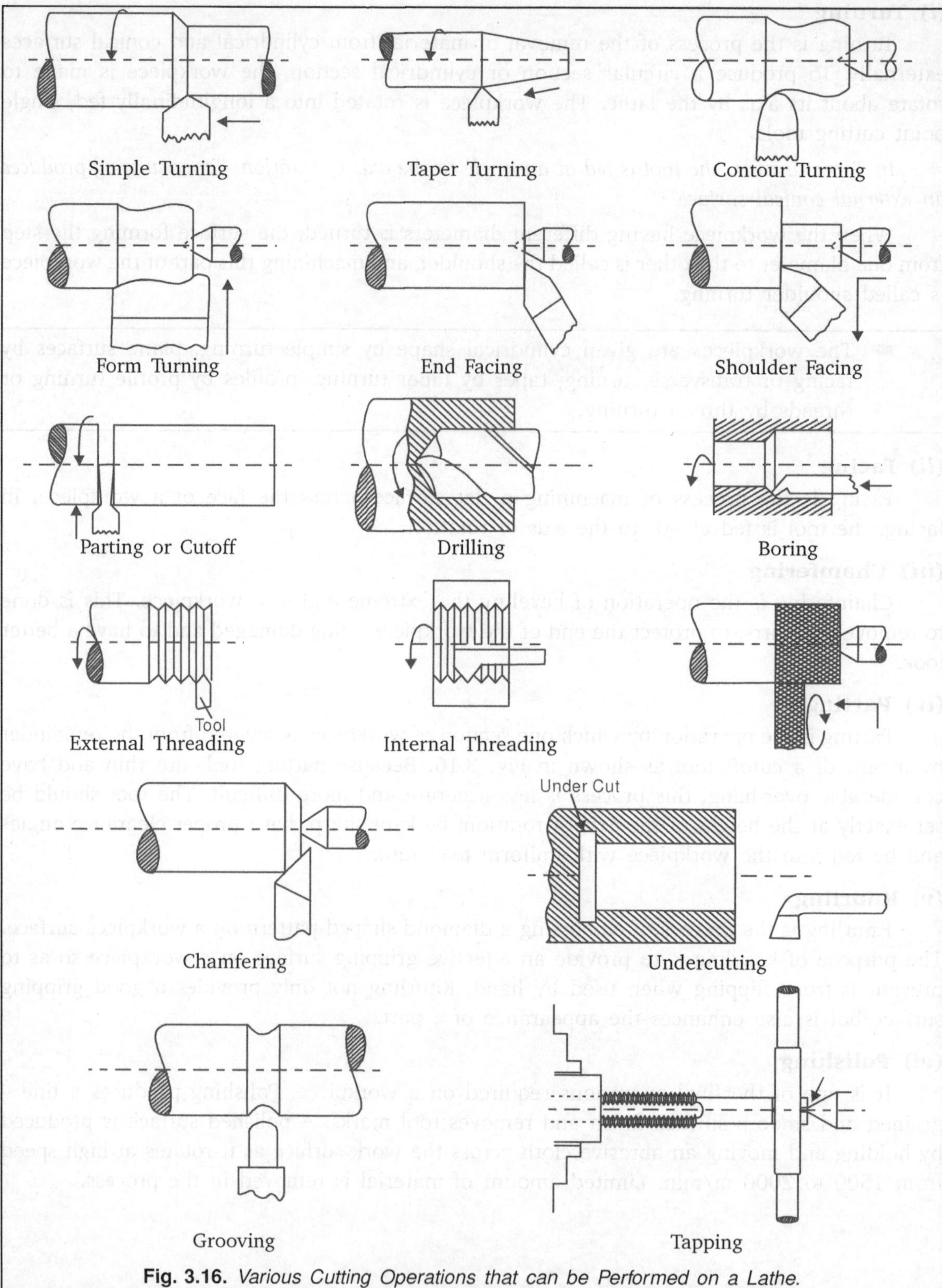

Fig. 3.16. *Various Cutting Operations that can be Performed on a Lathe.*

(*vii*) Grooving

It is the process of reducing the diameter of a workpiece over a very narrow surface.

(*viii*) Tapping

Tapping is the operation of cutting internal threads of small diameter using a multi point cutting tool called the tap. In a lathe, the work is mounted on a chuck or on a faceplate and is revolved slowly. A tap of required size held on a special fixture is mounted on the tail stock spindle. The axis of both the work and tool must coincide. The tap will automatically fed into the work to cut internal threads with the help of specially designed fixtures.

(*ix*) Drilling

Drilling is the operation of producing a cylindrical hole in a workpiece by a tool known as the drill. Drilling in a lathe is performed by any one of the following methods:

- Most drilling on lathes is done with a drill held in the tailstock quill and fed against a workpiece that is rotated in a chuck.
- It is also possible to drill on a lathe with the cutting tool (drill bit) mounted and rotated in the spindle while the work remains stationary and supported on the tailstock. Workpieces of irregular shape those cannot be accommodated on a chuck or faceplate are drilled by this method. Coolants should be used during the drilling.

(*x*) Boring

Boring is one of the machining operations, which involves enlarging of an existing hole that may have been produced by a drill or may be the result of a core in a casting. Boring is also used to make the previously drilled hole concentric with the axis of rotation of the workpiece. When boring is done in a lathe, the work is usually held in a chuck or on a faceplate.

(*xi*) Counter boring

Counter boring is the operation of enlarging a hole through a certain distance from one end instead of enlarging the whole drilled surface.

(*xii*) Reaming

It is the operation of finishing and sizing a hole, which has been previously drilled or bored. A tool called reamer is held on the tail stock spindle. The workpiece is rotated slowly against the stationary reamer. The required feed varies from 0.5 to 2 mm/min.

(*xiii*) Milling

Milling is a machining operation in which the material is removed by feeding the workpiece against a rotating cutter having multiple cutting edges.

(*xiv*) Spinning

Spinning Fig. (3.17) is the process of forming a thin sheet of metal by revolving the job at high speed and pressing it against a former attached to the headstock spindle. A support can also be given from the tailstock end.

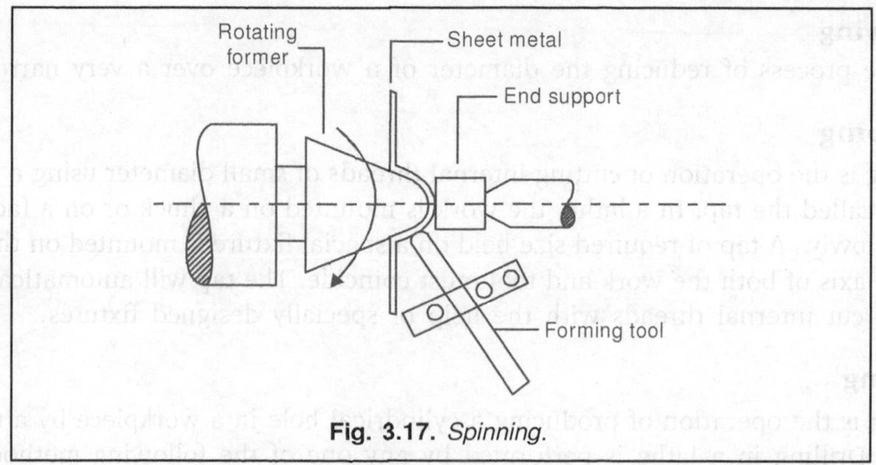

Fig. 3.17. *Spinning.*

Fig. (3.18) shows the different types of tools used on a lathe machine.

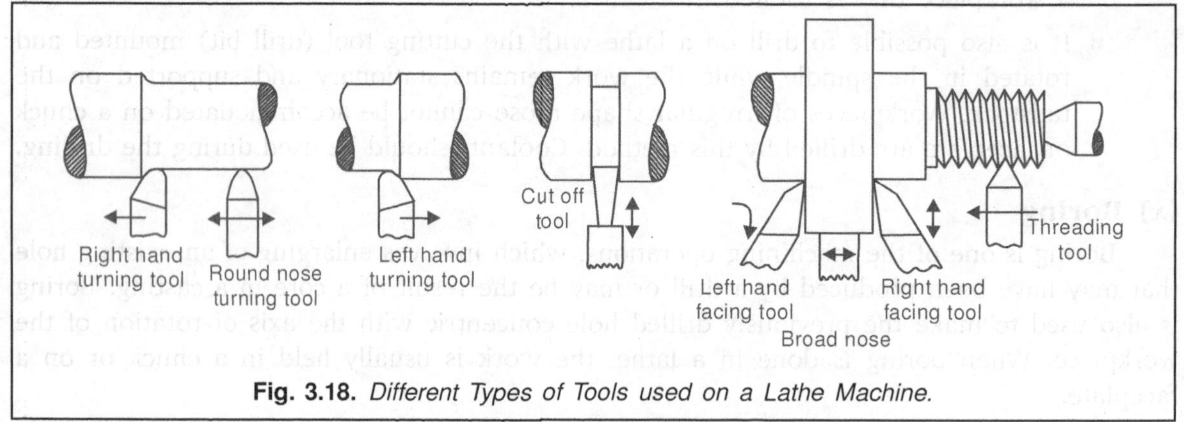

Fig. 3.18. *Different Types of Tools used on a Lathe Machine.*

3.6.7 Machining Parameters (Lathe)

☞ **Cutting speed**

It is defined as the speed at which the material is removed from the workpiece by a cutting tool. It is given by

$$V = \frac{\pi d N}{1000} \text{ m/min.}$$

where, d = diameter of the workpiece in mm.

N = spindle RPM.

The proper cutting speed should be selected in order to carry out the satisfactory material removal from the workpiece. The cutting speed can not be chosen at random, if the cutting speed is too low, the machining time will be too long, if the cutting speed is high, the cutting edge may loose its hardness due to high temperature rise.

☞ **Feed**

Feed is the distance, the cutting tool advances in each revolution of the work. It is expressed in mm/min. It is also expressed as mm/revolution of a workpiece.

☞ **Depth of cut**

It is the distance measured perpendicular from the machined surface to the uncut surface of the workpiece.

☞ **Machining time**

It is the time during which operations are performed which contributes directly to the completion of the work order.

$$\text{Machining time} = \frac{\text{Turning length (or the length of a workpiece)}}{\text{Feed}}$$

☞ **Material removal rate**

It is given by

$$\boxed{\text{MRR} = \pi D_{avg}\, tfN}$$

where,

$$D_{avg} = (D_I + D_f)/2$$
$$D_I = \text{initial diameter of the workpiece}$$
$$D_f = \text{final diameter of the workpiece}$$
$$t = \text{depth of cut} = (D_I - D_f)/2$$
$$f = \text{feed in mm/rev}$$
$$N = \text{work speed}$$

3.6.8 How to Carry-out the Centring of a Workpiece on a Lathe?

To carry out the turning operation, the workpiece is usually held between centres, fastened in a chuck, clamped to the face plate or clamped to the saddle. It is very important to carefully centre the workpiece to be turned. A properly centred workpiece saves the material from wastage in case if the work material does not have enough stock to be removed. Even in case of carelessly centred workpiece having enough stock, still the chances of spoiling the workpiece are more. This is due to the fact that the centres if out of true will result in more thick chip on one side of diameter and a small chip of low thickness on the other side of the diameter. This unevenness of the cut increases the machining time and may render the work useless after turning operation.

There are various methods available on the shop floor to accurately locate the centre holes, which must be drilled in each end of the work before it can be mounted on the lathe centres for turning. The correct centre can be located on the workpiece by using any one of the techniques mentioned below before turning the work.

(*i*) By using divider

(*ii*) By using hermaphrodite-caliper

(*iii*) By centre head method

(*iv*) By using centre gauge.

(i) By Using Divider

First mark the ends of the workpiece with chalk or copper sulphate, and then set the divider to ½ the diameter of the workpiece and then scribe four lines across each end as shown in Fig. (3.19). The centre of the square in the piece will be the centre of the work.

(ii) Hermaphrodite-caliper Method

The centre is found when four short circular arcs are drawn from the periphery having set the caliper equal to the radius of the workpiece. The four arcs will form an approximate square, the centre of which is the centre required as shown in Fig. (3.20).

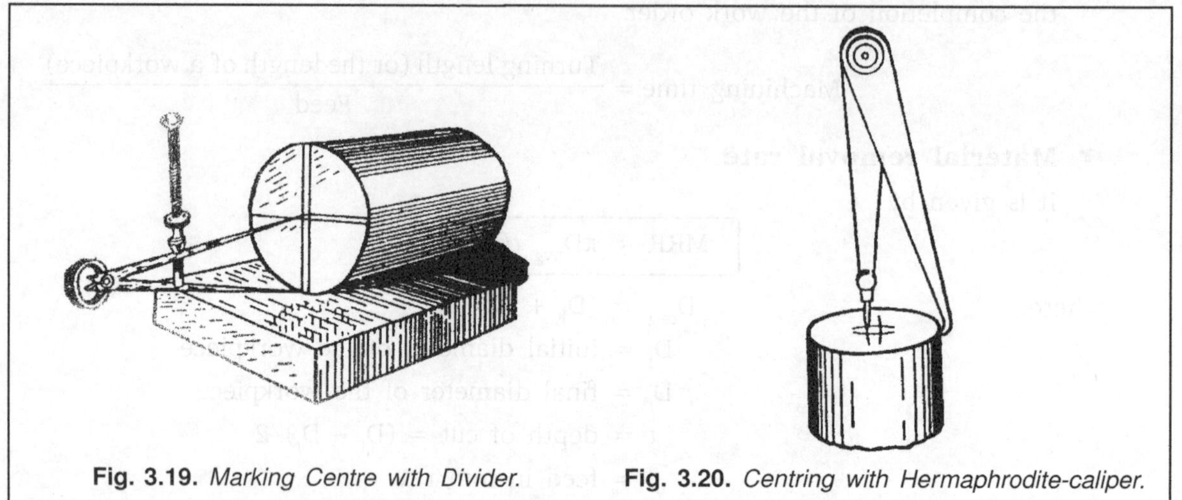

Fig. 3.19. *Marking Centre with Divider.* **Fig. 3.20.** *Centring with Hermaphrodite-caliper.*

(iii) Centre Head Method

In this method the workpiece is placed in V-block on a marking plate. Hold both limbs of the centre head of a combination square against the workpiece surface and draw two lines at right angle to each other with a scriber. The required centre is the point of intersection of the two lines drawn as shown in Fig. (3.21).

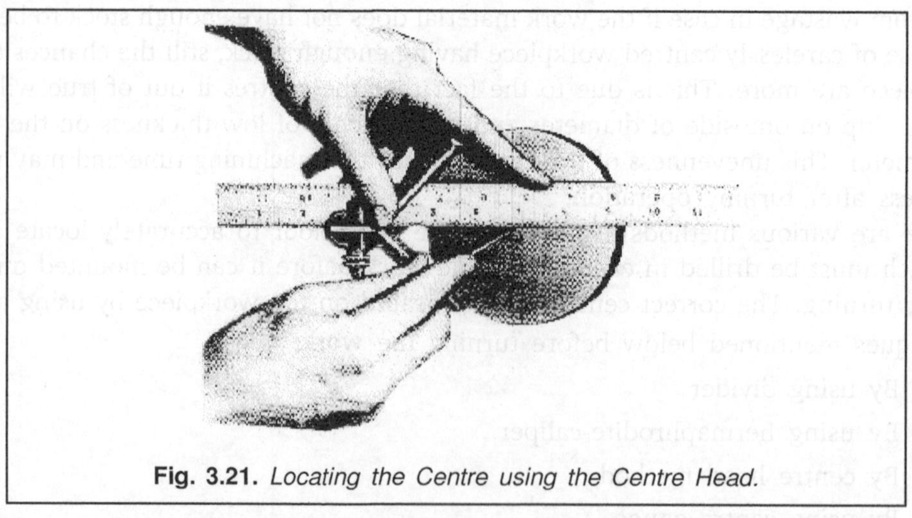

Fig. 3.21. *Locating the Centre using the Centre Head.*

(*iv*) Centre Gauge Method

The centre gauge method is also used to mark the centre. The centre gauge divides the angle between the two side rests. A line drawn along the centre rule always runs through the centre of the circle. The point of intersection of the two lines is the required centre of the job.

3.6.9 Taper Turning

Taper on a workpiece is required for different purposes *e.g.*, for fastening and seating of the lathe spindle, set screw etc. A taper angle is the angle included between the tapering sides of the job when extended to meet a point as shown Fig (3.22)

$$\tan \alpha = \frac{D-d}{2L}$$

where,

α = half taper angle

2α = full taper angle

D = large diameter of taper in mm.

d = small diameter of taper in mm.

L = length of taper in mm.

The amount of taper in a workpiece is the ratio of the difference in diameters of the taper to its length. This can be expressed by a letter T*a*.

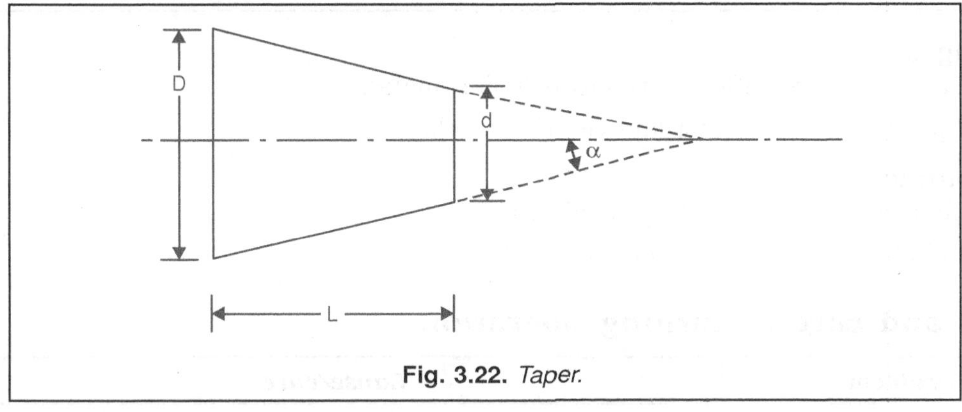

Fig. 3.22. *Taper.*

$$Ta = (D - d)/L$$

where

D = large diameter of taper in mm.

d = small diameter of taper in mm.

L = length of taper.

The tapers on a workpiece can be produced by (*i*) Using compound rest, (*ii*) offsetting the tail stock, (*iii*) using a taper turning attachment.

Compound rest method

This is the most commonly used method of producing tapers on a workpiece at the shop floor. This method is used for turning steep and short tapers. The tool mounted on the compound rest is attached to a circular base, which is graduated in degrees. It can be swiveled at any angle from the centre line of the lathe centres as shown in Fig. (3.23). While turning a taper, the base of the compound rest is swivelled through half taper angle. The feed is controlled manually by the compound rest feed handle.

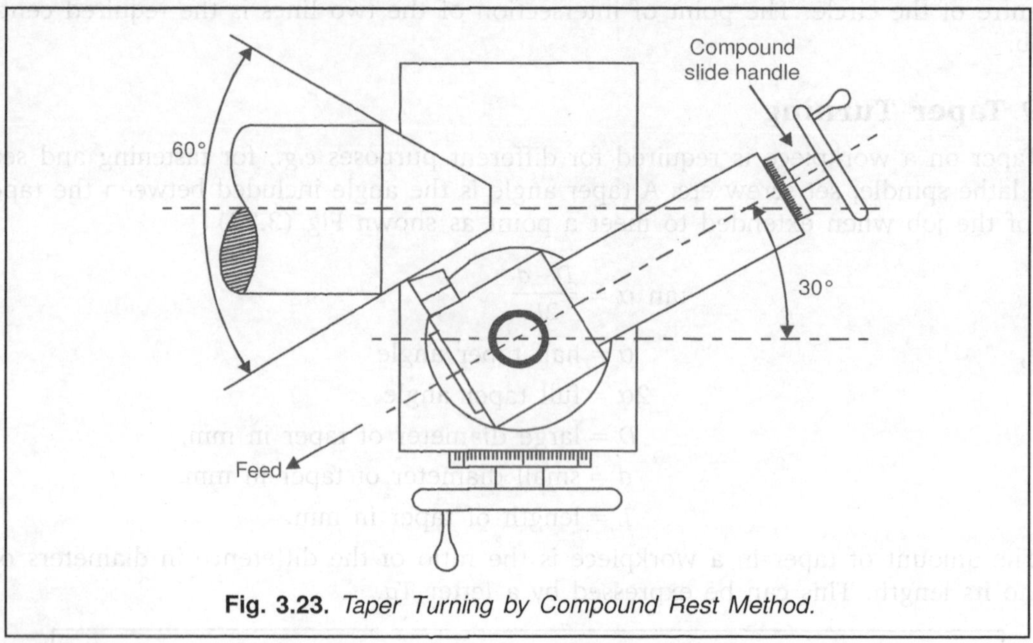

Fig. 3.23. *Taper Turning by Compound Rest Method.*

Advantages

- Easy to set for different values of taper angles.
- Applicable to chucking and centring work.

Disadvantages

- Tool feed is controlled by hand only.
- Suitable to small cuts (50–100 mm).

Causes and cure in turning operation

Problem	***Cause/cure***
Tool breakage	Lack of rigidity (poor tool material toughness), improper tool angle, too high cutting parameters.
Excessive tool wear	Too high cutting parameters, poor quality of tool material, ineffective cutting fluid, improper tool angle.
Rough surface finish	Built up edge (BUE) on the tool, incorrect tool geometry, incorrect cutting conditions.
Variation in dimensions	Excessive temperature rise, tool wear, lack of stiffness.
Tool chatter	Lack of stiffness, workpiece not supported rigidly, excessive tool overhang.

3.7 DRILLING

3.7.1 Introduction

When you see different products in your daily life, you will realize that many items have various holes in them. Holes are required for assembling one part with another with the help of temporary fasteners such as bolts, screws and rivets etc. All fasteners require a hole to serve their purpose. A motor cycle or a scooter is the best example to illustrate the importance of holes in the vehicle body.

Ancient man particularly Egyptians around 1200 BC had drilled holes in bone and wood with a bow string wrapped around and arrow tipped with a specially shaped flint. A wooden block was used to put pressure on the arrow top. This old idea of applying pressure on a rotating tool into work is still used in modern drilling machines.

Hole making is one of the most important operation in the manufacturing. Drilling is the process of making holes. *"Drilling is a machining process by which a hole is produced with the help of a rotating cutting tool called 'drill' "*. In drilling, the cutting action takes place inside the workpiece.

3.7.2 Types of Drilling Machines

A drilling machine or a drill press is the type of machine where the drill is held and it has the necessary mechanism to rotate the cutting tool (drill) at the desired rotating speed and feed.

Different types of drilling machines are:

(*i*) Sensitive drilling machine
(*ii*) Upright drilling machine
(*iii*) Radial drilling machine
(*iv*) Portable drilling machine
(*v*) Gang drilling machine
(*vi*) Multiple spindle drilling machine
(*vii*) Automatic drilling machine
(*viii*) Deep hole drilling machine
(*ix*) Numerically controlled drilling machine.

The three most commonly used drilling machines for general drilling operations are discussed here. These three drilling machines are: (*i*) sensitive drilling machine (*ii*) upright drilling machine (*iii*) radial drilling machine.

(*i*) Sensitive Drilling Machine

As the name implies, sensitive drilling machine allows the operator to "feel" the cutting action of the drill as the operator feeds the tool into a workpiece by purely hand control. The base of the machine is either mounted on a bench or floor. The major parts of a sensitive drilling machine are:

(*a*) The head

(*b*) Column

(*c*) Table

(*d*) Base

(*e*) A vertical spindle for driving and rotating the drill.

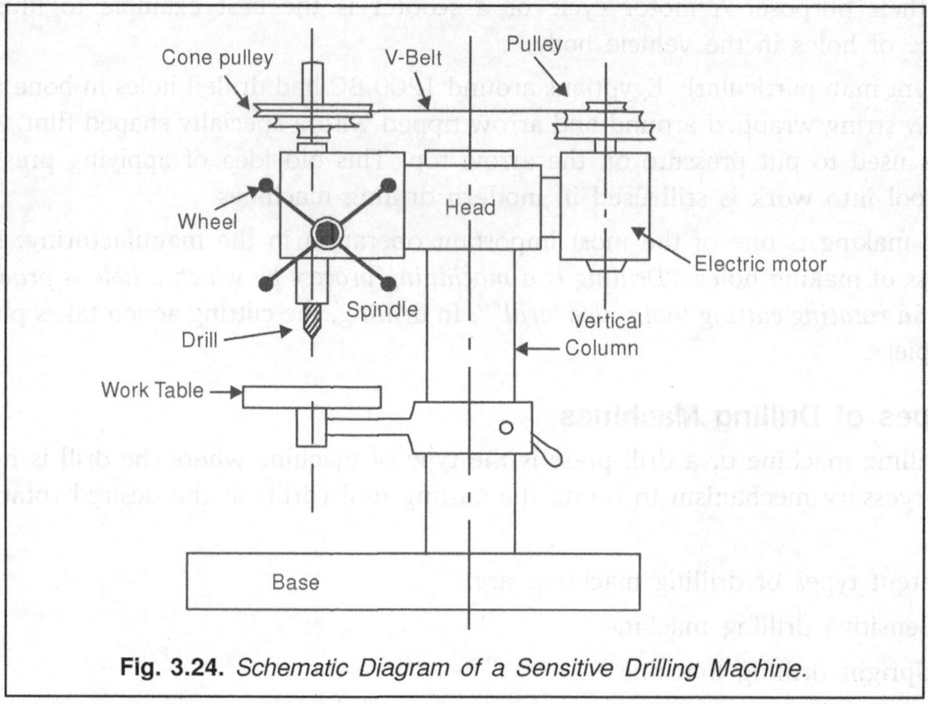

Fig. 3.24. *Schematic Diagram of a Sensitive Drilling Machine.*

Fig. (3.24) shows the schematic diagram of a sensitive drilling machine. The sensitive drilling machine is used for light duty applications. This machine is used for tool and die work and for drilling very small holes. This machine permits the maximum drill size of 1/2″ diameter. The machine capacity is usually measured by the hole diameter that can be drilled.

The spindle rotates within the quill, which does not rotate but carries the spindle up and down. The spindle shaft is driven by a stepped V-pulley and belt or by a variable speed drive. Sensitive drilling machines are capable of operating at higher spindle speeds up to 30000 rpm. The high spindle speeds are required to obtain proper cutting speed by small sized drill. Hand feed allows the operator to feel or sense the progress of the drill or hole into the work so in case of any trouble as sensed by the hand, the pressure on the drill is released to prevent the drill from breaking.

(*ii*) Upright Drilling Machine

It is similar to a sensitive drilling machine. It is used for handling much heavier work as compared to sensitive drilling machine. This machine has the necessary capability of using different spindle speeds and feeds for drilling different types of work. The spindle speeds ranges from 60 to 3500 rpm. Modern machines use a single speed motor and a gear drive to provide different range of speeds and feeds. Being gear driven these machines are capable of drilling holes of the order of 2″ diameter or more. The motor is required to be stopped while changing speeds on a machine.

The power feed mechanism is provided in upright drilling machines so as to drill large size holes in different workpieces. The operator may either feed manually with a lever or hand wheel or he may engage the power feed. Work tables on most upright drilling machines contain holes and slots for use in clamping work and have a channel around the edges to collect the cutting fluid when it is used.

(*iii*) Radial Drilling Machine

This is the most versatile drilling machine. Its size is determined by the diameter of the column and the length of the arm measured from the centre of the spindle to the outer edge of the column. Radial drilling machines are used for drilling holes in large size workpieces that can not be easily handled manually. It finds extensive use for drilling holes in large sized castings that are too heavy as it is difficult to reposition the casting every time for drilling each hole. This machine has a heavy round vertical column supported on a large base.

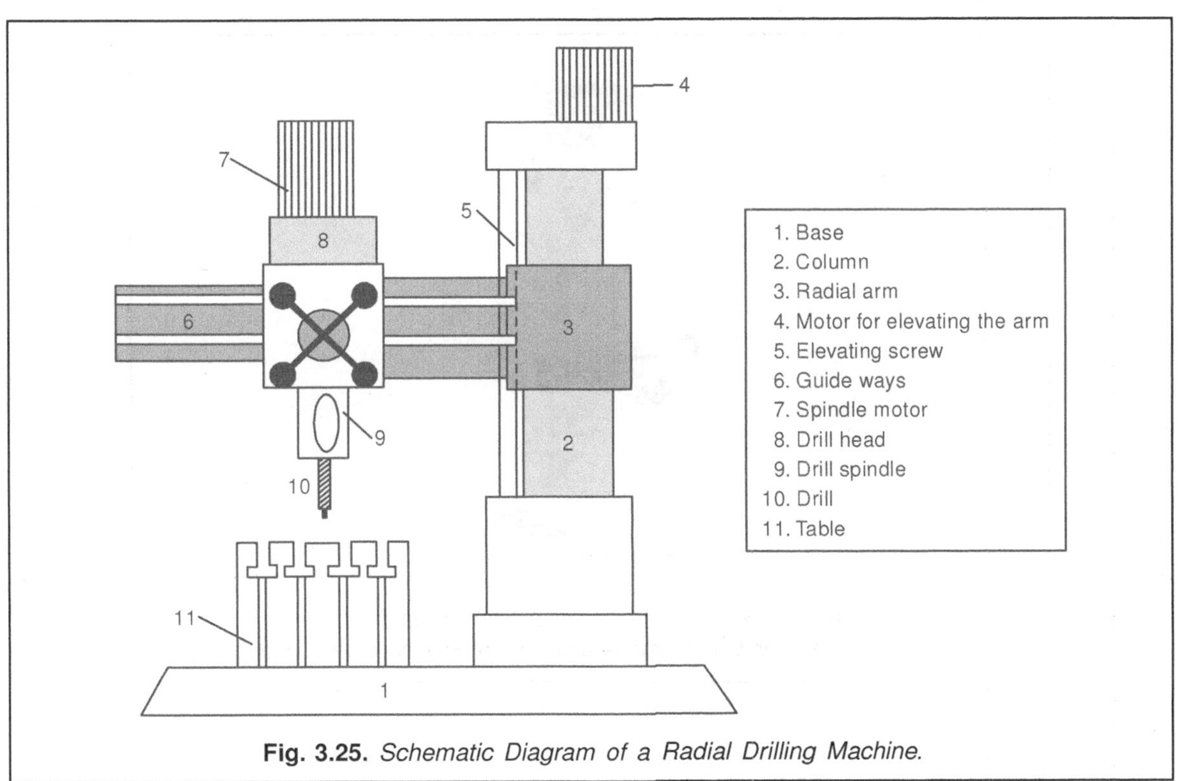

Fig. 3.25. *Schematic Diagram of a Radial Drilling Machine.*

The column supports a radial arm that can be raised and lowered by power and rotated over the base. The drilling head or the spindle head with its speed and feed changing mechanism is mounted on the radial arm. It can be moved horizontally to any desired position on the arm. The workpiece is clamped to the table or base and the drill head can then be positioned where it is needed by swinging the arm and moving the drill head along the arm.

Fig. (3.25) shows a line diagram of a radial drilling machine. The different parts of a radial drilling machine are:

- Base
- Column
- Radial arm
- Drill head (or spindle head)
- Spindle speed and feed mechanism.

Radial drilling machines have a wide range of speeds and feeds. These machines carry the necessary provision to perform the boring (enlarging the previous drilled hole) and tapping (cutting internal threads) operations. Spindle speeds ranging from 20 to 2000 rpm and feed from 0.05 to 3 mm per revolution are available. Figure 3.26. Shows a schematic of a drilling machine.

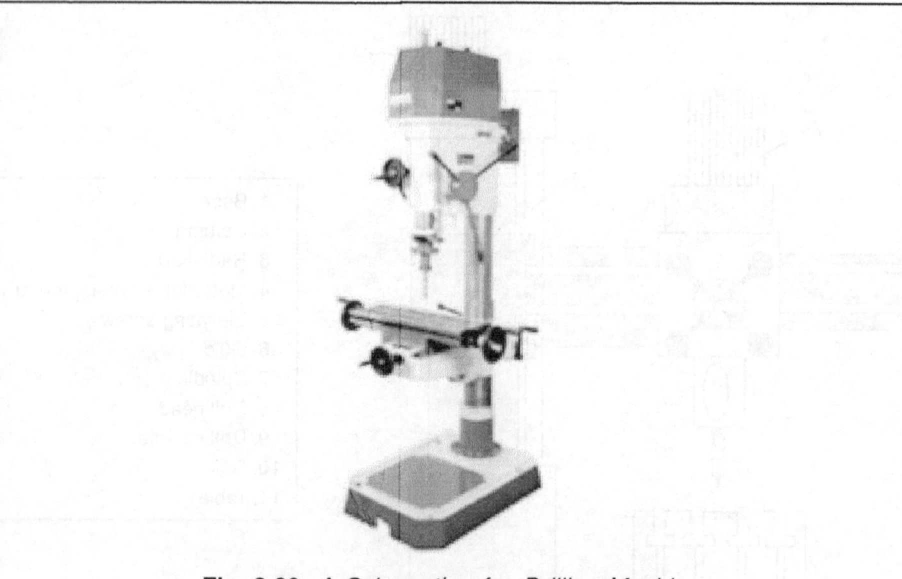

Fig. 3.26. *A Schematic of a Drilling Machine (Courtesy Premier Lathe Manufacturing Co. Rajkot, Gujrat, India).*

3.7.3 Drilling Machine Operations

Following operations as shown in Fig. (3.27) can be performed on the drilling machines.

- Drilling
- Boring
- Reaming
- Counterboring
- Countersinking
- Spot facing
- Trepanning
- Tapping.

Boring

It is the operation of enlarging a previously drilled hole. Boring is done with cutting tools that are similar to those used in turning.

Reaming

It is a machining operation used to make an existing hole dimensionally more accurate then the one obtained on a drilling machine and to improve its surface finish with the help of a tool called reamer. A reamer is a multiple cutting edge tool with straight or helical fluted edges that removes very little material.

Counterboring

It is the operation of enlarging the end of a hole cylindrically so as to accommodate a bolt head or a screw head in such a way that the entire bolt head or screw head will be below the surface of the part.

Countersinking

It is the operation of producing a tapered enlargement at the opening of a hole so as to provide a proper seat for a flat head screw or rivet. The most common angles are 60° and 90°.

Spot facing

It is the operation of producing a flat seat for a bolt head, washer, nut or a similar machine element at the opening of a drilled hole.

Trepanning

It is the operation of producing a hole by removing a disk shaped piece usually from flat plates. This produces the hole without reducing the entire material in the form of chips.

Tapping

It is the operation of cutting internal threads in the work.

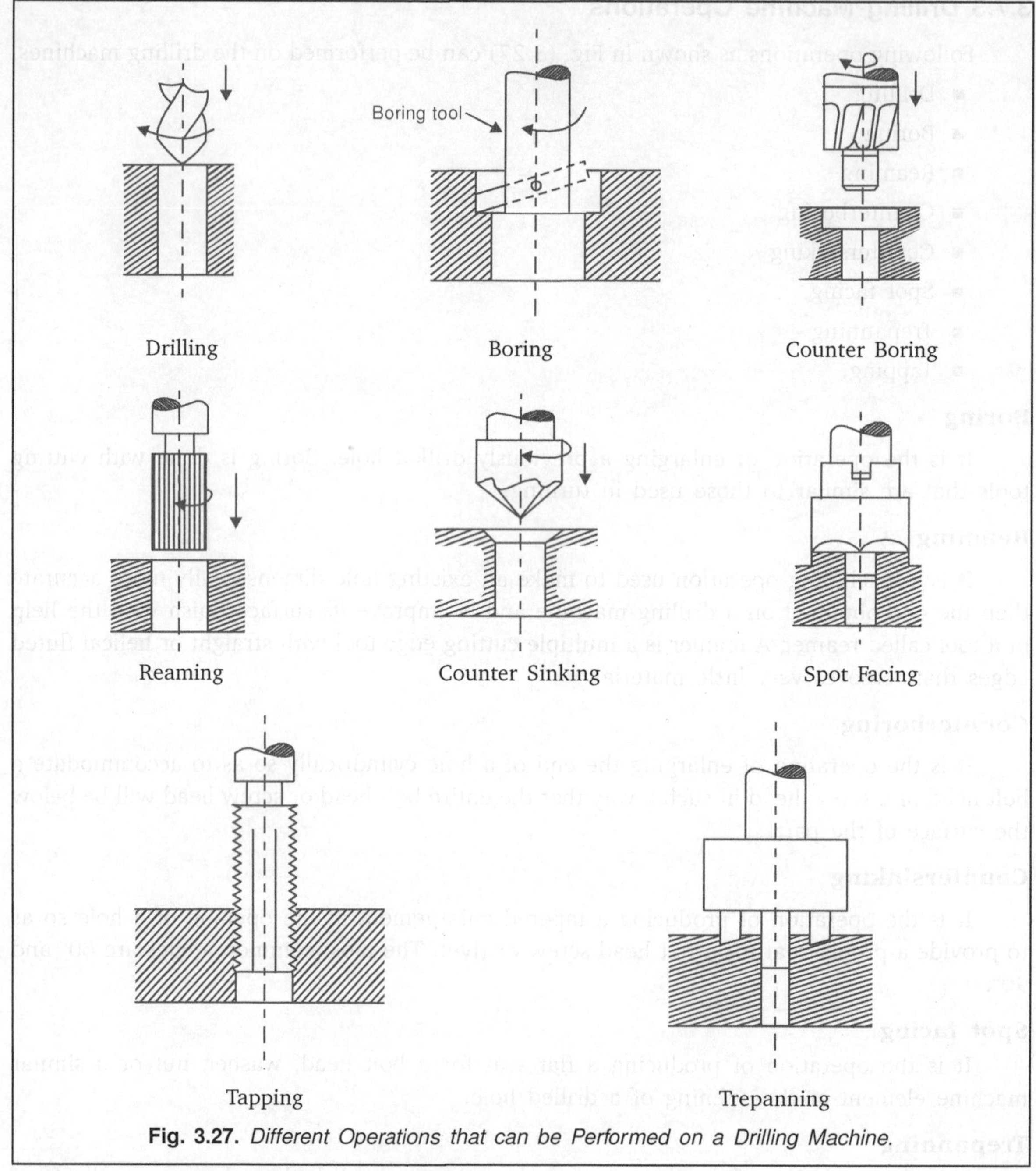

Fig. 3.27. *Different Operations that can be Performed on a Drilling Machine.*

3.7.4 Machining Parameters (Drilling)

☞ **Cutting speed**

The cutting speed for drilling operation is the speed of the periphery of the cutting edge of the drill (cutting tool) in m/min.

It is calculated as

$$V = \frac{\pi DN}{1000} \text{ m/min.}$$

where, D = diameter of the drill in mm.

N = spindle RPM.

☞ Feed

It is the distance moved per revolution by the drill into the work. Feed is expressed in mm/rev. of drill. Thickness of chip, necessary feed power and surface quality of hole depends on the feed. The permissible feed depends on the workpiece material, drill diameter, hole depth and type of the finish required.

☞ Depth of cut

It is equal to one half of the drill diameter.

☞ Material removal rate

The material removal rate in drilling is the volume of material removed by the drill per unit time. It is given by

$$MRR = \frac{\pi D^2}{4} fN \text{ mm}^3\text{/min.}$$

where, D = diameter of the drill in mm.

f = feed in mm/rev.

N = spindle RPM.

☞ Machining time

It is the time during which the drill removes material from the workpiece in the form of chips.

Let h_l = depth of the hole

L = feed distance of the drill = depth of hole + drill point

$L = h_l + 0.29\text{ D}$

D = diameter of the drill in mm.

Feed/min = feed/rev. × Number of revolutions/min. = $f \times N$

$$\therefore \quad \text{Machining time} = \frac{\text{feed distance}}{\text{Feed / min.}} = \frac{L}{f \times N} \text{ minutes.}$$

3.7.5 Specification of a Drilling Machine

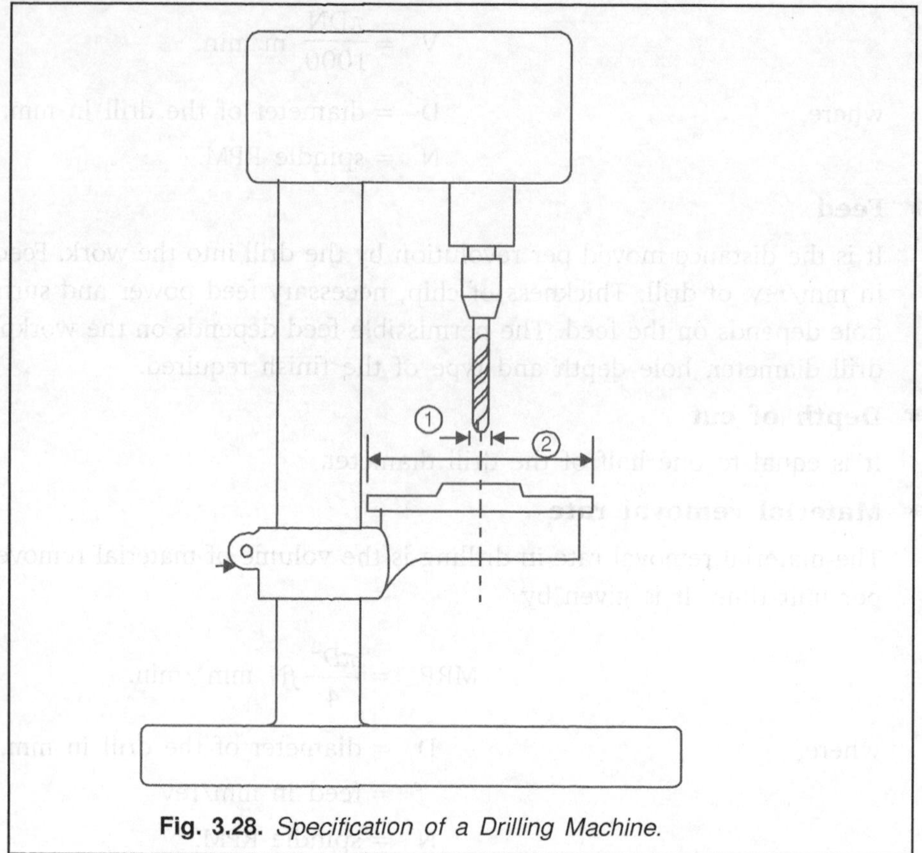

Fig. 3.28. *Specification of a Drilling Machine.*

The principal specifications as shown in Fig. (3.28) of a drilling machine are:

- Size of the maximum drill (or hole) that can be done in a given material.
- The maximum radial dimension of a workpiece that can be accommodated between the column and the tool (*i.e.,* drill).
- Power consumption of the motor (in HP or in Kilowatts)
- Range of spindle speeds
- Range of feeds.

Important note:

The drill does not possess the centring capability. It tends to walk on the workpiece surface at the beginning of the operation. This problem is very pronounced in the use of small diameter drills. To start a hole properly the drill should be guided, using fixtures such as a bushing to keep it from deflecting sideways.

3.7.6 Technical Specification of a Radial Drilling Machine

Capacity

Drilling capacity	: 40 mm

Drill Head

Spindle nose	: MT-4
Spindle travel	: 175 mm
Number of spindle speeds	: 8
Range of spindle speeds	: 100–1000 RPM
No. of Power feed	: 2
Power feed range	: 0-20–0.10 mm/rev

Working Range

Drilling radius (Max./Min)	: 1125–505
Distance from column to spindle centre (Max./Min)	: 1020–400
Distance from spindle to base (Max./Min.)	: 1180–380
Column Diameter	: 210 mm
Vertical Power Movement of the Arm	: 800 mm
Horizontal movement of drilling head	: 635 mm
Box Table size (L×W×H)	: 510 × 380 × 180

Base Plate

Base size (L×W×H)	: 1625 × 850 × 180
Working surface	: 1010 × 690
Number of T slot	: 16

Power

Drilling Motor	: 1.5/1400 KW/RPM
Elevating motor for arm	: 0.75/1400 KW/RPM
Shipping Size (approx L×W×H)	: 1750 × 1000 × 200

3.7.7 Drill Tools

A drill is a fluted cutting tool. Drills are used to produce or enlarge the holes. Drills are capable of producing deep holes into the work material as they have high length to diameter ratio. The commonly used drills types are as follows:

- Twist drill
- Step drill
- Straight flute drill
- Flat or Spade drill
- Gun drill.

Twist Drill

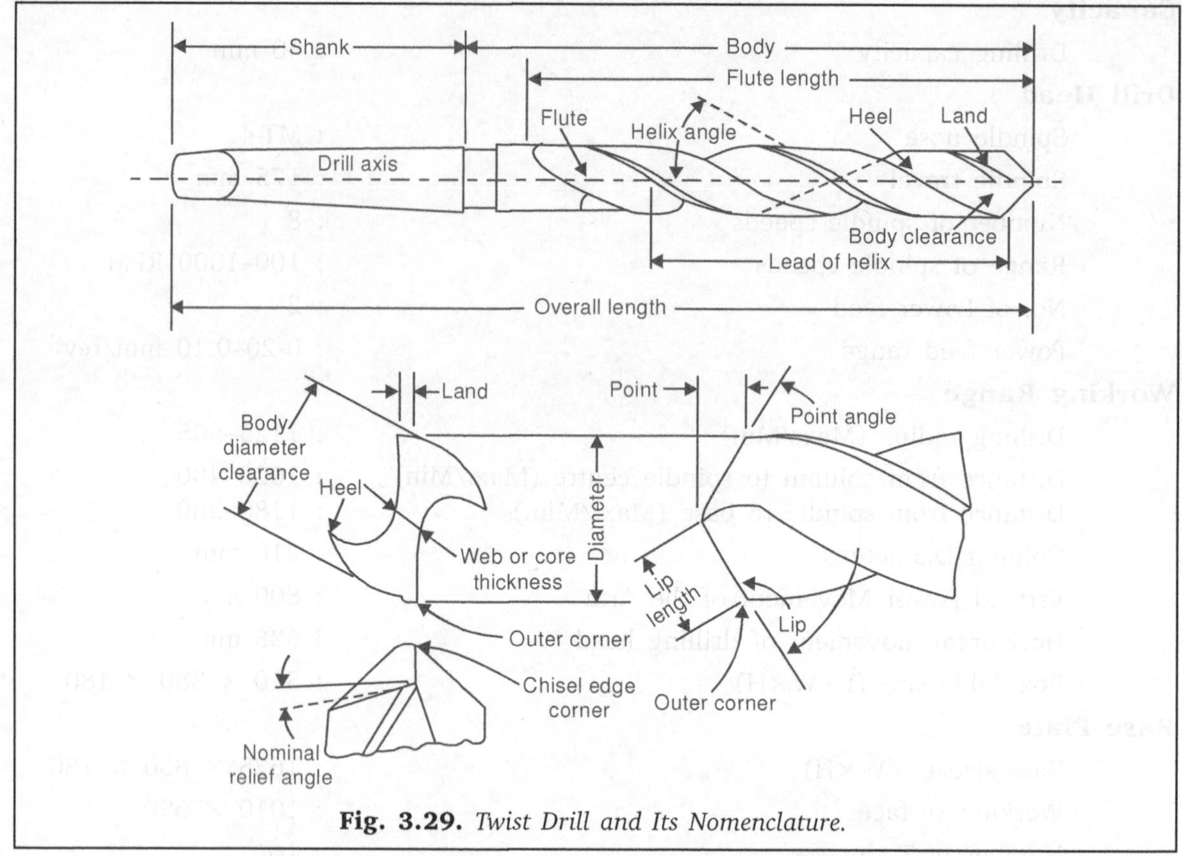

Fig. 3.29. *Twist Drill and Its Nomenclature.*

The most commonly used drill is the conventional standard point twist drill. The twist drill has three basic parts: the body, the point and the shank as shown in Fig. (3.29). Generally two or more spiral or helical grooves, called flutes, run the length of the drill, and the chips produced are guided upwards through these grooves. The cutting fluid is directed through the grooves on the drill in the cutting region from where the material is removed. The rake angle of a drill varies along the cutting edges or lips. Its value is zero near to the point and equal to the helix angle at the lip. The helix angle is built into the drill. The commonly used value of helix angle on the twist drill is 24° but drills with large helix angle above 30° are preferred for fast drilling and high material removal rate. Drills with helix angle less than 20° are used for softer materials.

The point angle as shown in Fig. (3.29) affects the direction of chip flow across the tool face and into the flute. For general purpose drilling, the point angle is 118°. The recommended drill point angles for various materials are:

Workpiece material	***Point angle***
Steel, cast iron, copper alloys	118°
Wood, plastics	90°
Aluminum alloys	118–135°

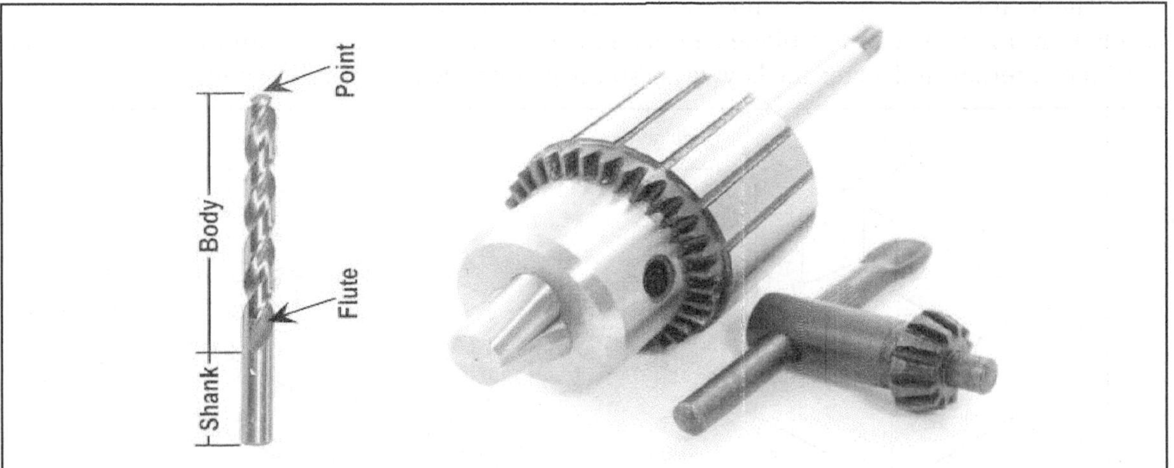

Fig. 3.30. *Shows a Schematic of a Twist Drill and Drill Chuck Commonly used with a Drilling Machine.*

Causes and cure in drilling operation

Problem	***Cause/cure***
Drill breakage	Point improperly ground, heavy feed, drill seizing in hole because of chips clogging flutes, dull drill, insecure clamping of work/drill.
Excessive drill wear	Excessive speed, ineffective cutting fluid, hard spots in material, too high rake angle.
Oversize hole	Loose machine spindle, unequal length or angle of the cutting edges, side pressure on the workpiece.
Hole with poor surface finish	Dull drill, ineffective cutting fluid, improperly ground drill, improper alignment.

3.8 SHAPER

Shaping and planing machining operations are used to produce straight flat surfaces. Both shaping and planing operations are used for removal of material from workpieces. A single point cutting tool is used in material removal action. This single point cutting tool is supported by a ram that reciprocates against workpiece.

The primary difference between the shaping and planing operations is that in shaping the tool is given cutting motion and feed is given to the workpiece whereas in planing cutting motion is provided to the workpiece and the feed is given to the tool. Figure 3.31 (*a*) and (*b*) clearly shows the difference between the two material removal operations. Shaping is preferred on small sized workpieces whereas planing is done on long workpieces.

The shaping is done on a machine called shaper as shown in Fig. (3.31) and planing is done on a machine called planer. By using special clamping devices and tools shaper can machine internal and external keyways, spiral groves, gear racks, dovetails, T-slots etc.

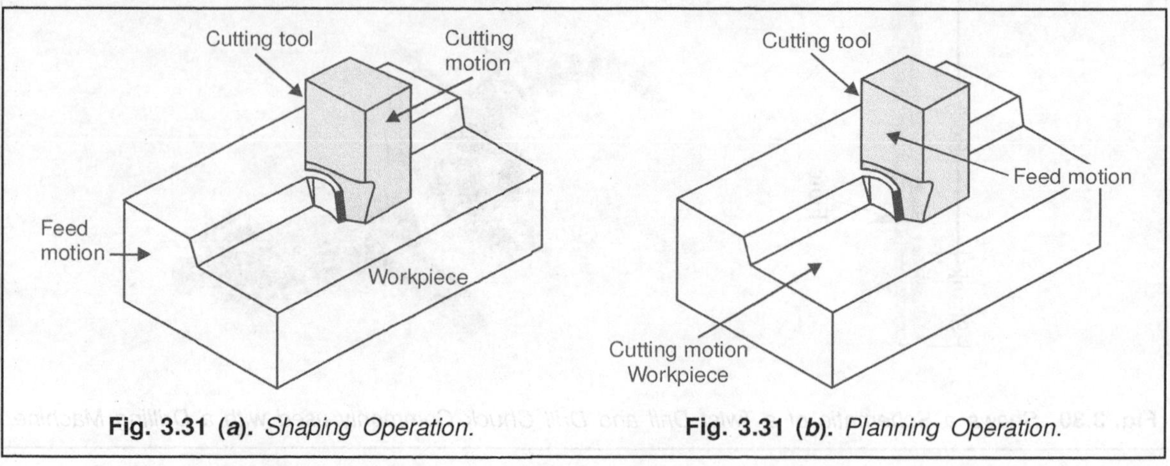

Fig. 3.31 (*a*). *Shaping Operation.* **Fig. 3.31 (*b*).** *Planning Operation.*

3.8.1 Description of a Shaper

Shaping operation is performed on a machine tool called a shaper. The straight line reciprocating motion between the cutting tool and the work is responsible for producing the cutting motion.

The main components of a shaper are:

(*i*) Base
(*ii*) Column
(*iii*) Ram
(*iv*) Worktable
(*v*) Tool head
(*vi*) Cross rail.

Base

It provides support to other parts of a shaper. It is generally made of cast iron.

Column

It is a box type structure which is mounted on the base. It is also made of cast iron. Column as shown in Fig. 3.32 contains the driving mechanism. It carries two machined guide ways on the top of it on which the ram reciprocates.

Ram

Ram moves with reference to a column that produces cutting motion. The ram has reciprocating motion and in most of the shapers, the material removal takes place during the forward stroke of the ram. It is also known as 'push cut'. In few shapers, the material removal takes place during the return stroke of the ram where it is called 'draw cut'.

Worktable

The worktable holds the workpieces and is responsible for feeding the workpiece. Standard T-slots are provided on the worktable to directly mount some of the workpieces with the help of bolts, blocks and clamp plates. But still vises are provided on worktables to clamp the workpieces.

Tool Head

As shown in Fig. 3.32, the tool head is an important part of a shaper. The clapper box has twin roles. Firstly, during the cutting stroke, it holds the tools tightly and secondly during non-cutting or return stroke, it allows the cutting tool to tilt upward so as to pass over the finished work surface.

Cross Rail

It is attached with the column front portion. Its twin tasks are (*i*) to lift the working table in up direction and (*ii*) the table can be moved perpendicular to ram axis over the cross rail.

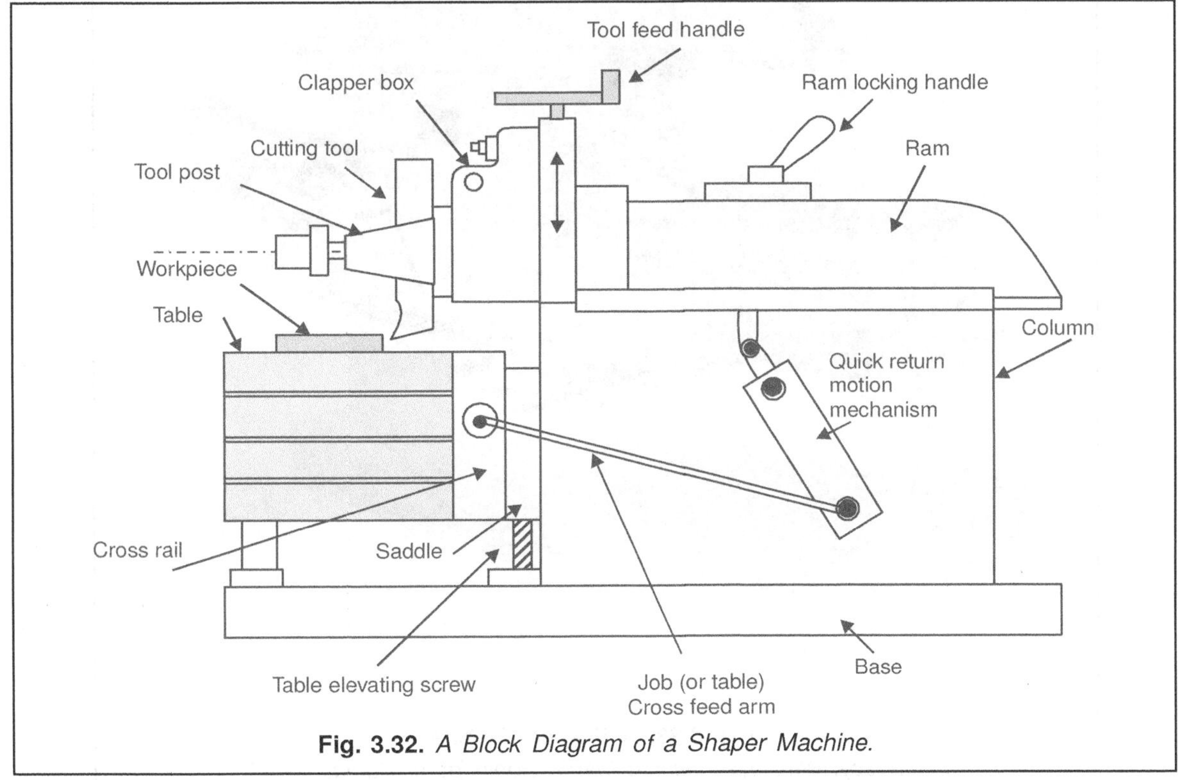

Fig. 3.32. *A Block Diagram of a Shaper Machine.*

In most of the shapers, ram drives the cutting tool in a horizontal direction but in few shapers, ram moves the cutting tool in vertical direction. In both these cases the workpiece as usual rests on a worktable which provides feed to workpieces by moving these towards the cutting tool. A shaper with a vertical cutting motion is called as 'Slotter'. A slotter is mainly used to machine notches, keyways, and dies.

There are two ways of operating a shaper either by (*a*) mechanically *i.e.*, crank driven system or by (*b*) a hydraulically driven system. A hydraulic shaper uses a piston and cylinder to move the ram. The motion of the ram consists of a forward stroke to cut the material and a return stroke during which the cutting tool is lifted slightly to clear the workpiece and it is then reset for the next cut stroke. On completion of each return stroke, the worktable is advanced laterally relative to the ram motion so as to provide feed to the workpiece. In shaper, feed is specified in mm/stroke. However the hydraulic drive is more costly than a mechanical drive but with hydraulic drive, reasonable flexibility in adjusting the stroke length and uniform speeds during the cut stroke can be achieved. Figure (3.33) shows a schematic of a shaper machine.

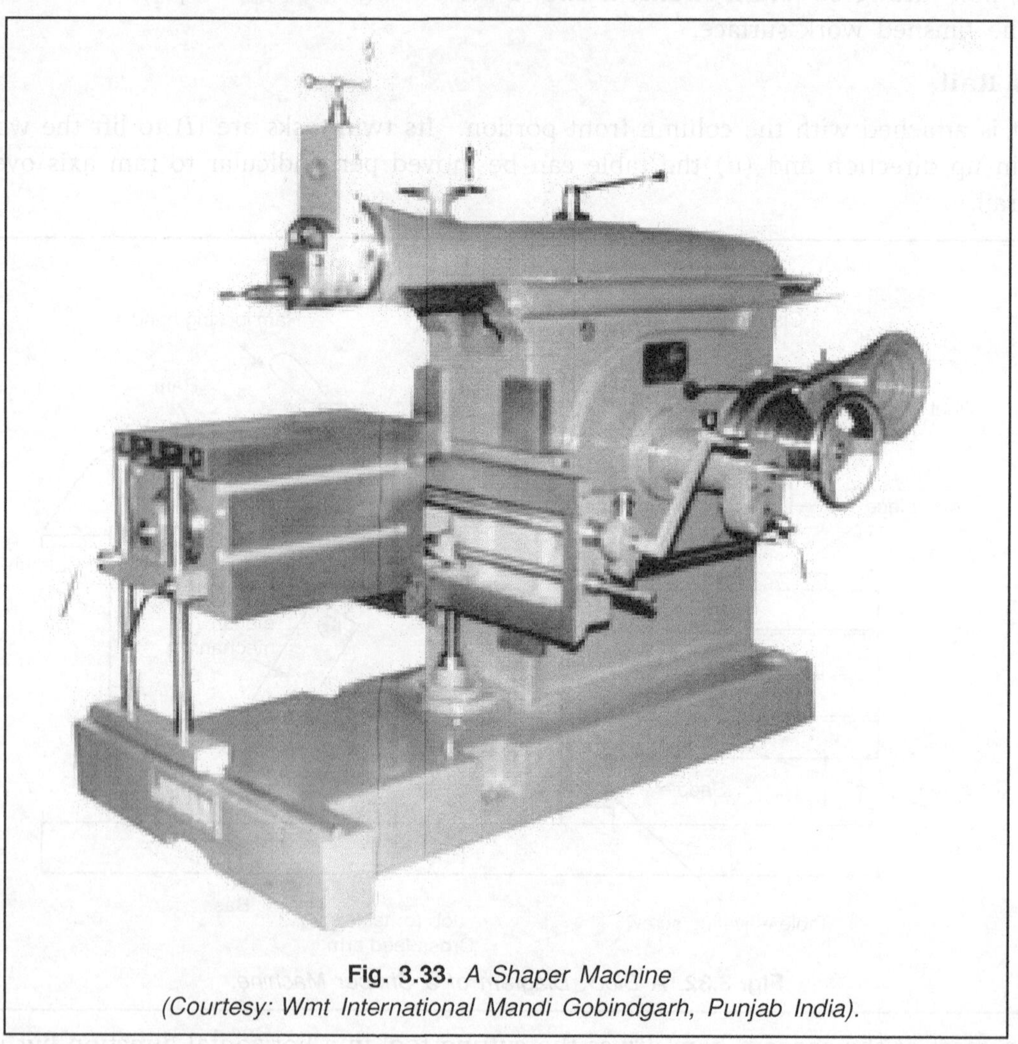

Fig. 3.33. *A Shaper Machine*
(Courtesy: Wmt International Mandi Gobindgarh, Punjab India).

A quick return motion mechanism for driving the ram is used in shaper so that the return stroke of shaper is executed at a faster rate in comparison with the cutting stroke. This reduces the non-machining time.

3.8.2 Specification of a Shaper

A shaper (or shaping machine) is usually specified by:

- The maximum length of the cutting stroke.
- Power consumption of the motor
- Size of the working table
- Number of cutting strokes per minute
- Type of quick return motion mechanism
- Floor area needed for its installation.

Horizontal shapers are commercially available in both bench and full size models. Small bench size models have a maximum stroke length of 150 mm. the large size shapers have a stroke length of about 915 mm. For vertical shapers, stroke length varies from 150 to 300 mm.

The workpiece size to be shaped or slotted is limited by the maximum length of stroke. The standard workpiece size is about 915 mm. For producing flat surfaces larger than 915 mm, usually such parts are then machined by a planer.

3.8.3 Technical Specification of a Shaping Machine

A typical specification of a shaper is as follows

Ram slide

Length of stroke of ram slide	: 610 mm
Length of ram slide	: 1275 mm
Max. and Min. distance from Table to ram slide	: 457 × 76

Table

Working surface of Table	: 610 × 356
Maximum horizontal table travel	: 610 mm
Maximum vertical table travel	: 381 mm
Swivel of either slide	: 60°

Tool Slide

Maximum vertical travel	: 153 mm
Max. Swivel of tool slide on either side	: 70°
Swivel of Clapper box on either side	: 15°

Speeds and auto feeds

Number of speeds	: 4
Range of speeds (strokes/min)	: 18-31-52-88
No. of auto feeds to table (Horizontal)	: 3
Feed to tool slide	: Manual

3.8.4 Shaping Parameters

For machining a workpiece of width B, the machining time is given by

$$t_m = \frac{B}{f \times N_c}$$

where, t_m is the machining time in sec.

f is the feed in mm/min

N_c is the number of cutting strokes

B is the width of the workpiece

The material removal rate during shaping operation is given by

$$MRR = L\, d\, f\, Nc$$

where, L is the length of the job in mm

d is the depth of cut in mm

f is the feed in mm/min

N_c is the number of cutting strokes

3.9 MILLING MACHINE

3.9.1 What is Milling?

Milling is a machining process in which metal is removed by a rotating multi tooth tool. The cutting tool in a milling machine is called a milling cutter and cutting edges are called teeth. The shaft on which this cutter is mounted is called arbor. The machine tool which performs this operation is called a milling machine. Figure 3.34 shows the commonly used types of milling cutters.

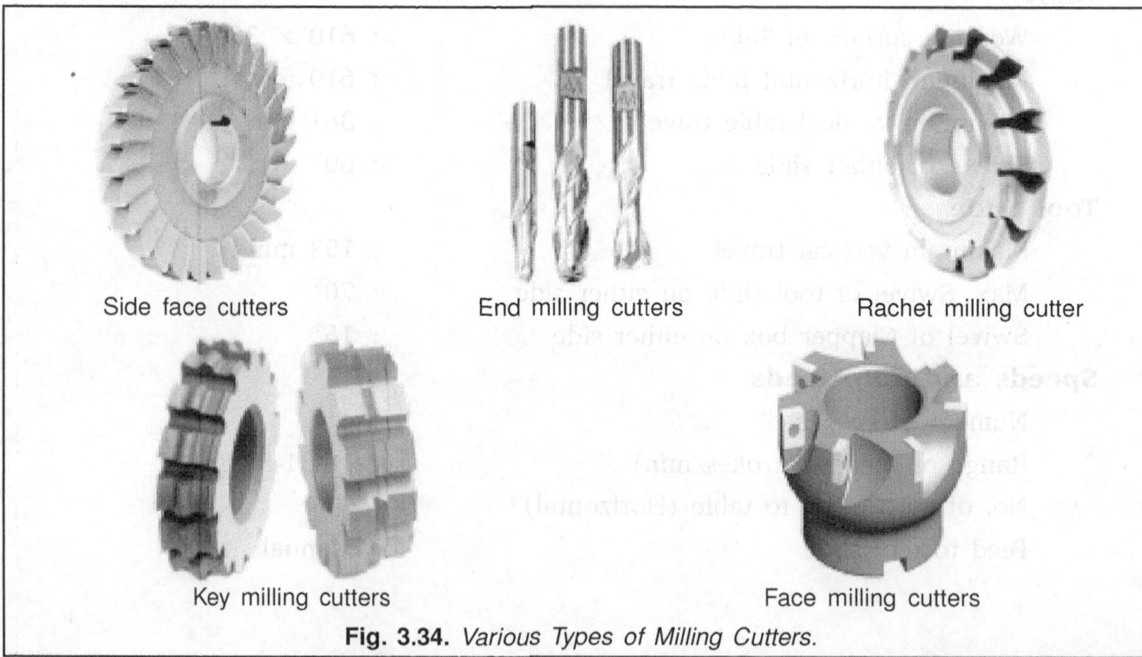

Fig. 3.34. *Various Types of Milling Cutters.*

Some of the cutting tool shapes used in milling resembles the drilling tools due to which confusion is created in the mind of the unskilled users. This confusion can be cleared by understanding the milling operation.

3.9.2 Milling Operations

Basically in milling operation, the axis of rotation of the cutting tool is perpendicular to the direction of feed. This orientation between the tool axis and the feed direction is one of the features that distinguish milling from drilling. In drilling cutting tool is fed in a direction parallel to its axis of rotation.

There two types of milling operations:

(*a*) peripheral milling

(*b*) Face Milling.

3.9.2.1 *Peripheral Milling*

In peripheral milling, the axis of cutter rotation is parallel to the workpiece surface to be machined. It is also called as plain milling. This operation is performed by cutting edges on the outside periphery of the cutter.

Depending on the direction of feed and cutting motion, peripheral milling can be grouped in two classes.

(*i*) Up milling (also called as conventional milling)

(*ii*) Down milling (also called as Climb milling).

Up Milling

It is generally referred as 'milling against the feed'. In up milling, the cutting direction is opposite to the feed motion as shown in Fig.(3.35).

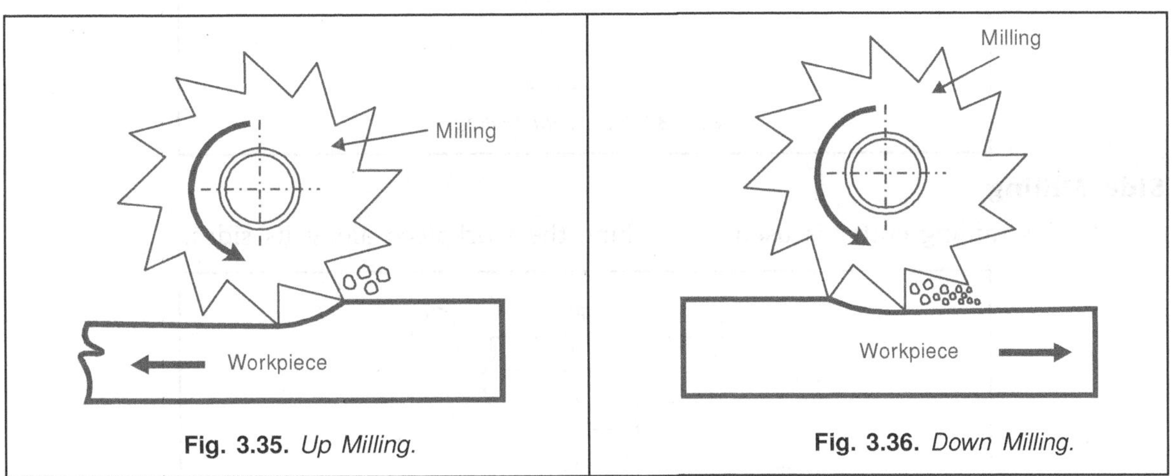

Fig. 3.35. *Up Milling.*

Fig. 3.36. *Down Milling.*

Down Milling

It is referred as 'milling with feed'. In down or climb milling as shown in Fig. (3.36) the direction of both the cutter motion and feed motion is same.

Up milling is considered safe in comparison to down milling. The reason is that in down milling the job has a tendency to move and dragged towards the cutter.

Different types of peripheral milling operations are shown in Fig. 3.37 given below:

Slab Milling

This is the basic form of peripheral milling. In this milling operation, the cutter width is more than the workpiece and it extends on both sides of the workpiece.

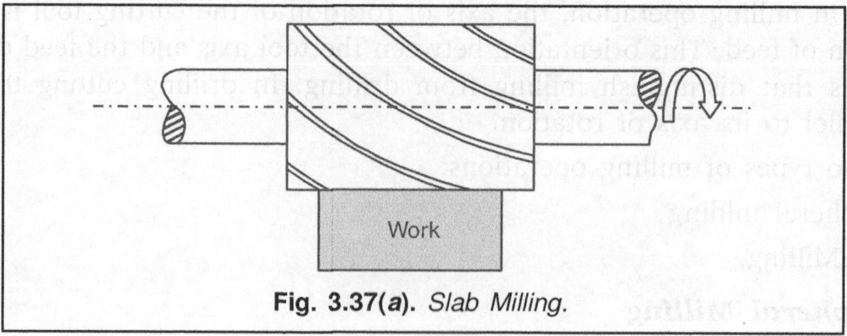

Fig. 3.37(*a*). *Slab Milling.*

Slot Milling

In slot milling, the width of the milling cutter is less than the workpiece width. As name suggests, it is used to mill a slot in the workpiece. It is also used to cut a piece in two parts.

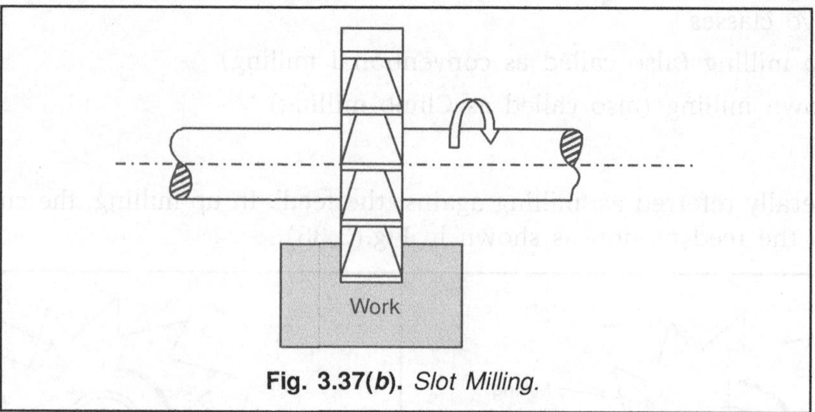

Fig. 3.37(*b*). *Slot Milling.*

Side Milling

In this milling cutter is used to machine the workpiece along its sides.

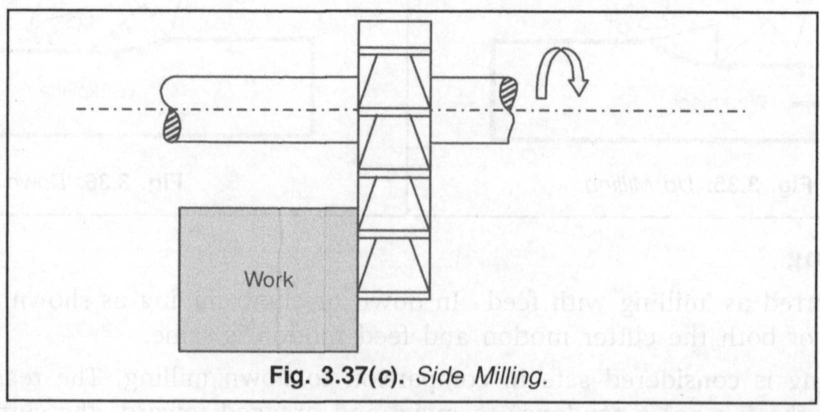

Fig. 3.37(*c*). *Side Milling.*

Straddle Milling

This is similar to side milling but machining takes place on both sides of the work.

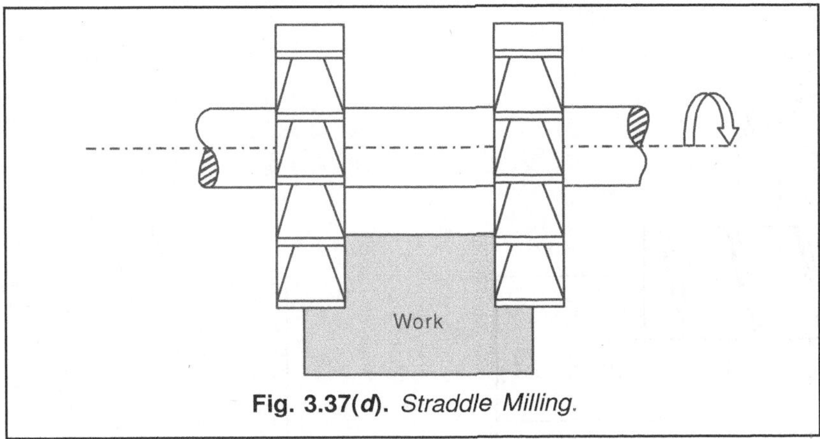

Fig. 3.37(*d*). *Straddle Milling.*

3.9.2.2 *Face Milling*

Face milling is used for machining flat work surfaces by means of milling cutter. In face milling, the axis of cutter is perpendicular to the surface being machined. In this milling cutters have teeth on their peripheral faces.

Different types of face milling operations are shown in Fig.(3.38) given below:

Traditional Face Milling

In traditional face milling, the diameter of the milling cutter is more than the workpiece width. Due to this the milling cutter overhangs the workpiece on both sides.

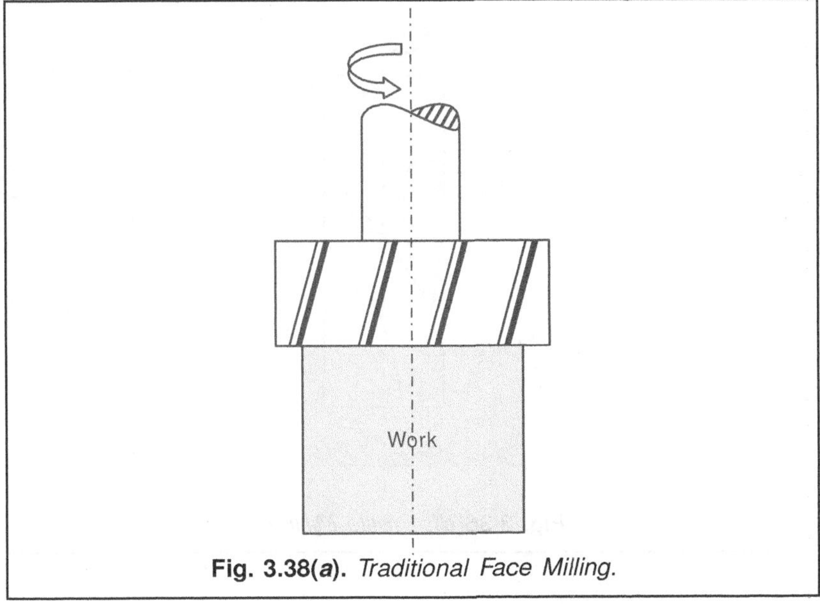

Fig. 3.38(*a*). *Traditional Face Milling.*

Partial Face Milling

In partial face milling, the cutting tool overhangs the workpiece from one side only.

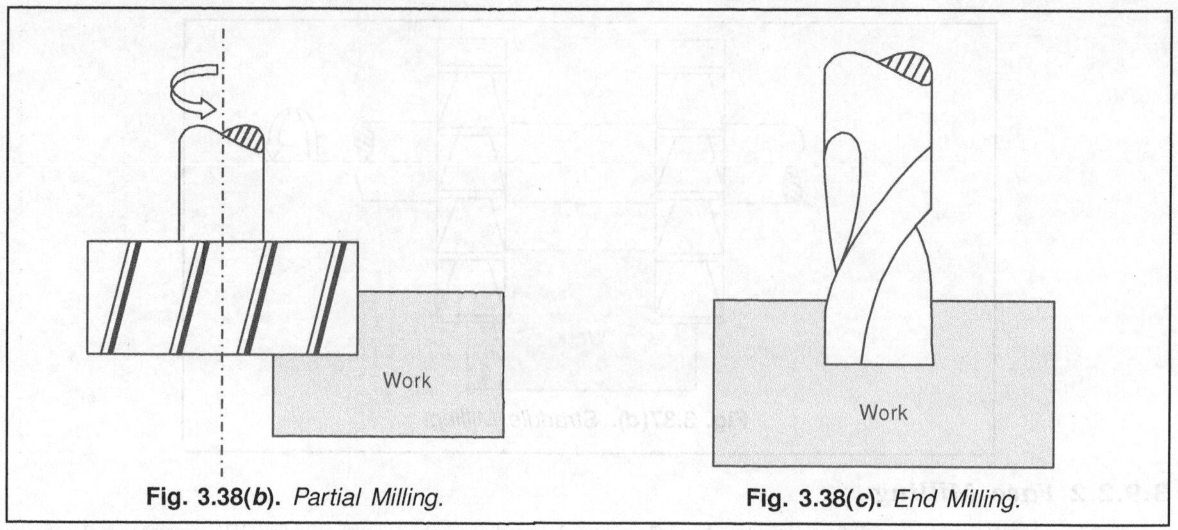

Fig. 3.38(*b*). *Partial Milling.*

Fig. 3.38(*c*). *End Milling.*

End Milling

In end milling, diameter of a chosen milling cutter is less than the workpiece width. This allows the cutter to mill a slot on the workpiece.

Profile Milling

In profile milling, outside profile of a workpiece is machined. It is similar to end milling operation.

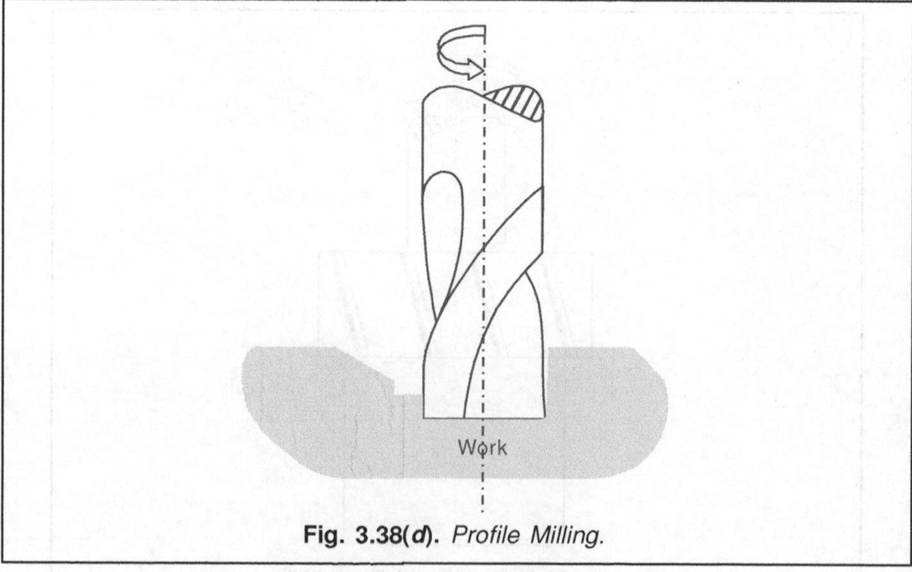

Fig. 3.38(*d*). *Profile Milling.*

Pocket Milling

It is one of the ends milling operation which is used to machine shallow pockets into flat workpieces.

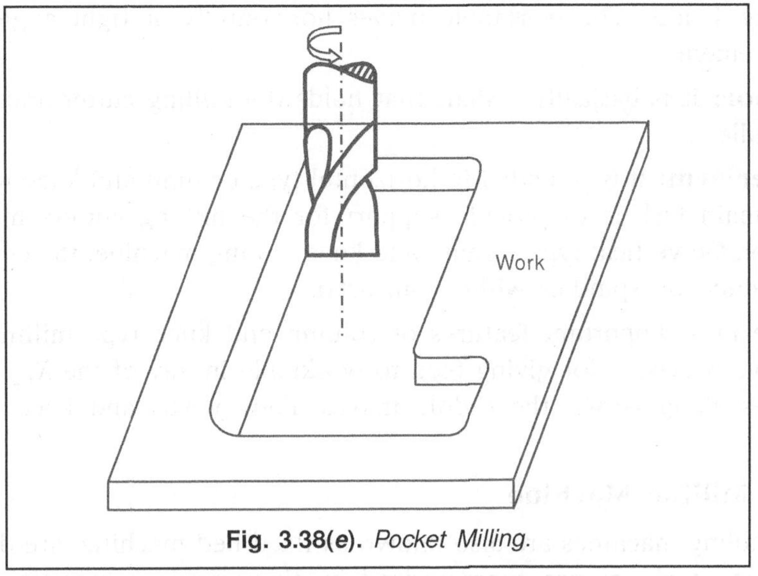

Fig. 3.38(*e*). *Pocket Milling.*

3.9.3 Milling Machine

Mostly milling is done on machines which are specifically designed to carryout milling. But milling can also be done on machines which can rigidly hold the workpiece and rotate a cutter and enable a workpiece to be fed into the cutter or vice versa.

On the basis of construction, the milling machines can be classified in four categories.

(A) Column and knee type milling machine

(B) Bed type milling machine

(C) Planer type milling machine

(D) Special purpose milling machine

(A) Column and Knee Type Milling Machine

Out of these four categories, the column and knee type milling machines are used for general purpose milling operations. The spindle on which milling cutter is mounted may be horizontal for slab milling or vertical for face and end milling, boring and drilling operations. It is available either as a horizontal or vertical machine as shown in Figs. 3.39(*a*) and 3.39 (*b*).

The basic components of column and knee type milling machine are:

(*a*) **Base:** It provides support to other machine components.

(*b*) **Column:** It contains the spindle and its driving mechanism.

(*c*) **Knee:** It is a separate section of machine which is attached to the column and moves vertically on the column.

(*d*) **Saddle:** this part as shown in Fig. 3.39(*a*) is supported by knee and moves horizontally on the knee.

(*e*) **Worktable:** It is an important part of a machine on which workpiece is clamped using T-slots. The worktable moves horizontally at right angles to the saddle movement.

(*f*) **Arbor:** it is basically a shaft that holds the milling cutter and is driven by the spindle.

(*g*) **Over arm:** It is provided in horizontal type column and knee milling machines. Its main task is to provide support for the milling cutters mounted over the arbor. On vertical type column and knee milling machine, the cutters can directly mounted on spindles without an arbor.

One of the most important features of column and knee type milling machine is its capability to have provision for giving feed to worktable in any of the *X*, *y* and *z*-axes. The worktable moves along *x*-axis, the saddle moves along *y*-axis and knee moves vertically along *z*-axis.

(B) Bed Type Milling Machine

Bed type milling machines are also known as fixed bed machine, are designed for high production. These machines are characterized by their rigid construction. The extremely rigid construction permits the use of high feed rates and depth of cuts and thus higher material removal rates. Bed types of machines are available in market in horizontal, vertical and multi spindle versions.

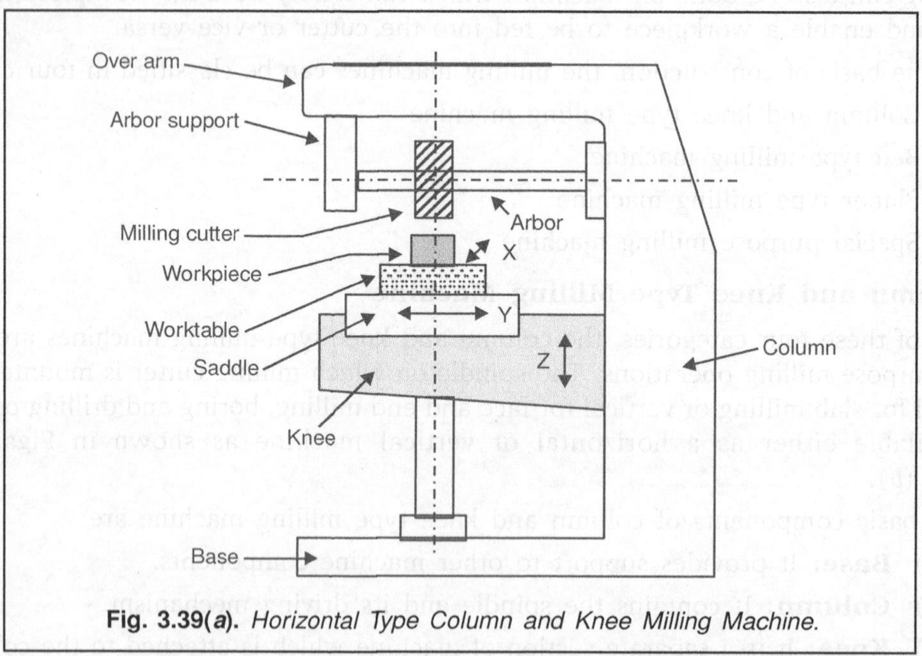

Fig. 3.39(*a*). *Horizontal Type Column and Knee Milling Machine.*

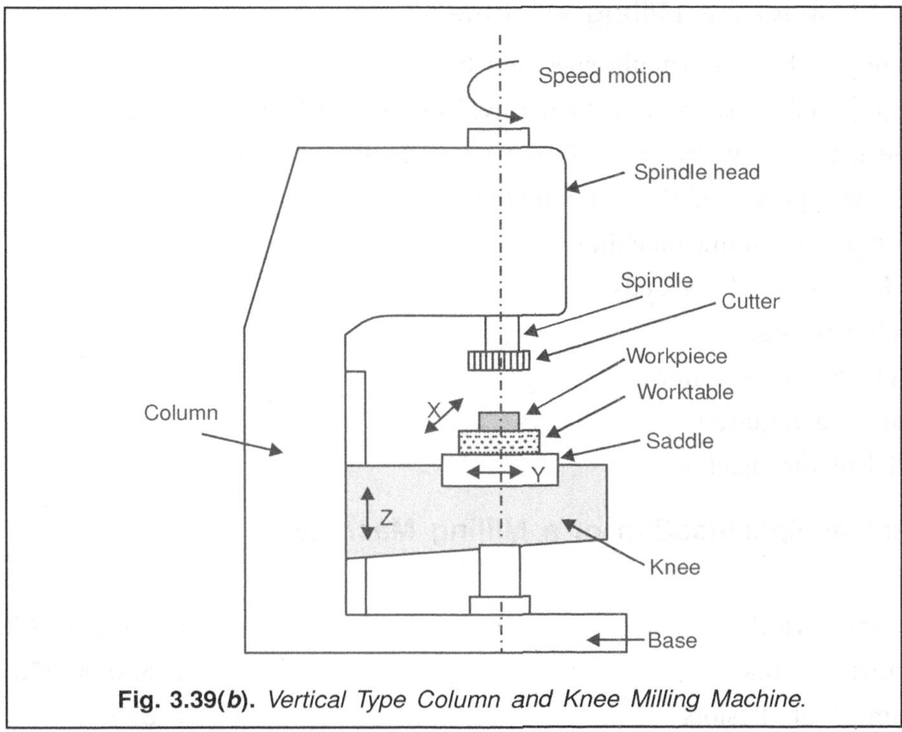

Fig. 3.39(*b*). *Vertical Type Column and Knee Milling Machine.*

(C) Planer Type Milling Machine

This particular type of milling machine has structural resemblance to a large planer. Planer type milling machines are built to machine very large parts. These machines can perform several milling and boring operations simultaneously. These machines utilize several milling heads to achieve high material removal rates.

(D) Special Purpose Milling Machine

Special purpose milling machines are designed and built for the special types of workpiece configuration. Special purpose machines include the following:

- Profiling mill or tracer mill
- Duplicating mill.

A profiling mill is used to reproduce an irregular internal or external part geometry that has been created on a template. These mills find extensive applications in the machining of moulds and dies. Now-a-days profiling mills have been replaced by the CNC (Computer Numerically Controlled) milling machines.

Duplicating mills are used to produce three dimensional part geometry. These are mostly used in aerospace industry. Presently these mills are also replaced by the CNC milling machines.

3.9.4 How to Specify a Milling Machine?

A milling machine is usually specified by:

- Longitudinal, cross and vertical travel of the worktable
  Table length × width = 1100 mm × 310 mm
- The horse power of the main motor
- The type of milling machine
- Number of spindle speeds
- Number of feeds
- Weight of the machine
- Floor area required
- Model of the machine.

3.9.5 Technical Specification of a Milling Machine

Table	
Length×Width	: 1050 × 250 mm
Clamping area	: 850 × 250 mm
Number of 'T' slots	: 3
Centre distance of 'T' slots	: 75 mm
Axes travel	
Longitudinal travel	: 500 mm
Vertical travel	: 250 mm
Spindle centre to table face	: 300 mm
Transverse travel	: 225 mm
Spindle	
Spindle nose taper	: ISO-40
Number of spindle speeds	: 6
Speed range	: 50 to 1000 RPM
Arbor Diameter	: 25.4 mm
Feed	
Number of feed	: 3
Power	
Main Motor	: 3HP/1440 rpm
Working range	
Swivel of table	: 45°
Distance between spindle axis to table surface	: 300 mm
Distance of spindle axis to bottom surface of over arm	: 150 mm
Coolant tank capacity	: 25 litres

Weight of machine : 2000 kgs
Overall height of the machine : 2450 mm

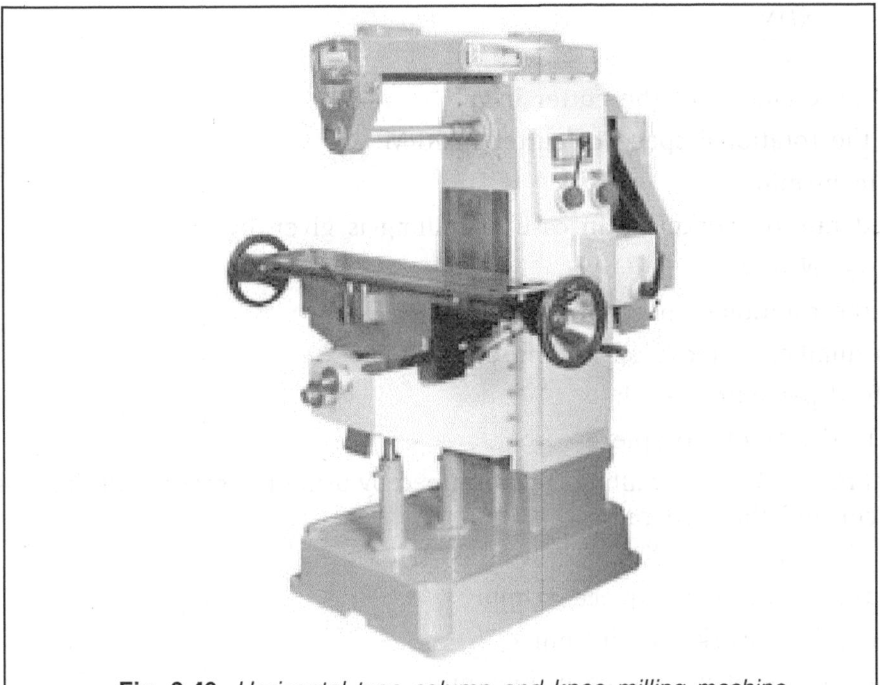

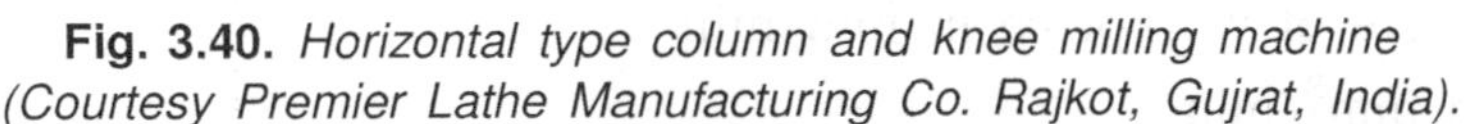

Fig. 3.40. *Horizontal type column and knee milling machine (Courtesy Premier Lathe Manufacturing Co. Rajkot, Gujrat, India).*

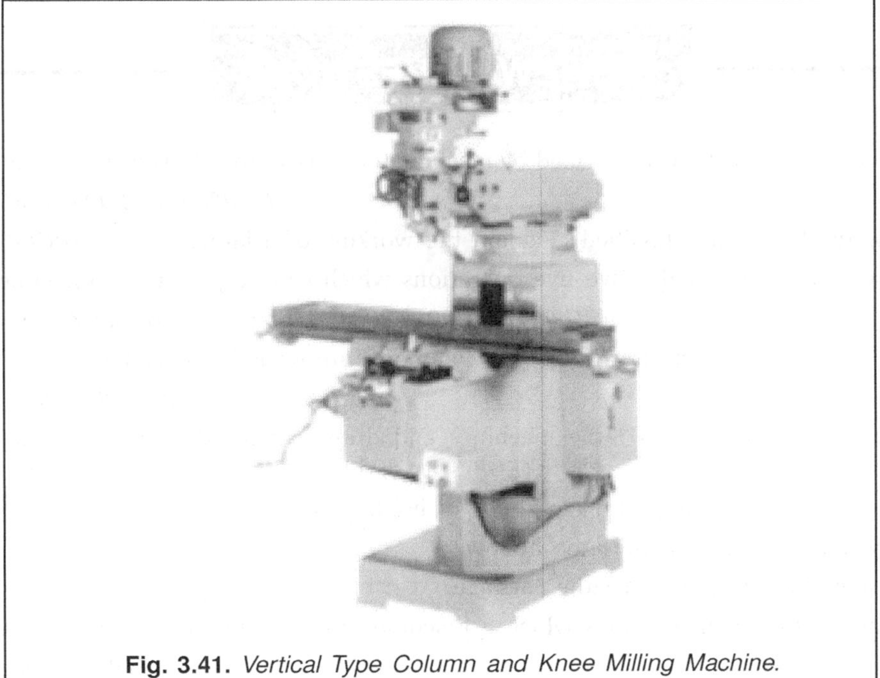

Fig. 3.41. *Vertical Type Column and Knee Milling Machine.*

3.9.6 Milling Parameters

Cutting speed is determined at the outside diameter of a milling cutter. It is expressed by

$$v = \frac{\pi DN}{60}$$

Where, D is the diameter of the cutter in m

N is the rotational speed of cutter in RPM

v is in m/min

The feed rate of workpiece in case of milling is given by

$$f_r = N\, n_1\, f$$

Where, N is the rotational speed of cutter in RPM

n_1 is number of teeth on milling cutter

f is feed per cutter tooth

f_r is feed rate of workpiece

Material removal rate in milling is determined by using the product of the cross sectional area of the cut and the feed rate.

$$MRR = wdf_r$$

Where, w is the width of workpiece in mm

d is depth of workpiece in mm

f_r is feed rate of workpiece in mm/min

MRR is in mm^3/min.

REVIEW QUESTIONS

1. What are the functions performed by machine tools? Give the classification of machine tools. *(RGPV June 2004, Nov./Dec. 2007)*
2. How the lathes are classified? Explain the working of a lathe. *(RGPV June 2002)*
3. How lathe is specified? Give five operations which can be performed on lathe? *(RGPV Feb. 2005, 2001, Jan./Feb. 2006, 2007, Nov./Dec. 2007)*
4. Write the detailed specifications of a common purpose lathe machine. *(RGPV June/July 2006, June 2007)*
5. Name different types of lathe machines and draw a block diagram of a 'Centre Length' showing its main parts. *(RGPV Dec.2003, Jan./Feb. 2007, June 2007, June 2008)*
6. Describe any three important methods of holding work in a lathe.
7. Why is the spindle of the lathe hollow?
8. What is the function of a lathe carriage?
9. What are the main functions which a machine tool is generally required to perform? *(RGPV Jan./Feb. 2006)*

10. How will you specify the feed on a lathe?

11. What will happen to the work piece while turning if it is held between centres and the centres are not exactly in line?

12. What is the purpose of a mandrel? List the types of mandrels used in a lathe?

13. What are the advantages and disadvantages of a four jaw independent chuck over a three-jaw chuck.

14. Describe the difference between a steady rest and a follower rest. Give an application of each.

15. Explain the working of a Drilling Machine. Name the various parts. *(RGPV-2001)*

16. Give the constructional details of a drilling machine with a neat sketch. *(RGPV-June 2002, June/July 2006)*.

17. Write typical specifications for a drilling machine. *(RGPV Dec. 2005)*

18. How drilling machines are classified? Draw block diagram of a drilling machine and describe it in detail? *(RGPV Dec. 2006)*

19. What is the difference between up milling and down milling?

20. What is the difference between the peripheral milling and face milling?

21. What machines are used for machining a flat surface and a round surface? *(RGPV Dec. 2005)*

22. What do you understand by cutting tool nomenclature? Sketch and label tool angles/ tool nomenclature?

OBJECTIVE TYPE QUESTIONS

Select the most appropriate answer:

1. The material used in the construction of a lathe bed is

(A) HSS (B) Stainless steel

(C) Cast iron (D) Plastic.

2. The include angle of lathe centre is

(A) 30° (B) 45°

(C) 60° (D) 90°

3. The lathe size is usually specified by

(A) height of lathe centres (B) length between centres

(C) swing diameter over the bed (D) swing diameter over the carriage

(E) all of the above.

4. Swing diameter over bed as compared to swing diameter over carriage is

(A) more (B) less

(C) same (D) none of the above.

5. .... is provided on the lathe spindle.

(A) external taper (B) internal taper

(C) internal and external taper (D) none of the above.

6. Knurling is an operation of
(A) cutting threads (B) making holes
(C) undercutting
(D) roughening the surface for the proper grip.

7. Reducing the diameter of workpiece over a very narrow surface is known as
(A) chamfering (B) grooving
(C) facing (D) knurling.

8. Operation of bevelling the extreme ends of a workpiece is called as
(A) turning (B) taper turning
(C) chamfering (D) facing.

9. Operation to be performed first on a job is
(A) boring (B) drilling
(C) counter boring (D) tapping.

10. Centreing can be done most accurately on
(A) four jaw chuck (B) three jaw chuck
(C) collet chuck (D) magnetic chuck.

11. The slowest speed is used for .... operation.
(A) taper turning (B) knurling
(C) facing (D) thread cutting.

12. The twist drill is
(A) front cutting tool (B) side cutting tool
(C) end cutting tool (D) none of the above.

13. The cutting edges of a twist drill are known as
(A) lips (B) flutes
(C) wedges (D) flanks.

14. Half nut is used to
(A) give desired direction of movement to the lathe carriage
(B) lock the lathe carriage to the lead screw for thread cutting
(C) drill the holes
(D) reduce spindle speed.

15. .... is preferred for quick setting and accurate centreing of the workpiece on a lathe.
(A) four jaw chuck (B) three jaw chuck
(C) combination chuck (D) magnetic chuck.

16. The taper on lathe spindle is
(A) 1:10 (B) 1:20
(C) 1:30 (D) 1:15.

17. ....is not provided with a power feed.
(A) compound rest (B) carriage
(C) cross slide (D) lead screw.

18. The included angle between lathe centres is
(A) 15° (B) 30°
(C) 45° (D) 60°

19. Which operation requires slowest speed on lathe machine

(A) taper turning (B) shoulder turning

(C) thread cutting (D) knurling.

20. Size of a shaper is specified by

(A) size of table (B) length of stroke

(C) size of tool (D) horse power of motor.

21. Size of a planer is specified by

(A) size of table and height of cross rail

(B) length of stroke

(C) size of table

(D) all the above.

22. In case of a shaper, the job is fed by

(A) ram movement (B) table movement

(C) tool movement (D) clapper box movement.

23. The difference between the planer and shaper is that in the former case

(A) tool and job both reciprocates

(B) job reciprocates while tool is stationary

(C) tool moves over the stationary work

(D) all the above.

24. Match List I (Machine tools) with list II (Machine tool parts) and select the correct answer using the codes given below:

List I (Machine tools)	***List II (Machine tool parts)***
A. Lathe	1. Lead screw
B. Milling machine	2. Rocker arm
C. Shaper	3. Universal indexing
D. Drilling machine	4. Flute

(A) A4-B2-C3-D1 (B) A1-B3-C2-D4

(C) A4-B3-C2-D1 (D) A1-B2-C3-D4.

25. Milling cutters are mounted on a part called

(A) Arbor (B) Dividing head

(C) Bracket (D) Shaft.

26. Milling machines are classified as horizontal type or vertical type, depending on the position of

(A) column (B) knee

(C) spindle (D) workpiece.

27. Consider the following forces acting on a finish turning tool:

1. Feed force
2. Thrust force
3. Cutting force.

The correct sequence of the decreasing order of the magnitudes of these forces is

(A) 1, 2, 3 (B) 2, 3, 1

(C) 3, 1, 2 (D) 3, 2, 1.

28. Match List 1 (Machine tools) with list II (Features) and select the correct answer using the codes given below the lists:

List I (Machine tools)	*List II (Features)*
A. Lathe	1. Push or pull tool
B.Drilling machine	2. Ratchet and pawl mechanism
C. Shaper	3. Dividing head
D. Broaching machine	4. Hollow tapered spindle
	5. Face plate

(A) A2-B4-C5-D1 (B) A5-B3-C2-D4

(C) A2-B3-C5-D4 (D) A5-B4-C2-D1.

29. In a shaper machine, the mechanism for tool feed is

(A) Geneva mechanism (B) Whitworth mechanism

(C) Ratchet and Pawl mechanism (D) Ward-Leonard system.

30. Consider the following operations

1. Under cutting
2. Plain turning
3. Taper turning
4. Thread turning.

The correct sequence of these operations in machining a product is

(A) 2, 3, 4, 1 (B) 3, 2, 4, 1

(C) 2, 3, 1, 4 (D) 3, 2, 1, 4.

31. Match List I with list II and select the correct answer using the codes given below the lists:

List I	*List II*
A. Reaming	1. smoothing and squaring surface around the hole for proper seating
B. Counter-boring	2. sizing and finishing the hole
C. Counter-sinking	3. enlarging the end of the hole
D. Spot facing	4. making a conical enlargement at the end of the hole

(A) A3-B2-C4-D1 (B) A2-B3-C1-D4

(C) A3-B2-C1-D4 (D) A2-B3-C4-D1.

32. Which one of the following processes results in the best accuracy of the hole made?

(A) Drilling (B) Reaming

(C) Broaching (D) Boring.

33. Cast iron is used for making machine beds because of its

(A) tensile strength (B) endurance strength

(C) damping capacity (D) compressive strength.

34. Low feed rate in a milling operation leads to

(A) low power consumption (B) improvement in surface finish

(C) better milling of harder materials

(D) dullness of the milling cutter as it will rub and scrap the workpiece surface.

35. Which of the following is not a part of a shaper

(A) clapper box (B) ram

(C) cross slide (D) tool head.

36. A twist is specified by

(A) its shank and diameter (B) shank, material and diameter

(C) its thickness (D) its lip angle.

37. Which type of milling causes the maximum friction

(A) up milling (B) down milling

(C) climb milling (D) all the above.

38. ....milling operation, the cutting tool tends to lift the work piece

(A) down (B) up

(C) inclined (D) climb.

39. In a milling machine, the arbor is used to hold...............

(A) Mandrel (B) Face plate

(C) Cutting tool (D) Job.

40. The milling operation in which the material is removed simultaneously from two sides of a workpiece is called ....

(A) Slab milling (B) Slot milling

(C) side milling (D) Straddle milling.

ANSWERS

1. (C)	**2.** (C)	**3.** (E)	**4.** (A)	**5.** (B)	**6.** (D)
7. (B)	**8.** (C)	**9.** (A)	**10.** (A)	**11.** (D)	**12.** (C)
13. (A)	**14.** (B)	**15.** (C)	**16.** (B)	**17.** (A)	**18.** (D)
19. (C)	**20.** (B)	**21.** (A)	**22.** (B)	**23.** (B)	**24.** (B)
25. (A)	**26.** (C)	**27.** (C)	**28.** (D)	**29.** (C)	**30.** (C)
31. (D)	**32.** (B)	**33.** (D)	**34.** (D)	**35.** (C)	**36.** (B)
37. (A)	**38.** (B)	**39.** (C)	**40.** (D).		

Fluids

4.1 INTRODUCTION

Fluid Mechanics: Deals with study of all fluids under static and dynamic conditions.

Fluid Statics: Deals with study of fluids at rest.

Fluid Dynamics: Deals with study of fluids in motion.

As we know, there are three states of matter. These are: (*i*) Solid, (*ii*) Liquid and (*iii*) Gases. The term fluid refers to both liquids and gases. Since we know the meaning of a solid so let see what is fluid?

Liquid: For a given mass, a liquid can be treated as a fluid that will have a definite volume independent of the shape of a container. It means that the liquid will assume the shape of the container and fill the container up to that level which is equal to the volume of the liquid.

☞ ***An Important Tip:***

Liquids are said to be incompressible so that their volume does not change with change in pressure. This is not exactly true. The change in volume due to pressure changes comes out to be very small that is why it is ignored in most of the fluid engineering problems.

Gases: Gases are fluids that are readily compressible and their volume will vary to fill the container. Liquids have almost fix volume for a given mass whereas the volume of gas used to change as per the shape of the container. On the other hand, the pressure has been found to have a significant effect on gases. An increase in pressure leads to decrease in the volume of the gas and decrease in pressure causes the volume of the gas to increase.

☞ The main difference between the liquid and gas state is that the gas will occupy the whole volume whereas liquid has definite volume.

A fluid is a collection of randomly arranged molecules held together by weak cohesive forces between these molecules and forces exerted by the container walls.

4.2 PROPERTIES OF FLUID

(*i*) **Density:** Density of a fluid is defined as the mass per unit volume. It is represented by symbol ρ (rho). It is also called as mass density.

Mathematically, we can write

$$\text{Density} = \frac{\text{mass}}{\text{volume}}$$

Its dimensions are $\frac{\text{kg}}{\text{m}^3}$. The density of water is 1000 $\frac{\text{kg}}{\text{m}^3}$.

(*ii*) **Specific weight:** The specific weight of a fluid is defined as the weight per unit volume. It is represented by symbol γ (gamma).

$$\text{Specific weight } \gamma = \rho g$$

Its dimensions are $\frac{\text{N}}{\text{m}^3}$.

(*iii*) **Specific Gravity:** The specific gravity (S_g) of a given fluid is defined as the ratio of the weight density of a fluid to the weight density of pure water at standard conditions. It is also called as relative density. It is a dimensionless parameter.

$$\textit{Specific gravity} = \frac{\textit{Weight density of liquid}}{\textit{weight density of water}}$$

(*iv*) **Specific Volume:** The specific volume is defined as the volume per unit mass of the fluid.

$$\textit{Specific volume} = \frac{\textit{volume of fluid}}{\textit{mass of fluid}} = \frac{1}{\frac{\textit{mass of fluid}}{\textit{volume of fluid}}} = \frac{1}{\rho}$$

Thus specific volume is the reciprocal of mass density. Its dimension is $\frac{\text{m}^3}{\text{kg}}$.

(*v*) **Pressure:** Pressure is one of the basic properties of all fluids. Pressure is defined as the force per unit area. In all cases of fluid statics which deals with the study of fluids at rest, the fluid molecules are in the state of compression. The degree of compression is measured by pressure. Thus the compressive stress acting on a fluid is also a representative of pressure.

Mathematically pressure can be expressed as the force per unit area.

$$\textit{Pressure} = \frac{\textit{Force}}{\textit{Area}}$$

$$p = \frac{F}{A}$$

The basic unit of pressure is Pascal and denoted by 'Pa'. When a fluid applies a force of 1N over an area of 1 m^2, then the pressure is equivalent to 1 Pascal.

i.e., $$1\ Pa = 1\frac{N}{m^2}$$

Since Pascal is a very small unit, so for the purpose of calculation and other derivation, its higher value is generally expressed in *kPa* (kilopascal) or *MPa* (megapascal).

$$1\ kPa = 1 \text{ kilopascal} = 1000\ Pa = 1000\ \frac{N}{m^2}$$

$$1\ bar = 100\ kPa = 100 \times 1000\ Pa = 10^5\ Pa = 10^5\ \frac{N}{m^2}$$

As we know that the atmosphere around us has air at some pressure therefore zero pressure point can only be defined with reference to some datum. Atmospheric pressure provides most suitable reference for pressure measurement. The standard value of atmospheric pressure has been taken as $P_{atm} = 101.3\ kPa$.

In real life applications, the pressure is usually expressed with reference to three scales. These are:

(*a*) **Gauge scale:** On this scale, atmospheric pressure is marked as zero. Pressure above the atmospheric value is called gauge pressure. Readings taken on this scale are termed as gauge pressure.

(*b*) **Absolute scale:** It starts at the point of no pressure *i.e.,* the absolute zero pressure. Readings taken on this scale are termed as absolute pressure and one has to use suffix 'abs' with the value. Absolute pressure is defined as the sum of the atmospheric and gauge pressure.

$$P_{abs} = P_{gauge} + P_{atm}$$

(c) **Vacuum scale:** Vacuum scale is used for expressing the pressure below atmospheric level. Its zero point is set at the standard value of atmospheric pressure. Vacuum is the difference between the atmospheric pressure and the absolute pressure.

$$P_{vacuum} = P_{atm} - P_{gauge}$$

(*vi*) **Viscosity:** Viscosity is the most valuable property of a fluid. Viscosity represents the fluid resistance to flow. The low viscosity means fluid flows easily and a fluid that flows with difficulty has a high viscosity. The thin and thick appearances of a fluid are representative of low and high viscosity respectively.

4.3 NEWTON'S LAW OF VISCOSITY

To derive a relationship for viscosity and to understand its concept, let us consider a fluid layer confined between the two plates. One plate is stationary and other is moving with velocity *V* with respect to the first plate due to applied force *F* on it as shown in Fig. 4.1. The distance between the two plates is *h*. The fluid adheres to both the plate surfaces due to viscosity. The Force *F* is also responsible to overcome the friction caused by the layers of molecules sliding over one another. Thus the velocity of fluid layer in contact with stationary plate will be zero and the velocity of the fluid layer in contact with the top layer will be V. This leads to a linearly varying velocity profile with a slope of V/h. Thus the fluid will be subjected to shearing forces and the velocity will vary across the gap.

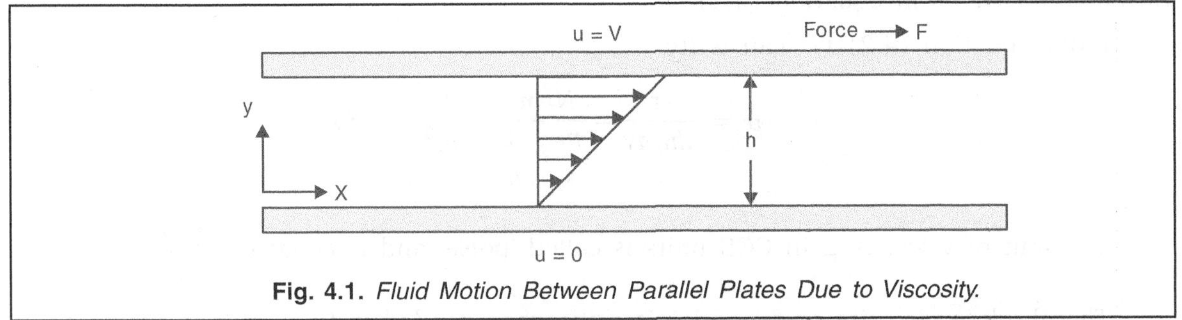

Fig. 4.1. *Fluid Motion Between Parallel Plates Due to Viscosity.*

Experiments have confirmed that the force per unit area required to move the upper plate is proportional to the velocity V of this plate and inversely proportional to the distance h between these two plates.

$$\frac{F}{A} \propto \frac{V}{h}$$

As we know that the force per unit area is called shear stress which is designated by τ.

$$\tau \propto \frac{V}{h}$$

This can be further simplified as

$$\tau = \mu \frac{V}{h} \qquad \text{...(4.1)}$$

Where μ is called the constant of proportionality and is known as coefficient of viscosity. It has also been referred to as dynamic viscosity or absolute viscosity or simply as viscosity at some places.

The equation (4.1) may be generalized for the case of two adjacent fluid layers separated by a distance dy, both moving in the x-direction. If the difference between velocities of the two layers is du, then equation (4.1) becomes

$$\tau = \mu \frac{du}{dy} \qquad \text{...(4.2)}$$

The equation (4.2) is known as Newton's Law of viscosity, and fluid satisfying the equation (4.2) is called a Newtonian fluid. From equation (4.2) it is clear that in one-directional flow, the shear stress in a Newtonian fluid is directly proportional to the rate of shear deformation. The fluids which fail to satisfy equation (4.2) are known as Non-Newtonian fluids.

Ironically, the three commonly used fluids *e.g.,* air, water and petroleum etc., very closely satisfy the Newton's law of viscosity. On the other hand the fluids like blood, paints and polymer solutions do not satisfy the equation (4.2) and are called as Non-Newtonian fluids.

In SI system, the unit of viscosity is **Pascal.Second.** It is obtained by substituting various dimensions in equation (4.2).

From Equation (4.2) we can write

$$\mu = \frac{\tau}{du/dy} = \frac{N/m^2}{\frac{m}{s} \times \frac{1}{m}} = \frac{Ns}{m^2} = Pa{\cdot}s$$

The unit of viscosity μ in CGS units is called 'poise' and is equal to $\frac{dyne \cdot sec}{cm^2}$

The absolute viscosity or dynamic viscosity when needed in CGS units is expressed in 'poise', designated by letter 'P' or 'centipoises', which is designated by letter 'cP'. Its equivalent in SI unit is given below.

$$1\ poise = 1\frac{gm}{cm \cdot s} = 1\frac{dyne \cdot sec}{cm^2} = \frac{10^{-5}\ N \cdot s}{\left(10^{-2}\right)^2 m^2} = \frac{1}{10}\ Pa{\cdot}s$$

$$1\ centipoise = \frac{1}{100}\ poise = \frac{1}{1000}\ Pa{\cdot}s$$

4.4 KINEMATIC VISCOSITY

Kinematic viscosity of a fluid is defined as the ratio of the coefficient of dynamic viscosity to the density of fluid. It is important in cases in which significant viscous and gravitational forces coexist.

It is denoted by **ν** (called 'nu'). In SI system, the unit of Kinematic Viscosity is m^2/s. It can be obtained from its definition as mentioned below.

$$Kinematic\ Viscosity = \nu = \frac{\mu}{\rho} = \frac{N \cdot s/m^2}{kg/m^3} = \frac{m^2}{s}$$

In CGS units, the kinematic viscosity **ν** is measured in stoke or centistokes.

$$1\ stoke = 1\frac{cm^2}{s} = 10^{-4}\frac{m^2}{s}$$

$$1\ centistoke = \frac{1}{100}\ stoke = 10^{-6}\frac{m^2}{s}$$

The coefficient of viscosity μ depends upon the temperature. It has been found that for liquids, with an increase in temperature, the value of μ starts decreasing whereas for gases an increase in temperature leads to an increase in the value of μ.

4.5 PRESSURE VARIATION WITH DEPTH (or Hydrostatic law)

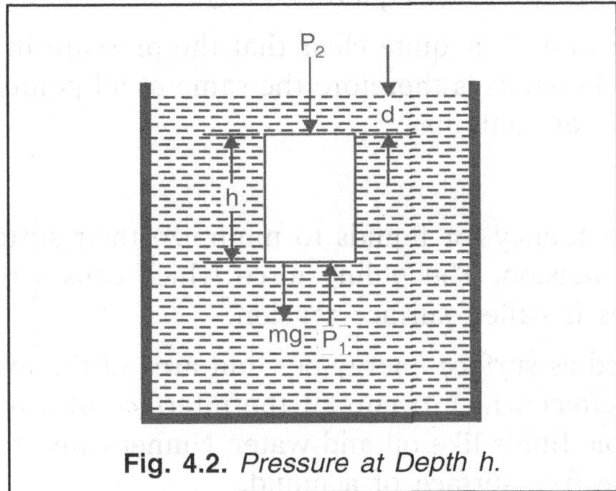

Fig. 4.2. *Pressure at Depth h.*

In order to know the effect of depth on pressure let us consider a fluid of density ρ is at rest in a container (Fig. 4.2). A fluid sample of imaginary cross sectional A is at a distance '*d*' from the free surface of the fluid. The height of the fluid sample is '*h*'. Pressures act normal to the surface. This sample of fluid is in equilibrium and at rest. From the Newton's second law of motion, the net force on the sample must be equal to zero.

At the top of the sample, the force due to atmospheric pressure is P_2A. At the bottom, the force due to fluid pressure is P_1A. Being symmetrical the pressure acting on the sides (in horizontal direction) will cancel out. In addition to this a gravitational force also acts on the sample.

As the sample is in equilibrium, the net summation of the forces in the vertical direction must be equal to zero.

$$F_y = 0$$

$$P_1A - P_2A - mg = 0 \qquad \text{...(4.3)}$$

Where m is mass of the fluid sample.

As we know $$\text{Density} = \frac{\text{mass}}{\text{volume}}$$

$\because$ $$\rho = \frac{m}{v}$$

$\therefore$ $$m = \rho\, v = \rho\, A\, h$$

Substituting the value of m in equation (4.3), we get

$$P_1A - P_2A - \rho\, A\, h\, g = 0$$

$$P_1A = P_2A + \rho\, g\, h\, A$$

$$P_1 = P_2 + \rho\, g\, h \qquad \text{...(4.4)}$$

If the top surface of sample is exposed (or open) to the atmospheric pressure, then P_2 becomes the atmospheric pressure in equation (4.4).

Thus from equation (4.4) it is quite clear that the pressure in a fluid depends only on depth h within the fluid. Pressure is therefore the same at all points having the same depth, independent of the shape of container.

(*i*) **Surface tension**

There is a natural tendency for liquids to minimize their surface area like a stretched elastic membrane under tension. The puling force which causes this due to the attractive forces between molecules is called surface tension.

It can also be defined as surface energy per unit area of the interface. It is the property of the apparent tension effect which occurs at the interface of a liquid and a gas or at the interface of two immiscible fluids like oil and water. Numerically It is expressed as force per unit length drawn on the free surface of a liquid.

(*ii*) **Capillary effect**

The molecular attraction in liquids due to cohesion and adhesion is the cause of capillary effect. It is related to the rise and fall of a liquid in a small diameter tube inserted into the liquid.

Example: Capillary effect is responsible for the rise of kerosene by a cotton wick inserted into the tank of a kerosene lamp.

The curved free surface of the liquid as shown in Fig. 4.3 in the tube is called the 'meniscus'.

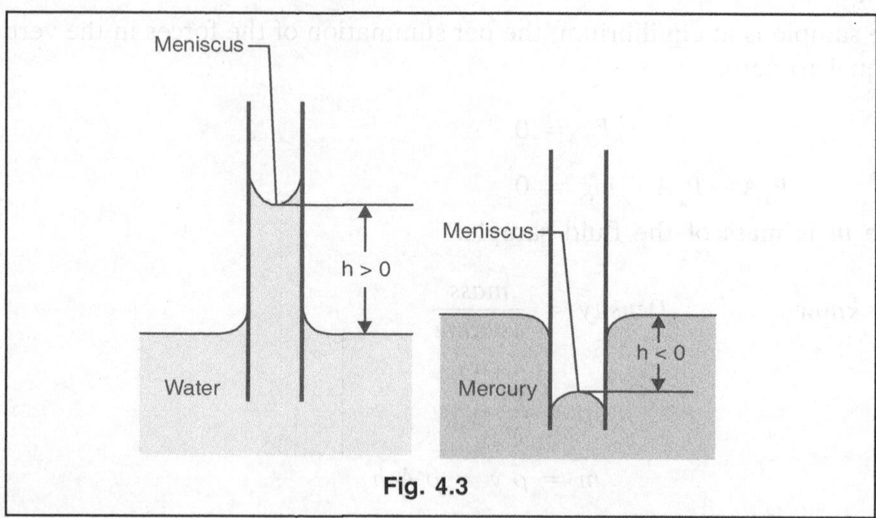

Fig. 4.3

4.6 CLASSIFICATION OF FLUID

Fluid can be classified in four categories:

(*i*) **Ideal or perfect fluid:** A fluid having zero viscosity (or no viscosity) and incompressibility is called an ideal fluid. In such a fluid there will no frictional effects between moving fluid layers or between the layers and its boundary walls and thus no energy loss due to friction and turbulence. Ideal fluid is of academic interest only as all the real fluids which exist in nature have some viscosity.

(*ii*) **Real fluid:** A fluid having viscosity is known as real fluid. All the fluid available and used in practical applications are real fluids.

(*iii*) **Newtonian fluid:** A fluid which obeys the Newton's linear law of viscosity is called a Newtonian fluid. In these fluids, the shear stress is directly proportional to velocity gradient.

Examples are: air, water and petroleum fluids.

(*iv*) **Non-Newtonian fluid:** A fluid which does not obey the Newton's linear law of viscosity is called Non-Newtonian fluid.

Examples: Paints, polymer solutions, blood, liquid food items like soup, jellies etc.

☞ ***Important note:***

The words compressible and non-compressible represent the type of flow not the type of fluid.

4.7 TYPES OF FLOW

When fluid in motion, the fluid can be classified as:

(*i*) **Steady and Unsteady flow:** In steady flow, the fluid variables such as velocity, pressure and density etc., do not change at a point with time. If any variable changes with time, the condition of unsteady flow prevails.

(*ii*) **Laminar and turbulent flow:** The fluid flow is said to be laminar if each particle of a fluid follows a smooth path so that the paths of other particles do not cross each other. A flow in which each fluid particle does not have specific path and characterized by small whirlpool like region is called turbulent flow.

(*iii*) **Uniform and non-uniform flow:** A flow in which velocity of all fluid particles is identical through out the flow field is called uniform flow. At a particular instant of time, if the velocity components at different location in a flow field are different then the flow is called non-uniform flow.

(*iv*) **Compressible and incompressible flow:** A flow in which density of the fluid changes during flow is called compressible flow. The fluids flow during which the density of fluid does not change is called incompressible flow.

(*v*) **Rotational and irrotational flow:** A flow in which the fluid particles rotate about their own axes while flowing along streamlines is called a rotational flow. A flow is said to be irrotational if while flowing its fluid particles do not rotate about its axis.

4.8 FLOW PATTERNS

(*i*) Streamlines

The path taken by a particle of the fluid under steady flow is called a streamline. No two streamlines can cross each other. If they cross each other then the flow under consideration will not be steady flow.

Streamline sketches help in visualizing the fluid flows. These are extremely helpful in finding the low and high velocity and pressure areas during fluid flows.

(*ii*) Stream surface

When a number of closely drawn streamlines pass through an arbitrary curve XY as shown in Fig. 4.4, these streamlines are said to have generated a stream surface.

(*iii*) Streamtube

When a number of streamlines are drawn through a closed curve, these streamlines form boundaries across which the fluid particles can not cross. The space created between the streamlines forms a tube called "streamtube".

The concept of stremtubes comes in handy when different fluid equations need to be derived.

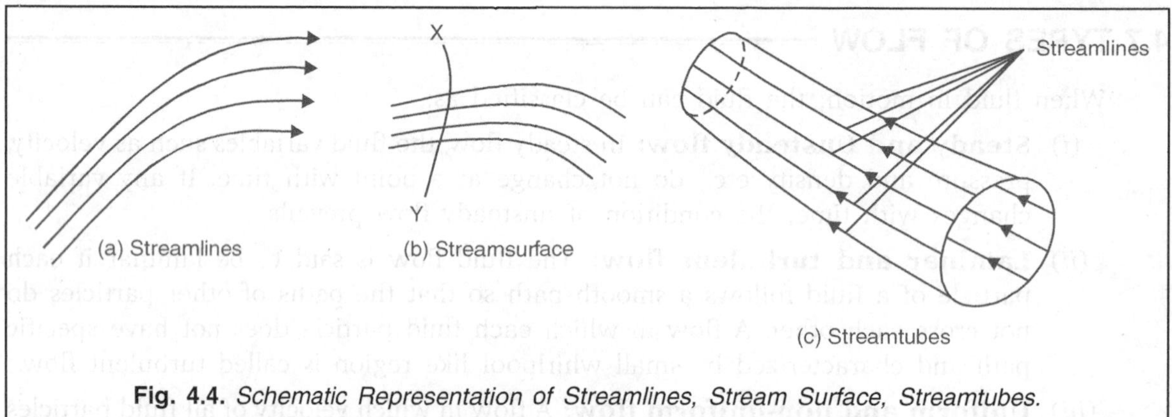

Fig. 4.4. *Schematic Representation of Streamlines, Stream Surface, Streamtubes.*

For steady flow, the velocity of a particle moving along a streamline is given by

$$v = \frac{ds}{dt} \qquad \text{...(4.5)}$$

Where *ds* is the distance traveled by fluid particle in time *dt*. If the velocity changes during the distance ds, then acceleration of the fluid particle is expressed by

$$a = \frac{dv}{dt} = \frac{ds}{dt}\frac{dv}{ds} = v\frac{dv}{ds}$$

However if flow is not steady then velocity used to get changed with point to point.

If this change is designated by $\frac{dv}{dt}$ then the total acceleration is given by

$$\text{Total acceleration} \qquad a_t = v\frac{dv}{ds} + \frac{dv}{dt} \qquad \text{...(4.6)}$$

4.9 THE EQUATION OF CONTINUITY

The equation based on the principle of conservation of mass is called continuity equation. Let us apply this principle to fluid flow in a streamtube. Along the streamtube the continuity of flow from point to point can be ensured for the ideal flow.

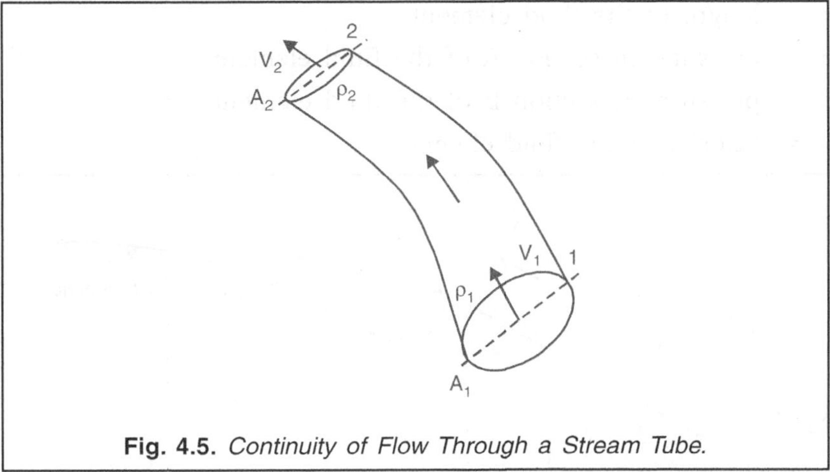

Fig. 4.5. *Continuity of Flow Through a Stream Tube.*

Let us consider two cross-sections at point 1 and 2 of a streamtube.

Let

A_1 = the area of streamtube at section 1
V_1 = average velocity at cross section 1
ρ_1 = density at section 1
A_2 = the area of streamtube at section 2
V_2 = average velocity at cross section 2
ρ_2 = density at section 2

The quantity of fluid passing section 1 per unit time = A_1V_1 and the mass of fluid passing section 1 per unit time = $A_1V_1\rho_1$. The mass of fluid passing per unit is also known as flow rate. Similarly the mass of fluid passing section 2 will be $A_2V_2\rho_2$.

From the law of the conservation of mass, no fluid mass is being created or destroyed between sections 1 and 2. Therefore,

Mass of fluid passing section 1 per unit time = Mass of fluid passing section 2 per unit time

$$A_1V_1\rho_1 = A_2V_2\rho_2 \qquad \text{...(4.7)}$$

Equation (4.7) is called continuity equation which is applicable to both the compressible and incompressible fluids.

For the fluids of constant density ($\rho_1 = \rho_2$) the equation (4.7) can be written as

$$Q = A_1V_1 = A_2V_2 \quad \text{...(4.8)}$$

Where Q is called rate of flow or discharge. Its unit is m^3/s.

4.10 EULER'S EQUATION

Euler's equation for steady flow of ideal fluids is derived from the Newton's law of motion. The fluid motion is considered along a streamtube. Let us consider a differentially small section AB of the stream tubes as shown in Fig. 4.6.

Let dA = cross sectional area of fluid element

ds = length of the fluid element

p = pressure on section A of the fluid element

$p + dp$ = pressure on section B of the fluid element

v = velocity of the fluid element.

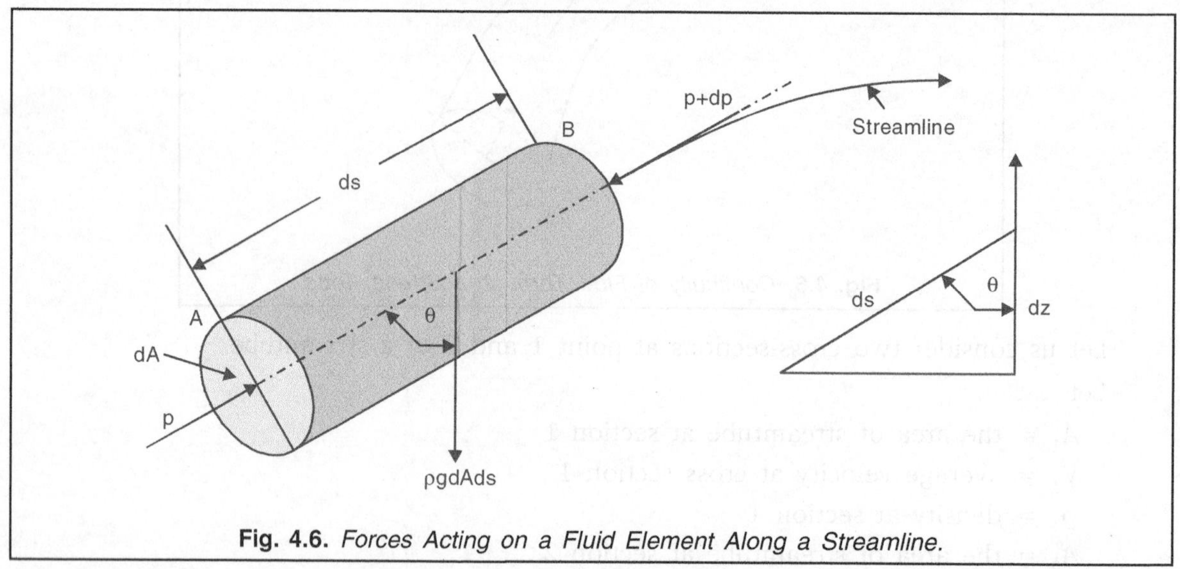

Fig. 4.6. *Forces Acting on a Fluid Element Along a Streamline.*

From the Fig. 4.6 the forces responsible for accelerating the fluid mass are:

1. Component of weight of fluid element in the direction of motion.
2. The forces acting due to pressure on the ends of the fluid element in the direction of motion.

Assuming that the fluid motion is in an upward direction and therefore pressure and velocity will also increase in upward direction.

Let us write the expression for the forces acting on the fluid element due to pressure at both ends

$$dF_p = pdA - \left(p + \frac{\partial p}{\partial s} ds\right) dA$$

$$= -\, dp\, dA \quad \text{...(4.9)}$$

Now let us write the expression for the weight of the fluid element

$$dF_w = \rho\ g\ dA\ ds \qquad ...(4.10)$$

From the Fig.4.6 it is clear that the cosine component of this force will help in accelerating the motion in up direction.

$$= -\ \rho\ g\ dA\ ds\ \text{Cos}\theta$$

$$= -\ \rho\ g\ dA\ ds\left(\frac{dz}{ds}\right)$$

$$= -\ \rho\ g\ dA\ dz$$

The mass dM of the fluid element

$$dM = \rho\ dA\ ds$$

As we know that total acceleration of a fluid element across a streamtube is given by equation (4.6) which is rewritten as equation (4.11) for unsteady flow and equation (4.12) for steady flow.

For unsteady flow $$a_t = v\frac{dv}{ds} + \frac{dv}{dt} \qquad ...(4.11)$$

For steady flow $$a_t = v\frac{dv}{ds} \qquad ...(4.12)$$

From the Newton's second law of motion we know that

$$F = M\ A\ (\text{Force} = \text{mass} \times \text{acceleration})$$

Summation of the forces due to weight and pressure is $dF_p + dW_w$

Substituting the values of total force, mass and acceleration in Newton's equation we get

$$dF_p + dF_w = dM\left(v\frac{dv}{ds}\right)$$

$$dF_p + dF_w = dM\left(v\frac{dv}{ds}\right)$$

$$-\ dp\ dA + (-\ \rho\ g\ dA\ dz) = \rho\ dA\ ds\left(v\frac{dv}{ds}\right)$$

On dividing both sides of the equation by $-\ \rho\ dA$, we get

$$\frac{dp}{\rho} + g\ dz = -\ v\ dv$$

$$\frac{dp}{\rho} + g\ dz + v\ dv = 0 \qquad ...(4.13)$$

Equation (4.13) is the Euler's differential equation of motion for the steady flow. One can also get the equation of motion for the unsteady flow by using equation (4.11).

Bernoulli's Equation

By integrating the Euler's equation along the streamtube we can get the Bernoulli's equation.

$$\int \frac{dp}{\rho} + \int g\, dz + \int v\, dv = 0$$

$$\frac{p}{\rho} + \frac{v^2}{2} + gz = \text{constant} \qquad \ldots(4.14)$$

4.11 BERNOULLI'S EQUATION

The Bernoulli's equation gives the relationship between fluid velocity, pressure and elevation for a fluid flowing under certain conditions. It was first derived by Daniel Bernoulli in 1738.

Assumptions made in the derivation of Bernoulli's equation are:

- The fluid should be ideal
- The fluid flow should be steady
- The fluid is incompressible
- The fluid flow is irrotational.

Let us consider a sample of an ideal fluid flowing through a non uniform pipe in a small time of Δt as shown in Fig. 4.7.

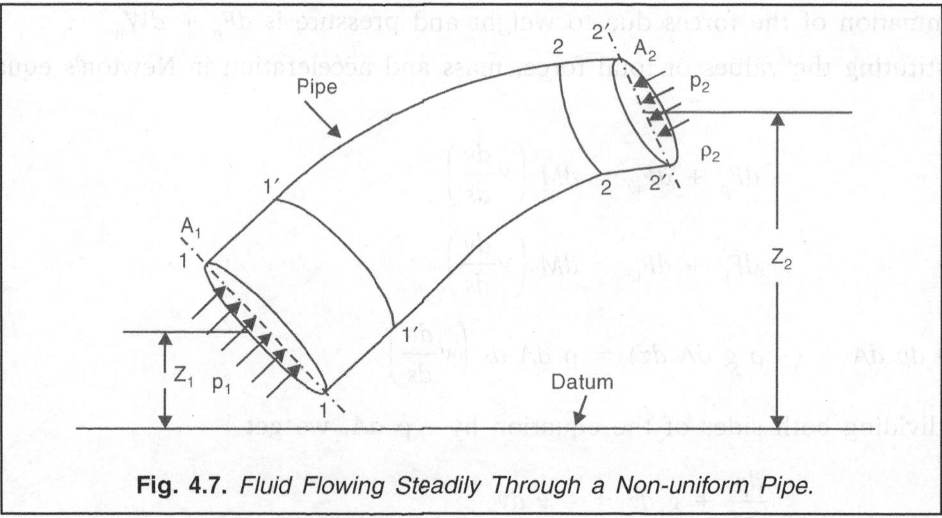

Fig. 4.7. *Fluid Flowing Steadily Through a Non-uniform Pipe.*

Let us consider some fluid at section 1–1 at the left end of the pipe and some at section 2–2 at the right end of the pipe. During the small time interval Δt, the left end of fluid moves from section 1–1 to 1′–1′ towards right by a small distance ds_1 and the fluid from the right end of the section 2–2 moves towards right 2′–2′ by a small distance ds_2.

Now let us consider the forces exerted by on the fluid on its left and right end segments.

Force on the left end of the segment is $p_1 A_1$.

The work done on left end segment is

$$F_1 \Delta x_1 = p_1 A_1 \Delta x_1 \qquad ...(4.15)$$

This is equal to $p_1 V$ where V is the volume of fluid at section 1–1.

Similarly the work done on right segment to displace the fluid from 2–2 to 2′–2′ is

$$F_2 \Delta x_2 = -p_2 A_2 \Delta x_2 = p_2 V \qquad ...(4.16)$$

The negative sign is taken as the direction of p_2 is opposite to that of p_1.

The total work done by the pressure acting on the movement of the fluid is

$$WD = (p_1 - p_2)V \qquad ...(4.17)$$

Part of this work goes into changing the kinetic energy of the segment of fluid and part of that goes into changing the potential energy of the segment.

Change in kinetic energy of the segment of fluid is

$$\Delta KE = \frac{1}{2}mv_2^2 - \frac{1}{2}mv_1^2 \qquad ...(4.18)$$

Where m is mass of the fluid at section 1–1 as well as at section 2–2 because the volume at both the section is same therefore the mass will also be the same.

Let us now find out potential energy due to elevation.

Change in potential energy (PE) is given by

$$PE = mgz_2 - mgz_1$$

$$PE = mg(z_2 - z_1) \qquad ...(4.19)$$

Total work done on the fluid segment is given by WD = KE + PE

$$(P_1 - P_2)V = \frac{1}{2}mv_2^2 - \frac{1}{2}mv_1^2 + mg\ (z_2 - z_1) \qquad ...(4.20)$$

Dividing each term of the above equation by V and substituting for $\rho = \frac{m}{V}$ we get

$$P_1 - P_2 = \frac{1}{2}\rho v_2^2 - \frac{1}{2}\rho v_1^2 + \rho g z_2 - \rho g z_1 \qquad ...(4.21)$$

Rearranging the terms we get

$$p_1 + \frac{1}{2}\rho v_1^2 + \rho g z_1 = p_2 + \frac{1}{2}\rho v_2^2 + \rho g z_2$$

Now dividing the both sides of the equation by ρ, we get the Bernoulli's equation.

$$\frac{p_1}{\rho} + \frac{v_1^2}{2} + gz_1 = \frac{p_2}{\rho} + \frac{v_2^2}{2} + gz_2 \qquad ...(4.22)$$

Limitations of Bernoulli's Equation

- It is applicable to steady flow.
- It is applicable along a streamline in the core region of the flow but not along a streamline near to surface.
- It is not applicable to those devices such as pump, turbine, compressor which destroy the streamlines and carryout the interaction with the fluid particles.
- It treats flow as incompressible. When Mach number is less than 0.3 then liquid and gas flow can be treated as incompressible.
- It is not applicable for flow sections that involve significant temperature variations.
- If the flow is irrotational Bernoulli's equation can be used across streamlines. However if flow is rotational use of Bernoulli's equation is restricted to flow along a streamline.

Applications of Bernoulli's Equation

- Free jets (spraying water into air)
- Flow rate, discharge
- Flow rate measurement
- Velocity measurement by the Pitot tube
- Velocity measurement by the Stagnation tube.

Problem 1. *The water is flowing through a pipe having diameters 20 cm at inlet section 1 and 10 cm at outlet section 2. The rate of flow through pipe is 35 litres/sec. The section 1 is 6 m above datum and section 2 is 4 m above datum. If pressure at section 1 is 39.24 N/cm², find the intensity of pressure at section 2.* *(RGTU April 2009)*

Solution.

First of all draw a diagram as per the problem and convert different parameters given in SI units.

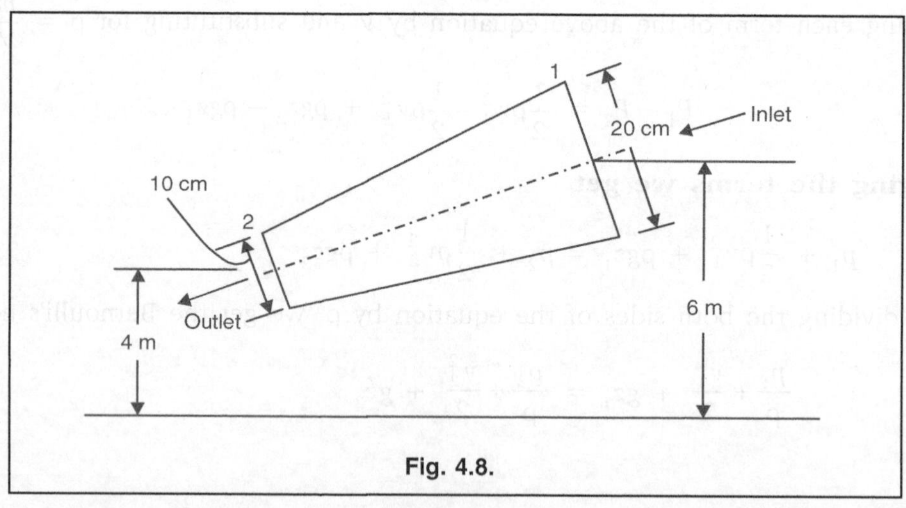

Fig. 4.8.

Given:

Rate of flow through the pipe $Q = 35$ litre/sec $= \dfrac{35}{1000}$ m^3/sec $= 0.035$ m^3/sec

$$d_1 = 20 \text{ cm} = 0.2 \text{ m}$$

At inlet section:

$$p_1 = 39.24 \text{ N/cm}^2 = \frac{39.24}{(100)^2} \text{ N/m}^2$$

$$z_1 = 6.0 \text{ m}$$

$$A_1 = \frac{\pi}{4}(d_1)^2 = \frac{\pi}{4}(0.2)^2 = 0.0314 \text{ m}^2$$

$$d_2 = 10 \text{ cm} = 0.1 \text{ m}$$

At outlet section:

$$P_2 = ?$$

$$z_2 = 4.0 \text{ m}$$

$$A_2 = \frac{\pi}{4}(d_2)^2 = \frac{\pi}{4}(0.1)^2 = 0.00785 \text{ m}^2$$

We know that density of water is $\rho = 1000$ kg/m^3

Using continuity equation $Q = A_1 v_1 = A_2 v_2$

$$\therefore \qquad v_1 = \frac{Q}{A_1} = \frac{0.035}{0.0314} = 1.114 \text{ m/s}$$

and

$$v_2 = \frac{Q}{A_2} = \frac{0.035}{0.00785} = 4.456 \text{ m/s}$$

Now applying Bernoulli's equation at sections 1 and 2, we get

$$\frac{p_1}{\rho g} + \frac{v_1^2}{2g} + z_1 = \frac{p_2}{\rho g} + \frac{v_2^2}{2g} + z_2$$

$$\frac{39.24 \times 10^4}{1000 \times 9.81} + \frac{(1.114)^2}{2 \times 9.81} + 6 = \frac{p_2}{1000 \times 9.81} + \frac{4.456^2}{2 \times 9.81} + 4$$

$$40 + 0.063 + 6 = \frac{p_2}{9810} + 1.012 + 4$$

$$46.036 = \frac{p_2}{9810} + 5.012$$

$$p_2 = 402710.3 \text{ N/m}^2 = 40.27 \text{ N/cm}^2$$

$$p_2 = 40.27 \text{ N/cm}^2 \textbf{ Ans.}$$

Problem 2: *The water is flowing at 1.56 m/sec through a pipe AB of uniform diameter 25 cm. The pressure of 45 kPa was observed at end A, which was at elevation of 12 m. At the*

B end of the pipe the elevation was 14 m and pressure was 15 kPa. Determine the direction of flow and energy loss across the ends of pipe AB.

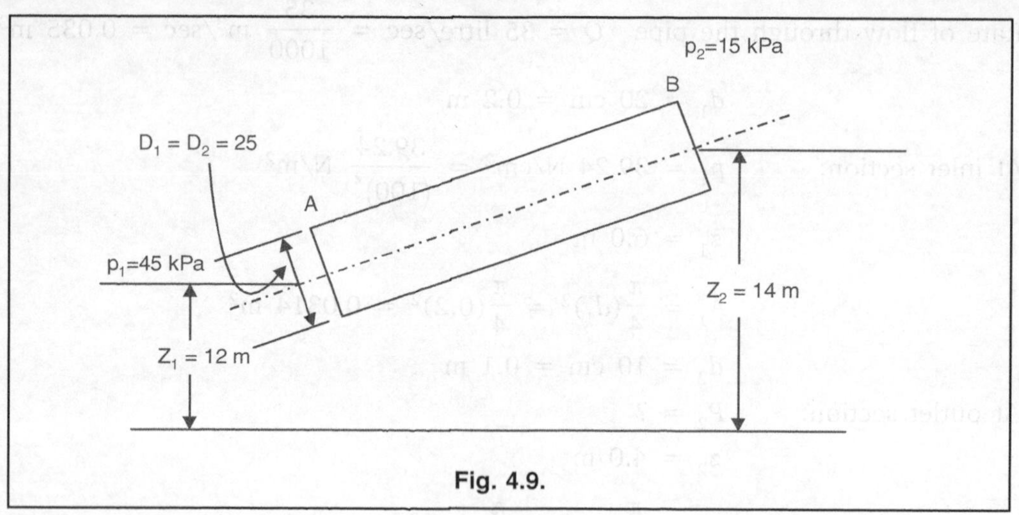

Fig. 4.9.

Solution. Given:

$d_1 = 25 \text{ cm} = 0.25 \text{ m}$

At pipe end A $\quad P_1 = 45 \text{ kPa} = 45000 \text{ N/m}^2$

$z_1 = 12 \text{ m}$

$d_2 = 25 \text{ cm} = 0.25 \text{ m}$

At pipe end B $\quad P_2 = 15 \text{ kPa} = 15000 \text{ N/m}^2$

$z_2 = 14 \text{ m}$

Using continuity equation $Q = A_1 v_1 = A_2 v_2$

$$\therefore \quad Q = \frac{\pi}{4}(0.25)^2\ 1.56 = 0.0765 \text{ m}^3/\text{s}$$

Total energy at section A is

$$E_1 = \frac{p_1}{\rho g} + \frac{v_1^2}{2g} + z_1$$

$$= \frac{45000}{1000 \times 9.81} + \frac{(1.56)^2}{2 \times 9.81} + 12$$

$$= 4.587 + 0.124 + 12$$

$$= 16.711$$

Total energy at section B is

$$E_2 = \frac{p_2}{\rho g} + \frac{v_2^2}{2g} + z_2$$

$$= \frac{15000}{1000 \times 9.81} + \frac{(1.56)^2}{2 \times 9.81} + 14$$

$$= 1.529 + 0.124 + 14$$

$$= 15.653$$

As $E_1 > E_2$, the flow is from end A to end B **Ans.**

Problem 3. *Water is flowing through a pipe having 250 mm diameter at section 1 which is the lower end of the pipe and 150 mm diameter at section 2 which is the upper end of the pipe. The intensity of pressure at the lower end is 24.54 Newton per square cm and the pressure at the upper end is 10.56 Newton per square cm. Find out the difference in datum head if the flow rate through pipe is 50 litres/sec.*

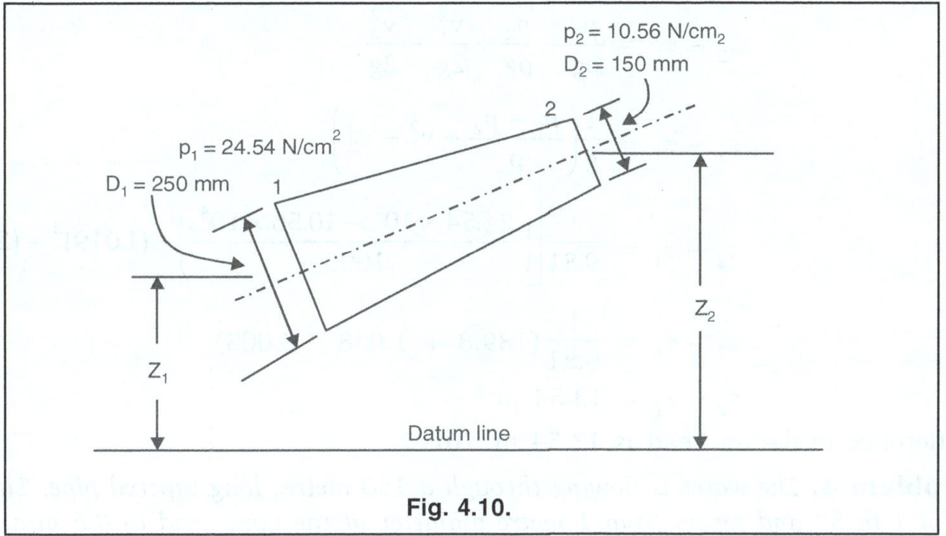

Fig. 4.10.

First of all draw a diagram as per the problem and convert different parameters given in SI units.

Solution. Given:

$$D_1 = 250 \text{ mm} = \frac{250}{1000} \text{ m} = 0.250 \text{ m}$$

$$D_2 = 150 \text{ mm} = \frac{150}{1000} \text{ m} = 0.150 \text{ m}$$

$$P_1 = 24.54 \text{ N/cm}^2 = 24.54 \frac{\text{N}}{\left(\frac{1}{100}\right)^2 \text{m}^2} = 24.54 \times 10^4 \text{ N/m}^2$$

Similarly, $$P_2 = 10.56 \text{ N/cm}^2 = 10.56 \times 10^4 \text{ N/m}^2$$

$$Q = 50\frac{\text{litres}}{\text{s}} = \frac{50}{1000} \text{ m}^3/\text{s} = 0.05 \text{ m}^3/\text{s}$$

Using continuity equation $Q = A_1 v_1 = A_2 v_2$

$$\therefore \quad v_1 = \frac{Q}{A_1} = \frac{0.05}{\frac{\pi}{4}(0.250)^2} = 1.019 \text{ m}^3/\text{s}$$

$$v_2 = \frac{Q}{A_2} = \frac{0.05}{\frac{\pi}{4}(0.150)^2} = 2.829 \text{ m}^3/\text{s}$$

Now using Bernoulli's equation at lower and upper end of the pipe, we get

$$\frac{p_1}{\rho g} + \frac{v_1^2}{2g} + z_1 = \frac{p_2}{\rho g} + \frac{v_2^2}{2g} + z_2$$

Difference in datum head is given by

$$z_2 - z_1 = \frac{p_1}{\rho g} - \frac{p_2}{\rho g} + \frac{v_1^2}{2g} - \frac{v_2^2}{2g}$$

$$z_2 - z_1 = \frac{1}{g}\left(\frac{p_1 - p_2}{\rho} + v_1^2 - v_2^2\right)$$

$$z_2 - z_1 = \frac{1}{9.81}\left[\left(\frac{24.54 \times 10^4 - 10.56 \times 10^4}{1000}\right) + (1.019)^2 - (2.829)^2\right]$$

$$z_2 - z_1 = \frac{1}{9.81}(139.8 + 1.038 - 8.003)$$

$$z_2 - z_1 = 13.54 \text{ m}$$

Difference in datum head is 13.54 m. **Ans.**

Problem 4. *The water is flowing through a 150 metres long tapered pipe. The pipe has a slope of 1 in 50 and tapers from 1 metre diameter at the upper end to 0.6 m at the lower end. The flow rate through the pipe is 50 Litres/sec. find out the pressure at the lower end if the pressure at the upper end is 18.63 N/cm².*

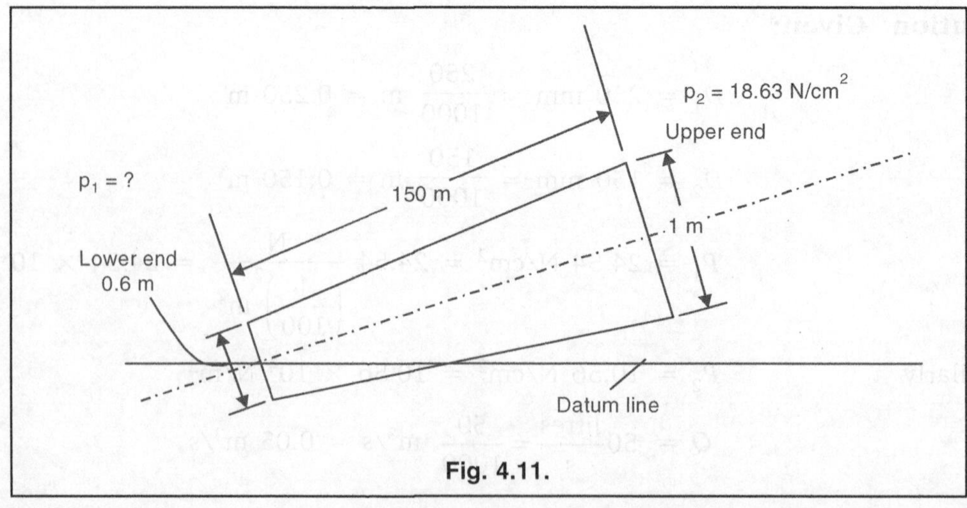

Fig. 4.11.

First of all draw a diagram as per the problem and convert different parameters given in SI units.

Flow rate of the pipe = $Q = 50\dfrac{\text{litres}}{\text{sec}} = \dfrac{50}{1000}\ \text{m}^3/\text{s}$

Diameter at the lower end $D_1 = 0.6$ m

Diameter at the upper end $D_2 = 1.0$ m

Pressure at the upper end $P_2 = 18.63\ \text{N/cm}^2 = 18.63\dfrac{1}{\left(\dfrac{1}{100}\right)^2}\ \text{N/cm}^2$

$= 18.63 \times 10^4\ \text{N/m}^2$

Using continuity equation $Q = A_1 v_1 = A_2 v_2$

We can find out the values of v_1 and v_2

Area at the lower and upper end can be calculated as

$$A_1 = \frac{\pi}{4}(0.6)^2 = 0.283\ \text{m}^2$$

$$A_2 = \frac{\pi}{4}(1.0)^2 = 0.785\ \text{m}^2$$

Now,

$$v_1 = \frac{Q}{A_1} = \frac{0.05}{0.283} = 0.177\ \text{m/s}$$

$$v_2 = \frac{Q}{A_2} = \frac{0.05}{0.785} = 0.064\ \text{m/s}$$

As datum line is assumed to pass through the centre of lower end of the pipe $\therefore z_1 = 0$

As slope is given 1 in 50 it means $z_2 = \dfrac{1}{50} \times 150 = 3$ m

Now using Bernoulli's equation at lower and upper end of the pipe, we get

$$\frac{p_1}{\rho g} + \frac{v_1^2}{2g} + z_1 = \frac{p_2}{\rho g} + \frac{v_2^2}{2g} + z_2$$

$$\frac{p_1}{\rho g} = \frac{P_2}{\rho g} + \frac{v_2^2}{2g} + z_2 - \frac{v_1^2}{2g} - z_1$$

$$p_1 = \rho g\left(\frac{p_2}{\rho g} + \frac{v_2^2}{2g} + z_2 - \frac{v_1^2}{2g} - z_1\right)$$

$$p_1 = 1000 \times 9.81\left[\left(\frac{18.63\times 10^4}{1000\times 9.81}\right) + \frac{(0.064)^2}{2\times 9.81} + 3 - \frac{(0.177)^2}{2\times 9.81} - 0\right]$$

$$p_1 = 9810[18.99 + 2.08 \times 10^{-4} - 1.59 \times 10^{-3}]$$

$$p_1 = 9810 \times 29.98 = 215708.34 \text{ N/m}^2$$

$$p_1 = 21.57 \text{ N/m}^2 \textbf{ Ans.}$$

4.12 TURBOMACHINES

Turbomachines are the machines that either extract energy from a fluid (*e.g.*, turbine) or add energy to a fluid (*e.g.*, pump) by the dynamic action of one or more moving blade rows.

A turbomachine or a fluid machine which converts fluid power into mechanical energy is called a turbine and a turbo machine which converts mechanical energy into fluid power is called a pump or a compressor.

A turbomachine consists of two main parts

(*i*) **Rotating Element:** It is known as **runner** if it is a part of turbine and the **impeller** if it is a part of a pump.

(*ii*) **Casing:** It encloses the rotating element.

The rotating element has a number of vanes or curved blades mounted on it and here takes place the transfer of energy between the fluid and rotating element.

4.13 CLASSIFICATION OF TURBOMACHINES

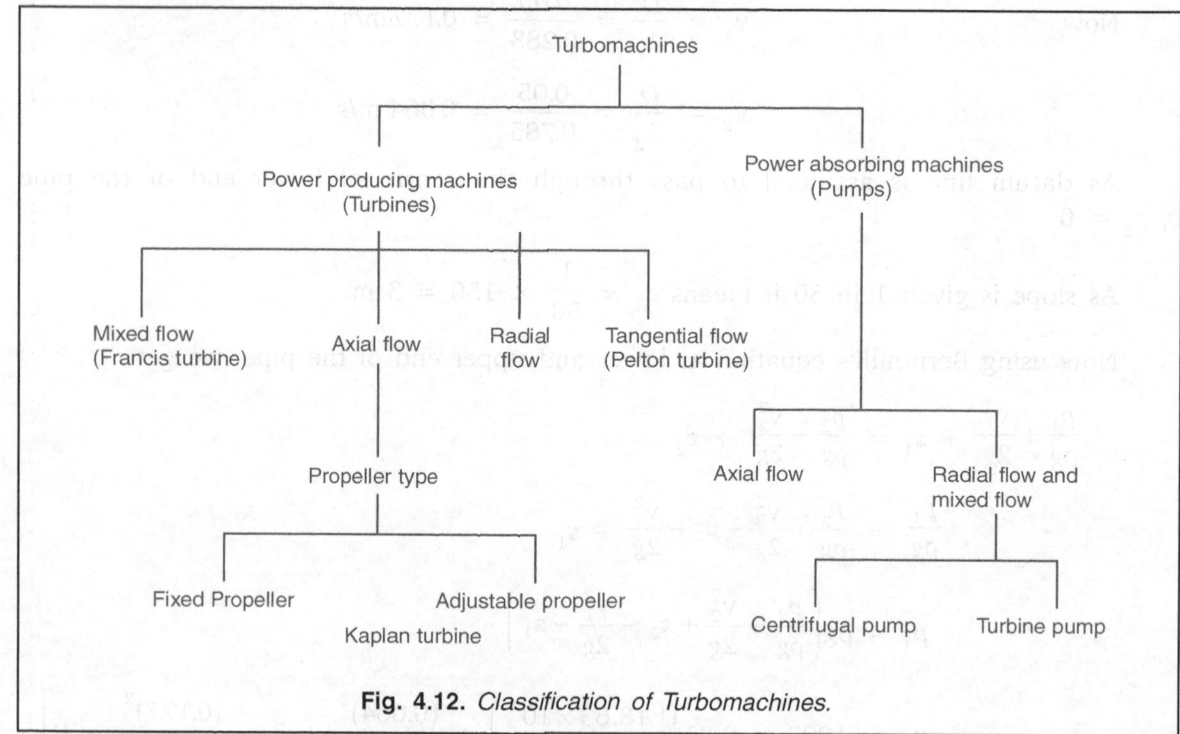

Fig. 4.12. *Classification of Turbomachines.*

(A) Broadly speaking, turbomachines are of two types

1. Power producing machines
2. Power absorbing machines.

Power producing machines are turbines which are usually classified on the basis of the working fluid used.

Hydraulic turbine	–	Water
Steam turbine	–	Steam
Gas turbine	–	Air or gas

Power absorbing machines are pumps and compressors and on the basis of working fluid these are classified as:

Pumps	–	water, oils, and other fluids
Compressors	–	air or gas
Fans, propellers	–	air or gas

(B) On the basis of their geometrical features

The turbomachines can be categorized into four types:

(*i*) **Axial flow machines:** An axial flow machine is one in which direction of water flow is parallel to the axis of rotation of the runner.

Examples: Kaplan turbine, axial flow compressors,

(*ii*) **Radial flow machines:** A radial flow machine is one in which water flows in the radial direction through the runner.

Examples: Centrifugal pump

(*iii*) **Mixed flow machines:** A mixed flow machine is one in which water flows through the runner in the radial direction but leaves in the direction parallel to the axis of rotation of the runner.

Examples: Francis Turbine

(*iv*) **Tangential flow machines:** Pelton turbine is a tangential flow machine in which high velocity water jets strikes the buckets mounted on runner tangential to the path of rotation.

4.14 PUMPS

A pump is a device which converts mechanical energy into hydraulic energy. An electric motor is used to deliver mechanical energy to a pump. Due to mechanical action, partial vacuum is developed at the pump inlet. This allows atmospheric pressure to push the fluid through the pump inlet. From there the pump then forces the fluid into the hydraulic system.

In hydraulic systems, incompressible fluids like oils are used to provide flow. Here hydraulic system means that the fluids (or liquids) are used as the medium to cause change in energy. On the other side, in pneumatic system, compressible fluid like air is used to provide fluid flow. A device which is used to produce fluid flow in pneumatic system is called a compressor.

Pumps can be classified in two categories.

1. Hydrodynamic pumps
2. Hydrostatic pumps.

1. Hydrodynamic pumps

Hydrodynamic pumps are also known as 'nonpositive displacement pump'. These pumps are used for low pressure and high–volume flow applications. These are mainly used for transporting fluids from one location to another.

Examples of hydrodynamic pumps are:

(*i*) Centrifugal pump

(*ii*) Axial pump.

Hydrodynamic pumps are not self-priming due to large clearance between the rotating and stationary pump elements. The displacement between inlet and outlet is not a positive one as there is lot of clearance space to seal against atmospheric pressure.

2. Hydrostatic pumps

These pumps are also known as 'positive displacement pump'. These type of pumps have nearly constant output flow as these pump used to eject a fixed quantity of fluid per revolution of the pump shaft. A pressure relief valve is used to protect the pump against very high pressure by diverting the pump flow back to the reservoir or sump.

On the basis of the motion of internal elements, positive displacement pumps can be classified in three categories.

(*i*) Gear pumps (*ii*) Vane pumps

(*iii*) Piston pumps

4.14.1 Centrifugal Pump

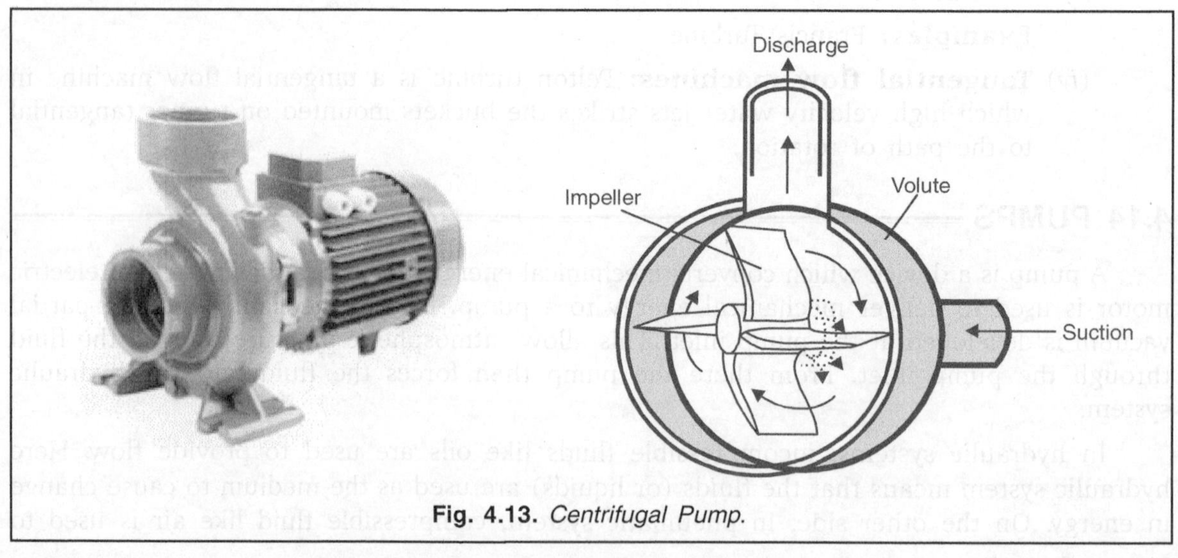

Fig. 4.13. *Centrifugal Pump.*

The working of a centrifugal pump is very simple. The fluid enters (Fig. 4.13) at the centre of the impeller called 'eye' and is picked up by the rotating impeller. As the fluid rotates with the impeller, the centrifugal force pushes the fluid radially outwards. The fluid that comes out from impeller is collected in the casing (also known as volute) where its velocity is converted to pressure before it leaves the pump's discharge. This causes the fluid to flow through the discharge pipe to the desired location. It runs at a much higher speed and is coupled directly to an electric motor. A typical layout of a centrifugal pump is shown in Fig. 4.14.

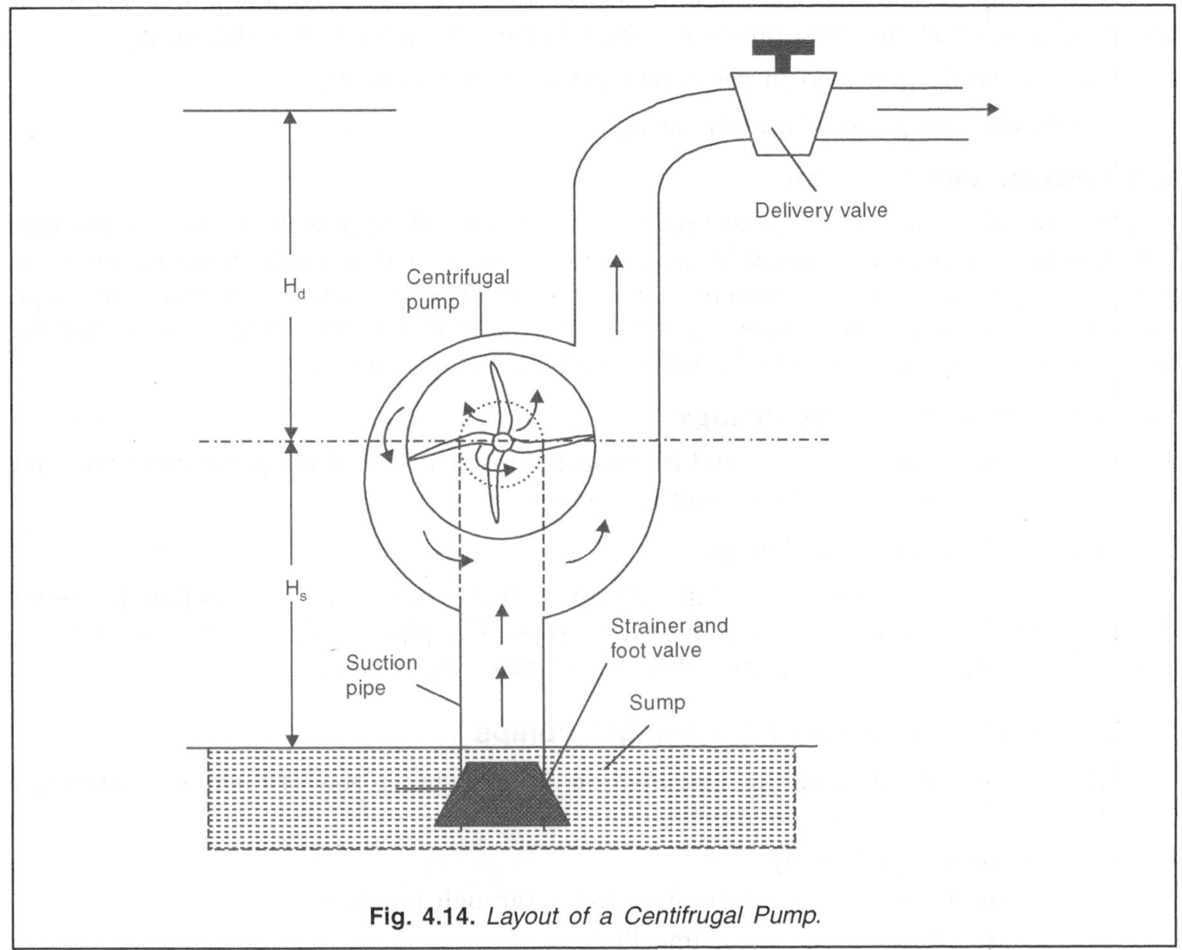

Fig. 4.14. *Layout of a Centifrugal Pump.*

The main parts of centrifugal pump are:

(*i*) The rotary element called 'Impeller',

(*ii*) The stationary element called 'Casing or volute',

(*iii*) Strainer and foot valve, and

(*iv*) Delivery pipe and its fittings

(*v*) Suction pipe and its fittings.

(*i*) **Rotary element called 'impeller'**

The rotating element of a centrifugal pump is called 'impeller'. It carries the backward curved blades alongside its outer periphery. The straight blades are used when heavy fluids like abrasive slurries are need to be pumped. 4 to 16 blades are used in curved blade impellers. The impeller is mounted on a shaft which is also connected to the electric motor for providing desired rotation to the shaft.

(*ii*) **Casing or volute**

The casing for a centrifugal pump is designed in such a way that the kinetic energy of the water gets converted into pressure energy before the water leaves the casing.

Casing usually preferred in centrifugal pump is of two types:

(*a*) volute casing (*b*) diffusion casing

(*iii*) **Strainer and foot valve**

It is fitted at one end of the suction pipe and is submerged in water in such a way that it is always a few metres deep inside the water in a sump. The water from the sump enters the suction pipe through the strainer and foot valve. The foot valve is a non-return type valve and this prevents the downward flow of the water. Foot valve is necessary so that the water poured in for priming can be retained in the pump casing.

(*iv*) **Delivery pipe and its fittings**

It is a simple pipe whose one end is connected to the outlet of the pump and other end is free to deliver the water at the required height.

(*v*) **Suction pipe and its fittings**

The suction should be arranged in such a way that water should always flow in the up direction. There should be no loop in suction pipe. The pipe fittings should be air tight. Otherwise a pump will not run smoothly if it contains air pockets.

4.14.2 The Classification of Centrifugal Pumps

The centrifugal pumps can be classified on different basis. Few of them are mentioned below:

(*i*) on the basis of casing
(*ii*) on the basis of flow direction of water through impeller
(*iii*) on the basis of number of impellers.

(*i*) On the basis of casing

On the basis of casing centrifugal pump can be classified as volute type and diffuser type.

(*a*) **Volute casing:** This is the simplest design of pump casing as shown in Fig. 4.15. It has a spiral shape and its cross sectional area is gradually increased upto the discharge pipe. It helps in decreasing the velocity of water and increases its pressure and its increasing cross-sectional area helps in keeping more and more water coming out of the impeller vanes.

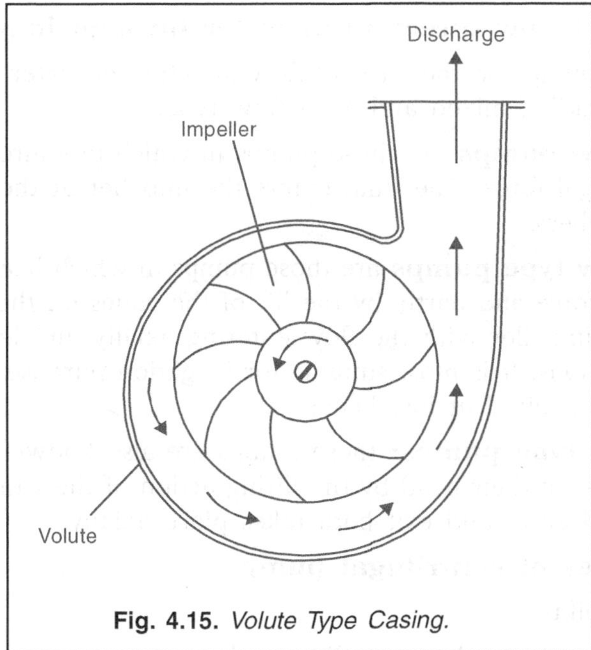

Fig. 4.15. *Volute Type Casing.*

(*b*) **Diffuser type:** It is an improved design of casing in which guide vanes are provided at the outlet of impeller vanes as shown in Fig. 4.16. When water comes out from the impeller vanes and passes through the guide vanes, this causes decrease in water velocity and increase its pressure energy which helps in lifting the water to great heights. When guide vanes are installed in a centrifugal pump, it looks like a Francis turbine therefore diffuser type centrifugal pump is also called a 'Turbine pump'.

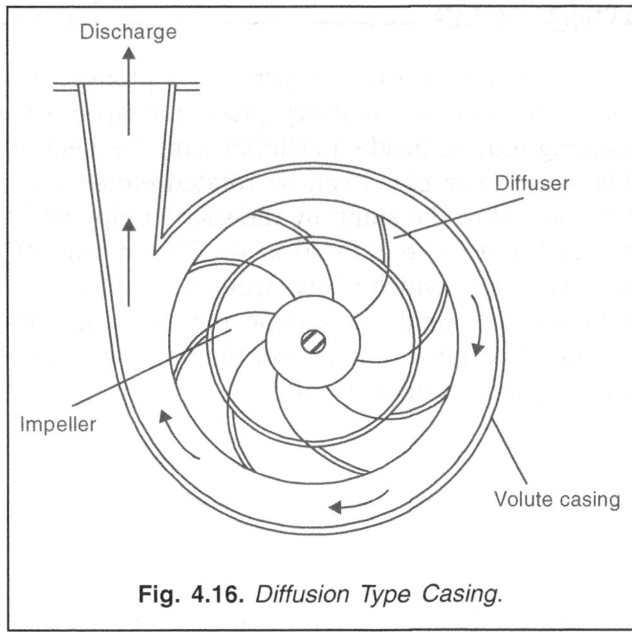

Fig. 4.16. *Diffusion Type Casing.*

(*ii*) **On the basis of flow direction of water through impeller**

The centrifugal pumps on the basis of flow direction of water through impeller can be further classified as radial, mixed and axial flow type.

(*a*) **Radial flow pumps** are those pumps in which pressure is mainly developed by the centrifugal force. The fluid enters the impeller at the hub and flows radially to the periphery.

(*b*) **Mixed flow type pumps** are those pumps in which head is developed partly by centrifugal force and partly by the lift of the vanes on the fluid. This pump has a single-inlet impeller with the flow entering axially and leaving in both axial and radial directions. It is most suitable for irrigation purposes where a large quantity of water is required at low heads.

(*c*) **Axial flow type pumps:** these pumps are also known as propeller pumps and develop most of their head by the lifting action of the vanes on the fluid. In these pumps, fluid entry and exit both takes place axially.

The advantages of centrifugal pump:

(*i*) High reliability
(*ii*) Most commonly used for pipeline applications
(*iii*) Uniform discharge
(*iv*) High volumetric efficiency
(*v*) Easy installation
(*vi*) Little maintenance
(*vii*) Low initial cost.

4.15 RECIPROCATING PUMP

A reciprocating pump is one which uses a piston or a plunger to forcibly push the fluid out of the pump cylinder and into the delivery pipe. A reciprocating pump consists of a piston having a reciprocating motion inside a cylinder with the help of a connecting rod and a crank as shown in Fig. 4.17. The crank can be rotated either manually or by an electric motor. The cylinder is connected to the sump by the suction pipe and to the reservoir by the delivery pipe. A strainer and foot valve is fitted at the bottom end of the suction pipe. The foot valve is a non-return type valve which allows water to flow only in the upward direction. Before starting the reciprocating pump, its suction pipe and the clearance volume of the cylinder is filled with water. In reciprocating pump, the pressure is generated by pushing the fluid with front face of the pump's moving part.

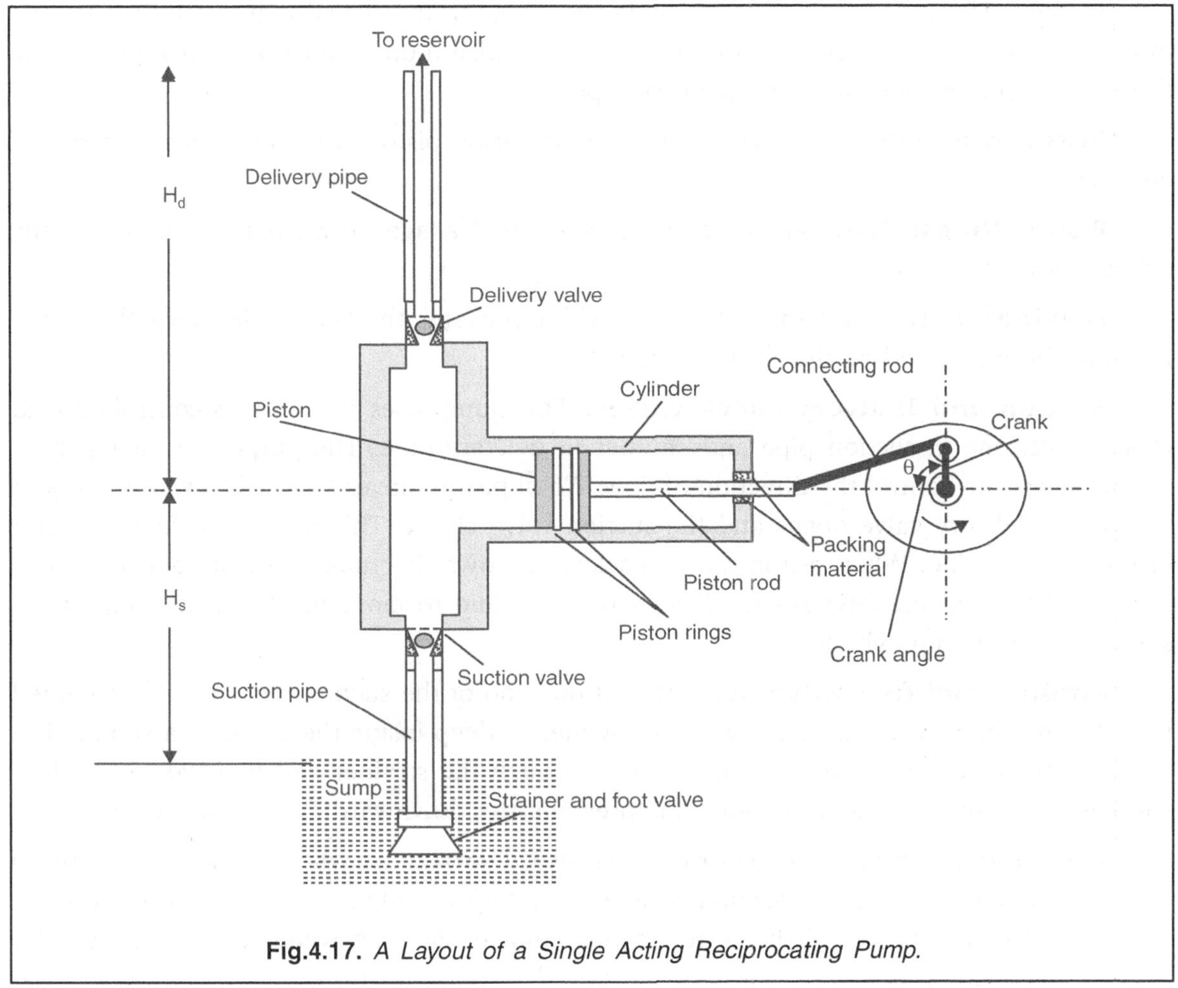

Fig.4.17. *A Layout of a Single Acting Reciprocating Pump.*

The main parts of a reciprocating pump are:

(*i*) Cylinder

(*ii*) Piston

(*iii*) Piston rod

(*iv*) Piston rings

(*v*) Packing for piston rod

(*vi*) Suction and Delivery valves

(*vii*) Strainer and foot valve.

Cylinder: In reciprocating pump, the cylinder acts as the casing of the pump that contains a piston and the fluid in the pump. It has two openings, one for the suction pipe and other for the delivery pipe.

Piston: The piston reciprocates inside the cylinder. It is usually driven by an electric motor or an engine. The reciprocating motion of the piston on its forward stroke pushes the fluid out of the cylinder into the delivery pipe.

Piston rod: Piston rod connects the reciprocating piston with the connecting rod of an engine.

Piston Rings: These are used to prevent the leakage between the piston and the cylinder walls.

Packing for the piston rod: This packing prevents the leakage between the piston rod and the crank end of the cylinder side.

Suction and Delivery check valves: This pump uses two springs controlled feed check valve, one in suction pipe and another in delivery pipe. The purpose of using these check valves is to control the direction of flow. When piston moves forward (towards cylinder head), the delivery valve opens and the suction valve closes. This causes the fluid to enter in the delivery pipe. When piston moves backward (towards crank end), the delivery valve closes and the suction valve opens. This causes the fluid to move up the suction pipe from sump into the pump cylinder.

Strainer and foot valve: It is fitted at one end of the suction pipe and is submerged in water in such a way that it is always a few metres deep inside the water in a sump. The water from the sump enters the suction pipe through the strainer and foot valve. The foot valve is a non-return type valve and this prevents the downward flow of the water.

Working: When piston moves on its backward stroke *i.e.,* when crank rotates from 0 to 180°, this creates a partial vacuum inside the pump cylinder. The atmospheric pressure acting on the free surface of the water in the sump pushes the water into pump cylinder through the suction pipe. The filling of the cylinder continues till the crank reaches at the outermost position (*i.e.,* till crank angle θ becomes 180°). Piston moves on its forward stroke when crank rotates from 180° to 360°. On its forward stroke the piston pushes the water upwards in the delivery pipe and finally from the delivery pipe to the reservoir.

4.16 ELECTRICITY GENERATION THROUGH RENEWABLE RESOURCES

Since last two decades, lot of emphasis has been put on producing electricity from renewable energy resources as the conventional sources are depleting very fast. In year 2007, a total of **18971 TWh** (Terra watt hour) electricity was generated worldwide from all the available resources present on this planet. Out of 18971 TWh of electricity produced, a total of 3604 TWh electricity was produced from renewable energy resources. The category wise breakup of the power produced globally from the renewable energy resources are given in Table 4.1.

Table 4.1: Power produced from Renewable Energy Resources

S. No.	Renewable Resource	Electricity Generated Globally in Year 2007 (in TWh)	
			% contribution
1	Biomass	217.9	6.0
2	Geothermal	62.6	1.7
3	Hydraulic	3145.5	87.3
4	Marine energies	0.6	0.02
5	Solar	8.5	0.2
6	Wind	169.3	4.7
	Total	**3604.4**	**100%**

From the Table 4.1 it is clear that among the renewable resources, the majority of the power *i.e.*, 87.3% has been generated by hydraulic means *i.e.*, by using water. This underlines the importance attached to this topic.

4.17 HYDROELECTRIC POWER

Hydroelectric power is one of the ancient systems of extracting energy from nature for industrial use. The underlying principle of this is to convert the potential energy of water into electrical energy. This is being made possible by using a hydraulic turbine which is directly connected with the electrical generator shaft. To achieve this, the water needs to exert its power on a turbine to produce mechanical energy. This mechanical energy is then converted into electricity in a generator mounted on the same shaft as the turbine.

4.17.1 Hydroelectric Power Plants

Hydroelectric power plants generate electricity by using flowing water's kinetic and potential energies. Desired water flow can be obtained either from a natural river or by erecting a dam or reservoir. The water in the dam is at a pressure, called hydrostatic pressure, which is responsible in creating the pressure potential energy from water flow. Hydrostatic pressure is proportional to water depth which is also known as head.

4.17.2 Classification of Hydroelectric Power Plants

(*i*) On the Basis of Operation

According to its operation in meeting the demand, a hydroelectric plant can be classified as:

(*a*) Base load plant

(*b*) Peak load plant.

A hydroelectric power plant is called a base load plant if it generates continuous power. If the conditions prevailing at the power station permit regulated release of water, plant can be used to generate peak power. Due to its quick starting and easy adaptability in picking up the load, hydroelectric power plant is quite suitable for peak load operation. A base load power plant usually operates at a constant load. The examples of these are coal based thermal power station and nuclear power station. Further Run-of-river plants come under the category of base load plants whereas pumped-storage plants function as peak load plants.

(*ii*) **On the Basis of Hydraulic Features**

On the basis of hydraulic features the hydraulic power plants can be categorized into three types.

(*a*) Conventional hydroelectric plants

(*b*) Pumped storage plants

(*c*) Tidal power plants.

(*a*) **Conventional hydroelectric plants:** These plants utilize the fluid energy of flowing river water.

(*b*) **Pumped storage plants:** These plants use a dam to create an upper reservoir and release water into lower reservoir. Pumping activities take place during night when power demand is low. These plant generate electricity during the peak load hours, called generating phase and pump water back from the lower reservoir to upper reservoir during off peak hours, called pumping phase.

Advantages

- Pumped storage plant is economical
- High reliability
- Pickup load easily
- No pollution
- Flexible operation
- Permits other units to run as base load and thus improve the overall efficiency of the system.

Disadvantages

- Small operational range
- Remain idle for long hours
- Need some time (about 6 to 8 hours) before picking up the full load

(*c*) The tidal power plant uses the tidal energy of sea waves.

(*iii*) On the Basis of Storage

The hydroelectric power units can also be classified on the basis of availability of storge reservoir whether they have storage reservoir or not. It becomes mandatory for the hydropower units to have storage facilities to overcome the fluctuations in the availability of water supply from the natural flowing rivers.

(*iv*) On The Basis of Plant Capacity

On the basis of the plant capacity, the hydroelectric power plants can be categorized in three types:

Low capacity plant	:	upto 1 MW
Medium capacity plant	:	upto 10 MW
High capacity plant	:	more than 10 MW

(*v*) On The Basis of Water Head

On the basis of the head, the hydroelectric power plants can be categorized in four types:

Low head plant	:	< 15 m
Medium head plant	:	15–70 m
High head plant	:	71 to 250 m
Very high head plant	:	>250

4.17.3 Main Parts of a Hydroelectric Power Plant

The main parts of a typical hydroelectric power plant as shown in Fig. 4.18 are:

1. Dam or a reservoir constructed across a river to supply desired quantity of water as per the requirement.
2. Penstock: It is the large pipe that conveys the water from the reservoir to the turbine.
3. Gate: to control the water flow through the penstock.
4. Hydraulic turbine (impulse or reaction type) to drive a generator.
5. Draft tube which carries the water from the turbine to a lower reservoir after the water has done the work on the turbine blades.
6. A tailrace also known as the tailing pond, which is a deep pool of water for dissipating the kinetic energy of the discharge from the draft tube.

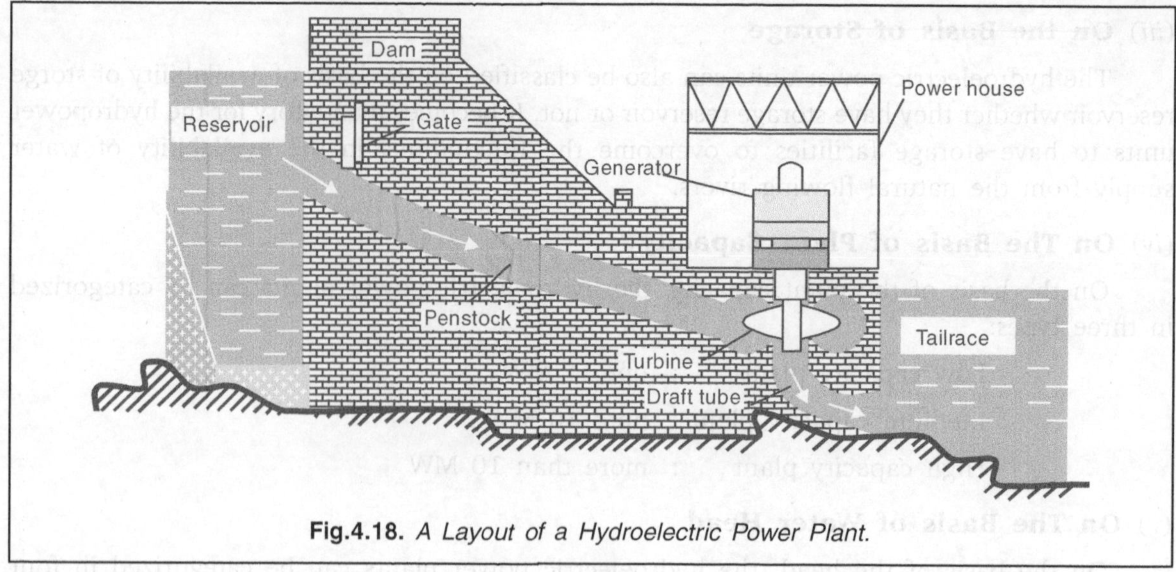

Fig.4.18. *A Layout of a Hydroelectric Power Plant.*

Advantages of hydroelectric power plants

These are:

- Long useful life
- Simple and well tested technology
- High efficiency (about 90%)
- Extremely useful for irrigation, flood protection, and drinking water supply.
- Pollution free operation
- Low operating costs.

Disadvantages of hydroelectric power plants

- Shortage of suitable sites
- High initial investment
- Disturbance in biological equilibrium
- Loss of agriculture land
- Displacement of the village population (Example Sardar Sarovar Project on Narmada)
- Large distance between the power generating plant and utility regions (city).

4.18 WATER TURBINES

Water turbines also known as hydraulic turbines are used in hydroelectric power projects to convert the potential and kinetic energy of water to electrical energy.

A water turbine is a mechanical device that converts the potential energy of water head into mechanical energy. This mechanical energy is used in running an electric generator which is directly coupled to the turbine shaft. Thus the main function of a water turbine is to rotate the generator coupled to it for producing electricity.

On the basis of action of fluid, water turbines are of two types:

(A) Impulse Turbines

(B) Reaction Turbines.

(A) Impulse Turbine

In impulse turbine, high velocity water jet impinges on the buckets (or blades) mounted on the runner (or rotor). The impulsive force applied by the water jets on the buckets (or blades) causes the runner blades to move which in turn rotates the turbine shaft. This action results in converting all the pressure energy of water into kinetic energy. The water enters and leaves the runner blades at atmospheric pressure. As impact of water jet is responsible for the rotation of the runner, therefore the turbine is called impulse turbine.

Examples: Pelton wheel turbine
Cross flow turbine
Turgo wheel turbine.

(B) Reaction Turbine

Reaction turbines are used where the energy of water flow is moderate to low, and volume of water is high. At the inlet of the turbine, the water has both kinetic as well as pressure energy. Water enters the turbine at high pressure and low velocity. As the water flows through the runner, its pressure energy is gradually converted into kinetic energy. The reaction force applied by the water on the runner blades causes the runner to rotate. In a reaction turbine water is under pressure and it can be admitted from all the direction over the runner and this keeps turbine full with water during running.

Examples: Francis turbine
Kaplan turbine
Propeller turbine
Deriaz turbine.

4.18.1 Pelton Turbine

Pelton turbine is suitable for very high heads (>500 m). Pelton turbine was developed by L.A. Pelton, an American engineer in 1880. In this turbine, entire available water head is converted into kinetic energy at the atmospheric pressure by means of one or more nozzles before the water strikes the buckets. The high velocity water jets coming from the nozzle impinges on the buckets and causes the runner rotation.

Fig. 4.19. *Pelton Turbine Runner.*

The main parts of a Pelton turbine are:

- Nozzle
- Vanes
- Runner
- Penstock
- Casing
- Spear rod

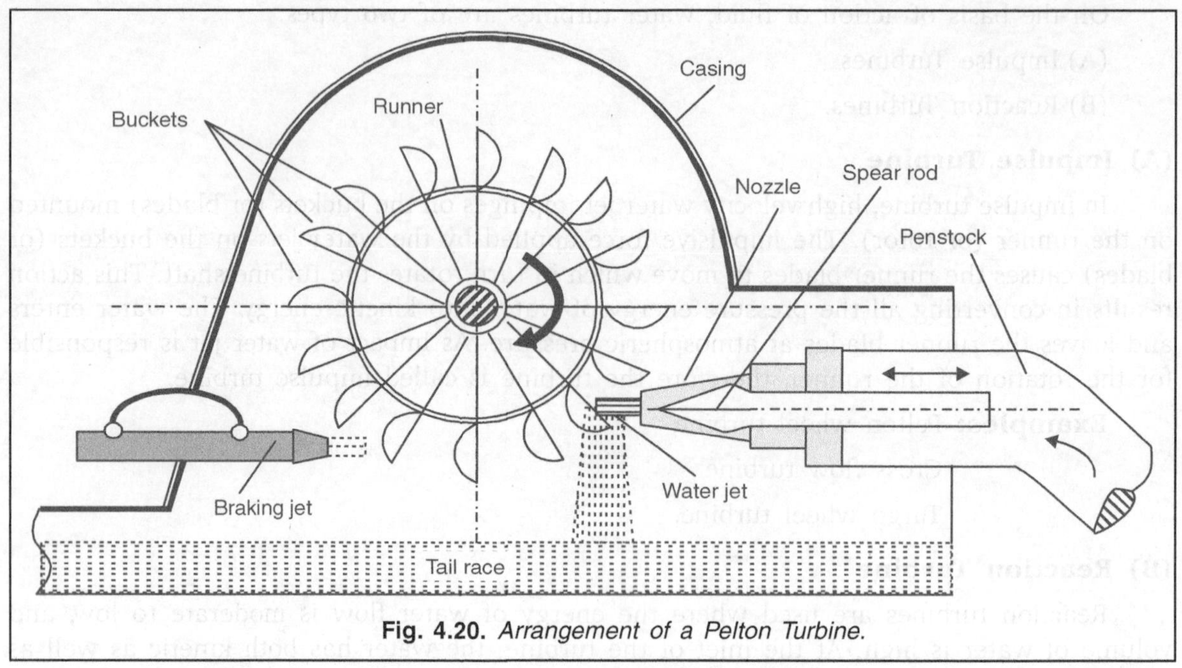

Fig. 4.20. *Arrangement of a Pelton Turbine.*

A schematic diagram of a Pelton turbine is shown in Fig. 4.20. It consists of a penstock which supplies water from the high head reservoir through one or more nozzles at its exit. The runner consists of a circular disc with a number of spoon shaped buckets evenly spaced around its periphery. Nozzle is attached to penstock. The number of the nozzles used in one setup depends on the power requirement. Due to physical constraints on orderly outflow from the unit, the number of nozzles is generally limited to six or less. The nozzle controls the discharge to the runner as per the power demand.

Fig. 4.20 shows a schematic of a Pelton turbine runner. The buckets are made of from steel or bronze for high head applications and iron is used for low heads. The buckets have highly polished surface with a dividing wall at its centre. This diving wall is called as splitter edge. It divides the striking water jet into two equal parts. The buckets are so designed that water jet after impinging on the buckets surface leaves with a relative velocity in opposite direction of the impinging water jet.

Buckets are usually bolted to the runner disc. But in few applications, the buckets and disc is cast as a single unit. 16 to 24 buckets are evenly spaced around the Pelton turbine runner periphery.

There are two methods of regulating the speed of the Pelton turbine runner. In first method, a spear rod is used inside the nozzle as shown in Fig. 4.20. The spear rod is fitted axially in the nozzle. To regulate the speed of the runner, spear rod is pushed towards the nozzle exit. This action reduces the quantity of water coming out of the nozzle. By moving spear rod backwards, the quantity of water can be restored to its original condition. In second method, a deflector plate is placed in the path of the high velocity water jets which deflect the water jet and prevent it from striking the bucket.

Casing: The entire turbine machinery is assembled in a casing to prevent the splashing of water and to safeguard the runner from possible accident. It helps in leading the water to tail race. It does not serve any role in the dynamics of the turbine.

Braking jet: Braking jet is used to stop the runner movement as quickly as possible. For this reason, arrangement along the turbine casing is made to direct the water jet at the back of the buckets. This provides the braking force for stopping the runner rotation in short time.

4.18.2 Francis Turbine

Francis turbine is suitable for medium head (15 to 500 m) installations. It is developed by an American engineer James B. Francis in 1849. Francis turbine as shown in Fig. 4.21 is a mixed flow type turbine in which fluid enters the runner along the radial direction and leaves axially. The turbine runner as shown in Fig. 4.22 has fixed blades on its periphery.

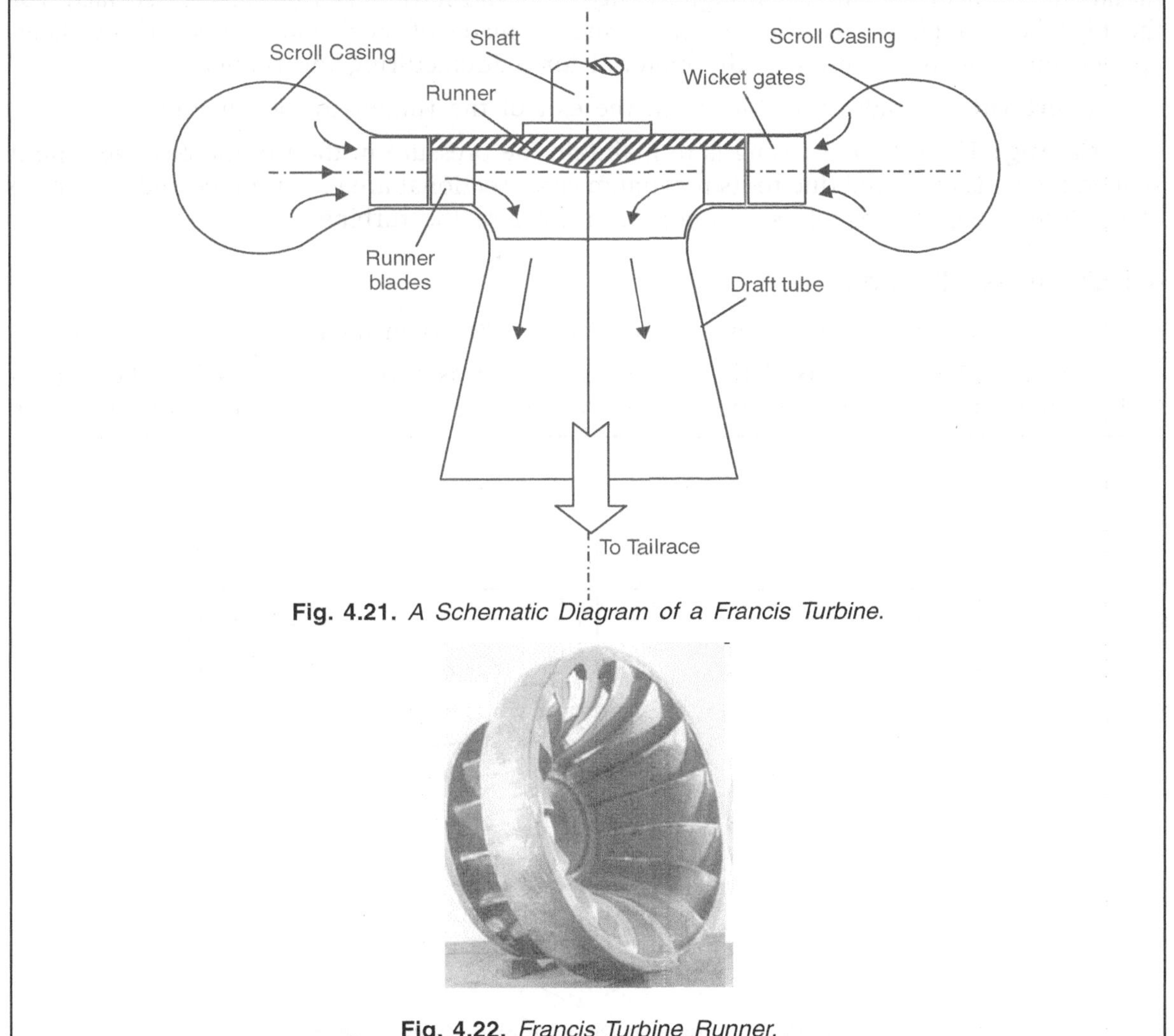

Fig. 4.21. *A Schematic Diagram of a Francis Turbine.*

Fig. 4.22. *Francis Turbine Runner.*

The main parts of a Francis turbine are:

- Turbine casing
- Penstock
- Guide vane
- Runner
- Draft tube

Wicket gates:

Wicket gates also known as stator gates are used to control the power of a reaction turbine. Wicket gates as shown in Fig. 4.21 are the pivoted guide vanes that allow or control the magnitude and direction of the inlet flow to the turbine.

Runner: The runner of a reaction turbine consists of a circular disc mounted on a shaft which carries a series of radial curved blades (or vanes) fixed on its periphery. The radial curved blades are made of high quality steel and have very fine finished surface. For the high head applications, the turbine runner is made of steel and in case of low head application; cast iron is used as the material for manufacturing the runner.

Draft tube: Guides the fluid from the exit of the runner to the tail race.

Casing: The role of a casing is to maintain the pressure in the turbine passage. Spiral shaped casing is preferred due to its maximum cross-sectional area at entrance and minimum at tip. The casing completely surrounds the runner of the turbine.

4.18.3 Kalpan Turbine

Kaplan turbine is suitable for Low head (5 to 70 m) installations.

Kaplan turbine is an axial flow reaction turbine as shown in Fig. 4.23. The Kaplan turbine, named after Victor Kaplan, a German professor, has adjustable angles on both the

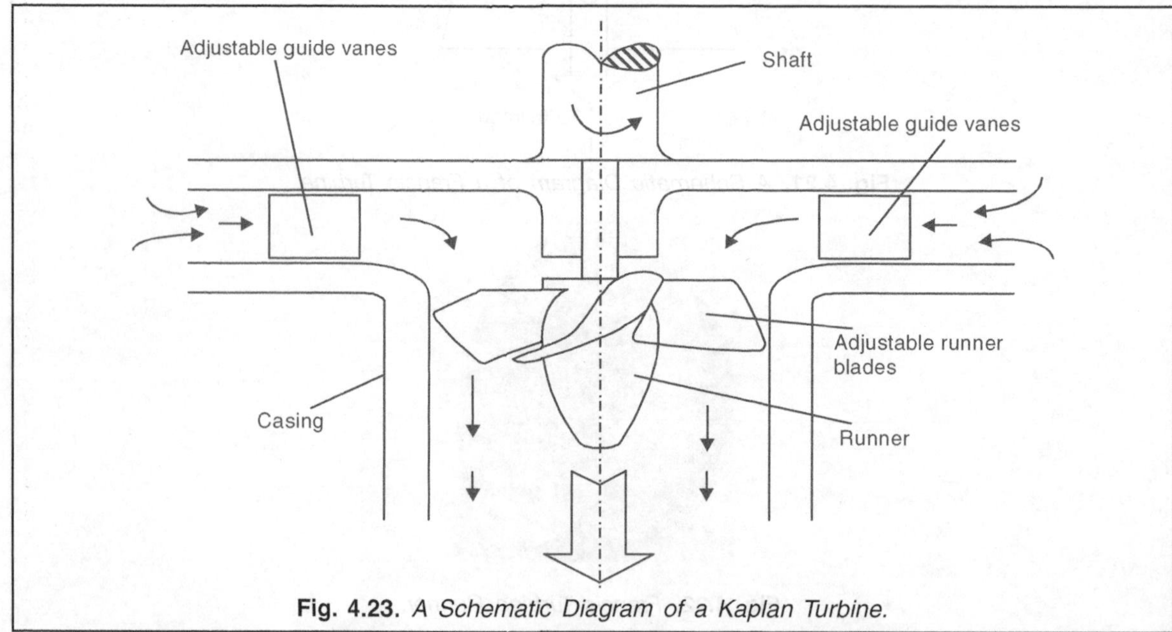

Fig. 4.23. *A Schematic Diagram of a Kaplan Turbine.*

guide vanes and runner blades. Four to six rotor blades are provided on runner as shown in Fig. 4.24. The adjustable feature of guide and runner blades makes it possible to maintain high efficiency over widely varying flow rates and heads. The propeller turbine has similar construction like that of Kaplan turbine but has fixed angle blades on its runner.

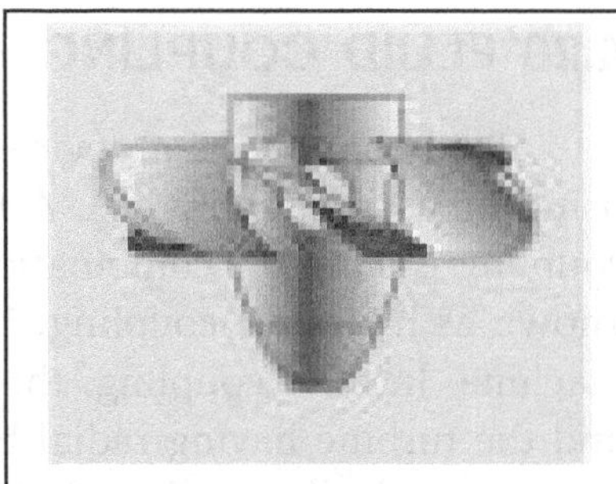

Fig. 4.24. *Kaplan Turbine Rotor.*

The main parts of a Kaplan turbine are:

- Turbine casing
- Penstock
- Adjustable guide vane
- Runner
- Adjustable blades on runner
- Draft tube.

Similar to a Francis turbine, water flows from the penstock to the turbine casing. The plane of the runner blades is below the level of the casing. From the turbine casing, water flows to the runner space through adjustable guide vanes. After passing through guide vanes water takes a 90° turn and then moves past the runner blades in a direction parallel to the turbine axis.

4.19 COMPARISON OF IMPULSE AND REACTION TURBINES

No.	*Impulse Turbine*	*Reaction Turbine*
1.	The total available energy of water is converted into kinetic energy by nozzles and water jets thus produced strike the buckets or vanes on a runner.	Only a fraction of available fluid energy is converted into kinetic energy before fluid (water) enters the runner.
2.	The high velocity water jet impinges on the buckets.	The water under pressure flows over the moving vanes.
3.	Flow control is done by fitting axially a spear or a needle valve in nozzle or by using a servo controlled deflector plate.	Here the flow control is achieved through the adjustable guide vane assembly.
4.	Working fluid operates at atmo-spheric pressure.	Working fluid pressure is reduced after it moves over the vanes.
5.	No need of a draft tube.	Draft tube is necessary.
6.	Total work is done by virtue of the kinetic energy of the impinging water jets.	Work is done partly by change in kinetic energy and rest by change in pressure energy of water.
7.	Suitable for high heads.	Suitable for low to medium heads.

4.20 FLUID COUPLING

Fluid coupling is a device which transmits power from driving shaft to driven shaft by using fluid (preferably oil). A line diagram of fluid coupling is shown in Fig. 4.25 A fluid coupling is simply a combination of a centrifugal pump and a turbine. Fluid coupling is also known as hydraulic coupling. It is neither a power producing nor a power absorbing turbo-machine. In fluid coupling, the radial impeller of a pump is mounted on driving shaft (D_R) and the turbine having radial blades is mounted on driven shaft (D_N). Both these shafts are not mechanically connected. They are independent of each other. Both the pump impeller and turbine runner being same in shape are enclosed in a casing completely filled with oil. The oil viscosity should be around Saybolt 130 seconds at 82.5°C. This oil helps in transmitting the torque from the pump impeller to the turbine runner. The runners of the turbine should be set in such a way that the flow from the pump impeller is tangential to the runner blades at inlet of the turbine.

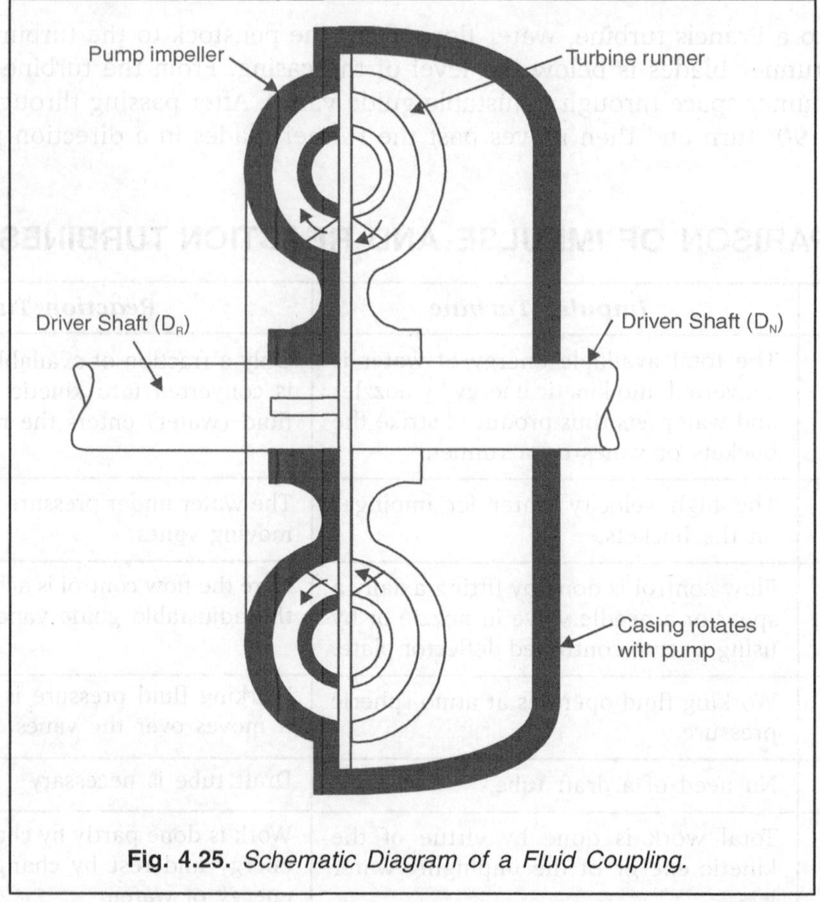

Fig. 4.25. *Schematic Diagram of a Fluid Coupling.*

Working

When the driver shaft (D_R) is rotated by a prime mover or an engine, the pump impeller also starts rotating. Due to centrifugal action, the rotating impeller forces the oil out of the pump impeller and strikes the turbine runner blades. The oil from the runner then flows back into the pump impeller, thus having a continuous circulation. The turbine shaft starts rotating as soon as sufficient head is created by increasing the speed of the driver shaft.

Applications

- Fluid or hydraulic coupling is used to transmit power in automobiles, sky ropeway, marine engines etc.
- It is very useful in cases where initially high load acts on driver shaft(D_R) and vibration and shock free operation is needed.

4.21 AIR COMPRESSOR

Air compressor is a machine which compresses the air and delivers it at high pressure to a storage tank. The process of compressing the air requires work to be done on it. Therefore a primer mover is required to drive a compressor.

Applications of compressed air

Compressed air is used for many industrial applications as mention below:

1. Compressed air is used in coal mining for driving the air motors.
2. It is used to inject fuel in air injection diesel engines.
3. It is used to operate pneumatic hand tools, drills, hammers, crane hoist, and for paint spraying.
4. It is used for applying brakes in locomotives and railway carriages.
5. It is used to produce blast of air in blast furnace.
6. It is used in the running of lifts, rams, pumps and pneumatic conveyors.
7. It is used in gas turbine plants.
8. It is used to cool aircraft cabins.

4.22 CLASSIFICATION OF COMPRESSORS

Compressors can be classified on the basis of following considerations.

(*i*) On the Basis of the Action of Air

(*a*) **Single acting:** A compressor is considered single acting when the total compression is done using only one side of a piston.

(*b*) **Double acting:** A compressor which uses both sides of a piston is called double acting.

(*ii*) On the Basis of the Number of Stages

(*a*) **Single stage:** In a single stage compressor, the total compression takes place in one cylinder. The delivery pressure of air in single stage compressor is limited up to 10 bar.

(*b*) **Multi stage:** In multi-stage compressor, the compression takes place in two or more than two cylinders combined in series. The delivery pressure of air in multi-stage compressor is more than 10 bar.

(*iii*) **On the Basis of the Capacity of Working Substance**

(*a*) Low capacity compressor: Air volume delivered is less than 0.15 m^3/s

(*b*) Medium capacity compressor: air volume delivered is in the range of 0.15 to 5.0 m^3/s

(*c*) High capacity compressor: air volume delivered is more than 5 m^3/s.

(*iv*) **On the Basis of the Number of Cylinders**

(*a*) Single cylinder

(*b*) Multi cylinder.

(*v*) **On the Basis of Working Principle**

(*a*) Reciprocating compressor

(*b*) Rotary compressor.

(*vi*) **On the Basis of the Method of Cooling**

(*a*) Air cooled

(*b*) Water cooled

(*vii*) **On the Basis of Pressure Range**

(*a*) Low pressure

(*b*) Medium pressure

(*c*) High pressure

(*viii*) **On the Basis of Running Drive Used**

(*a*) Steam engine drive

(*b*) Steam turbine drive

(*c*) IC engine drive

(*d*) Electric motor drive

Important terms and their definitions related to air compressor

(*i*) **Inlet pressure:** It is the absolute air pressure available at the inlet of a compressor.

(*ii*) **Deliver pressure:** It is the absolute air pressure available at the outlet of a compressor.

(*iii*) **Compression ratio:** It is the ratio between delivery pressure to inlet pressure.

(*iv*) **Swept volume:** It is the volume of air taken by a compressor during its suction stroke.

(*v*) **Work done:** It is the quantity of work required to compress the particular volume of air from suction pressure conditions to delivery pressure conditions.

(*vi*) **Mean effective pressure:** It is the ratio of work done to swept volume.

(*vii*) **Capacity:** It is the volume of air delivered by a compressor under specified conditions of pressure, temperature and composition prevailing at the compressor inlet.

4.23 RECIPROCATING COMPRESSOR

Reciprocating compressor finds extensive use in air and refrigerant compression. Its working principle is similar to that of a bicycle pump. It consists of a piston which reciprocates in a cylinder driven by a connecting rod and crank shaft mounted in a crank case.

An air compressor has a number of inlet and delivery valves on its cylinder. These are automated spring controlled valves and operated only when there is a pressure difference across the sides of these valves. Fig. 4.26 shows a schematic diagram of an air compressor. When the piston moves towards bottom dead centre, this causes drop in air pressure inside the cylinder and inlet valve opens. Delivery valve of compressor opens when the pressure in the cylinder is more than the delivery pressure. When the piston moves from top end of the cylinder towards the bottom end *i.e.,* on its downward stroke, air enters into the cylinder. The pressure in the cylinder starts increasing when piston moves from bottom end towards top end of the cylinder. When this air pressure become slightly greater than the pressure in delivery chamber, the delivery valve opens and pressurized air is delivered into the delivery chamber. When the piston reaches the top end of the cylinder *i.e.,* it completes its upward stroke then a small volume of pressurized air is left in the clearance space of the cylinder. When the piston moves from bottom end of the cylinder to top end, compression of air takes place. When piston reaches top end compression stroke gets completed. At the end of this compression stroke, piston again moves towards bottom end of the cylinder. This again causes drop in air pressure inside the cylinder and due to pressure difference the delivery valves closes and inlet valve gets opened and the cycle is repeated.

From the above discussion it is clear that in a single acting air compressor, in one revolution of crank shaft one cycle of operation is completed. This one cycle of operation is completed in two strokes of piston *i.e.,* one is downward and other is upward.

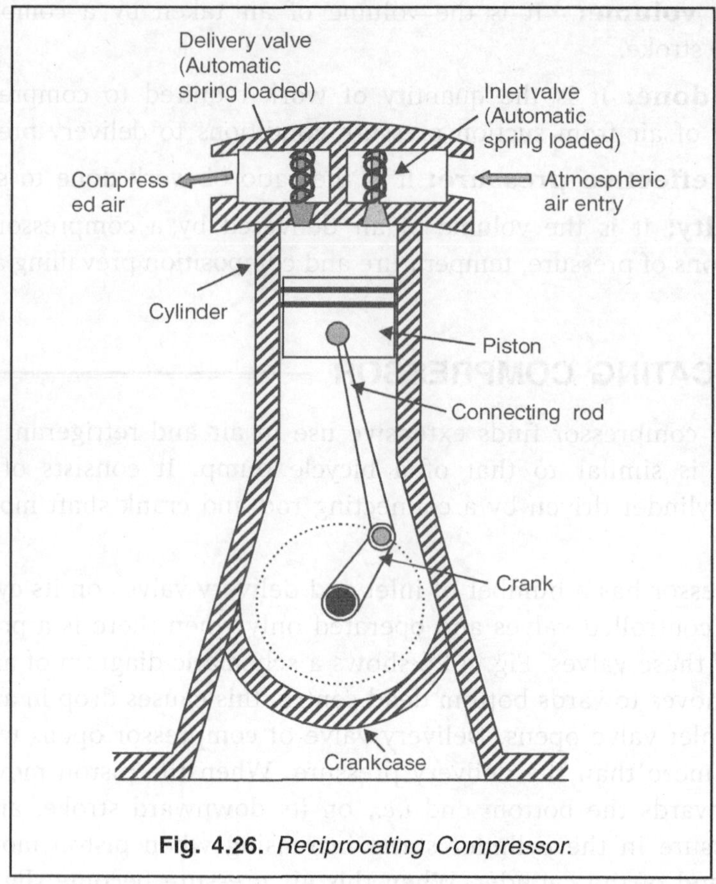

Fig. 4.26. *Reciprocating Compressor.*

4.24 ROTARY COMPRESSORS

In these compressors, air is compressed and delivered by means of rotating elements. Rotary compressors run at high speeds and generally are used for supplying large quantities of compressed air at low pressure. Rotary compressors find extensive application in supercharging the internal combustion engines and these are also used as vacuum pump.

Rotary Compressors are of Two Types

(*i*) **Positive displacement compressor**

These are of two types:

(*a*) Roots blower compressor

(*b*) Vane blower compressor.

(*ii*) **Non-positive displacement compressor**

These are also known as dynamic compressors. These are also of two types:

(*a*) Centrifugal compressor

(*b*) Radial compressor.

Either air or water cooling may be used for rotary compressors. Since the cooling takes place inside the compressor, its rotating parts do not face high operating temperatures.

4.25 ROOTS BLOWER COMPRESSOR

Figure 4.27 shows a schematic diagram of a roots blower compressor. Its main parts consist of two or more than two lobed rotors with their profiles generated by epicycloids and hypocycloids. These lobes rotor rotates in a airtight casing. The two lobed rotors mesh together similar to a gear drive. For any angular position of lobed rotors, the high pressure region is sealed from the low pressure region due to their geometrical shapes. Out of the two rotors as shown in Fig. 4.27, one rotor is coupled to the source of power and other is driven by an external gear drive.

Working

A prime mover is used to drive one of the lobed rotors and the other is a driven lobe rotor. When rotor starts rotating, air from the inlet port enters the empty space between the lobes and casing. The air gets compressed between interfaces of the rotors. This compressed air then moves towards the delivery port. There is one more reason for this compression. As we know that entry port of the compressor is at atmospheric pressure and on the other hand delivery port is exposed to delivery pipe line where pressure is more than atmospheric pressure. When the delivery port opens usually some compressed air from the delivery port enters into the region between the casing and lobed rotors. This entry of pressurized air is called as back flow of air. This back flow of air ensures that the entrapped air in the blower gets compressed at constant volume and this continues till the blower pressure become same at all regions. Lastly the blower delivers the compressed air.

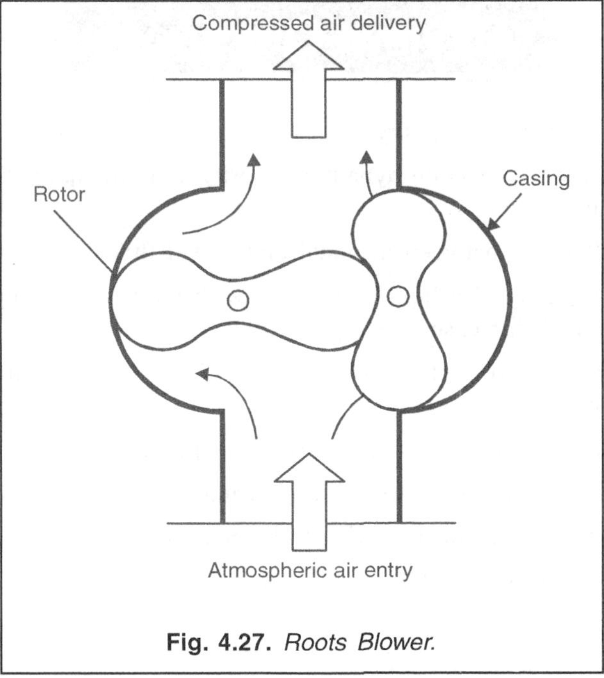

Fig. 4.27. *Roots Blower.*

4.26 COMPARISON BETWEEN RECIPROCATING AND ROTARY COMPRESSOR

No.	*Reciprocating Compressor*	*Rotary Compressor*
1.	In this case supply of compressed air is intermittent.	In case of rotary compressor, supply of compressed air is continuous.
2.	It is used to handle low volumes of air.	It can handle large volumes of air.
3.	It delivers compressed air at high pressure.	It is used to deliver compressed air at low pressure.
4.	It is suitable for low delivery rates.	It is used where high delivery rates are needed.
5.	Reciprocating compressor has low speeds.	Rotary compressor has high speeds.
6.	Compressed air delivered gets dirty due to contact of air with the lubricating oil of reciprocating surfaces of the piston.	In this case air delivered is clean and free from dirt.
7.	It is bulky in size.	Being compact, its size is small.
8.	It requires high capital costs.	It needs low capital costs

REVIEW QUESTIONS

1. What is Newton's law of viscosity? *(RGTU April 2009)*
2. What is the difference between dynamic viscosity and kinematic viscosity? State their units of measurements.
3. What is the difference between an ideal and a real fluid?
4. Why does the viscosity of a liquid decreases with increase in temperature whereas that of a gas increases with increase in temperature?
5. Distinguish clearly between the Newtonian and Non-Newtonian fluids. Give three examples of each category.
6. State and explain Bernoulli's equation for incompressible fluids. *(RGTU June 2008)*
7. Derive the Bernoulli's equation from Euler's equation for an ideal flow. *(RGTU April 2009)*
8. Enlist the assumptions made in deriving the Bernoulli's equation.
9. Enumerate the applications of Bernoulli's theorem.

10. Explain the construction and working and uses of the following:

(*i*) Venturimeter,

(*ii*) Rotameter *(RGTU Jan.-Feb. 2008)*

11. Give a comprehensive classification of water turbines.

12. Compare the impulse and reaction turbines.

13. Describe the construction and working principle of the following:

(*i*) Pelton wheel,

(*ii*) Kaplan turbine *(RGTU Jan.-Feb. 2008)*

14. Explain the working principle of a fluid coupling. *(RGTU June 2008)*

15. How pumps are classified? Explain. *(RGTU June 2008)*

16. Sketch the layout of a centrifugal pump and explain the functions of its major parts.

17. Compare the reciprocating and rotary compressors.

18. Explain the following in brief:

(*i*) Base load plants

(*ii*) Peak load plants

(*iii*) Pumped storage plants.

19. Explain briefly the different types of casing for pumps.

20. Explain the construction and working of a Kaplan turbine with the help of a neat sketch.

OBJECTIVE TYPE QUESTIONS

Select the most appropriate answer:

1. A flow is said to be steady when

(A) Fluid variables do not change with time at some points

(B) Fluid variables change steadily with time

(C) Fluid variables do not change with time at any point

(D) None of the above

2. Dimension of absolute viscosity is....................

(A) L^3T^{-1} (B) $ML^{-1}T^{-2}$

(C) L^2T^{-1} (D) $ML^{-1}T^{-1}$

3. Newton's law of viscosity defines the relationship between

(A) Fluid pressure and shear stress

(B) The shear stress and rate of shear strain

(C) Shear stress, pressure, and temperature

(D) All the above

4. The viscosity of
   (A) Gases increase with increase in temperature
   (B) Fluids increase with increase in temperature
   (C) Liquids increase with increase in temperature
   (D) Fluid decreases with increase in temperature
5. Local atmospheric pressure is measured by
   (A) Altimeter (B) Bourdon's pressure gauge
   (C) Barometer (D) Hydrometer
6. Assumptions not made in the derivation of Bernoulli's theorem is
   (A) The flow is steady (B) The flow is irrotational
   (C) The fluid is real (D) The flow is incompressible
7. The equation of continuity in fluid mechanics is based on
   (A) Newton's second law of motion (B) Hooke's law
   (C) Einstein's law (D) Law of the conservation of mass
8. An ideal fluid is
   (A) Incompressible and frictionless (B) Compressive and gaseous
   (C) The one which follows the perfect gas law
   (D) All of the above
9. The kinematic viscosity ν is related to the dynamic viscosity μ and density ρ as $\nu =$
   (A) μ/ρ (B) $\mu \rho$
   (C) ρ/μ (D) $\mu/\rho g$
10. Bernoulli's equation is applicable between any two points
    (A) In steady, rotational flow of an incompressible fluid
    (B) In any rotational flow of an incompressible fluid
    (C) In any type of compressible flow
    (D) In steady, irrotational flow of an incompressible fluid
11. Poise is the unit of
    (A) Shear stress (B) Viscosity
    (C) Mass density (D) Pressure
12. Bernoulli's theorem deals with the law of conservation of
    (A) Mass (B) Both mass and energy
    (C) Both mass and momentum (D) Energy
13. Venturi meter is used to measure
    (A) Discharge (B) Instantaneous velocity
    (C) Velocity at a point (D) Pressure at a point
    (E) Both discharge and temperature
14. Pitot tube is used to measure
    (A) Average velocity (B) Instantaneous velocity
    (C) Velocity at a point (D) Fluid pressure at a point
15. Francis turbine is
    (A) An axial flow turbine (B) An tangential flow turbine
    (C) A radial flow reaction turbine (D) A radial flow impulse turbine

16. Kaplan turbine is

(A) An axial flow impulse turbine
(B) An axial flow reaction turbine
(C) A radial flow impulse turbine
(D) A radial flow reaction turbine

17. Kaplan is a propeller type axial flow turbine in which the blades fixed on the runner are:

(A) Adjustable
(B) Fixed
(C) Partially fixed
(D) All of the above

18. ....is suitable for high head and low discharge applications.

(A) Pelton
(B) Francis
(C) Deriaz
(D) Kaplan

19. In a centrifugal pump casing, the flow of water leaving the impeller is.............

(A) Centrifugal flow
(B) Free flow
(C) Forced vortex flow
(D) Free vortex flow.

20. Draft tube is used in case of.......................turbine to discharge the water to tailrace.

(A) Pelton turbine
(B) Fluid coupling
(C) Reciprocating pump
(D) Francis turbine.

ANSWERS

1. (C)	**2.** (D)	**3.** (B)	**4.** (A)	**5.** (C)	**6.** (C)
7. (D)	**8.** (A)	**9.** (A)	**10.** (D)	**11.** (B)	**12.** (D)
13. (A)	**14.** (C)	**15.** (C)	**16.** (B)	**17.** (A)	**18.** (A)
19. (D)	**20.** (D).				

Basic Concepts of Thermodynamics

5.1 INTRODUCTION

Since ages it has been the human tendency to search for more and more power. In early days men used animals for producing power. However the industrial revolution of the nineteenth century had motivated the engineers to search for the more effective and reliable sources of power. Being closer to nature, man first came in contact with fossil fuel followed by petroleum based fuel and then nuclear fuel and now lot of emphasis is being put on renewable resources. The continuous efforts made by the man in the direction of getting mechanical work resulted in the invention of a heat engine so that he can convert the energy stored in these fuels into useful energy capable of doing some useful work. Thermodynamics is the fundamental science that governs the processes taking place in a heat engine. Historically, the origin of thermodynamics is traced to a publication in 1824 by a French engineer, N.L.S. Carnot about his early studies on the performance of steam engine.

Thermodynamics is one of the important branches of science and engineering, which deals with energy and its transformation and relationship between the heat, work and properties of the system. Thermodynamics deals only with macroscopic structure of the matter and does not concern itself with events happening at the molecular level.

Every thermodynamics process involves the use of a working substance which has the ability to receive, store and reject energy as per the requirement of a particular application. The properties of the working substance should suit the specific process.

Name of Heat Engine	***Working Substance***
Steam power plant	Water (liquid and vapour form)
I.C. engine	Air and fuel mixture
Rockets	Solid or liquid fuel
Ice plant	Ammonia or any other standard refrigerant

5.2 MACROSCOPIC AND MICROSCOPIC VIEWPOINT

There are two view points to describe a system (*i*) Macroscopic and (*ii*) Microscopic. A few measurable properties are needed to describe a system from the macroscopic viewpoint. For example consider the contents of a piston cylinder arrangement as a system. At any moment the system occupies a certain volume depending on the position of the piston. Pressure of the gases inside the cylinder can be another useful quantity, which can describe the system. A pressure gauge may be used to measure the pressure changes as the engine operates. Other important measurable quantities are the temperature and the chemical composition. A system may be described by means of the above quantities which we call the properties of the system. These are called macroscopic coordinates and provide a macroscopic description of the system. The macroscopic approach is used in **engineering thermodynamics** and therefore in the study of heat engines.

From the microscopic point of view, matter in the cylinder *i.e.,* gas is composed of a large number of molecules. Each molecule at a given moment has certain position, velocity and energy, and for each molecule these changes vary frequently. The behaviour of the system is described by summing up the behaviour of each molecule. This type of approach is used in **statistical thermodynamics.**

5.3 THERMODYNAMIC SYSTEM

A **'system'** is defined as a quantity of matter or region in space on which our attention is focused. The mass or region outside the system is called the ***surroundings.*** The system is separated from the surroundings by the system boundary as shown in Fig. 5.1. The system boundary may be real or imaginary, fixed or movable. From the mathematics point of view, the boundary has zero thickness and thus it can neither contain any mass nor occupy any volume in space. A *system* and its *surrounding* together comprise a ***universe.***

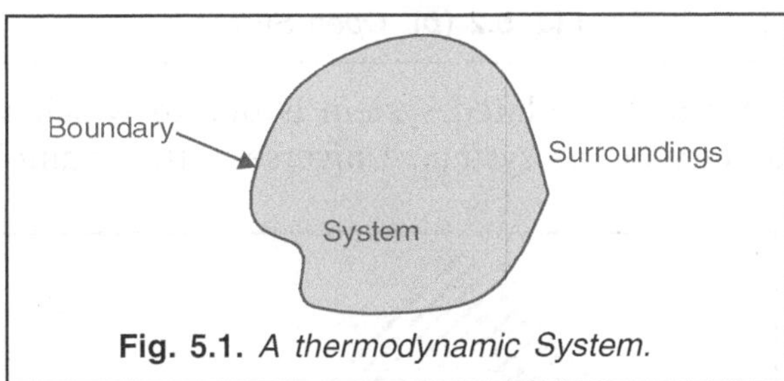

Fig. 5.1. *A thermodynamic System.*

There are three types of thermodynamic systems:

(*a*) Closed system

(*b*) Open system

(*c*) Isolated system.

(*a*) Closed system: It is also known as control mass. In a closed system there is no transfer of mass but transfer of energy across the system boundaries does take place [Fig. 5.2 (*a*)]. Heating of water in a closed vessel, bomb calorimeter, matter bounded by a piston in cylinder, refrigerator etc are the examples of closed system.

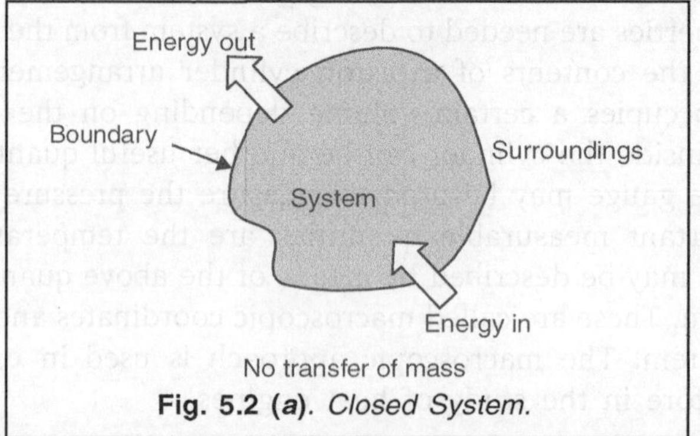

Fig. 5.2 (*a*). *Closed System.*

(*b*) Open system: In an open system both mass and energy may across the system boundary. Open system is also referred to as **control volume** Fig. 5.2 (*b*). Steam turbines, gas turbines, air compressors, and boilers delivering steam, a car radiator are the few examples of an open system.

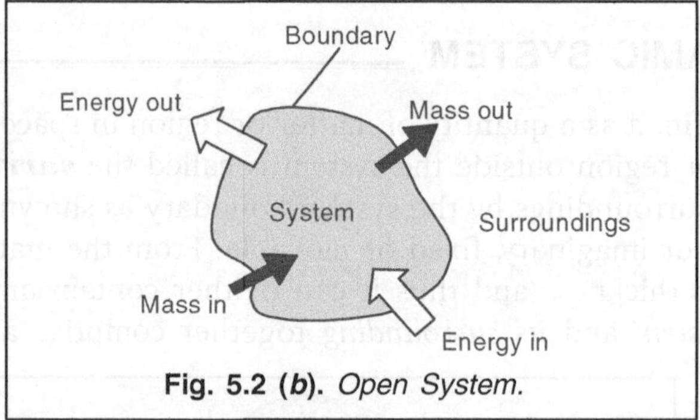

Fig. 5.2 (*b*). *Open System.*

(c) Isolated system: An isolated system is one in which neither mass nor energy cross the boundary of the system. Universe is the example of an isolated system Fig. 5.2 (*c*).

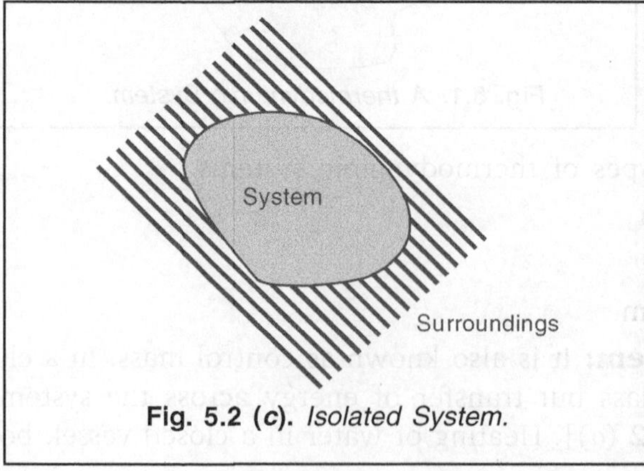

Fig. 5.2 (*c*). *Isolated System.*

5.4 CONTROL VOLUME

The concept of control volume has been used in modelling various engineering problems which involves mass flow in and out of a system. A fixed region in space is called a 'control volume' through which mass, momentum, and energy may flow (Fig. 5.3). The boundaries of control volume are called as control surfaces. The control volume may have either fixed or moving boundaries as in case of a piston cylinder arrangement. Control volume is equivalent to an open system when there is a change in size and position of the control volume. The control volume becomes synonymous with closed system if no mass transfer takes place across the control surface.

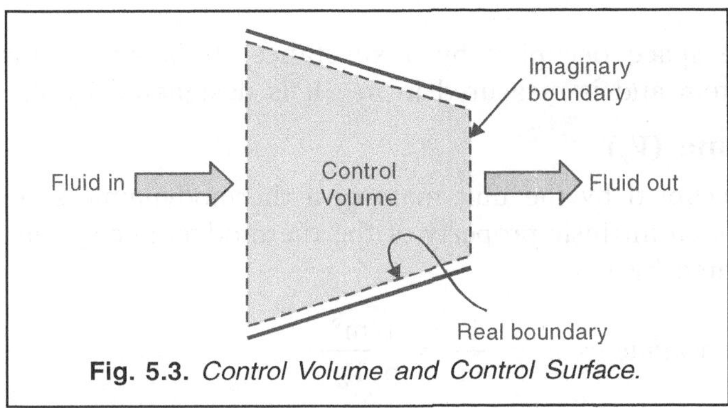

Fig. 5.3. *Control Volume and Control Surface.*

5.5 THERMODYNAMIC EQUILIBRIUM

A system is said to be in thermodynamic equilibrium, if conditions for the following three types of equilibrium are satisfied:

(*a*) Mechanical equilibrium

(*b*) Thermal equilibrium

(*c*) Chemical equilibrium.

Mechanical equilibrium is related to pressure or elastic stress and a system is said to be in mechanical equilibrium if there is no change in pressure at any point of the system with time.

A system is said to be in **thermal equilibrium** if the temperature remains same throughout the entire system. If there are variations in temperature from point to point of an isolated system, one may observe that the temperature at every point, at first changes with time. This rate of change of temperature keeps on decreasing and eventually this change will cease. The system will be in thermal equilibrium when no further changes in temperature are noticed.

The system is said to be in **chemical equilibrium** if there is no chemical reaction between the system and the surrounding.

The science of **classical thermodynamics** deals mainly with the systems in thermodynamics equilibrium or in which the departure from the equilibrium is negligibly small. Such types of the systems are more appropriately described as *Thermostatics* whereas leaving the term 'Thermodynamics' reserved for the study of systems not in thermodynamic equilibrium. When the conditions for any one of the three types of equilibrium are not satisfied, a system is said to be in a non-equilibrium state.

5.6 FEW THERMODYNAMIC PROPERTIES

(a) Density

It is defined as the mass per unit volume of a substance. It is measured in kg/m³. It is designated by ρ. It is always equal to the reciprocal of the specific volume.

$$\rho = m/V = 1/V_s \qquad ...(5.1)$$

(b) Volume (V)

Volume is the space occupied by a substance. It is an extrinsic property of the thermodynamic system and is measured in m³. It is designated by V.

(c) Specific volume (V_s)

The volume occupied by the unit mass of a thermodynamic system is known as the specific volume. It is an intrinsic property of the thermodynamic system and is measured in m³/kg. It is designated by V_s.

$$\textit{specific volume} = v = \frac{V}{m} = \frac{1}{\rho}\frac{\text{m}^3}{\text{kg}} \qquad ...(5.2)$$

(d) Specific weight

It is defined as the force of gravity acting on a unit volume of the substance. It is measured in N/m³ and designated by W_s.

(e) Specific gravity or relative density

Sometimes the density of a substance is mentioned relative to the density of a well-known substance. Then it is called specific gravity or relative density and is defined as the ratio of the density of a substance to the density of some standard substance at a specified temperature.

Usually water at 4°C is taken as standard substance whose density is taken as 1000 kg/m³.

$$\textit{specific gravity} = \frac{\rho}{\rho_{water}} = \frac{\rho}{1000} \qquad ...(5.3)$$

(f) Pressure

Pressure is defined as the force exerted by a fluid per unit area, measured over a given point or over a surface. The basic unit of pressure is Pascal and denoted by 'Pa'. When a fluid applies a force of 1N over an area of 1m², then the pressure is equivalent to 1 Pascal.

i.e.,

$$1Pa = 1\frac{\text{N}}{\text{m}^2} \qquad ...(5.4)$$

Since Pascal is a very small unit, so for the purpose of calculation and other derivation, its higher value is generally expressed in kPa (kilopascal) or MPa (megapascal).

$$1\ kPa = 1 \text{ kilopascal} = 1000\ Pa = 1000\ N/m^2$$

$$1\ bar = 100\ kPa = 10^5 \text{ kPa} = 10^5\ Pa = 10^5\ N/m^2$$

The other important pressure units are:

$$1\frac{kgf}{\text{cm}^2} = 9.807 \text{ N/cm}^2 = 9.807 \times 10^4 \text{ N/m}^2 = 9.81 \times 10^4 \text{ Pa}$$

$$1\frac{kgf}{\text{cm}^2} = 14.223\ psi$$

As we know that the atmosphere around us has air at some pressure therefore zero pressure point can only be defined with reference to some datum. Atmospheric pressure provides most suitable reference for pressure measurement. The atmospheric pressure is due to the weight of the earth's atmosphere. The standard value of atmospheric pressure has been taken as $P_{atm} = 101.3\ kPa$. The pressure of atmosphere does not remain constant at a given location due to changes in climatic conditions.

Atmospheric pressure is equivalent to the height of air column at that place.

$$\text{Atmospheric pressure} = \frac{\text{pressure exerted due to height of air column}}{\text{area of air column}}$$

$$= \frac{\text{mass of air column} \times \text{acceleration due to gravity}}{\text{area of air column}}$$

$$= \frac{\text{density of air column} \times \text{volume of air column} \times \text{acceleration due to gravity}}{\text{area of air column}}$$

As we know that **volume = area × height**

$$= \frac{\text{density} \times \text{area} \times \text{height} \times \text{acceleration due to gravity}}{\text{area}}$$

$$\therefore \text{ atmospheric pressure} = \rho g h\ \frac{\text{kg}}{\text{ms}^2}. \qquad ...(5.5)$$

The unit of atmospheric pressure may be derived by dividing and multiplying its unit $\frac{\text{kg}}{\text{ms}^2}$ by m^2

we get

$$= \frac{\text{kg}}{\text{ms}^2}\frac{\text{m}^2}{\text{m}^2} \qquad (\text{since } 1 \text{ N} = 1 \text{ kg m/s}^2)$$

$$= \frac{\text{N}}{\text{m}^2}$$

There are three categories of pressure measurements namely:

(i) Absolute pressure

(ii) Gauge pressure

(iii) Differential pressure.

(i) Absolute Pressure

Absolute pressure is the difference between the pressure at a particular point and the absolute zero of pressure. Absolute zero of pressure will exist only in a completely evacuated space (perfect vacuum).

(ii) Gauge Pressure

Mostly pressure measuring devices are calibrated to read zero in the atmosphere and so these devices indicate the difference between the absolute pressure and local atmospheric pressure. This difference is called the gauge pressure. Simply speaking pressure above the atmospheric value is called gauge pressure. Pressure measured below the atmospheric pressure is called vacuum pressure. Vacuum gauges are used to measure vacuum pressures that indicate the difference between the atmospheric pressure and absolute zero pressure. Absolute, gauge and vacuum pressures are all positive quantities. Relationship between pressures stated for different references is shown in Fig. 5.4.

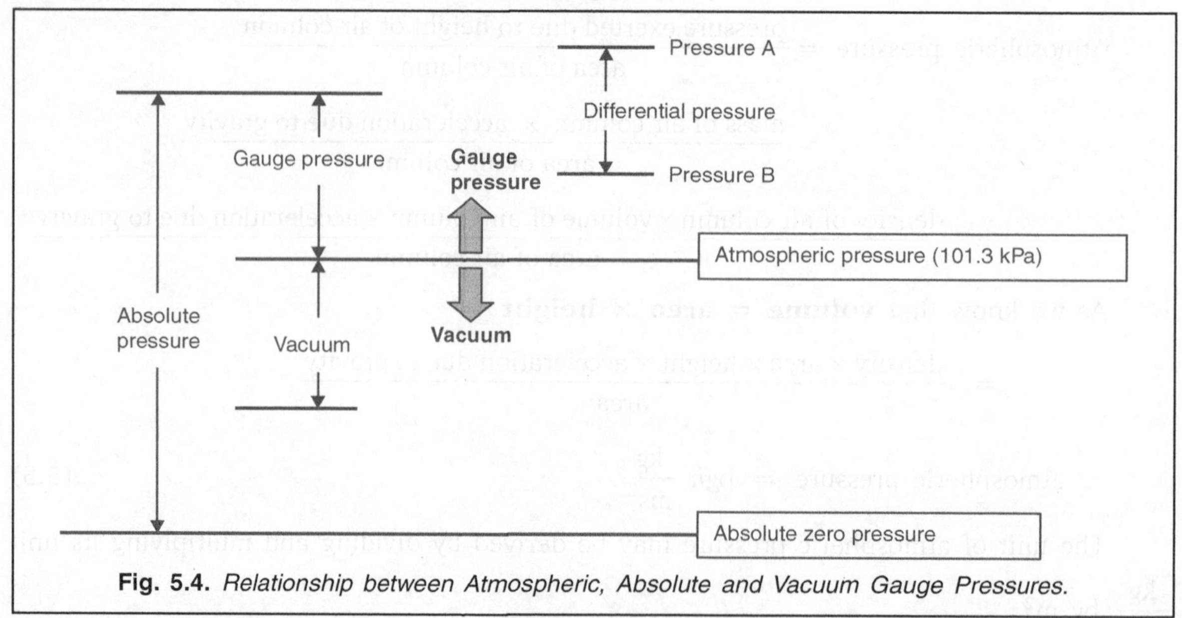

Fig. 5.4. *Relationship between Atmospheric, Absolute and Vacuum Gauge Pressures.*

$$P_{gauge} = P_{abs} - P_{atm}$$

$$P_{vacuum} = P_{atm} - P_{gauge} \qquad \ldots(5.6)$$

(iii) Differential Pressure

When the pressure measuring device measures the difference between the two unknown pressures, neither of which atmospheric pressure, then measurement is known as the differential pressure.

Problem 1. *A manometer is used to measure the pressure in a tank. The fluid used has a specific gravity of 0.90 and the manometer column height is 40 cm as shown in Fig. 5.5. If the local atmospheric pressure is 100 kPa, determine the absolute pressure in the tank.*

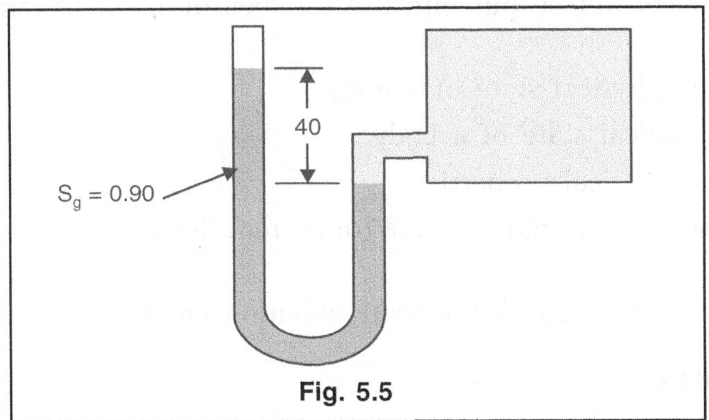

Fig. 5.5

Solution. Let us first find out the density of the fluid

As we know

$$\text{Specific gravity } S_g = \frac{\rho_{fluid}}{\rho_{water}} = \frac{\rho_{fluid}}{1000}$$

$$\therefore \quad 0.90 = \frac{\rho}{1000}$$

$$\therefore \quad \rho = 0.90 \times 1000 = 900 \text{ kg/m}^3$$

As we know

$$p_{abs} = p_{gauge} + p_{atm}$$

$$p = \rho g h + p_{atm}$$

$$p = 900 \times 9.81 \times 0.40 + 100$$

$$p = 103.53 \; kPa$$

Problem 2. *A pressure gauge fitted on a boiler shows the pressure of steam inside the boiler as 1.35 MPa. The barometer reads the atmospheric pressure as 775 mm of mercury. Find out the absolute pressure of steam inside the boiler?*

Solution. Given:

Pressure gauge reading p_{gauge} = 1.35 *MPa*

Atmospheric pressure p_{atm} = 775 mm of H_g

Absolute pressure p_{abs} = ?

As we know

$$p_{abs} = p_{atm} + p_{gauge}$$

$$p_{atm} = \rho g h = 13.6 \times 9.81 \times \frac{775}{1000}$$

$$p_{atm} = 103.4 \; kPa = 0.1034 \; MPa$$

$$\therefore \quad p_{abs} = 1.35 + 0.1034 = 1.4534 \; MPa \; \textbf{Ans.}$$

(g) Temperature

Temperature may be defined as the degree of hotness or coldness of a body measured on a definite scale. When a body is heated or cooled down one can observe different effects and one of these effects can be put into use for measurement of temperature. The major effects include:

- Change in the physical state of a body
- Change in chemical state of a body
- Variations in electrical properties
- Generation of an electromotive force (or current flow) at the junction of two dissimilar metals.
- Changes in the intensity of the total radiation emitted.

Temperature Scales

Temperature scales provide us a common basis for measuring temperatures. All temperature scales are based on some easily reproducible states such as freezing and boiling point of water. Freezing point is also known as ice point and boiling point as steam point.

☞ A mixture of ice and water that is in equilibrium with air saturated with vapour at 1 atm pressure is termed as ice point.

A mixture of water and its vapour in equilibrium at 1 atm pressure is referred as steam point.

(a) Celsius Scale

It is also known as centigrade scale. In this scale, the ice and boiling points of water are treated as fixed points. In Celsius scale, the mercury in a glass tube is calibrated to read 0°C as ice point and 100°C as boiling point of water. The interval between these two points is divided in 100 equal parts.

(b) Fahrenheit Scale

In this scale, the ice and boiling points of water are used as fixed points. In Fahrenheit scale, mercury in a glass tube is calibrated to read 32°F as ice point and 212°F as boiling point of water. This interval between these two points is divided in 180 equal parts.

(c) Thermodynamic Scale

It is really desirable to have a temperature scale that is independent of the properties of any substance or matter. Such type of a temperature scale is known as thermodynamic temperature scale.

Kelvin scale: In SI units, the thermodynamic scale is known as Kelvin scale. The temperature unit on this scale is the 'Kelvin' which is denoted by 'K'. The lowest temperature on the Kelvin scale is 0 K. No molecule activity takes place at absolute zero Kelvin and therefore body does not possess any heat content. The temperature at this condition is the lowest temperature possible and is referred to as absolute zero. Kelvin scale has a value of 273.16 K and a value of 373.16 K at 100°C.

Rankine Scale: The thermodynamic scale in FPS or English system is known as Rankine cycle. The temperature unit on this scale is the Rankine which is denoted by 'R'. In this scale absolute zero is marked at 460°F.

The following expressions are used to convert one scale reading into another.

$$°C = \frac{5}{9}(°F - 32)$$

$$K = (°C + 273.16) \qquad \text{...(5.7)}$$

$$R = (°F + 460)$$

Figure 5.6 shows the comparison of various temperature scales.

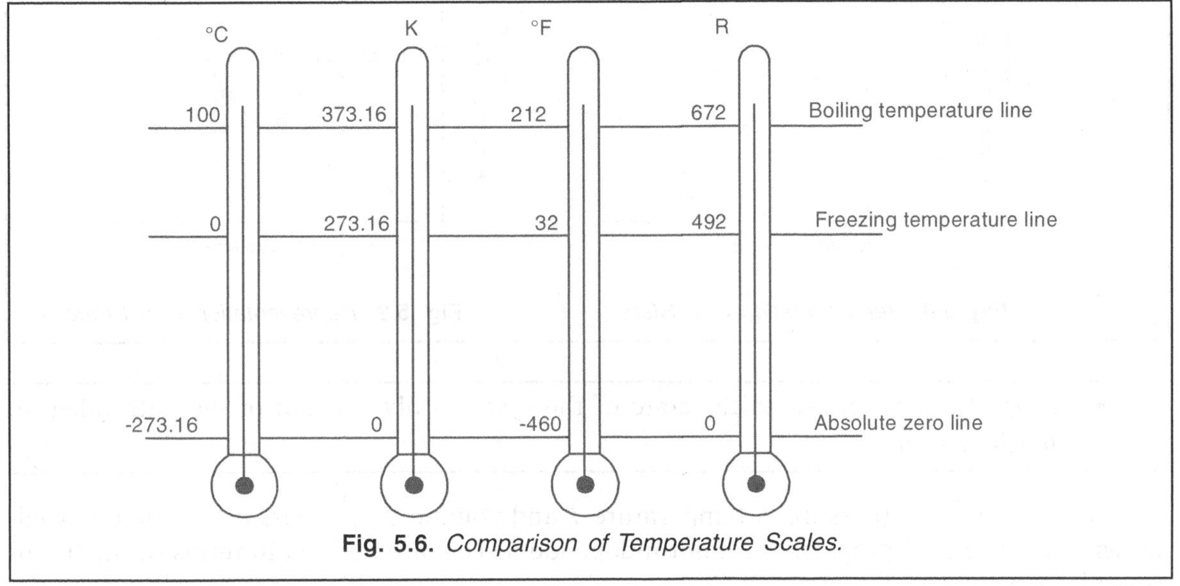

Fig. 5.6. *Comparison of Temperature Scales.*

5.7 ZEROTH LAW OF THERMODYNAMICS

This law states that "if two systems *A* and *B* are separately in thermal equilibrium with a third system *C*, then the two systems *A* and *B* will also be in thermal equilibrium with each other".

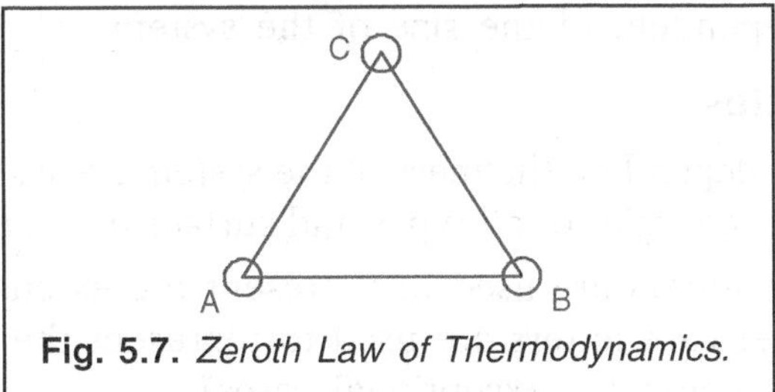

Fig. 5.7. *Zeroth Law of Thermodynamics.*

This law enables a comparison between the temperatures of the two bodies with the help of a third taken as reference. The zeroth law is, in fact, an axiom since it is derived from past experience and does not have mathematical proof. This law provides the basis for temperature measurements.

5.8 STATE AND PROPERTIES OF A SYSTEM

The condition of a system at a given time is called the '**state**' of the system (Fig. 5.8) and the state of the system is described by the '**properties**' of the system. The number of the properties required to define the state of the system depends on the complexity of the system. When a system changes its state from one equilibrium condition to another, it is said to have undergone a process (Fig. 5.9). The series of states through which a system passes during a process is called the '**path**' of the process as shown in Fig. 5.9.

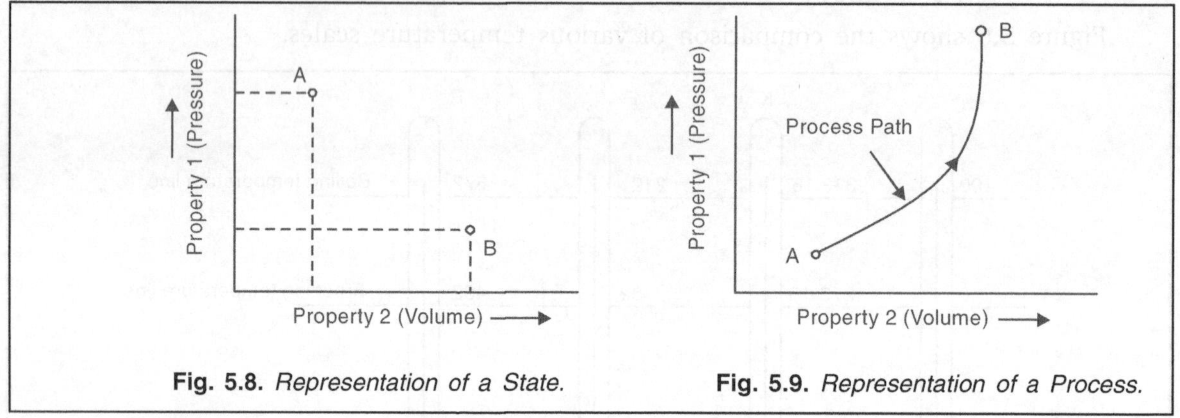

Fig. 5.8. *Representation of a State.*

Fig. 5.9. *Representation of a Process.*

☞ Properties are related to the state of the system only and not of the path taken to reach that state.

As we know that pressure P, temperature T and volume V and mass m are the few well-known properties. All properties are not independent. Few are defined in terms of others for example density.

Properties of a system can be of two types:

(i) Intensive Properties

The properties which do not depend on the mass of the system are called as intensive properties. For examples pressure, temperature, velocity, viscosity etc are intensive properties. These properties are independent of the size of the system.

(ii) Extensive Properties

The properties which depend on the mass of the system are called as extensive properties. For examples total volume, energies of all types and surface area are the extensive properties.

Generally upper case letters are used to represent the extensive properties (with mass being exception) and lower case letters are used to represent the intensive properties (with pressure and temperature being the exceptional cases).

The properties can also classified as

(*a*) **Fundamental property:** The fundamental property is one which can be measured directly. For example pressure, temperature and volume are the fundamental properties.

(*b*) **Thermodynamic Property:** The thermodynamic property is one which can not be measured directly but its value is determined through calculation. For example: work, energy, enthalpy, entropy and internal energy are the few thermodynamic properties.

The prefix–**iso** is often used to designate a process for which a particular property remains constant.

Name of the Process	***Property which Remains Constant During the Process***
Isothermal process	temperature
Isobaric process	pressure
Isochoric or isometric process	specific volume
Isentropic process	entropy

The important feature of a property is that it has a definite numerical value when the system is in a particular state and that this value can be determined by measurement without knowing the past or future history of the system. Since the value of a property is independent of the process through which the system has passed in reaching a particular state, it follows that a change in the value of a property depends only on the initial and final states of the system.

Mathematically, this is stated as follows:

If ψ is a quantity which represents a property of a system, then $d\psi$ is an exact differential and the total change between states 1 and 2 of the system is given by

$$\int_1^2 d\psi = \psi_2 - \psi_1 \qquad ...(5.8)$$

5.9 CONTINUUM

Matter is made up of the atoms that are widely spaced in the gas phase. Yet it is very convenient to disregard the atomic nature of a substance and view it as a continuous homogeneous matter with no holes. The continuum idealization enables us to treat properties as point functions and to assume the properties to vary continually in space with no jump discontinuities.

5.10 PROCESSES AND CYCLES

Any change that a system undergoes from one equilibrium state to another is called a process. In order to describe a process completely, one should specify the initial and final states of the process, as well as the path it follows and the interactions with the surroundings.

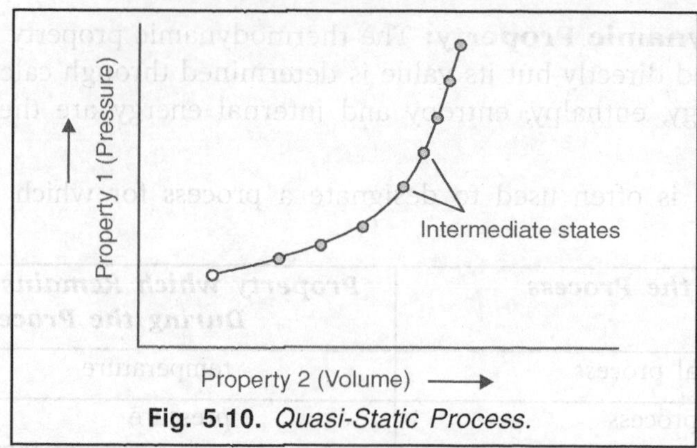

Fig. 5.10. *Quasi-Static Process.*

When a process proceeds in such a manner that the system remain infinitesimally close to an equilibrium state at all times, it is called a quasi-static process (Fig. 5.10). Being a slow process, it allows the system to adjust itself internally so that properties in one part of the system do not change quickly than those at other parts. One should keep in mind that a quasi-static process is an idealized process and is not a true representation of an actual process. But many actual processes closely approximate it and they can be modeled as quasi-static process.

Quasi-static process is of great interest to engineers related to thermal branch for two reasons.

(*i*) It is easy to analyze.

(*ii*) Devices operating on this process produce more work.

These reasons make a quasi-static process to serve as standard to which real life processes can be compared.

5.10.1 Cyclic Process

In cyclic process, the state of the system at the end of the process is the same as it was at the beginning.

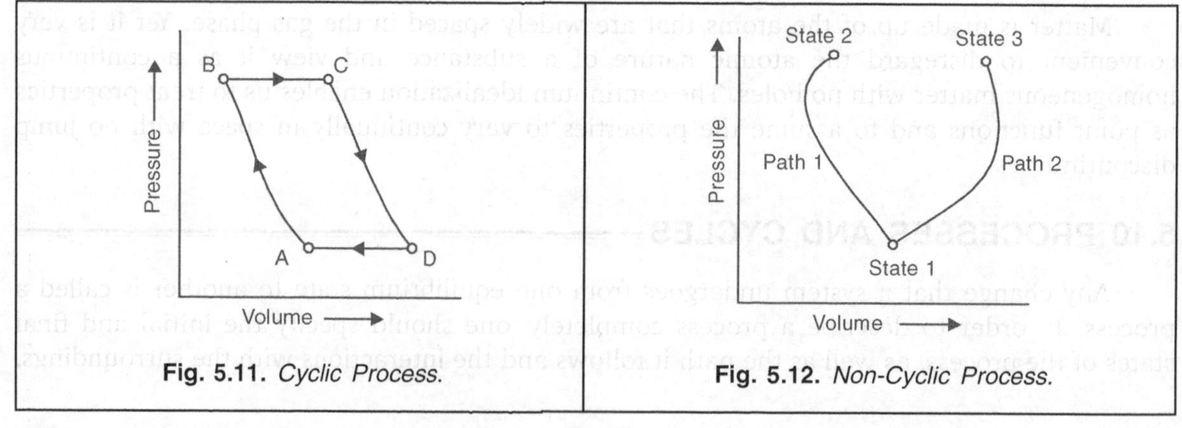

Fig. 5.11. *Cyclic Process.*

Fig. 5.12. *Non-Cyclic Process.*

5.10.2 Non-cyclic Process

Non-cyclic process as shown in Fig. 5.12 is one in which the state of the system at the end is different from the starting one.

5.10.3 Reversible Process

A reversible process may be defined as one during which no portion of the system or its surroundings ever departs from a state of equilibrium by more than an infinitesimal amount and no friction of any kind occurs.

No after-effects or changes will be evident in the system or in the environment when a truly reversible process occurs in forward and then in reverse direction.

5.10.4 Irreversible Process

An **irreversible** process is a process that is not reversible.

Any energy transferred, such as mechanical, thermal, and chemical, would have to be transformed reversibly in case of reversible process. Thus processes involving dissipative effects such as fluid flow, in which friction losses occur and convective heat transfer in which there is turbulent flow cannot be reversible. For a heat transfer to occur there must be temperature difference. Consider any process in which heat is transferred from system A to system B. It means that system A is at higher temperature than system B. The process will be a reversible process if heat also flows from system B to system A. But this is impossible since system B is at lower temperature than A and heat never flows in that direction without making a change in the universe. Therefore it can be stated that actual processes involving heat interaction are irreversible.

5.10.5 Steady Flow Process

A steady flow process is defined as a process during which a fluid flows through a control volume steadily. That is the fluid properties can change from point to point within the control volume but at any fixed point they remain the same during the entire process.

5.11 WORKING SUBSTANCE

The matter contained within the boundaries of the system can be liquid, Vapour or gas and is referred to as the working substance or working fluid. A substance that has a fixed chemical composition throughout is called a pure substance. Water, nitrogen, helium and carbon dioxide for example are all pure substances. Clean atmospheric air is a pure substance as long as it is in gaseous state. The concept of pure substance is useful in determining the properties of the working substance at various conditions of temperature and pressure.

In most of the mechanical devices like heat engines, heat pumps, and refrigerators, the working substance is gas or vapour or vapour and liquid in equilibrium. In order to understand their working it is essential to know the behaviour and properties of working substances. For example steam power generating plant uses water vapour as working substance and refrigerator uses ammonia or hydrocarbons (HC-600a) as the working substance.

5.12 HEAT AND WORK

Energy may be transformed into and out of a system because of a temperature difference between the system and the surroundings. This mode of energy transfer is called the heat or heat transfer. Heat is defined as the action of energy transfer that occurs on the boundary of a system due to temperature differences between the system and its surroundings.

Energy may also be transformed in and out of a system without involving temperature difference between it and its surroundings. For example, in case of the forging of the metal, the energy is transformed by a force. This type of energy transfer is called work or work transfer.

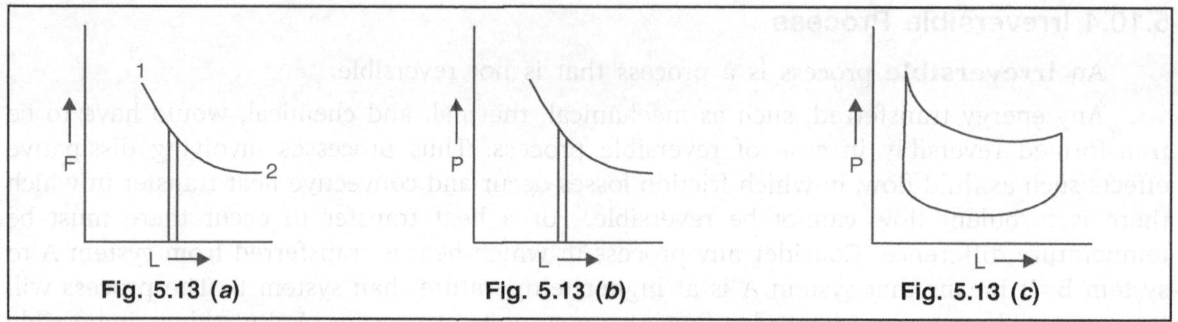

Fig. 5.13 (*a*) **Fig. 5.13 (*b*)** **Fig. 5.13 (*c*)**

Figure 5.13 (*a*) represents work in terms of force and distance whereas Fig. 5.13 (*b*) in terms of pressure and volume and Fig. 5.13 (*c*) depicts a reversible cycle on p-V diagram.

The concept of work as studied in a course on mechanics defines the mechanical work as the product of force and the distance moved in the direction of application of force. But from the thermodynamics point of view, work can also arise from other effects.

In Fig [5.13(*a*)] the force F is shown as varying with the distance moved. The work performed in this case is represented graphically by the area under the force curve 1–2 and its amount can be determined by integrating the areas of all the small segments *FdL*.

Mathematically it can be expressed as

$$\delta W = F.dL$$

$$= (pA)\ dL$$

$$= pdV \qquad \text{since } dV = AdL$$

$$_1W_2 = \int_{L_1}^{L_2} FdL \qquad \text{...(5.9)}$$

If the unit of force is Newton and that of distance in metre, then the unit of work will be N–m. In the study of thermodynamics it is easy to represent work in terms of pressure and volume coordinates rather than in terms of force and distance. The process as shown in Fig [5.13 (*b*)] involves the integration of the elements of the area *pdV* and the expression for the work becomes

$$_1W_2 = \int_{V_1}^{V_2} pdV \qquad \text{...(5.10)}$$

Since the unit of pressure is in N/m^2 and that of volume is m^3 therefore the unit of work will be N–m.

When a reversible process is represented on a diagram of properties it forms a closed figure and on pressure volume diagram, the area of the diagram gives the net work as shown in Fig [5.13 (*c*)].

5.13 WORK AS A PATH FUNCTION

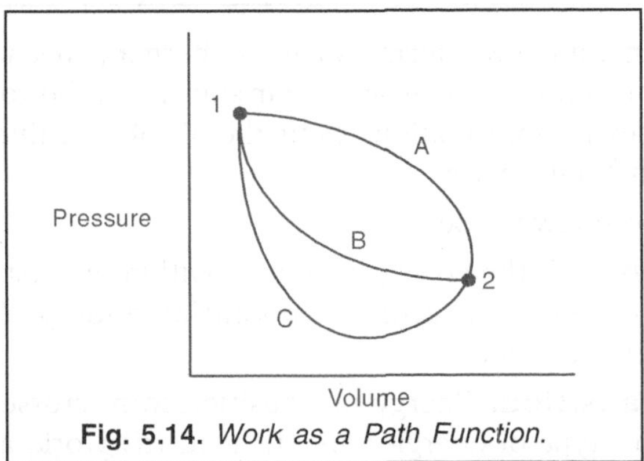

Fig. 5.14. *Work as a Path Function.*

Different reversible processes leading from state 1 to state 2 are shown in Fig. (5.14). The area underneath the corresponding path represents the work interaction in each case. The workdone is obviously different in each case because the path depends on the nature of the process. This means that work is not a property or state function, rather it is a path function. Therefore system does not possess work, instead work is a mode of transfer of energy. This transfer occurs only at the boundaries of the system as the system undergoes a change of state. That is why work is expressed by

$$\int_1^2 dW = W_{1-2} \text{ and not by } W_2 - W_1 \qquad \text{...(5.11)}$$

5.14 SPECIFIC HEATS

The specific heat of a substance is defined as the energy required to raise the temperature of a unit mass of the substance by one degree. Its unit is J/kg K. It is usually measured at either constant volume or constant pressure.

5.14.1 Specific Heat at Constant Volume (C_v)

C_v is defined as the quantity of heat energy required to raise the temperature of a unit mass of a substance by one degree with the volume remaining constant.

5.14.2 Specific Heat at Constant Pressure (C_p)

C_p is defined as the quantity of heat energy required to raise the temperature of a unit mass of a substance by one degree with the pressure remaining constant.

5.15 ENERGY

Engineering processes involve the conversion of energy from one form to another, transfer of energy from place to place and the storage of energy in various forms using a working substance.

☞ Energy is defined as the capacity to do the work. Changes in the properties of a system occur whenever energy is added or subtracted from a system.

Energy can exist in numerous forms such as thermal, mechanical, kinetic, potential, electric, magnetic, chemical and nuclear and their sum gives the total energy E of a system. Thermodynamics provides no information about the absolute value of total energy. It deals only with the change in total energy.

The energy may be of two types:

(*a*) **Stored energy:** It is the energy contained within the system boundaries. Examples of this type of energies are: (*i*) potential energy, (*ii*) kinetic energy and (*iii*) Internal energy, etc.

(*b*) **Energy in transition:** Energy in transition form crosses the system boundaries. Examples of this type of energies are (*i*) heat, (*ii*) work, and (*iii*) electrical energy, etc.

Further there are two different modes of energy namely macroscopic and microscopic. A system which possesses energy with respect to some outside source belongs to the macroscopic form of energy. Examples of this type of energy are (*i*) kinetic energy, and (*ii*) potential energy. The microscopic form of energy is not dependent on the outside reference source but it is related to the molecular structure of a system and degree of the molecular activity. The sum of all the microscopic forms of energy is called the internal energy of a system.

5.15.1 Potential Energy (PE)

It is defined as the energy possessed by a body by virtue of its position.

$$PE = mgh$$

Where m is mass of the body in kg

h is height of the body in metres at a particular datum line in m

g is acceleration due to gravity in m/s^2

$$PE = mgh \quad \text{...(5.12)}$$

☞ The unit of PE in SI units is Nm or joule.

One can get this unit directly from its equation as described below.

$$PE = mgh = kg \times m \times \frac{m}{s^2} = \text{Nm (as we know 1 N = 1 kg m/s}^2\text{)}$$

5.15.2 Kinetic Energy (KE)

It is defined as the energy possessed by a body by virtue of its motion.

Kinetic energy of a body of mass m moving with velocity v is given by

$$KE = \frac{1}{2}mv^2 \quad ...(5.13)$$

It should be noted here that the potential and kinetic energies are external forms of energy.

5.15.3 Internal Energy (IE)

The internal energy U of a substance may be defined as the algebraic sum of internal kinetic energy and the internal potential energy of its molecules. It is usually represented by U or E.

Internal energy of a substance is the energy stored within the substance and it is due to the movement and arrangement of its molecules. As we know that if the temperature of a gas is increased, the molecular activity increases. Therefore, the internal energy is a function of temperature and its value can be increased or decreased by adding or subtracting heat to or from the substance.

The absolute value of internal energy cannot be determined but one can find out the changes in internal energy from changes of the state of the system.

The change in internal energy can be expressed in a simple way when the substance passes from state 1 to state 2.

i.e.,
$$\Delta U = U_2 - U_1$$

The change in internal energy of a fluid (or a gas) of mass m when its temperature changes from T_1 to T_2 is given by the equation

$$\Delta U = mC_v(T_2 - T_1) \text{ kJ} \quad ...(5.14)$$

where

C_v is the specific heat at constant volume of the gas.

5.15.4 Total Energy (E)

Total Energy possessed by a mass is the sum of all forms of internal and external energy. It is denoted by letter 'E' and its unit is Joule.

$$\text{Total Energy} = PE + KE + IE \quad ...(5.15)$$

$$E = mgh + \frac{1}{2}mv^2 + U$$

Problem 3. *A fighter plane is flying at an altitude of 7000 m at a speed of 350 km per hour. Find out the kinetic and potential energy of the fighter plane if the mass of the plane is 5150 kg?*

Solution. Given

Mass of the plane $m = 5150\ kg$

Speed of the plane $v = 350\ km/hr = 350 \times \frac{1000}{3600}\ m/s = 97.22\ m/s$

$$\text{Kinetic energy} = KE = \frac{1}{2}mv^2 = \frac{1}{2} \times 5150 \times (97.22)^2 = 24.33\ MJ$$

$$\text{Potential energy} = PE = mgh = 5150 \times 9.81 \times 7000 = 353.65\ MJ$$

5.16 POWER

☞ Power is defined as rate of doing work.

The unit of power is joule/s or watt (W).

$$1\ \text{W} = 1\ \text{J/s} = 1\ \text{Nm/s}$$

$$1\ \text{kW} = 1000\ \text{W}$$

$$1\ \text{MW} = 10^6\ \text{W}$$

5.17 ENTHALPY

Enthalpy is defined as the sum of internal energy and flow work. It is denoted by H.

$$H = U + pV \quad \text{...(5.16)}$$

Where

U is the internal energy

p is the absolute pressure

V is the volume

It is not possible to measure the absolute enthalpy of a substance, as absolute value of internal energy cannot be determined. Therefore the change in enthalpy can be determined.

$$\therefore \text{ Change of enthalpy, } \Delta H = H_2 - H_1 = m\ C_p\ (T_2 - T_1)\ \text{kJ} \quad \text{...(5.17)}$$

Where C_p is the specific heat at the constant pressure of the gas.

5.18 LAW OF CONSERVATION OF ENERGY

It states that energy can neither be created nor destroyed. Energy can be changed from one form into another. For example: when brake is suddenly applied in a bicycle, kinetic energy of the wheel is converted into heat as the wheel rim temperature increases due to friction between wheel rim and brake shoe materials.

5.19 LAW OF CONSERVATION OF MASS

This law states that the mass can neither be created nor destroyed but can only be changed from one form to another form. For example, a substance can be changed from solid to liquid and liquid to gas or vice-versa.

5.20 GAS LAWS

5.20.1 Boyle's Law

It states that "when mass and temperature of a perfect gas remain constant the volume of a given mass of a perfect gas is inversely proportional to the absolute pressure".

Let p = absolute pressure of gas

V = Volume of gas at pressure p

T = Absolute temperature of gas

Expressing Boyle's law mathematically we can write

$$V \; \alpha \; \frac{1}{p}$$

$$pV = \text{constant} \qquad ...(5.18)$$

or $$p_1V_1 = p_2V_2$$

The subscript 1 denotes the original values and subscript 2 the final values. The graph obtained from the experimental results conducted for verifying this law gives the rectangular hyperbola and because of this it is also known as hyperbolic law.

5.20.2 Charles Law

This law states that if mass and pressure of a perfect gas remain constant, "the volume of a given mass of gas is directly proportional to the absolute temperature".

Expressing Charles's law mathematically we can write

$$V \; \alpha \; T$$

or $$\frac{V}{T} = \text{Constant} \qquad ...(5.19)$$

If the process occurs from state 1 to state 2, we can write

$$\frac{V_1}{T_1} = \frac{V_2}{T_2}$$

5.20.3 Characteristic Equation of Gas

Characteristic equation of a gas can be obtained by combining the Boyle's and Charle's laws. This law provides a general equation between pressure, temperature and volume of a given mass of gas.

$$V \propto \frac{1}{p} \text{ when } T \text{ is constant} \qquad ...(5.20)$$

$$V \propto T \text{ when } p \text{ is constant} \qquad ...(5.21)$$

$$\therefore \quad V \propto \frac{T}{p} \text{ when both } T \text{ and } p \text{ vary} \qquad ...(5.22)$$

For 1 kg of a gas

$$\frac{pV}{T} = R \text{ where } R \text{ is a constant}$$

$$pV = RT \text{ for unit mass}$$

when m kg of gas is considered then

$$pV = mRT \qquad ...(5.23)$$

where **R** is called the characteristic gas constant and equation (5.23) is called the characteristic equation of a gas.

The unit of **R** is $\frac{Nm}{kg\ K}$ or $\frac{kJ}{kg\ K}$

$$R = 287\ J/kg\ K \text{ for air.}$$

REVIEW QUESTIONS

1. What is meant by system and surrounding? Explain
   (*i*) closed system (*ii*) open system (*iii*) isolated system.
2. Define the following:
   (*i*) state (*ii*) property (*iii*) path
   (*iv*) process (*v*) cycle.
3. Explain the concept of thermodynamic equilibrium.
4. What is energy? Describe its various forms.
5. Define enthalpy. Derive the expression for the heat supplied in a constant pressure process.
6. Define internal energy and prove that it is a property of system.
7. Give the thermodynamic definition of work. Differentiate between heat and work.
8. What do you mean by reversible and irreversible process? Give one example of each.
9. State and explain the Zeroth law of thermodynamics.
10. Why does the enthalpy of an ideal gas depend only on temperature?
11. Explain the concept of control volume.
12. Define a quasi-static process. Explain its physical significance.

■■■

Laws of Thermodynamics and Entropy

6.1 THE FIRST LAW OF THERMODYNAMICS

The first law of thermodynamics is same as the law of conservation of energy, which states that "*energy can neither be created nor destroyed*". Energy however can be converted from one form into another form. The first law of thermodynamics is based on the relationship between heat and work.

> *The first Law of Thermodynamics:*
>
> ☞ For a closed system under a cyclic process, the algebraic sum of work done is proportional to the algebraic sum of heat transferred.

Mathematically

$$\oint dW \; \alpha \oint dQ$$

$$\oint dW = J\oint dQ$$

$$\oint dW - J\oint dQ = 0$$

Where,

J is mechanical equivalent of heat

$$J = 427 \; kgm/kcal \qquad \text{(Mks units)}$$

In SI units both heat and work are measured in same unit, that is joules.

For a non-cyclic process for example if heat transferred is more than the work done than this may bring a change in some of the properties of the system and the system would go under a change of state. The difference between these transferred energies Q and W has close impact on the internal energy of the system.

Mathematically

$$Q - W = \Delta U$$

or

$$Q = W + \Delta U$$

where, ΔU is the change in internal energy of the system.

In differential form this can be written as

$$dQ = dU + dW \qquad \text{...(6.1)}$$

The first law of thermodynamics as applied to closed system processes.

6.1.1 Corollary–1

☞ There exists a property of a system called energy such that a change in its value is equal to the difference between the heat supplied and the workdone during any change of state.

$$Q - W = \Delta E = E_2 - E_1 \quad ...(6.2)$$

Where,

E_1 Energy of the system at state 1.

E_2 Energy of the system at state 2.

6.1.2 Energy is a Property

Let us consider a system which is undergoing a cyclic process changing through state 1 to state 2 by the process A and returns from state 2 to state 1 by the process B or C as shown in Fig. 6.1.

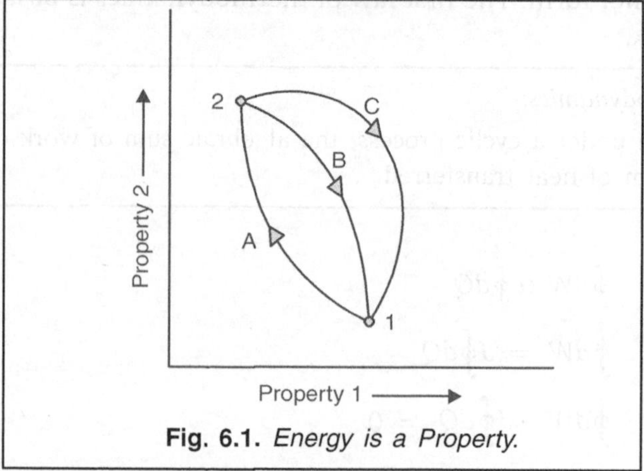

Fig. 6.1. *Energy is a Property.*

As we know from the first law of thermodynamics

$$\oint dW = \oint dQ$$

Applying this law to process 1A2B1 we can write

$$\oint_{A1}^{2} dQ + \oint_{B2}^{1} dQ = \oint_{A1}^{2} dW + \oint_{B2}^{1} dW \quad ...(6.3)$$

Next considering the cyclic process 1A2C1 we can write

$$\oint_{A1}^{2} dQ + \oint_{C2}^{1} dQ = \oint_{A1}^{2} dW + \oint_{C2}^{1} dW \quad ...(6.4)$$

Subtracting equation (6.3) from (6.4) we get

$$\oint_{B2}^{1} dQ + \oint_{C2}^{1} dQ = \oint_{B2}^{1} dW - \oint_{C2}^{1} dW$$

or

$$\oint_{B2}^{1} (dQ - dW) = \oint_{C2}^{1} (dQ - dW) \quad ...(6.5)$$

The quantity $(dQ - dW)$ is the same for any process between state 1 and state 2. The value of the integral $(dQ - dW)$ depends on the end states and is independent of the process path. Hence it is a point function and is an exact differential and thus a thermodynamic property.

It is generally designated by letter 'E'.

The energy E is the sum of potential energy (PE), kinetic energy (KE), and internal energy (IE), Internal energy is designated by letter 'U'.

$$E = PE + KE + IE$$

$$= PE + KE + U$$

The internal energy cannot be measured directly. It is determined by inference.

$$dE = d(PE) + d(KE) + dU$$

The *PE* and *KE* terms are zero for a closed or non-flow thermodynamic system. Therefore energy is equal to internal energy.

Hence equation (6.2) becomes

$$Q - W = \Delta U$$

$$Q - W = U_2 - U_1 \qquad \text{...(6.6)}$$

Where $U_2 - U_1$ represents the change in internal energy

6.1.3 Corollary–2

☞ In an isolated system the energy of the system remains constant.

For an isolated system both Q and W are zero, hence change in energy is zero.

when

$$Q = 0$$

$$W = 0$$

Hence, $\Delta E = 0$

or E is Constant.

6.1.4 Corollary–3

☞ A perpetual motion machine of the first kind (PMM1) is impossible.

A perpetual motion machine of the first kind *i.e.,* PMM1 is a hypothetical machine producing a continuous supply of work without consuming energy from the surroundings.

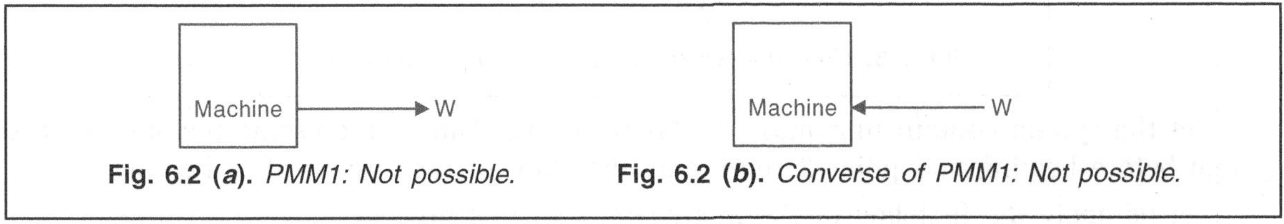

Fig. 6.2 (*a*). *PMM1: Not possible.* **Fig. 6.2 (*b*).** *Converse of PMM1: Not possible.*

6.2 APPLICATION OF FIRST LAW TO NON-FLOW PROCESSES

What do you mean by a non-flow process?

The laws concerned with the relationships between pressure, volume and temperature and discussed below are applicable only to an ideal or perfect gas.

☞ An ideal and perfect gas is one in which there is no interaction between the molecules of a gas. Examples of the perfect gases are air, N_2, H_2 and H_e.

In a non-flow process, the substance remains in a closed system. As we know that in a closed system mass of substance does not leave the system but only energy crosses the boundary in the form of heat and work.

Non-flow processes may be reversible and irreversible.

Under this category following processes are discussed.

6.2.1 Constant volume process (Isochoric/Isometric)

6.2.2 Constant pressure process (Isobaric)

6.2.3 Constant temperature process (Isothermal)

6.2.4 Adiabatic or constant entropy process (Isentropic)

6.2.5 Polytropic process.

6.2.1 Constant Volume Process

A process in which the volume of the system remains constant during its execution is known as a constant volume process (Fig. 6.3).

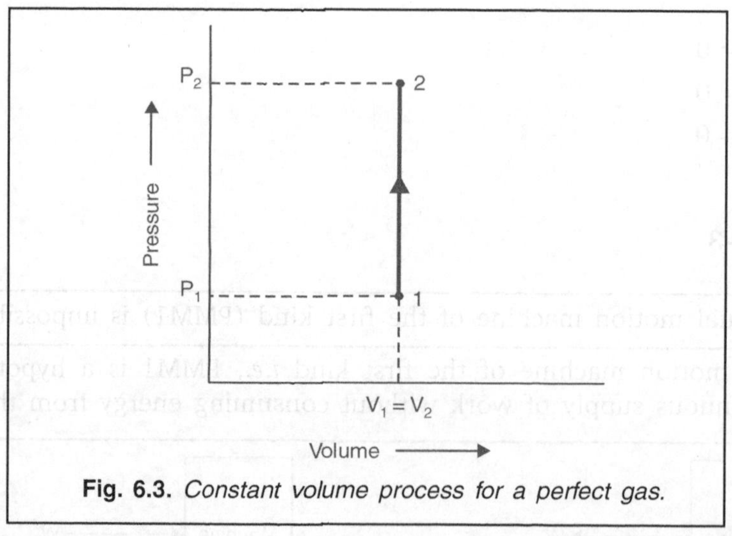

Fig. 6.3. *Constant volume process for a perfect gas.*

Let the system contain unit mass of a perfect gas. Point 1 represents the state of the system before heating and point 2 represents the state of the system after heating.

Let us apply the first law of thermodynamics to this process.

$$dQ = dW + dU$$

In the constant volume process, the volume does not change [no area under the curve in Fig. 6.3] and hence no work is done *i.e.,* $dW = 0$.

As $\quad dV = 0$

$$\therefore \quad dW = \int p dV = p(V_2 - V_1) = 0$$

$$\therefore \quad dQ = dU \qquad \text{...(6.7)}$$

From equation (6.7) it is clear that during constant volume process, the heat transfer equals the change in internal energy of the system. Thus the addition of heat increases the internal energy and loss of heat decreases the internal energy of the system.

$$\therefore \quad dQ = dU = C_v dT \qquad \text{...(6.8)}$$

$$\int_1^2 dU = \int_1^2 C_v dT$$

Heat Supplied

$$Q_{1-2} = m\, C_v[(T_2 - T_1)] \qquad \text{...(6.9)}$$

6.2.2 Constant Pressure Process

A process in which the pressure of the system remains constant is known as constant pressure process. The process is shown in Fig. 6.4.

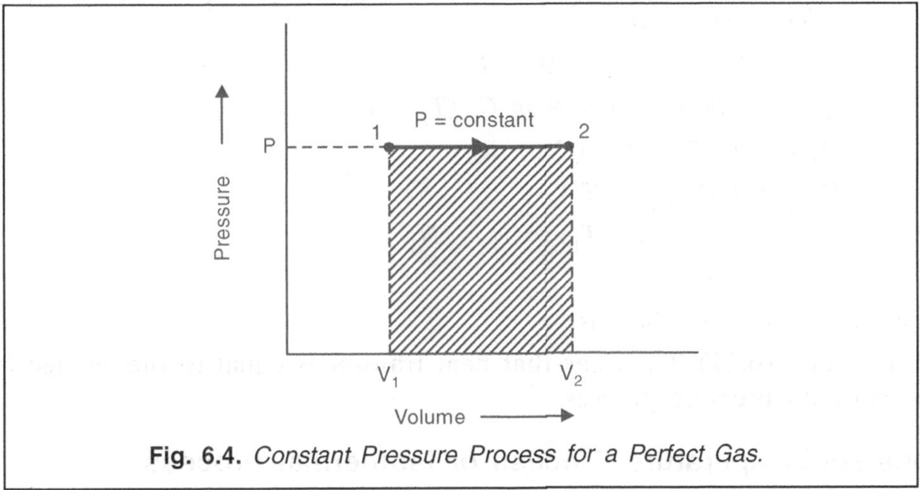

Fig. 6.4. *Constant Pressure Process for a Perfect Gas.*

The constant pressure or isobaric process is carried out in such a way that the pressure remains constant throughout the process. The work done is represented by the shaded area as shown in Fig. 6.4.

Let a system contains unit mass of a perfect gas. Point 1 represents the state of the system before heating and point 2 represents the state of the system after heating. If heat is supplied to a gas under constant pressure, the volume of the gas will increase in direct proportion to the change in absolute temperature.

Applying the characteristic gas equation to this process, we can write

$$\frac{p_1 V_1}{T_1} = \frac{p_2 V_2}{T_2}$$

Since $p_1 = p_2$

We get

$$\frac{V_1}{T_1} = \frac{V_2}{T_2}$$

or $$\frac{V}{T} = \text{Constant}$$

Let us apply the first law of thermodynamics to this process

$$dQ = \delta W + dU$$

Let Q be the amount of heat to be supplied to the system during the process from state 1 to 2.

Work done by the system

$$\delta W = \int pdV$$

or $$W_{1-2} = \int_1^2 pdV = p(V_2 - V_1)\text{kJ} \qquad ...(6.10)$$

Heat transfer

$$dQ = \delta W + dU$$

$$Q_{1-2} = p(V_2 - V_1) + U_2 - U_1$$

$$Q_{1-2} = mR(T_2 - T_1) + m\,C_v\,(T_2 - T_1)$$

$$Q_{1-2} = m(T_2 - T_1)\,(C_v + R)$$

$$Q_{1-2} = mC_p\,(T_2 - T_1) \qquad ...(6.11)$$

$$Q_{1-2} = mC_p\,(T_2 - T_1) = H_2 - H_1$$

$$dQ = dH$$

Where H is the enthalpy of the gas.

From equation (6.11) it is clear that heat transfer is equal to the change in enthalpy during the constant pressure process.

6.2.3 Constant Temperature Process or Isothermal Process

A process in which temperature of the system remains constant is known as constant temperature or isothermal process. It is very difficult to execute the isothermal process in practice.

An isothermal process is expressed by $pV = C$ where C is a constant. When this process is plotted on PV diagram (Fig. 6.5) it takes the form a rectangular hyperbola.

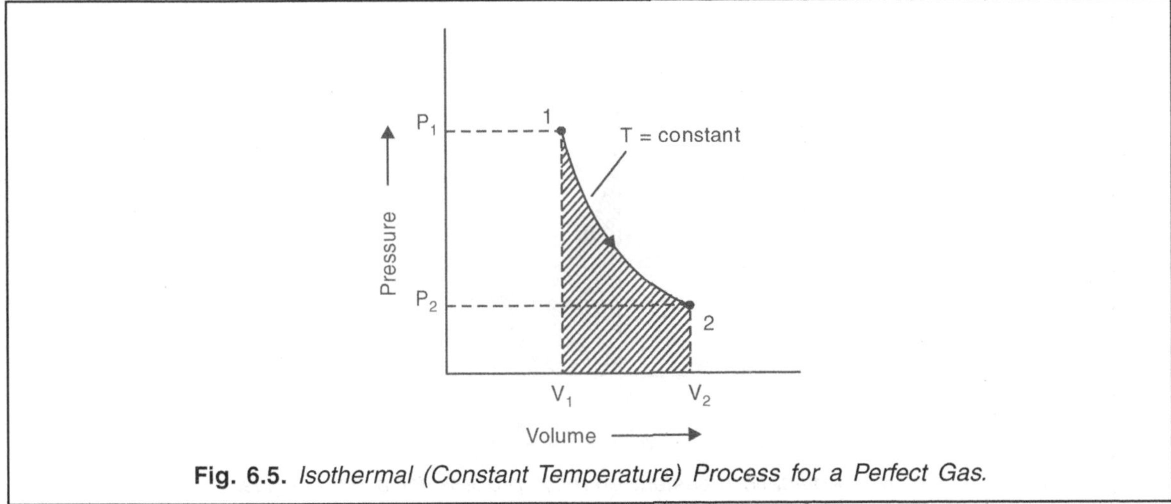

Fig. 6.5. *Isothermal (Constant Temperature) Process for a Perfect Gas.*

As there is no change in the temperature during the process, hence there is no change in the internal energy of the system.

Work done during the process

Let us apply the first law of thermodynamics to this process

$$dQ = \delta W + dU$$

$\because \quad dU = 0$ as temperature during the process is constant and $\delta W = \int pdV$ on substituting $p = \dfrac{C}{V}$

$$\therefore \quad W_{1-2} = C\int_1^2 \frac{dV}{V} = C\left[\ln V\right]_{V_1}^{V_2} = C \ln\left[\frac{V_2}{V_1}\right]$$

Let us now apply the ideal gas equation

$$pV = mRT$$

$$\therefore \quad W_{1-2} = pV \ln\left[\frac{V_2}{V_1}\right] = mRT \ln\left(\frac{V_2}{V_1}\right) \qquad ...(6.12)$$

Heat transferred

$$dQ = \delta W + dU$$

$$\because \quad dU = 0$$

$$\because \quad dQ = \delta W$$

$$\therefore \quad Q_{1-2} = W_{1-2} = pV \ln\left[\frac{V_2}{V_1}\right] = mRT \ln\left(\frac{V_2}{V_1}\right) \qquad ...(6.13)$$

6.2.4 Constant Entropy Process or Adiabatic Process or (Isentropic Process)

An adiabatic process as shown in Fig. (6.6) is one in which the heat transfer to or from the system is zero.

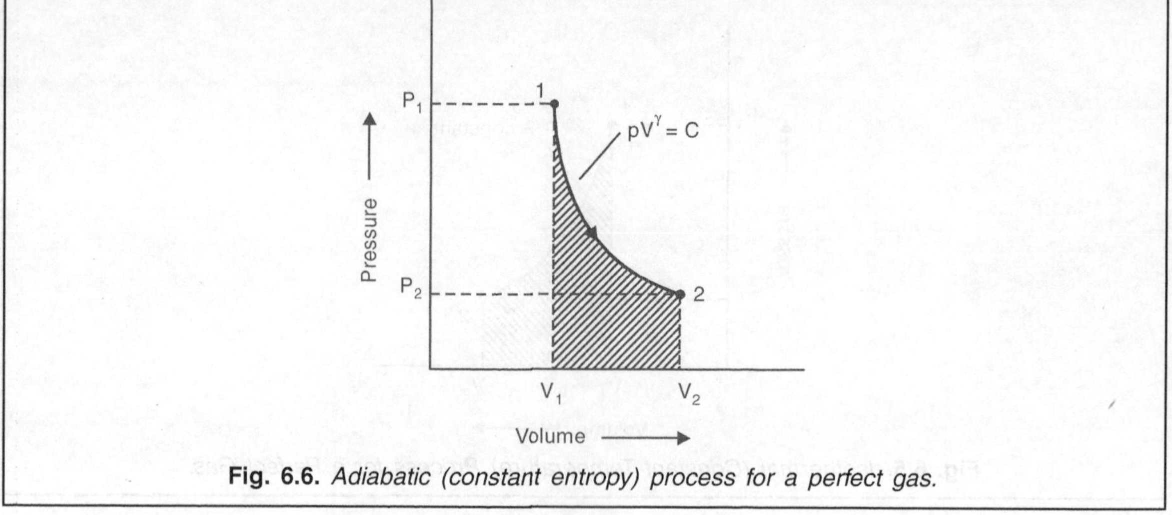

Fig. 6.6. *Adiabatic (constant entropy) process for a perfect gas.*

Let us apply the first law of thermodynamics to this process

$$dQ = \delta W + dU.$$

Since $dQ = 0$

$\therefore$ $\delta W = - dU$

Thus work is done by the system at the cost of internal energy

$\therefore$ $\delta W = - dU$

$$pdV = - dU$$

$$pdV + dU = 0$$

$$pdV + mC_v dT = 0 \quad ...(6.14)$$

Using perfect gas equation $pV = mRT$ in differential from, we get

$$pdV + Vdp = mRdT \quad ...(6.15)$$

It should be noted here that pdV, Vdp and RdT are expressed in the same unit.

or

$$dT = \frac{pdV + Vdp}{mR} \quad ...(6.16)$$

Substituting the value of dT from equation (6.16) into equation (6.14), we get

$$pdV + mC_v \frac{pdV + Vdp}{mR} = 0 \quad ...(6.17)$$

$$pdV \times R + C_v(pdV + Vdp) = 0 \quad ...(6.18)$$

As we know $R = C_p - C_v$ and substituting this value in (6.18).

We get

$$pdV \times (C_p - C_v) + C_v (pdV + Vdp) = 0$$

$$C_p\, pdV - C_v\, pdV + C_v\, pdV + C_v\, Vdp = 0$$

$$C_p\, pdV + C_v\, Vdp = 0 \quad ...(6.19)$$

Dividing both sides of the equation (6.19) by $C_v\ pV$, we get

$$\frac{C_p p\ dV}{C_v\ pV} + \frac{C_v V\ dp}{C_v\ pV} = 0$$

$$\frac{C_p}{C_V}\frac{dV}{V} + \frac{dp}{p} = 0$$

$$\frac{dp}{p} + \gamma\frac{dV}{V} = 0 \quad (\text{as } C_p/C_v = r) \qquad ...(6.20)$$

Integrating equation (6.20), we get

$$\log_e P + \gamma \log_e V = \text{Constant}$$

$$\therefore \qquad pV^{\gamma} = \text{Constant} \qquad ...(6.21)$$

Therefore adiabatic process follows the law $\therefore\ pV^{\gamma} = C$

The equation (6.21) may also be written in the following forms for the two state points 1 and 2.

$$p_1V_1^{\gamma} = p_2V_2^{\gamma} = C$$

or

$$\frac{p_1}{p_2} = \left(\frac{V_2}{V_1}\right)^{\gamma}$$

or

$$\frac{V_2}{V_1} = \left(\frac{p_1}{p_2}\right)^{\frac{1}{\gamma}}$$

For a perfect gas we can write

$$\frac{p_1V_1}{T_1} = \frac{p_2V_2}{T_2}$$

$$\frac{T_2}{T_1} = \frac{p_2}{p_1}\frac{V_2}{V_1} \qquad ...(6.22)$$

Substituting the value of $\frac{V_2}{V_1} = \left(\frac{p_1}{p_2}\right)^{\frac{1}{\gamma}}$ in equation (6.22) we can write

$$\frac{T_2}{T_1} = \frac{P_2}{P_1}\left(\frac{p_1}{p_2}\right)^{\frac{1}{\gamma}}$$

$$\frac{T_2}{T_1} = \left(\frac{P_2}{P_1}\right)^{\gamma-1/\gamma} \qquad ...(6.23)$$

Substituting the value of $\frac{p_1}{p_2} = \left(\frac{V_2}{V_1}\right)^{\gamma}$ in equation (6.22) we can write

$$\frac{T_2}{T_1} = \left(\frac{V_1}{V_2}\right)^{\gamma} \frac{V_2}{V_1}$$

$$\frac{T_2}{T_1} = \left(\frac{V_1}{V_2}\right)^{\gamma-1} \quad ...(6.24)$$

Work done during the adiabatic process

$$W_{1-2} = \int_{V_1}^{V_2} pdV \quad ...(6.25)$$

but $$pV^{\gamma} = C$$

$\therefore$ substituting $p = \frac{C}{V^{\gamma}}$ in equation (6.25), we get

$$W_{1-2} = \int_{V_1}^{V_2} \frac{C}{V^{\gamma}} dV = C\int_{V_1}^{V_2} V^{-\gamma} dV = C\left[\frac{V_2^{-\gamma+1} - V_1^{-\gamma+1}}{1-\gamma}\right]$$

$$W_{1-2} = \left[\frac{CV_2^{-\gamma+1} - CV_1^{-\gamma+1}}{1-\gamma}\right] \quad ...(6.26)$$

on substituting the value of C from equation (6.25) $p_1V_1^{\gamma} = p_2V_2^{\gamma} = C$ in equation (6.26).

$$W_{1-2} = \left[\frac{p_2V_2^{\gamma}V_2^{-\gamma+1} - p_1V_1^{\gamma}V_1^{-\gamma+1}}{1-\gamma}\right]$$

$$W_{1-2} = \left[\frac{p_2V_2 - p_1V_1}{1-\gamma}\right]$$

or $$W_{1-2} = \left[\frac{p_1V_1 - p_2V_2}{\gamma-1}\right] \text{ Nm/kg or J/kg} \quad ...(6.27)$$

if p is expressed in N/m^2.

Equation (6.27) can also be expressed as $W_{1-2} = \left[\frac{R(T_1 - T_2)}{\gamma-1}\right]$ Nm/kg or J/kg. If R is expressed in Nm/kg K.

Heat transferred and heat supplied

As we know that heat transferred or heat supplied in case of an adiabatic process is zero, therefore

$$Q_{1-2} = 0 \quad ...(6.28)$$

6.2.5. Polytropic Process

Polytropic process is one in which energy for doing work is supplied partly from an external source and partly from the gas itself. This process follows a path that falls between those of isothermal and adiabatic process as shown in Fig. 6.7.

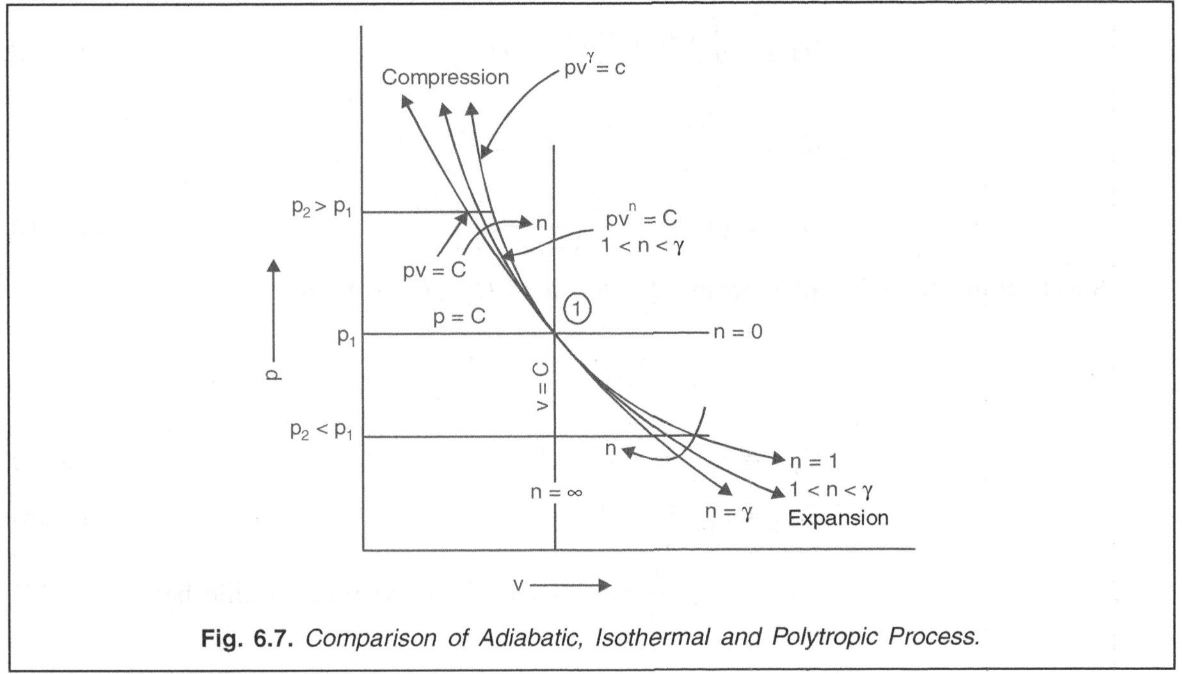

Fig. 6.7. *Comparison of Adiabatic, Isothermal and Polytropic Process.*

It is represented by

$$pV^n = C \quad ...(6.29)$$

Where n is called the polytropic index of compression or expansion.

By using the analogy with reversible adiabatic process presented before this process and replacing γ by n we can write following equations

$$p_1V_1^n = p_2V_2^n \quad ...(6.30)$$

$$\frac{T_2}{T_1} = \left(\frac{p_2}{p_1}\right)^{n-1/n} \quad ...(6.31)$$

$$\frac{T_2}{T_1} = \left(\frac{V_1}{V_2}\right)^{n-1} \quad ...(6.32)$$

Similarly work done during the polytropic process is given by

$$W_{1-2} = \left[\frac{p_1V_1 - p_2V_2}{n-1}\right] = \frac{mR(T_1 - T_2)}{n-1} = \frac{mR(T_2 - T_1)}{1-n} \quad ...(6.33)$$

Heat transferred or heat supplied during the polytropic process

From first law of thermodynamics

We can write

$$dQ = \delta W + dU \quad \text{...(6.34)}$$

$$Q_{1-2} = \int_1^2 p\,dV + C_v(T_2 - T_1) \quad \text{...(6.35)}$$

$$Q_{1-2} = \frac{R(T_2 - T_1)}{1-n} + C_v\,(T_2 - T_1)$$

$$Q_{1-2} = (T_2 - T_1)\left[\frac{R}{1-n} + C_v\right] \quad \text{...(6.36)}$$

Substituting the value of R from equation $R = C_p - C_v$, we get

$$Q_{1-2} = (T_2 - T_1)\left[\frac{C_p - C_v}{1-n} + C_v\right]$$

$$Q_{1-2} = \frac{C_p - nC_v}{1-n}(T_2 - T_1) \quad \text{...(6.37)}$$

$$Q_{1-2} = C_n\,(T_2 - T_1) \quad \text{...(6.38)}$$

where, $C_n = \dfrac{C_p - nC_v}{1-n}$ is called the polytropic specific heat ...(6.39)

On substituting $\gamma = \dfrac{C_p}{C_v}$, we get

$$C_n = \left(\frac{\gamma - n}{1-n}\right) C_v \quad \text{...(6.40)}$$

Heat transferred during the polytropic process is given by

$$Q_{1-2} = \left(\frac{\gamma - n}{1-n}\right) C_v\,(T_2 - T_1) \text{ kJ/kg} \quad \text{...(6.41)}$$

Problem 1. *A system while going through a cycle receives 260 kJ of heat at constant volume and rejects 270 kJ of heat at constant pressure and 70 kJ of work is done on the system [Fig. (6.8)]. The system cycle is brought to its initial state by an isentropic process. If the initial internal energy is 240 kJ then find out (a) the internal energy at all the points, (b) the isentropic work.*

Solution. Considering process A-B, we can find out internal energy at point B.

As mentioned in the problem the process A – B is a constant volume heat addition process. As we know during a constant volume process, the heat transfer equals the change in internal energy of the system.

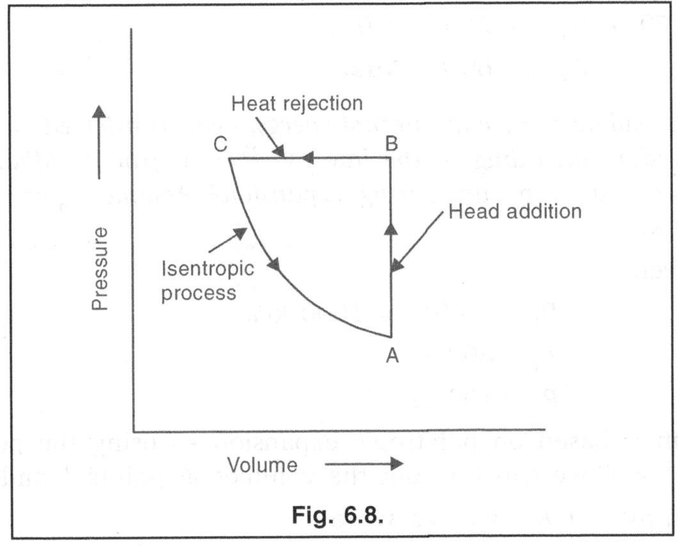

Fig. 6.8.

Initial internal energy is given as $U_A = 240$ kJ

$$dQ = du$$

$$Q_{A-B} = U_B - U_A$$

$$260 = U_B - 240$$

$$\therefore \quad U_B = 260 + 240 = 500 \text{ kJ } \textbf{Ans.}$$

⇒ Considering process $B - C$, we can find out internal energy at point C.

The process B–C is a constant pressure heat rejection process. As we know during constant pressure, the heat transfer is given by

$$dQ = \delta W + dU$$

Applying this equation to process $B - C$, we get

$$Q_{B-C} = W_{B-C} + U_C - U_B$$

As given in the problem

$$Q_{B-C} = -270 \text{ kJ}$$

$$W_{B-C} = -70 \text{ kJ}$$

$$-270 = -70 + U_C - 500$$

$$\therefore \quad U_C = 300 \text{ kJ } \textbf{Ans.}$$

⇒ Considering process C-A.

As process C–A is isentropic (or adiabatic) process $\therefore Q_{C-A} = 0$.

Applying the first law of thermodynamics we can write

$$\oint dW = \oint dQ$$

$$W_{A-B} + W_{B-C} + W_{C-A} = Q_{A-B} + Q_{B-C} + Q_{C-A}$$

$$0 - 70 + W_{C-A} = 260 - 270 + 0$$

$$W_{C-A} = 60\ kJ \textbf{ Ans.}$$

Problem 2. *Calculate change in internal energy, heat transferred and change in enthalpy for 1.5 kg air expanding according to the law $pV^{1.25} = C$ from 1 MPa, 600K to 100 kPa. What will be the work done by air during expansion? Assume $C_p = 1.005$ kJ/kg K and $C_v = 0.7818$ kJ/kg K.*

Solution: Given.

$$P_1 = 1 \text{ MPa} = 1000 \text{ kPa}$$

$$T_1 = 600 \text{ K}$$

$$P_2 = 100 \text{ kPa}$$

As the problem is based on polytropic expansion so using the perfect gas equation, $pV = mRT$ and $pV^n = C$ we can find out the volumes at points 1 and 2.

Using equation $pV = mRT$, we can write

$$1000 \times V_1 = 1.5 \times 0.287 \times 600$$

$$\therefore \quad V_1 = 0.2583 \text{ m}^3$$

$$p_1V_1^n = p_2V_2^n$$

$$1000 \times 0.2583^{1.25} = 100 \times V_2^{1.25}$$

$$V_2^{1.25} = \frac{1000 \times 0.2583^{1.25}}{100} = 1.8414$$

$$\therefore \quad V_2 = (1.8414)^{\frac{1}{1.25}} = 1.629 \text{ m}^3$$

$$\frac{p_1V_1}{T_1} = \frac{p_2V_2}{T_2}$$

$$T_2 = \frac{p_2V_2T_1}{p_1V_1} = \frac{100 \times 1.629 \times 600}{1000 \times 0.2583}$$

$$\therefore \quad T_2 = 378.4 \text{ K.}$$

(A) For polytropic process, change in internal energy is given by

$$U_2 - U_1 = mC_v\ (T_2 - T_1)$$

$$= 1.5 \times 0.718\ (378.4 - 600)$$

$$= -238.66 \text{ kJ}$$

(B) For polytropic process, workdone is given by

$$W_{1-2} = \frac{p_1V_1 - p_2V_2}{n-1}$$

$$W_{1-2} = \frac{1000 \times 0.2583 - 100 \times 1.629}{1.25 - 1}$$

$$W_{1-2} = 381.6 \text{ kJ.}$$

(C) For polytropic process, heat transferred is given by

$$dQ = \delta W + dU$$

$$Q_{1-2} = W_{1-2} + U_2 - U_1$$

$$= 381.6 + (-238.66)$$

$$= 142.94 \text{ kJ}$$

(D) Change in enthalpy is given by

$$H_2 - H_1 = mC_p \, (T_2 - T_1)$$

$$= 1.5 \times 1.005 \times (378.4 - 600)$$

$$= -334.06 \text{ kJ.}$$

6.3 APPLICATION OF FIRST LAW TO FLOW PROCESSES

The processes occurring in the open system which allow the transfer of mass to and from the system are known as flow processes. In a flow process, mass or working substance enters the system and leaves after doing work.

Flow processes are of two types:

(*i*) Steady flow processes

(*ii*) Unsteady flow processes.

The term 'steady' implies no change with time. The opposite of steady is unsteady or transient. The term uniform however implies no change with location over a specified region.

6.3.1. Steady Flow Processes

A steady flow process is one during which a fluid flows through a control volume steadily, in this mass flow rate is same at the inlet and outlet and properties of working substance do not change with time.

Fluid properties can change from point to point within the control volume but at any fixed point they remain the same during the entire process. Examples of steady flow processes are flow through turbines, pumps, boilers, nozzles, heat-exchangers and compressors etc.

A steady flow process is characterized by the following features:

- During a steady flow process, there is no change in intensive or extensive properties with time within the control volume. Thus the volume V, mass m and the total energy content E of the total volume remains constant.
- The properties at the boundaries of the control volume do not change with time.
- The work and heat interactions between the system and its surroundings do not change with time.

6.3.2 Unsteady Flow Processes

Unsteady flow processes are those processes which involve changes within the control volume with time. In this process, mass flowing into system is not equal to mass flowing out of the system in a given time and there is no change in storage energy of the system.

Let ΔE be the change in storage energy

$$\therefore \quad Q - W = \Delta E + \Delta E_{exit} - \Delta E_{inlet} \quad ...(6.42)$$

While analyzing an unsteady flow process, one should keep track of the mass and energy content of the control volume as well as the energy interactions across the boundary. Examples of unsteady flow processes include the followings:

- Cooking with an ordinary pressure cooker.
- Inflating tyre.
- Discharging a fluid from a pressurized vessel.
- Driving a gas turbine with pressurized air stored in a large container.

6.3.3 Control Volume

The space bounded by imaginary boundaries around a mechanical device like pump, compressor or nozzle etc., is called the control volume. It is a volume of fixed shape and position. The difference between system and control volume is that the boundary of the system may change but the boundary of control volume remains fixed. However, working substance may flow across the boundary of the control volume.

6.3.4 Flow Energy or Flow Work

Flow work or Flow energy is the energy required to push the working substance into or out of a control volume.

Let us consider a fluid flowing through a pipe of cross-sectional area A. Let one kg of working substance moves a distance ΔL. Pressure p and velocity v at the section X – X′ of a pipe under consideration are uniform. Let an imaginary piston behind the section X – X′ causes the working substance to move by a distance ΔL. If this work is being done without changing the kinetic, potential and internal energies then the entering working substance must possess some form of energy to account for this work. This particular energy is called flow energy or flow work. For an ideal process, the quantity of flow work must be equal to the work done by the piston.

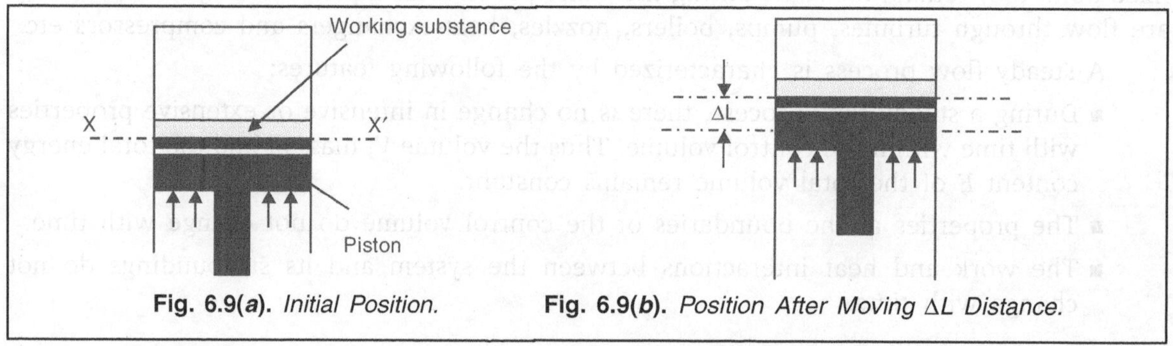

Fig. 6.9(*a*). *Initial Position.* **Fig. 6.9(*b*).** *Position After Moving ΔL Distance.*

$$\therefore \text{ Flow work} = p\,A\,\Delta L = pv_s \text{ kJ/kg} \quad ...(6.43)$$

Where p is the pressure at section $X - X'$ in kPa and

v_s is the specific volume in m^3/kg.

6.3.5 Total Energy in a Flow Process

The different types of energies involved with in a flow process are:

(*i*) Potential energy (PE). It is equal to mgZ

(*ii*) Kinetic energy (KE). It equals $\frac{1}{2}mv^2$

(*iii*) Internal energy (IE). It is represented by letter 'U'.

(*iv*) Flow energy. It is equal to pV_s.

Total energy accompanied per unit mass in a flow process is the sum of all the four types of energies listed above.

$$E_f = gZ + \frac{v^2}{2} + pv_s + U$$

$$E_f = gZ + \frac{v^2}{2} + H \qquad \text{...(6.44)}$$

Where $H = (U + pv_s)$ is the enthalpy of the system under consideration.

Equation (6.44) represents the energy of a fluid mass passing into and out of a control volume.

6.4 THE STEADY FLOW ENERGY EQUATION

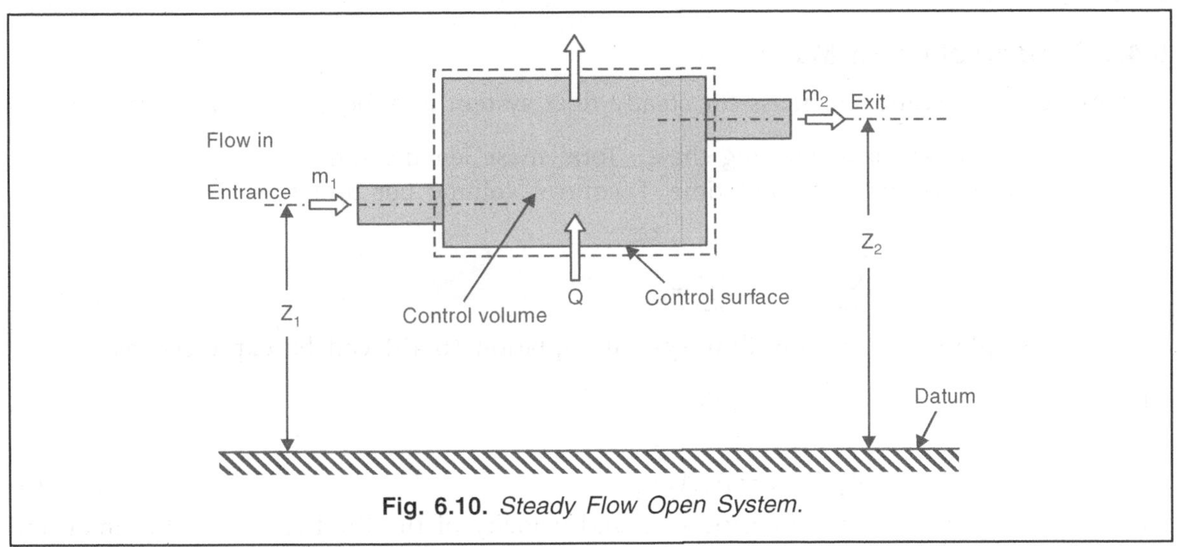

Fig. 6.10. *Steady Flow Open System.*

Let

m_1, m_2 = the mass flow rate in kg/s at entrance and exit of control volume.

p_1, p_2 = absolute pressures at entrance and exit

v_{s1}, v_{s2} = specific volume in m^3/kg at entrance and exit

v_1, v_2 = velocity in m/s at entrance and exit

Z_1, Z_2 = elevation above datum in m at entrance and exit
U_1, U_2 = specific internal energy at entrance and exit
Q = heat introduced into the system in J
W = amount of work available in J
$\dot{Q}$ = rate of heat flow from entrance to exit in J/s
$\dot{W}$ = rate of work done from entrance to exit in J/s.

Potential energy is proportional to the height Z above the datum level under consideration which is usually happens to be the earth. *i.e.*, $PE = mgZ$

For a unit mass this reduces to $PE = gZ$

Similarly the kinetic energy is $KE = \frac{1}{2}mv^2$ which for a unit mass reduces to

$$KE = \frac{1}{2}v^2.$$

In Fig. 6.10 a steady flow open system has been shown in which one stream of fluid enters the control volume and other fluid stream leaves it. The fluid steam at the inlet has pressure p_1, specific volume v_{s1}, internal energy U_1 and velocity v_1 and it enters the control volume at height Z_1. Inside the system, the transfer of heat energy Q and work done W take place across the system boundaries. The fluid stream leaves the system at pressure p_2, specific volume v_{s2}, internal energy U_2 and velocity v_2. The fluid exists at a height of Z_2 above the datum level.

6.4.1 Conservation of Mass

Principle of conservation of mass for steady flow system may be expressed in rate form as

$$\text{Total mass entering the control volume per unit time} = \text{Total mass leaving the control volume per unit time}$$

or

$$\sum \dot{m}_{inlet} = \sum \dot{m}_{exit} \qquad \text{...(6.45)}$$

For a single stream steady flow system, equation (6.45) can be expressed as

$$\dot{m}_{inlet} = \dot{m}_{exit}$$

or

$$\rho_1 A_1 v_1 = \rho_2 A_2 v_2 \qquad \text{...(6.46)}$$

where ρ_1, A_1, and v_1 are the density, area and velocity of the fluid stream at the inlet and ρ_2, A_2, and v_2 are the density, area and velocity of the fluid stream at the exit.

6.4.2 Conservation of Energy

When a steady state is reached in any system, the sum total of different forms of energy entering the system is equal to the sum total of different forms of energy leaving it. This assumes that no energy is used up by the system.

Using the principle of conservation of energy for a two flow open system can be expressed as

$$\text{Total energy entering the control volume per unit time} = \text{Total energy leaving the control volume per unit time}$$

$$\dot{Q} + \dot{m}_1 E_{f1} = \dot{W} + \dot{m}_2 E_{f2} \quad \text{...(6.47)}$$

where E_f is the total energy of the flowing fluid, including the flow work per unit mass.

Now writing equation (6.47) using equation (6.44), we get

$$\dot{Q} - \dot{W} = \sum \dot{m}_2 \left(g_2 Z_2 + \frac{v_2^2}{2} + H_2 \right) - \sum \dot{m}_1 \left(g_1 Z_1 + \frac{v_1^2}{2} + H_1 \right) \quad \text{...(6.48)}$$

$$\dot{Q} + \sum \dot{m}_1 \left(g_1 Z_1 + \frac{v_1^2}{2} + H_1 \right) = \sum \dot{m}_2 \left(g_2 Z_2 + \frac{v_2^2}{2} + H_2 \right) + \dot{W} \quad \text{...(6.49)}$$

In a steady flow system, the mass flow rate of the working substance is considered to be constant. That is $\dot{m}_{inlet} = \dot{m}_{exit}$

Now we can write the equation (6.49) as

$$\dot{Q} + \left(g_1 Z_1 + \frac{v_1^2}{2} + H_1 \right) = \left(g_2 Z_2 + \frac{v_2^2}{2} + H_2 \right) + \dot{W} \quad \text{...(6.50)}$$

Equation (6.50) is called the steady flow energy equation.

Now let us consider the flow of unit mass per second, through the system then we get

$$\boxed{Q + \left(g_1 Z_1 + \frac{v_1^2}{2} + h_1 \right) = \left(g_2 Z_2 + \frac{v_2^2}{2} + h_2 \right) + W} \quad \text{...(6.51)}$$

The mass flow rate is given by

$$\boxed{\dot{m} = \frac{\text{Total volume}}{\text{specific volume}} = \frac{v\,A}{v_s}} \quad \text{...(6.52)}$$

6.5 EXAMPLES OF STEADY FLOW PROCESSES

6.5.1 Nozzle and Diffuser

A nozzle is a device that increase velocity of the working substance (*i.e.*, a fluid) at the expense of pressure whereas a diffuser is a device that increases the pressure of a fluid by slowing it down.

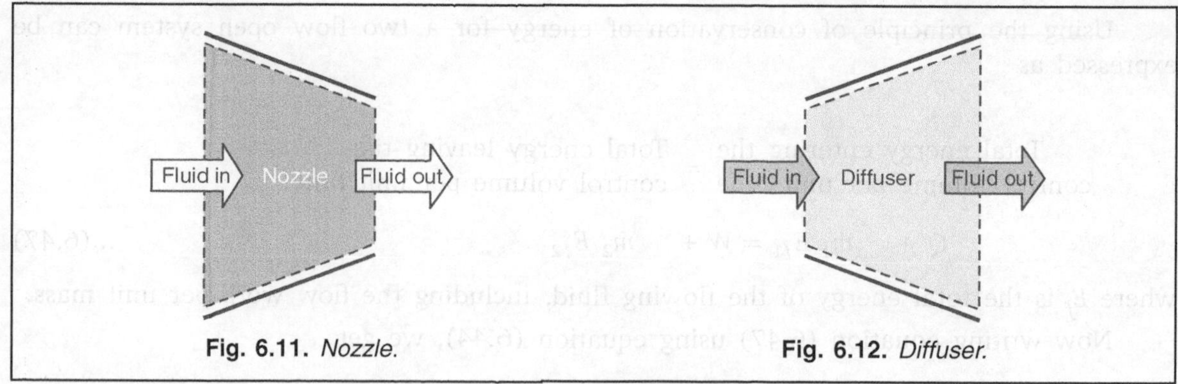

Fig. 6.11. *Nozzle.* **Fig. 6.12.** *Diffuser.*

The rate of heat transfer between fluid flowing through a nozzle and a diffuser and the surroundings is usually very small (Q ≈ 0) as the fluid used to move with high velocity and it does not stay in device for much time.

Both the nozzle and diffuser typically involve no work *i.e.*, $\dot{W} = 0$ and any change in potential energy (PE) is negligible. As the fluid moves through a nozzle or diffuser, it experiences significant changes in its velocities. Therefore, the changes in kinetic energy (KE) must be taken into consideration while analyzing the flow through these devices.

In this case

Net sum of energy entering the system = net sum of energy leaving the system

So using the equations.

$$\dot{Q} + \dot{m}_1 E_{f1} = \dot{m}_2 E_{f2} + \dot{W}$$

$$\dot{Q} + \sum \dot{m}_1 \left(g_1 Z_1 + \frac{v_1^2}{2} + H_1 \right) = \sum \dot{m}_2 \left(g_2 Z_2 + \frac{v_2^2}{2} + H_2 \right) + \dot{W}$$

Putting $\dot{Q} = 0$, $\dot{W} = 0$ and considering the same elevation $Z_1 = Z_2 = 0$ in equation, we get

$$\therefore \quad \frac{v_1^2}{2} + h_1 = \frac{v_2^2}{2} + h_2 \qquad ...(6.53)$$

6.5.2 Turbines

A turbine is a device which converts fluid energy into mechanical work. As fluid passes through the turbine, work is done against the blades, which are attached to the shaft. This causes the shaft to rotate and thus turbine produces work. The work done in a turbine is positive since it is done by the fluid. Heat transfer from a turbine is usually negligible ($\dot{Q} = 0$) as the turbine is well insulated. Further changes in potential energy are also negligible. In most turbines, fluid velocities encountered are very high and the fluid experiences a significant change in its kinetic energy. However, the change in enthalpy is more pronounced and so in comparison with this the change in kinetic energy is usually neglected.

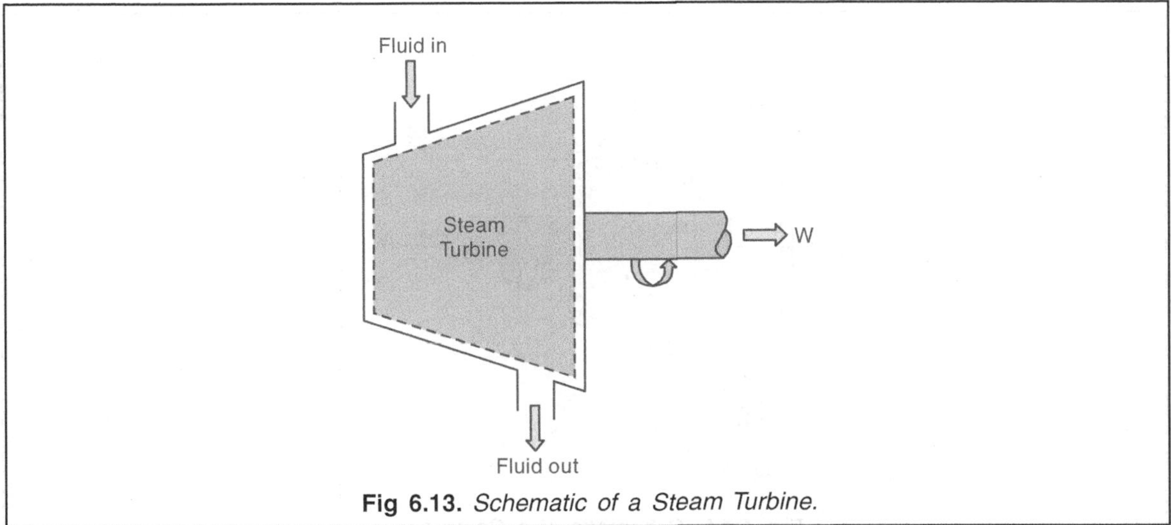

Fig 6.13. *Schematic of a Steam Turbine.*

In this case

Net sum of energy entering the system = net sum of energy leaving the system.

So using the equation.

$$\dot{Q} + \sum \dot{m}_1 E_{f_2} = \sum \dot{m}_2 E_{f_2} + \dot{W}$$

We can write

$$\dot{Q} + \sum \dot{m}_1 \left(g_1 Z_1 + \frac{v_1^2}{2} + H_1 \right) = \sum \dot{m}_2 \left(g_2 Z_2 + \frac{v_2^2}{2} + H_2 \right) + \dot{W}$$

Substituting $\dot{Q} = 0$, $\dot{W} = 0$, $v_1 = v_2 = 0$ and considering the same elevation $Z_1 = Z_2 = 0$ we get from equation.

$$\therefore \qquad W = h_1 - h_2 \qquad \text{...(6.54)}$$

6.5.3 Compressor

A compressor is a device which is used to increase the pressure of a fluid. Work is supplied to a compressor from an external source by a rotating shaft. As in case of a compressor work input is required from an outside source so W is negative and since heat Q is lost from the system so Q is also negative. Further changes in PE and KE can be disregarded.

On applying the principle of conservation of energy to a flow through a compressor we can write

Net sum of energy entering the system = net sum of energy leaving the system

$$\dot{Q} + \dot{m}_1 E_{f1} = \dot{m}_2 E_{f2} + \dot{W}$$

$$\dot{Q} + \dot{m}_1 \left(g_1 Z_1 + \frac{v_1^2}{2} + H_1 \right) = \dot{m}_2 \left(g_2 Z_2 + \frac{v_2^2}{2} + H_2 \right) + \dot{W}$$

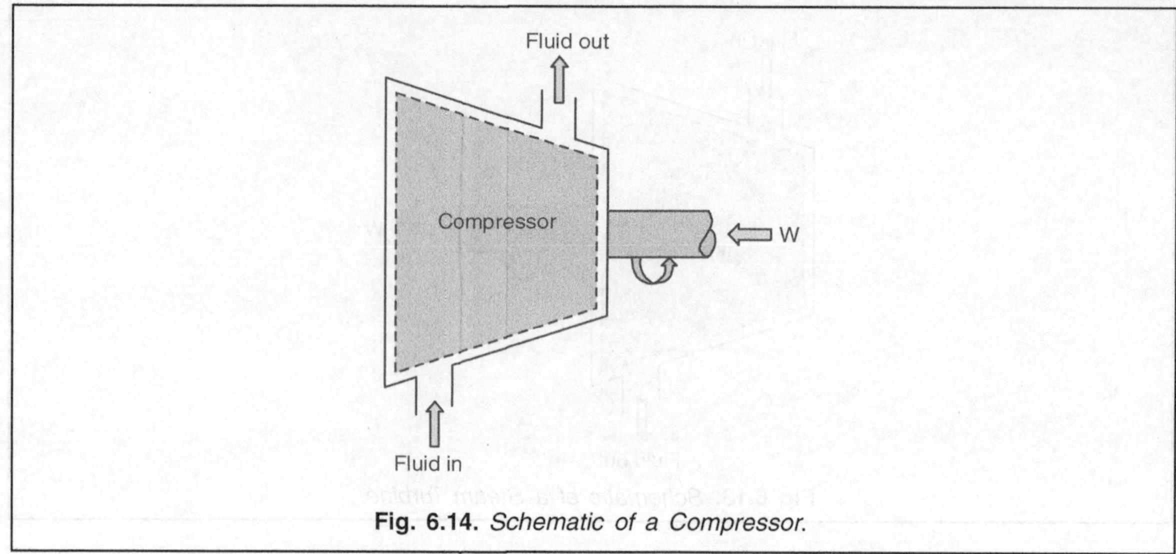

Fig. 6.14. *Schematic of a Compressor.*

Substituting $\dot{Q} = -\dot{Q}$, $\dot{W} = -\dot{W}$, $v_1 = v_2 = 0$ and considering the same elevation $gZ_1 = gZ_2 = 0$ we get from equation

$$-\dot{Q} + \dot{m}_1\left(g_1 Z_1 + \frac{v_1^2}{2} + H_1\right) = \dot{m}_2\left(g_2 Z_2 + \frac{v_2^2}{2} + H_2\right) - \dot{W} \quad ...(6.55)$$

Further solving the above equation, we get

$$\therefore \quad H_1 - \dot{Q} = H_2 - \dot{W}$$

$$\therefore \quad \dot{W} = H_2 - H_1 + \dot{Q} \quad ...(6.56)$$

6.5.4 Boiler

A boiler is a steam generating device in which water is heated by an external heat source and water gets converted into steam (Fig. 6.15). This steam is used to run a steam engine or a steam turbine. Since the boiler does not perform any work $\therefore W = 0$.

In this case PE and KE terms in energy equations can be neglected.

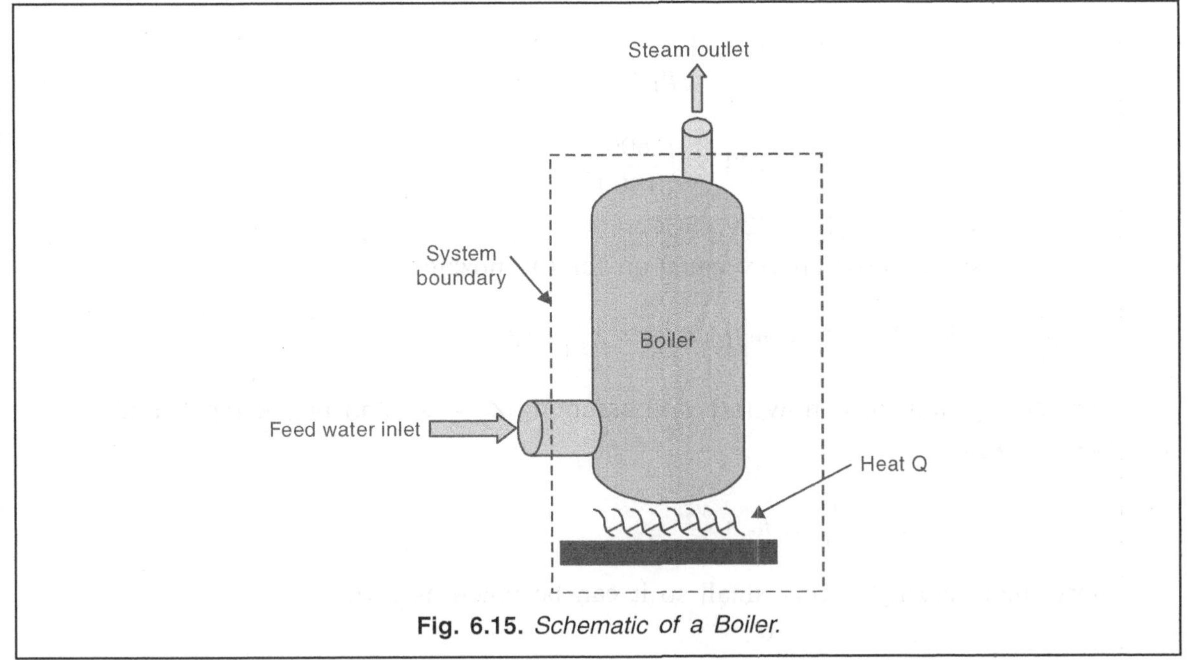

Fig. 6.15. *Schematic of a Boiler.*

On applying the principle of conservation of energy to a boiler we can write

Net sum of energy entering the system = net sum of energy leaving the system

$$\dot{Q} + \dot{m}_1 E_{f1} = \dot{m}_2 E_{f2} + \dot{W}$$

$$\dot{Q} + \dot{m}_1\left(g_1 Z_1 + \frac{v_1^2}{2} + H_1\right) = \dot{m}_2\left(g_2 Z_2 + \frac{v_2^2}{2} + H_2\right) + \dot{W}$$

Substituting $\dot{W} = 0$, $v_1 = v_2 = 0$ and considering the same elevation $gZ_1 = gZ_2 = 0$ we get from equation.

$$\therefore \quad \dot{Q} = H_2 - H_1 \qquad ...(6.57)$$

Problem 3. *Air is expanded reversibly and adiabatically in a nozzle from 13 bar and 150°C to a pressure of 6 bar. The inlet velocity of the nozzle is very small and the process occurs under steady flow conditions. Calculate the exit velocity of the nozzle.*

Solution.

The process under consideration is reversible adiabatic *i.e.,* isentropic.

$p_1 = 13\ bar = 1300\ kPa$

$T_1 = 150°C = 150 + 273 = 423$ K

$p_2 = 6\ bar = 600\ kPa$

Using the perfect gas equation and $pV^{\gamma} = C$ we can write

$$\frac{T_2}{T_1} = \left(\frac{p_2}{p_1}\right)^{\frac{\gamma-1}{\gamma}}$$

$$\therefore \quad T_2 = T_1 \times \left(\frac{p_2}{p_1}\right)^{\frac{\gamma-1}{\gamma}}$$

$$\therefore \quad T_2 = 423 \times \left(\frac{600}{1300}\right)^{\frac{1.4-1}{1.4}}$$

$$\therefore \quad T_2 = 339.15\ K$$

Using the steady flow energy equation for the nozzle

$$m\left(h_1 + \frac{v_1^2}{2} + Z_1\right) + Q = m\left(h_2 + \frac{v_2^2}{2} + Z_2\right) + W$$

Considering unit mass flow rate and assuming $Z_1 = Z_2$ and neglecting Q and W. We can write

$$h_1 + \frac{v_1^2}{2} = h_2 + \frac{v_2^2}{2}$$

Since inlet velocity is very small so it can be taken as zero.

$$v_1 = 0$$

$$h_1 + 0 = h_2 + \frac{v_2^2}{2}$$

$$\therefore \quad v_2^2 = 2\ (h_1 - h_2)$$

$$\therefore \quad v_2 = \sqrt{2(h_1 - h_2)} = \sqrt{2C_p(T_1 - T_2)}$$

$$\therefore \quad v_2 = \sqrt{2 \times 1.005\,(423 - 339.15)}$$

$$\therefore \quad v_2 = 12.98\ \text{m/s}.$$

Problem 4. *In a gas turbine, the gas enters at the rate of 5 kg/s with a velocity of 50 m/s and enthalpy of 900 kJ/kg and leaves the turbine with a velocity of 150 m/s and enthalpy of 400 kJ/kg. The loss of heat from the gases to the surroundings is 25 kJ/kg. Assume for gas R = 0.285 kJ/kg K and* C_p *= 1.004 kJ/kg K and the inlet conditions to be at 100 kPa and 27°C. Determine the power output of the turbine and the diameter of the inlet pipe.*

Solution. Given:

$$v_1 = 50\ \text{m/s}$$

$$h_1 = 900\ \text{kJ/kg}$$

$$v_2 = 150\ \text{m/s}$$

$$h_2 = 400\ \text{kJ/kg}$$

$$\dot{m} = 5\ \text{kg/s}$$

$$\text{heat loss} = -25\ \text{kJ/kg}$$

$$\text{Total } \dot{Q} = -25 \times 5 = -125\ \text{kJ/kg}$$

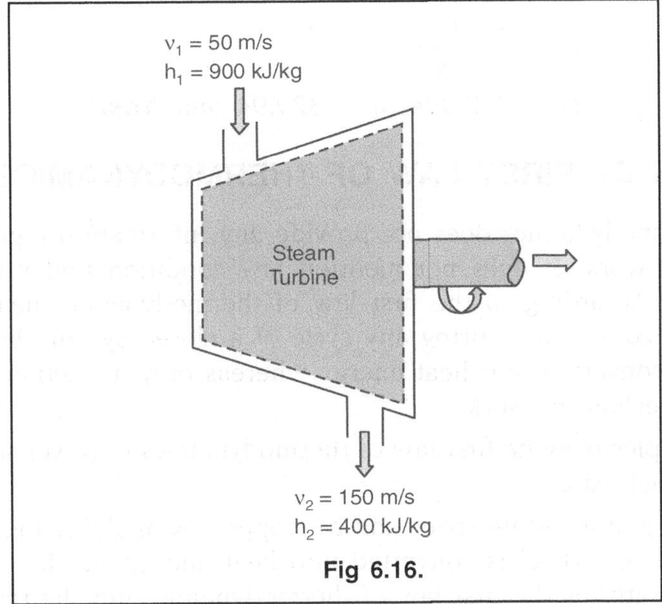

Fig 6.16.

Using the steady flow energy equation for the turbine

$$m\left(h_1+\frac{v_1^2}{2}+Z_1\right)+Q = m\left(h_2+\frac{v_2^2}{2}+Z_2\right)+W$$

Considering unit mass flow are and assuming $Z_1 = Z_2$, we can write

$$m\left(h_1+\frac{v_2^2}{2}\right)+Q = m\left(h_2+\frac{v_2^2}{2}\right)+W$$

$$W = m\left[(h_1-h_2)+\frac{1}{2}\left(v_1^2-v_2^2\right)\right]-Q$$

$\therefore$ $$W = 5[(900 - 400) + 0.5(50^2 - 150^2) \times 10^{-3}] -25 \times 5$$

$\therefore$ $$W = 2325 \text{ kJ/kg}$$

Using perfect gas equation at the turbine inlet we can write

$$p_1\dot{V}_1 = m_1RT_1$$

Where V is the volume flow rate at the turbine inlet

$$\dot{V}_1 = \frac{m_1RT_1}{p_1} = \frac{5\times 0.285\times 10^3\times 300}{100\times 10^3} = 4.275 \text{ m}^3/\text{s}$$

As we know that $\dot{V}_1 = A_1v_1$

$\therefore$ $$A_1 = \frac{\dot{V}_1}{v_1} = \frac{4.275}{50} = 0.0855 \text{ m}^2$$

$$A_1 = \frac{\pi}{4}\times D_1^2$$

$$\therefore \qquad D_1^2 = \frac{4 \times 0.0855}{\pi}$$

$$\therefore \qquad D_1 = 0.32994 \text{ m} = 329.94 \text{ mm } \textbf{Ans.}$$

6.6 LIMITATIONS OF FIRST LAW OF THERMODYNAMICS

The first law of thermodynamics does not provide any information regarding the direction of flow of heat and work. It does not mention any condition under which these energy transfers take place. According to the first law of thermodynamics heat energy and work energy are mutually convertible during any cycle of a closed system, but in real life, work energy can be fully converted into heat energy whereas only a fraction of heat energy can be converted into mechanical work.

Following examples obey he first law of thermodynamics however some limitations will be observed in all such cases.

A bicycle running at a certain speed can be stopped by applying the brakes. In this case the kinetic energy of the wheel is converted into heat energy and brakes get heated. This part of the example satisfies the first law of thermodynamics but the reverse of this should also obey the first law *i.e.,* if you cool down the hot brakes, the bicycle should have started moving but this will never happens in practice.

Similarly a hot pie lying on a plate in the Canteen will eventually cools down. This process satisfies the first law of thermodynamics as the amount of heat energy lost by a pie is equal to the amount of heat energy gained by the nearby surroundings. However the reverse of this process is not possible. Now let us consider the reverse process. Can a hot pie get even hotter in a cool surrounding due to heat transfer from cool surrounding air? As we are well aware of the fact that his process will not take place in real life even though the occurrence of this process would not have violated the first law of thermodynamics.

From these examples we can conclude that the first law of thermodynamics is a necessary but not a sufficient condition for a process to take place. This inadequacy of the first law to identify whether a process can take place is solved by the second law of thermodynamics.

6.7 SECOND LAW OF THERMODYNAMICS

The first law of thermodynamics is a law of conservation of energy only. It does not specify the direction of the process. The second law of thermodynamics enables us to determine the direction in which a spontaneous process proceeds. It is essential to understand the meaning of the following terms in order to discuss the second law of thermodynamics.

1. Thermal Energy Reservoir—is a hypothetical large body which can supply finite quantity of energy or which can absorb finite quantity of energy as heat without changing its temperature.
2. A reservoir that supplies energy in the form of heat is called a source.
3. A reservoir that absorbs energy in the form of heat is called a sink.

Thermal energy reservoirs are often referred to as heat reservoir since they supply or absorb energy in the form of heat. A typical example of a source is a constant temperature furnace where heat is continuously burnt and a typical sink is a river or a sea or the atmosphere itself.

6.7.1 Statements of Second Law of Thermodynamics

The second law of thermodynamics is the law of degradation of energy. A large number of statements have been put forth to formulate the second law of thermodynamics. The two most popular statements defining the major aspects of second law of thermodynamics are mentioned below:

☞ **Kelvin Planck Statement**

It is impossible to construct a heat engine working on a thermodynamic cycle, whose sole aim is to extract heat energy from a reservoir and convert it into an equivalent amount of work.

In other words, no actual engine operating in cycles can convert all the heat energy supplied to it into work. It means that some of the heat energy must be lost to the surroundings. As in real life conditions only a small part of the heat supplied to the engine gets converted into mechanical work whereas a large percentage of heat supplied is lost to the surroundings. The ratio of work to the heat supplied is known as the efficiency of the engine. Thus a 100% efficient engine is not possible. Figure 6.17 shows an impossible engine whereas Fig. 6.18 shows a schematic of a possible engine in which heat is supplied from the source and a part of the heat is rejected to sink. The work done on the surroundings by the engine is equal to the difference of heat supplied from source and heat rejected to sink.

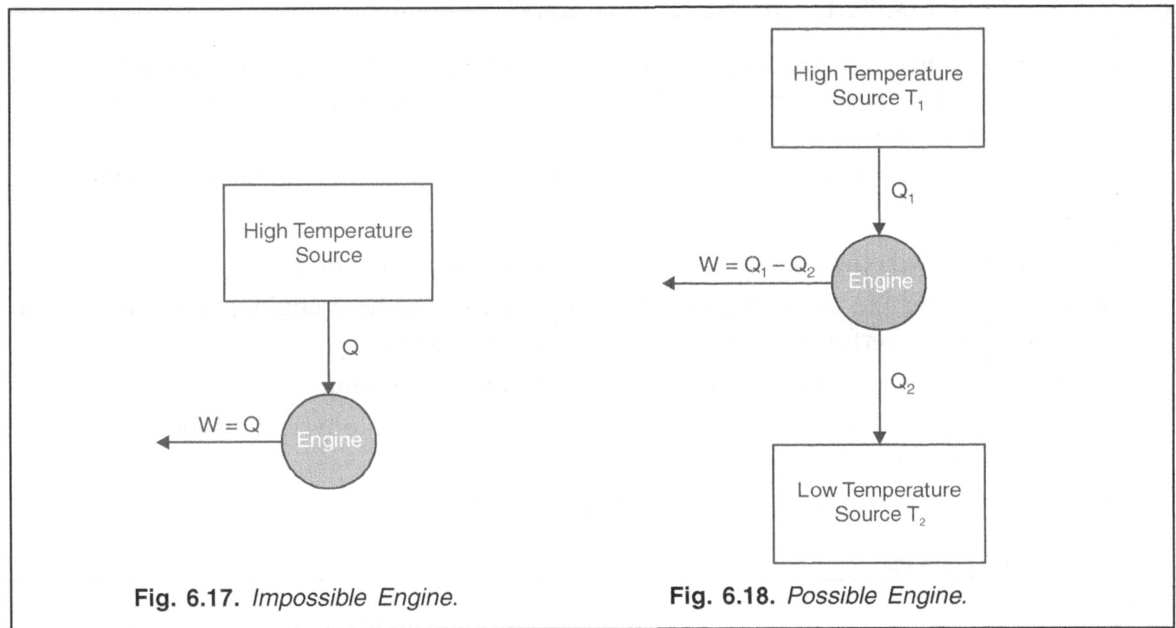

Fig. 6.17. *Impossible Engine.*

Fig. 6.18. *Possible Engine.*

☞ **Clausius Statement**

It is impossible for the heat energy to flow from a body at lower temperature to a body at higher temperature without the aid of an external agency.

According to this statement, if the source is connected to higher temperature T_1 and a sink is connected to a lower temperature T_2 then heat energy can not flow from sink to source without the help of an external agency like a heat pump as shown in Fig. 6.19 which takes mechanical work to deliver heat continuously from sink to source.

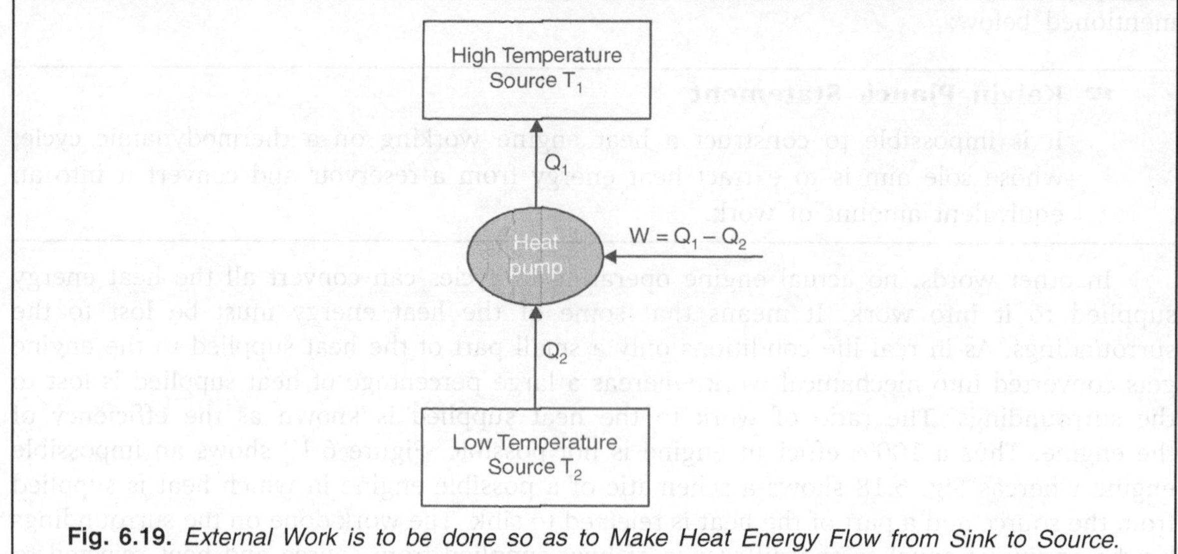

Fig. 6.19. *External Work is to be done so as to Make Heat Energy Flow from Sink to Source.*

6.8 THE IMPORTANCE OF SECOND LAW

The second law of thermodynamics may be expressed by different statements. In the beginning one would think that these statements do not have any relationship with each other but in reality all these statements are equivalent.

Second law of thermodynamics provides solution to a large number of problems. This law enables:

- To find out whether a particular process will take place or not?
- To determiner the maximum possible efficiency of the heat engine, or the maximum coefficient of performance of a refrigerating machine.
- To find out the direction and extent of chemical reaction.
- To develop a temperature scale, independent of the physical properties of any thermodynamic substance.
- To define an important property known as 'Entropy'.

6.9 HEAT ENGINE

Heat engine is a device used to convert heat energy into mechanical work in a cyclic process. It receives heat energy from a high temperature source, converts a part of it into work and rejects the rest to a low temperature sink.

Heat engines and other cyclic devices need a working substance (*i.e.*, a fluid) to and from which heat is transferred while completing a cycle. The term heat engine is often used in a broader sense to include work producing devices like IC engines and gas turbines which do not operate in a thermodynamic cycle.

The best example of a heat engine is the steam power plant (Fig. 6.20). The steam power plant comprises of a boiler (or heater), turbine (or expander), condenser (or cooler), compressor (or pump). Here the working substance is water. In a heat engine, the working substance returns to its initial state after going through a number of processes. The water flows through the boiler, turbine, condenser and feed pump. High pressure steam is generated in the boiler and supplied to the steam turbine. The work is developed in the turbine and the exhaust steam from turbine then flows to the condenser where it gets condensed and is returned to the boiler by means of a pump to complete the cycle.

Let

Q_{boiler} — heat transferred from the combustion of the fuel to the steam in boiler

$W_{Turbine}$ — work done by turbine

$Q_{condenser}$ — heat transferred from the condensing steam to the cooling water

W_{Pump} — work done on the pump

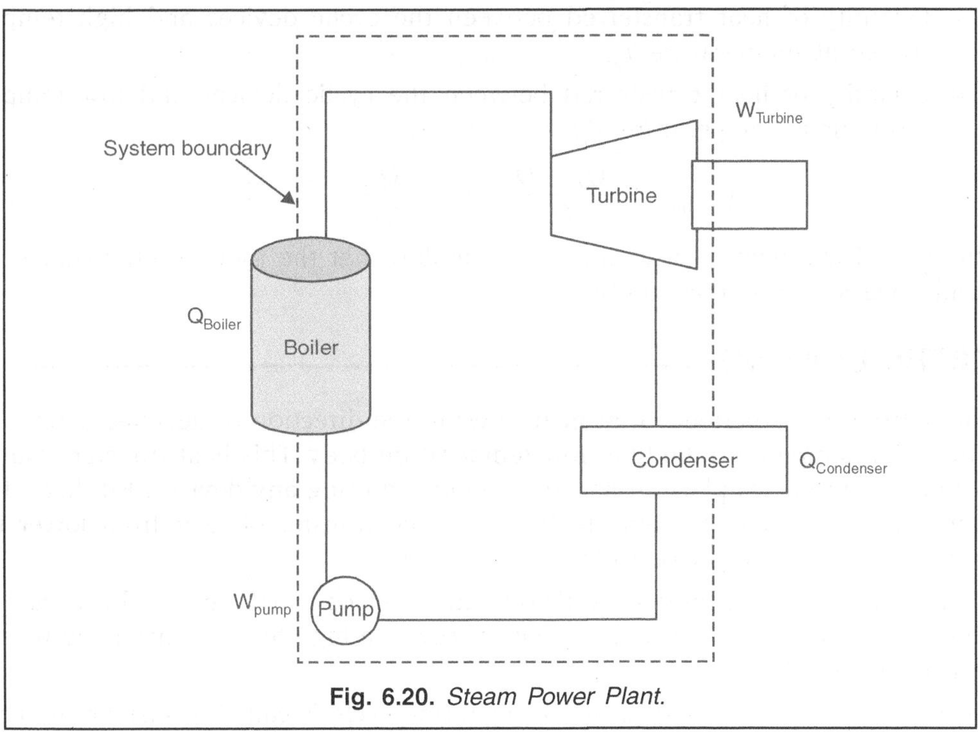

Fig. 6.20. *Steam Power Plant.*

From the first law of thermodynamics

$$\oint dQ = \oint dW$$

Using it to above case

$$Q_{boiler} - Q_{condenser} = W_{Turbine} - W_{Pump} \quad ...(6.58)$$

As we know that the efficiency of any engine is given by the ratio of the output to the input. In this case the net output is the work done by the turbine minus the work done on the pump whereas the input is the quantity of heat transferred to boiler for the generation of stream.

$$\therefore \quad \eta = \frac{W_{Turbine} - W_{Pump}}{Q_{boiler}} = \frac{Q_{boiler} - Q_{condenser}}{Q_{boiler}} \quad ...(6.59)$$

The fraction of the heat input that is converted to net work output is a measure of the performance of a heat engine and is called the thermal efficiency.

$$\eta_{thermal} = \frac{Net\ work\ output}{Total\ heat\ input} = \frac{W_{turbine} - W_{pump}}{Q_{boiler}} \quad ...(6.60)$$

Cyclic devices are of great practical interest such as heat engines, refrigerators and heat pump. These devices operate between a high temperature source at temperature T_H and a low temperature sink at temperature T_L. To bring uniformity to the treatment of heat engines, refrigerators and heat pumps, let us define following two quantities:

Q_H = Quantity of heat transferred between the cyclic devices and high temperature source at temperature T_H.

Q_L = Quantity of heat transferred between the cyclic devices and low temperature reservoir at temperature T_L.

$$\eta_{thermal} = \frac{Q_H - Q_L}{Q_H} = 1 - \frac{Q_L}{Q_H} \quad ...(6.61)$$

The Both Q_H and Q_L being positive quantities implies that the thermal efficiency of a heat engine will always be less than unity.

6.10 REFRIGERATORS

We all know from our experience that heat flows in the direction of decreasing temperature that is from high temperature body to low temperature body. This heat transfer from higher to lower temperature takes place in nature without requiring any devices. But heat does not flow from lower to higher temperature by itself. The transfer of heat from lower to high temperature needs special device called refrigerators.

Refrigerator like heat engine is cyclic operating device which absorbs heat energy from a low temperature body and rejects this heat energy to a high temperature body when work is performed on the device.

The working substance used in the refrigeration cycle is called a refrigerant. The most commonly used refrigeration cycle is the vapour compression refrigeration cycle. This cycle consists of four main components.

(*a*) Compressor

(*b*) Condenser

(*c*) Expansion valve and

(*d*) Evaporator.

The working fluid *i.e.,* refrigerant enters the compressor as a vapour and is compressed to the condenser pressure. It leaves the compressor at a relatively high temperature and cools down and condenses as it flows through the coils of the condenser by rejecting heat to the surrounding medium. It then enters the expansion valve where drop in its pressure and temperature takes place. This low temperature refrigerant then enters the evaporator where it absorbs heat from the space where cooling is desired. This space is referred to as refrigerated space. The cycle is completed as the refrigerant leaves the evaporator and re-enters the compressor.

6.10.1 Coefficient of Performance

Coefficient of performance (COP) is used to represent the efficiency of a refrigerator. The main purpose of a refrigerator is to remove heat Q_1 from the refrigerated space.

$$COP_{refrigerator} = \frac{\text{desired cooling effect produced}}{\text{work done in producing the cooling effect}}$$

$$COP_{refrigerator} = \frac{Q_L}{Q_H - Q_L}$$

$$COP_{refrigerator} = \frac{1}{\frac{Q_H}{Q_L} - 1} \qquad \text{...(6.62)}$$

The *COP* of a refrigerator may have a value greater than unity *i.e.,* the amount of heat removed from the refrigerated space can be more than the amount of work input.

6.11 HEAT PUMP

A heat pump is defined as a cyclically operating device which removes heat from a low temperature body and delivers it to a high temperature body. Usually the atmosphere acts as a low temperature body. This can be achieved by doing some external work on the heat pump. The performance of heat pump is also expressed in terms of *COP*.

Let heat pump takes Q_L heat out from a low temperature body and supply this heat to a room at high temperature T_H for the purpose of heating then the *COP* of the pump is given by

$$COP_{Pump} = \frac{\text{heating effect produced}}{\text{work input}}$$

$$COP_{Pump} = \frac{Q_H}{Q_H - Q_L}$$

$$COP_{Pump} = \frac{1}{1 - \frac{Q_L}{Q_H}} \qquad \text{...(6.63)}$$

Substituting the value of $\dfrac{Q_L}{Q_H} = \dfrac{COP_{refrigerator}}{1 + COP_{refrigerator}}$ from equation (6.62) in equation (6.63), we get

$$COP_{Pump} = \frac{1}{1 - \dfrac{COP_{refrigerator}}{1 + COP_{refrigerator}}}$$

$$COP_{Pump} = 1 + COP_{refrigerator} \qquad ...(6.64)$$

From equation (6.64) it is clear that *COP* of a heat pump is always greater than unity. Majority of the heat pump have a *COP* of 2 to 3.

6.12 EQUIVALENCE OF KELVIN PLANCK AND CLAUSIUS STATEMENTS

The equivalence of Kelvin Planck and Clausius statements can be proved by showing that violation of the Clausius statements implies the violation of the Kelvin Planck statement or vice-versa.

It appears that the Kelvin-Planck statement and the Clausius statement of second law of thermodynamics are altogether different, but they are equivalent. A violation of the Kelvin Planck statement leads to a violation of the Clausius statement and vice-versa.

To prove this fact let us consider a device A which acts as a heat pump and transfers Q_L amount of heat from a low temperature sink to a high temperature source without needing any external work. This device A violates the Clausius statement of second law of thermodynamics. Now let us consider another devices *B* that does not violate the Kelvin Planck statement. The device *B* acts as heat engine which extract energy Q_H from a high temperature reservoir and rejects Q_L to low temperature sink while producing work $W = Q_H - Q_L$.

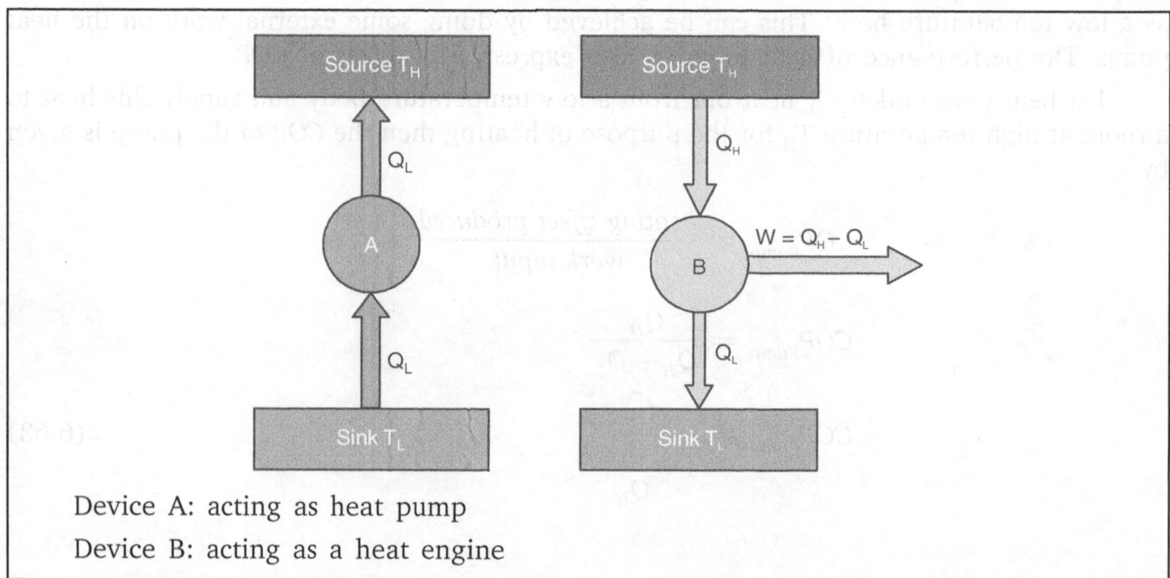

Device A: acting as heat pump

Device B: acting as a heat engine

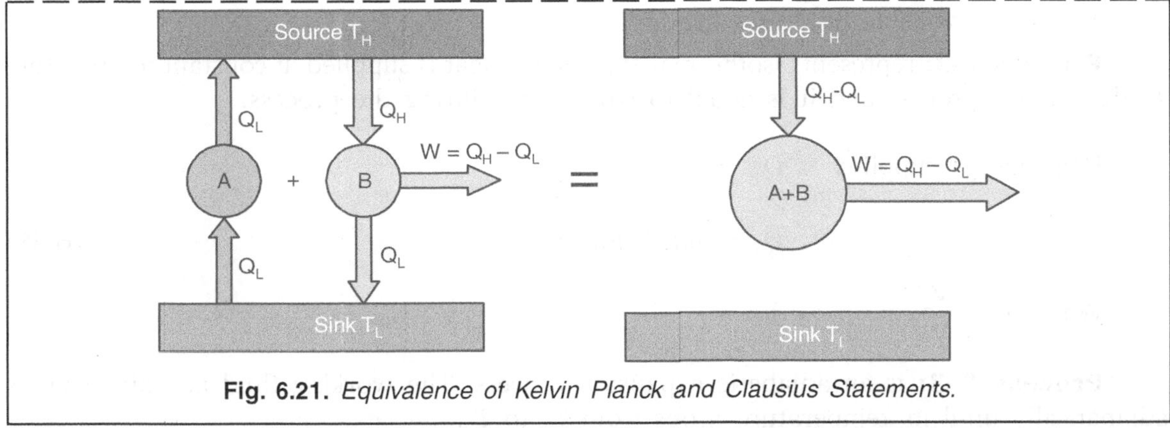

Fig. 6.21. *Equivalence of Kelvin Planck and Clausius Statements.*

Now let us combine these devices A and B as shown in Fig 6.21. Once these two devices are combined, no heat is transferred from lower temp sink and the combined device operates cyclically exchanging heat from a high temperature reservoir and produces no effect other than that of thermodynamic work. This combined device thus violates the Kelvin Planck statement. Therefore, it proves that violation of the Clausius statement leads to the violation of the Kelvin Planck statement also or vice-versa.

6.13 CARNOT CYCLE

In 1824, A French engineer Sadi Carnot first proposed an engine using perfect gas undergoing a cyclic process and consisting of reversible processes. An engine based on the working of a Carnot cycle is termed as Carnot engine. A Carnot engine is a hypothetical device and is not feasible to construct as it consists of reversible processes which proceeds at a very slow piston speed.

Carnot engine is used as a yardstick to measure the performance of any other engine. If the direction of Carnot cycle is reversed than it can be used for refrigeration/heat pump.

The Carnot cycle consists of four reversible processes which include two isothermal and two adiabatic processes. The PV and T–S diagram of the Carnot cycle is shown in Fig. 6.22.

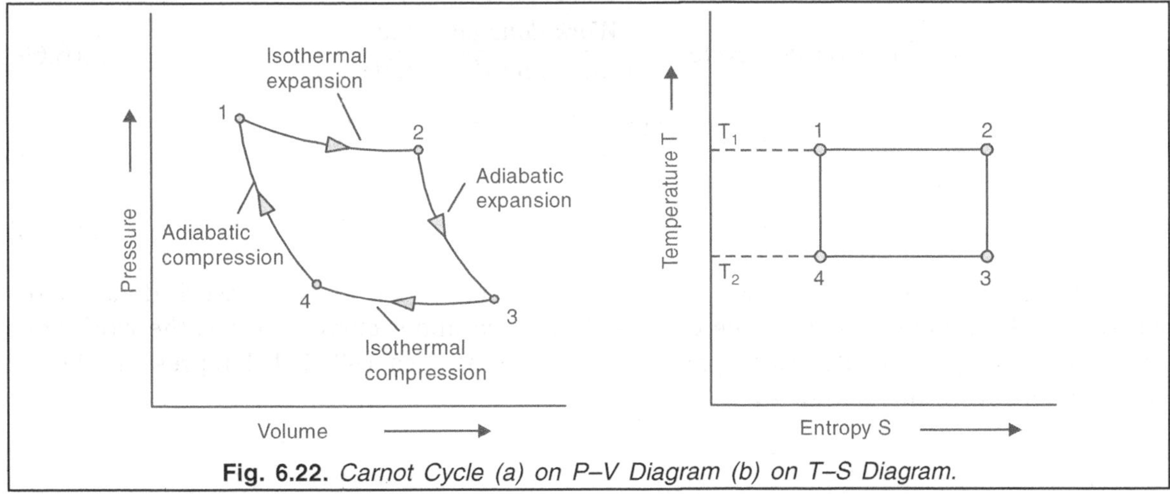

Fig. 6.22. *Carnot Cycle (a) on P–V Diagram (b) on T–S Diagram.*

The four reversible processes are as follows:

Process 1–2: represents isothermal expansion. Heat is supplied at constant temperature T_1 during this process and it is equal to work done during the process.

$$\text{Heat energy supplied} \quad Q_1 = p_1V_1 \log_e \frac{V_2}{V_1}$$

$$Q_1 = mRT_1 \log_e r \qquad ...(6.65)$$

Where $r = \frac{V_2}{V_1}$.

Process 2–3: is an adiabatic expansion process. The working fluid *i.e.,* air expands adiabatically until its temperature drops from T_1 to T_2.

$$\text{Work done during this process is } \frac{p_2V_2 - p_3V_3}{\gamma - 1} \qquad ...(6.66)$$

Process 3–4: during the process 3–4 heat rejection at constant temperature T_2 takes place. The working fluid is slowly compressed up to point 4. The heat is rejected by the air to the sink.

$$\text{Heat rejected} \quad Q_2 = p_3V_3 \log_e \frac{V_3}{V_4} \quad \text{as we know } \therefore\ r = \frac{V_3}{V_4} = \frac{V_2}{V_1}$$

$$Q_2 = mRT_2 \log_e r \qquad ...(6.67)$$

Process 4–1: the system returns back to the initial state by adiabatic compression. The temperature of air rises from T_2 to T_1 and work is done on the air.

$$\therefore \text{ Work done during compression } = \frac{p_4V_4 - p_1V_1}{\gamma - 1} \qquad ...(6.68)$$

$\therefore$ Net work done = Heat supplied – Heat rejected

$$= Q_1 - Q_2$$

$$Q_1 - Q_2 = mR(T_1 - T_2) \log_e r$$

$$\therefore \text{ Efficiency of the Carnot cycle } = \frac{\textit{Work done per cycle}}{\textit{heat supplied per cycle}} \qquad ...(6.69)$$

$$\eta = \frac{mR\,(T_1 - T_2) \log_e r}{mRT_1 \log_e r} = \frac{T_1 - T_2}{T_1}$$

$$\eta = \frac{T_1 - T_2}{T_1} \qquad ...(6.70)$$

As shown in Fig. 6.22 of the cycle, area under curve 1–2–3 is the work done by the gas during the expansion part of the cycle and the area under curve 3–4–1 is the work done on the gas during the compression part of the cycle. The area 1–2–3–4–1 represents the net work done during the Carnot cycle.

6.14 THE CARNOT PRINCIPLES

Kelvin-Planck and Clausius statements on second law of thermodynamics put some limits on the operation of cyclic devices. A heat engine cannot operate by exchanging heat with a single reservoir and a refrigerator cannot operate without a network input from an external source.

The Carnot principles are the conclusions pertaining to the thermal efficiency of reversible and irreversible engines.

1. The efficiency of irreversible heat engine is always less than the efficiency of a reversible engine operating between the same two reservoirs.
2. The efficiency of all reversible heat engines operating between the same two reservoirs are the same.
3. The efficiency of any reversible engine operating between two reservoirs does not dependent on the nature of the working fluid but depends solely on the temperature of the reservoirs.

These statements can be proved by demonstrating that the violation of these statement results in the violation of the second law of thermodynamics.

6.14.1 Corollary 1: Maximum Reversible Engine Efficiency

☞ The efficiency of irreversible heat engine is always less than the efficiency of a reversible engine operating between the same two reservoirs.

To prove Corollary 1 let us assume contrary to corollary 1 that the efficiency of the irreversible engine I_E is greater than the efficiency of a reversible engine R_E whereas both the engines are opening between the same temperature limits of heat source T_1 and heat sink T_2.

Let us assume that the 100 kJ heat is supplied to both the engines. As we have assumed that the irreversible engine is more efficient than the reversible engine. Therefore irreversible engine I_E produces more work (W_1 = 30 kJ) than the reversible engine (W_R = 20 kJ). The irreversible and reversible engines reject 70 kJ (100–30) and 80 kJ (100–20) heat respectively.

Now let both the engines I_E and R_E are combined mechanically such that the engine R_E is reversed and acts as a heat pump. It receives 80 kJ (100–20 = 80 kJ) heat from the low temperature reservoir and work W_R = 20 kJ from the irreversible engine and pumps 100 kJ heat to high temperature. The high temperature source supplies 100 kJ to the irreversible engine. The net results of this combination is shown in Fig. 6.23 (*c*).

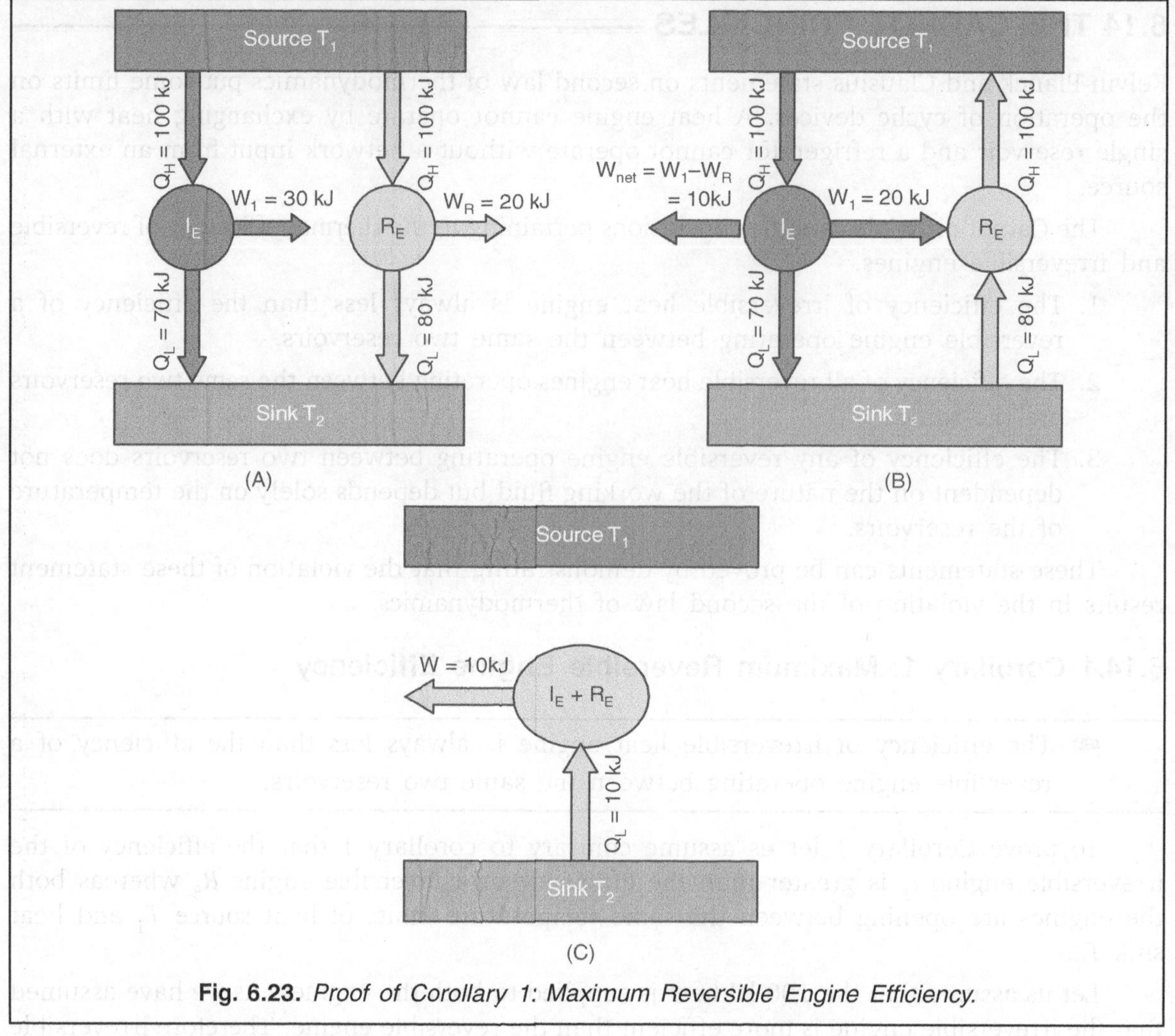

Fig. 6.23. *Proof of Corollary 1: Maximum Reversible Engine Efficiency.*

The combined device consisting of engines I_E and R_E draws 10 kJ (80–70) heat from low temperature reservoir and produces equal amount of work 10 kJ (30–20). This violates the second law of thermodynamics as it is impossible to develop a machine which takes heat from one source and produces an equal amount of work.

Thus our assumptions of irreversible engine having more efficiency than reversible engine is wrong. The only possibility is that the efficiency of any irreversible engine can be less than or equal to the efficiency of a reversible engine.

6.14.2 Corollary 2: The Efficiency of all Reversible Heat Engines Operating between the Same Two Reservoirs are the Same

The second corollary can also be proved in a similar manner. In this case, let us replace the irreversible engine I_E by another reversible engine R_{E_1} which is more efficient and thus delivers more work than the first reversible engine R_E.

The work produced by engine R_{E1} is more than the by engine R_E.

$$W_{RE_1} > W_R$$

$$\therefore \quad (Q_H - Q_{L_1}) > (Q_H - Q_L)$$

$$\therefore \quad Q_L > Q_{L_1}$$

Now engine R_E is reversed to run as a heat pump. The engine R_E absorbs heat Q_I from a low temperature sink and rejects heat Q_H to high temperature sources at temperature T_1. Since the heat energy required by R_{E_1} is also Q_H so we can eliminate the higher temperature source. The net heat energy taken from low temperature sink is $Q_L - Q_{L_1} = W_{net} = (W_{E1} - W_E)$.

This combination will end up in having an engine which exchanges the heat from a single reservoir and produces the equal amount of work. Thus it violates the Kelvin Planck statement on second law of thermodynamics.

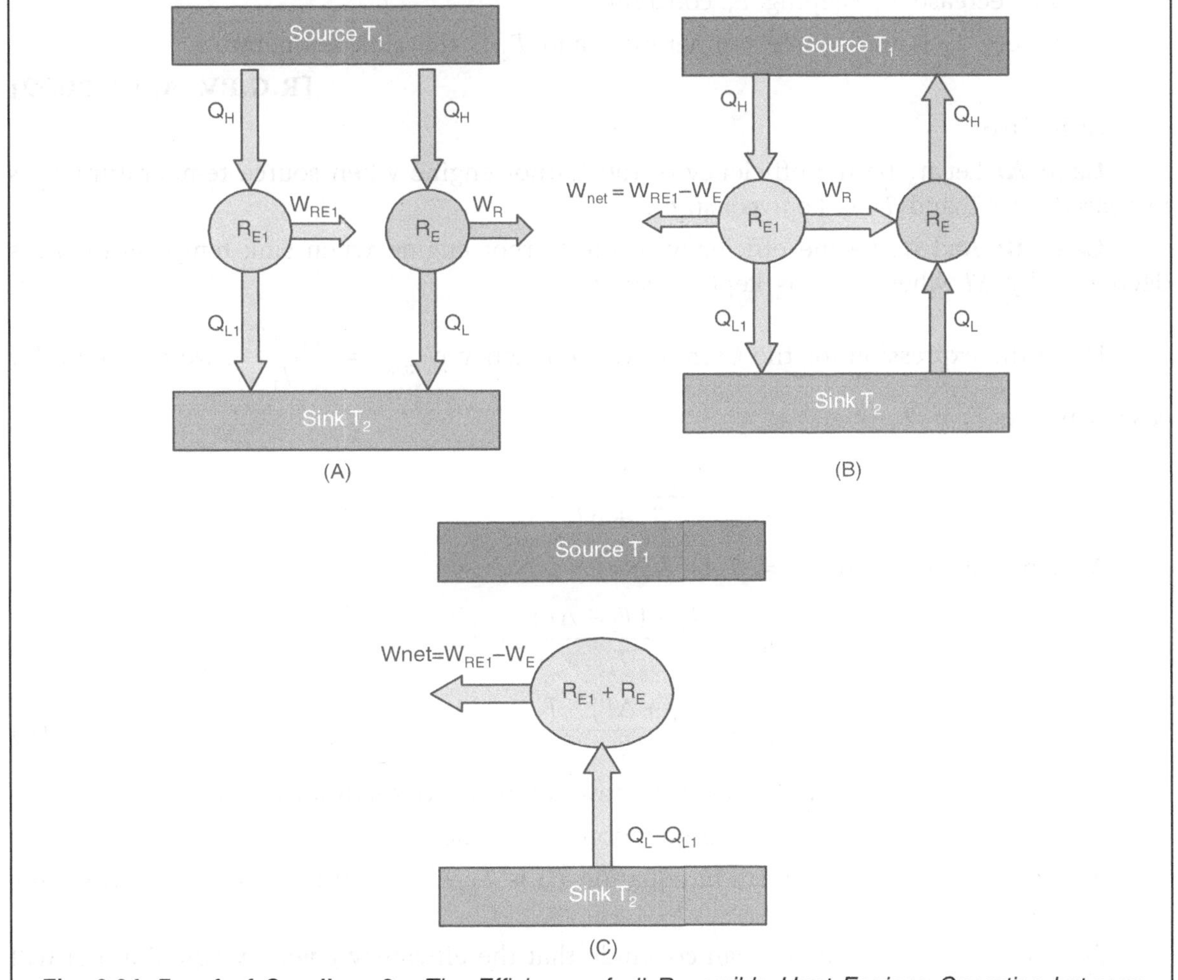

Fig. 6.24. Proof of Corollary 2 : *The Efficiency of all Reversible Heat Engines Operating between the Same Two Reservoirs are the Same.*

Problem 5. *A Carnot engine works between the temperature limits of 1127°C and 427°C. Find out the thermal efficiency of this engine.*

Solution:

Source temperature $T_1 = 1127 + 273 = 1400$ K

Sink temperature $T_2 = 427 + 273 = 700$

$$\eta_{Carnot} = \frac{T_1 - T_2}{T_1} = \frac{1400 - 700}{1700} = \frac{700}{1400} = 0.50$$

$$\eta_{Carnot} = 50\%$$

Problem 6. Which is the more effective way to increase the efficiency of a Carnot engine:

To increase T_1 keeping T_2 constant

or

To decrease T_2 keeping T_1 constant

Where T_1 is the source temperature and T_2 is the sink temperature

[R.G.P.V. April 2009]

Solution:

Case A: Let η_A be the efficiency of the Carnot engine when source temperature T_1 is increased by ΔT and T_2 is kept constant.

Case B: And η_B be the efficiency of the Carnot engine when sink temperature T_2 is decreased by ΔT whereas T_1 is kept constant.

Using the expression for the Carnot cycle efficiency $\eta_{Carnot} = \dfrac{T_1 - T_2}{T_1}$, we can write for case A where $T_1 = T_1 + \Delta T$:

$$\therefore \quad \eta_A = \frac{(T_1 + \Delta T) - T_2}{T_1 + \Delta T} \quad \text{...}(i)$$

And for case B where $T_2 = T_2 - \Delta T$:

$$\therefore \quad \eta_B = \frac{T_1 - (T_2 - \Delta T)}{T_1}$$

$$\therefore \quad \eta_B = \frac{(T_1 + \Delta T) - T_2}{T_1} \quad \text{...}(ii)$$

From equations (*i*) and (*ii*) we can draw following observations that

- Numerator terms in both the equations are same.
- Whereas denominator term in equation (*i*) is $T_1 + \Delta T$ and in equation (*ii*) it is only T_1.

From these observations we can conclude that the efficiency given by equation (*ii*) will be more than that by equation (*i*), it means

$$\therefore \quad \eta_B > \eta_A$$

Therefore we can conclude that decreasing the sink temperature T_2 is more effective way to increase the efficiency of the Carnot engine. Ans.

Problem 7. *A house is to be maintained at 25°C in summer as well as in winter. For this purpose it is proposed to use a reversible device as a refrigerator in summer and as a heat pump in winter. The ambient temperature is 40°C in summer and 3°C in winter. The energy losses as heat from the roof and the walls are estimated at 5 kW per degree Celsius temperature difference between the room and the ambient conditions. Calculate the power required to operate the device in summer and winter.*

Solution: For Summer:

$$T_L = 25 + 273 = 298 \text{ K}$$

$$T_H = 40 + 273 = 313 \text{ K}$$

$$\text{Energy losses} = 5 \times (40 - 25) = 75 \text{ kW}$$

$$COP = \frac{T_L}{T_H - T_L} = \frac{298}{313 - 298} = \frac{298}{15} = 19.86$$

Heat to be transferred in 75 kW

$$\text{Power required} = \frac{\textit{heat to be transferred}}{COP} = \frac{75}{19.86} = 3.77 \text{ kW}$$

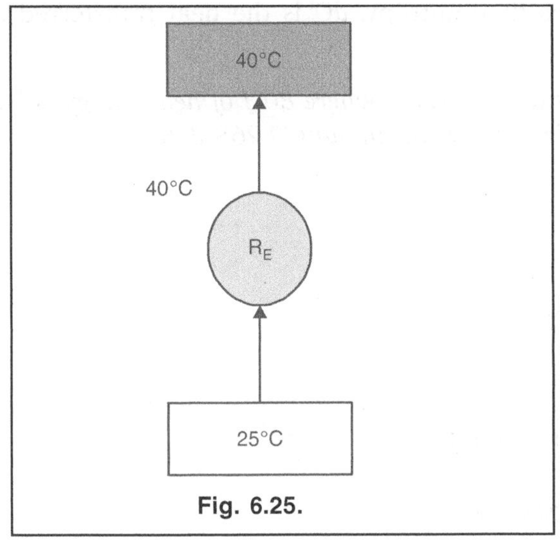

Fig. 6.25.

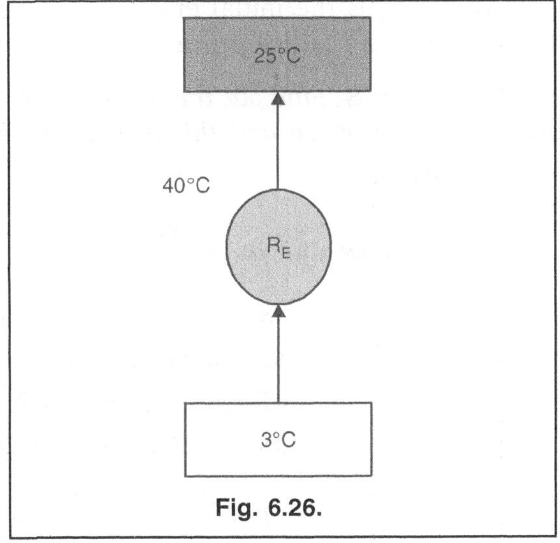

Fig. 6.26.

For winter:

$$T_L = 3 + 273 = 276 \text{ K}$$

$$T_H = 25 + 273 = 298 \text{ K}$$

$$\text{Heat energy to be being lost} = 5 \times (25 - 3) = 110 \text{ kW}$$

$$COP = \frac{T_H}{T_H - T_L} = \frac{298}{298 - 276} = \frac{298}{22} = 13.54$$

$$\text{Power required} = \frac{110}{13.54} = 8.12 \text{ kW}$$

6.15 ENTROPY

Entropy of a substance is a thermodynamic property. It increases with the addition of heat and decreases with the removal of heat, when the state of the substances changes in reversible manner. During such a change, the net heat transferred divided by the absolute temperature of the substance measures the change in entropy.

Let a small amount of heat dQ be added to a gas under reversible conditions and thus causing the entropy to increase by dS and let T be the absolute temperature during that particular time. Then as per the definition of entropy we can write

$$dS = \frac{dQ}{T} \qquad ...(6.71)$$

Mathematically change in entropy from state 1 to state 2 for a cyclic process can be expressed as

$$\int_1^2 dS = \int_1^2 \frac{dQ}{T}$$

or
$$S_2 - S_1 = \int_1^2 \frac{dQ}{T} \qquad ...(6.72)$$

Where S_1 is the initial entropy, and S_2 is final entropy, dQ is the heat transferred and T is the absolute temperature.

Problem 8. *Find out the temperature of a study room where 80 J of heat energy is being added to a cup of tea and this increase its entropy by an amount 0.268 J/K.*

Solution.

As we know that $dS = \frac{dQ}{T}$

Where dS = increase in entropy

dQ = heat added

T = temperature =?

Given dS = 0.268 J/K and dQ = 80 J

$$\therefore \quad T = \frac{dQ}{ds} = \frac{80}{0.268} = 298.5 \text{ K}$$

Temperature of the study room is 298.5 K. **Ans.**

6.15.1 Clausius Inequality

When a system undergoes a cycle, the cyclic integration of $\frac{\delta Q}{T}$ *is less than zero for an irreversible cycle and equal to zero for a reversible cycle.* This theorem is known as Clausius inequality.

Mathematically

$$\oint \frac{dQ}{T} = 0 \qquad \text{...(6.73)}$$

Clausius introduced the idea of entropy. He observed that if an amount of heat Q_1 is added reversibly at temperature T_1 to a substance and an amount of heat Q_2 is rejected reversibly at temperature T_2, then we can write

$$\frac{Q_1}{T_1} = -\frac{Q_2}{T_2}$$

As heat added is positive and heat rejected is negative.

or $$\frac{Q_1}{T_1} + \frac{Q_2}{T_2} = 0$$

or $$\oint \frac{dQ}{T} = 0$$

or $$cyclic \int \frac{dQ}{T} = 0 \qquad \text{...(6.74)}$$

From equation (6.74) it is clear that during a reversible cycle, the sum of the ratios of heat transfers to absolute temperatures equals zero. This result is applicable to all the reversible cycles.

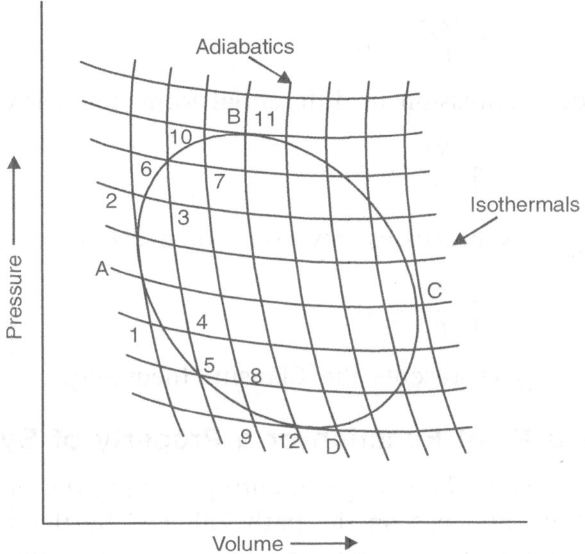

Fig. 6.27. *Clausius Inequality.*

Let us consider a reversible cycle ABCD as shown in Fig. 6.27. This cycle is divided into a large number of small Carnot cycles. Here each Carnot cycle consists of two reversible isothermals and two reversible adiabatic processes. From the Fig. 6.27 it is evident that the inside boundaries of these tiny Carnot cycles cancel out. For example, process 3–4 is an adiabatic expansion process and process 5–6 is consists of three adiabatic compression processes 5–4, 4–3 and 3–6. The process 3–4 and 4–3 will cancel out as 3–4 is expansion (energy gain) and other 4–3 is compression (energy loss) process. Therefore, one can approximate the cycle by *ABCD* and for all these Carnot cycles one can write

$$\frac{Q}{T} = 0$$

or $$\oint \frac{dQ}{T} = 0 \qquad ...(6.75)$$

This equation is applicable to reversible cycle. Let us now take the case of an irreversible cycle.

As we know that the thermal efficiency of a reversible cycle is given by

$$\eta_{rev} = \frac{Q_1 - Q_2}{Q_1} = \frac{T_1 - T_2}{T_1}$$

But thermal efficiency of an irreversible cycle is less.

$$\frac{Q_1 - Q_2}{Q_1} < \frac{T_1 - T_2}{T_1}$$

or $$\frac{Q_2}{Q_1} > \frac{T_2}{T_1}$$

or $$\frac{Q_2}{T_2} > \frac{Q_1}{T_1}$$

or $$\frac{Q_1}{T_1} - \frac{Q_2}{T_2} < 0$$

Writing the above expression in differential form, for a irreversible cycle engine

$$\oint \frac{dQ}{T} < 0 \qquad ...(6.76)$$

Thus for any cycle of processes, reversible or irreversible

$$\oint \frac{dQ}{T} \le 0 \qquad ...(6.77)$$

The equation (6.77) represents the Clausius Inequality.

6.15.2. Entropy is a Point Function or a Property of System

In a thermodynamic process, the change in entropy during the process depends only on the start and end conditions and not on the path followed by the process. Entropy is a point function, *i.e.,* it has got one and only one value for each point (pressure, volume and temperature). While solving numerical problems based on entropy, the change in entropy is needed and absolute value of entropy is not required. So it has been assumed that entropy of all substances is zero at the ice point. It means that if temperature is above 0°C then entropy is positive and if temperature is below 0°C then entropy is negative.

6.15.3. Change in Entropy of a Perfect Gas

Let us consider a closed system containing a certain amount of gas undergoing a reversible process during which there are heat and mechanical work interconversions take place.

Let m kg of gas at a pressure p_1, volume V_1, absolute temperature T_1 and entropy S_1 be heated initially such that its final conditions of pressure, volume, absolute temperature and entropy be p_2, V_2, T_2 and S_2 respectively.

Now from the first law of thermodynamics, we get

$$dQ = dW + dU \qquad \text{...(6.78)}$$

As we know $dW = pdV$ and $dU = mC_v\, dT$

Substituting these values in equation (6.78), we get

$$dQ = pdV + m\ C_v dT \qquad \text{...(6.79)}$$

Dividing both sides of the equation (6.79) by T, we get

$$\frac{dQ}{T} = \frac{P}{T}dV + m\,C_v\frac{dT}{T}$$

But

$$\frac{dQ}{T} = dS$$

$$\therefore \quad dS = \frac{P}{T}dV + m\,C_v\frac{dT}{T} \qquad \text{...(6.80)}$$

As we know $pV = mRT$

$$\therefore \quad \frac{P}{T} = \frac{mR}{V}$$

On substituting the above value of $\frac{P}{T}$ in equation (6.80), we get

$$\therefore \quad dS = mR\frac{dV}{V} + mC_v\frac{dT}{T}$$

Integrating the above equation between states 1 and 2, we get

$$\int_1^2 dS = mR\int_1^2 \frac{dV}{V} + m\,C_v\int_1^2 \frac{dT}{T}$$

$$\therefore \quad S_2 - S_1 = mR\ \log_e \frac{V_2}{V_1} + mC_v\ \log_e \frac{T_2}{T_1}$$

As we know $C_p - C_v = R$ and substituting for R in the above equation, we get

$$\therefore \quad S_2 - S_1 = m\ (C_p - C_v)\ \log_e \frac{V_2}{V_1} + m\ C_v\ \log_e \frac{T_2}{T_1} \qquad \text{...(6.81)}$$

Equation (6.81) represents the change in entropy in terms of volume and temperature.

To represent the entropy in terms of pressure and volume, let us use the general gas equation, we have

$$\frac{p_1V_1}{T_1} = \frac{p_2V_2}{T_2}$$

or

$$\frac{T_2}{T_1} = \frac{p_2V_2}{p_1V_1}$$

Substituting the value of $\frac{T_2}{T_1}$ in equation (6.81), we get

$$\therefore \quad S_2 - S_1 = mC_p \log_e \frac{V_2}{V_1} - mC_v \log_e \frac{V_2}{V_1} + mC_v \log_e \frac{P_2}{P_1} + mC_v \log_e \frac{V_2}{V_1}$$

$$\therefore \quad S_2 - S_1 = mC_p \log_e \frac{V_2}{V_1} + mC_v \log_e \frac{P_2}{P_1} \qquad \text{...(6.82)}$$

Equation (6.82) represents the change in entropy in terms of volume and pressure.

Similarly the change in entropy can be represented in terms of pressure and temperature. Using the perfect gas equation again, we can write

$$\frac{p_1V_1}{T_1} = \frac{p_2V_2}{T_2}$$

or

$$\frac{V_2}{V_1} = \frac{T_2}{T_1} \times \frac{p_1}{p_2}$$

Substituting the value of $\frac{V_2}{V_1}$ in equation (6.81), we get

$$\therefore \quad S_2 - S_1 = m(C_p - C_v) \log_e \frac{T_2}{T_1} + m(C_p - C_v) \log_e \frac{P_1}{p_2} + mCv \log_e \frac{T_2}{T_1}$$

$$\therefore \quad S_2 - S_1 = mC_p \log_e \frac{T_2}{T_1} + m(C_p - C_v) \log_e \frac{P_1}{p_2}$$

$$\therefore \quad S_2 - S_1 = mC_p \log_e \frac{T_2}{T_1} - m(C_p - C_v) \log_e \frac{p_2}{p_1} \qquad \text{...(6.83)}$$

6.15.4. Changes of Entropy for Ideal Gases During Different Reversible Processes and their Representation on Temperature-Entropy (T–S) Diagram

6.15.4.1 *Constant Volume Process*

Let m kg of an ideal gas is supplied with a small amount of heat dQ at constant volume which increases its temperature by dT.

As we know that in case of a constant volume process $V_2 - V_1$.

Therefore substituting this value in equation (6.81), we get

$$\therefore \quad S_2 - S_1 = m(C_p - C_v) \log_e \frac{V_2}{V_1} + mC_v \log_e \frac{T_2}{T_1}$$

$$\therefore \quad S_2 - S_1 = m(C_p - C_v) \log_e 1 + mC_v \log_e \frac{T_2}{T_1}$$

Since $\log_e 1 = 0$, change in entropy for a constant volume process is given by equation (6.84)

$$\therefore \quad S_2 - S_1 = mC_v \log_e \frac{T_2}{T_1} \qquad ...(6.84)$$

The constant volume process of heating is represented on T–S diagram as shown in Fig. 6.28.

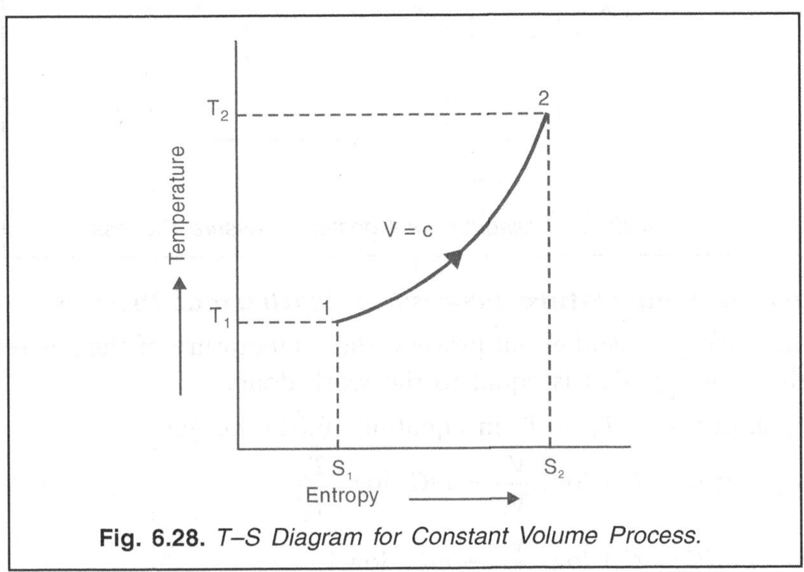

Fig. 6.28. *T–S Diagram for Constant Volume Process.*

6.15.4.2 *Constant Pressure Process*

As we know that when a gas is heated at constant pressure, a portion of the heat supplied is utilized in doing the external work and rest of the heat energy supplied is stored in the gas in the form of internal energy.

As we know that in case of a constant pressure process $p_2 = p_1$.

Substituting this value in equation (6.83), we get

$$\therefore \quad S_2 - S_1 = mC_p \log_e \frac{T_2}{T_1} - m(C_p - C_v) \log_e \frac{P_2}{P_1}$$

$$\therefore \quad S_2 - S_1 = mC_p \log_e \frac{T_2}{T_1} - m(C_p - C_v) \log_e 1$$

Since $\log_e 1 = 0$, change in entropy for a constant pressure process is given by

$$\therefore \quad S_2 - S_1 = mC_p \log_e \frac{T_2}{T_1} \qquad ...(6.85)$$

The constant pressure process of heating is represented on T–S diagram as shown in Fig. 6.29.

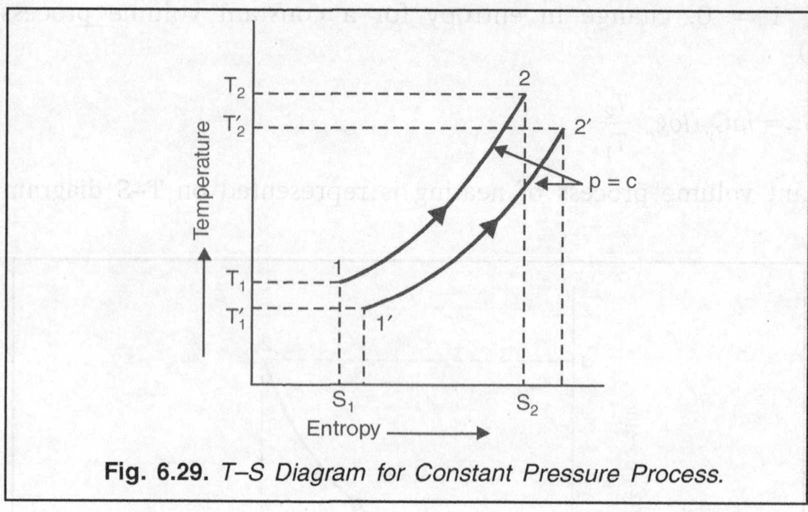

Fig. 6.29. *T–S Diagram for Constant Pressure Process.*

6.15.4.3 *Constant Temperature Process or Isothermal Process*

As we know that during an isothermal process, the temperature of the gas remains constant $T_1 = T_2$ and the heat supplied is equal to the work done.

Therefore, substituting $T_1 = T_2$ in equation (6.81) we get

$$\therefore \quad S_2 - S_1 = m(C_p - C_v) \log_e \frac{V_2}{V_1} + mC_v \log_e \frac{T_2}{T_1}$$

$$\therefore \quad S_2 - S_1 = m(C_p - C_v) \log_e \frac{V_2}{V_1} + mC_V \log_e 1$$

Since $\log_e 1 = 0$, change in entropy for an isothermal process is given by

$$\therefore \quad S_2 - S_1 = m(C_p - C_v) \log_e \frac{V_2}{V_1} \qquad \text{...(6.86)}$$

The constant temperature process or isothermal process is represented on T–S diagram as shown in Fig. 6.30.

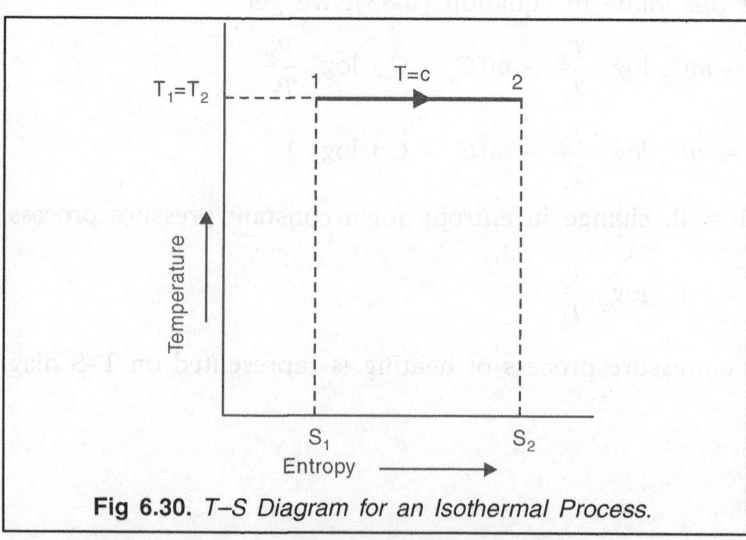

Fig 6.30. *T–S Diagram for an Isothermal Process.*

6.15.4.4 *Reversible Adiabatic Process or Isentropic Process*

As we know that during an adiabatic process, no heat is supplied or rejected by the system.

$$\therefore \quad dQ = 0$$

We know that

$$\therefore \quad dS = \frac{dQ}{T} \text{ but } dQ = 0$$

$$\therefore \quad dS = \frac{0}{T} = 0$$

or $$S = \text{Constant} \qquad \text{...(6.87)}$$

The isentropic process is represented by vertical lines on T–S diagram as shown in Fig. 6.31.

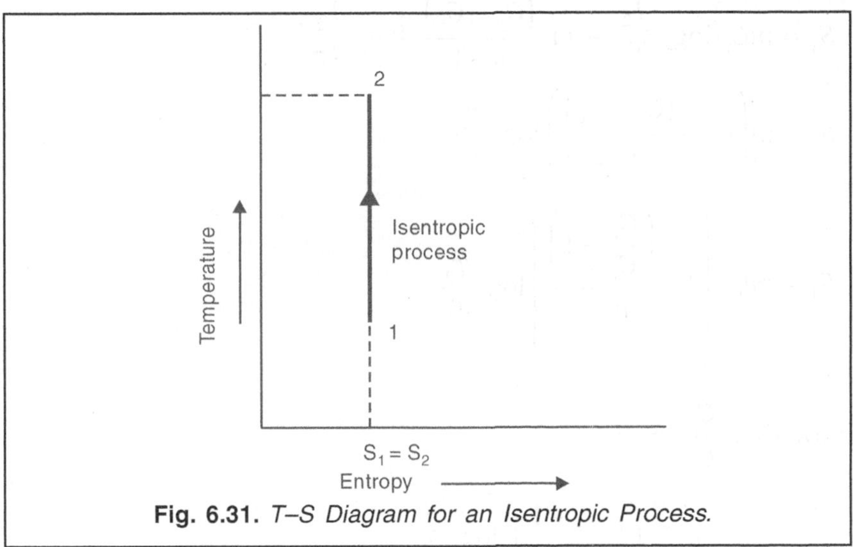

Fig. 6.31. *T–S Diagram for an Isentropic Process.*

6.15.4.5 *Polytropic Process*

The change in entropy for any process is given by equation (6.81).

$$\therefore \quad S_2 - S_1 = m(C_p - C_v) \log_e \frac{V_2}{V_1} + mC_v \log_e \frac{T_2}{T_1}$$

As we know that polytropic process follows the law $pV^n = C$

So for a polytropic process, we can write

$$p_1V_1^n = p_2V_2^n$$

or $$\frac{p_1}{p_2} = \left(\frac{V_2}{V_1}\right)^n$$

We also know that in a polytropic process

$$\frac{V_2}{V_1} = \left(\frac{T_1}{T_2}\right)^{\frac{1}{n-1}} \qquad \text{...(6.88)}$$

Substituting the value $\frac{V_2}{V_1}$ in equation (6.81), we get

$$\therefore \quad S_2 - S_1 = m(C_p - C_v)\log_e \frac{V_2}{V_1} + mC_v \log_e \frac{T_2}{T_1}$$

$$\therefore \quad S_2 - S_1 = m(C_p - C_v)\log_e \left(\frac{T_1}{T_2}\right)^{\frac{1}{n-1}} + mC_v \log_e \frac{T_2}{T_1}$$

$$\therefore \quad S_2 - S_1 = m\frac{(C_p - C_v)}{n-1}\log_e \frac{T_1}{T_2} + mC_v \log_e \frac{T_2}{T_1}$$

Now rearranging the terms we get

$$\therefore \quad S_2 - S_1 = mC_v \log_e \frac{T_2}{T_1} - m\frac{(C_p - C_v)}{n-1}\log_e \frac{T_2}{T_1}$$

$$\therefore \quad S_2 - S_1 = m\left(C_v - \frac{(C_p - C_v)}{n-1}\right)\log_e \frac{T_2}{T_1}$$

$$\therefore \quad S_2 - S_1 = mC_v\left(1 - \frac{\left(\frac{C_p}{C_v} - 1\right)}{n-1}\right)\log_e \frac{T_2}{T_1}$$

As we know that $\frac{C_p}{C_v} = \gamma$

$$\therefore \quad S_2 - S_1 = mC_v\left(1 - \frac{\gamma - 1}{n-1}\right)\log_e \frac{T_2}{T_1}$$

$$\therefore \quad S_2 - S_1 = mC_v\left(\frac{n-\gamma}{n-1}\right)\log_e \frac{T_2}{T_1} \qquad \text{...(6.89)}$$

By using perfect gas equation, the change in entropy can be expressed in terms of pressure and volume by equations (6.90) and (6.91) respectively.

$$\therefore \quad S_2 - S_1 = mC_v\left(\frac{n-\gamma}{n}\right)\log_e \frac{p_2}{p_1} \qquad \text{...(6.90)}$$

$$\therefore \quad S_2 - S_1 = mC_v(\gamma - n)\log_e \frac{V_2}{V_1} \qquad \text{...(6.91)}$$

Problem 9. *An ideal gas is heated from temperature T_1 and T_2 by keeping its volume constant. The gas is expanded back to its initial temperature according to law pV^n = constant. If the entropy change in the two processes are equal, find the value of 'n' in terms of adiabatic index 'γ'.*

Solution.

As we know that change in entropy during a constant volume process is given by

$$\therefore \quad S_2 - S_1 = mC_v \log_e \frac{T_2}{T_1} \quad \text{...}(i)$$

And change in entropy during polytropic process is given by

$$S_2 - S_1 = mC_v \left(\frac{\gamma - n}{n-1}\right) \log_e \frac{T_2}{T_1} \quad \text{...}(ii)$$

Since the entropy changes in the two processes are equal

$$\therefore \quad mC_v \log_e \frac{T_2}{T_1} = mC_v \left(\frac{\gamma - n}{n-1}\right) \log_e \frac{T_2}{T_1}$$

$$\therefore \quad 1 = \left(\frac{\gamma - n}{n-1}\right)$$

$$\therefore \quad n - 1 = \gamma - n$$

$$\therefore \quad n = \frac{\gamma + 1}{2}.$$ **Ans.**

Problem 10. *A heat engine operates between source at 500K and sink 250K. Heat supplied by source is 400 kcal/s. Evaluate feasibility of engine and nature of cycle for the following conditions. (i) Heat rejected being 150 kcal/s. (ii) heat rejected being 300 kcal/s, (iii) heat rejected being 200 kcal/s.*

Solution.

In this case Clausius inequality can be used to evaluate the feasibility of engine.

Here subscript 1 stands for source and 2 for sink.

As we know that

$$\oint \frac{dQ}{T} = \frac{Q_1}{T_1} - \frac{Q_2}{T_2}$$

Given $Q_1 = 400$ kcal/s

$T_1 = 500$ K

$T_2 = 250$ K

Condition (*i*), Heat rejected being 150 kcal/s

i.e., $Q_2 = 150$ kcal/s

$$\oint \frac{dQ}{T} = \frac{Q_1}{T_1} - \frac{Q_2}{T_2}$$

$$\oint \frac{dQ}{T} = \frac{400}{500} - \frac{150}{250}$$

$$\oint \frac{dQ}{T} = 0.2$$

Since $\oint \frac{dQ}{T} > 0$, therefore for condition (*i*), engine is not possible. **Ans.**

Condition (*ii*), Heat rejected being 300 kcal/s

$$Q_2 = 300 \text{ kcal/s}$$

$$\oint \frac{dQ}{T} = \frac{Q_1}{T_1} - \frac{Q_2}{T_2}$$

$$\oint \frac{dQ}{T} = \frac{400}{500} - \frac{300}{250}$$

$$\oint \frac{dQ}{T} = -0.4$$

In this case $\oint \frac{dQ}{T} < 0$ it means that engine is feasible and cycle is irreversible. **Ans.**

Condition (*iii*), Heat rejected being 200 kcal/s

$$Q_2 = 200 \text{ kcal/s}$$

$$\oint \frac{dQ}{T} = \frac{Q_1}{T_1} - \frac{Q_2}{T_2}$$

$$\oint \frac{dQ}{T} = \frac{400}{500} - \frac{200}{250}$$

$$\oint \frac{dQ}{T} = 0$$

In this case $\oint \frac{dQ}{T} = 0$ it means that engine is feasible and cycle is reversible. **Ans.**

Problem 11. *3 kg of air is expanded at constant pressure to two times its initial volume. The initial temperature of the air is 300°C. Calculate the change in entropy.*

Solution.

As we know that change in entropy is given by

$$\therefore \quad S_2 - S_1 = mC_p \log_e \frac{T_2}{T_1}$$

Given

$$m = 3 \text{ kg}$$

$$T_1 = 300°\text{C} = 300 + 273 = 573 \text{ K}$$

$$V_2 = 2V_1$$

$$T_2 = ?$$

Using perfect gas equation, the value of T_2 can be determined.

$$\frac{p_1 V_1}{T_1} = \frac{p_2 V_2}{T_2}$$

$$T_2 = \frac{p_2 V_2}{p_1 V_1} \times T_1$$

or $$T_2 = \frac{p_2}{p_1} \times \frac{V_2}{V_1} \times T_1$$

As it is a constant pressure process $\therefore\ p_2 = p_1$ and

Substituting the given values, we get

$$T_2 = \frac{p_1}{p_1} \times \frac{2V_1}{V_1} \times 573$$

$$T_2 = 2 \times 573 = 1146 \text{ K}$$

Now change in entropy can be calculated by using equation

$$\therefore \quad S_2 - S_1 = mC_p \log_e \frac{T_2}{T_1}$$

$$\therefore \quad S_2 - S_1 = 3 \times 1.005 \log_e \frac{1146}{573}$$

$$\therefore \quad S_2 - S_1 = 2.089 \text{ kJ/K}$$

Problem 12. *5 m³ of air at a pressure of 1000 kPa and a temperature of 27 °C expand reversibly in a cylinder to a pressure of 200 kPa. The final volume is 25 m³. Find out*

(i) Heat transferred during the process.

(ii) Change of entropy during the process.

Specify heat for this process is constant.

Solution. Given.

$V_1 = 5 \text{ m}^3$

$p_1 = 1000$ kPa

$T_1 = 27 + 273 = 300$ K

$p_2 = 200$ kPa

$v_2 = 25 \text{ m}^3$

Using perfect gas equation we can find out T_2

$$\frac{p_1 V_1}{T_1} = \frac{p_2 V_2}{T_2}$$

$$T_2 = \frac{p_2 V_2}{p_1 V_1} \times T_1$$

$$T_2 = \frac{p_2}{p_1} \times \frac{V_2}{V_1} \times T_1$$

$$T_2 = \frac{200}{1000} \times \frac{25}{5} \times T_1$$

$$T_2 = T_1$$

Since $T_2 = T_1$, it means that the process is a constant temperature or an isothermal process. The change in internal energy for this process is zero

$$\therefore \quad dU = U_2 - U_1 = 0$$

The heat transferred during the process is given by

$$\therefore \quad Q_{1-2} = W_{1-2} = p_1 V_1 \log_e \left[\frac{V_2}{V_1}\right]$$

$$\therefore \quad Q_{1-2} = W_{1-2} = 1000 \times 5 \log_e \left[\frac{25}{5}\right]$$

$$\therefore \quad Q_{1-2} = W_{1-2} = 5000 \times \log_e 5 = 8047.19 \text{ kJ. } \textbf{Ans.}$$

Change in entropy for an isothermal process is given by

$$\therefore \quad S_2 - S_1 = m(C_p - C_v) \log_e \frac{V_2}{V_1}$$

Change in entropy for an isothermal process is given by

$$\therefore \quad S_2 - S_1 = R \log_e \frac{V_2}{V_1}$$

$$\therefore \quad S_2 - S_1 = \frac{p_1 V_1}{T_1} \log_e \frac{V_2}{V_1} = \frac{Q_{1-2}}{T_1} = \frac{8047.19}{300} = 26.82 \text{ kJ/kg. } \textbf{Ans.}$$

REVIEW QUESTIONS

1. What is isentropic process? Prove that for an isentropic process pV^γ = C.
2. Show that for a polytropic process

$$Q = \left(\frac{\gamma - n}{\gamma - 1}\right) W$$

Where Q and W are heat and work interactions and n is the polytropic index?

3. State and explain the first law of thermodynamics. *(R.G.P.V. Nov./Dec. 2007)*
4. State and explain the law of conservation of energy. Does it differ from the first law of thermodynamics?
5. Derive the expression for the steady flow energy equation. State clearly the assumptions made.
6. Write down the steady flow energy equation for (*i*) Nozzle, (*ii*) Diffuser, (*iii*) Turbine.
7. Explain the limitations of first law of thermodynamics.
8. Define the following terms: (*i*) heat engine, (*ii*) heat pump, (*iii*) refrigerator.
9. State and explain second law of thermodynamics. *(R.G.P.V. June 2008)*
10. Give the two statements of the second law of thermodynamics. *(R.G.P.V. April 2009)*
11. Write down Kelvin Planck and Clausius statements of the second law of thermodynamics. *(R.G.P.V. Sep. 2009)*

12. Justify the following statement

"It is impossible to construct an engine to work between two heat reservoirs each having a fixed and uniform temperature, which will exceed in efficiency a reversible engine working between the same reservoir".

13. Show that coefficient of performance of heat pump and refrigerator can be related

$$COP_{REF} = COP_{HEAT\ PUMP} - 1$$

14. Name the operations involved in Carnot cycle. Does the efficiency of Carnot cycle depend upon the nature of its working substance?

15. During winter season a house to be kept at 21°C by using a heat pump. The ambient temperature is 0°C, find out the power required to run the heat pump if the heat transfer from the house takes place at 15 kJ/s. *(1.07 kW)*

16. A boiler generates steam at the rate of 1200 kg/hr. The water enters the boiler with an enthalpy of 190 kJ/kg and velocity of 10 m/s. Steam leaves the boiler with an enthalpy of 2780 kJ/kg and a velocity of 25 m/s. if the outlet pipe from the boiler is 9100 mm above the inlet pipe, then find out the heat transferred and the fuel consumption rate. The calorific value of the fuel is 31500 kJ/kg. *(51.8 MJ/min, 98.7 kg/hr)*

17. Water enters a boiler as saturated liquid at 8 MPa, h = 1316.7 kJ/kg and leaves as superheated steam with 6 MPa, 500°C having the enthalpy of 3422.2 kJ/kg. Find out the amount of heat energy required to generate one kg of steam under these conditions. Neglect the kinetic potential energy terms and assume negligible heat losses. *(2105.5 kJ/kg)*

18. Air at 17°C and 1.1 bar occupies 0.05 m^3. The air is first heated at constant volume until the pressure is 4.4 bar and then cooled at constant pressure back to original temperature. Find out (*a*) the net heat transfer to or from the air and (*b*) the net entropy change. Given that for air C_p = 1.005 and C_v = 0.718 kJ/kg K *(624.7 kJ, 1.39 kJ)*

19. A heat engine receives 85 kW of heat from a high temperature reservoir and rejects 50 kW to a low temperature sink. Determine the work produced and the thermal efficiency. *(35 kW, 41.2)*

20. If the network output of a cyclic process is 65 kNm and heat input is 135 kJ, find out the efficiency of the cycle. *(48.15%)*

21. In a steam turbine, steam enters at the rate of 20 kg/s. With a velocity of 60 m/s and enthalpy of 3300 kJ/kg and leaves the turbine with a velocity of 150 m/s and enthalpy of 2400 kJ/kg. The specific volume of steam at the inlet is 0.4 m^3/kg. The power developed by the turbine is 17600 kW. Inlet conditions are 100 kPa and 300 K. Find out the rate at which heat is rejected by the turbine and the diameter of the inlet pipe.

■■■

Properties of Steam (Properties of pure substance)

7.1 INTRODUCTION

A pure substance is homogenous in nature and retains constant chemical composition throughout its mass even though there may be a change of phase. It may exist in more than one phase. Its typical examples are water, atmospheric air, steam water-mixture etc. Water is an example of a pure substance which exists in all the three phases *i.e.,* solid (ice), liquid, and vapour (steam).

A working substance (or pure substance) is required for converting heat energy into mechanical work. It means a substance, which is capable of receiving heat energy from the combustion of the fuel can be used to run a prime mover (*e.g.,* steam engine or steam turbine) for producing mechanical work. Steam is used as a working substance in the running of the steam engines and steam turbines. Some notable features of the steam are:

- Steam possess the necessary capability to carry large amount of heat in liquid as well as in vapour form.
- Steam is produced from water, which is readily available and it costs a little.
- After serving its useful purpose in the steam engine/turbine, it can be used for heating purposes.

The perfect gas law is not applicable on steam as a vapour unless it is in totally dried condition. Steam in a completely dried state is called superheated steam and then it can be considered to behave like a perfect gas. First Regnault and than Prof. Callender carried out detailed experimentation for finding the properties of the steam. The important properties of the steam are : pressure, temperature, volume, internal energy, enthalpy, and entropy.

7.2 FORMATION OF STEAM

Let us just see how the water from 0°C changes into steam. Consider one kg of water initially at temperature 0°C is taken in a piston cylinder arrangement. The piston is frictionless which exerts a constant pressure P equal to atmospheric pressure on it. Heat is supplied to the water in the cylinder at constant pressure. As heat is given to water from and at 0°C,

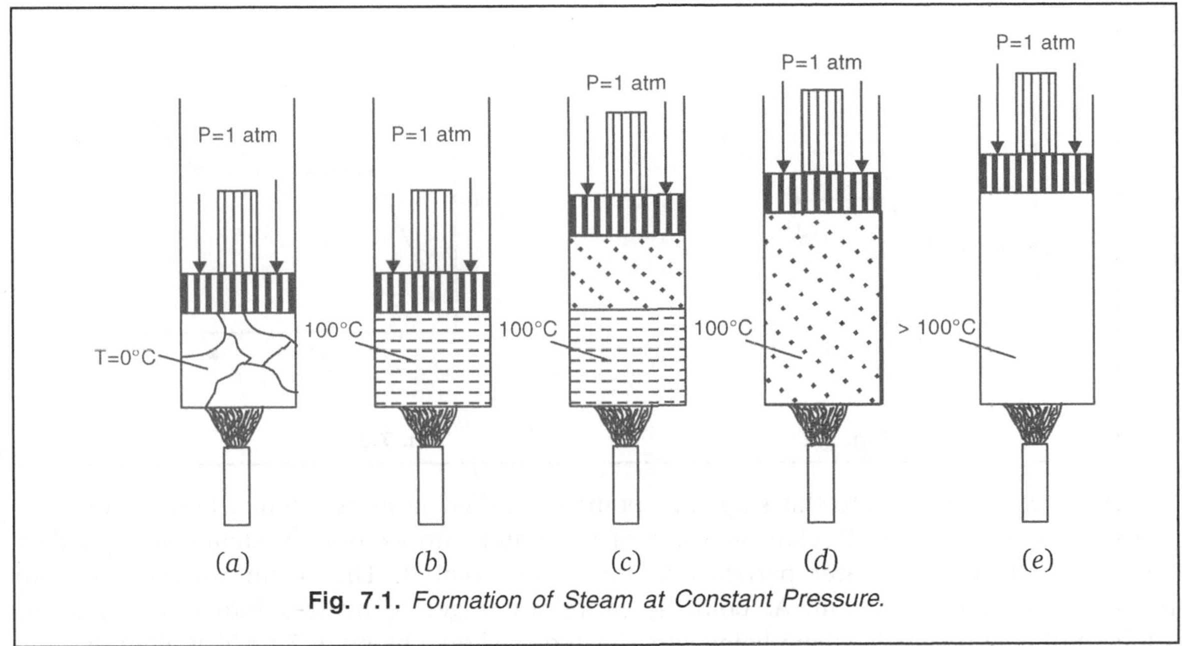

Fig. 7.1. *Formation of Steam at Constant Pressure.*

the water temperature rises until it reaches 100°C as shown in Fig. 7.1 (*b*) and the water starts boiling as shown in Fig. 7.1 (*c*). The temperature at which the water boils depends upon the applied pressure. There is one boiling point for any given pressure. The boiling point is also called as "saturation temperature" (t_S) or the temperature of steam formation. *Now what is this boiling process?* It is nothing but simply a phase transition from liquid to vapour (gaseous steam). The heat added in converting water to steam is known as evaporation enthalpy or latent heat. In Fig. 7.1 (*c*), only a portion of water is evaporated and at this stage the cylinder has a mixture of water and steam. *The mixture of water and steam is known as wet steam.* On further heat addition to the cylinder the entire water gets converted into steam (vapour) as shown in Fig. 7.1 (*d*). The steam produced corresponding to stage (*d*) is known as *dry saturated steam.* If further heat addition is continued at constant pressure at stage (*d*), the temperature of steam increases and this results in increase in the steam volume. Such a state of steam is known as superheated steam [Fig. 7.1 (*e*)].

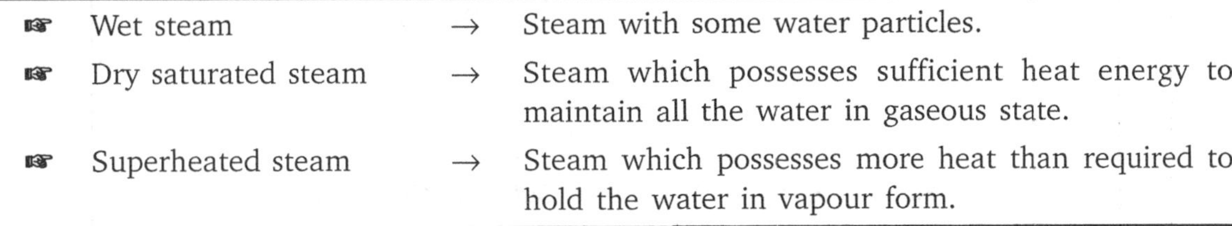

- ☞ Wet steam → Steam with some water particles.
- ☞ Dry saturated steam → Steam which possesses sufficient heat energy to maintain all the water in gaseous state.
- ☞ Superheated steam → Steam which possesses more heat than required to hold the water in vapour form.

Fig. (7.2) illustrates graphically the transformation of water from and at 0°C into steam where heat is added to it at constant pressure. The T–H (temperature-enthalpy) diagram is plotted for one atmospheric pressure. Fig. (7.3) shows a T–V (temperature-volume) diagram for the formation of steam under constant pressure from water at 0°C. In Fig. (7.3), point 1 represents the initial state of water. The water reaches point 2 on adding heat to it.

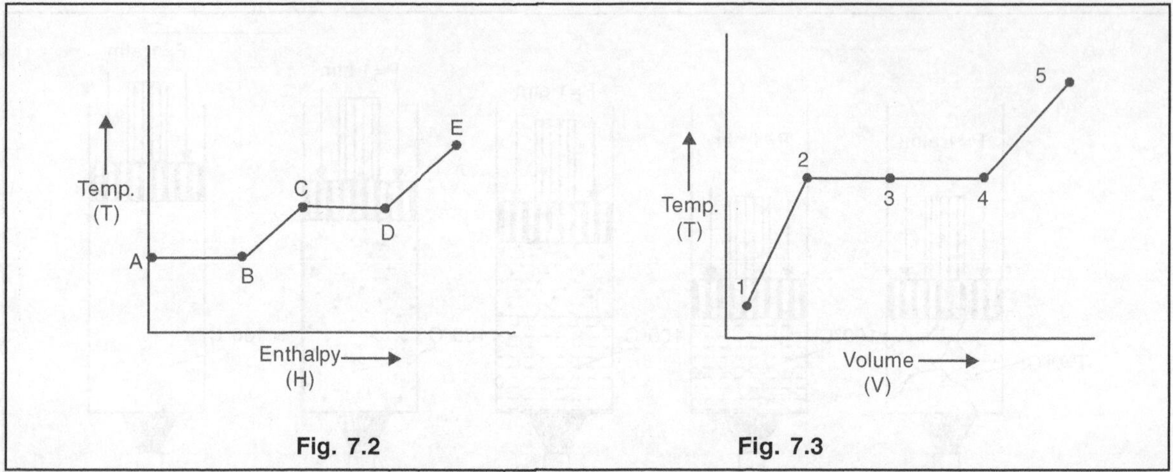

Fig. 7.2 Fig. 7.3

The water corresponding to this stage *i.e.,* point 2 is called saturated liquid. Further addition of heat to water at point 2, changes some of the water into vapour (or steam) and yields a mixture of steam and water particles as shown by point 3. The steam corresponding to point 3 is called wet steam. At point 4, entire water gets converted into steam. Further addition of heat results in steam being superheated as shown by point 5 on T–V diagram. The steam corresponding to point 5 is called super-heated steam.

7.3 EFFECT OF PRESSURE ON SATURATION TEMPERATURE

If the pressure on water is increased, the boiling point of water also gets increased. Because of this fact the water boils at low temperature in hill areas and at high temperature in plain areas. The boiling point of water at particular pressure is called saturation temperature and the corresponding pressure is called saturation pressure. The variation of saturation temperature of water with pressure is shown in Fig. (7.4).

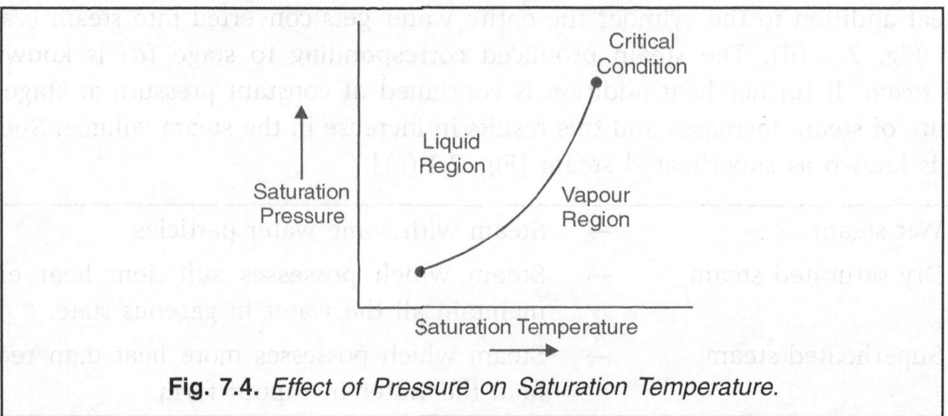

Fig. 7.4. *Effect of Pressure on Saturation Temperature.*

7.4 T-V DIAGRAM FOR A PURE SUBSTANCE (*i.e.,* STEAM) AT DIFFERENT PRESSURES

Fig. (7.5) shows T-V diagrams for the generation of steam at different pressures from water at 0°C. On connecting the locus of point 2 and point 3 one can get a curve as shown in Fig. (7.6). The line connecting the locus of point 2 is called saturated liquid line and the line connecting the locus of point 3 is called saturated vapour line.

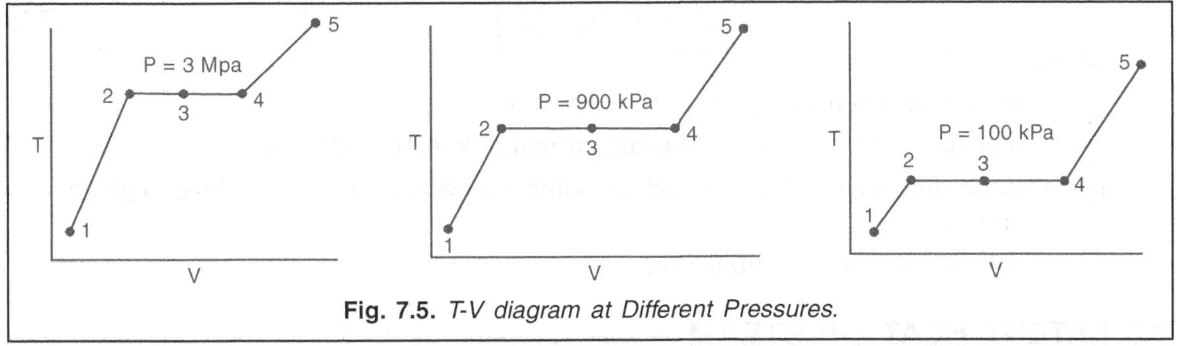

Fig. 7.5. *T-V diagram at Different Pressures.*

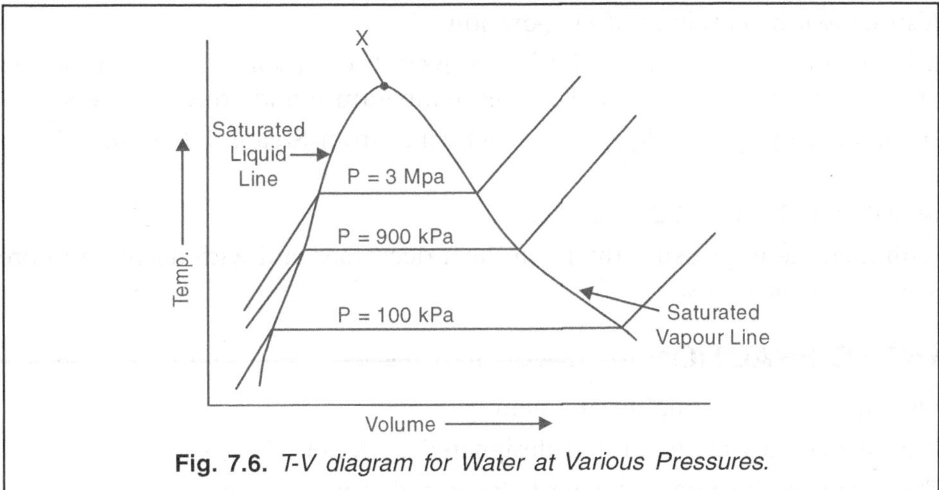

Fig. 7.6. *T-V diagram for Water at Various Pressures.*

The point X is called the critical point. The temperature, pressure and volume corresponding to this point are called critical temperature, critical pressure and critical volume. Above the critical point there is no difference in the proportion of liquid and vapour and there is no significant change in the density.

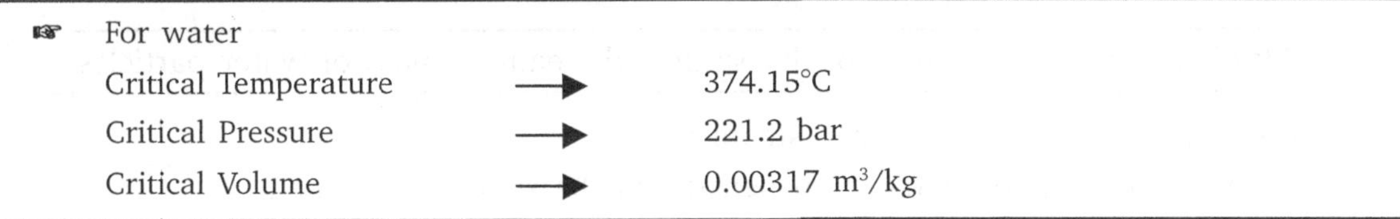

☞ For water

Critical Temperature	⟶	374.15°C
Critical Pressure	⟶	221.2 bar
Critical Volume	⟶	0.00317 m^3/kg

7.5 SENSIBLE HEAT OF WATER (ENTHALPY OF WATER)

It is defined as the amount of heat required to raise the temperature of 1 kg of water from 0°C to the boiling point temperature t_S (or saturation temperature).

- Also called enthalpy of water.
- Designated by symbol h_f. (To get its value from Steam Table refer column 5 of the table.)

- Mathematically enthalpy of water at any temperature t_s above 0°C (initial stage) is given by:

$$h_f = C_p (t_s - t_i) \quad ...(7.1)$$

where

h_f = total heat or enthalpy of saturated water

C_p = specific heat of water at constant pressure = 4.187 kJ/kg K

t_s = saturation temperature or boiling point temperature corresponding to given pressure

t_i = temperature at the initial stage.

7.6 LATENT HEAT OF STEAM

- Also known as enthalpy of evaporation.
- It is the amount of heat required to convert 1 kg of water at saturation temperature into dry saturated steam at the same temperature and constant pressure.
- Designated by symbol $\boldsymbol{h_{fg}}$. (To get its value from Steam Table see column 6 of the table.)
- Its value at 1 bar is 2258 kJ/kg.
- With increase in pressure the latent heat decreases and with decrease in pressure the latent heat increases.

7.7 DRYNESS FRACTION

- Also known as the quality of steam.
- It is related to wet steam and designated by letter 'x'.
- The steam with some water particles is called wet steam.
- Dryness fraction provides an idea about dryness of wet steam.
- It is defined as the ratio of mass of dry saturated steam actually present to the total mass of wet steam.
- Total mass of wet steam is the sum of the mass of the dry saturated steam and the mass of water particles in suspension in the mixture under consideration.

Mass of wet steam = mass of dry saturated steam + mass of water particles

For example: In a sample of wet steam let

m = mass of water particles in suspension in the steam considered.

m_s = mass of dry saturated steam in the steam considered.

Then

$$x = \frac{m_s}{m_s + m} \quad ... (7.2)$$

Thus if dryness fraction of wet steam is x =0.85, then one kg of wet steam contains 0.15 kg of moisture (water particles) in suspension and 0.85 kg of dry saturated steam.

Enthalpy of evaporation of 1 kg of wet steam = $x\, h_{fg}$ kJ/kg.

Enthalpy of 1 kg of wet steam = Enthalpy of saturated water + Enthalpy of evaporation.

$$h_{wet} = h_f + xh_{fg} \quad ... (7.3)$$

7.8 WETNESS FRACTION

- This term is also related to wet steam.
- It is defined as the ratio of the mass of water particles in suspension to the total mass of the wet steam.
- Wetness fraction when expressed in terms of percentage is known as priming.

Let in a sample of wet steam:

m = mass of water particles in suspension in the steam under consideration.

m_s = mass of the dry saturated steam in the mixture.

M = total mass of wet steam = $m_s + m$.

$$\text{Wetness fraction} = \frac{m}{m_s + m} \quad \text{... (7.4)}$$

By adding and subtracting 1 in equation (7.4), we get

$$= \frac{m}{m_s + m} + 1 - 1 = 1 - \left(1 - \frac{m}{m_s + m}\right) = 1 - x$$

Wetness fraction = 1 – Dryness fraction

or

$$\boxed{\text{Wetness fraction + Dryness fraction = 1}} \quad \text{...(7.5)}$$

7.9 ENTHALPY OF DRY SATURATED STEAM

- It is defined as the amount of heat required to raise the temperature of 1 kg of water from 0°C to the temperature of evaporation (t_s) and then convert it into dry saturated steam at the same temperature and pressure.
- Designated by symbol h_g.
- Enthalpy of 1 kg of dry saturated steam

$$\boxed{h_g = h_f + h_{fg}} \quad \text{...(7.6)}$$

7.10 ENTHALPY OF SUPERHEATED STEAM

- In practice steam is superheated by passing it through superheater.
- It is defined as the amount of heat required to convert 1 kg of water from its freezing point (0°C) into superheated steam.
- The amount of superheating is generally measured by the increase in the temperature of the steam above its saturation temperature.
- Heat added per kg of dry saturated steam during superheating is given by:

  $= C_p (t_{sup} - t_s)$

  (C_p varies from 2.0 to 2.5 kJ/kg K)

- The enthalpy of 1 kg of superheated steam is therefore the sum of the enthalpy of water, enthalpy of evaporation and the heat of superheat.
- It is designated by symbol h_{sup}.

$$h_{sup} = h_f + h_{fg} + C_p(t_{sup} - t_s)$$

$$\boxed{h_{sup} = h_g + C_p(t_{sup} - t_s)} \quad ...(7.7)$$

- ***Degree of superheat*** is the difference between the temperature of superheated steam and that of dry saturated steam *i.e.*, this equals $(t_{sup} - t_s)$.
- ***Heat of superheat*** is the amount of heat supplied during superheating of steam only *i.e.*, this equals $C_p(t_{sup} - t_s)$.

7.11 THERMODYNAMIC PROPERTIES OF STEAM

The main thermodynamic properties of steam which attach great importance to this course are:

1. Pressure
2. Temperature
3. Internal energy
4. Enthalpy
5. Entropy
6. Specific volume.

7.11.1 Specific Volume of Steam

The specific volume of dry saturated steam is the volume occupied by 1 kg of steam. It is designated by symbol V_g. Its unit is m³/kg.

As we know the density of steam is defined as mass per unit volume of steam.

$$\rho = 1/V$$

$$\rho_g = 1/V_g$$

$$v_g = 1/\rho_g \quad ...(7.8)$$

where ρ_g is the density of dry saturated steam and V_g is its specific volume.

If the steam is wet and its dryness fraction is x, then 1 kg of wet steam contains x kg of dry steam and $(1 - x)$ kg of water particles in suspension in the steam.

Specific volume of wet steam

= vol. of dry steam + vol. of water particles in suspension

$$v_g = x\,V_g + (1 - x)V_f \quad ...(7.9)$$

where V_f = specific volume of water

Since the volume of water V_f is very small as compared to volume of dry saturated steam v_g at low pressure. Hence the quantity $(1 - x)V_f$ in equation (7.9) can be neglected.

Specific volume of wet steam $= x\,V_g$

$$\text{Density of wet steam} = \frac{1}{xv_g}$$

In case of superheated steam, the specific volume can be determined by using Charles law as superheated steam is considered to behave like a perfect gas.

$$\frac{V_g}{t_s} = \frac{V_{sup}}{t_{sup}}$$

$$V_{sup} = \frac{V_g\, t_{sup}}{t_s} \qquad \text{... (7.9)}$$

where

V_g = specific volume of dry saturated steam

t_s = saturation temperature

V_{sup} = specific volume of superheated steam.

7.11.2 External Work Done During Evaporation

The amount of heat absorbed due to increase in volume when water is converted into steam is called external work done during evaporation.

1 kg of water from and at 0°C is converted into dry saturated steam at constant pressure. V_f is the volume of 1 kg of water in m^3.

∴ External work done during evaporation per kg of dry saturated steam

$$= \frac{10^5 p\,(V_g - V_f)}{10^3} \text{ kJ/kg} \quad \text{where } p \text{ is pressure (in bar)} \qquad \text{... (7.10)}$$

as we know the V_f is the volume of 1 kg of water is very small as compared to V_g which is the volume of 1 kg of dry saturated steam. Hence neglecting V_f we get

∴ External work done during evaporation per kg of dry saturated steam

$$= \frac{10^5 p\,(V_g)}{10^3} \text{ kJ/kg} \quad \text{where } p \text{ is pressure (in bar)} \qquad \text{...(7.11)}$$

$$= 100\, pv_g \text{ kJ/kg}$$

Similarly external work done per kg of wet steam = $100\, px\, V_g$ kJ/kg

and external work done per kg of superheated steam = $100\, p\, V_{sup}$ kJ/kg

where p is the pressure of steam in bar and V_g is the volume of dry saturated steam in m^3 at pressure p.

7.11.3 Internal Latent Heat of Steam

The total heat supplied h_{fg} is the sum of internal latent heat and the external work done during evaporation. The internal latent heat h_{fgi} is defined as the difference between the total latent heat h_{fg} and the external work done during evaporation.

For dry saturated steam

$$\therefore \quad h_{fg} = h_{fgi} + 100p\, V_g \text{ kJ/kg} \qquad \text{...(7.12)}$$

$$h_{fgi} = h_{fg} - 100p\, V_g \text{ kJ/kg}$$

Similarly for wet steam

$$h_{fgi} = x.h_{fg} - 100px\, V_g \text{ kJ/kg}$$

Where p is the pressure of steam in bar, V_g the volume of dry saturated in m^3 at pressure p, x is the dryness fraction of steam and h_{fg} is the enthalphy of evaporation in kJ/kg.

7.11.4 Internal Energy of Steam

As we know the total enthalpy of steam is the sum of internal energy of the steam and the external work done during evaporation.

So the internal energy of steam is defined as the difference between the total enthalpy of steam and the external work done during evaporation. It is designated by letter 'U'.

Work done during evaporation of 1 kg of dry steam $= 100\ p\ V_g$ ...(7.13)

Enthalpy of dry saturated steam $\rightarrow$ $h_g = h_f + h_{fg}$

For dry saturated steam $U = h_g - 100\ p\ V_g$ kJ/kg ...(7.14)

Similarly,

For wet steam $U = h_g - 100\ pxV_g$ kJ/kg

For superheated steam $U = h_{sup} - 100\ pV_g$ kJ/kg

7.11.5 Entropy of Water

Let 1 kg of water is heated from initial temperature T_o to the saturation temperature T_S at constant pressure. Then from the definition of enthalpy, change in entropy is given by:

$$dS = \frac{\delta q}{T} = \frac{C_p dT}{T} \qquad \text{... (7.15)}$$

$$\text{Therefore,} \quad S_2 - S_1 = \int_{T_o}^{T_s} C_p \frac{d_T}{T_o} = C_p \log_e \frac{T_s}{T_o} \qquad \text{... (7.16)}$$

But the entropy of water is generally reckoned above 0°C or 273 K.

∴ Entropy of saturated water (S_f) is given by

$$S_f = C_p \log_e \frac{T_s}{273.16} \qquad \text{... (7.17)}$$

as we know $C_p = 1$ for water

$$\therefore \quad \boxed{S_f = \log_e \frac{T_s}{273.16}}$$

7.11.6 Entropy of Steam

Change in entropy of steam during evaporation (S_{fg}) is given by

$$S_{fg} = h_{fg} / T_S \qquad \text{... (7.18)}$$

7.12 STEAM PROPERTY CHARTS

Steam property charts are used to find out the various steam properties. The commonly used charts are:

- Temperature—Enthalpy chart (T–S diagram)
- Enthalpy—Entropy chart (H–S diagram) or Mollier chart

7.12.1 Temperature Entropy Chart (T–S diagram)

This diagram is very useful in determining the properties of steam. In this diagram absolute temperature is plotted as ordinate (along Y-axis) with origin as absolute zero and entropy is plotted about the abscissa (X-axis). The T–S diagram is constructed by plotting the saturation temperature of water and steam against the corresponding values of the entropy at the same temperature corresponding to various values of absolute pressure. The properties of steam can be read directly from the chart. The curve line OA is called the *liquid line or water line*. The line OA represents the relationship between entropy and temperature before the start of steam formation. The curve BD represents *dry saturated steam line*. The curve BD represents that stage of relationship between the temperature and entropy when all the water has been transformed into dry saturated steam.

The area between water line and dry saturated steam line is known as *wet steam region* whereas the region to the right of curve BD is called the *superheated steam* region. On further adding heat at point B will cause the temperature and entropy of steam to increase further as shown by curve BC in Fig. (7.7).

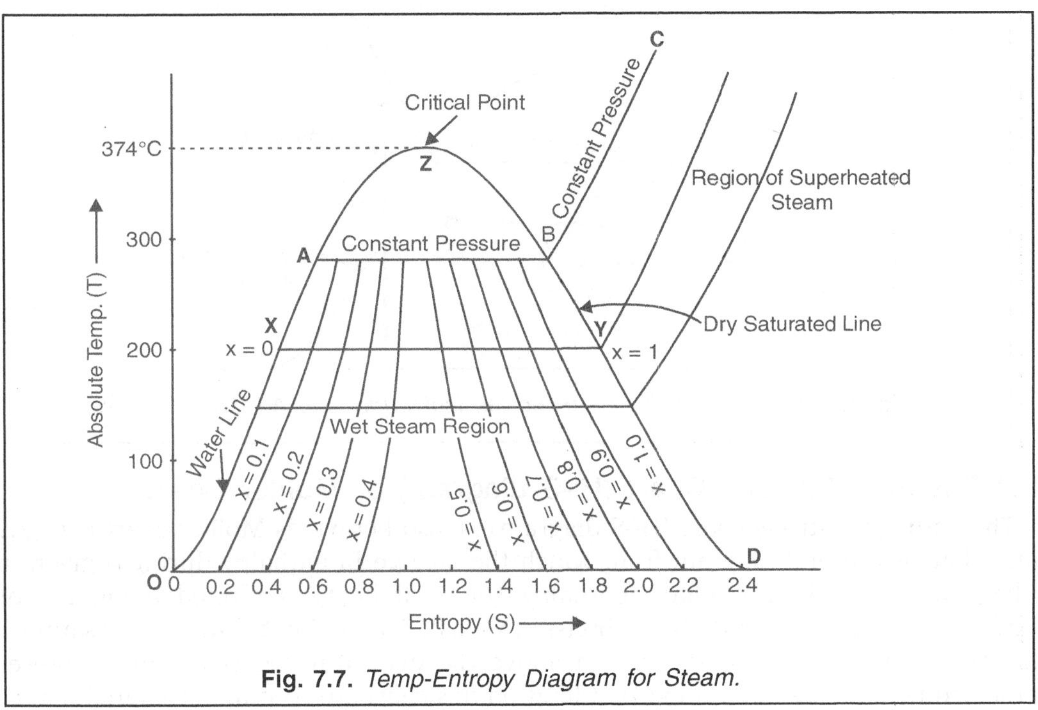

Fig. 7.7. *Temp-Entropy Diagram for Steam.*

The lines of *dryness fraction* can also be plotted in T–S diagram. A horizontal line between the water line and dry saturated steam line represents the process of evaporation of 1 kg of water at a definite temperature. For example point X (dryness fraction $x = 0$) represents 1 kg of water at the particular temperature whereas the point Y (dryness fraction $y = 1.0$) represents 1 kg of dry saturated steam at the same temperature. Any other point on this line XY represents 1 kg of wet steam. Corresponding to different temperatures, different horizontal lines can be plotted between the liquid line and dry saturated steam line. If these lines are equally divided say into 10 parts (*i.e.,* 0.1, 0.2, 0.3, 0.4, 0.5, 0.6, 0.7, 0.8, 0.9, 1.0) and then join the corresponding points on the other horizontal lines, the lines obtained thus represent 1 kg of wet steam at different temperatures and different values of dryness fractions. Any point on water line represents zero dryness fraction and at any point on the dry saturated steam line the value of dryness fraction is equal to 1. The point Z is the critical point. The temperature and pressure corresponding to this point is called critical temperature (374°C) and critical pressure (225 kg/cm²). The water gets transformed into dry saturated steam without absorbing any latent heat at the critical point.

- ☞ An isothermal process (1–2) is represented in T–S diagram Fig. (7.8) by a horizontal line because the temperature during the process remains constant.
- ☞ An adiabatic process (3–4) is represented by a vertical line in the T–S diagram Fig. (7.8). The entropy remains constant during the adiabatic process, as there is no heat addition or rejection takes place during the process.
- ☞ A throtting process cannot be truly represented in the T–S diagram as it is an irreversible process.

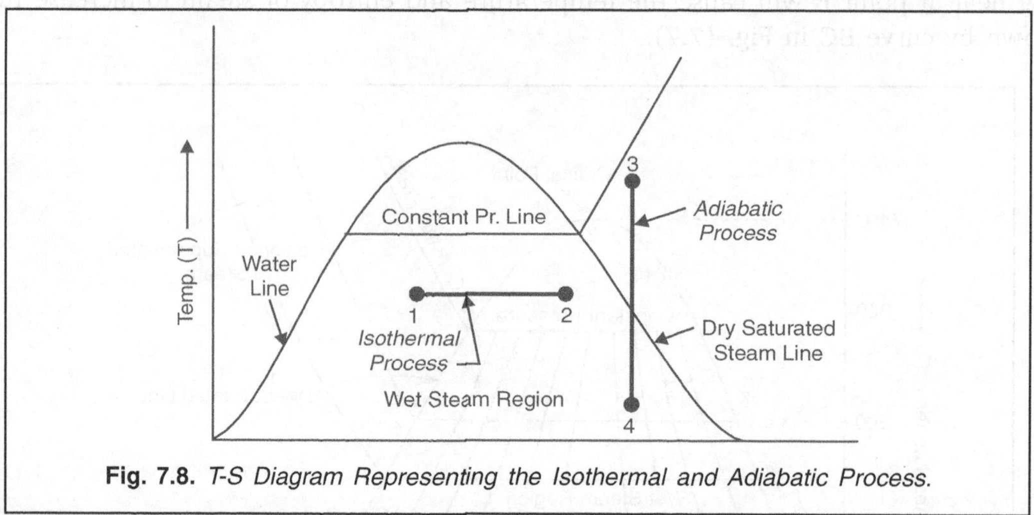

Fig. 7.8. *T-S Diagram Representing the Isothermal and Adiabatic Process.*

7.12.2 Enthalpy–Entropy Chart (H–S diagram) or Mollier Chart

The enthalpy-entropy chart (H–S diagram) is also known as Mollier chart Fig. (7.9). The Mollier chart is a useful diagram from which the change in enthalpy during a given process can be determined. In this chart, enthalpy (or total heat) is plotted along the ordinate (*Y*-axis) and entropy is plotted along abscissa (X-axis). The region below the dry saturation line is the wet steam region and the region above the dry saturated steam line represents the superheated region. The other important lines on this chart are constant pressure lines, constant temperature lines, constant volume lines and constant dryness fraction lines. The lines of constant dryness fraction are shown in the wet steam region where as the lines of constant temperatures are shown in the region of superheat. The constant pressure lines are straight in the wet steam region and curved in the superheated steam region.

- ☞ The throttling process is represented by a horizontal line on Mollier chart, as it is a constant enthalpy process.
- ☞ The reversible adiabatic process is represented by a vertical line on Mollier chart, as it is a constant entropy process.

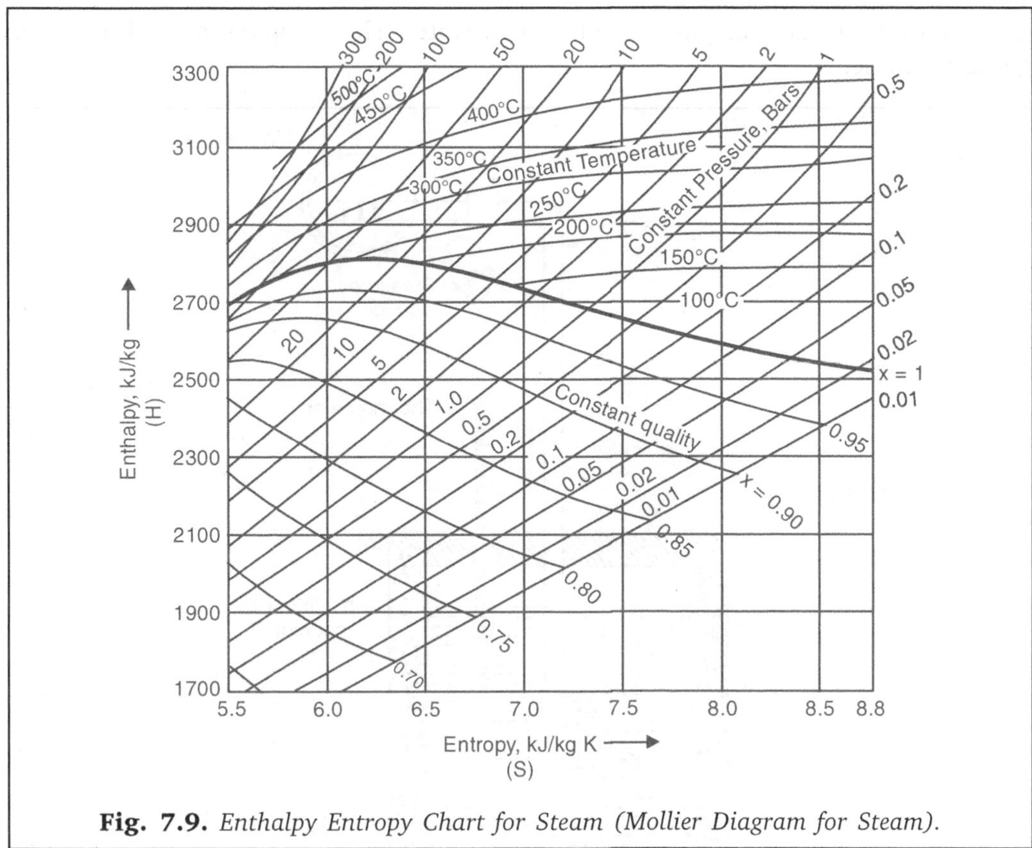

Fig. 7.9. *Enthalpy Entropy Chart for Steam (Mollier Diagram for Steam).*

One of the benefits of H–S chart is that the drop in total heat during adiabatic process can be directly read from it. If the initial and final conditions of the steam of a particular process are known then in such cases H–S chart can be used to find out all the changes in properties during a process.

7.13 MEASUREMENT OF DRYNESS FRACTION

It is very important to know the actual state of wet steam before using steam for any application. To know the actual state of wet steam, one has to find out the dryness fraction of wet steam.

The dryness fraction of steam can be determined experimentally with the help of steam calorimeters. Four calorimeters used for measuring the dryness fraction of the wet steam are discussed below.

7.13.1 Barrel Calorimeter

It is very easy to determine the dryness fraction of the steam by a barrel or tank calorimeter. It consists of a copper calorimeter which is placed in a wooden block with a wooden cover as shown in Fig. (7.10). An outer vessel surrounds the copper calorimeter and some air gap is kept between the two vessels to prevent the heat transfer from the copper calorimeter to the outside. The value of water equivalent of the calorimeter is stamped on it. The calorimeter contains a fixed quantity of cold water. The calorimeter is covered with a wooden cover, through which a thermometer passes and dips in water. The complete assembly is placed on a weighing machine.

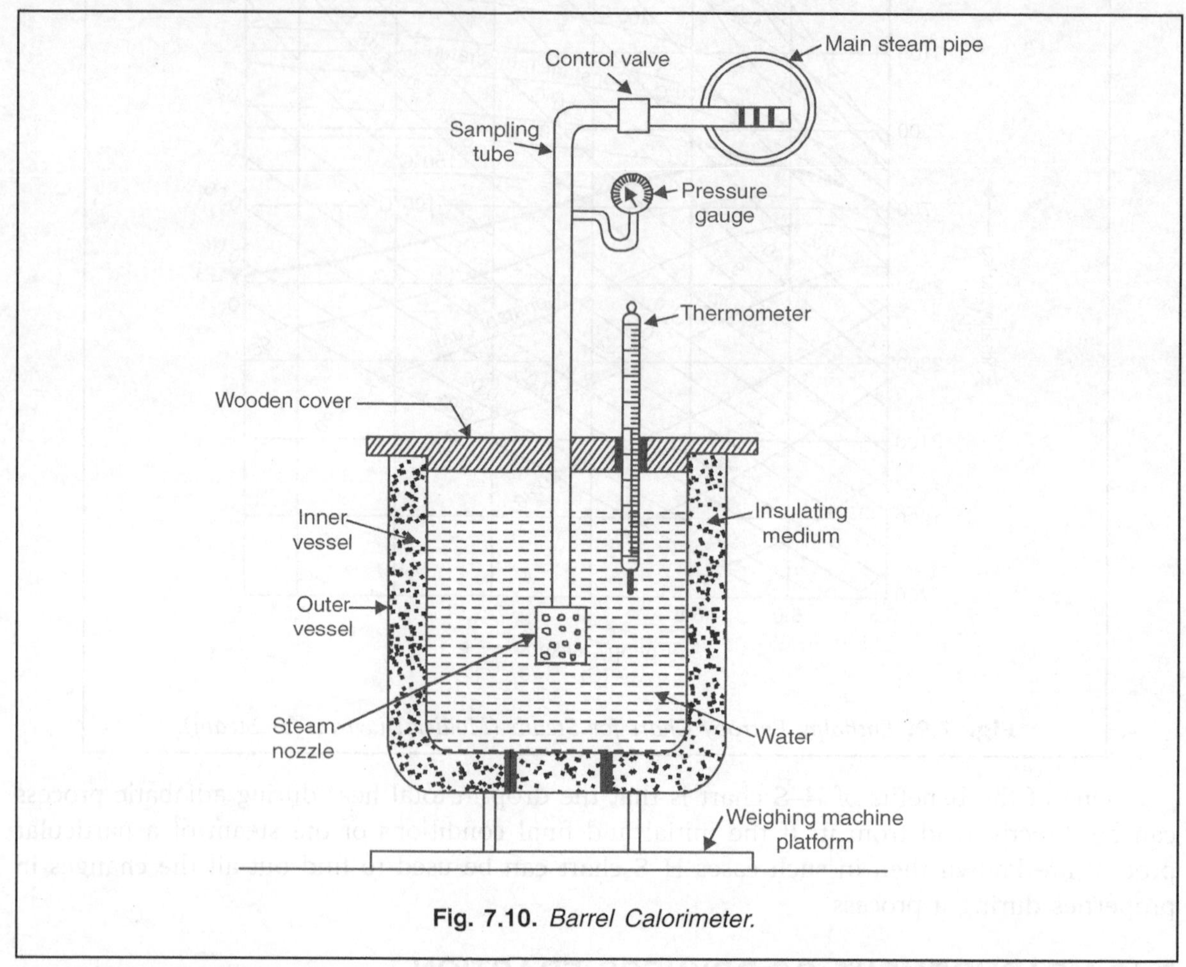

Fig. 7.10. *Barrel Calorimeter.*

A known quantity of steam whose dryness fraction is to be determined, is passed through the sampling tube to water in the calorimeter. The steam on entering in the calorimeter condenses and raises the temperature of the water. The mass of the steam condensed is measured by a weighing machine from the difference of readings of the weighing machine before and after the condensation. An ordinary thermometer is used to find the increase in temperature.

Let

t_1 = initial temperature of cold water in °C in the calorimeter before condensation

t_2 = temperature of water in °C in the calorimeter after condensation

x = dryness fraction of steam

p = pressure of steam

h_{fg} = enthalpy of evaporation corresponding to pressure p.

t_s = saturation temperature corresponding to pressure p.

m = mass of steam condensed.

M_1 = mass of cold water in the calorimeter before the steam supply.

M_2 = water equivalent of calorimeter in kg.

C_p = specific heat at constant pressure.

$$M = M_1 + M_2$$

From the law of conservation of energy, we can write

Heat lost by wet steam = heat gained by cold water

$$m\ \{x\ h_{fg} + C_p\ (t_s - t_2)\} = C_p\ M\ (t_2 - t_1)$$

$$\therefore \qquad x = \frac{\left\{C_p \frac{M}{m}(t_2 - t_1)\right\} - C_p(t_s - t_2)}{h_{fg}} \qquad ...(7.19)$$

It gives an approximate value of dryness fraction. In order to get the correct results there should be a significant rise in temperature but by the same time one should also keep an eye on the excessive rise in temperature otherwise error will creep in on account of heat losses due to conduction and convection.

7.13.2 Separating Calorimeter

This calorimeter is used to test steam of practically any degree of wetness. This calorimeter is based on the principle of separating the water mechanically from the steam by suddenly changing the direction of steam flow and water is collected in a separate chamber where its mass can be measured with the help of a scale which reads in kg.

Figure (7.11) shows a schematic diagram of a separating calorimeter. The separating calorimeter consists of two chambers, the inner chamber and outer chamber. The space between the chambers forms a jacket. The inner chamber is connected to a glass tube water gauge. The outer chamber is connected to steam gauge, which indicates the pressure in the outer chamber. The steam under test is passed through a sampling tube from the steam pipe into outer chamber. The moisture in the steam is separated mechanically by changing the direction of wet steam as indicated by arrows in the Fig. (7.12). Steam then passes through the perforated tubes where the water separates due to inertia of the droplets. The steam leaving the outer chamber is condensed in the bucket calorimeter.

Let

m = mass of water separated in a specified time say 10 minutes.

M = mass of steam measured in the same interval of time.

x = dryness fraction of the steam.

then the dryness fraction of steam by definition is

$$\therefore \qquad x = \frac{M}{M + m} \qquad ...(7.20)$$

In practice it is very difficult to separate out all the suspended water particles from wet steam so the dryness fraction calculated from this calorimeter may not be accurate. Its main benefit lies in its ability to quickly determine the dryness fraction of the highly wet steam.

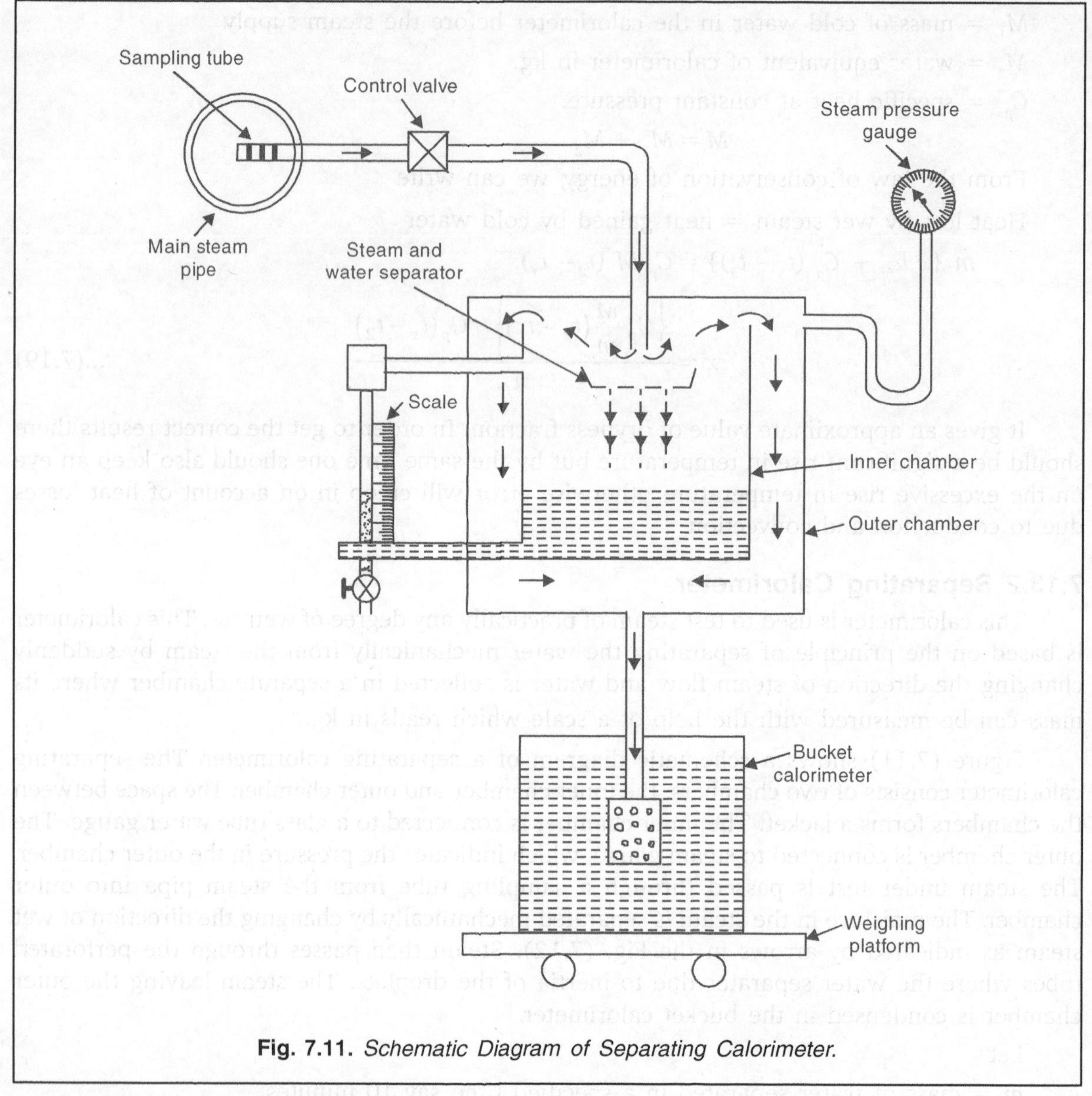

Fig. 7.11. *Schematic Diagram of Separating Calorimeter.*

7.13.3 Throttling Calorimeter

This calorimeter gives very good results. The working principle of this calorimeter is based on the fact that the steam at medium or high pressure has more enthalpy than that at low-pressure.

A schematic diagram of throttling calorimeter is shown in Fig. (7.12.) The pressure of steam drops when it passes through the small opening. Since no external work is done by the steam when it passes through a small opening (orifice) in the throttle valve, the enthalpy of steam remains constant before and after the throttling process. This fact ensures that if the steam is initially wet, the pressure drop during throttling will produce enough heat to raise the temperature of wet steam *i.e.*, the steam gets superheated. It consists of an inner chamber having provision for holding a thermometer at its centre. The inner chamber has a throttle valve (orifice) at the top through which steam enters the calorimeter. The outer chamber, which is placed around the inner chamber is insulated to prevent the heat losses.

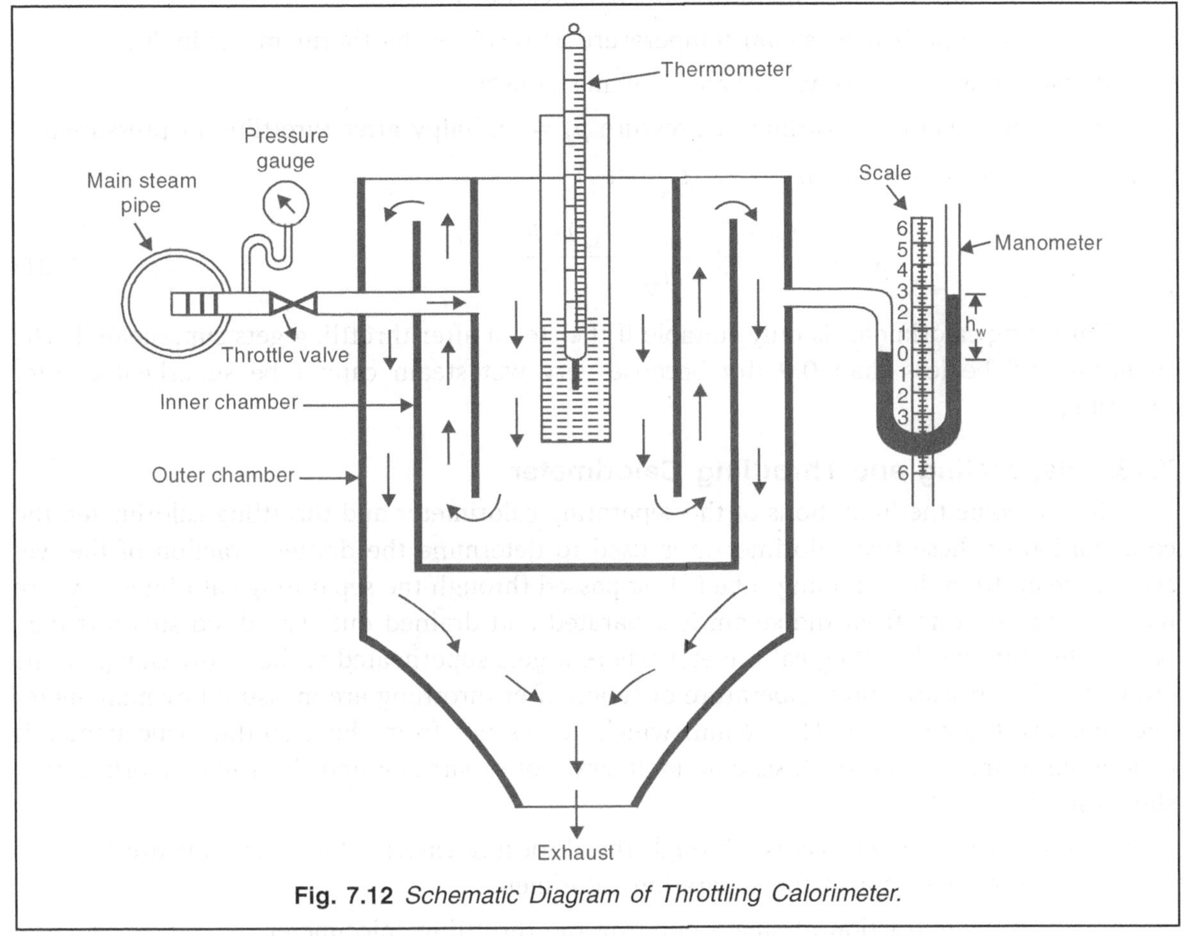

Fig. 7.12 *Schematic Diagram of Throttling Calorimeter.*

The pressure of steam before entering the throttle valve is measured by a steam pressure gauge connected to main steam pipe. The steam expands down to a lower pressure until it becomes superheated (usually by 5°C of superheat). This superheated steam then enters into the inner chamber of the calorimeter. First the steam moves down and then moves upward and then come out from the inner chamber and enters in the outer chamber. The temperature of throttled steam is measured by a centrally located thermometer in the inner chamber and pressure of steam after throttling is measured by means of a manometer gauge.

Let p_1 = initial pressure of wet steam or dry steam before throttling in kPa.

x_1 = dryness fraction of wet steam

h_{f1} = enthalpy of water at p_1.

h_{fg1} = enthalpy of vaporization at p_1 in kJ.

p_2 = final pressure in kPa.

C_p = specific heat of superheated steam after throttling in kJ/kg.

h_{g2} = enthalpy of saturated steam at p_2 in kJ/kg.

t_{S2} = saturation temperature at final pressure in °C.

t_{sup2} = superheated steam temperature as recorded by thermometer in °C.

As we know in throttling enthalpy remains constant

∴ Enthalpy before throttling at pressure p_1 = enthalpy after throttling at pressure p_2

$$h_{f1} + x_1 h_{fg1} = h_{g2} + C_p (t_{sup2} - t_{S2})$$

$$x_1 = \frac{h_{g2} + C_p (t_{sup2} - t_{s2}) - h_{f_1}}{h_{fg1}} \quad \text{... (7.21)}$$

Throttling calorimeter is only suitable if the steam after throttling gets superheated. The steam should be less than 0.9 dry because very wet steam cannot be superheated after throttling.

7.13.4 Separating and Throttling Calorimeter

To overcome the limitations of the separating calorimeter and throttling calorimeter, the combination of these two calorimeters is used to determine the dryness fraction of the wet steam. Steam from the sampling tube is first passed through the separating calorimeter where most of the moisture from the steam is separated and drained out. The dried steam is next passed through the throttling calorimeter where it gets superheated without any change in its enthalpy. The pressure and temperature of steam after throttling are measured by manometer and mercury thermometer. The steam, which comes out from the throttling calorimeter is condensed in the condenser. A schematic diagram of separating and throttling calorimeter is shown in Fig. (7.13).

Let M = mass of steam passed through the throttling calorimeter after moisture is separated out in the separating calorimeter.

x_2 = dryness fraction of steam entering the throttling calorimeter.

m = mass of water particles separated in the separating calorimeter.

x = dryness fraction of the sample steam whose dryness fraction is to be determined.

Total mass of the dry steam entering the throttling calorimeter = Mx_2

Total mass of steam supplied = $M + m$

As we know $$x = \frac{\textit{Mass of dry steam present in the steam mixture}}{\textit{Total mass of steam mixture}}$$

$$= \frac{Mx_2}{M+m} = \frac{M}{M+m} x_2 \quad \text{...(7.22)}$$

∴ $x = x_1 x_2$ where $x_1 = M/(M + m)$ (from equation 7.20)

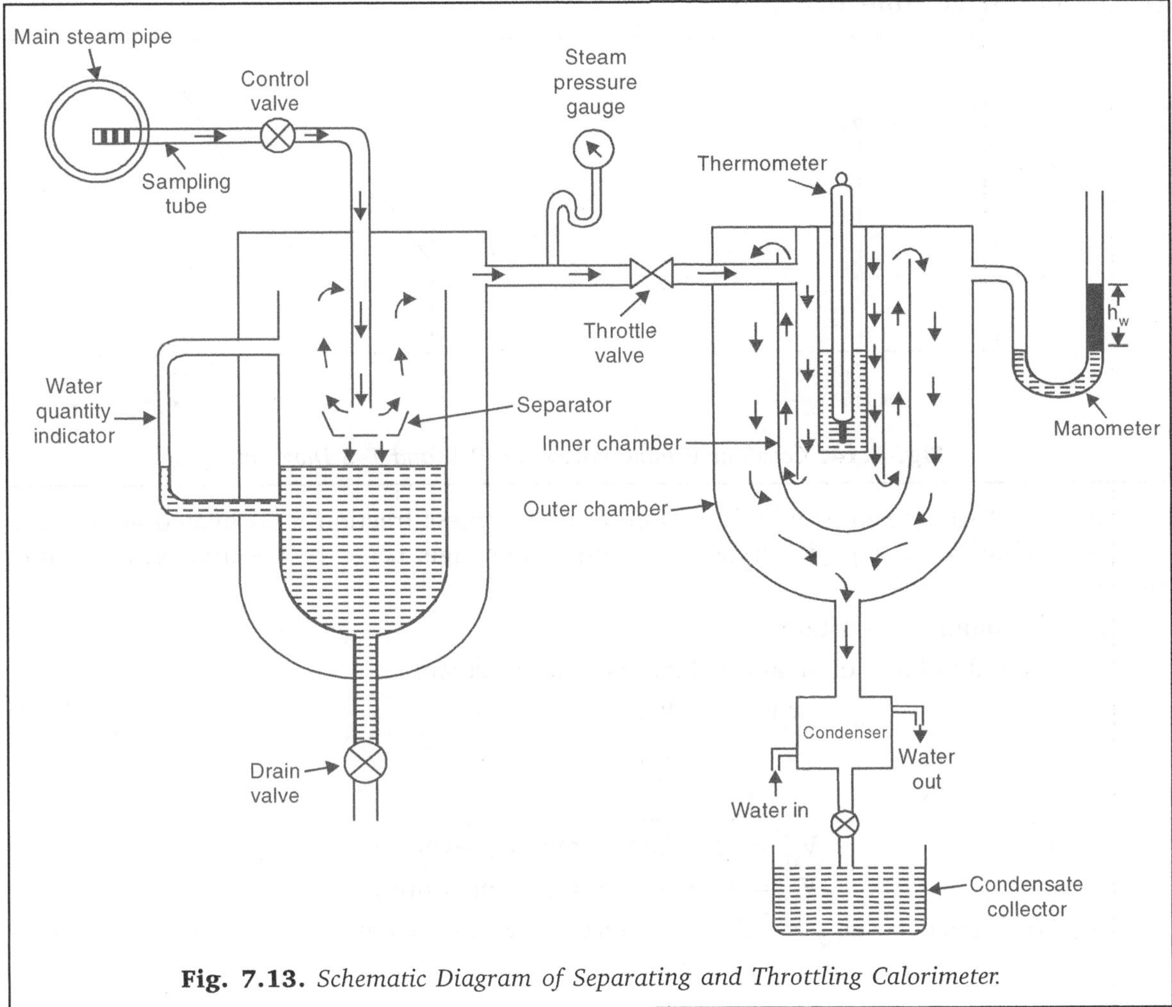

Fig. 7.13. *Schematic Diagram of Separating and Throttling Calorimeter.*

7.14 VAPOUR PROCESSES

The equations derived from ideal gas equation ($PV = mRT$) are not applicable to vapours (steam) whereas the basic equations derived from the first and second law of thermodynamics and different processes involving ways of heating, cooling and expanding the gases hold good to vapours (steam). The equations for work done, heat supplied, internal energy and enthalpy can be applied to the thermodynamic processes (vapours) with minor changes. Following are the important thermodynamic processes:

- Constant volume process
- Constant pressure process
- Constant temperature process (also known as *"isothermal process"*)
- Polytropic process
- Isentropic process (frictionless adiabatic or constant entropy process)
- Constant enthalpy process (also known as *"Throttling process"*).

(A) Constant volume process

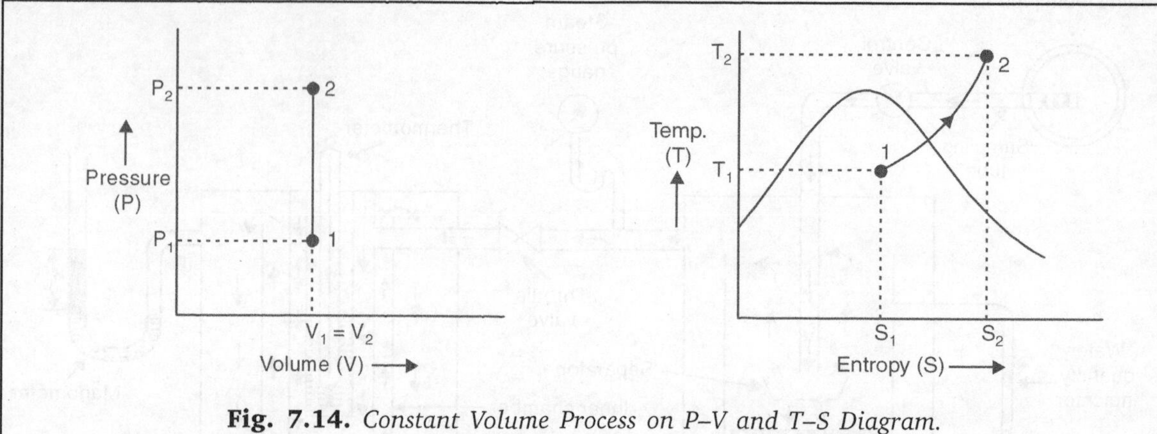

Fig. 7.14. *Constant Volume Process on P–V and T–S Diagram.*

If 1 kg of wet steam at initial pressure p_1 and dryness fraction x_1, is heated at constant volume to final pressure p_2, its dryness fraction will change. Let x_2 be the dryness fraction of steam in final condition.

As the volume is constant

∴ Initial volume of steam = Final volume of steam

$$x_1 V_{g1} = x_2 V_{g2} \quad \text{... (7.23)}$$

$$\therefore \quad x_2 = \frac{x_1 V_{g1}}{V_{g2}}$$

where, V_{g1} = specific volume at pressure p_1

V_{g2} = specific volume at pressure p_2

(*i*) Work done = $w_{1-2} = \int pdV = 0$, since there is no change in volume ∴ $dV = 0$.

(*ii*) Change in internal energy (ΔU)

As we know

Enthalpy (h) = Internal energy (U) + External work of evaporation (pV)

$$h = U + pV \quad \text{... (7.24)}$$

Change in internal energy,

$$\Delta U = U_2 - U_1$$

$$\therefore \quad U_2 = h_2 - p_2 V_2$$

$$\therefore \quad U_2 = h_2 - p_2 V_{sup} \quad \text{... (7.25)}$$

(as in this case $V_2 = V_{sup}$)

where, U_1 = internal energy of 1 kg of steam at pressure p_1 bar

h_1 = Enthalpy of wet steam at p_1 bar in kJ/kg

U_2 = Internal energy of 1 kg of steam at pressure p_2 bar

h_2 = Enthalpy of steam at pressure p_2 bar

The internal energy of steam before heating,

$$U_1 = h_1 - p_1 V_1$$

As we know $V_1 = V_{f_1} + x V_{g_1}$ (neglecting the volume of water V_{f1})

$\therefore$ $V_1 = x V_{g_1}$

$\therefore$ $U_1 = h_1 - p_1 x V_{g_1}$ ... (7.26)

Similarly, internal energy of steam after heating

$$U_2 = h_2 - p_2 V_2 = h_2 - p_2 V_{g2} = h_2 - p_2 V_{sup}$$

where V_{sup} is the volume of the superheated stem in m³/kg

and $h_1 = hf_1 + xV_{g_1}$ ... (7.27)

$h_2 = h_{g_2} + C_p(t_{sup2} - t_{s2})$ ... (7.28)

$V_1 = xV_{g_1}$, neglecting V_{f1} and $V_2 = V_{sup}$ (in this case)

$\therefore$ Change in internal energy $\Delta U = U_2 - U_1$

$$= (h_2 - p_2V_2) - (h_1 - p_1V_1)$$

$$= [(h_2 - h_1) - (p_2V_2 - p_1V_1)]$$

$$= [(h_2 - h_1) - (p_2V_{sup} - p_1xV_{g1})] \quad \text{... (7.29)}$$

(*iii*) Heat transferred

From the first law of thermodynamics as we know

$$\delta q = du + pdV$$

$$_1Q_2 = U_2 - U_1 + \int_1^2 pdV$$

since workdone $= \int_1^2 pdV = 0$

$$\therefore \quad _1Q_2 = U_2 - U_1 = [(h_2 - h_1) - (p_2V_{sup} - p_1xV_{g_1})] \quad \text{...(7.30)}$$

(B) Constant pressure process

The generation of steam in a boiler is an example of constant pressure process heating.

Let us consider the heating of 1 kg of wet steam at point 1 at the constant pressure to become superheated at point 2.

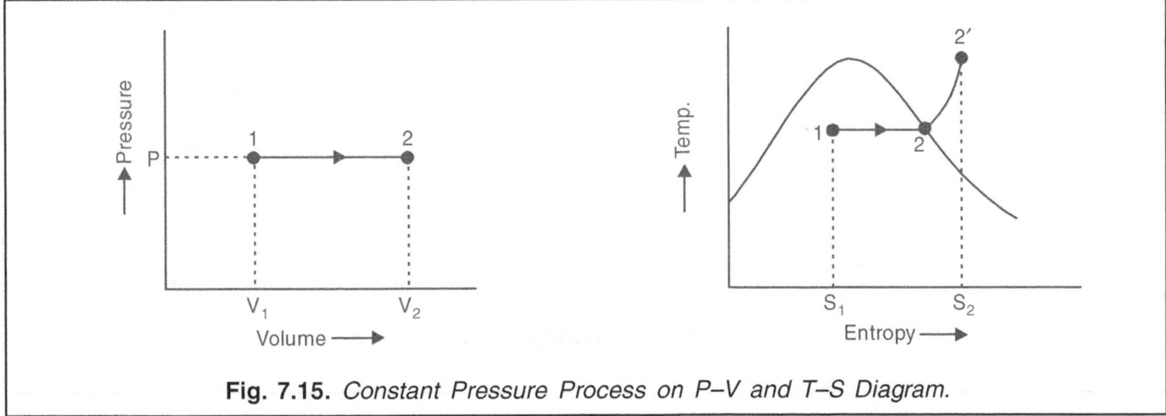

Fig. 7.15. *Constant Pressure Process on P–V and T–S Diagram.*

(*i*) Work done $\int_1^2 pdV = p_1(V_2 - V_1)$ ... (7.31)

where $V_1 = V_{f_1} + xVg_1$ (corresponding to pressure p_1)

and $V_2 = V_{sup}$ (corresponding to temp t_2)

(*ii*) Change in internal energy

$$\Delta U = U_2 - U_1$$
$$= (h_2 - p_2V_2) - (h_1 - p_1V_1)$$
$$= [(h_2 - h_1) - (p_2V_2 - p_1V_1)] \text{ kJ/kg} \quad \text{... (7.32)}$$

(*iii*) Heat transferred

From the first law of thermodynamics as we know

$$\delta q = du + pdV$$

$$_1Q_2 = U_2 - U_1 + \int_1^2 pdV \quad \text{...(7.33)}$$

substituting for $U_2 - U_1$, we get,

$$_1Q_2 = (h_2 - h_1) - \{(p_2V_2 - p_1V_1)\} + p_1(V_2 - V_1)$$

as $p_1 = p_2$ ∴ $_1Q_2 = (h_2 - h_1)$ kJ/kg

where $h_2 = h_{sup}$ and $h_1 = h_{f1} + x\, h_{fg1}$

Thus heat transfer at constant pressure equals the change in internal energy.

(C) Constant enthalpy process

Constant enthalpy process is also known as throttling process. (Fig. 7.16)

In this process steam expands from higher to lower pressure while passing through a small opening (throttle valve or orifice). During the process of expansion no work is done and no heat is supplied. The enthalpy or total heat of steam remains constant before and after the expansion of steam.

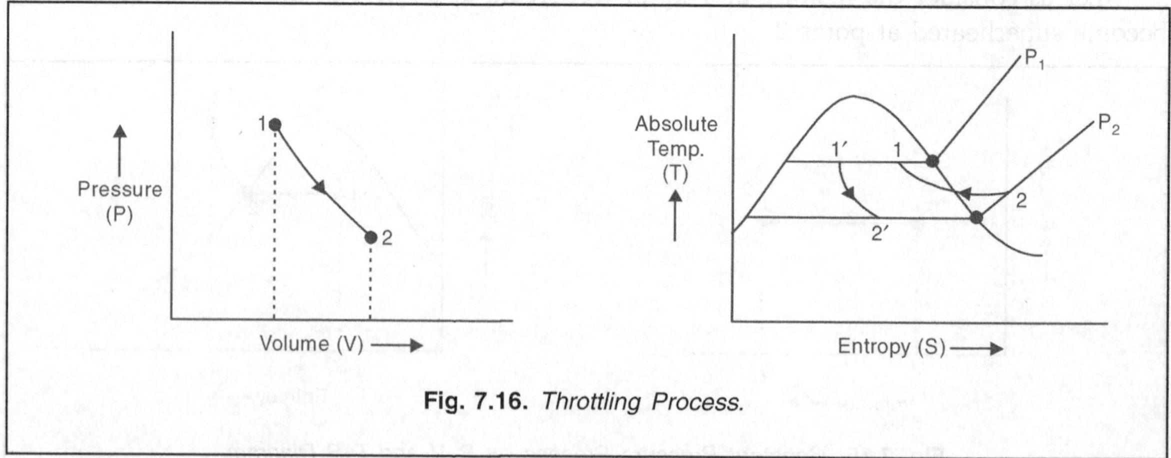

Fig. 7.16. *Throttling Process.*

Let us consider 1 kg of steam at a pressure p_1 bar having dryness fraction x is expanded by throttling process to p_2 bar.

$\therefore$ Enthalpy before expansion (throttling) = Enthalpy after expansion (throttling)

Enthalpy of wet steam = enthalpy of super heated steam

$$h_{f1} + x\, h_{fg1} = h_{f2} + h_{fg2} + C_p(t_{sup2} - t_{s2})$$

$$\therefore \quad x = \frac{h_{f2} + h_{fg2} + C_p\,(t_{sup2} - t_{s2}) - h_{f1}}{h_{fg1}} \quad \text{... (7.34)}$$

SOLVED PROBLEMS

Problem 1. *Find out the heat required to convert 5 kg of water at 40°C into 90% dry steam at 10 bar. Take specific heat of water as 4.187 kJ/kg K.*

Solution. From steam table at 10 bar

$$h_f = 762.81 \text{ kJ/kg}$$
$$h_{fg} = 2015.3 \text{ kJ/kg}$$

Enthalpy of 1 kg of wet steam at 10 bar and 0.9 dry:

$$h_{wet} = 762.81 + 0.90(2015.3)$$
$$= 2576.58 \text{ kJ/kg}$$

Enthalpy of 1 kg of feed water at 40°C

$$h_{feed} = mC_p\,(t_{final} - t_{ini})$$
$$h_{feed} = 1\times 4.187\times(40 - 0) = 167.48 \text{ kJ/kg}$$

Quantity of heat required to convert 1 kg of feed water from and at 40°C into 90% dry steam (*i.e.*, steam is wet and its dryness fraction is 0.90) at 10 bar

$$= 2576.58 - 167.48$$
$$= 2409.1 \text{ kJ/kg}$$

Quantity of heat required to convert 5 kg of water from and at 40°C into 90% dry steam (*i.e.*, steam is wet and its dryness fraction is 0.90) at 10 bar

$$= 2409.1 \times 5$$
$$= \mathbf{12045.5 \text{ kJ.}}$$

Problem 2. *A pressure cooker has 2 kg of steam at 5 bar pressure and 0.9 dry. What quantity of heat be rejected so as to get 60% dry steam from the pressure cooker.*

Solution. The cooling in a pressure cooker is a constant volume process. As we know in a constant volume process the quantity of heat transferred is equal to change in internal energy. So let us first find out the internal energy at the initial stage and then at the final stage.

Initially the condition of steam is wet, hence first determine the internal energy of wet steam by using equation:

$$U = h_{wet} - 100\ pxV_g$$

and $$h_{wet} = h_f + xh_{fg}$$

From steam table at 5 bar

$$h_f = 640.23 \text{ kJ/kg}$$
$$h_{fg} = 2107.5 \text{ kJ/kg}$$
$$V_g = 0.3749 \text{ m}^3\text{/kg}$$

Dryness fraction $x = 0.90$ (given)

∴ Initial internal energy of steam *i.e.*,

$$U_1 = 640.23 + 0.90 \times 2107.5 - 100 \times 5 \times 0.90 \times 0.3749$$
$$= 2369.18 \text{ kJ/kg}$$

In the pressure cooker the volume of steam before and after the heat rejection remains the same.

∴ Volume before heat rejection = Volume after the heat rejection

$$x_1 V_{g_1} = x_2 V_{g_2}$$
$$x_1 = 0.90$$
$$x_2 = 0.60$$
$$V_{g_1} = 0.3749 \text{ m}^3\text{/kg}$$

∴ $$V_{g_1} = \frac{x_1 V_{g_1}}{x_2} = \frac{0.90 \times 0.3749}{0.60}$$

$$V_{g2} = 0.562 \text{ m}^3\text{/kg}$$

Now corresponding to this value of specific volume (*i.e.*, $V_{g_2} = 0.562$ m³/kg), the pressure from the steam table can be read directly *i.e.*, $p_2 = 3.25$ bar.

Final value of internal energy is given by

$$U = h_{wet} - 100\ pxVg$$

and $$h_{wet} = h_f + xh_{fg}$$

From steam table at 3.25 bar

$$h_f = 573.25 \text{ kJ/kg}$$
$$h_{fg} = 2155.8 \text{ kJ/kg}$$

Dryness fraction $x = 0.60$ (given)

$$U_2 = 573.25 + 0.60 \times 2155.8 - 100 \times 3.25 \times 0.60 \times 0.562$$
$$U_2 = 1757.14 \text{ kJ/kg}$$

As the cooling in a pressure cooker is a constant volume process therefore

Heat transferred = change in internal energy

$$_1Q_2 = m(U_2 - U_1)$$
$$= 2(1757.14 - 2369.18)$$
$$= \mathbf{-1224.08\ kJ}$$

Negative sign means the heat is rejected from the cooker.

Problem 3. *Calculate the heat supplied to the boiler and volume of 1 kg of steam at a pressure of 15 bar. The feed water temperature is 40°C. When*

(i) Steam is dry saturated

(ii) Steam is wet and dryness fraction is 0.90

(iii) Steam is superheated to 275°C.

Take C_p of superheated steam as 2.1 kJ/kg K and specific heat of water as 4.187 kJ/kg K.

Solution.

(*i*) When steam is dry saturated

From steam table at 15 bar

$$h_f = 844.89 \text{ kJ/kg}$$

$$h_{fg} = 1947.3 \text{ kJ/kg}$$

$$V_g = 0.13177 \text{ m}^3\text{/kg}$$

$$t_s = 198.32°\text{C}$$

Enthalpy (total heat) of dry saturated steam

$$h_g = h_f + h_{fg}$$

$$h_g = 844.89 + 1947.3 = 2792.2 \text{ kJ/kg}$$

Enthalpy of 1 kg of feed water

$$h_{feed} = mC_p \; (t_{final} - t_{ini})$$

$$h_{feed} = 1 \times 4.187 \times (40 - 0) = 167.48 \text{ kJ/kg}$$

∴ Heat supplied to boiler per kg of steam

$$= 2792.2 - 167.48$$

$$= \mathbf{2624.72 \text{ kJ/kg.}}$$

From steam table specific volume of dry saturated steam at 15 bar is

$$= \mathbf{0.13177 \text{ m}^3\text{/kg.}}$$

(*ii*) When steam is wet and dryness fraction is 0.9

Enthalpy of wet steam is $h_{wet} = h_f + x\,h_{fg}$

$$h_{wet} = 844.89 + 0.90 \times 1947.3$$

$$h_{wet} = 2597.46 \text{ kJ/kg.}$$

Enthalpy of 1 kg of feed water

$$h_{feed} = mC_p \; (t_{final} - t_{ini})$$

$$h_{feed} = 1 \times 4.187 \times (40 - 0) = 167.48 \text{ kJ/kg}$$

∴ Heat supplied to boiler per kg of steam

$$= 2597.46 - 167.48 = \mathbf{2430 \text{ kJ/kg.}}$$

Specific volume of wet steam of dryness fraction x

$$= xV_g$$

$$= 0.90 \times 0.13177$$

$$= \mathbf{0.1186 \text{ m}^3\text{/kg.}}$$

(*iii*) When steam is superheated to 275°C

Enthalpy of superheated steam

$$h_{sup} = h_f + h_{fg} + C_p\,(t_{final} - t_s)$$

$$\therefore \quad h_{sup} = 844.89 + 1947.3 + 2.1(275 - 198.32)$$

$$= 2953.22 \text{ kJ/kg}$$

Enthalpy of 1 kg of feed water

$$h_{feed} = mC_p\,(t_{final} - t_{ini})$$

$$h_{feed} = 1 \times 4.187 \times (40 - 0) = 167.48 \text{ kJ/kg}$$

Heat supplied to boiler per kg of Steam

$$= 2953.22 - 167.48 = \mathbf{2785.74\ kJ/kg.}$$

Absolute temperature of saturated steam $T_s = 198.32 + 273 = 471.32$ K

Absolute temperature of superheated steam $T_{sup} = 275 + 273 = 548$ K

$\therefore$ Specific volume of superheated steam $V_{sup} = V_g \times \dfrac{T_{sup}}{T_s} = 0.13177 \times \dfrac{548}{471.32}$

$$= \mathbf{0.1532\ m^3/kg.}$$

Problem 4. *A steam boiler contains 1.5 m³ of dry saturated steam at 14 bar. Find out the temperature and mass of steam.*

Solution. From steam table at 14 bar,

The saturation temperature of steam $t_s = 195.07$°C.

V_g is the specific volume of steam at 14 bar from steam table = 0.1408 m³/kg.

$$\therefore \quad \text{Mass of steam} = \frac{\textit{Volume of Steam}}{\textit{Specific volume of Steam}}$$

$$= \frac{1.5}{0.14084}$$

$$= \mathbf{10.65\ kg.}$$

Problem 5. *Two boilers X and Y are discharging steam in equal proportion into the main steam pipe both at a pressure of 14 bar. The steam from boiler X is at 14 bar and 310⁰C and from the other at 14 bar and 90% dry. What will be the resulting quality of steam after mixing in the main pipe. Take C_p of superheated steam as 2.1 kJ/kg K.*

Solution. Given,

p_1 is pressure of boiler X = 14 bar

t_{sup1} is temperature of superheated steam from boiler A = 310°C

p_2 is the pressure of boiler Y = 14 bar

x_2 is the dryness fraction of steam from boiler Y = 0.90

C_p is specific heat of superheated steam = 2.1 kJ/kg K.

From steam table at 14 bar

$$t_s = 195.07\text{°C}$$

$$V_g = 0.1408 \text{ m}^3\text{/kg}$$

$$h_f = 830.3 \text{ kJ/kg}$$
$$h_{fg} = 1959.7 \text{ kJ/kg}$$

Enthalpy of superheated steam (Boiler X)

$$h_{sup} = h_f + h_{fg} + C_p (t_{sup} - t_s)$$
$$= 830.3 + 1959.7 + 2.1(310 - 195.07)$$
$$= 3031.35 \text{ kJ/kg}$$

Enthalpy of wet steam (Boiler *Y*)

$$h_{wet} = h_f + xh_{fg}$$
$$= 830.3 + 0.90\ (1959.7)$$
$$= 2594.03 \text{ kJ/kg}$$

When these two qualities of steam mix up in the main pipe, then the total heat possessed per kg of steam will be the average of the two enthalpies (total heats).

$$\therefore \quad \text{Average total heat} = \frac{3031.35 + 2594.03}{2}$$
$$= 2812.69 \text{ kJ/kg}$$

Let us assume that resulting steam is wet, then

Total heat (or Enthalpy) of wet steam $= h_{wet} = h_f + xh_{fg}$

$$2812.69 = 830.3 + x\ (1959.7)$$
$$x = \frac{2812.69 - 830.3}{1959.7}$$
$$x = 1.012$$

Since the dryness fraction of the mixed steam from the two boilers X and Y comes out to be more than unity. It means that the steam in the main pipe will be superheated.

Thus enthalpy (or total heat) of this superheated steam is 2812.69 kJ/kg

$$2812.69 = h_f + h_{fg} + C_p\ (t_{sup} - t_s)$$
$$2812.69 = 830.3 + 1959.7 + 2.1\ (t_{sup} - 195.07)$$
$$\therefore \quad 2.1\ (t_{sup} - 195.07) = 2812.69 - 830.30 - 1959.7$$
$$(t_{sup} - 195.07) = 22.69/2.1 = 10.80$$
$$\therefore \quad t_{sup} = 195.07 + 10.80$$
$$= \mathbf{205.87°C.}$$

Hence the resulting steam will be superheated to **205.87°C.**

Problem 6. *Find out the external workdone during evaporation and the internal energy of 1kg of steam at a pressure of 10 bar. When*

(i) Steam is 0.9 dry

(ii) Steam is dry saturated

(iii) Steam is superheated (t_s = 230°C).

Neglect the volume of water.

Solution. From steam table at 10 bar:

$$h_f = 762.81 \text{ kJ/kg}$$
$$h_{fg} = 2015.3 \text{ kJ/kg}$$
$$V_g = 0.1944 \text{ m}^3\text{/kg}$$
$$t_s = 179.91°\text{C}$$

First enthalpy is to be calculated in order to find out the internal energy because as we know

Internal Energy of Steam = Enthalpy of Steam – External Workdone during Evaporation

(*i*) When steam is 0.9 dry

Total heat (or Enthalpy) of wet steam $= h_{wet} = h_f + xh_{fg}$

$$h_{wet} = 762.81 + 0.9 \times 2015.3 = 2576.58 \text{ kJ/kg}$$

External workdone per kg of wet steam $= 100\, pxV_g$

$$= 100 \times 10 \times 0.90 \times 0.1944$$
$$= \mathbf{175\ kJ.}$$

Internal Energy of wet steam U_{wet}

$$= h_{wet} - \text{External workdone during evaporation}$$
$$U_{wet} = 2576.6 - 175 = \mathbf{2401.6\ kJ/kg.}$$

(*ii*) When steam is dry saturated

Enthalpy (total heat) of dry saturated steam $= h_g = h_f + h_{fg}$

$$h_g = 762.81 + 2015.3 = 2778.11 \text{ kJ/kg}$$

External workdone during evaporation per kg of dry saturated steam $= 100\, pV_g$

$$= 100 \times 10 \times 0.1944$$
$$= \mathbf{194.4\ kJ/kg.}$$

Internal Energy of dry saturated steam $= U_{dry} = h_g - 100\, pV_g$

$$U_{dry} = 2778.11 - 194.44$$
$$= \mathbf{2583.67\ kJ/kg.}$$

(*iii*) When steam is superheated to 230°C

Enthalpy of superheated steam $h_{sup} = h_f + h_{fg} + C_p\,(t_{sup} - t_s)$

$$h_{sup} = 762.81 + 2015.3 + 2.1\,(230 - 179.91)$$
$$= 2883.3 \text{ kJ/kg}$$

To find out the external workdone per kg of superheated steam, the specific volume of superheated steam is to be calculated.

$$V_{sup} = V_g \frac{T_{sup}}{T_s} = 0.1944 \times \left(\frac{230+273}{179.91+273}\right) = 0.216 \text{ m}^3\text{/kg.}$$

$\therefore$ External workdone per kg of superheated steam $= 100\ pV_{sup}$

$$= 100 \times 10 \times 0.216$$

$$= \mathbf{216\ kJ/kg.}$$

$\therefore$ Internal Energy of superheated steam $U_{sup} = h_{sup} - 100\ pV_{sup}$

$$U_{sup} = 2883.3 - 216$$

$$= \mathbf{2667.3\ kJ/kg.}$$

Problem 7. *1000 kg of steam at a pressure of 16 bar and 0.9 dry is generated by a boiler per hour. The steam passes through a superheater via boiler stop valve where its temparature is raised to 380°C. If the temperature of feed water is 30°C, determine*

(i) total heat supplied to feed water per hour to produce wet steam

(ii) total heat absorbed in the superheater

Take Specific heat of superheater steam as 2.2 kJ/kg K **(R.G.P.V. Dec. 2006)**

Solution.

Mass of steam generated $m = 1000$ kg/hr

Pressue of steam $p = 16$ bar

Dryness fraction of steam $x = 0.9$

Temperature of superheated steam $t_{sup} = 380°C = 380 + 273 = 653$ K

From steam table at 16 bar

$$h_f = 858.8 \text{ kJ/kg}$$

$$h_{fg} = 1935.2 \text{ kJ/kg}$$

$$t_s = 201.40°C$$

Enthalpy of wet steam is

$$h_{wet} = h_f + x\, h_{fg}$$

$$= 858.8 + 0.9 \times 1935.2$$

$$= 2600.48 \text{ kJ/kg}$$

Enthalpy of 1 kg of feed water

$$h_{feed} = mC_p(t_{final} - t_{Initial})$$

$$= 1 \times 4.187 \times (30 - 0)$$

$$= 125.61 \text{ kJ/kg}$$

Total heat supplied to boiler $= m \times (h_{wet} - h_{feed})$

$$= 1000 \times (2600.48 - 125.61) = 2.47 \times 10^6 \text{ kJ/kg}$$

Heat absorbed by superheater per hour $= m[(1-x)\, h_{fg} + C_p\,(t_{sup} - t_s)]$

$$= 1000[(1-0.9)1935.2 + 2.2(380-201.4)]$$

$$= 1000[0.1 \times 1935.2 + 2.2 \times 178.6]$$

$$= 1000[193.52 + 392.92]$$

$$= 5.86 \times 10^5 \text{ kJ. } \textbf{Ans.}$$

Problem 8. *A boiler is supplying steam at pressure of 12 bar and 95% dry to a steam engine. As the steam flows through the pipeline, it loses 20 kJ/kg of heat due to poor pipe insulation. However during the steam flow its pressure remains constant. What would be the condition of steam at the engine end of the pipeline?*

Solution.

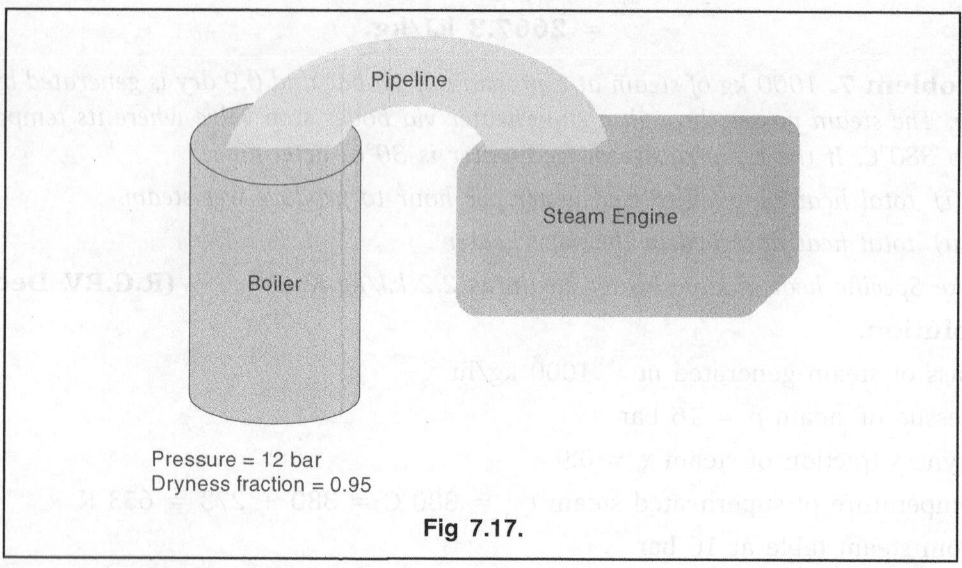

Fig 7.17.

From steam table at 12 bar

$$t_s = 187.97°\text{C}$$
$$V_g = 0.1633 \text{ m}^3/\text{kg}$$
$$h_f = 798.6 \text{ kJ/kg}$$
$$h_{fg} = 1986.2 \text{ kJ/kg}$$

Let enthalpy of steam at boiler end is h_B and that of engine end is h_E

Using equation (7.3) we get

Enthalpy at boiler end

$$h_B = 798.6 + 0.95 \times 1986.2$$
$$h_B = 2685.5 \text{ kJ/kg}$$

Enthalpy at engine end is

$$h_E = 798.6 + x_E \times 1986.2$$

Now as mentioned in the problem there is a loss of 20 kJ/kg of heat during steam flow from boiler end to engine end.

$$\therefore \quad h_B - 20 = h_E$$
$$2685.5 - 20 = 798.6 + x_E\, 1986.2$$
$$x_E = \frac{2685.5 - 20 - 798.6}{1986.2} = 0.94$$

Steam at the engine end is 0.94 dry and its saturation temp is 189.97°C **Ans.**

Note: saturation temp is taken from steam table at 12 bar.

Problem 9. *Calculate the internal energy per kg of superheated steam at a pressure of 10 bar and a temperature of 300°C. Find the change of internal energy if this steam expands to 1.4 bar and dryness fraction 0.8.* *(R.G.P.V. Jan./Feb. 2007)*

Solution.

From steam table at 10 bar

$$h_f = 762.81 \text{ kJ/kg}$$
$$h_{fg} = 2015.3 \text{ kJ/kg}$$
$$V_g = 0.1944 \text{ m}^3\text{/kg}$$
$$t_s = 179.91°\text{C}$$

First enthalpy is to be calculated in order to find out the internal energy because as we know

Internal energy of steam = enthalpy of steam – external work done during evaporation

Enthalpy of superheated steam is given by

$$h_{sup} = h_f + h_{fg} + C_p\,(t_{sup} - t_s)$$
$$h_{sup} = 762.81 + 2015.3 + 2.1\,(300 - 179.91)$$
$$= 3030.3 \text{ kJ/kg}$$

To find out the external work done per kg of superheated steam, the specific volume of superheated steam is to be calculated.

$$V_{sup} = \frac{V_g T_{sup}}{T_s} = \frac{0.1944 \times (300 + 273)}{(179.91 + 273)}$$
$$= 0.1944 \times 1.265 = 0.246 \text{ m}^3\text{/kg}$$

External work done per kg of superheated steam

$$100\,pV_{sup} = 100 \times 10 \times 0.246$$
$$= 246 \text{ kJ/kg}$$

Internal energy of superheated steam $U_{sup} = h_{sup} - 100\,pV_{sup}$

$$U_{sup} = 3030.3 - 246$$
$$= 2784.3 \text{ kJ/kg}$$

From steam table at 1.4 bar

$$h_f = 458.4 \text{ kJ/kg}$$
$$h_{fg} = 2231.9 \text{ kJ/kg}$$
$$V_g = 1.2363 \text{ m}^3\text{/kg}$$
$$t_s = 109.32°\text{C}$$

Enthalpy of wet steam is $h_{wet} = h_f + x\,h_{fg}$

$$h_{wet} = 458.4 + 0.8\,(2231.9) = 2243.92 \text{ kJ/kg}$$

External work done per kg of wet steam

$$= 100\,p\,x\,V_g$$
$$= 100 \times 1.4 \times 0.8 \times 1.2363 = 138.46 \text{ kJ/kg}$$

Internal energy of wet steam

$$U_{wet} = 2243.92 - 138.46 = 2105.45 \text{ kJ/kg}$$

Therefore Change in internal energy per kg is $= U_{sup} - U_{wet}$

$$= 2105.45 - 2784.3$$

$$= -678.84 \text{ kJ}$$

Since change in internal energy comes out in negative, it means that there is decrease in internal energy. **Ans.**

Problem 10. *Two boilers one with superheater and other without superheater are delivering equal quantities of steam into a common main. The pressure in the boilers and the main is 15 bar. The temperature of the steam from a boiler with superheater is 300°C and the temperature of the steam in the main is 200°C. Determine the quality of steam supplied by the other boiler.*

(RGPV Jan./Feb. 2007)

Solution.

Let boiler A is operating with a superheater and boiler B is without superheater.

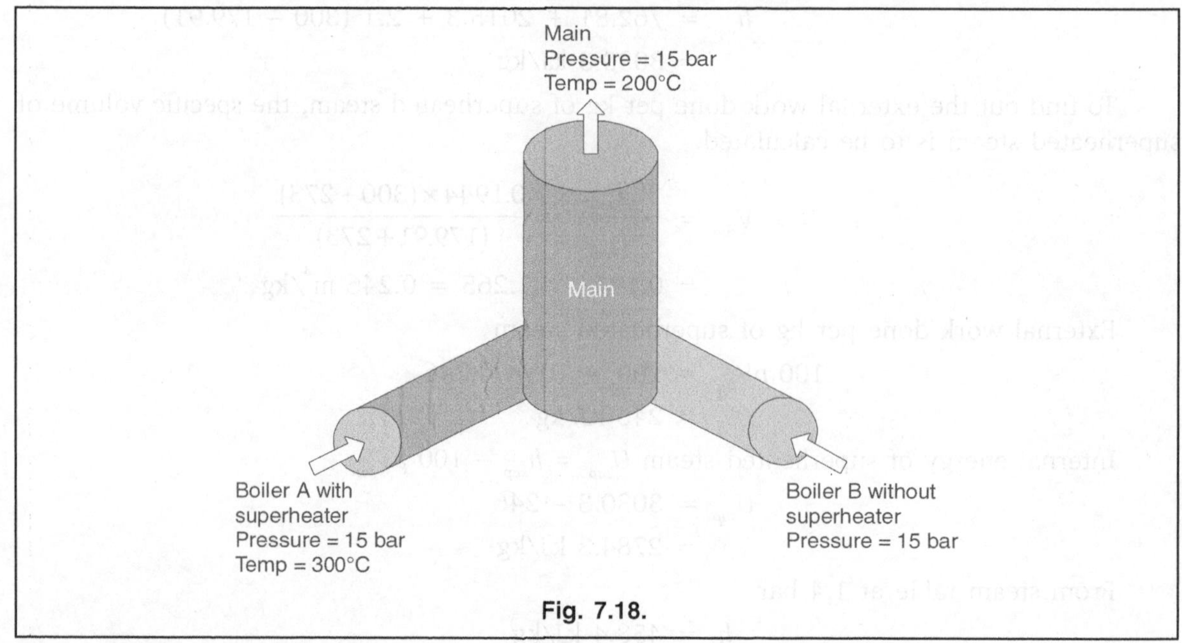

Fig. 7.18.

For boiler A

From steam table at 15 bar

We get

$$h_f = 844.9 \text{ kJ/kg}$$

$$h_{fg} = 1947.3 \text{ kJ/kg}$$

$$V_g = 0.1317 \text{ m}^3\text{/kg}$$

$$t_s = 198.31°\text{C}$$

Enthalpy of superheated steam in boiler A is given by

$$h_A = h_{sup} = h_f + h_{fg} + C_p\,(t_{sup} - t_s)$$

$$h_{sup} = 844.9 + 1947.3 + 2.1(300 - 198.31) \qquad ...(1)$$

$$h_A = h_{sup} = 3005.75 \text{ kJ/kg}$$

For boiler B

Let x_B be the dryness fraction of steam in boiler B.

Enthalpy of steam in boiler B we get

$$h_B = h_{f2} + x_B h_{fg2}$$

$$h_B = 844.9 + x_B \times 1947.3 \qquad (2)$$

In main both the boiler A and B are delivering 1 kg of steam thus total 2 kg of steam is coming in the main.

Total heat in main at 15 bar is given by

$$h_{main} = 2[h_f + h_{fg} + C_p\,(t_{sup} - t_s)]$$

$$= 2[844.9 + 1947.3 + 2.1(200 - 198.31)] \qquad ...(3)$$

$$= 5591.5 \text{ kJ}$$

Now combining equation (1) and (2) and equating this with equation (3) we get

$$3005.75 + 844.9 + x_B \times 1947.3 = 5591.5$$

$$x_B = \frac{5591.5 - 3005.75 - 844.9}{1947.3} = 0.894$$

$\therefore$ Quality of steam supplied by Boiler B is 0.894. **Ans.**

Problem 11. *A boiler is generating steam of 0.87 dryness fraction at a pressure of 10 bar. This steam is then passed to a superheater where steam gets heated at constant pressure and its temperature increases upto 310°C. Find out (a) the quantity of heat added to steam in superheater and (b) change in internal energy of the steam.*

Solution.

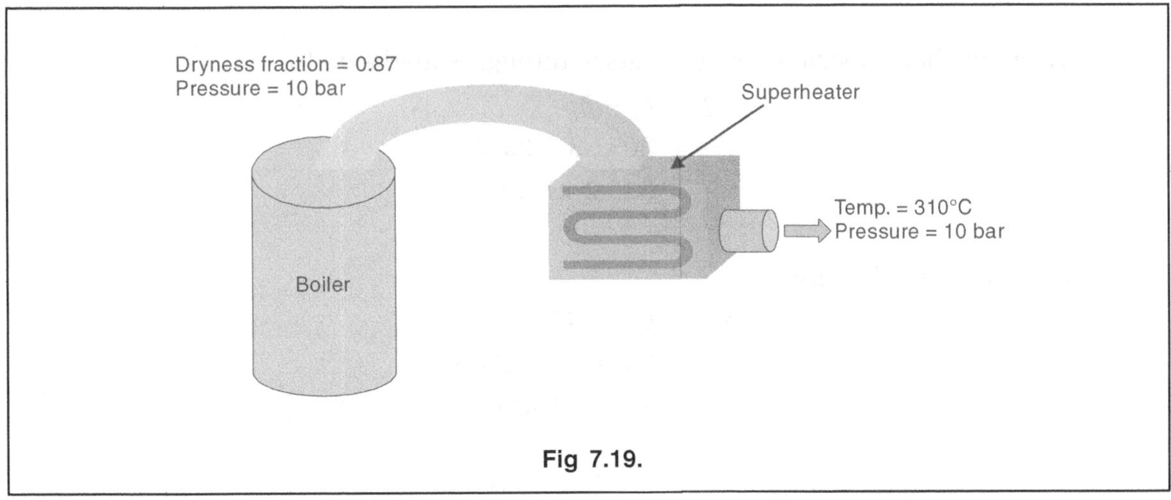

Fig 7.19.

From steam table at 10 bar

$$h_f = 762.81 \text{ kJ/kg}$$
$$h_{fg} = 2015.3 \text{ kJ/kg}$$
$$V_g = 0.1944 \text{ m}^3\text{/kg}$$
$$t_s = 179.91°\text{C}$$

Let h_B is the enthalpy of steam at the boiler end and h_s is the enthalpy of steam after coming out from the superheater.

Now

$$h_B = 762.81 + 0.87 \times 2015.3 = 2516.1 \text{ kJ/kg}$$

External work done per kg of wet steam

$= 100\ pxV_g = 100 \times 10 \times 0.87 \times 0.1944 = 169.13$ kJ

Internal energy of steam at boiler end is

$$U_B = h_B - 100\ pxV_g$$
$$= 2516.1 - 169.13$$
$$= 2346.97 \text{ kJ/kg}$$

Steam when comes out from the superheater will be in superheated state.

$$h_s = h_{sup} = 762.81 + 2015.3 + 2.1\ (310 - 179.71)$$
$$= 3051.3 \text{ kJ/kg}$$

External work done per kg of superheated steam $= 100\ pV_{sup}$

And $$V_{sup} = \frac{V_g T_{sup}}{T_s} = \frac{0.1944 \times (310 + 273)}{(179.91 + 273)}$$
$$= 0.1944 \times 1.287 = 0.2502 \text{ m}^3\text{/kg}$$

external work done $= 100 \times 10 \times 0.25 = 250.24$ kJ/kg

Internal energy of superheated steam $= U_{sup} = 3051.3 - 250.24 = 2801.06$ kJ/kg

(*a*)

∴ Heat supplied to steam when it passes through superheated is

$$Q = h_s - h_B$$
$$= 3051.3 - 2516.1$$
$$= 535.2 \text{ kJ/kg}$$

(*b*)

Change in internal energy

$$\Delta U = U_{sup} - U_B$$
$$= 2801.06 - 2346.97$$
$$= 454.09 \text{ kJ/kg}$$

Problem 12. *Calculate the internal energy of 1 kg of steam at a pressure of 10 bar when steam is*

(i) 0.8 dry

(ii) Dry and saturated

(iii) Superheated to 250°C (R.G.R.V. Nov./Dec. 2007)

Solution.

From steam table at 10 bar

$$h_f = 762.81 \text{ kJ/kg}$$
$$h_{fg} = 2015.3 \text{ kJ/kg}$$
$$V_g = 0.1944 \text{ m}^3\text{/kg}$$
$$t_s = 179.91°\text{C}$$

First enthalpy is to be calculated in order to find out the internal energy because as we know

Internal energy of steam = Enthalpy of steam – external work done during evaporation

(*i*) When steam is 0.8 dry.

Total heat or enthalpy of wet steam

$$h_{wet} = h_f + x\, h_{fg}$$
$$h_{wet} = 762.81 + 0.8 \times 2015.3$$
$$h_{wet} = 2375.05 \text{ kJ/kg}$$

External work done per kg of wet steam $= 100\, p\, x\, V_g$

$$= 100 \times 10 \times 0.8 \times 0.1944$$
$$= \mathbf{155.52 \text{ kJ.}}$$

Internal Energy of wet steam U_{wet}

$$= 2375.05 - 155.52 = \mathbf{2219.53 \text{ kJ/kg Ans.}}$$

(*ii*) When steam is dry saturated

Enthalpy of dry saturated steam

$$h_g = h_f + h_{fg}$$
$$h_g = 762.81 + 2015.3$$
$$= 2778.11 \text{ kJ/kg}$$

External work done during evaporation per kg of dry saturated steam $= 100\, pV_g$

$$= 100 \times 10 \times 0.1944$$
$$= \mathbf{194.4 \text{ kJ/kg.}}$$

Internal Energy of dry saturated steam is

$$U_{dry} = h_g - 100\, pV_g$$
$$= 2778.11 - 194.44$$
$$= \mathbf{2583.71 \text{ kJ/kg Ans.}}$$

***(iii)* When steam is superheated to 250°C**

Enthalpy of superheated steam is given by

$$h_{sup} = h_f + h_{fg} + C_p (T_{sup} - T_s)$$

$$h_{sup} = 762.81 + 2015.3 + 2.1 (523 - 452.91)$$

$$h_{sup} = 2925.3 \text{ kJ/kg}$$

To find out the external work done per kg of superheated steam, the specific volume of superheated steam is to be calculated. It is given by

$$V_{sup} = V_g \times \left(\frac{T_{sup}}{T_s}\right)$$

$$V_{sup} = 0.1944 \times \frac{523}{452.91} = 0.2245 \text{ m}^3\text{/kg}$$

External work done per kg of superheated steam $= 100\, pV_{sup}$

$$= 100 \times 10 \times 0.2245$$

$$= \mathbf{224.5 \text{ kJ}}$$

∴ Internal Energy of superheated steam $U_{sup} = h_{sup} - 100\, pV_{sup}$

$$= 2925.3 - 224.5$$

$$= \mathbf{2700.8 \text{ kJ/kg Ans.}}$$

Problem 13. *A rigid vessel is divided into two parts A and B by means of a frictionless, perfectly conducting piston. Initially part A contains 0.4 m^3 of air (ideal gas) at 10 bars pressure and part B containes 0.4 m^3 of wet steam at 10 bars. Heat is now added to both parts until all the water in part B is evaporated. At this condition the pressure in part B is 15 bars. Determine the initial quality of steam in part B and the total amount of heat added to both the parts.*

[U.P.S.C. 1992].

Solution.

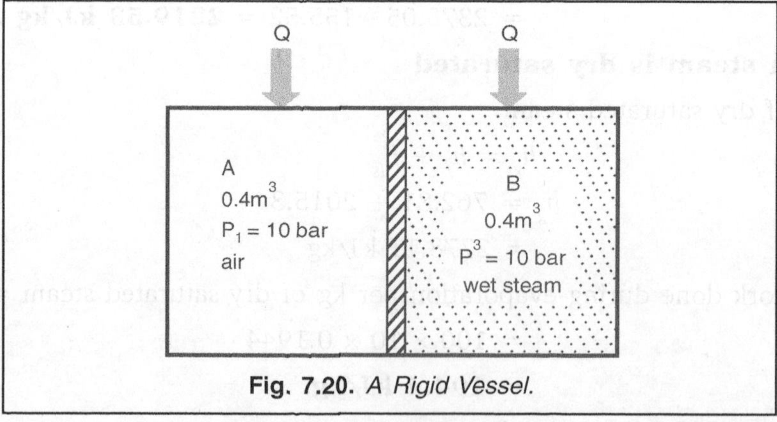

Fig. 7.20. *A Rigid Vessel.*

Heat is being added to both the parts A and B. after heat addition all the water in part B gets evaporated. The part B now contains dry saturated steam and its pressure at this point is 15 bar.

Given:

$$p_1 = 10 \text{ bar} = 1000 \text{ kPa}$$

$$p_2 = 15 \text{ bar} = 1500 \text{ kPa}$$

As piston is perferctly conducting so for balancing pressure in part A will rise to 15 bar and for this heat addition at constant volume will take place.

∴ Heat addition at constant volume in Part A is given by

$$Q = V(p_2 - p_1) = 0.4 \times (1500 - 1000) = 200 \text{ kJ}$$

Let x be the dryness fraction of steam in part B.

∴ Total heat of wet stream at 10 bar + Q = total heat of dry and saturated steam at 15 bar

From steam table at 10 bar

$$h_f = 762.81 \text{ kJ/kg}$$

$$h_{fg} = 2015.3 \text{ kJ/kg}$$

And at 15 bar

$$h_g = 2792.2 \text{ kJ/kg}$$

$$\therefore \quad 762.81 + x\ 2015.3 + 200 = 2792.2$$

$$\therefore \quad x = \frac{2792.2 - 762.81 - 200}{2015.3}$$

$$\therefore \quad x = 0.907$$ **Ans.**

Unsolved Problems

1. A pressure cooker is containing 1.5 kg of dry saturated steam at a pressure of 5 bar. Find out the quantity of heat, which must be rejected so that the quality of steam is reduced to 60% dry. **[Ans.** – 1218.43 kJ]
2. A steam boiler is generating steam at 13 bar. Find out the quantity of heat required to produce 1 kg of steam if the steam produced is
   (*i*) 0.97 dry
   (*ii*) dry saturated
   (*iii*) degree of superheat is 40°C **[Ans.** (*i*) 2519.2 kJ, (*ii*) 2578.3 kJ, (*iii*) 2662.3 kJ]
3. Steam escapes to atmosphere through a leak in a main steam pipe. The steam pressure and atmospheric pressure is 14 and 1 bar respectively. The observed steam temperature in the leakage region is 150°C. Find out initial dryness fraction of the steam in the main steam pipe. **[Ans.** 0.993]
4. 0.25 kg of wet steam is heated at constant pressure of 7 bar till its volume becomes twice the initial volume. The dryness fraction of steam is 0.85. Find out the work done and the heat transferred. **[Ans.** 40.6 kJ, 213.34 kJ]

5. A closed vessel of 0.2 m^3 capacity contains steam at a pressure of 8 bar and 200°C.
   (*i*) What is the weight of two steam contained in the vessel?
   (*ii*) The vessel is cooled so as the steam becomes dry and saturated, what will be pressure and temperature of steam at this state. *(R.G.P.V. Dec. 2003)*

   [Ans. 0.7 kJ, 6.76 bar, 163.5°C]

6. A sample of steam is available at 8 bar and 90% dry.
   Calculate (*i*) the external work of evaporation and the internal latent heat, (*ii*) internal energy (*iii*) the heat required to superheat two steam upto 250°C. Take C_p = 2.25 kJ/kg K. *(R.G.P.V. Dec. 2003)*

   [Ans. 172.78 kJ/kg, 1668.86 kJ/kg, 2389.8 kJ/kg, 179 kJ/kg]

7. A vessel of volume 0.04m^3 contains a mixture of saturated water and saturated steam at a temperature of 250°C. The mass of liquid present is 9 kg. Find the pressure, the mass, the specific volume, the enthalpy and the internal energy of the mixture.

   (R.G.P.V.-June 2004)

   [Ans. 39.73 bar, 9.56 kg, 0.0042 m^3/kg, 1188.33 kJ/kg, 1171.72 kJ/kg]

8. A closed vessel contains 1 kg of steam at a pressure of 6.4 bar and dryness fraction 0.9. The vessel is cooled until the dryness fraction falls to 0.6. Find out the quantity of heat extracted while neglecting the volume of water. **[Ans.** 0.6MJ/kg]

9. A cylinder contains a mixture of wet steam having dryness fraction 0.90 and air at 104.8°C, the pressure being 2.56 bar. This is compressed to one-sixth of its original volume, the final temperature being 140.8°C. Find out (*i*) the pressure in the cylinder, (*ii*) the dryness fraction of steam after compression. **[Ans.** 8.94 bar, 0.43]

10. Steam enters a surface condenser at 35.5°C at the rate of 4000 kg/hr. the dryness fraction of steam is 0.9. Determine the rate at which cooling water is to be supplied to the condenser if the rise in temperature of cooling water is 10°C. Take specific heat of water as 4.187 kJ/kg K. **[Ans.** 2.218 × 10^5 kg/hr]

REVIEW QUESTIONS

1. Explain the difference between wet steam, dry saturated steam and superheated steam.
2. Define the following terms
   (*i*) Sensible heat
   (*ii*) Latent heat
   (*iii*) Dryness fraction
   (*iv*) Wetness fraction
   (*v*) Degree of superheat
   (*vi*) Heat of superheat
   (*vii*) Enthalpy
   (*viii*) Internal Energy

   (R.G.P.V.-June 2002, Dec. 2002)
3. Distinguish clearly between the enthalpy and internal energy of steam.
4. Discuss mollier chart with the help of a neat sketch. Explain isentropic and throttling processes on mollier diagram.
5. Explain the principle of throttling calorimeter used for determining dryness fraction of steam. What is its limitation? What essential conditions must be fulfilled when using a throttling calorimeter. *(R.G.P.V.-June 2004)*

6. Sketch diagrammatically a combined separating and throttling calorimeter.
7. Explain on the following on T–S and H–S diagram
   (*i*) Constant volume process
   (*ii*) Constant pressure process
   (*iii*) Isentropic process.
8. Explain the process of generation of steam with the help of the following:
   (*i*) Temperature—entropy diagram
   (*ii*) Pressure—volume diagram *(R.G.P.V.-Dec. 2004)*

OBJECTIVE TYPE QUESTIONS

Select the most appropriate answer:

1. The perfect gas law is applicable to
   (A) dry steam (B) wet steam
   (C) superheated steam (D) all of the above.
2. The unit of pressure in SI units is
   (A) pascal (B) kg/cm^2
   (C) $dyne/cm^2$ (D) mm of Hg.
3. Specific heat of water is
   (A) 1 (B) 0.2
   (C) 0.3 (D) 0.1
4. Specific heat is least for
   (A) water (B) air
   (C) steam (D) ice.
5. Atmospheric pressure is equal to
   (A) 1.013 bar (B) $101.3 kN/m^2$
   (C) 760 mm of Hg (D) all of the above.
6. Barrel calorimeter is used for
   (A) for measuring temperature of steam
   (B) for measuring pressure of steam
   (C) for measuring dryness fraction of steam
   (D) none of the above.
7. Degree of superheat is
   (A) difference between temperature of superheated steam and that of dry saturated steam.
   (B) difference between temperature of superheated steam and that of wet steam.
   (C) difference between pressure of superheated steam and that of dry saturated steam.
   (D) none of the above.

8. During throttling process

(A) entropy remains unchanged (B) enthalpy remains unchanged

(C) pressure remains unchanged (D) internal energy remains unchanged.

9. Sensible heat is

(A) the heat which causes change of state.

(B) The heat which does not allow temperature to rise.

(C) The heat which brings about change in temperature.

(D) All of the above.

10. Enthalpy of a substance is defined as

(A) the sum of internal energy and the work done.

(B) The sum of sensible heat and latent heat of substance.

(C) The difference of sensible heat and latent heat of substance.

(D) None of the above.

11. The value of latent heat of steam at 1 bar is

(A) 1000 kJ/kg (B) 1258 kJ/kg

(C) 2258 kJ/kg (D) 3000 kJ/kg.

12. Triple point of a pure substance is a point at which

(A) liquid and vapour phases exist together

(B) liquid and solid phases exist together

(C) solid and vapour phases exist together

(D) solid, liquid and vapour phases exist together.

13. Enthalpy of water at 0°C is assumed to be

(A) 1 (B) 0

(C) –1 (D) –100

14. The specific heat of water with rise of temperature.................... .

(A) increases (B) decreases

(C) remains constant

(D) first decreases to a minimum then increases.

15. Dryness fraction represents

(A) pressure of steam. (B) temperature of steam.

(C) volume of steam. (D) quality of steam.

16. Wetness fraction is

(A) 1–dryness fraction (B) 1+dryness fraction

(C) 1/dryness fraction (D) none of the above.

ANSWERS

1. (C) **2.** (A) **3.** (A) **4.** (B) **5.** (D) **6.** (C)
7. (A) **8.** (B) **9.** (C) **10.** (A) **11.** (C) **12.** (D)
13. (B) **14.** (D) **15.** (D) **16.** (A).

Boilers

8.1 INTRODUCTION

The boilers are used for the purpose of heating and power generation. Steam is the main working substance for the power generation in steam engine and steam turbine. Boiler is one of the main part of the steam power plant. The main task of a boiler is to produce steam at required pressure, which is then transformed into electrical energy in the steam power plant.

- ☞ A steam boiler is a closed vessel made of high quality steel in which steam is generated from water by the application of heat. It is also known as steam generator.
- ☞ The function of a boiler is to convert chemical energy of fuel into heat and than this heat is used to produce steam from water at the required pressure.

8.2 BOILER NOMENCLATURE

Shell

A boiler shell consists of several steel plates bent into cylindrical form and riveted together. The ends of the shell are closed by means of boiler heads which are made of flat, convex or concave plates having flanged edges to permit riveting to the shell. The shell together with the heads forms the drum.

Setting

The boiler setting is made of brick. It not only provides the support to the boilers but also forms the furnace wall and combustion chamber.

Grate

The grate is made of cast iron bars or plates where the fuel is burned. The area of the surface upon which the fire rests is the grate surface.

Furnace

The furnace is also known as the firebox. It is the space above the grate and below the boiler shell and is enclosed by the side and front wall of the setting. It is the space in which the fuel is burnt.

Water level

It is the level at which water is stored in the boiler shell.

Steam space

It is the space above the water level in the boiler shell.

Mountings

The different types of valves and gauges, which are compulsory for the safety of the boiler, are known as mountings. The mountings include the following:

- Safety valve
- Stop valve
- Feed check valve
- Pressure gauge
- Water level indicator
- Blow off cock
- Fusible plug.

Accessories

These devices are used to improve the overall efficiency of a boiler. The important accessories are:

- Economiser
- Superheater
- Air-preheater
- Feed pump.

8.3 CLASSIFICATION OF BOILERS

The classification of boilers is difficult owing to the large number of variations in the design and working of the boilers. The commonly accepted form of boilers classification is given below:

(i) According to the Relative Position of Water and Hot Gases

(A) Water tube boiler

In a water tube boiler, water passes inside the tubes and flue gases (hot gases or flue gases) pass over the tubes. Examples of water tube boilers are Babcock and Wilcox boiler, Stirling boiler and high-pressure boilers.

(B) Fire tube boiler

A fire tube boiler is one in which the flue gases or the hot gases pass inside the tubes and water is outside the tubes. Examples of the fire tube boilers are simple vertical boiler, Cochran boiler, Lancashire boiler, Cornish boiler and Locomotive boiler.

(C) High pressure boiler

The commonly used high-pressure boilers are:

- LA Mont boiler
- Benson boiler
- Loeffler boiler
- Schmidt-Hartzmann boiler
- Velox boiler.

(*ii*) According to the Forms

On the basis of forms, the boilers are classified as:

- Plain cylindrical boilers
- Tubular boilers
- Sectional boilers.

(*iii*) According to the Location of the Furnace

On the basis of the location of the furnace, the boilers are of two types:

- Externally fired boiler: In which the furnace is outside the shell.
- Internally fired boiler: In which the furnace is inside the boiler shell.

(*iv*) According to the Use

On the basis of use, the boilers can be classified as:

- Stationary boilers
- Portable boilers
- Locomotive boilers
- Marine boilers.

(*v*) According to the Axis of the Shell

On the basis of the axis of the shell, the boilers can be classified as:

- Horizontal boilers
- Vertical boilers
- Inclined boilers.

(*vi*) According to the Method of Circulation of Water

On the basis of the method of water circulation, the boilers can be classified as:

- Natural circulation boilers
- Forced circulation boilers.

(*vii*) According to the Type of Fuel Used by the Boiler

On the basis of the type of fuel used, the boilers can be classified as:

- Solid fuel based boilers
- Liquid fuel based boilers
- Gas fuel based boilers.

(*viii*) According to the Pressure

On the basis of the pressure, the boilers can be classified as:

- Low pressure boilers
- Medium pressure boilers
- High pressure boilers.

(ix) According to the Capacity

On the basis of the capacity, the boilers can be classified as:

- Low capacity boilers
- Medium capacity boilers
- High capacity boilers.

8.4 REQUIREMENTS OF A GOOD BOILER

1. It should have high reliability and capability of producing desired quality of steam economically at a required pressure.
2. A boiler should have the minimum fuel consumption.
3. It should be able to adjust the fluctuations in steam pressure due to variations in loads.
4. It should occupy less floor area.
5. It should able to start quickly.
6. It should be constructed in such a way so that all its interior parts are easily accessible for inspection, cleaning and repair etc.
7. Its design should be based on the optimum use of the flue gases.

8.5 TYPES OF BOILERS

8.5.1 Simple Vertical Boiler

It consists of a cylindrical shell of steel. It has got its name due to its position, (as its axis is vertical). A cylindrical shell surrounds a cylindrical firebox. To increase the heating surface and to improve the circulation of water, cross tubes are fitted in the fire-box. These cross tubes are kept slightly inclined to ensure the effective circulation of the water. An uptake tube passes from the crown of the fire-box to the crown of the shell and then at the top where the uptake tube is joined with the chimney.

Hand holes are provided to clean the cross tubes, which are placed at the bottom of the water space surrounding the fire-box. A manhole is provided for cleaning and inspection of the boiler. The fuel (coal wood or charcoal) is burnt on the grate from where the hot gases travel through uptake to the chimney while heating the water that is stored in the shell. The steam is collected in the space above the water level and can be taken through the stop valve. A schematic diagram of a simple vertical boiler is shown in Fig. (8.1).

Steam capacity

- 50 Kg to 1000 kg of steam/hr

Pressure

- 2 to 10 bar

Applications

- Small industries
- Water supply plant

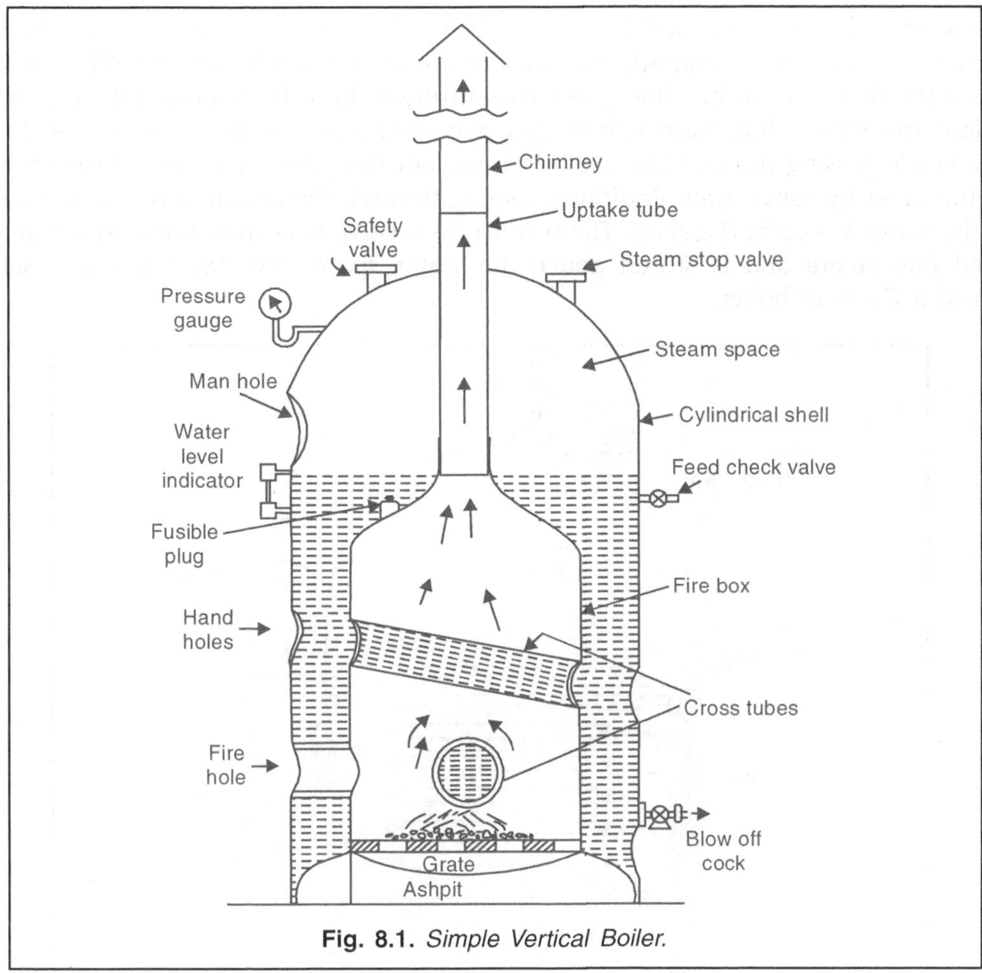

Fig. 8.1. *Simple Vertical Boiler.*

Advantages

- Less floor area
- Portable
- Cheap
- Easy to start.

Disadvantages

- Wastage of fuel
- Small heating surface.

8.5.2 Cochran Boiler

Cochran boiler is a multi tubular boiler. It has better efficiency due to its large heating surface as compared to simple vertical boiler. Its cylindrical shell is made of high quality steel. The furnace is of hemispherical shape. The hemispherical shape of furnace permits maximum absorption of radiation heat. This type of crown shape is very effective in handling high heat of the combustion of fuel.

The grate is placed at the bottom of the furnace and the ash pit is located below the grate. The furnace is connected to the combustion chamber by a flue pipe. The back of the combustion chamber is lined with firebricks. About 150 to 170 fire tubes (generally 6.25 cm in diameter) are placed horizontally and these connect the combustion chamber and smoke box. From the furnace, the hot flue gases pass through these horizontal fire tubes and then moves into the smoke box from where they are discharged to the atmosphere through a chimney. While passing through the fire tubes, the hot flue gases heat the surrounding water outside the tubes by convection. Radiation coming through the crown of the fire-box also add heat to the water stored in the shell. The water thus receive heat from these two sources, gets converted into steam and is stored above the water space. Fig. (8.2) shows a schematic diagram of a Cochran boiler.

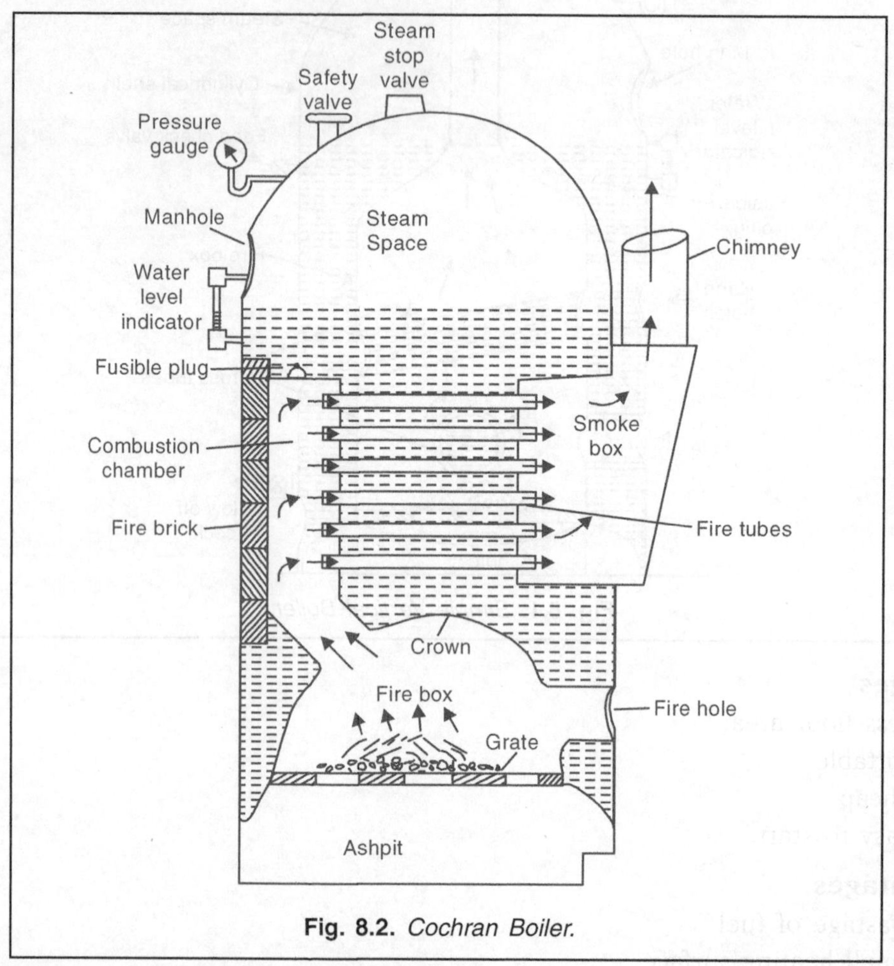

Fig. 8.2. *Cochran Boiler.*

Steam capacity

- 150 to 3000 kg of steam/hr.

Pressure

- 20 bar.

Thermal efficiency

- 70% with coal
- 75% with oil.

Applications

- In small and medium sized industries.
- In mines.

Advantages

- Need less floor space.
- Portable.
- Need no elaborate foundation.
- Can be used in limited space.
- Can use any type of fuel.

8.5.3 Lancashire Boiler

Lancashire boiler is a horizontal fire tube, internally fired, natural circulation boiler. Lancashire boiler is used as a stationary boiler. It generates good quality steam and can use coal of inferior quality. Lancashire boiler has two (internally fired) tubes having diameter between 75 to 100 cm. A fire grate is arranged inside the fire tube at the front. The brickwork setting of this boiler is shown in Fig. (8.3). The flue gas passage is shown in the brickwork setting. The Lancashire boiler consists of a cylindrical shell with flat ends. Its two fire tubes pass through the entire length of the boiler. The brickwork setting forms one bottom flue (marked A) and two side flues (marked as B and C).

Fuel is burnt on the fire grate and hot flue gases pass through the two internal fire tubes and reaches the back end of the boiler, where they dip and enter in to the bottom flue passage 'A'. After coming out from passage 'A' at the front of the boiler, the hot flue gases then divide in two and pass through the two side flues 'B' and 'C' in backward direction and then discharged to atmosphere through chimney. In this boiler, the heat is transferred to water in three ways through

(*a*) two internal flue tubes.

(*b*) bottom flue, and

(*c*) two side flues.

This arrangement utilizes the heating power of the hot flue gases more effectively. The presence of a fire bridge at the end of the grate prevents the flow of coal and ash particles into interior of the furnace tubes. The ash and coal particles carried with the flue gases may form deposits on interior of the flue tubes and may prevent the heat transfer to water. By deflecting the hot flue gases upwards, the fire bridge helps in providing better heat transfer. Air space is provided in the brickwork setting to prevent cracks in the brickwork due to expansion of the walls because of heat. Air space also reduces the heat loss through the brick walls.

The cross tubes called "*Galloway Tubes*" are sometimes fitted in Lancashire boiler. These tubes are fitted inside the fire tubes. These cross tubes increase the heating area and circulation of water but this also makes the flue tubes weak.

Steam capacity

- 8000–10000kg/hr.

Steam pressure

- 15–20 bar.

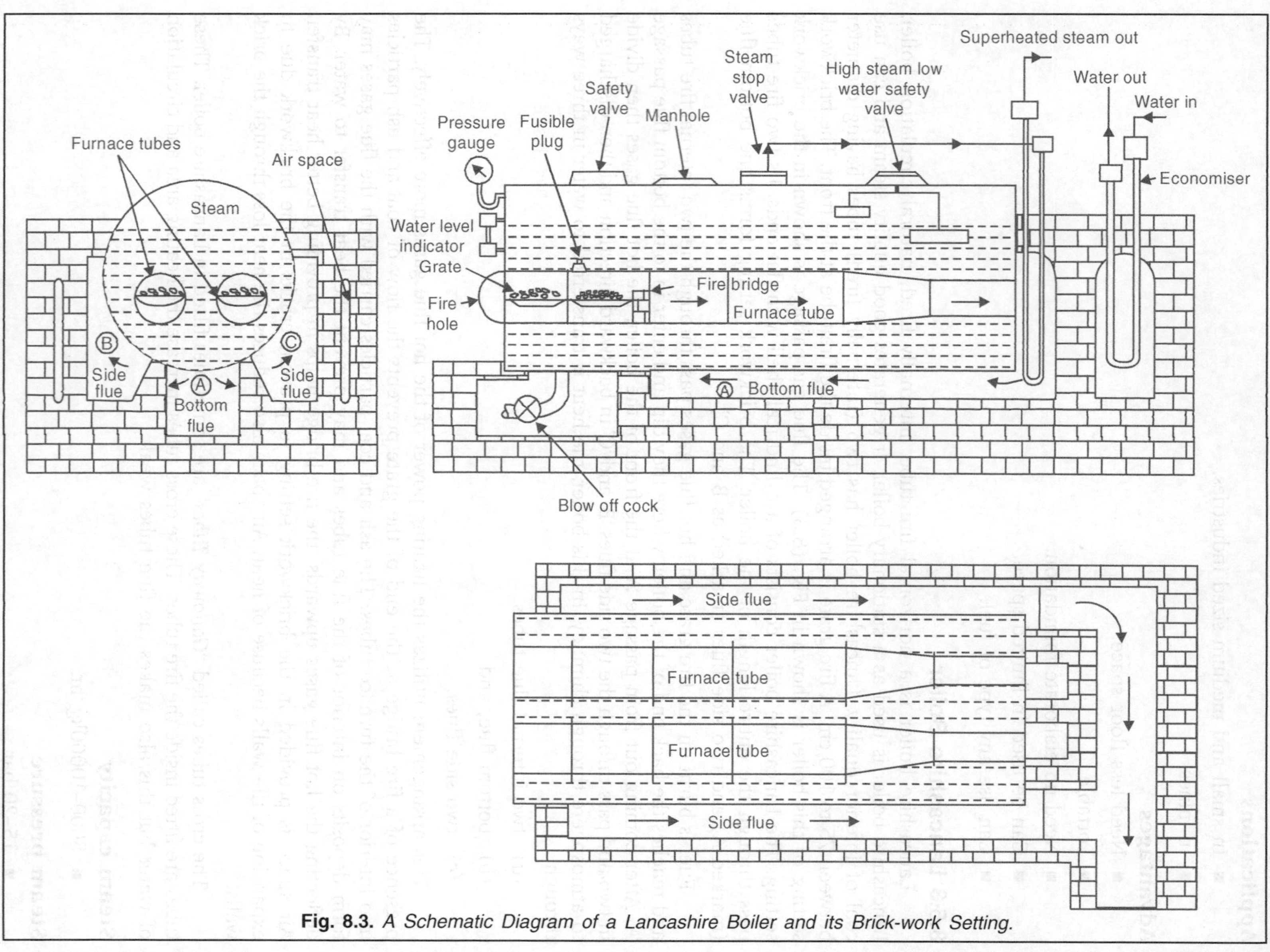

Fig. 8.3. *A Schematic Diagram of a Lancashire Boiler and its Brick-work Setting.*

Efficiency

- 80-85%(with superheater and economiser).
- Diameter of shell: 2–4 m.
- Length: 8–10 m.

Advantages

- Simple design.
- High availability of steam.
- Low maintenance cost.
- Load fluctuation can be met easily due to its large reserve capacity of water and steam.

Disadvantages

- Limited maximum pressure (up to 20 bar).
- More floor area due to brickwork setting.
- Restricted grate area.
- Need more time to raise pressure due to large water capacity.

8.5.4 Cornish Boiler

Cornish boiler is one of the earliest and the simplest types of fire tube boiler. The Cornish boiler has only one fire tube (also known as *flue tube* or *furnace tube*). The fire tube is usually centred on the vertical centre line of the boiler shell. To make internal cleaning easy, the fire tube is sometimes placed few centimetres away from the centre line of boiler shell. The furnace grate is arranged at the front end of the furnace tube. The main air for combustion enters below the grate. The coal is fed to the grate through the fire door. The brickwork setting of this boiler increases the heating surface by providing two side flues and one bottom flue as shown in Fig. (8.4) (see on next page). Fuel is burnt on the fire grate and the hot flue gases pass through the only fire tube and reaches the back end of the boiler, where it divides in two and pass through the two side flues B and C towards the front end of the boiler. At the front end of the boiler these flue gases combine and enter the bottom flue A through which they pass to the back end of the boiler and then to chimney. The overheating of the boiler plates due to scale and sludge deposits should be avoided.

The use of cross tubes (also called Galloway tubes) inside the fire tube improves the circulation of the water in the boiler and break the flue gas stream. These tubes (Fig. 8.5) are conical and placed with larger end at the upside.

Steam capacity

- 3000 Kg/hr

Pressure

- 10–12 bar

Shell diameter

- 1 to 2 m.

Length

- 4 to 8 m.

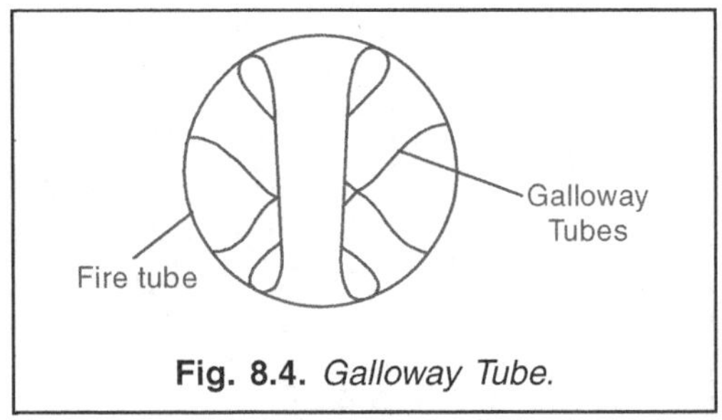

Fig. 8.4. *Galloway Tube.*

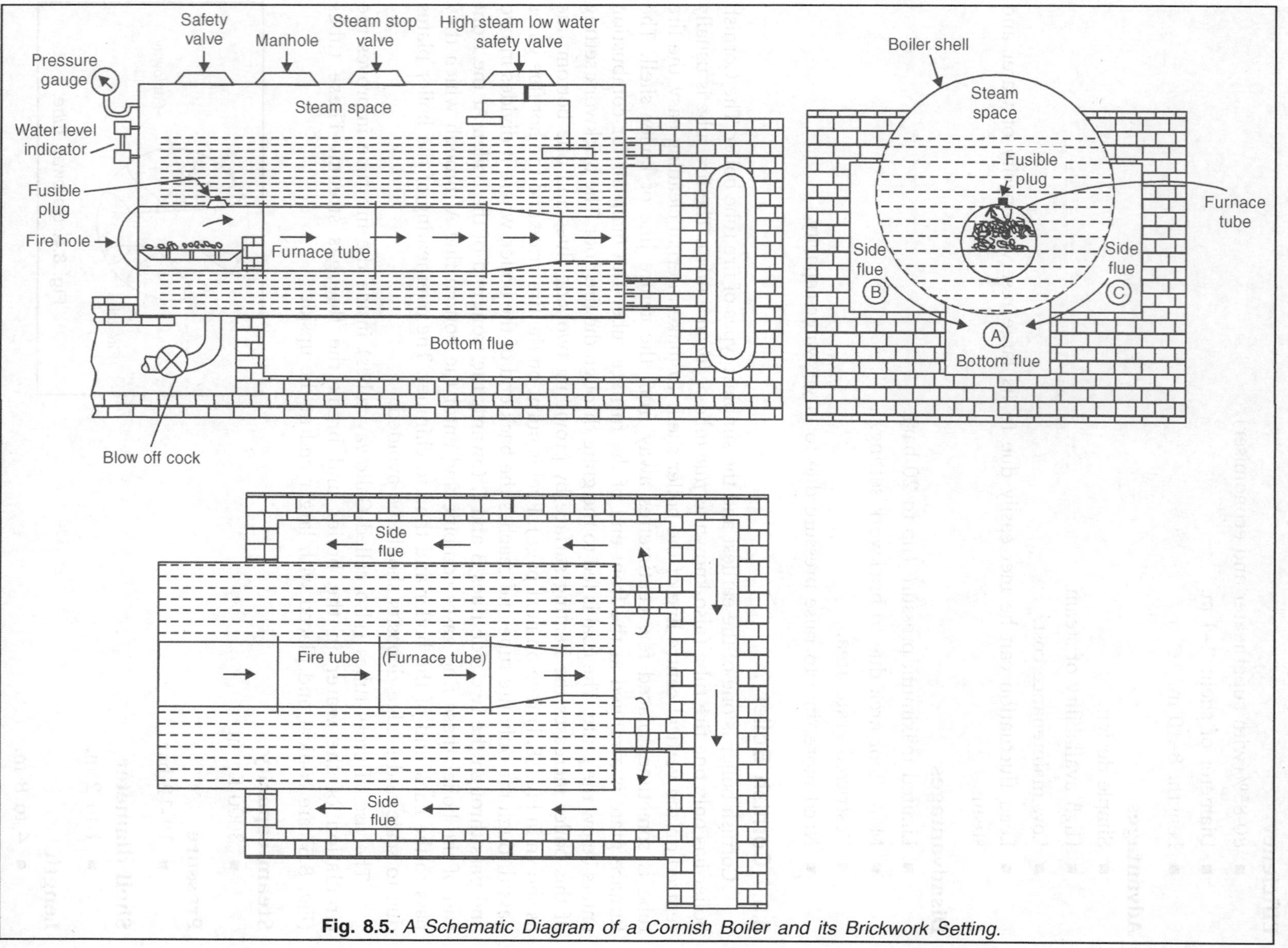

Fig. 8.5. *A Schematic Diagram of a Cornish Boiler and its Brickwork Setting.*

The fundamental difference between the Cornish and Lancashire boiler is given in Table 8.1.

Table 8.1: Fundamental difference between the Cornish and Lancashire Boiler

	Cornish Boiler	***Lancashire Boiler***
Number of fire tubes	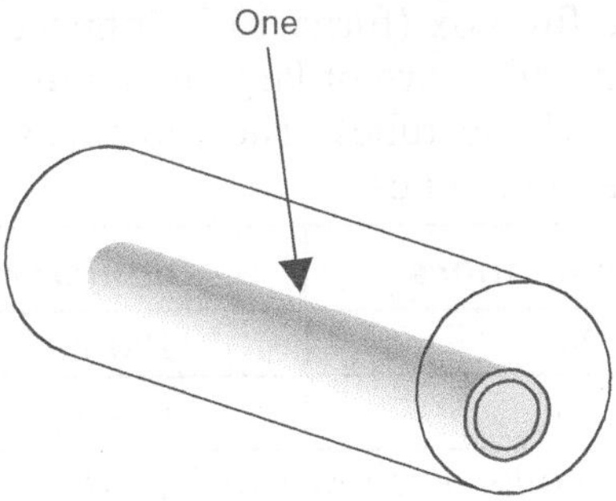	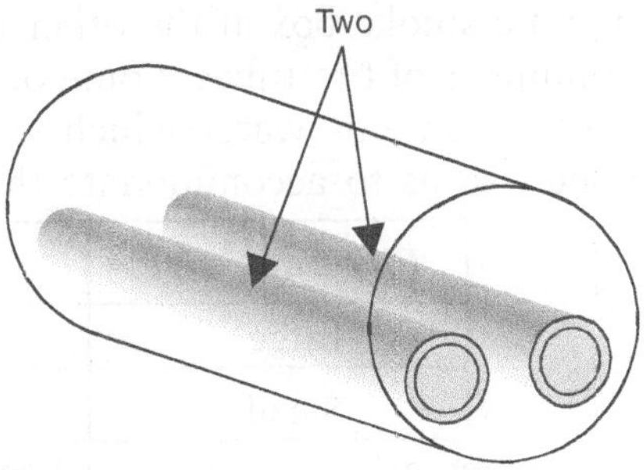
Direction of flue gases	Flue gases after moving through the single fire tube, ↓ first enters the two side flues at the back end of the boiler ↓ come out at the front end ↓ then they combine and pass through the bottom flue ↓ and come out at the back end of the boiler ↓ and then discharged to chimney	The flue gases after moving through the two fire tubes ↓ reaches the back end of the boiler and enters the bottom flue ↓ come out at the front end and divide in two end ↓ then pass through the two side flues ↓ and come out at the back end of the boiler ↓ and then discharged to chimney

8.5.5 Locomotive Boiler

Locomotive boiler is a horizontal fire tube type mobile boiler. The main purpose of a locomotive boiler is to produce steam at a fast rate. Therefore this boiler needs a large heating surface and a large grate area upon which coal is burnt at a rapid rate. This can be achieved by providing a large number of fire tubes of different diameters and by creating strong induced draught by using exhaust steam from the steam engine.

The Locomotive boiler consists of a cylindrical shell having a rectangular fire-box at one end and a smoke box at the other end. The fire-box (furnace) is connected to the smoke box by a number of fire tubes. Some of these fire tubes are of large diameter and some are small. The shell contains water which surrounds all the tubes. The top tubes are of large size in diameter so as to accommodate the superheater tubes.

Fire tube size	*Number of tubes*	*Diameter*
Large	24	130 mm
Small	160	47.5 mm

A schematic diagram of a Locomotive boiler is shown in Fig. (8.6).

Fuel (coal) is fed manually into fire-box through the fire hole. The fuel burns on the grate and resulting hot flue gases are deflected by a fire brick arc at the top. Hot flue gases pass through the fire tubes to the smoke-box and from there finally discharged to the atmosphere through a short chimney.

The heating of water inside the boiler takes place in two ways:

1. Water surrounding the fire-box gets heat from walls of the fire-box (by radiation).
2. Water gets heated as the hot flue gases while moving through the fire tubes pass their heat energy to surrounding water.

The slope of grate in the furnace is downward towards the front. This helps in deflecting the products of combustion, causing them to come in contact with heating surface of the fire-box before moving through the fire tubes. The supply of air for combustion is obtained from the atmosphere. The chimney being short in length is not able to produce the desired suction effect (called natural draft). To overcome this limitation, a blast pipe is placed below the chimney, through which the exhaust steam from the steam engine is discharged. The steam while passing through the nozzle placed in the smoke box produces enough suction (partial vacuum in smoke box) to draw out the flue gases from the fire tubes. This action results in the fresh air rushing into the fire-box for the rapid combustion of the fuel on the grate. The steam generated in the shell is collected above the water space and in the steam dome. The height of chimney is kept low in order to facilitate the locomotive to pass through tunnels and bridges.

The artificial draught is created with the help of exhaust steam when locomotive is moving and with the help of live steam when locomotive is stationary.

Steam capacity

- 10,000 kg of steam/hr.

Pressure

- 0.15–0.25 MPa.

Advantages

- Compact

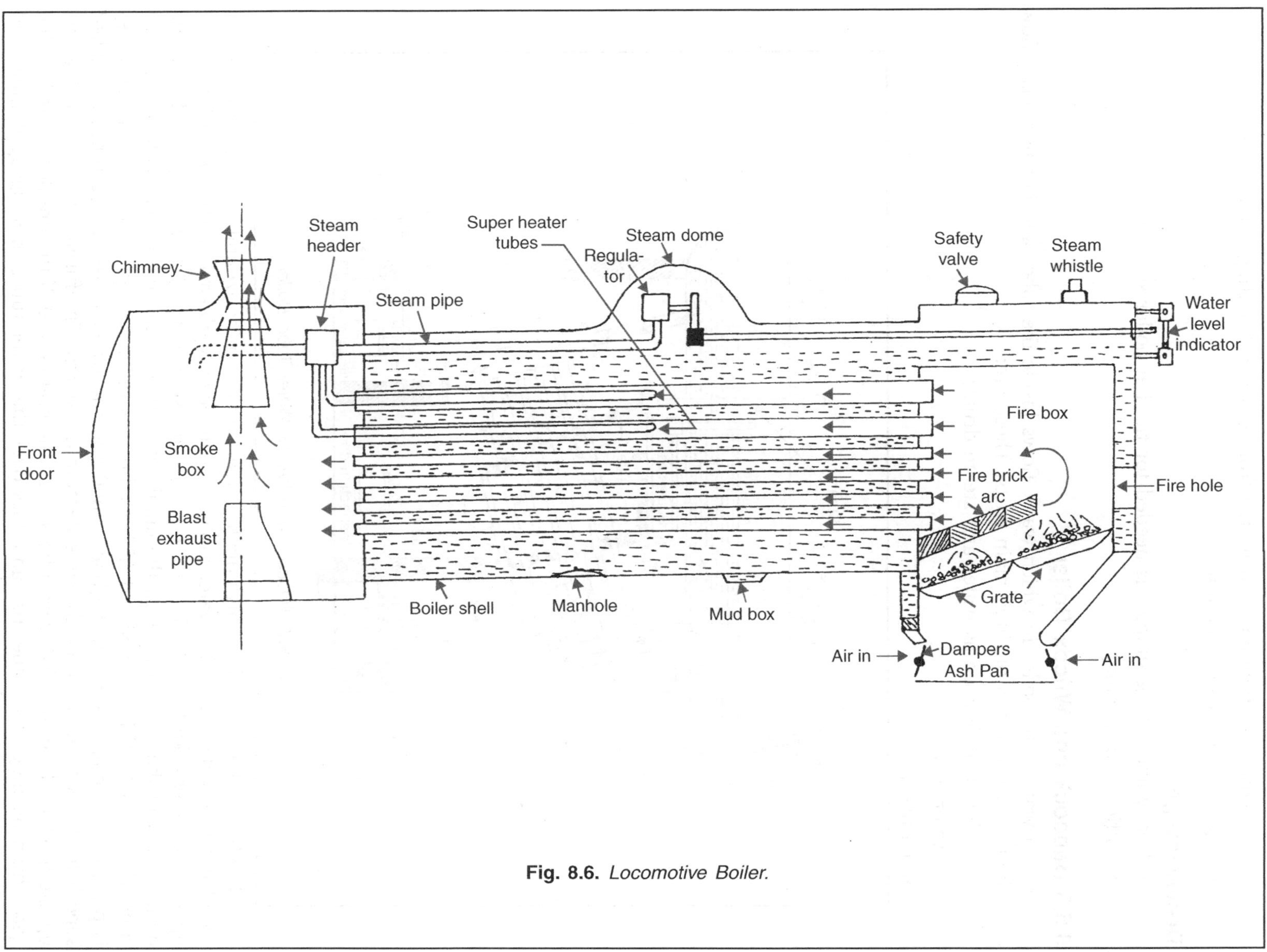

Fig. 8.6. *Locomotive Boiler.*

- Low cost of installation as it is free from brickwork setting, chimney etc.
- High rate of steam generation.

Disadvantages

- Heavy and strong support is needed for the flat boiler surface.
- Difficulty in cleaning.

8.5.6 Babcock and Wilcox Boiler

It is a water tube boiler in which water flows inside the tubes and hot flue gases flow over the tubes. It consists of three main parts. These are:

- Boiler shell (known as water and steam drum)
- Water tubes.
- Furnace.

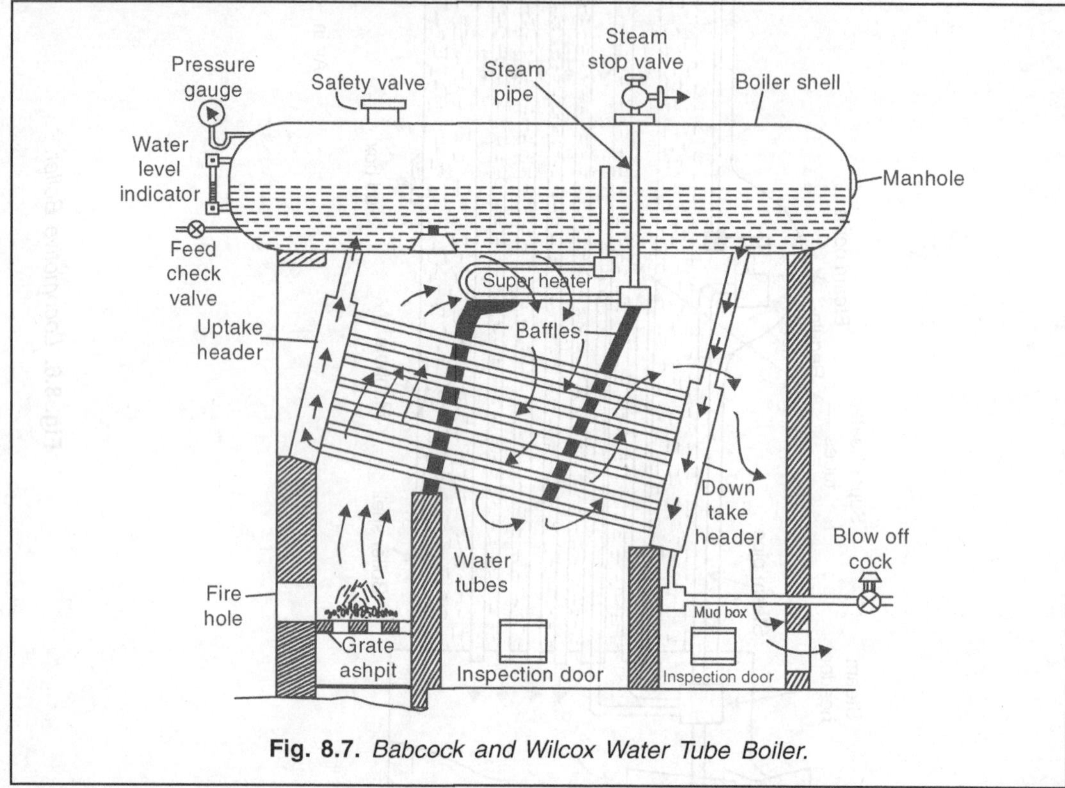

Fig. 8.7. *Babcock and Wilcox Water Tube Boiler.*

The boiler shell is made of high quality steel and lies in horizontal position. Water tubes are placed below the shell. The water tubes are connected with each other and with the shell uptake and downtake headers. The water tubes are kept inclined at an angle of 15° to the horizontal. The short tube by which the uptake header is connected to the shell is known as riser. The water level in the boiler shell is slightly above the centre. A set of U shaped tubes is placed horizontally between the boiler shell and the water tubes. These tubes are called superheater tubes and are used for superheating the steam. The furnace is placed below the uptake header. The coal is fed from a hopper to the grate through the fire door. The hot flue gases from burning of the fuel rise and pass over the water tubes in a zigzag manner and transferring the heat to the water tubes.

The zigzag movement of the hot flue gases is due to the staggered arrangement of the water tubes in the boiler and two firebrick baffles. The staggered arrangement of the water tubes exposes the surface of every tube to the hot flue gases. The fire brick baffles are arranged in such a manner that the hot gases from the grate are first guided to move up and then down and then again in up direction before escaping to the chimney. The water in the water tubes on heating becomes lighter and its density decreases. Because of decrease in density, the hot water moves up through the uptake header and enters the boiler shell. The water from the back side of the boiler drum enters in water tubes through the downtake header. The water then moves upward through the inclined tube and finally rises through uptake header to the boiler drum.

The circulation of water is produced due to the difference in the temperature of the front and the rear end of the furnace. This effect is known as thermo-siphon. It results in continuous circulation of water through the tubes. The steam from the steam space enters the superheater at the upper part and leaves it at the lower part from there it is carried by pipe to the stop valve. When the steam is to be raised from the cold boiler, the superheater is filled with water to boiler drum water level. This is essential to prevent the overheating of the superheater tubes. The superheater remains full with water until the steam reaches the working pressure.

The entire boiler with all the fittings is suspended from horizontal beams by steel slings or straps. The horizontal beams in turn are supported on cast iron columns. This arrangement makes the boiler unit independent of the brickwork which eliminates expansion problems. The feed water entering through the feed check valve is directed backwards to help thermo-siphon effect.

Steam capacity

- 20,000 to 40,000 kg of steam/hour

Pressure

- 11.5 to 17.5 kgf/cm^2

Advantages

- High evaporative capacity
- Less draught loss as compared to the other boilers,
- Easy replacement of defective parts,
- Good thermal efficiency.

8.5.7 Striling Boiler

This is an example of a water tube boiler in which bent tubes have been used instead of straight tubes. Striling boiler Fig. (8.8) may have two or three steam water drums at the top and a mud drum at the bottom.

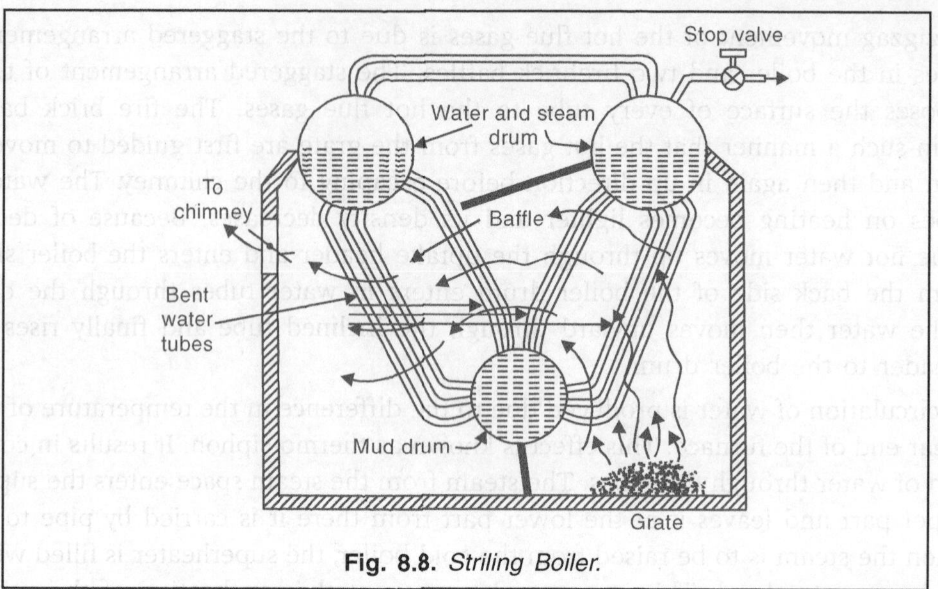

Fig. 8.8. *Striling Boiler.*

The bent tubes are advantageous as absence of header makes the boiler lighter.

Disadvantages

- Difficult to clean and inspect.
- Difficult to seal a tube in case of its rupture.

8.6 ADVANTAGES AND DISADVANTAGES OF FIRE AND WATER TUBE BOILERS

Fire Tube Boiler

Advantages	*Disadvantages*
1. Low initial cost. 2. Easy replacement of ruptured tubes. 3. Large water storage capacity.	1. Small steam space. 2. Fluctuations in steam pressure. 3. Difficult to clean and examine. 4. Leakage at the ends of tubes at stays and corner of the fire-box. 5. Limited capacity of steam generation.

Water Tube Boiler

Advantages	*Disadvantages*
1. Steam can be raised quickly. 2. Suitable for high pressure. 3. Safe operation. 4. Low weight for large capacity.	1. More sensitive to operational changes. 2. Difficult to deal with tubes having leakage.

8.7 HIGH PRESSURE BOILERS

The development of the high pressure boilers have been made possible because of the development of high strength and high temperature resistant materials. Modern high-pressure boilers used for power generation have steam capacities varying from 50 to 700 tons/hour, working pressure up to 200 bar and maximum steam temperature about 600°C. The layout of a high-pressure boiler is very compact due to the maximum utilization of heat of the flue gases. The steam above the critical pressure of water (221.2 bar) can be generated in the high-pressure boilers. The design and operation of high-pressure boilers is different then that of the low-pressure boilers. The reason being the different nature of the properties of steam in critical range.

The high-pressure boilers are usually characterized by the following:

(*i*) Forced circulation of water,

(*ii*) Arrangement of drum and tubing,

(*iii*) Improved method of heating.

Advantages of high pressure boilers

1. Light weight.
2. High evaporative capacity.
3. Reduction in number of drums.
4. Smaller bore tubes reduces the boiler weight.
5. Freedom in disposing the heating surfaces.
6. Capable of meeting the fluctuations in the demand of steam consumption.
7. With external power supply, quick start from cold is possible. Hence these boilers are suitable for handling peak loads and are used for standby purposes in hydraulic stations.
8. Uniform heating of the parts eliminates the danger of overheating and reduces the thermal stresses.
9. Absence from scaling troubles especially if storage drums are provided.

8.7.1 La Mont Boiler

This forced circulation boiler was introduced by La Mont in1925. The circulation diagram of a La Mont boiler is shown in Fig. (8.9). In La Mont boiler, water is supplied through an economiser to a separating and storage drum, which contains a feed regulator that controls the speed of the feed pump. The economiser provides sensible heat to the feed water. The economiser is placed in the path of the hot combustion gases.

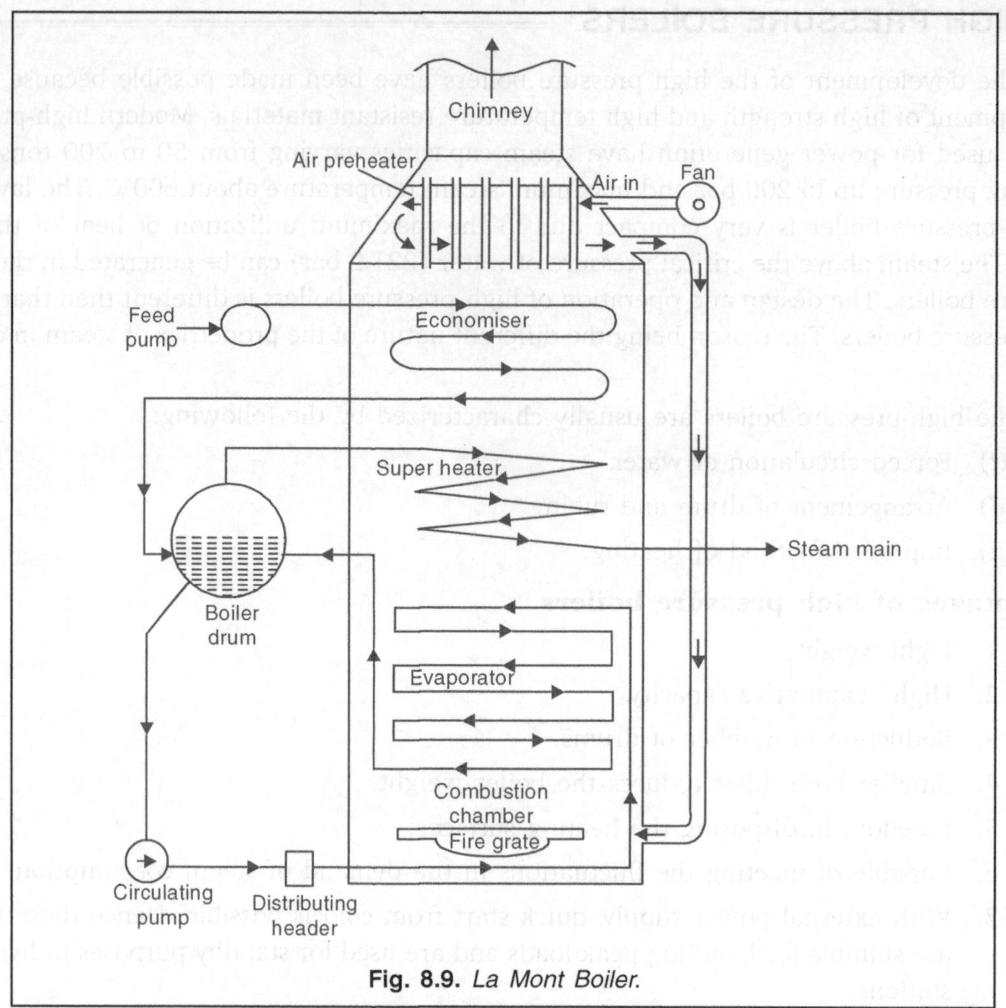

Fig. 8.9. *La Mont Boiler.*

From this drum a centrifugal pump circulates about 8–10 times water evaporated into the parallel circuits of water walls through distributing header at 2.5 bar. The distributing headers, as the names suggest, distribute water through nozzles to the parallel generating tubes. The impurities or sludge if any, settle down in the headers and are removed regularly from the bottom. The water from the distributing header flows through small diameter tubes positioned in the walls and roof of the furnace to the drum where the steam is separated and water comes back to the pump.

Inside the water tubes, some water gets converted into steam and a mix of water and steam is pushed into the boiler drum. In the boiler drum the steam produced is separated and collected in the steam space and then superheated by passing it through a superheater. The superheated steam is next send through the steam stop valve for the commercial use. Overheating of the water tubes is prevented by circulating water through them. The quantity of water required for this purpose is about 10 times the amount of the steam to be generated. About 0.5–0.6 percent of the boiler output is generally consumed by the pump. To ensure a uniform flow of fluid through each of the parallel circuits, a choke is fitted at the entrance to each circuit. The diameter of the choke is chosen in such a way so as to get the same discharge from each circuit. For the economic combustion, the air is preheated by the flue gases.

Steam capacity

- 300,000 kg of steam per hour

Pressure

- 120–130 bar

Temperature

- 500°C

8.7.2 Benson Boiler

Benson boiler is first super critical pressure boiler and was first built in England by Mark Benson in 1923 and later on commercially developed by Siemens Co. (Germany). As shown in Fig. (8.10), it is an once through boiler which takes in feed water at one end and discharges superheated steam at the other. It consists of the following:

(*i*) *Air preheater:* It preheats the air for economical combustion.

(*ii*) *Economiser:* It supplies sensible heat to water.

(*iii*) *Evaporative:* Surface where the major quantity of water is evaporated.

(*iv*) *Superheater:* It provides superheat to the steam before tapping it for external use.

(*v*) Radiant surface of tubes in which radiant heat is supplied by combustion.

The main feature of this boiler is the absence of the steam separating drum. The steam bubble formation in earlier boilers used to reduce the rate of heat transfer from flue gas to water and causes the pulsation in the water flow. This problem of steam bubble formation of earlier boilers was one of the crucial factor in Mark Benson's idea of raising the boiler pressure to critical level in Benson boiler. At critical pressure no steam bubble formation results as the density of steam and water is same at the critical pressure. Looking at the flow circuit of Benson boiler, it appears as a single tube of great length but it consists of many parallel circuits.

Advantages

- Pressure can be raised to super critical level.
- The initial cost of the boiler is low, as no drum is needed.
- Light in weight with reasonably good output.
- Early starting is possible (It can start within 15 minutes time).

Steam capacity

- 135,000 kg of steam per hour

Pressure

- 210 bar

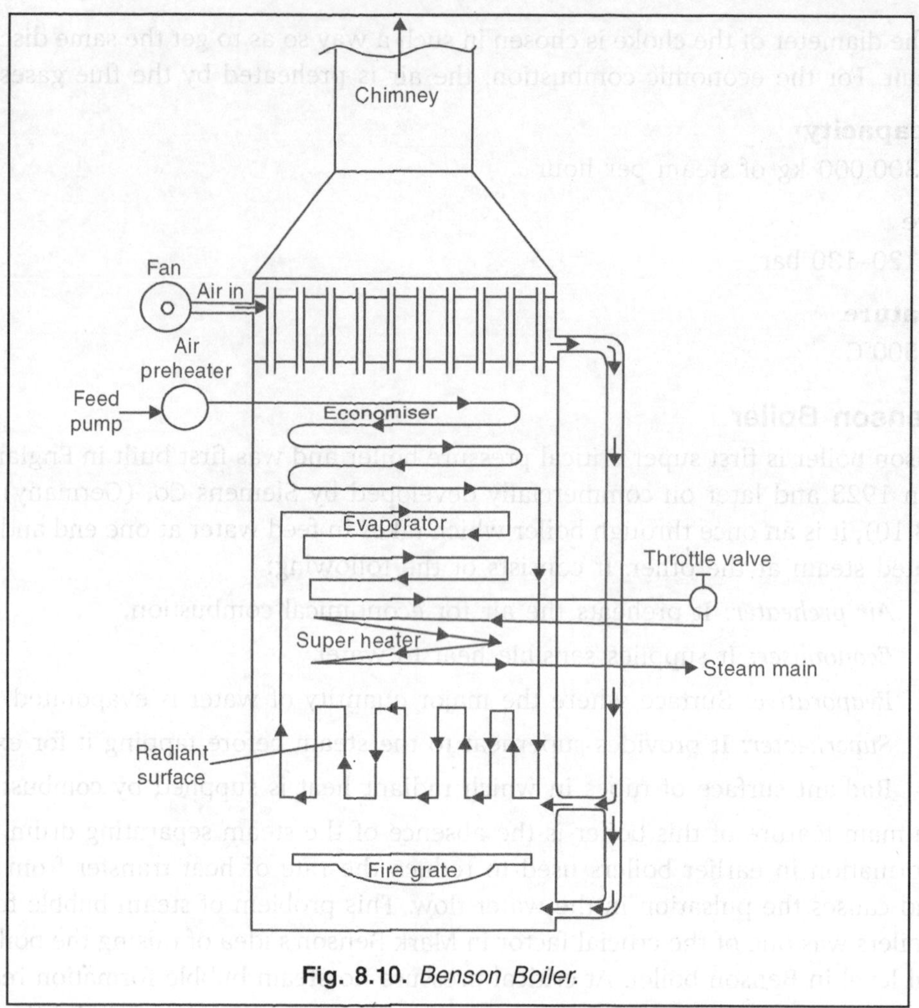

Fig. 8.10. *Benson Boiler.*

Temperature

- 405°C

Thermal efficiency

- 90%

8.7.3 Loeffler Boiler

It is a high-pressure water tube boiler, which has the twin advantages of forced circulation and indirect heating. The indirect heating reduces the problem associated with undesirable solid deposition on the heating surfaces. The other notable feature of this boiler is that it utilizes the steam as the heat absorbing and heat carrying medium. A schematic diagram of this boiler is shown in Fig. (8.11).

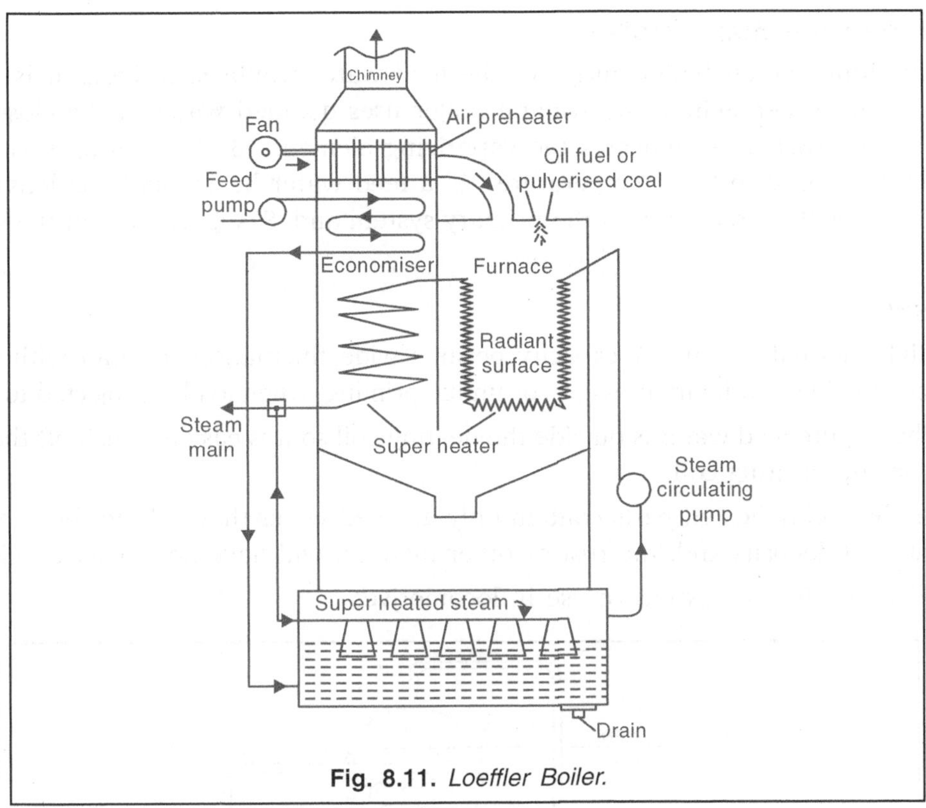

Fig. 8.11. *Loeffler Boiler.*

In Loeffler boiler, the feed water from the economiser is forced to mix with superheated steam coming from the superheater in the evaporative drum. The pressure and temperature of this steam is about 120 bar and 500°C respectively. The steam circulating pump draws the steam from the evaporator drum and guides it to pass through radiant and convective superheaters. A portion of this superheated steam (about 1/3) is tapped off for the external use. The remaining (about 2/3) superheated steam is made to pass through the evaporative drum where it gives up its superheat to the water contained in the drum and results in generating an amount of steam equal to that tapped off. The nozzles, which distribute the superheated steam throughout the water are of special design to avoid priming and noise. This enables the Loeffler boiler to carry higher salt concentrations as compared to any other boiler types.

Steam capacity

- 95,000 kg of steam per hour

Pressure

- 140 bar

Temperature

- 500°C

Applications: suitable for land and sea transport power generation.

8.7.4 Schmidt-Hartmann Boiler

This multiple circuit boiler mitigates the feed water troubles. In looks it is similar to an electric transformer. Schmidt-Hartmann boiler uses distilled water in the closed system as shown in Fig. (8.12). 'A' indicates the vaporizing element, 'B' the heating surface of the closed circuit exposed to high temperature, 'C' a feed water heater and condensate cooler which accelerates the circulation in the primary system and 'S' superheater in the secondary system.

Advantages

- High thermal and water capacity permits wide fluctuations of load without undue increase in the primary pressure, or undue priming, when load is subjected to variation.
- The impure feed water is outside the heating coil so it is easy to brush off the deposits from the heating coil.
- As the highly heated parts contain only distilled water they will not burn out due to internal deposits and, the rust or other material will have no impact on circulation.

Application: It finds extensive use in locomotives

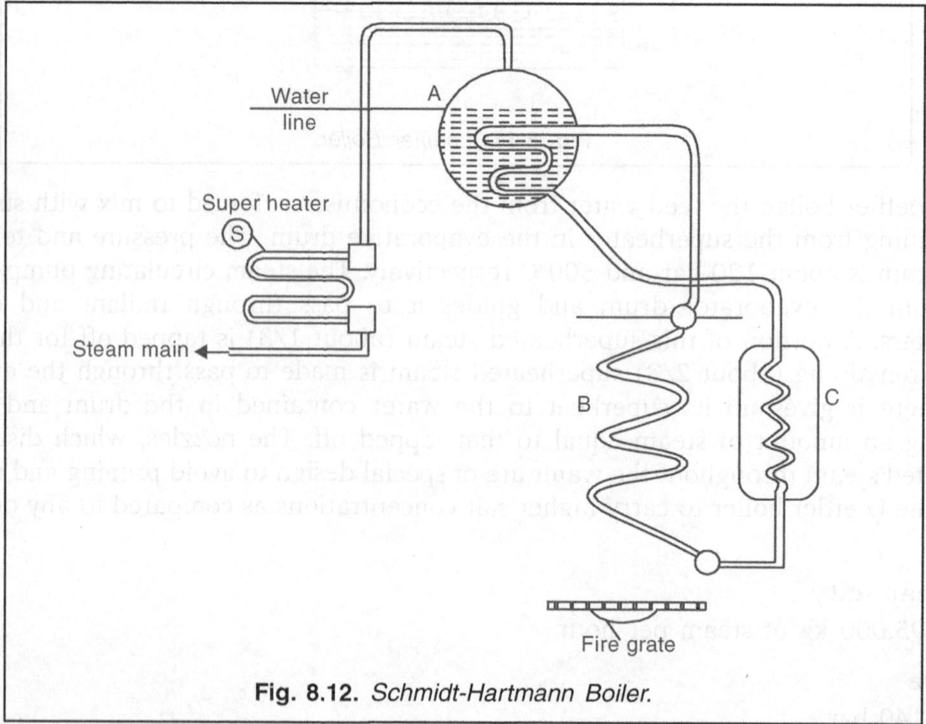

Fig. 8.12. *Schmidt-Hartmann Boiler.*

8.8 BOILER MOUNTINGS

For the safe working of a boiler different fitting and devices are used. The fittings and devices, which are necessary for the proper functioning of a boiler are known as boiler mountings.

The devices, which are essential from the safety point of view are:

- Safety valves
- Water level indicator
- Fusible plug
- Combined high steam and low water safety valve.

Control fittings required for the safety of a boiler are:

- Steam pressure gauge
- Stop valve
- Feed check valve
- Blow off cock
- Manhole.

8.8.1 Safety Valve

Safety valve is the most important valve on a boiler. The main function of a safety valve is to prevent steam pressure in a boiler exceeding the predetermined maximum pressure limit. This is done automatically by opening the safety valve and discharging the steam to the atmosphere. This brings down the pressure within the specified limits. Since the discharged steam is wasted, hence these valves should operate occasionally. To keep a safety valve in good working condition. It should be tested from time-to-time. Safety valves are usually classified on the basis of the method of loading.

The commonly used safety valves are:

(A) Dead weight safety valve

(B) Lever loaded safety valve

(C) Spring loaded safety valve.

(A) Dead weight safety valve

This safety valve as shown in Fig. (8.13) is loaded by applying the weights above the valve. It consists of a large vertical pipe having a flange at the bottom for fixing it on the boiler. The valve seat is fixed at the pipe top and the valve rests on it. A large cast iron weight carrier is suspended from top of the valve, which acts as a weight carrier. The dead weight of the cast iron rings enclosed in cast iron cover is placed on the weight carrier. The load on the valve consists of weights, the weight of the weight carrier and cover, and the weight of the valve itself. The combined load of all these balances the pressure of steam on the valve.

The required dead weight can be calculated from a given valve area.

Dead weight = valve area × steam pressure

Dead weight may be reduced for a given steam pressure, by reducing the valve area. In case of sudden rise in steam pressure, the valve area should be sufficient enough to release the steam quickly and bring back the steam pressure within the set limits. Whenever the steam pressure rises beyond the set limits, the valve and the weight carrier will lift and excess steam will escape to the enclosed discharge casing from which it is carried to the waste pipe. The blowing off the valve is prevented by a ring cast on the inside of the weight carrier and stop screws fitted on the discharge casing. A drain pipe is placed at the bottom of the discharge casing for draining the condensed steam.

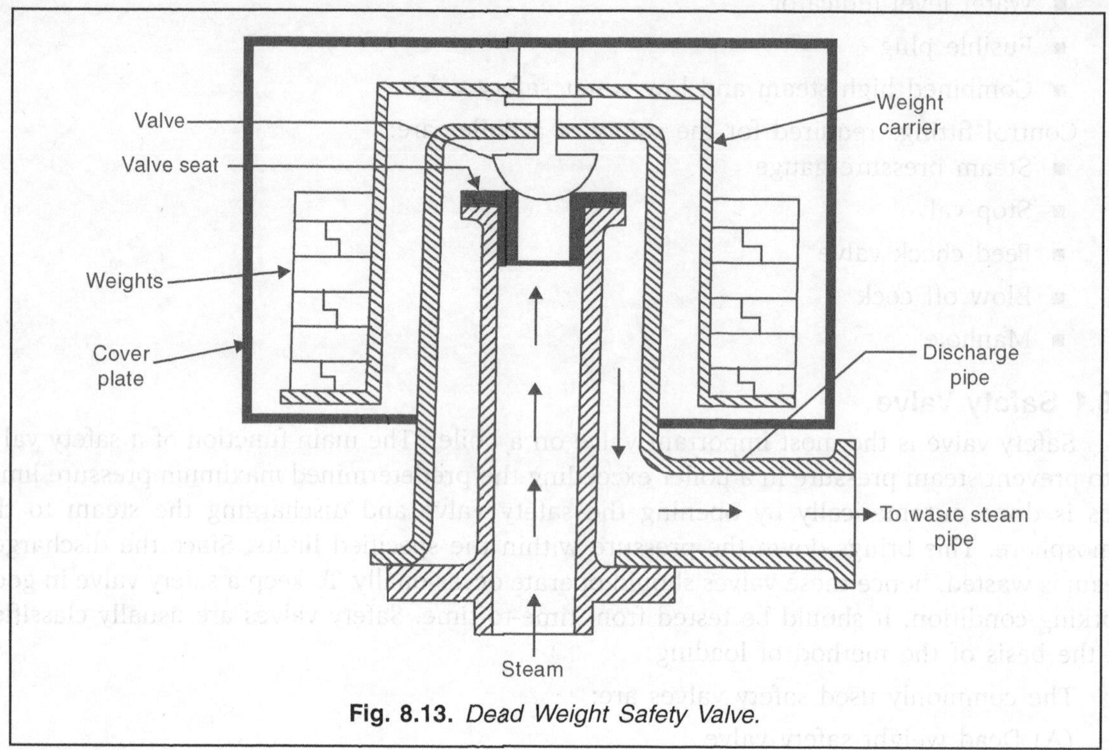

Fig. 8.13. *Dead Weight Safety Valve.*

Applications

- For stationary boilers

Limitations

- Not suitable to locomotive and marine boilers as (*i*) more weights are required, and (*ii*) the effective weight changes when the boiler is on the slope.

Advantages

- Reliable operation.

(B) Lever loaded safety valve

A schematic diagram of a lever loaded safety valve is shown in Fig. (8.14). The valve rests on the valve seat, which is screwed on the body. The valve is held by a short strut hanging from the lever. One end of the lever is pivoted at the fulcrum and the other end carries the weight. The guide not only prevents the lateral movement of the valve but also controls its blowing off. The weight on the lever can be moved to the desired position according to the boiler pressure.

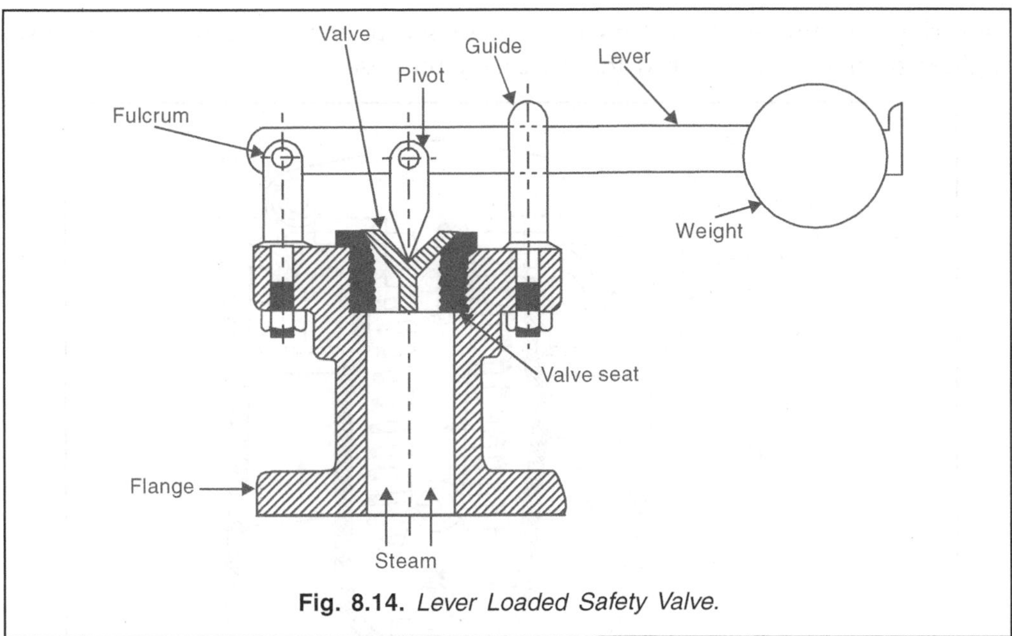

Fig. 8.14. *Lever Loaded Safety Valve.*

Limitations

- Valve friction due to the presence of two pivot points (one at the fulcrum and other at the valve spindle) which in extreme cases may prevent the valve from opening when the pressure exceeds the normal working pressure.
- Not suitable for mobile boilers.

Applications

- Suitable for stationary boilers.

(C) Spring loaded safety valve

It is the most commonly used safety valve on both stationary and mobile boilers. It is loaded by means of a spring instead of weight. The spring in helical form is made from a square steel rod. It consists of two valves, each is placed on a valve seat over a branch pipe. These two branch pipes are connected to a common block. The valves are held down on their seats by the spring and lever pivots. The lever has two pivots. One pivot is forged on the lever and other is connected by a pin. The spring upper end is attached to the lever while lower end of the spring is hooked to the shackle. The shackle is screwed to the body of the valve by the studs and nuts. The valve can be tested manually by pressing down or raising the lever. Two links on the either side of the liver is provided to prevent the valves being blown away in the event of spring failure. Figure (8.15) shows a schematic diagram of a spring loaded safety valve.

Applications

- Suitable for mobile boilers such as locomotive and marine type.

Disadvantages

- The valve sometimes may lift under less pressure as compared to that required to open it fully.

The spring pressure on the valve can be increased by increasing the spring tension. The spring tightening nut is kept locked for the safety purpose.

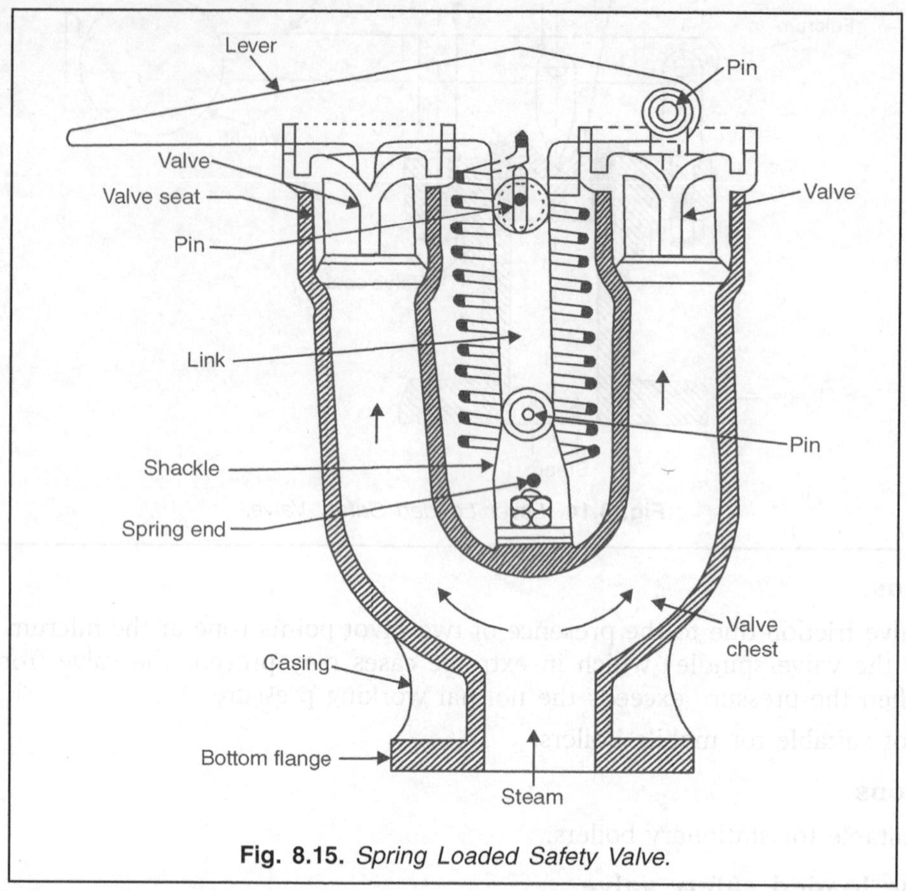

Fig. 8.15. *Spring Loaded Safety Valve.*

8.8.2 Water Level Indicator

Water level indicator ensures safe operation of a boiler by allowing the operator to see how much water is inside the boiler. Water level indicator shows the exact level of water in the boiler at any instant. The level of water inside the boiler should remain nearly constant and in no way it should be allowed to fall below a certain minimum level; otherwise the boiler plates may burn. Two water level indicators are provided on each boiler for the safety reasons. The one indicator is placed at the left-hand side of the boiler front and the other at the right hand side. When the boiler drum is situated at considerable height from the floor, the water level indicator is often inclined so as to make the water level visible from any position.

The schematic diagram of a water level indicator is shown in Fig. (8.16). AA is the front end plate of the boiler to which the water level indicator is attached and WW is the water level. G is a gauge glass tube, which is very tough and hard to withstand the boiler pressure. It is connected to the boiler plate through stuffing boxes assembly B and C. B is connected to the steam space and C is connected to the water space in the boiler. Two cocks D and E are provided for controlling the passages of steam and water respectively from the boiler. The cocks open in the vertical position. When the cocks are open, the water level is indicated in the gauge glass. The safe water level in the boiler is marked on the gauge glass. A third cock F called a blow through cock is provided for blow off purposes. This cock F closes in the vertical Position and opens in the horizontal position.

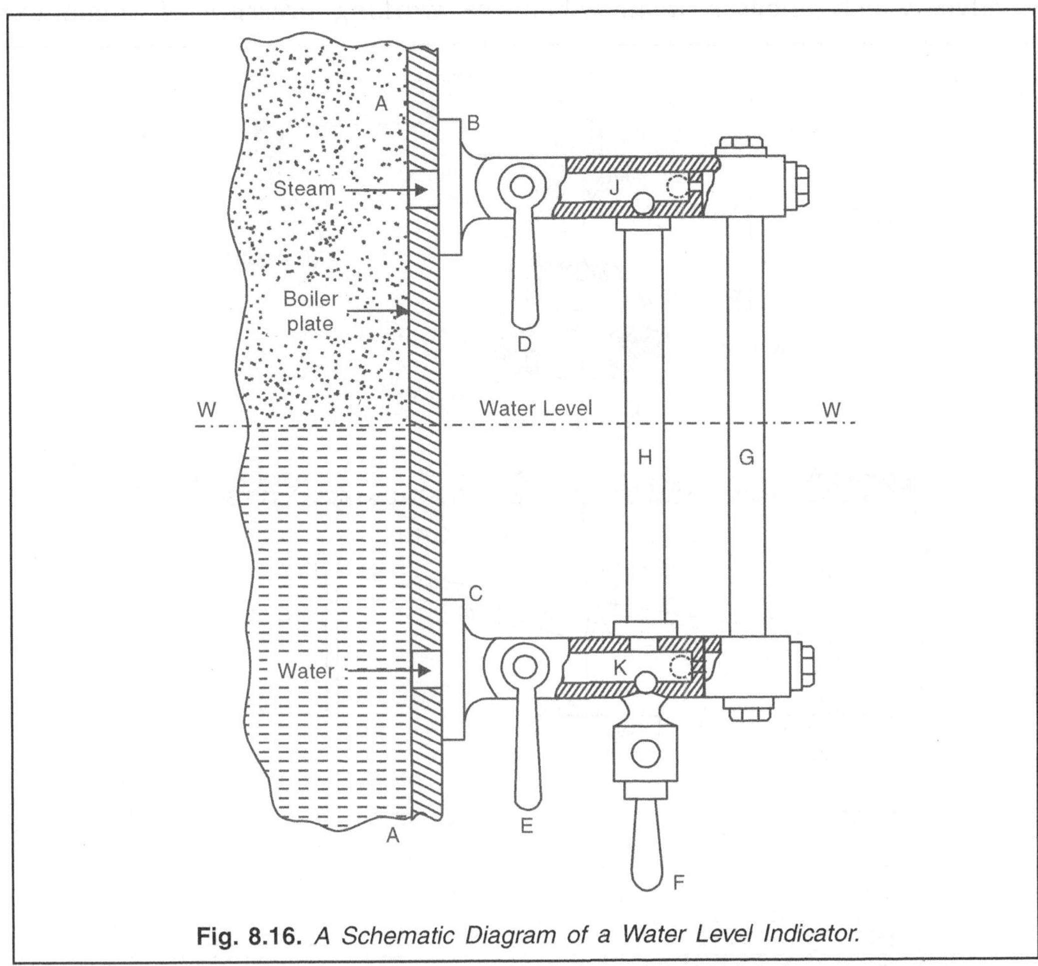

Fig. 8.16. *A Schematic Diagram of a Water Level Indicator.*

The water level indicator is manufactured with the provision of automatically shutting off the steam and water supply to the glass tube if in case it breaks. Upper and lower stuffing box assembly is also connected by hollow metal column H. The level of water in this column corresponds to the level of water in the glass tube though it is not visible. Under normal working, balls J and K rest in the position as shown in Fig. (8.16). In case the glass tube breaks, the rush of water from bottom passage and steam from top passage pushes the balls J and K in the dotted position and disconnects the guage glass with the boiler. At this moment the boiler operator can safely close the cocks D and E and replace the broken glass tube.

Every manual on boiler operation, under all circumstances incorporates the following instruction: the first duty of a boiler operator on taking over the operation is to check the level of water in the gauge glass. The water level indicator is a simple device, but a through understanding of its function and the means to ensure its accuracy is essential in boiler room operation.

8.8.3 High Steam and Low Water Safety Valve

This safety valve ensures the safety of a boiler against high steam pressure and low water level in the boiler. It is the combination of two valves, one of which acts as a lever loaded safety valve, which releases steam when the working pressure of steam exceeds the pre-determined value. The second valve releases the steam when the water level falls below the danger level (Pre-decided level). It raises an emergency alarm so that the fresh water can be pumped in to the boiler. It is used with stationary boilers such as Lancashire and Cornish boilers.

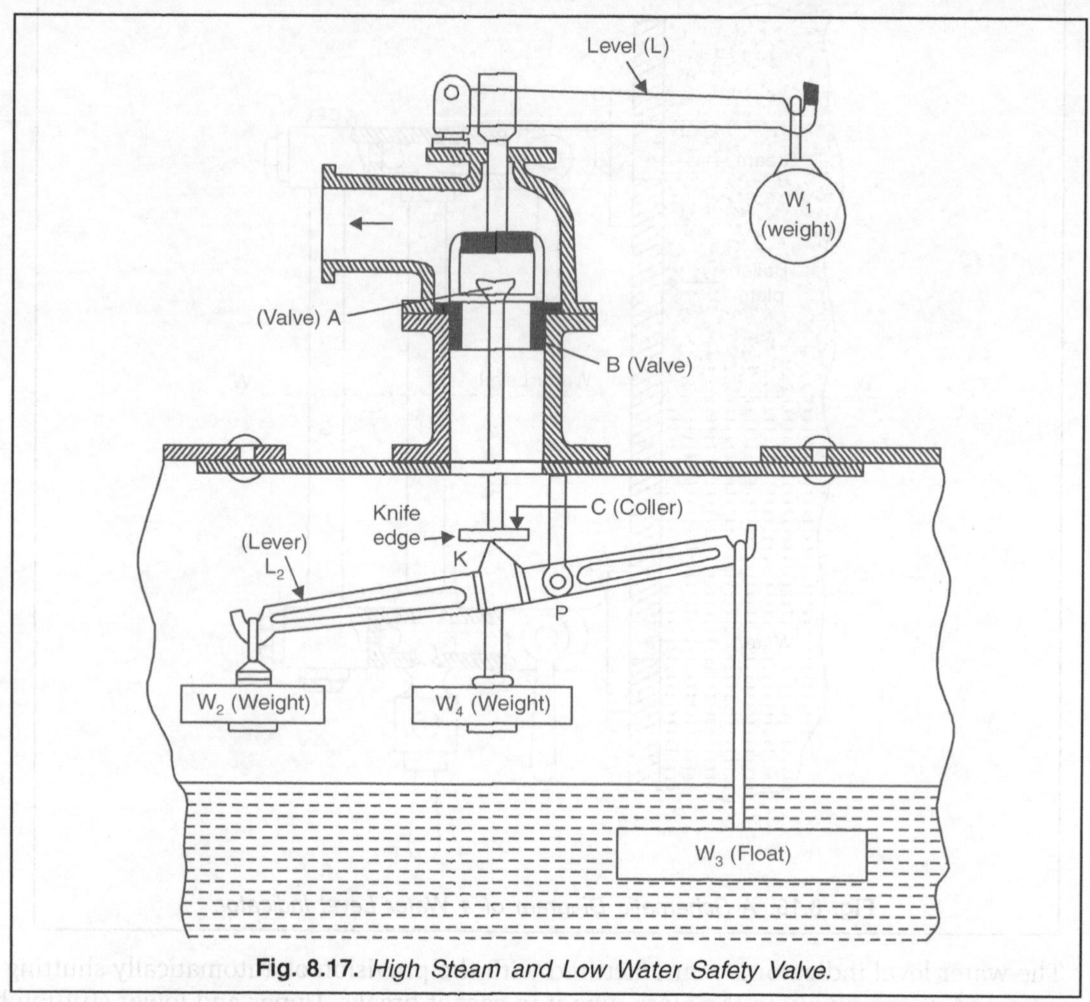

Fig. 8.17. *High Steam and Low Water Safety Valve.*

The Fig. (8.17) depicts a schematic diagram of high steam and low water safety valve. Safety valve is outside the boiler shell to protect the boiler from sudden increase in steam pressure. It consists of two valves A and B. The valve B acts as a valve seat for valve A. To maintain the desired pressure in the boiler it acts like a simple lever safety valve loaded by two weights, one attached to the lever and other to valve A. The lever is hinged at one of its end and the weight is at the other. A short pivot attached with the lever is placed on the valve A, which keeps it in position under normal working pressure. When the pressure exceeds the normal limit, the valve A is lifted along with lever L. This releases the steam through the passage between the valve seat and valve B. The low water safety device is placed inside the boiler shell. The lever L_2 is hinged to the boiler shell at point P. The weight W_2 is connected to one end of the lever and a large weight W_3 (float) at the other. The valve A is connected with a spindle at the lower end of which a weight W_4 is attached. The knife edge K provided on the lever L_2 touches a collar C on the spindle under normal pressure. Under the normal circumstances, the weight W_3 remains in water and the complete system is under balance.

When the water level falls below the set limit, the weight W_3 (float) is uncovered It causes an increase in its weight (according to Archimedes's Principle) which inturn swings the lever. This causes the knife edge to push the collar C which opens the valve A thus allowing the steam to escape between the valve A and valve B. The escaping steam makes a loud noise as it passes through the narrow opening between the valves A and B. This loud noise draws the attention of the boiler operator about the falling level of water in the boiler shell.

8.8.4 Fusible Plug

The main function of a fusible plug is to put off the fire in the furnace of a boiler when the water level in the boiler falls below the safe minimum level. Fusible plug is provided at the crown of the furnace or over the combustion chamber. Figure (8.18) shows a commonly used fusible plug. 'A' is a hollow gunmetal plug body screwed into the cross of the furnace. 'B' is a second hollow gunmetal plug screwed into the plug A. The third hollow gunmetal plug C with conical top and rounded bottom separated from plug B by the fusible metal. The normally preferred fusible metal is lead. The inner surface of B and the outer surface of C are grooved as shown in Fig. (8.18) so as to lock the plugs B and C when the fusible metal is poured.

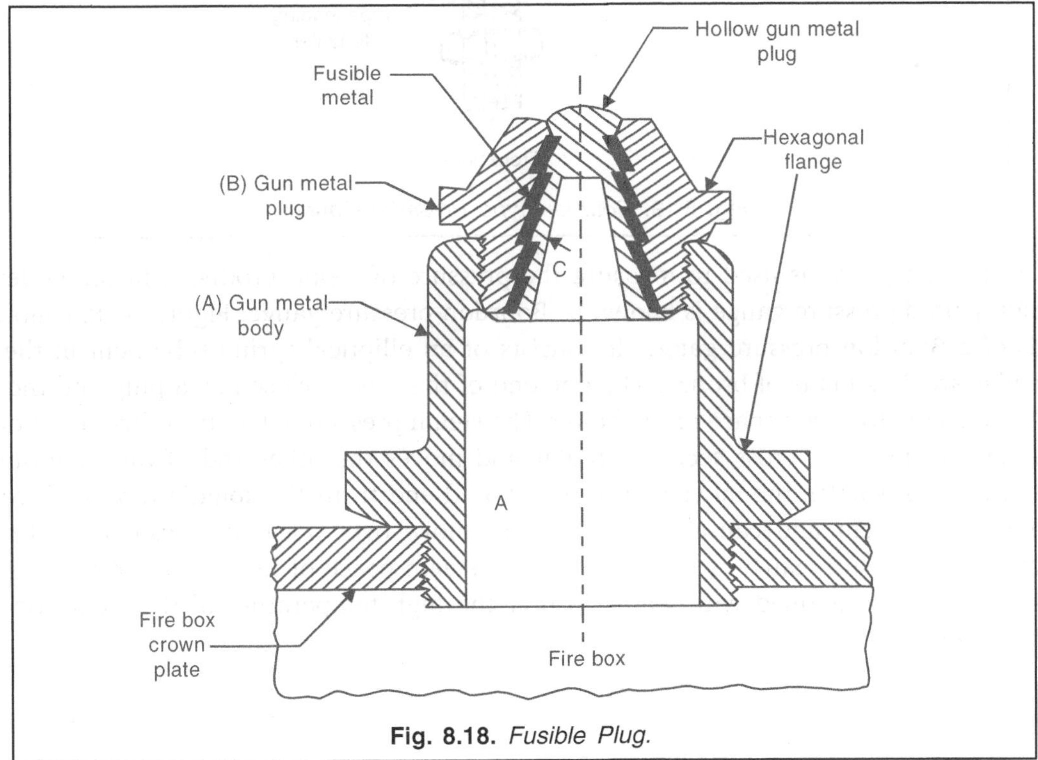

Fig. 8.18. *Fusible Plug.*

During normal running the fusible plug remains under water inside a boiler which keeps the temperature of the fusible metal below its melting point. When the water level in the boiler falls low enough to uncover the top of the plug, the fusible metal melts by the heat of the furnace as it gets exposed to steam space instead of water which is not able to keep it cool. The plug C drops out and the steam rushes in to the furnace through the opening created by falling of plug C and extinguishes the fire.

8.8.5 Steam Pressure Gauge

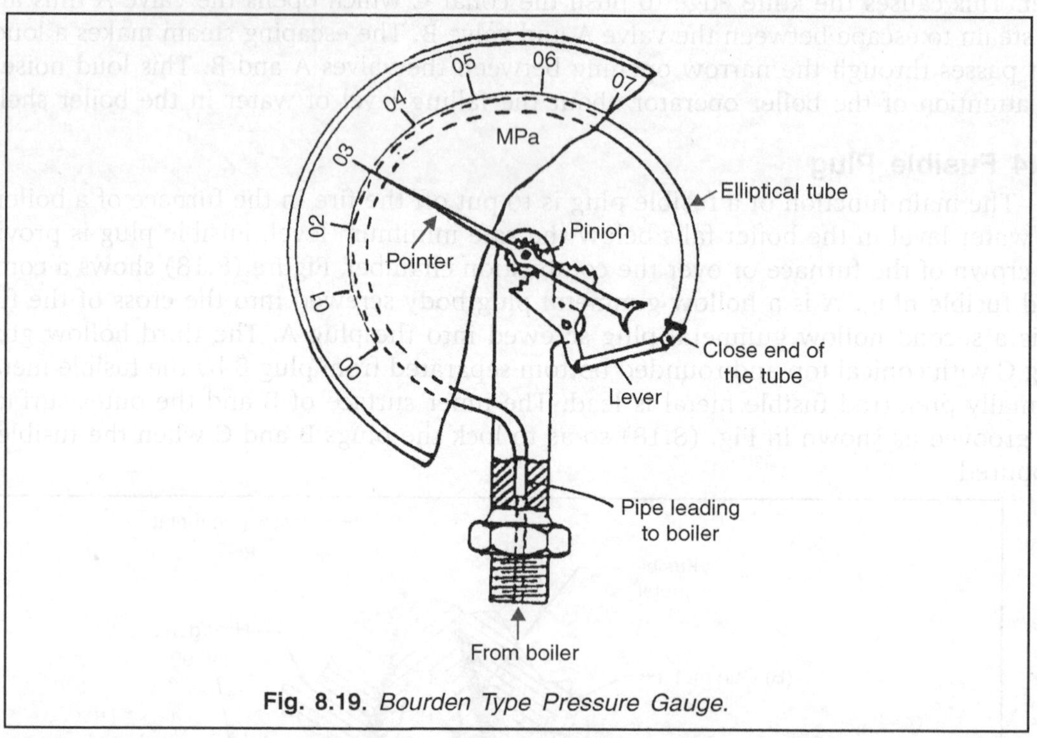

Fig. 8.19. *Bourden Type Pressure Gauge.*

A pressure gauge is used to measure the pressure of steam produced in the boiler. The commonly used pressure gauge is known as Bourdon pressure gauge. Figure (8.19) shows the details of a Bourdon pressure gauge. It consists of an elliptical spring tube bent in the form of circular arc. It is made of bronze. The one end of the tube is closed by a plug and the other end is connected to steam space of the boiler. The steam pressure when transferred to Bourdon tube, tends to make its cross section circular and moves the other end of the tube outward as shown in Fig. (8.19). the movement of the closed end turns the spindle through lever and gearing arrangement and causes the pointer to move over a circular scale indicating the pressure. The pressure gauge is connected to the steam peace of the boiler through a U-tube siphon. The U-tube is filled with water so that the high temperature of the steam does not affect the gauge.

8.8.6 Stop Valve

The main task of a stop valve (popularly known as junction valve) is to regulate the steam flow from the boiler and shut off the steam flow when not required. Figure 8.20 shows a simple steam stop valve. The stop valve is generally made of good quality steel. The valve, valve seat and the nut through which the valve spindle works are all made of brass for smooth working. The valve body has two flanges. The stop valve is connected to the boiler through bottom flange and the other is connected to the outlet.

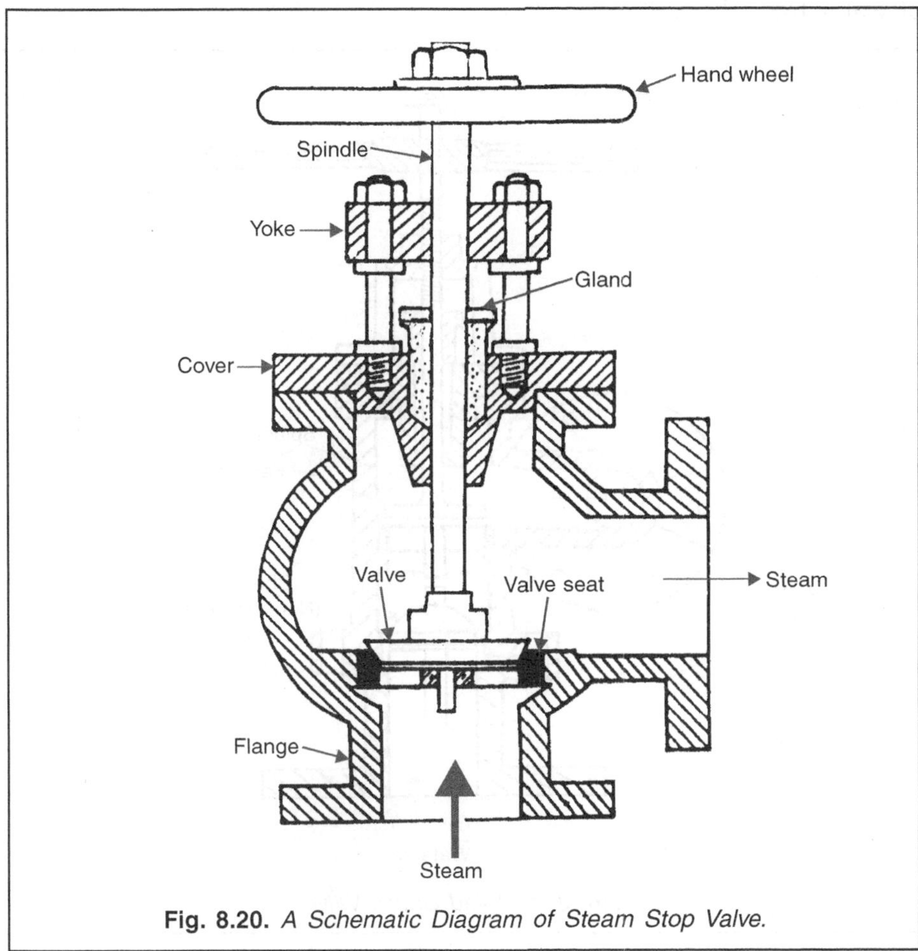

Fig. 8.20. *A Schematic Diagram of Steam Stop Valve.*

A hand wheel is provided with the valve spindle. The valve is connected with the spindle by nuts. The valve rests on valve seat, which is fixed on the valve body. The spindle passes through a gland to prevent the steam leakage. The spindle is rotated by means of a hand wheel. Due to the rotation of hand wheel, the valve may move up and down and it may close or open the passage fully or partially for the steam flow. The diameter of the steam pipe determines the size of the steam stop valve.

8.8.7 Feed Check Valve

The function of a feed check valve is to regulate the flow of water into the boiler at high pressure and prevents the back flow of water from the boiler when the feed water pump is not working or pump pressure is less than the boiler pressure. It is placed on top of the boiler shell.

It is a non-return valve, which permits flow of water in one direction only. The valve opening can be regulated by hand so the water level in the boiler remains nearly constant. A schematic diagram of a feed check valve is shown in Fig. (8.21). It consists of two valves namely the feed valve and check valve. The check valve permits the water flow in one direction only *i.e.,* from feed pump to the boiler side. When the feed pump fails, the check valve is closed automatically due to the pressure of the water from the boiler and prevents the backflow of water from the boiler.

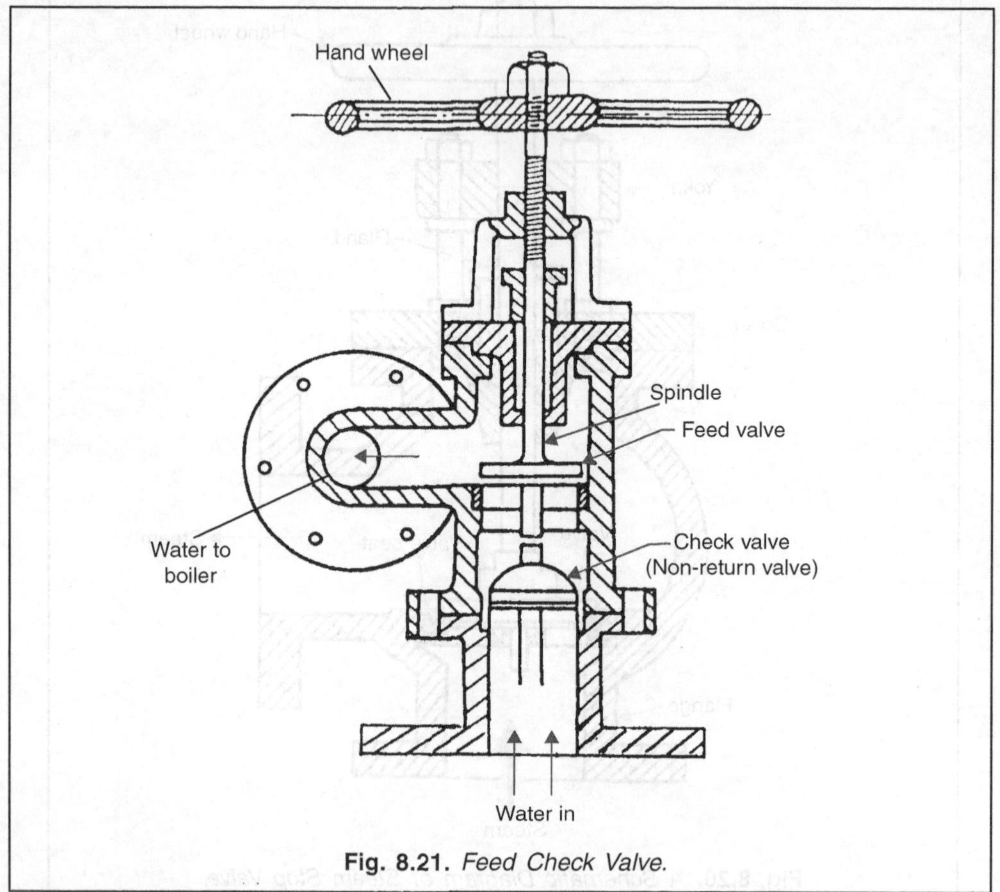

Fig. 8.21. *Feed Check Valve.*

The feed check valve should be kept in good running conditions. It can be inspected and cleaned even when the boiler is working.

8.8.8 Blow Off Cock

It is one of the important fittings on a boiler. Its functions are:

1. To empty the boiler for cleaning and inspection.
2. To discharge the loose mud, sediments and heavy deposits which settles down at the bottom of the boiler.

A schematic diagram of blow off cock is shown in Fig. (8.22). It is fitted to the lowest part of the boiler either directly with the boiler shell or to a pipe connected with the boiler. The blow off cock consists of a conical gunmetal plug fitted accurately in a conical casing. The plug has a hole or slot in it. The casing is comprised of two flanges; one is connected to the boiler and the other one to the pipe, which takes the discharge out of the boiler.

The shank of the plug passes through gland to avoid leakage. The slot or hole in the plug is brought in line with the passage of casting by rotating the plug; this causes the discharge of water from the boiler. Rotating the plug again can stop this discharge of water. At the bottom of the plug a set screw is provided which is used in case the plug is jammed.

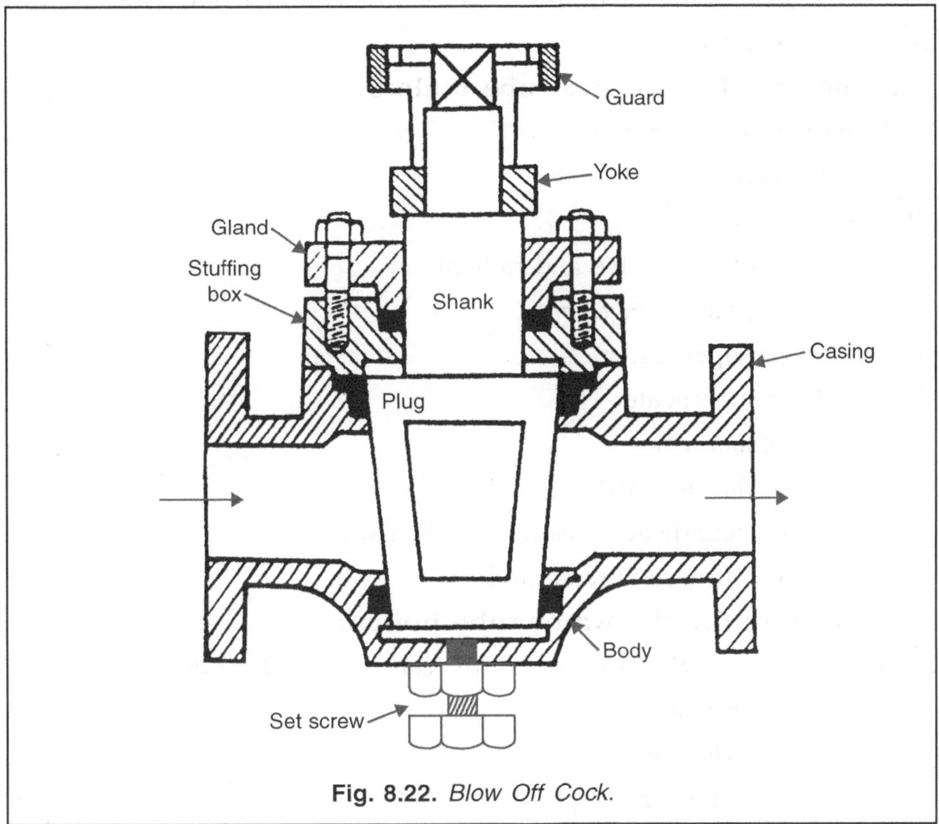

Fig. 8.22. *Blow Off Cock.*

8.8.9 Manhole

The manhole is an elliptical shaped opening provided on a boiler shell at an appropriate location for a man to enter inside the boiler for inspection, cleaning and repairing.

8.8.10 Mud Box

It is provided at the bottom of the boiler to collect mud. Generally this is provided along with the blow off cock.

8.9 BOILER ACCESSORIES

The function of the boiler accessories is to increase the overall efficiency of the boiler plant. The main accessories used on a boiler are:

8.9.1 Superheater

The function of a superheater is to dry the wet steam and then raise the temperature of the steam above its saturation temperature. In superheaters, steam pressure does not change but its volume increases as its temperature is raised. Superheaters are generally placed in the path of the furnace gases so as to utilize the heat from the hot flue gases.

The types of superheater are:

(*a*) **Superheaters for the fire tube boilers**

⇒ *On the basis of heat received*

- Convective superheater
- Radiant superheater
- Combination of convective and radiant

⇒ *On the basis of flow arrangement*

- Counter flow superheater
- Parallel flow superheater
- Mixed flow superheater

⇒ *On the basis of furnace used*

- Portable (*i.e.,* superheater has its own furnace)
- Non-portable (furnace is separated).

(*b*) **Superheaters for the water tube boilers**

⇒ *On the basis of superheater's position in relation to water tubes.*

- Over deck superheater
- Inter deck superheater
- Inter bank superheater.

Convective superheater

This type of superheater is placed in the path of the furnace or flue gases. The surface of superheater receives bulk of heat from the flue gases by convection.

Radiant superheater

This type of superheater is fitted in the walls of the boiler furnace where superheater tubes receive heat by direct radiation from the fire. It is used where high degree of superheat is required.

Combination type

This type of superheater receives heat by both the convection as well as radiation modes of heat transfer.

Parallel flow superheater

In this type of superheater, the direction of flow of steam in superheater tubes and the direction of flue gases is same.

Counter flow superheater

In this type of superheater, the direction of flow of steam in superheater tubes and the direction of flue gases is opposite to each other.

Mixed type

This category of superheater utilizes both types of flow arrangement while raising the temperature of steam.

Portable superheater

The superheater with separate furnace is called portable superheater.

Non portable superheater

It is the superheater without the separate furnace.

Over deck superheater

The over deck superheater is one which is placed in the space over the water tubes.

Inter-deck superheater

If the superheater is placed between the water tubes located near the furnace it is called inter-deck superheater.

Inter-bank superheater

Inter-bank superheater is one, which is placed between the rows of the water tubes.

Superheater tubes must be protected from over heating when no steam is passing through them. There are two ways of preventing the overheating of the superheater tubes.

1. By filling the superheater with water from the boiler
2. By diverting the direction of the flow of hot flue gases from the superheater.

The Fig. (8.23) illustrates the superheater commonly used in the Lancashire boiler. This superheater consists of two headers and a set of superheater tubes made of high quality steel in the form of U tube. Superheater is placed in the path of furnace gases where the temperature of the gases is not less than 560°C.

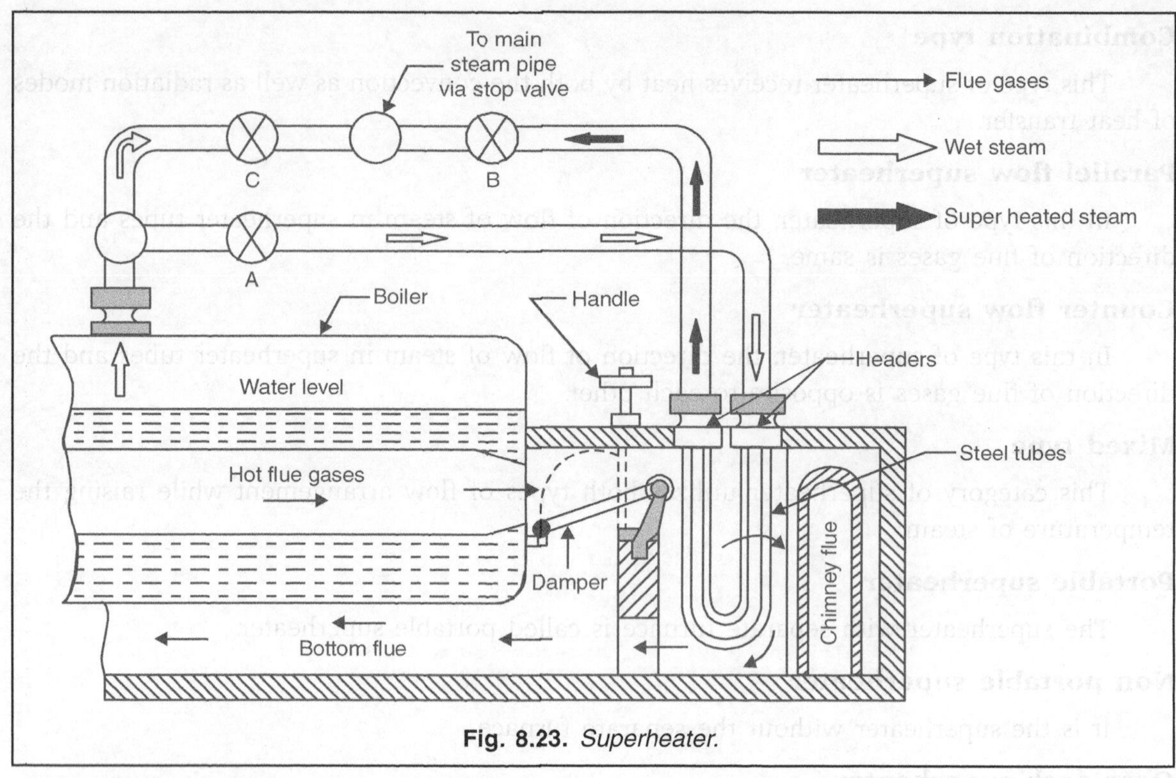

Fig. 8.23. *Superheater.*

When the superheater is in operation, valve C is closed, wet steam from the boiler moves to intake header from valve A, passes through the U tube, gets superheated and then passes through uptake header and valve B and then enters the stop valve for release to main steam pipe. When no superheated steam is required, the superheater is not exposed to hot flue gases by closing the damper, keeping valve A and B closed and valve C open. In this case wet steam directly comes out from the boiler through the valve C to the stop valve.

8.9.2 Economiser

It is also known as feed water heater. The function of an economiser is to utilize the heat of the waste furnace gases in heating feed water before it is supplied to the boiler. It is placed between the boiler and chimney. One of the major losses in a boiler plant is the heat carried away by the flue gases. An economiser consists of a number of horizontal and vertical tubes through which passes the feed water from the pump on its way to the boiler whereas the hot flue gases pass over the outside surface of the water tubes. Heating of feed water by 1°C reduces the temperature of the flue gases by 2 to 3°C.

The economisers are of two types:

(A) *Independent type:* This type of economiser is not the part of the boiler setting

(B) *Integral type:* This type of economiser is installed within the boiler setting.

Figure (8.24) represents an independent type vertical tube economiser popularly known as Green's economiser. It is commonly used in Lancashire boiler. The economiser consists of vertical cast iron tubes, which are pressed hydraulically into the top and bottom header.

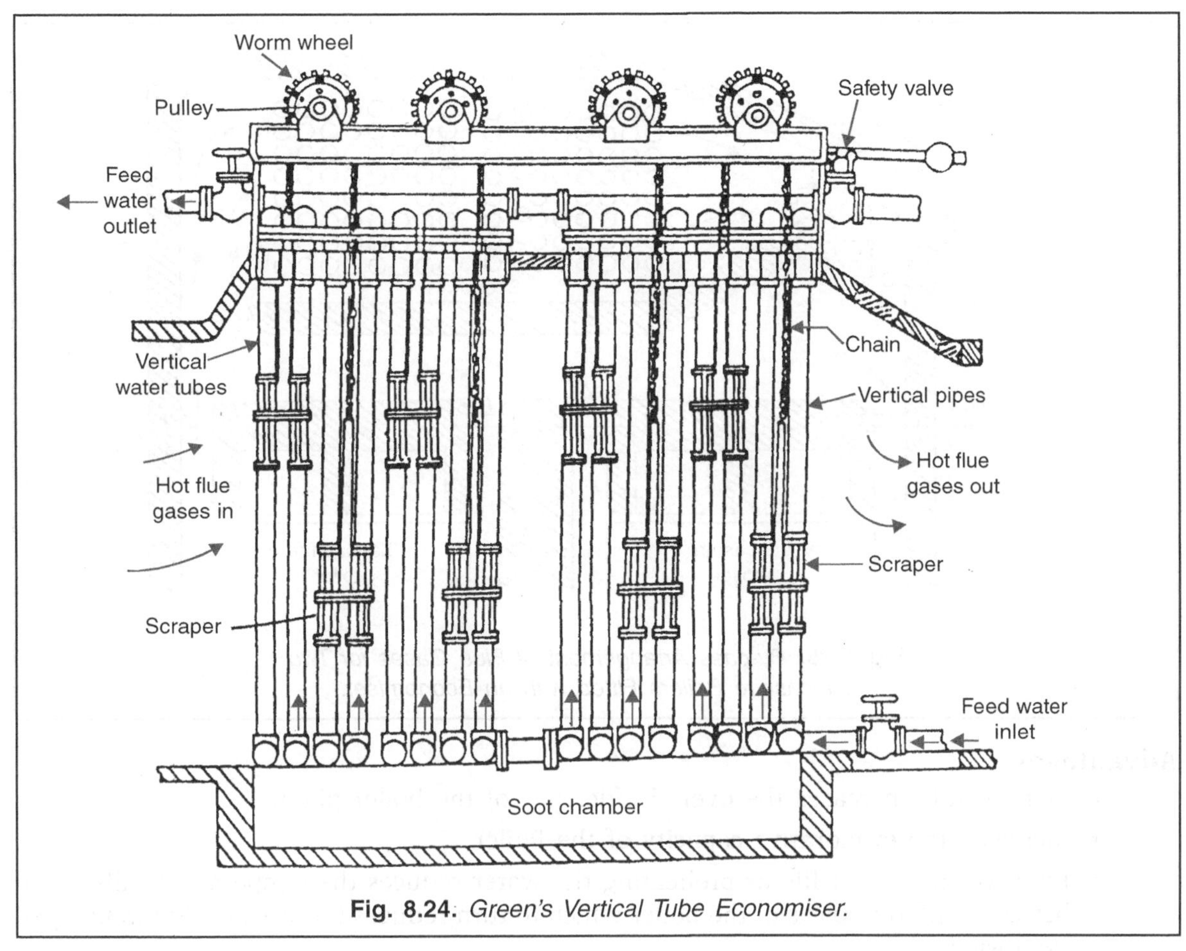

Fig. 8.24: *Green's Vertical Tube Economiser.*

The feed water from the feed pump enters bottom header, passes through the vertical tubes, then to the top header and then enters into the boiler through the feed check valve. The hot flue gases pass over the external surface of the tubes in the direction opposite to the flow of water thus giving its heat to feed water in the tubes.

For efficient heat transfer, the surface of the tubes has to be kept clean from soot and volatile ash deposits. To achieve this, each pipe is provided with a scrapper, which moves up and down with the help of a chain and gear arrangement and keeps the surfaces clean. The scrapping action on the tubes is in opposite direction *i.e.,* when one scrapper moves up, the other goes down. The soot removed from the pipes(tubes) is collected in soot chamber located below the bottom header. Being a pressure vessel, economiser is equipped with a safety valve which functions when the pressure in the economiser due to any reasons exceeds 10 to 15% of the boiler pressure.

In case of the failure of an economiser, by pass arrangement is provided for diverting the hot flue gases straight to chimney by closing the dampers as shown in Fig. (8.25).

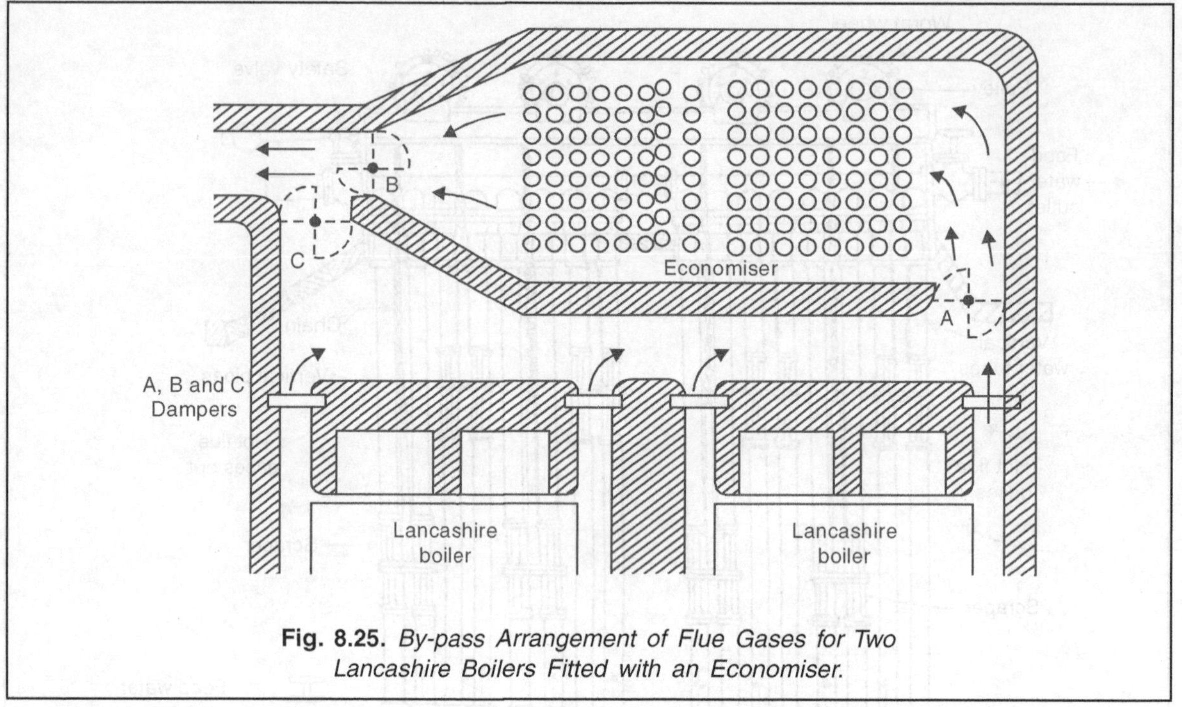

Fig. 8.25. *By-pass Arrangement of Flue Gases for Two Lancashire Boilers Fitted with an Economiser.*

Advantages

- Significantly increases the overall efficiency of the boiler plant.
- Increases the evaporative capacity of the boiler.
- Improves the boiler life as preheating the water reduces the temperature difference between different parts of the boiler which in turn reduces the stresses due to unequal expansion.

Disadvantages

- Costly
- Loss of draught for the flue gases.

8.9.3 Air Preheater

The function of an air preheater is to recover some portion of the heat from the waste furnace gases or exhaust gases (outgoing hot flue gases through chimney) and transfers the heat to the air before it passes in to the furnace for combustion. It is installed between the economiser and the chimney. The air is passed through the tubes of the preheater internally, whereas the hot flue gases are passed over the tubes.

The two commonly used types of air preheaters are:

(A) **Recuperative type air preheater**

In this type, the passage of air and hot gases are different, both being separated by a metallic wall. Figure (8.26) depicts a schematic diagram of a recuperative type air-preheater.

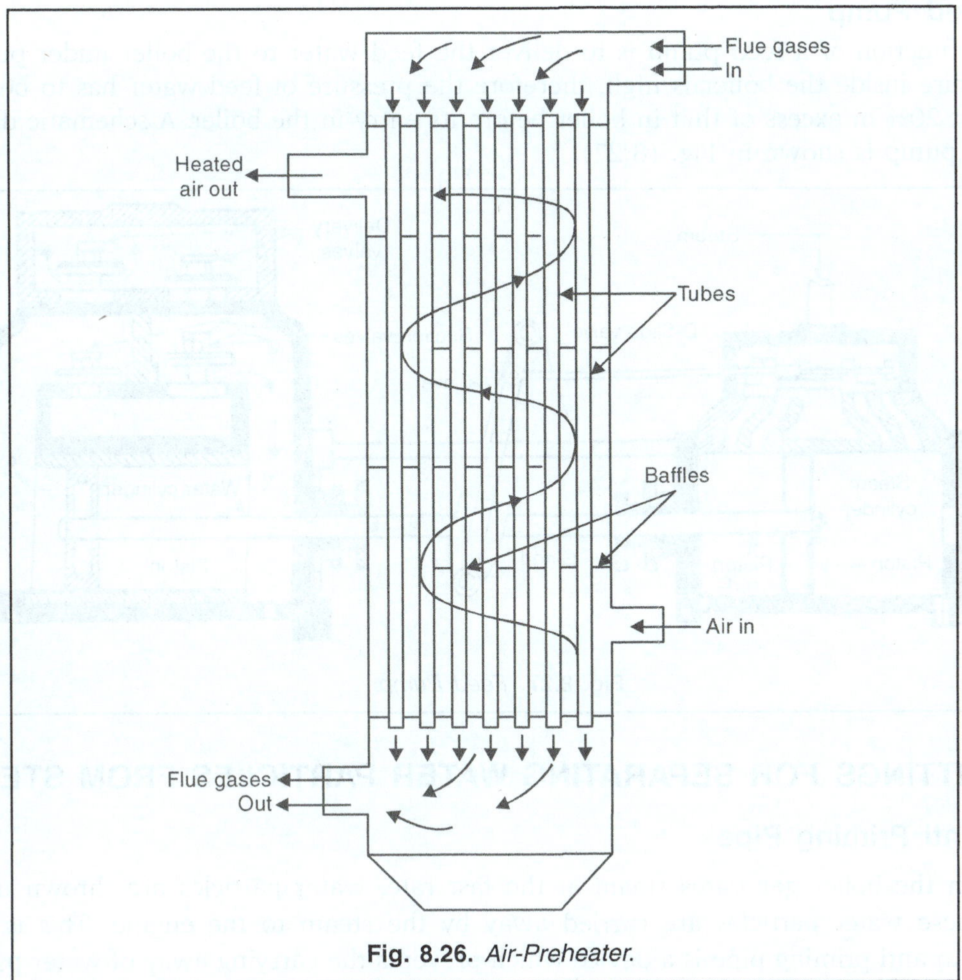

Fig. 8.26. *Air-Preheater.*

Figure (8.26) shows a recuperative type air preheater. Hot flue gases are passed through the tubes and air circulates around them. Air is forced to deflect by baffles as shown in Fig. (8.26) and compelled to move in a zigzag path several times. This increases the contact time between the air and hot surfaces and results in effective heating of the air. The flue gases impurities are collected in the hopper at the bottom and removed periodically through the soot gate.

(B) **Regenerative type air preheater**

In this type of air-preheater, the air and hot flue gases are made to pass alternatively.

Advantages

- Increases boiler efficiency (by about 2%).
- Improves evaporative rate.
- Better combustion with less soot, smoke and ash.
- Reduced fuel consumption.
- Inferior grade coal can also be used.

8.9.4 Feed Pump

The function of a feed pump is to deliver the feed water to the boiler under pressure. The pressure inside the boiler is high, therefore the pressure of feed water has to be raised by at least 20% in excess of that in boiler before its entry in the boiler. A schematic diagram of a feed pump is shown in Fig. (8.27).

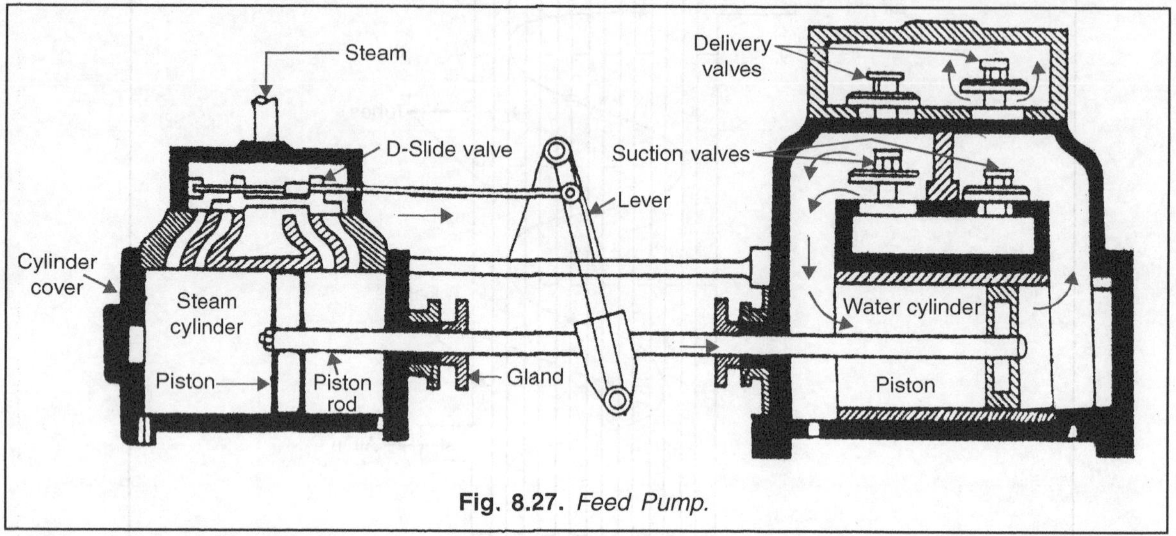

Fig. 8.27. *Feed Pump.*

8.10 FITTINGS FOR SEPARATING WATER PARTICLES FROM STEAM

8.10.1 Anti Priming Pipe

When the boiler generates steam at the fast rate, water particles are thrown up into steam. These water particles are carried away by the steam to the engine. This is called priming. An anti priming pipe is a device which prevents the carrying away of water particles with the steam. It is provided in the boiler shell just above the steam space and underneath the steam stop valve. It looks like an iron box with closed ends. Its upper half is perforated with several rectangular holes through which the steam passes on its way to the main steam pipe. When it passes through the slots, heavy water particles are left out and fall back in the boiler through a small hole provided at bottom of the anti priming pipe.

8.10.2 Steam Trap

The function of a steam trap is to collect and automatically discharge the water resulting from the partial condensation of steam without allowing any steam to escape.

8.10.3 Steam Drier or Separator

The function of a steam drier (Fig. 8.28) is to remove fine water particles carried along with the steam on its way from boiler to the engine. It is placed on the main steam pipeline near to the engine.

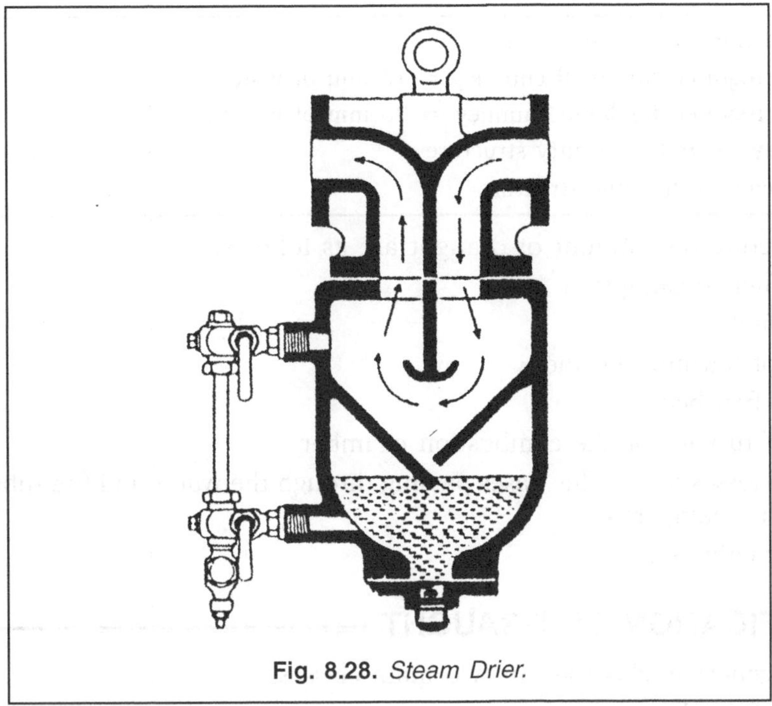

Fig. 8.28. *Steam Drier.*

8.10.4 Pressure Reducing Valve

The function of pressure reducing valve is to maintain a constant pressure on its delivery side with fluctuating boiler pressure.

8.11 DRAUGHT

The rate of steam generation in a boiler depends upon the rate at which the fuel is burnt. To burn the fuel rapidly, the continuous supply of fresh air through the grate is required which in turn discharges the products of combustion through the chimney to the atmosphere. *Draught is a small pressure difference that causes the flow of air through the bed of fuel on the furnace grate to the chimney.*

Objects of draught

1. To supply sufficient air for the combustion of fuel.
2. To push out the hot flue gases from the combustion chamber of the boiler to move through the system.
3. To discharge the used (spent) flue gases to the atmosphere through chimney.

The draught is usually measured in mm of water rather than in bar by a U-tube manometer. It is measured by noting the difference between the surface of two columns of water in the two legs of the manometer, one leg being connected to chimney and the other one is open to atmosphere.

- ☞ 1 mm of water = 9.80665 N/mm^2
- ☞ Draught required for small chimney = 12 mm of water
- ☞ Draught required for large chimney = 20 mm of water
- ☞ **Chimney** refers to masonry structure
- ☞ **Stack** refers to metallic structure

Factors affecting the amount of draught are as follows:

- Rate of fuel consumption
- Type of fuel used
- Method of burning the fuel
- Depth of fuel bed
- Design of furnace or the combustion chamber
- Frictional losses due to flue gases passing through the water and fire tubes, superheater, economiser, dampers etc.
- Height of chimney.

8.12 CLASSIFICATION OF DRAUGHT

Draught is generally classified in two main classes:

(*a*) Natural Draught

(*b*) Artificial Draught.

A detailed classification of draught is shown in Fig. (8.29).

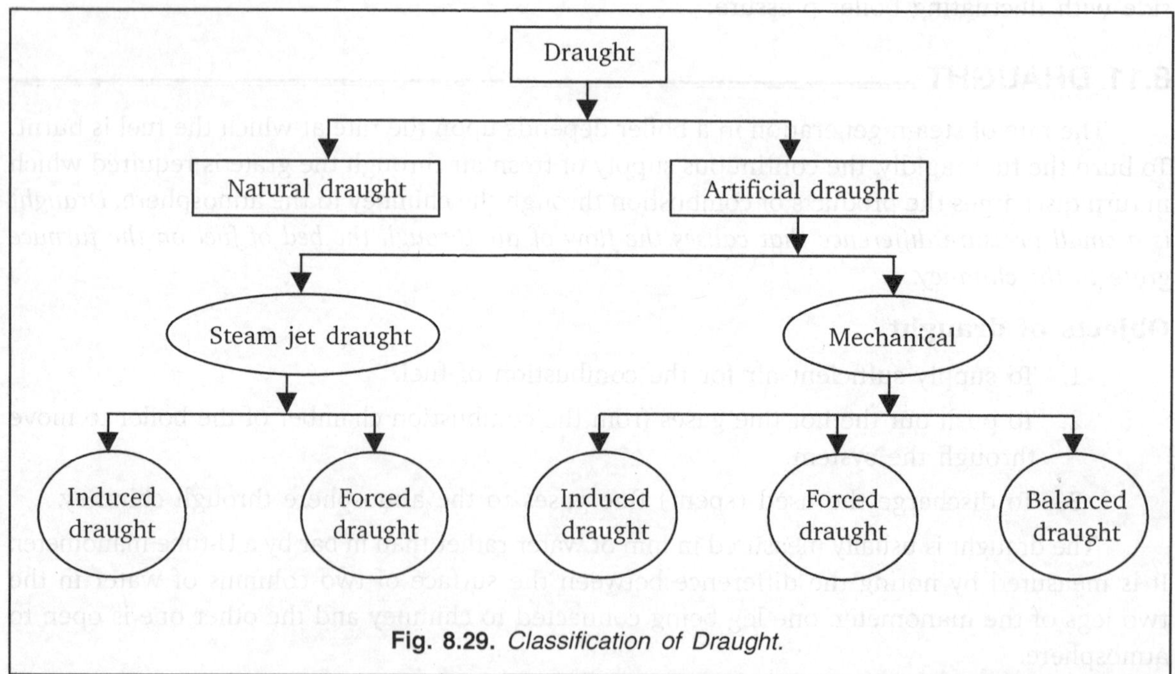

Fig. 8.29. *Classification of Draught.*

8.12.1 Natural Draught

Natural draught is produced by the boiler chimney hence it is also referred to as chimney draught. It is produced due to the density difference between the column of the hot gases inside the chimney and an equal column of cold air of same area outside the chimney.

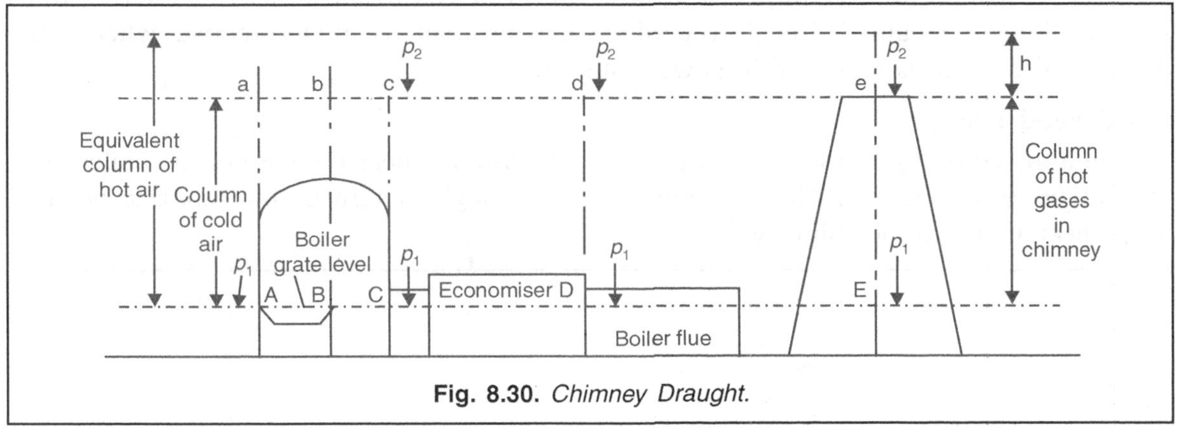

Fig. 8.30. *Chimney Draught.*

As shown in Fig. (8.30) when the boiler is not working (*i.e.,* the fuel is not ignited). The pressure at the grate level (*i.e.,* at points ABCDE) is atmospheric pressure say p_1. Let p_2 is the atmospheric pressure at all the points in a horizontal plane passing through the chimney top (*i.e.,* abcde). Since $p_2 < p_1$ as we know with altitude pressure keep on decreasing. When the boiler is working (*i.e.,* the fuel is burning) the chimney is filled with hot gases. The hot flue gases have lower density than outside air. This condition does not change the atmospheric pressure at the chimney top, which is the same p_2 but this condition changes the pressure at the base of the chimney, which is now p_2+H (hot gases). The pressure at the grate remain unchanged *i.e.,* $p_1 = p_2$+H (cold gases) which is higher than the pressure at the base of the chimney.

$\therefore\ p_1 > p_2$

p_2+ pressure due to cold column H > p_2 + pressure due to hot column H

Therefore,

Pressure at the grate due to cold column H > pressure at the chimney base due to hot column H

This pressure difference is called the static draught and is responsible for the flow of gases in the boiler. This static pressure difference decreases with the higher level of chimney and at the top of the chimney there is no static pressure and the gases are discharged in the atmosphere.

Advantages

- Draught is easily produced
- Low cost
- Negligible maintenance

8.12.2 Artificial Draught

The draught produced by some artificial means (other then chimney) is called artificial draught. When draught required is more than 40 mm of water column or when the rate of fuel burning is very high, artificial draught is provided. Artificial draught is usually produced by mechanical means such as fans, blower or steam jet. If the draught is produced by a fan, it is called *mechanical draught* or *fan draught* and if produced by steam jet it is referred to as *steam jet draught*. Steam jet draught is used in small installations and locomotives whereas the mechanical draught is preferred in power stations.

(A) Forced draught

In a forced draught system a fan or blower is installed near the base of the boiler or at the entrance of air-preheater. The air is forced to pass through the furnace, flue tubes, economiser, air-preheater, and to the chimney.

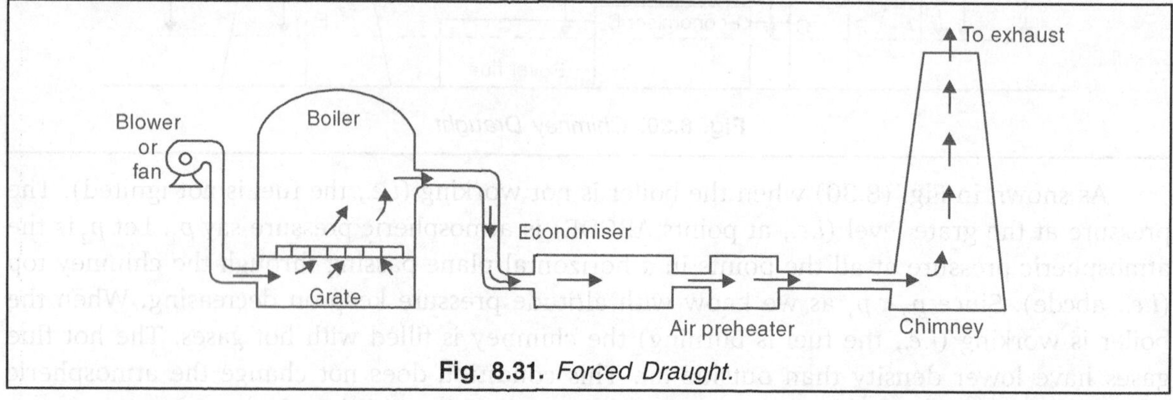

Fig. 8.31. *Forced Draught.*

This draught system as shown in Fig. (8.31) is known as positive draught system or forced draught system because the pressure of air throughout the system is above the atmospheric pressure and air is forced to flow through the system. Chimney used in this system is not for producing draught but just to dispose off the hot gases into the atmosphere.

(B) Induced draught

In the induced draught system a fan or blower is placed at the outlet of an economiser or air-preheater. In this draught system, the air pressure in the furnace is reduced below that of the atmosphere by means of a fan placed at (or near) the bottom of the chimney.

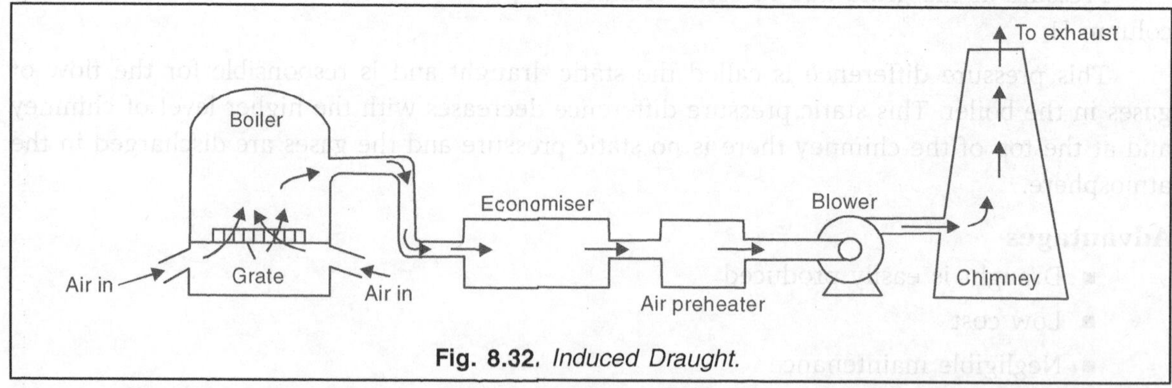

Fig. 8.32. *Induced Draught.*

It draws the product of combustion from the furnace and forces them to pass through the chimney. The action of induced draught is similar to the action of the chimney. The arrangement of this system is shown in Fig. (8.32).

(C) Balanced draught

Balanced draught is a combination of forced and induced draught. It is always better to use a combination of forced and induced draught instead of using either forced or induced draught alone. Here an air fan supplies air at a moderate pressure through the grate and also through the air-preheater. A comparison of forced and induced draught is given in Table 8.2.

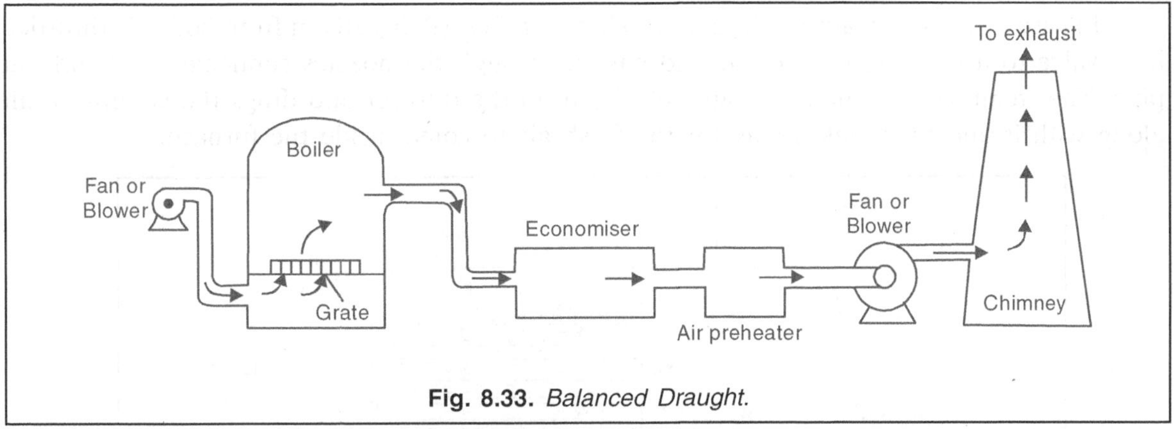

Fig. 8.33. *Balanced Draught.*

An induced draught fan is placed at the bottom of the chimney to draw the flue gases through the flue tubes, economiser air-preheater and discharges them to atmosphere through chimney.

Table 8.2 Comparison of Forced and Induced Draught

Factor	*Forced Draught*	*Induced Draught*
Air factor	Less *(handles only cold air)*	More *(handles air and fuel at elevated temp.)*
Volume of gas	Low	High
Rate of burning	Good	Satisfactory
Air leakage in to the furnace	No leakage *(pressure inside the furnace is above the atmosphere)*	Leakage is possible *(pressure inside the furnace is less than atmosphere)*
Fan life	Satisfactory	Poor

(D) Steam jet draught

The steam jet draught comes under the category of artificial draught. It can also be classified in two categories:

- Forced steam jet draught
- Induced steam jet draught

In the forced steam jet draught, the draught is produced by placing the steam jet in the ash pit which is placed under the fire grate of the furnace. In this case, the air will be forced into the furnace by the steam jet and the draught produced will be the forced draught.

The induced steam jet draught is produced when steam jet is directed into chimney from a nozzle placed in chimney. This causes the flow of gases through the fire tubes, ash pit and grate. Induced type steam jet draught is used in a locomotive boiler in which the exhaust steam from steam engine is passed through the nozzle to induce the air to pass through the fuel bed, furnace, flues and smoke box, thus creating a partial vacuum in the smoke box.

In case of forced steam jet draught as shown in Fig. (8.34) steam from boiler is throttled by a valve to a pressure of two bar and passed through the nozzles connected to a diffuser pipe. The steam comes out with high velocity from the diffuser and drags the column of air along with it and thus makes way for the fresh air to come inside the furnace.

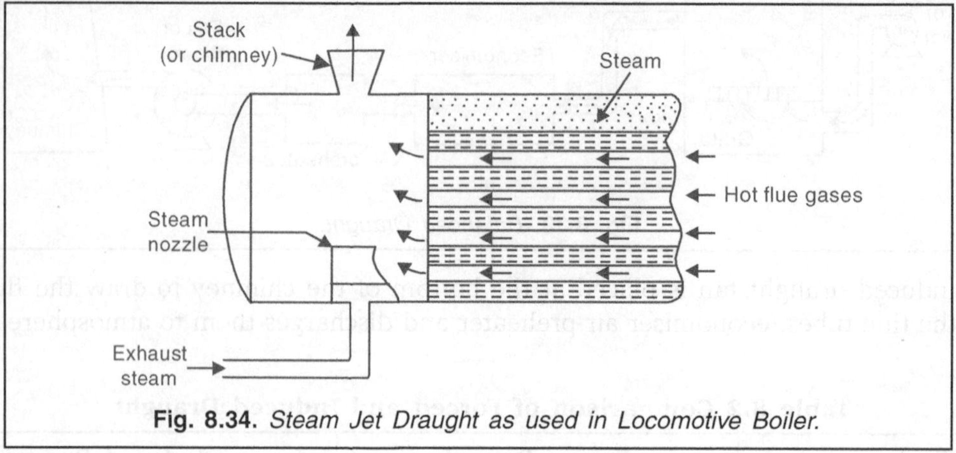

Fig. 8.34. *Steam Jet Draught as used in Locomotive Boiler.*

Advantages of steam jet draught

- Simple and economical
- Occupies little space
- Permits burning of inferior fuel
- No maintenance.

Disadvantages

- It cannot start until pressurized steam is available.

8.13 ADVANTAGES OF ARTIFICIAL DRAUGHT OVER THE NATURAL DRAUGHT

The artificial draught has the following advantages over the natural draught:

1. Effective control over the combustion.
2. High rate of combustion.
3. Permits burning of low grade fuel.
4. Improves overall thermal efficiency of the plant.

5. Reduces the height of chimney as it is only required to discharge the gases high in the atmosphere.
6. Low consumption of fuel.
7. Prevents smoke formation.
8. Independent of atmospheric temperature.
9. Enhances the evaporative capacity of boiler.
10. It can be regulated as per the requirement of the furnace.

Disadvantages

1. High installation cost.
2. High running and maintenance cost.

8.14 CHIMNEY HEIGHT

The natural draught produced in a boiler depends on the height of the chimney and the difference between the temperature of the hot flue gases leaving the boiler and that of the outside cold air.

An expression showing the relationship between the chimney height and draught produced by it in terms of temperature of outside air and temperature of the flue gases is deduced below:

Let

m = mass of air in kg per kg of fuel (or mass of air used in kg to burn 1 kg of fuel)

T = absolute temperature of the flue gases inside the chimney in K

T_1 = absolute temperature of air outside the chimney in K

Therefore,

Mass of chimney gases = $(m+1)$ kg per kg of fuel.

The volume of flue gases produced may be taken equal to the volume of air supplied.

Volume per kg of flue gases at 0°C (273K) = volume per kg of air at 273 K

As we know

$$pV = RT \qquad \text{where value of } R \text{ (for air)} = 287 \text{ kJ/kg K}$$

$$\therefore \quad V = \frac{RT}{p} = \frac{287 \times 273}{1.01325 \times 10^3} = 0.7734 \text{ m}^3/\text{kg}.$$

The pressure difference being very small, the pressure can be treated as constant at furnace and chimney base for the purpose of volume calculation at higher temperature.

As we know from Charles's law:

"Volume of a gas is proportional to its absolute temperature"

$$V \propto T$$

$$\frac{V}{T} = \frac{V_1}{T_1}$$

$$\therefore \quad V_1 = \frac{VT_1}{T}$$

The volume per kg of air at temperature $T_1 = \frac{0.7734\, T_1}{273}$

$\therefore$ Volume of m kg of air at temperature $T_1 = \frac{0.7734\, mT_1}{273}$

and volume of chimney gases at temp. $T = \frac{0.7734\, mT}{273}$ per kg of fuel burnt ...(8.1)

Therefore the density of air at T_1 = mass/volume

$$= \frac{m}{(0.7734\, mT_1)/273}$$

$$= 1.293 \left(\frac{273}{T_1}\right) \text{kg/m}^3 \quad ...(8.2)$$

$$\text{Mass density of chimney gases at } T = \frac{\text{Mass of chimney gases}}{\text{Volume of chimney gases}}$$

$$\rho = \frac{m+1}{(0.7734\, mT)/273}$$

$$\rho = 1.293\ \frac{m+1}{m} \times \frac{273}{T} \quad ...(8.3)$$

Let H be the height of the chimney required in metres measured from the grate level.

The pressure exerted per square metre at the furnace grate level by a column of hot gas of 1 metre in height = density of the gas;

$\therefore$ Pressure exerted by a column of hot air of H metre height

$$= \text{density} \times \text{height} \times g$$

$$= \rho g H \text{ N/m}^2 \quad ...(8.4)$$

Substituting ρ from (8.3) in (8.4), we get

$$= 1.293 \frac{m+1}{m} \times \frac{273}{T} \times gH \text{ N/m}^2 \quad ...(8.5)$$

Similarly pressure exerted by a column of cool air of H metre height from equation (8.3)

$$= 1.293 \times \frac{273}{T_1} \times gH \text{ N/m}^2 \quad ...(8.6)$$

Since the pressure causing the draught is due to the difference of pressure between the hot gas column in chimney of height H and cool air column outside the chimney of height H.

$$\therefore \quad p = \left\{1.293 \times \frac{273}{T_1} \times gH - \{1.293 \frac{m+1}{m} \times \frac{273}{T} \times gH\right\} \text{ N/m}^2$$

$$p = 1.293 \times 273 \times gH \left\{\frac{1}{T_1} - \left[\frac{m+1}{m}\right] \times \frac{1}{T}\right\} \text{N/m}^2$$

$$p = 353\ gH \left\{\frac{1}{T_1} - \left[\frac{m+1}{m}\right] \times \frac{1}{T}\right\} \text{N/m}^2 \qquad \text{...(8.7)}$$

$$p = 3462\ H \left\{\frac{1}{T_1} - \left[\frac{m+1}{m}\right] \times \frac{1}{T}\right\} \text{N/m}^2$$

The pressure of 1 N/m^2 is equivalent to 0.101972 mm of water column.

The draught pressure is generally expressed in terms of mm of water column as shown by the manometer.

Let h be the height of the column corresponding to the draught pressure in mm of water. Therefore the draught pressure p can be expressed as

$\therefore \quad p = \rho g h \qquad$ where ρ = density of water = 1000 kg/m^3

$g = 9.81\ \text{N/m}^2$

$$\therefore \quad p = 1000 \times 9.81 \times h/1000 = 9.81h\ \text{N/m}^2 \qquad \text{...(8.8)}$$

$$\therefore \quad h = \frac{p}{9.81} \text{ mm of water}$$

Substituting value of p from equation (8.7) in equation (8.8)

We get

$$\therefore \quad h = 353\ H \left\{\frac{1}{T_1} - \left[\frac{m+1}{m}\right] \times \frac{1}{T}\right\} \text{mm of water} \qquad \text{...(8.9)}$$

The equation (8.9) represents the theoretical draught which is also known as static draught. The actual draught is less than the static draught due to the following reasons:

- Resistance to the flow of the flue gases in its path.
- Losses in bends and curves in the the flue gases path.
- Energy required to impart velocity to the flue gases.

8.14.1 Condition for the Maximum Discharge Through the Chimney

Let h_G is the height of a column of hot gases which produces draught pressure p.

$$\therefore \quad p = \textit{density} \times h_G \times g$$

$$\therefore \quad h_G = p/\rho g$$

Substituting ρ from equation (8.3) and p from equation (8.8), we get

$$h_G = \frac{353gH\left\{\frac{1}{T_1} - \left[\frac{m+1}{m}\right] \times \frac{1}{T}\right\}}{1.293\frac{m+1}{m} \times \frac{273}{T} \times g}$$

On simplifying

$$h_G = H\left\{\left[\frac{m}{m+1}\right]\frac{T}{T_1} - 1\right\} \qquad \text{...(8.10)}$$

The mass of hot flue gases discharged per second through a cross section A and velocity V is given by

$$M_f = A \times V \times \rho \qquad ...(8.11)$$

Thus mass of hot gases discharged in a given time is proportional to the product of its density and velocity of its discharge.

i.e., Mass of hot gases discharged (M_f) $\propto$ *velocity (V)* $\times$ *density* (ρ)

As we know $\mathbf{V^2 = 2g\, h_G}$ and *density = mass/volume*

$\therefore$ *Volume per unit mass = 1/density* $\rightarrow V = 1/\rho$ and by using perfect gas law $pV = RT$ we get

$\therefore p/\rho = RT$

$$\therefore \quad \rho = p/RT = \frac{p/R}{T} = \frac{\text{constant}}{T}$$

$\therefore \quad \rho \propto \mathbf{1/T}$

From the equations given in the box it is clear that the velocity is proportional to square root of h_G and density is inversely proportional to temperature T

i.e., velocty $\propto \sqrt{h_G}$ and density $\propto$ 1/T

Thus, mass of hot flue gases discharged for a given height of chimney $\propto \sqrt{h_G} \propto \frac{1}{T}$

$$M_f \propto \sqrt{H\left\{\left[\frac{m}{m+1}\right]\frac{T}{T_1} - 1\right\}} \propto \frac{I}{T}$$

$$M_f = \text{constant}\sqrt{H\left\{\left[\frac{m}{m+1}\right]\frac{1}{TT_1} - \frac{1}{T^2}\right\}} \qquad ...(8.12)$$

The only variable in equation (8.12) are M_f and T whereas atmospheric temperature T_1, height of chimney H and cross section area of chimney A are constants.

To find out the maximum discharge, differentiate equation (8.12) with respect to variable temperature T and equating it to zero, we get

$$\frac{dM_f}{dT} = \text{constant} \times \frac{1}{2}\left[\frac{-\left[\frac{m}{m+1}\right]\left\{\frac{1}{T_1} \times \frac{1}{T^2} + \frac{1}{T^3}\right\}}{\sqrt{\frac{m}{m+1}\frac{1}{T^2 T_1} - \frac{1}{T^2}}}\right] = 0$$

$$\therefore \qquad \left[-\left[\frac{m}{m+1}\right]\left\{\frac{1}{T_1} \times \frac{1}{T^2} + \frac{2}{T^3}\right\}\right] = 0$$

$$\therefore \qquad T = 2\left[\frac{m+1}{m}\right]T_1 \qquad ...(8.13)$$

On substituting for T in equation 8.10, we get

$$h_G = H\left\{\left[\frac{m}{m+1}\right]\frac{T}{T_1} - 1\right\}$$

$$h_G = H\left\{\left(\frac{m}{m+1}\right)\frac{2\left(\frac{m}{m+1}\right)T_1}{T_1} - 1\right\}$$

$$\therefore \qquad \boxed{h_G = H} \text{ for maximum discharge.}$$

Thus for maximum discharge, the height of hot gas column producing the draught is equal to the height of the chimney.

Now substituting the value T in equation (8.9) from equation (8.13)

As we know from equation (8.9), $h = 353\ H\left\{\frac{1}{T_1}\left[\frac{m+1}{m}\right]\right\}$

and for max. discharge the temp. T from equation (8.13) is $T = 2\left(\frac{m+1}{m}\right)T_1$

$$\therefore h = 353\ H\left[\frac{1}{T_1} - \left(\frac{m+1}{m}\right)\left(\frac{m}{m+1}\right)\frac{1}{2T_1}\right] = 353\ H\left[\frac{1}{T_1} - \frac{1}{2T_1}\right]$$

$\therefore$ for maximum discharge

$$\boxed{h = \frac{353H}{2T_1} = \frac{176.5H}{T_1}} \qquad ...(8.14)$$

8.14.2 Diameter of Chimney

Let h_f be the frictional losses due to h_G height of column of hot gases, then the velocity of flue gases in the chimney is given by equation (8.15*a*)

$$V = \sqrt{2g(h_G - h_f)} = 4.43\sqrt{(h_G - h_f)} = K\sqrt{h} \qquad ...(8.15a)$$

Where

$$K = 4.43\sqrt{1 - \frac{h_f}{h_G}} \qquad ...(8.15b)$$

The value of constant K can be determined experimentally. Its value for steel chimney is 1.1 whereas for the brick chimney its value is 0.875.

Considering the steel chimney in use and substituting K = 1.1. in equation (8.15*b*), we get

$$1.1 = 4.43\sqrt{1 - \frac{h_f}{h_G}}$$

$$\therefore \quad \frac{h_f}{h_G} = 0.938 \qquad \text{...(8.16)}$$

Thus it means that about 93.8% of the draught is lost in friction in case of steel chimney and only 6.2% of the draught is available for creating the velocity of flow.

Total mass flow of hot gases $M_f = A \times V \times \rho$

$$\therefore \quad A = \frac{M_f}{V \times \rho} = \frac{M_f}{4.43\rho\sqrt{(h_G - h_f)}} \qquad \text{... (8.17)}$$

where, A = cross sectional area of the chimney.

8.14.3 Efficiency of Chimney

In the age of natural draught system, the temperature of the flue gases leaving the chimney is more as compared to artificial draught system as a certain minimum flue gas temperature is required to produce a given draught for a given height of chimney. This results in the loss of heat through the flue gases leading to the reduction in overall efficiency of the steam plant.

Let T_2 be the temperature of the flue gases in the case of artificial draught

∴ Extra heat carried away by 1 kg of flue gases $= C_p\,(T - T_2)$ kJ/kg

The draught pressure produced by the natural draught system in terms of the hot gas column height h_G is given by equation (8.10).

$$h_G = H\left\{\left(\frac{m}{m+1}\right)\frac{T}{T_1} - 1\right\}$$

Thus the energy obtained by 1 kg of air

$$= \frac{1 \times g \times H}{1000}\left\{\left(\frac{m}{m+1}\right)\frac{T}{T_1} - 1\right\} \text{kNm}$$

$$\therefore \quad \text{Efficiency of chimney} = \frac{gH\left\{\left(\frac{m}{m+1}\right)\frac{T}{T_1} - 1\right\}}{1000 \times C_p(T - T_2)} \qquad \text{...(8.18)}$$

where, $g = 9.81$ N/m², $C_p = 1.005$ kJ/kg K

8.15 PERFORMANCE OF BOILERS

The performance of a steam boiler is judged by calculating the quantity of heat produced by using the heat generated from combustion of the fuel. Steam generated by a boiler is usually expressed in kg per hour per kg of fuel. This is also known as evaporation rate or evaporative capacity.

Evaporation rate can also be determined by taking into consideration the grate area, heating surface, and storage capacity of the boiler. Evaporation rate plays an important role in comparing the performance of the boiler. Evaporation rate can be determined in any one of the three ways:

$$\text{Evaporation Rate} = \frac{\text{amount of steam generated per hour}}{\text{grate area}} \text{ kg/m}^2\text{/hr} \quad ...(8.19)$$

$$= \frac{\text{amount of steam generated per hour}}{\text{fuel consumed}} \text{ kg/hr/kg of fuel}$$

$$= \frac{\text{amount of steam generated per hour}}{\text{volume of furnce}} \text{ kg/m}^3\text{/hr}$$

Different types of boilers generate different quantities of steam at different pressures according to the requirements from the feed water at different temperatures and pressures. For example one boiler generates 10 kg of wet steam at 7 bar whereas the other boiler generates 15 kg of dry saturated steam at 12.5 bar.

In order to compare the performance of the different boilers, some standard conditions have been selected. These standard conditions selected are:

"From the feed water at 100°C, the dry saturated steam is produced at 100°C".

Under these conditions, the amount of heat required to evaporate 1 kg of water at 100°C requires 2257 kJ for converting the water into dry saturated steam at 100°C at a pressure of 1.01325 bar (1.0332 kgf/cm^2).

A term "equivalent evaporation" is used to compare the performance of the boilers generating steam at different evaporation rates at different pressures and temperatures from feed water at different temperatures. Equivalent evaporation is defined as the evaporation of feed water at 100°C into dry saturated steam at 100°C. It is expressed in kg/hr/kg of fuel.

Under actual working conditions of the boiler:

Let

m_a = actual mass of the water evaporated in kg per kg of fuel.

h_g = enthalpy of 1 kg of steam produced.

h_{feed} = enthalpy of 1 kg of feed water.

L_S = enthalpy of evaporation of 1 kg of steam at 100°C (2257 kJ)

m_e = equivalent evaporation in kg of water from and at 100°C per kg of fuel.

Heat required to produce 1 kg of steam = $(h_g - h_{feed})$ kJ

Heat required to produce m_a kg of steam = $m_a\,(h_g - h_{feed})$ kJ

$$\therefore \quad \text{Equivalent evaporation} = \frac{m_a\,(h_g - h_{feed})}{L_s} = \frac{m_a\,(h_g - h_{feed})k_g}{2257} \quad ...(8.20)$$

$$\text{For wet steam} \quad m_e = \frac{m_a\,(h_{wet} - h_{feed})k_g}{2257} \quad ...(8.21)$$

8.15.1 Factor of Evaporation

- It is defined as the ratio of the amount of heat absorbed by 1 kg of feed water under working conditions of the boiler to that required by 1 kg of feed water from and at 100°C for evaporation.

$$\text{Factor of evaporation} = \frac{h_g - h_{feed}}{L_s} \text{ kJ} \quad \text{...(8.22)}$$

8.15.2 Boiler Efficiency

- Boiler efficiency is the measure of the ability of a boiler to transfer the heat given to it by the furnace to the water and steam.
- It is defined as the ratio of the heat utilized for generating the steam in a specified period of time to the total heat generated by the combustion of fuel in the furnace in the same time.

$$\text{Boiler efficiency} = \frac{m_a\,(h_g - h_{feed})}{CV} \quad \text{...(8.23)}$$

Where

m_a = actual evaporation in kg/kg of fuel burnt

h_g = Enthalpy of 1 kg of steam.

h_{feed} = enthalpy of 1 kg of feed water.

CV = calorific value of fuel in kJ/kg.

$\eta_{thermal}$ (Water tube boiler) = 80%
$\eta_{thermal}$ (Fire tube boiler) = 70%

- The commonly used boiler accessories such as economiser, superheater have their own efficiency. If these accessories happens to be the part of a boiler then the efficiency in that case is known as overall efficiency of the boiler plant.

8.15.3 Boiler Horse Power

- It is the amount of water which a boiler can evaporate per hour from and at 100°C.
- ASME (American Society of Mechanical Engineers) defines one boiler horsepower as equivalent to the evaporation of 15.65 kg of water per hour from and at 100°C into dry saturated steam.

8.15.4 Heat Losses in Boiler

- It is not possible for a boiler to run at 100% efficiency. In practice a significant amount of heat produced by the combustion of fuel is not utilized in raising steam from water but is lost in the following ways:

(A) Heat lost to flue gases

- The flue gases consists of (*i*) heat carried away by dry flue gases (*ii*) heat carried away by steam generated due to combustion of hydrogen present in the fuel.

(B) Heat lost due to incomplete combustion (burning of carbon to CO)

- CO is formed due to incomplete combustion of fuel. When carbon burns to CO instead of CO_2 this condition is referred to as incomplete combustion.

  1 kg of carbon burns to $CO_2 \rightarrow$ 33830 kJ

  1 kg of carbon burns to CO $\rightarrow$ 10130 kJ

- Thus energy loss per kg of carbon burning to CO instead of CO_2

$$= (33830 - 10130) \text{ kJ} = 23700 \text{ kJ}$$

(C) Heat lost due to unburnt fuel

- This type of heat loss occurs when some of the fuel falls through the grate bars and is carried away with fly ash and clinkers.
- The heat loss is calculated by multiplying mass of unburnt fuel lost through the grate bars by the calorific value of the fuel.

(D) Heat lost due to moisture present in the fuel and air.

(E) Heat lost due to convection and radiation etc.

SOLVED PROBLEMS

Problem 1. *Determine the amount of heat required to generate 5 kg of steam at a pressure of 10 bar and temperature of 250°C from water at 25°C. Assume specific heat for superheated steam as 2.1 kJ/kg K.* (RGPV June 2002)

Solution:

Enthalpy of 1kg of feed water h_{feed}

(Specific heat of water is 4.187 kJ/kg K)

$$1 \times 4.187 \times (25 - 0) = 104.7 \text{ kJ/kg}$$

At 10 bar from steam table:

$$h_f = 762.81 \text{ kJ/kg}$$
$$h_{fg} = 2015.3 \text{ kJ/kg}$$
$$t_s = 179.91°\text{C}$$
$$h_g = h_f + h_{fg} = 2778.11 \text{ kJ/kg}$$

Enthalpy of 1 kg of superheated steam

$$h_{sup} = h_g + C_p \ (t_{sup} - t_s)$$
$$= 2778.11 + 2.1 \ (250 - 179.91)$$
$$= 2925.3 \text{ kJ/kg.}$$

Amount of heat required to produce 5 kg of steam

$$= m \ (h_{sup} - h_{feed})$$
$$= 5 \ (2925.3 - 104.7)$$
$$= \mathbf{14103 \ kJ}$$

Problem 2. *A boiler generates 900 kg of steam per hour at a pressure of 11 bar and with 50°C superheat. The coal burnt per hour in the furnace is 150 kg and the feed water temperature is 40°C. The calorific value of the coal is 30000 kJ/kg. Take C_p of supherhead steam as 2.1 kJ/kg K and specific heat of water as 4.187 kJ/kg K. calculate:*

(a) The factor of equivalent evaporation.

(b) The actual evaporation per kg of coal.

(c) The equivalent evaporation from and at 100°C per kg of coal.

(d) The boiler efficiency.

Solution:

(*a*) At 11 bar from steam table :

h_g = 2781.7 kJ/kg

Enthalpy of superheated steam produced by the boiler is given by

$$h_{sup} = h_g + C_p \ (t_{sup} - t_s)$$
$$= 2781.7 + 2.1 \times 50$$
$$= 2886.7 \text{ kJ/kg.}$$

Enthalpy of 1 kg of feed water is given by

$$h_{feed} = mC_p \ (t_{final} - t_{ini})$$

(*where specific heat of water is 4.187 kJ/kg*)

$$h_{feed} = 4.187 \times (40 - 0)$$
$$= 167.48 \text{ kJ/kg}$$

∴ Heat required to produce 1 kg of superheated steam from feed water at 40°C

$$= h_{sup} - h_{feed}$$
$$= 2886.7 - 168.48$$
$$= 2719.22 \text{ kJ/kg}$$

$$\text{Factor of equivalent evaporation} = \frac{h_{sup} - h_{feed}}{2257} \text{ kg} = \frac{2719.22}{2257} = \mathbf{1.2048}$$

(*b*) Actual evaporation = m_a = 900/150 = **6kg/kg of coal.**

(*c*) Equivalent evaporation in kg of water from and at 100°C per kg of coal.

$$m_e = \frac{m_a \ (h_{sup} - h_{feed})}{2257}$$
$$= \frac{6 \times 2719.22}{2257}$$
$$= \mathbf{7.23 \ kg/kg \ of \ coal.}$$

(*d*)

$$\text{Boiler efficiency} = \frac{m_a \ (h_{sup} - h_{feed})}{\text{C.V.}}$$
$$= \frac{6 \times 2719.22}{30000}$$
$$= 0.54238 = \mathbf{54.38\%}$$

Problem 3. *Calculate the equivalent evaporation of the boiler per kg of coal fired, if the boiler produces 50,000 kg of wet steam per hour with a dryness fraction of 0.95 and operating at 10 bar. The coal burnt per hour in the furnace is 5500 kg and feed water temperature is 40°C.* *(RGPV 2001)*

Solution:

The enthalpy of wet steam of dryness fraction (x = 0.95) is

$$h_{wet} = h_f + xh_{fg}$$

From steam table at 10 bar-

$$h_f = 762.8$$
$$h_{fg} = 2015.3$$

$\therefore$ Enthalpy of steam produced by the boiler is

$$h_{wet} = 762.81 + 0.95 \times 2015.3 = 2677.34 \text{ kJ/kg}$$

Enthalpy of 1 kg of feed water at 40°C is $= mC_p\,(t_{final} - t_{ini})$

$$= 1 \times 4.187 \times (40 - 0)$$
$$= 167.48 \text{ kJ/kg}$$

Heat required to produce 1 kg of steam $= h_{wet} - h_{feed} = 2677.34 - 167.48$

$$= 2509.86 \text{ kJ/kg}$$

$$\text{Actual Evaporation } m_a = \frac{50000}{5500} = 9.09 \text{ kg/kg of fuel}$$

$\therefore$ Equivalent evaporation in kg of water per kg of coal $= \dfrac{m_a\,(h_{wet} - h_{feed})}{2257}$

$$= \frac{9.09 \times 2509.86}{2257}$$

$$= \mathbf{10.10 \text{ kg/kg of coal}}$$

Problem 4. *A boiler produces steam at 12 bar and 0.90 dry. The steam thus produced by the boiler then passes through the superheater where its temperature is raised to 300°C at constant pressure. If the boiler generates 900 kg of superheated steam per hour, find out the amount of heat received by the superheater per hour. Take C_p of superheated steam as 2.1 kJ/kg K.*

Solution:

From the problem statement it is clear that initially the boiler is generating wet steam and thereafter the steam is superheated in the superheater.

From steam table at 12 bar:

$$h_f = 798.65 \text{ kJ/kg}$$
$$h_{fg} = 1986.2 \text{ kJ/kg}$$
$$t_s = 187.99°\text{C}$$

Enthalpy of wet steam

$$h_{wet} = h_f + xh_{fg} = 798.65 + 0.90 \times 1986.2 = 2586.23 \text{ kJ/kg.}$$

Enthalpy of superheated steam is given by

$$h_{sup} = h_f + h_{fg} + C_p\,(t_{sup} - t_s)$$
$$= 798.65 + 1986.2 + 2.1(300\text{-}187.99)$$
$$= 3020 \text{ kJ/kg}$$

Now heat received by 1 kg of wet steam in the superheater = 3020–2586.23

$$= 433.7 \text{ kJ/kg}$$

Heat received by steam per hour in the superheater = 900×433.77 = **390393 kJ/hr.**

Problem 5. *A steam turbine requires 6.5 kg of steam per kWh at a pressure of 40 bar and at 295°C. The temperature of feed water was 145°C and the efficiency of boiler is 75%. If the cost of the coal consumed in producing the steam is ₹ 1050 per tonne. The calorific value of coal is 30000 kJ. Calculate:*

(i) Cost of the fuel/kWh.

(ii) Mass of the coal required in kg/kWh.

Solution:

(*i*) As we know the boiler efficiency is given by

$$\eta = \frac{m_a\,(h_{sup} - h_{feed})}{m_f \times CV}$$

From steam table at 40 bar

$$h_f = 1087.31 \text{ kJ/kg}$$
$$h_{fg} = 1714.1 \text{ kJ/kg}$$
$$t_s = 250.40°\text{C}$$

Enthalpy of superheated steam

$$h_{sup} = h_f + h_{fg} + C_p\,(t_{sup} - t_s)$$
$$= 1087.31 + 1714.1 + 2.1\,(295 - 250.40)$$
$$= 2895 \text{ kJ/kg}$$

Enthalpy of feed water $h_{feed} = mC_p\,(t_{final} - t_{ini}) = 1 \times 4.187 \times (145 - 0)$

$$= 607.12 \text{ kJ/kg}$$

Now from the boiler efficiency equation, the mass of the fuel can be calculated

$$\eta = \frac{m_a\,(h_{sup} - h_{feed})}{m_f \times CV}$$

$$0.75 = \frac{6.5\,(2895 - 607.12)}{m_f \times 30000}$$

$$m_f = \mathbf{0.661 \text{ kg/kWh}}$$

(*ii*) The cost of fuel/kWh=m_f in tonne/kWh × cost/tonne

$$= \frac{0.661 \times 1050}{1000} \times 100$$

= **₹ 69.41 P/kWh.**

Problem 6. *Estimate the temperature of the flue gases leaving the chimney 30 m high to produce a draught of 15mm of water column if 17 kg of air is required per kg of fuel burnt. The temperature of atmospheric air is 28°C.*

Solution:

Given:

Height of chimney H = 30 m

Height of water column h = 15 mm of water column

Mass of air used in burning per kg of fuel m = 17 kg

Outside air temperature is = 28°C = 28+273 = 301 K

As we know

$$h = 353H\left\{\frac{1}{T_1} - \left(\frac{m+1}{m}\right) \times \frac{1}{T}\right\} \text{mm of water}$$

Where

h = height of the column

T = absolute temperature of flue gases inside the chimney

T_1 = absolute temperature of air outside the chimney

m = mass of air used in kg to burn one kg of fuel.

$$15 = 353 \times 30\left[\frac{1}{301} - \frac{17+1}{17} \times \frac{1}{T}\right]$$

$$15 = 10590\left[\frac{1}{301} - \frac{18}{17T}\right]$$

$$15 = \left[\frac{10590}{301} - \frac{10590 \times 18}{17\,T}\right] = 35.18 - \frac{11212.94}{T}$$

on rearranging the terms $= \dfrac{11212.94}{T} = 35.18 - 15 = 20.18$

$$T = \frac{11212.94}{20.18} = 555.65 \text{ K}$$

To get the temperature in degree Celsius deduct 273 from the above value.

= (555.65 − 273)

= **282.65°C.**

Problem 7. *A boiler is equipped with a chimney of 30m height. The ambient temperature is 25°C. The temperature of flue gases passing through chimney is 300°C. If the airflow is 20 kg/kg of fuel burnt, find*

(i) *Draught produced*

(ii) *The velocity of the flue gases passing through the chimney if 50% of the theoretical draught is lost in friction.*

Solution:

Given that:

Height of chimney $H = 30\text{m}$

Mass of air used in burning per kg of fuel $m = 20$ kg

Outside air temperature T_1 is $= 25°\text{C} = 25 + 273 = 298$ K

Flue gases temperature inside the chimney is $T = 300°\text{C} = 300 + 273 = 573$ K

Height of water column (or the draught produced) $h = ?$

Velocity of the flue gases = ?

(*i*) As we know $h = 353\,H\left\{\frac{1}{T_1} - \left(\frac{m+1}{m}\right) \times \frac{1}{T}\right\}$ mm of water

Where

h = height of the water column

T = absolute temperature of flue gases inside the chimney

T_1 = absolute temperature of air outside the chimney

m = mass of air used in kg to burn one kg of fuel.

$$h = 353 \times 30\left[\frac{1}{298} - \frac{20+1}{20} \times \frac{1}{573}\right]$$

$\boldsymbol{h}$ **= 16.13 mm of water column**

(*ii*) As 50% of the draught is lost in friction.

$\therefore$ Available draught = 50% of h_G

Where h_G is height of column of hot gases which produces draught

We know that

$$h_G = H\left\{\left(\frac{m}{m+1}\right)\frac{T}{T_1} - 1\right\}$$

$$h_G = 30\left[\left(\frac{20}{20+1}\right) \times \frac{573}{298}\right]$$

$$h_G = 24.94 \text{ m of hot gas column.}$$

Available draught is = 0.50 × 24.94 = 12.47 ***m***

Let V is the velocity of flue gases passing through chimney

$\therefore$ V $= \sqrt{2g\,h_G} = \sqrt{2 \times 9.81 \times 12.47}$ = **15.64 m/sec.**

Problem 8. *A 35 m high chimney deals with flue gases at 310°C, when the ambient (outside) temperature is 10°C. The air supplied for combustion is 19 kg per kg of fuel burnt. The specific heat and the minimum temperature of the waste gases is 1.004 kJ/kg K and 125°C. Find out the following:*

(*i*) *Draught produced in mm of water.*

(*ii*) *Efficiency of the chimney.*

Solution : Given that:

Height of chimney H = 35 m

Mass of air used in burning per kg of fuel m = 19 kg

Outside air temperature T_1 is = 10°C = 10 + 273 = 283 K

Flue gases temperature inside the chimney is T = 310°C = 310+ 273 = 583 K

Height of water column (or the draught produced) h =?

Efficiency of the chimney = ?

(*i*) As we know

$$h = 353\,H\left\{\frac{1}{T_1}-\left(\frac{m+1}{m}\right)\times\frac{1}{T}\right\}\text{ mm of water}$$

Where

h = height of the water column

T = absolute temperature of flue gases inside the chimney

T_1 = absolute temperature of air outside the chimney

m = mass of air used in kg to burn one kg of fuel.

$$h = 353\times 35\left\{\frac{1}{283}-\left(\frac{19+1}{19}\right)\times\frac{1}{583}\right\}$$

$$h = 12355\times 1.73\times 10^{-3}$$

$$\boldsymbol{h = 21.35}\textbf{ mm of water column.}$$

(*ii*) Draught in terms of height of hot gas column is expressed by equation (8.10)

$$h_G = H\left\{\left(\frac{m}{m+1}\right)\frac{T}{T_1}-1\right\}\text{ mm of water}$$

$$h_G = 35\left\{\left(\frac{19}{19+1}\right)\frac{583}{283}-1\right\}$$

$$h_G = 33.5\text{ m.}$$

Energy given to 1 kg of air = 1 × (9.81/1000) × 33.5 = 0.3286 kJ

Heat carried away by 1 kg of flue gases = 1 × 1.004 × (310 – 125) = 185.74 kJ

Efficiency of chimney can be obtained by using equation (8.18)

Efficiency of chimney = 0.3286 × 100/185.74 = **0.177%.**

Problem 9. *With a chimney of height of 45 m, the temperature of flue gases with natural draught was 370°C. The same draught was developed by induced draught fan and the temperature of the flue gases was 150°C. Mass of the flue gases formed is 25 kg per kg of coal fired. The boiler house temperature is 35°C. Assuming* C_p = *1.004 kJ/kg K for the flue gases, determine the efficiency of the chimney.* *(RGPV Jan./Feb. 2006)*

Solution:

Given: Height of chimney H = 45 m

Temperature of flue gases with natural draught, T = 370°C
= 370 + 273
= 643 K

Temperature of flue gases with induced draught fan, T_2 = 150°C
= 150 + 273
= 423 K

Mass of flue gases formed per kg of coal fired $(m + 1)$ = 25 kg

$\therefore \quad m = 25 - 1 = 24$ kg

where m is the mass of air used in burning 1 kg of fuel

Boiler house temperature T_1 = 35° C = 35 + 273 = 308 K

Specific heat of flue gases C_p = 1.004 kJ/kg K

Efficiency of chimney η = ?

Then using equation (8.18)

$$\eta = \frac{gH\left\{\left(\frac{m}{m+1}\right)\frac{T}{T_1} - 1\right\}}{1000 \times C_p \times (T - T_2)}$$

$$\eta = \frac{9.81 \times 45\left\{\left(\frac{24}{24+1}\right)\frac{643}{308} - 1\right\}}{1000 \times 1.004 \times (643 - 423)}$$

$$\eta = \frac{9.81 \times 45.19}{1000 \times 1.004 \times 220}$$

$$\eta = 2.00 \times 10^{-3}$$

$$\eta = 0.2\%$$

Unsolved Problems

1. Find out the mass of air per kg of coal burnt in a boiler having a chimney of 30 m height to produce a draught of 16 mm of water when the temperature of the flue gases in the chimney is 317°C and the outside air temperature is 30°C.
   [Ans. 17.8 kg/kg of fuel burnt]
2. Find out the minimum height of a chimney required to produce a draught of 16 mm of water. The atmospheric temperature is 30°C, and the average flue gas temperature is 330°C. Air supplied per kg of fuel burnt is equal to 19 kg. **[Ans.** 29.1m]
3. A boiler chimney of 40m height deals with flue gases at 300°C, when the ambient temperature is 27°C. Air supplied for the combustion is 19 kg per kg of fuel burnt in the furnace. Find out:

(*i*) draught in mm of water column

(*ii*) draught in terms of height of hot flue gas column

(*iii*) velocity of the flue gases passing through chimney if 60% of the theoretical draught is lost in friction.

(*iv*) draught produced in mm of water, and the temperature of flue gases for the maximum discharge conditions.

[**Ans.** (*i*) 21.2 mm, (*ii*) 32.6 m, (*iii*) 16 m/sec, (*iv*) 23.5 mm, 358.6°C]

4. The following data were recorded during a test on a steam boiler:

The temperature of feed water 20°C, the pressure of the steam generated 10 bar, mean steam dryness fraction 0.9, fuel burnt per hour 250 kg, water supplied to boiler per hour 2260 kg. The mass of water in the boiler at the end of test was less than that at the start of test by 240 kg. Calculate:

(*i*) Actual evaporation per kg of fuel

(*ii*) Equivalent evaporation. [**Ans.** (*i*) 10 kg (*ii*) 11.05 kg]

5. Calculate the height of chimney required to produce a draught equivalent to 1.7 cm of water if the flue gases temperature is 270°C and ambient temperature is 22°C and minimum amount of air per kg of fuel is 17 kg. *(RGPV-Dec. 2003)* [**Ans.** 33.45 m]

REVIEW QUESTIONS

1. What is a steam boiler?

2. What is the function of a boiler? *(RGPV-Dec. 2003)*

3. Enlist the detailed classification of boilers.

4. Classify the boilers on the basis of various factors? Write down the names of at least one boiler in each category.

5. What are the uses of steam generated by a boiler?

6. Write down the essential features of a good boiler.

7. List the factors, which affect the selection of a boiler.

8. Explain the difference between a water tube and a fire tube boiler.

Or

Give the comparison between fire tube and water tube boiler. *(RGPV-June 2004)*

9. Explain the following terms used in connections with boilers

(*a*) boiler shell (*b*) grate (*c*) setting (*d*) water space (*e*) steam space (*f*) mountings (*g*) accessories.

10. Describe with the help of a neat sketch the construction and working of a simple vertical boiler.

11. List the main parts of the Cochran boiler. Explain the working of a Cochran boiler with the help of a neat sketch. Show the path of the flue gases. *(RGPV-Dec. 2002)*

12. Explain the working of a Lancashire boiler with the help of a neat sketch clearly showing (*i*) the path of the flue gases and (*ii*) path of water and steam, (*iii*) main fittings.

13. What are the benefits of a Lancashire boiler? List down the shortcomings of this boiler.

14. Explain with the help of a sketch the construction and working of a locomotive boiler. Describe the method of obtaining draught in a locomotive boiler. What are the salient features of this boiler?

15. Why the water tubes are kept inclined in Babcock and Wilcox boiler?

16. Explain with the help of a neat sketch the construction and working of a Babcock & Wilcox boiler. Indicate on it the path of the flue gases and water circulation.

17. Distinguish between the mounting and accessory fitted in a boiler. Explain with neat sketch any two types of the boiler accessories. *(RGPV-Dec. 2003)*

18. Name all the mounting and accessories of a steam boiler and describe an economiser. *(RGPV-Dec. 2001)*

19. Sketch and describe the working of any one type of the safety valve. Give its advantages and disadvantages.

20. What is the purpose of using the fusible plug in a boiler? Where is its location in the boiler?

21. Describe advantages from incorporating economiser, air pre-heater and superheater in a steam plant. *(RGPV-June 2002)*

22. Explain with the help of a neat sketch the working of a mounting that gives warning of the low water level in a boiler.

23. What is a manhole?

24. Draw a schematic diagram of a water tube boiler. Clearly mark the position of the economiser and superheater in the boiler and indicate the path of the flue gases.

25. What is boiler draught? Mention the types of draught used in practice. *(RGPV-Dec. 2003)*

26. Distinguish between

(*a*) Natural draught and artificial draught

(*b*) Forced draught and induced draught.

27. Establish a condition for maximum discharge of flue gases through a chimney of given height. *(RGPV-June 2002)*

28. What is the function of a boiler chimney? Explain why there is no chimney in the case of a locomotive boiler?

29. Prove that the height of a hot gas column is equal to the height of the chimney at a maximum mass flow rate of the gas through the chimney. *(RGPV 2001).*

30. List the advantages of artificial draught over natural draught.

31. What is the function of boiler mounting? Can a boiler work without mounting? Explain the any two of the boiler mountings. *(RGPV-June 2004)*

32. What do you mean by evaporative capacity of a boiler? How is it expressed? Define and explain equivalent evaporation. *(RGPV-June 2004)*

OBJECTIVE TYPE QUESTIONS

Select the most appropriate answer:

1. The device which is used for the generation of steam and is made of closed vessel of steel is known as

(A) Steam Condenser (C) Steam Turbine

(B) Steam Boiler (D) Steam Injector

2. The ratio of heat utilized to produce steam and the heat released in the furnace is known as

(A) Boiler effectiveness (C) Boiler horsepower

(B) Boiler efficiency (D) Boiler evaporative capacity

3. The number of the fire tubes in Lancashire boiler is

(A) Nil (B) One
(C) Two (D) Four
(E) Five (F) 150

4. Boiler rating is usually defined in terms of

(A) Maximum temperature of steam in Kelvin.
(B) Heat transfer rate in kJ/hr
(C) Heat transfer area in m^2
(D) Steam output in kg/hr

5. Lancashire boiler is of

(A) Horizontal type (B) Natural circulation type
(C) Stationary fire tube type (D) Internally fired type
(E) All of the above

6. The .... coal has the highest calorific value.

(A) Wood (B) Peat
(C) Anthracite (D) Bituminous
(E) Lignite

7. The high pressure boiler generates steam at a pressure more then

(A) Atmospheric pressure (B) 0.5 MPa
(C) 1.0 MPa (D) 7.0–8.0 MPa
(E) 4.0 MPa

8. The evaporation of 15.653 kg of water per hour from and at 100°C into dry saturated steam is known as

(A) Boiler efficiency (B) One boiler horse power
(C) Factor of evaporation (D) Equivalent evaporation
(E) None of the above

9. The small difference between the pressure of air outside and that of gases within a chimney which causes the flow of gas to take place is known as

(A) Boiler efficiency (B) Equivalent evaporation
(C) Boiler performance (D) Boiler draught
(E) None of the above

10. The boiler draught is used

(A) To draw resulting hot gases through the system.
(B) To force air through the fuel bed to aid in the proper combustion of fuel.
(C) To discharge the products of combustion to the atmosphere.
(D) All of the above.

11. Which one of the following is a boiler mounting?

(A) Steam injector (B) Blow-off-cock
(C) Air pre-heater (D) Economiser

12. Which one of the following is a boiler accessory?

(A) Manhole (B) Blow-off-cock
(C) Air pre-heater (D) Stop valve

13. The Economiser in a boiler
(A) Increases the life of a boiler (B) Increases steam pressure
(C) Reduces steam pressure (D) Reduces fuel consumption

14. The .... of tubes specify the size of boiler tubes.
(A) Outside diameter (OD) and thickness (B) Inside diameter (ID) and thickness
(C) OD and ID (D) OD, ID and thickness
(E) Mean diameter, and thickness

15. Where is the fusible plug located in small boilers
(A) Near the steam space of the boiler shell
(B) Inside the fire tubes
(C) At the chimney entrance
(D) Over the combustion chamber

16. Steam in water tube boilers as compared to fire tube boilers can be raised
(A) Quickly (B) Slowly
(C) At the same rate (D) None of the above

17. The position of water tubes in a Babcock and Wilcox boiler are
(A) Horizontal (B) Vertical
(C) Inclined (D) None of the above

18. Water tube boiler is one in which
(A) Steam generation takes place outside the boiler shell.
(B) Flue gases move through (inside) the tubes and water is outside the tubes.
(C) Water passes through (inside) the tubes and flue gases are outside the tubes.
(D) Water and flue gases both pass through the tubes.

19. The fire tube boiler is one in which
(A) Steam generation takes place outside the boiler shell.
(B) Flue gases move through the tubes and water is outside the tubes.
(C) Water passes through the tubes and flue gases are outside the tubes.
(D) Water and flue gases both pass through the tubes.

20. The condition of steam in a boiler drum is always ....
(A) Wet (B) Dry
(C) Superheated (D) None of the above

21. The draught produced by steel chimney as compared to that produced by brick chimney for the same height is
(A) Less (B) More
(C) Same (D) None of the above.

ANSWERS

1. (B)	**2.** (B)	**3.** (D)	**4.** (D)	**5.** (E)	**6.** (C)
7. (D)	**8.** (B)	**9.** (D)	**10.** (D)	**11.** (B)	**12.** (C)
13. (D)	**14.** (A)	**15.** (D)	**16.** (A)	**17.** (C)	**18.** (C)
19. (B)	**20.** (A)	**21.** (B).			

Steam Engine

9.1 INTRODUCTION

A steam engine is a device in which heat energy of fuel is converted into mechanical work through some working substance (steam). It is a reciprocating type of basic *prime mover* as it is used to drive other machines. Since steam engine converts energy directly from natural source into mechanical work therefore, it is also known as 'prime mover of first order'.

Engines can be classified in two categories namely:

(*i*) External Combustion Engine
Examples: *Steam Turbine*
Steam Engine

(*ii*) Internal Combustion Engine
Examples: *Petrol Engine*
Diesel Engine

An external combustion engine is one in which combustion of fuel takes place outside the engine cylinder and heat energy generated by combustion of the fuel is transferred to a working substance, which in turn does the mechanical work. Steam engine is the oldest known heat engine to be used for producing mechanical power since 1930. The external combustion engines produce mechanical work either using reciprocating motion or the rotary motion. The steam engine does the mechanical work by using reciprocation motion of the piston whereas the steam turbine does its work based on rotary motion of the rotor. A reciprocating steam engine is one in which piston reciprocates in a cylinder.

Advantages

- Any type of fuel can be used
- Low cost of machinery
- Ease of repair and maintenance.

Disadvantages

- Low efficiency.
- large surface area

Applications

- Steam engine is used for driving pumps, fans, blowers, compressors, stokers, locomotives and electric generators etc.

An internal combustion engine is one in which combustion of fuel takes place inside the engine cylinder. *(For details see next chapter)*

9.2 CLASSIFICATION OF STEAM ENGINES

Steam engines can be classified depending upon the different operations and constructional features.

1. As per the action of steam on sides of the piston

(*i*) Single acting steam engine

(*ii*) Double acting steam engine.

In single acting steam engine, the steam acts on one side of the piston and the other side is open to the atmosphere. There is only one working stroke per revolution of crankshaft. A double acting steam engine has two working strokes per revolution of the crankshaft and here steam acts on both sides of the piston.

2. As per the position of axes of the engine cylinders

(*i*) *Horizontal steam engine*: A horizontal steam engine is one in which axis of the cylinder is horizontal.

(*ii*) *Vertical steam engine:* If axis of the cylinder is vertical it is called vertical steam engine.

(*iii*) *Inclined steam engine*: If axis of the cylinder is inclined it is called inclined steam engine.

3. As per the field of application of engine

(*i*) *Stationary steam engine*: If the foundation of the steam engine is fixed and it is used to run other machinery such as rolling mills etc., it is called stationary steam engine.

(*ii*) *Portable steam engine*: If the steam engine runs along side the machinery it is called portable steam engine. Example: Locomotive steam engine.

(*iii*) *Marine steam engine*: If the steam is working at or under the surface of water, it is called marine steam engine.

4. As per the method of removal of exhaust gases

(*i*) *Condensing steam engine*: In condensing steam engine, the steam after serving its purpose, is discharged (exhausted) into a condenser where the pressure maintained is much lower than atmospheric. Steam is condensed here and the condensate is sent back to boiler by means of a feed pump.

(*ii*) *Non-condensing steam engine*: In a non-condensing steam engine, the steam after expansion is discharged into the atmosphere.

5. As per the nature of expansion

(*i*) *Expansive steam engine:* If the steam is supplied for a portion of the stroke only and then it is allowed to expand to release pressure, then the engine is said to be expansive engine. Expansive steam engines are used for Industrial purposes.

(*ii*) *Non-expansive steam engine:* If the steam is supplied throughout the stroke *i.e.,* the cutoff is unity, then the engine is said to be non-expansive steam engine. It is used in boiler feed pumps, rolling mills etc.

6. As per the number of stages in which expansion of steam takes place

(*i*) *Simple steam engine:* A simple steam engine is one in which complete expansion of steam takes place in one cylinder of the engine.

(*ii*) *Compound steam engine:* A compound steam engine is one in which the complete expansion of steam takes place in more than one cylinder (*i.e.,* two or three or four cylinders).

7. As per the method of governing

(*i*) *Throttled governing:* In a throttled governed steam engine, a control valve placed in the steam pipe controls the speed of the engine. The function of this valve is to regulate the pressure of steam entering the engine cylinder and to control its speed.

(*ii*) *Cut-off governing:* In the cut-off governed engine, a mechanical governor regulates the speed of the engine by regulating the steam quantity entering in the engine cylinder.

8. As per the speed of the engine

(*i*) *Low speed engine*: Steam engine having speed less than 100 RPM comes under the category of low speed engine.

(*ii*) *Medium speed engine*: Steam engine having speed more than 100 RPM but less than 200 RPM comes under the category of medium speed engine.

(*iii*) *High-speed engine*: Steam engine having speed over 200 RPM comes under the category of high-speed engine.

9. Types of valve used

(*i*) Slide valve steam engine

(*ii*) Corlis valve steam engine

(*iii*) Drop valve steam engine.

9.3 STEAM ENGINE PARTS

A line diagram of a single cylinder double acting non-condensing, horizontal type reciprocating steam engine is shown in Fig. (9.1). The different parts of a steam engine can be grouped in two classes namely stationary parts and moving parts.

(1) Stationary parts

- Engine frame
- Cylinder
- Stuffing box
- Crosshead guides
- Main bearings

(2) Moving parts

- Piston
- Piston rod
- Crosshead
- Connecting rod
- Crankshaft
- Flywheel
- Slide valve
- Valve rod
- Eccentric
- Governor

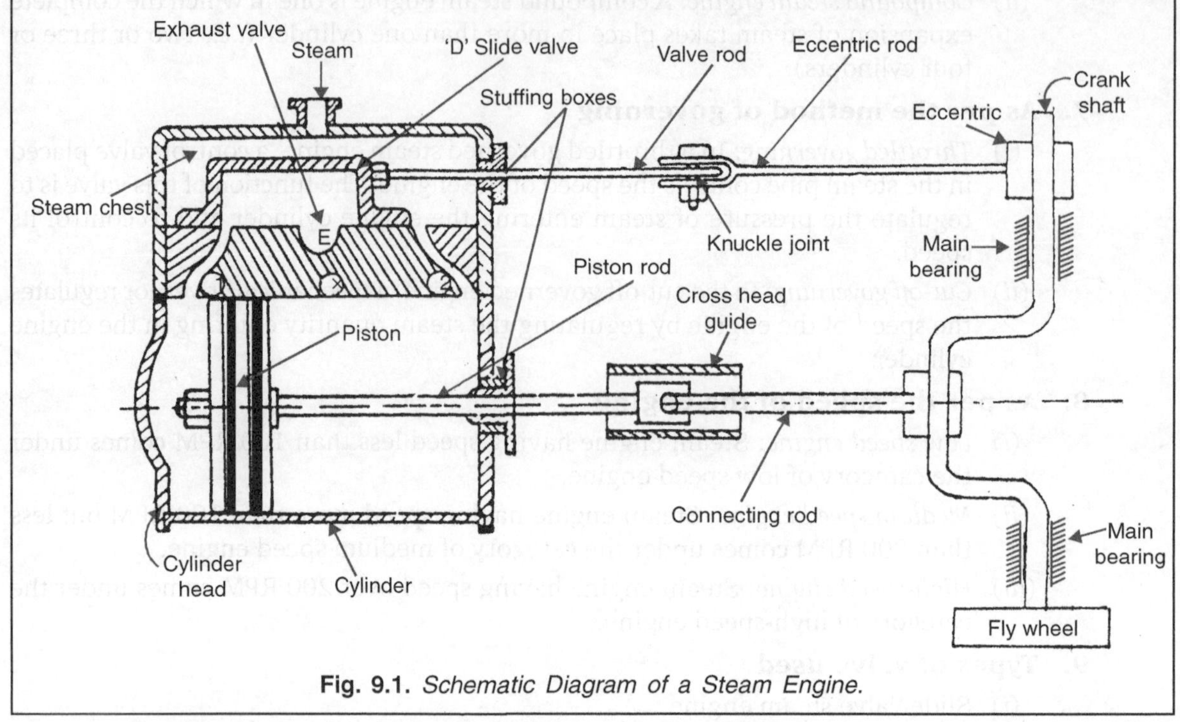

Fig. 9.1. *Schematic Diagram of a Steam Engine.*

(*i*) **Engine frame:** It supports all the stationary and moving parts of an engine and holds them in proper alignment. It is made of cast iron and has significant weight.

(*ii*) **Cylinder:** The engine cylinder is made of cast iron. It is a cylindrical hollow vessel in which piston reciprocates.

(*iii*) **Steam chest:** The steam chest is generally cast integral with the cylinder. It is a closed chamber. Steam chest supplies the steam to the cylinder through steam ports when the slide valve moves. The cylinder is connected at each of its end with the steam chest by means of steam ports.

(*iv*) **Stuffing box:** Its function is to prevent leakage of fluid when a sliding or rotating rod passes through a vessel containing fluid under pressure. In reciprocating steam engines, stuffing boxes are used at two places (*a*) where the piston rod passes through the end of the cylinder or cylinder cover, (*b*) where the valve rod passes out of the valve chest. A good packing such as asbestos is necessary for preventing leakage between the stuffing box and a gland. The gland is usually made of either gunmetal or cast iron. It is tightened by means of studs and nuts in order to keep packing in compressed position.

(*v*) **Piston:** The function of a piston is to convert the heat energy of steam into its reciprocating movement. It is made of cast iron or cast steel or forged steel. It makes to and fro movement in the cylinder. The piston rings usually made of cast iron, are provided on the piston peripherry so as to prevent the leakage of steam. The piston rings reduce the friction between the piston and cylinder walls.

(*vi*) **Crosshead and guide:** The crosshead connects the outer end of the piston rod to the small end of the connecting rod. It guides the motion of the piston rod and prevents it from bending.

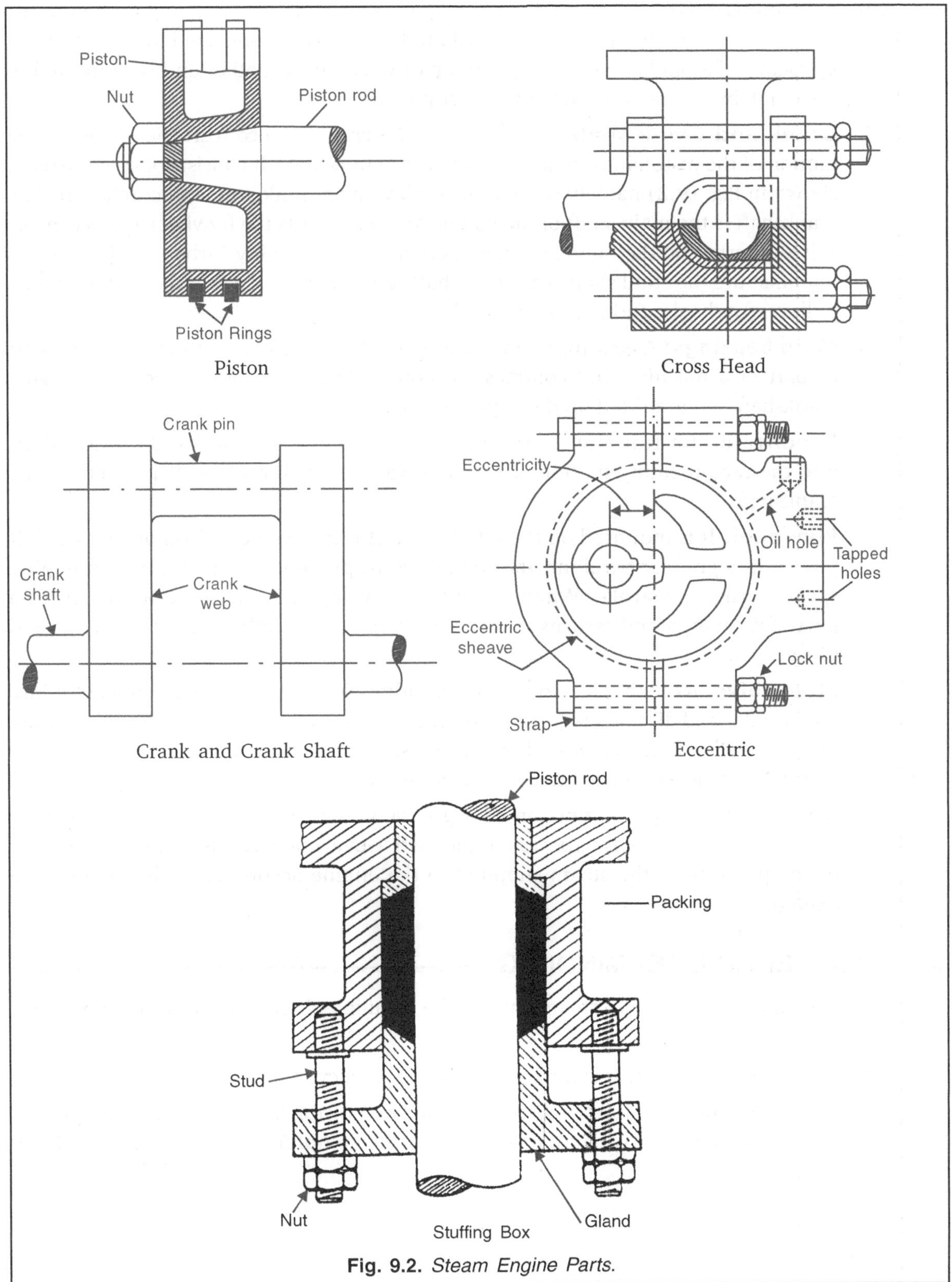

Fig. 9.2. *Steam Engine Parts.*

(*vii*) **Connecting rod:** It's function is to help in converting reciprocating motion of the piston into rotary motion of the crankshaft. Its one end is connected to the crosshead by means of a pin known as gudgeon pin or wrist pin and other end is connected to the crank by means of a pin known as crank pin.

(*viii*) **Crank and crank shaft:** The function of a crank is to convert the reciprocating motion of the piston into rotary motion of the crankshaft. It consists of three parts as shown in Fig. (9.2) namely crank pin, crank web or crank arm, and crankshaft. The crankshaft is the main shaft of the engine and it supports the flywheel and eccentric. It is supported on the main bearings. Generally good quality carbon steel is used to manufacture the crankshaft. The crankshaft runs a variety of systems *e.g.,* generators, pulleys, flywheels, and pumps etc.

(*ix*) **Main bearings:** A bearing is a machine element, which supports the revolving shaft or part of a machine and confines its motion. Main bearings support the engine crankshaft and are fitted on the engine frame.

(*x*) **Eccentric:** Its function is to convert the rotary motion of the crankshaft into the reciprocating motion of the slide valve or piston valve. It is fitted on the crankshaft.

(*xi*) **Flywheel:** It is mounted on the crankshaft. It is made of cast iron or cast steel. It possesses significant weight. Its function is to prevent the fluctuation in engine speed during the stroke. When the engine develops power, the flywheel stores a part of its energy and returns this energy to the crankshaft during the idle period of the cycle.

(*xii*) **Slide valve:** Its function is to control the admission of steam from steam chest into the cylinder and discharges (exhausts) the steam from the cylinder at the appropriate time. The slide valve is located inside the steam chest. Slide valve gets to and fro motion from the eccentric fitted on the crankshaft.

(*xiii*) **Governor:** The main function of a governor is to regulate the mean speed of an engine when there is variation in the load. Governor controls either the quantity or the pressure of the steam supplied to the engine according to the load on the engine.

9.4 STEAM ENGINE TERMINOLOGY

To understand the working of steam engine, the meaning of terms mentioned below must be understood.

(*i*) **Cylinder bore (D):** It is the inside diameter of a cylinder.

(*ii*) **Piston Stroke (L):** It is the distance moved by a piston from one end of the cylinder to the other end. Two stroke of the piston completes one revolution of the crankshaft.

(*iii*) **Forward stroke:** It means the movement of the piston from the head end to the crank end.

(*iv*) **Return stroke:** It means the movement of the piston from the crank end to the head end.

(*v*) **Crank radius:** It is the distance between the centre of the crankshaft and the centre of the crank pin.

$$\text{Crank radius} = \text{Piston stroke}/2$$

(*vi*) **Average piston speed:** It is the linear distance travelled by the piston per minute and can be calculated by equation (9.1)

$$\begin{aligned}\text{Piston speed} &= 2 \times \text{piston stroke} \times \text{crank speed}\\ &= 2 \times L \times N = 2\,LN \text{ m/min.}\end{aligned} \quad \ldots (9.1)$$

(*vii*) **Dead centres:** The dead centres define the positions of the piston at the end of the stroke when the centre lines of piston rod, connecting rod and crank are in the same straight line. There are two dead centres.

For horizontal steam engine	*For vertical steam engine*
Inner Dead Centre (IDC)	Top Dead Centre (TDC)
Outer Dead Centre (ODC)	Bottom Dead Centre (BDC)

(*viii*) **Swept volume (V_S):** The volume swept by the piston when it moves from one dead centre position to another dead centre position is called swept volume. It is also known as piston displacement. It is given by:

V_S = Area of piston × Length of stroke

(*ix*) **Clearance volume:** It is the volume of the space between the piston and the cylinder cover, when the piston is at the end of the stroke plus the volume of the steam port leading to this space.

(*x*) **Eccentricity:** It is the distance between the centre of the eccentric and the centre of the crankshaft.

(*xi*) **Valve travel:** It is the distance traveled by valve in one direction. It is equal to twice the throw of the eccentric.

(*xii*) **Back pressure:** The steam pressure which acts on the exhaust side of the piston is known as back pressure. It reduces the net work done on the piston during a stroke.

9.5 WORKING OF A STEAM ENGINE

The function of a steam engine is to convert the heat energy of steam into mechanical work. The piston moves to and fro inside the cylinder under the pressure of steam. This to and fro motion of the piston is converted into rotary motion with the help of the connecting rod and crank of the steam engine Fig. (9.1). The high pressure steam is first admitted to the cover end side of the cylinder from the steam chest when steam port (1) is opened by the D-slide valve whereas the steam on the crank end of the cylinder after doing useful work passes through the steam port (2) to the exhaust valve and from there it either goes to the condenser or thrown in the open atmosphere.

The steam entered from the cover end through port (1) exerts pressure on the surface of the piston and pushes it towards the crank end of the cylinder. The crankshaft provides the movement to the D-slide valve through eccentric and valve rod. D-slide valve covers port (1) when piston moves towards the crank end and cuts off the supply of steam from steam chest to the engine cylinder. Position marked by point (X) at which the supply of steam to engine cylinder is cutoff, is referred to as point of cutoff and the piston stroke is called charging stroke. At this position both the steam ports are covered.

At the end of this stroke as shown in Fig. (9.1) fresh steam enters in the cylinder through the port (2) from the crank end of side of the cylinder whereas the used steam from the cover end of the cylinder passes through the steam port (1) to the exhaust port. Thus the steam at the cover end is pushed out from the cylinder. While the fresh admitted steam pushes the piston from the crank end to the cover end. This completes the two piston strokes and the crankshaft completes one revolution when the engine is double acting. This cycle is repeated several times.

9.6 HYPOTHETICAL INDICATOR DIAGRAM

An indicator diagram is a plot of steam pressure and volume of steam in the cylinder during a cycle of operations. Such a plot is known as theoretical or hypothetical indicator diagram as it is based up on the following assumptions:

(*i*) The opening and closing of the ports is instantaneous.

(*ii*) There is no pressure drop due to condensation of steam.

(*iii*) The steam is admitted to engine cylinder at the boiler pressure and discharged (exhausted) at condenser pressure.

(*iv*) The expansion of steam is hyperbolic, *i.e.,* it follows the law PV= constant.

(*v*) The clearance volume is negligible.

(*vi*) There is no compression after exhaust.

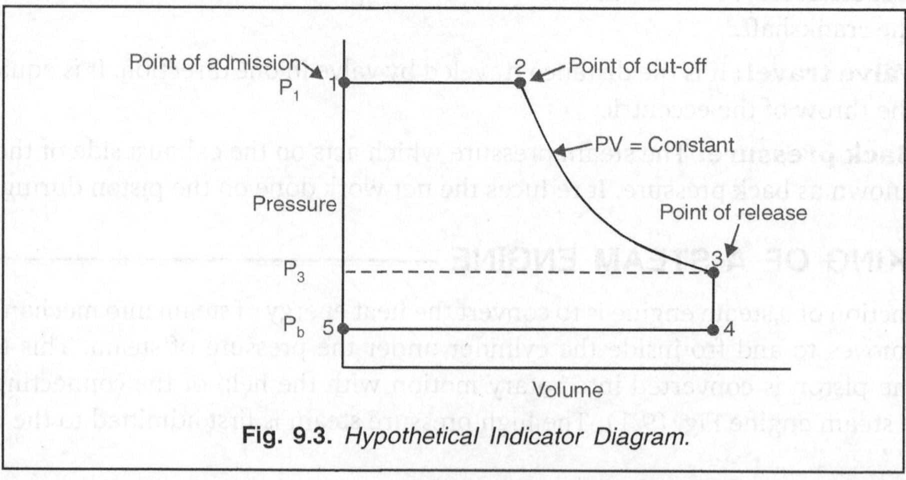

Fig. 9.3. *Hypothetical Indicator Diagram.*

Figure (9.3) shows a hypothetical indicator diagram. As shown in Fig. (9.3), the steam is admitted at boiler pressure in the engine cylinder at point 1. The point 1 is called the point of admission. From point 1 to point 2, the steam is supplied at constant pressure. The point 2 is called the point of cut-off. Next from 2→3 the expansion of steam takes place in the cylinder till end of the stroke. The expansion of steam follows the law PV = constant. At point 3 steam is released and the pressure falls instantaneously to exhaust pressure (or back pressure (P_b) as shown by line 3→4. Exhaust takes place at constant pressure P_b and line 4→5 represents the exhaust stroke. The exhaust of steam takes place at constant pressure *i.e.,* condenser pressure or back pressure p_b. At point 5, the piston reaches its extreme position enclosing the clearance volume. At this stage the engine cylinder is connected with the boiler and fresh steam at pressure P_1 is admitted in the engine cylinder during operation 5→1. This completes one cycle of operations and similarly next cycle repeats.

Thermodynamically the expansion of steam should be adiabatic but in real life such conditions never occurs as some quantity of the steam condenses before the cut-off and quite often the steam is not superheated at the entry level. Condensation of steam takes place during the expansion and at the end of the expansion, certain amount of re-evaporation results. That is why the expansion process 2→3 follows the hyperbolic law PV = C and not adiabatic. It is found its shape is quite different from that of the actual indicator diagram. But this diagram is very useful in making the calculations for designing different components of a steam engine.

9.7 HYPOTHETICAL MEAN EFFECTIVE PRESSURE

9.7.1 Hypothetical Indicator Diagram without Considering Clearance and Compression

Mean effective pressure is defined as the net uniform constant pressure that acts upon the piston face of an ideal engine during its length of stroke, and produces the same effect as is produced by a real engine in one complete cycle. Figure (9.4) shows a hypothetical indicator diagram without considering the clearance and compression. As shown earlier if the pressure is in kPa and volume in m^3 then the area of PV diagram represents the work done in kJ.

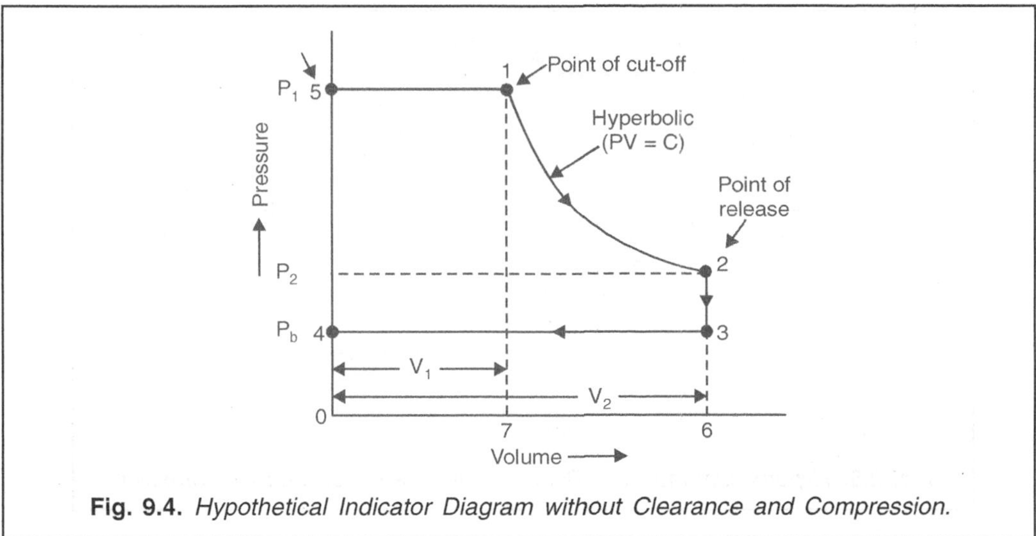

Fig. 9.4. *Hypothetical Indicator Diagram without Clearance and Compression.*

Let P_1 = Steam pressure at the time of entering in the engine cylinder in kPa

P_b = Back pressure in kPa

V_1 = Volume of steam in the cylinder at the cut-off in m^3.

V_2 = Swept volume in m^3

$\therefore$ Work done per cycle in kJ = area of the hypothetical indicator diagram

= area 512345

= Area 5170 + area 1267 – area 4360

$$= P_1V_1\left\{1+\log_e\frac{V_2}{V_1}\right\} - P_bV_2 \quad \text{...(9.2)}$$

$$\therefore \text{Hypothetical mean effective pressure} = \frac{\text{Area of hypothetical indicator diagram}}{\text{Length of the base of the hypothetical diagram}}$$

$$= \frac{\text{Work done per cycle in kJ}}{\text{Swept volume in m}^3}$$

$$= \frac{P_1V_1\{1+\log_e\left(\frac{V_2}{V_1}\right) - P_bV_2}{V_2}$$

$$= P_1\frac{V_1}{V_2}\{1+\log_e(V_2/V_1)\} - P_bV_2$$

But as we know $V_2/V_1 = r$ (expansion ratio)

And $V_1/V_2 = 1/r$ = cut-off ratio

$$\therefore \text{Hypothetical mean effective pressure} = \frac{P_1}{r}\{1+\log_e(r)\} - P_b \text{ kPa} \quad \text{... (9.3)}$$

9.7.2 Hypothetical Indicator Diagram with Clearance and Compression

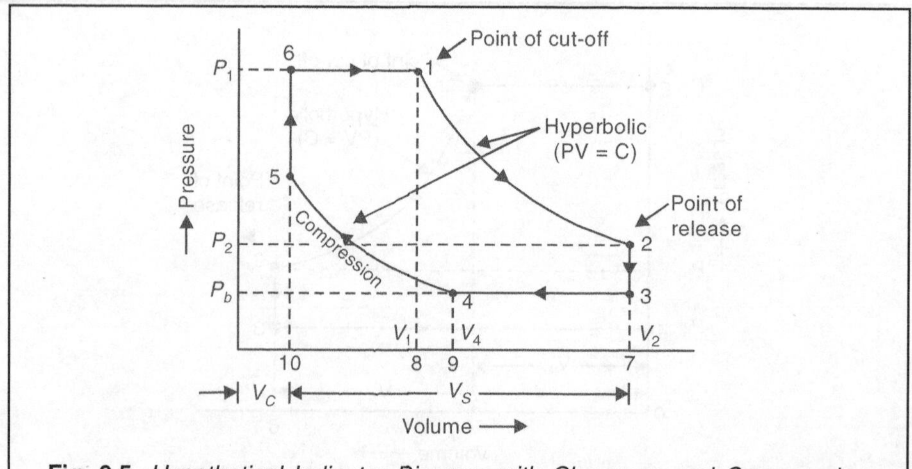

Fig. 9.5. *Hypothetical Indicator Diagram with Clearance and Compression.*

With reference to above Fig. (9.5). The compression starts at 4 and finishes at 5 which is also point of admission. The admission is continued upto 1. The expansion curve is represented by $1 \rightarrow 2$. 2 is the point of release and $3 \rightarrow 4$ is the exhaust line. The compression curve is represented by $4 \rightarrow 5$. The expansion and compression curve are assumed to be hyporbolic.

Let V_s = Volume swept by the piston in one stroke (m^3)

V_c = Clearance volume (m^3)

α = Ratio of the volume between points of compression and admission to the swept volume $= \dfrac{V_4 - V_c}{V_s}$

C = Fractional ratio of clearance volume to swept volume (V_s) $= \dfrac{V_c}{V_s}$

P_1 = Initial pressure of steam or boiler pressure (kPa)

P_b = Back pressure of steam or condenser pressure (kPa)

$1/r$ = Cut-off ratio = ratio of volume between the points of admission and cut-off to the swept volume $= \dfrac{V_1 - V_c}{V_s}$

Net workdone per cycle = area of the indicator diagram = Area 1234561

= Area 6-1-8-10-6 + area 1-2-7-8-1–area 3-4-9-7-3-area –area 4-5-10-9-4

$$\text{Workdone} = P_1(V_1 - V_c) + P_1 V_1 \log_e \frac{V_2}{V_1} - p_b(V_2 - V_4) - p_b[(V_4 - V_5) + V_c] \log_e \left\{\frac{(V_4 - V_5) + V_c}{V_c}\right\} \quad \ldots (9.4)$$

As we know $C = \dfrac{V_c}{V_s}$ $\quad \therefore V_c = CV_s$

and $V_2 = V_s + V_c$ $\quad \therefore V_2 = V_s(1 + C)$

and $\dfrac{1}{r} = \dfrac{V_1 - V_c}{V_s}$ $\quad \therefore V_1 = V_s\left(\dfrac{1}{r} + C\right)$

and $\alpha = \dfrac{V_4 - V_c}{V_s}$ $\quad \therefore V_4 = \alpha V_s + V_c = \alpha V_s + CV_s = V_s(\alpha + C)$

and $V_5 = V_c$

on substituting these values in equation (9.4)

we get net work done per cycle

$$= V_s\left[P_1\left\{\frac{1}{r} + \left(\frac{1}{r} + C\right)\log_e\left(\frac{1 + C}{\frac{1}{r} + C}\right)\right\} - P_b\left\{(1 - \alpha) + (\alpha + C)\log_e\left(\frac{\alpha + C}{C}\right)\right\}\right] \quad \ldots (9.5)$$

$$\therefore \text{Hypothetical mean effective pressure} = \frac{\text{Work done per cycle in kJ}}{\text{Swept volume in m}^3}$$

∴ Hypothetical mean effective pressure

$$= V_s\left[P_1\left\{\frac{1}{r}+\left(\frac{1}{r}+C\right)\log_e\left(\frac{1+C}{\frac{1}{r}+C}\right)\right\}-P_b\left\{(1-\alpha)+(\alpha+C)\log_e\left(\frac{\alpha+C}{C}\right)\right\}\right] V_s \quad \text{... (9.6)}$$

∴ Hypothetical mean effective pressure

$$= P_1\left\{\frac{1}{r}+\left(\frac{1}{r}+C\right)\log_e\left(\frac{1+C}{\frac{1}{r}+C}\right)\right\}-P_b\left\{(1-\alpha)+(\alpha+C)\log_e\left(\frac{\alpha+C}{C}\right)\right\} \quad \text{... (9.7)}$$

If compression effect is neglected *i.e.*, $\alpha = 0$, then Hypothetical mean effective pressure

$$= P_1\left\{\frac{1}{r}+\left(\frac{1}{r}+C\right)\log_e\left(\frac{1+C}{\frac{1}{r}+C}\right)\right\}-P_b \quad \text{... (9.8)}$$

If the effect of clearance volume is neglected *i.e.*, $C = 0$, we get

∴ Hypothetical mean effective pressure $= \frac{P_1}{r}\{1 + \log_e(r)\} - P_b$ ... (9.9)

Equation 9.9 is same as equation 9.3, which was derived earlier.

9.8 ACTUAL INDICATOR DIAGRAM

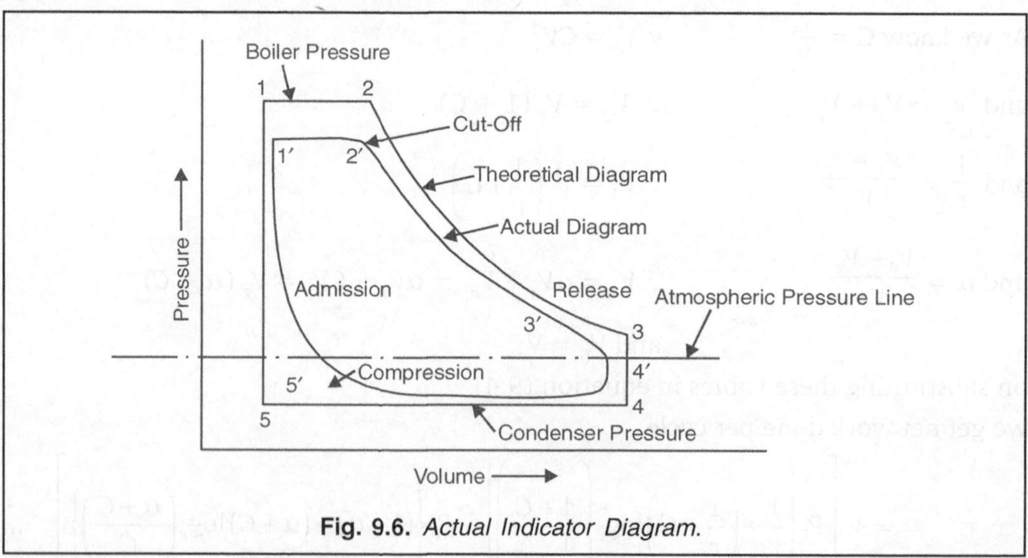

Fig. 9.6. *Actual Indicator Diagram.*

An actual indicator diagram of a steam engine is a P–V diagram, which is obtained from an indicator while performing a test on the engine. Figure (9.6) depicts the actual diagram in thick lines. The area of actual indicator diagram is smaller than the theoretical indicator diagram.

The actual indicator diagram is different from the theoretical diagram due to the following reasons as these reasons were not considered while plotting theoretical indicator diagram.

- When the high-pressure steam from the boiler is supplied to the steam engine through steam pipe via control valves, there is always a drop in steam pressure due to
  - Condensation caused by loss of heat in steam pipe,
  - Frictional losses in the steam supply line and admission valve
  - Wire drawing effect of the valves.

  The fresh steam when mixes with steam in the clearance volume in the engine cylinder causes further drop in the steam pressure. As shown in the Fig. (9.6) the line 1′2′ lies below line 12.
- The opening and closing of the inlet and exhaust ports is not instantaneous, this being a gradual process requires some time in opening and closing the ports thus causing a drop in the pressure. Because of this, there is rounding off the actual indicator diagram at the beginning of admission, cut-off and release points.
- When the high-pressure steam comes in contact with comparatively cool walls of the engine cylinder, it condenses causing a drop in pressure and due to this also, the line 1′2′ lies below 12.
- The actual expansion curve is not hyperbolic due to the high temperature of the steam in early stage of expansion as compared to the temperature of the cylinder walls. This results in the condensation of steam and consequently the volume of steam decreases. At the end of the expansion stroke, temperature of steam becomes lower than the temperature of the cylinder walls. Heat then flows from the cylinder walls to steam, which tends to re-dry the steam.
- The pressure of exhaust steam in the cylinder during the exhaust stroke is higher than the pressure in the condenser despite the quick opening of the exhaust port. It is represented by line 3′4′.
- The clearance volume in steam engine is provided so as to prevent the piston from striking the cylinder head and simultaneously it ensures the cushioning effect on the piston. During the exhaust stroke some steam is left in the clearance space, as the entire steam can not be exhausted in the condenser. This steam gets compressed by the end of the exhaust stroke, which in turn produces cushioning effect on the piston. This is shown by rounding off the line 4′5′ of the P–V-diagram.

Diagram factor is the ratio of the area of actual indicator diagram to the area of hypothetical diagram of a steam engine. It is designated by letter 'K_{DF}'.

$$\text{Diagram factor } K_{DF} = \frac{\text{Area of actual indicator diagram}}{\text{Area of hypothetical indicator diagram}} \quad \text{... (9.10)}$$

Since the area of indicator diagram gives the work done per stroke

$$\therefore \quad K_{DF} = \frac{\text{Actual work done per stroke}}{\text{Theoretical work done per stroke}} \quad \text{... (9.11)}$$

If the two diagrams are drawn to the same scale then

$$K_{DF} = \frac{\text{Mean height of actual indicator diagram}}{\text{Mean height of hypothetical diagram}}$$

$$K_{DF} = \frac{\text{m.e.p. from actual indicator diagram}}{\text{m.e.p. from hypothetical indicator diagram}} \quad \text{... (9.12)}$$

where m.e.p. is mean effective pressure.

The value of the diagram factor depends upon the type of the engines. Its average value is about 0.75.

9.9 INDICATED POWER (IP)

The indicated power is the power developed in the engine cylinder. The indicated power is the rate of doing work on the moving piston. It is called indicated power because an instrument known as indicator is used to obtain the P–V diagram. In SI units, the power developed in the engine cylinder is known as indicated power (IP) whereas in MKS units the power developed in the engine cylinder is known as indicated horsepower (IHP).

For a single acting steam engine, the power developed in any engine cylinder can be determined if the engine cylinder diameter, piston stroke, and speed are known and if an indicator diagram is available. First mean effective pressure is to be determined. The mean effective pressure is the average effective pressure on the piston.

$$\therefore \text{Mean height of indicator diagram in mm, } h = \frac{\text{Area of indicator diagram in mm}^2}{\text{Length of indicator diagram in mm}}$$

Indicated mean effective pressure = mean height × spring scale (or spring number)

For a single acting steam engine, let

p_{m_1} = actual mean effective pressure acting on the piston on cover end side of the engine (N/m^2).

A_1 = area of the piston on cover end side (m^2).

L = length of stroke (m).

N = engine speed (crankshaft speed in RPM)

Force on the piston during a stroke $= p_m \times A$

piston area on crank end side (m^2) $= p_m$ AL

Work done per minute $= p_{m1} \times A_1 LN$ (N-m/min)

$$IP = \frac{p_{m1} A_1 LN}{60} \text{ Watts}$$

For a single cylinder indicated power

$$IP = \frac{p_{m1}A_1LN}{60 \times 1000} \text{ kW} \quad \text{... (9.13)}$$

IP developed on the crank end of the engine cylinder

$$IP = \frac{p_{m2}A_2LN}{60 \times 1000} \quad \text{... (9.14)}$$

Where,

p_{m2} = mean effective pressure acting on crank end side of the piston (N/m^2)

A_2 = piston area on crank end side (m^2) = II $(D^2–d^2)$

D = Diameter of the piston (m^2)

d = Diameter of the piston rod (m^2)

$$\text{Total Indicated Power} = \frac{p_{m1}A_1LN}{60 \times 1000} + \frac{p_{m2}A_2LN}{60 \times 1000} \text{ kW}$$

Assuming mean effective pressure equal on both sides of the piston ($p_{m2} = p_{m1}$) and neglecting the piston rod area, we get

$$\text{Total Indicated Power} = \frac{2p_mALN}{60 \times 1000} \text{ kW (for double acting steam engine)}$$

If the steam engine has n cylinders then

$$\text{Indicated Power} = \frac{2p_mALNn}{60 \times 1000} \text{ kW.}$$

9.10 BRAKE POWER

Brake power is the actual power available from an engine for doing useful work. The brake power is always less than the indicated power since a portion of the power developed in the engine cylinder is used to overcome the frictional losses of different moving parts of the engine. The difference between the indicated power and brake power is called friction power

∴ Indicated power = brake power + frictional power

The brake power is usually determined by means of some brake mechanism. The rope brake dynamometer is used widely for determining the brake power of the engine. A schematic diagram of the rope brake dynamometer is shown in Fig. (9.7). It consists of one or more ropes wound round the brake drum or flywheel of the engine whose brake power is to be determined. The one end of the rope is connected to the spring balance suspended from overhead whereas the other end carries the load W. The load W is called dead load. A number of U-shaped wooden blocks are attached to these ropes at different points so as to maintain a tight grip over the flywheel rim. Since considerable heat is produced in this arrangement, the wheel rim must be water-cooled. It is clear from Fig. (9.7) that load W is opposing the wheel rotation whereas the pulls S indicated by the spring balance is helping to rotate the wheel. Therefore the net brake load which opposes the rotation is (W – S).

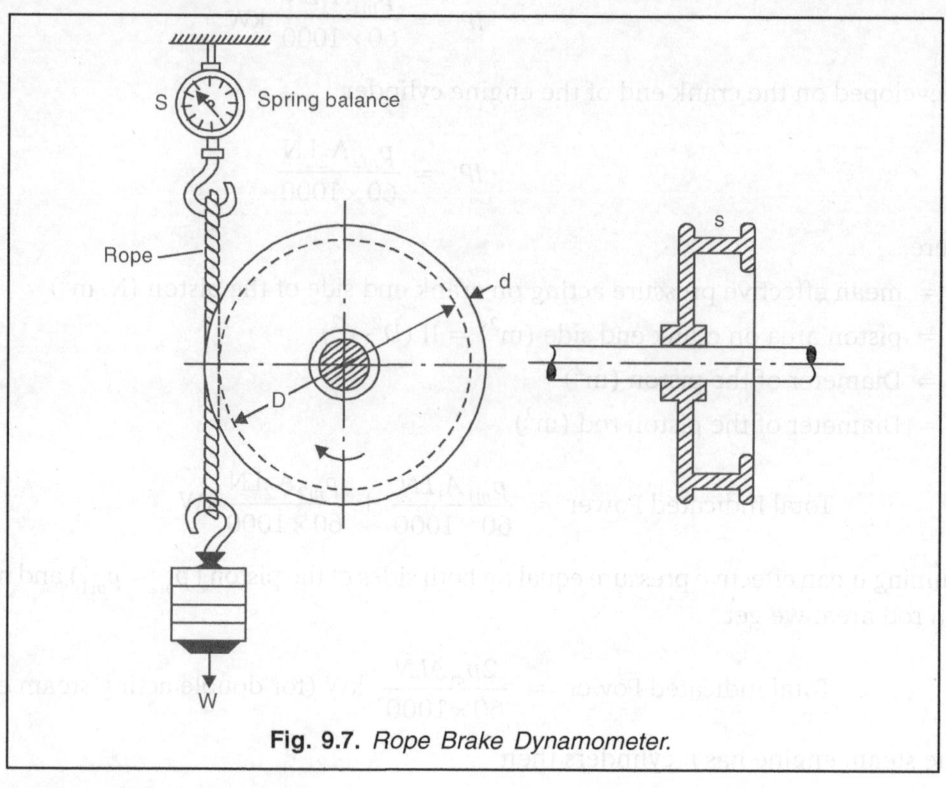

Fig. 9.7. *Rope Brake Dynamometer.*

Let, W = dead weight on the rope (N)

S = spring balance reading (N)

D = diameter of the brake drum (m)

d = diameter of the rope (m)

N = speed of the engine (RPM)

Net braking load or frictional force on the brake drum = $(W - S)$ N

Mean effective radius at which the braking force acts = $(D + d)/2$ m

$\therefore$ Braking torque T = frictional force × radius

$$= (W - S)\left(\frac{D+d}{2}\right) \text{ Nm}$$

$\therefore$ Power absorbed = Braking force × Angle turned in one minute

$$= T \times 2\pi N$$

$$= (W - S)\left(\frac{D+d}{2}\right) \times 2\pi N \quad \text{Nm/min}$$

$$= \frac{(W - S)\,(D + d)\pi \text{N}}{60} \quad \text{Nm/sec. or J}$$

$$\text{Brake Power (BP)} = \frac{(W - S)\,(D + d)\pi N}{60 \times 1000} \text{ kW} \quad \text{... (9.15)}$$

If rope diameter 'd' is neglected, then

$$\text{Brake Power (BP)} = \frac{(W - S)\,\pi DN}{60 \times 1000} \text{kW} \quad \text{... (9.16)}$$

9.11 EFFICIENCIES OF A STEAM ENGINE

The performance of a steam engine is usually expressed in terms of the following:

(*i*) Mechanical efficiency

(*ii*) Thermal efficiency

(*iii*) Overall efficiency.

Mechanical efficiency: It is the ratio of brake power to the indicated power. The mechanical efficiency of high speed steam engine at full load generally lies between 80 to 90%.

$$\eta_m = \frac{\text{Brake power}}{\text{Indicated power}} \quad \text{... (9.17)}$$

A part of the energy developed in an engine cylinder is always lost in overcoming the friction of the moving parts of the engine. The power lost in friction is the difference between the indicated power and the brake power.

$$\therefore \quad FP = IP - BP$$

$$\therefore \quad \eta_m = BP/IP = BP/(BP + FP)$$

or

$$\eta_m = \frac{IP - FP}{IP} \quad \text{... (9.18)}$$

Thermal efficiency: The thermal efficiency of a steam engine is the basis which shows how efficiently the thermal energy (heat energy) carried by steam is converted into mechanical work by the engine. It is defined as the ratio of the heat converted into useful work to the heat supplied. The useful work obtained can be determined either from the indicated power or brake power. If the heat equivalent of indicated power developed by the engine is considered as the output, the efficiency is known as indicated thermal efficiency.

Let, m_S = mass of steam supplied to the engine cylinder in kg/sec.

h_1 = enthalpy of steam supplied to the engine in kJ.

h_{f2} = enthalpy of water (of exhaust steam) in kJ.

$$\therefore \quad \text{Indicated thermal efficiency } \eta_I = \frac{\text{heat equivalent of indicated power in kJ / sec.}}{\text{heat energy supplied in steam in kJ / sec.}}$$

$$\eta_I = \frac{\text{Indicated power}}{m_s\,(h_1 - h_{f2})} \quad \text{... (9.19)}$$

where indicated power is in kW

Brake thermal efficiency: It is the ratio of useful work available at the crankshaft to the heat carried by the steam supplied to the engine.

$$\text{Brake thermal efficiency} = \frac{\text{heat equivalent of useful shaft work}}{\text{Heat energy supplied in steam}}$$

$$\eta_b = \frac{\text{Brake power}}{m_s\ (h_1 - h_{f2})} \quad \text{... (9.20)}$$

where brake power is in kW

$$\text{Mechanical efficiency} = \frac{\text{Brake thermal efficiency}}{\text{Indicated thermal efficiency}} \quad \text{... (9.21)}$$

Overall efficiency: It is defined as the ratio of the net output of an engine to the heat energy supplied by the fuel to the boiler.

$$\eta_{overall} = \frac{\text{Net output or brake power}}{\text{Heat supplied by fuel to the boiler}}$$

$$\eta_{overall} = \frac{\text{Brake power in kW}}{m_f \times CV} \quad \text{... (9.22)}$$

where,

m_f = mass of fuel consumed in boiler in kg/sec.

CV = calorific value of fuel used in the boiler in kJ/kg.

9.12 GOVERNING OF STEAM ENGINE

The function of a governor is to keep the engine speed constant by either controlling the quantity of steam or the pressure of steam supplied to the engine as per the load requirements of an engine.

The steam engines are usually governed by the following two methods:

9.12.1 Throttle Governing

In throttle governing, the quantity of steam entering the engine cylinder is controlled by opening and closing the throttle valve under the action of a governor. In throttle governing, initial steam pressure is reduced due to throlling (reduction in in-take pressure) as shown in Fig. (9.8).

Advantages

- Initial condensation is minimized.

Disadvantages:

The throttling governing is not effective due to:

- loss of available energy
- wasteful in consumption of steam.

In throttle governing, the steam consumption is directly proportional to the indicated power. Therefore the plot between steam consumption and indicated power is a straight line Fig. (9.9). This is known as Willan's law. It holds good only for a throttled governed engine running at constant speed.

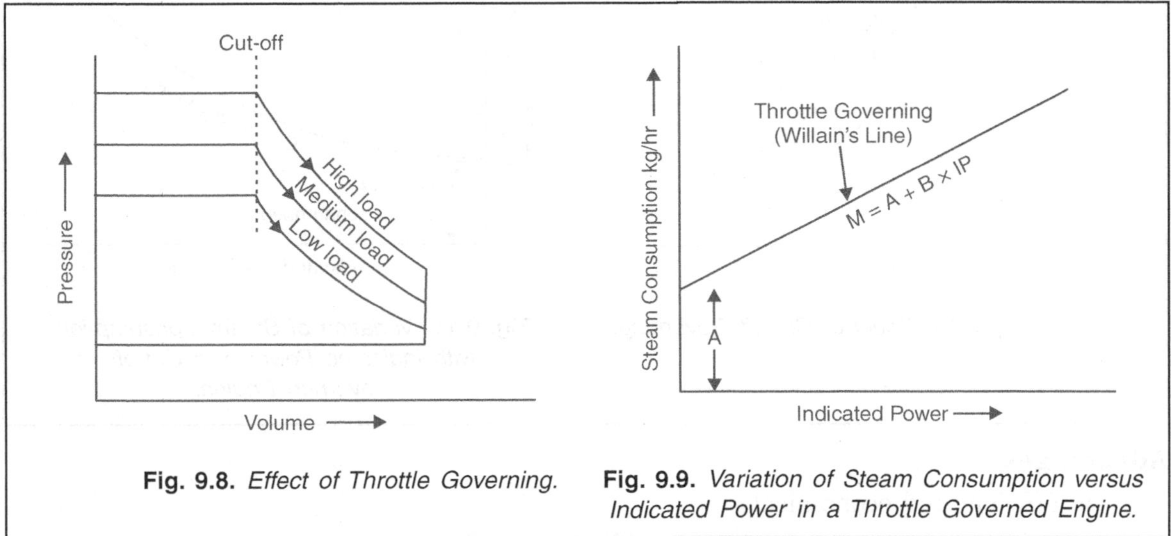

Fig. 9.8. *Effect of Throttle Governing.*

Fig. 9.9. *Variation of Steam Consumption versus Indicated Power in a Throttle Governed Engine.*

Willan's line is a straight line and it is expressed as

$$m = A + B \times IP \qquad \ldots (9.23)$$

where,
m = steam consumption in kg/hr
A = steam consumption at no load in kg/hr
B = slope of the curve (Willian's line)
IP = indicated power.

9.12.2 Cut-off Governing

In cut-off governing Fig. (9.10), the volume of steam supplied to the engine is varied by changing the point of cut-off by a special side valve working under the control of a centrifugal governor. In cut-off governing, the governor changes the point of cut-off but does not change the inlet pressure. Figure (9.11) shows the variation of steam consumption with indicated power in a cut-off governed engine.

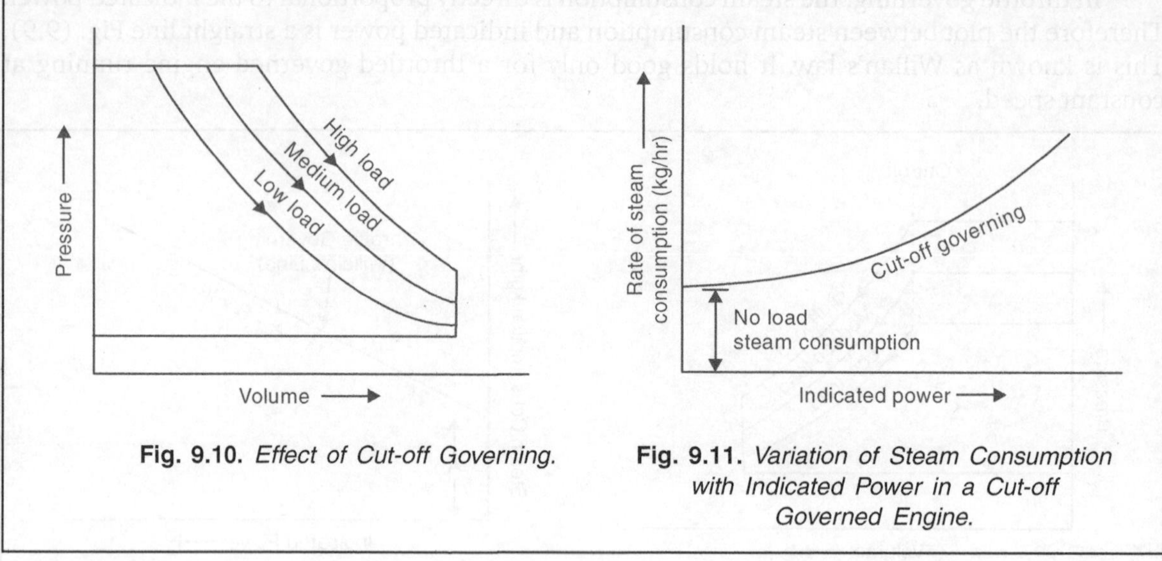

Fig. 9.10. *Effect of Cut-off Governing.*

Fig. 9.11. *Variation of Steam Consumption with Indicated Power in a Cut-off Governed Engine.*

Advantages

- Efficient and economical
- Effective steam consumption.

Disadvantage

- Loss of heat energy due to initial condensation.

Application

- Locomotives

Comparison of throttle and cut-off governing

S.No.	*Factor*	*Throttle governing*	*Cut-off governing*
1.	Thermal efficiency at part load	Low	High
2.	Design	Simple	Intricate
3.	Cost	Low	High
4.	Steam consumption (at part load)	High	Low

9.13 COMPOUND STEAM ENGINE

The thermal efficiency of a steam engine increases by increasing the pressure and temperature of the steam supplied to it. In order to meet the requirement for the high thermal efficiency, the steam is required to be generated at high pressure. The expansion of high pressure steam to that of condenser pressure (or atmospheric pressure) in one engine cylinder of steam engine leads to the following troubles:

1. High expansion ratio will require large stroke length.
2. Condensation of steam will increase.
3. The cylinder has to be of considerable weight and large volume to withstand the high pressure and temperature.
4. Large size flywheel will be required to balance out variations in turning moment.
5. Increased leakage loss because of large pressure difference between the inlet and exhaust.

In order to overcome these problems it is required to expand high pressure steam in two or more successive cylinders arranged in such a way that exhaust steam from the first cylinder is passed to a second cylinder to do further work before discharging it into condenser. This is known as compounding of steam engine and such engines are called compound steam engines.

☞ A compound steam engine is one in which complete expansion of steam takes place in two or more than two cylinders in succession

---

The cylinder in which the expansion first occurs is called high-pressure cylinder (HP cylinder) and the cylinder in which the expansion takes place later is called low- pressure cylinder (LP cylinder). The cylinders (if any) lying in between these two are called intermediate pressure cylinders.

Advantages of compounding

1. Increased efficiency due to use of high-pressure steam.
2. Low pressure and temperature range in each cylinder reduces the condensation of steam.
3. High-pressure cylinder is of small size because of low expansion ratio. HP cylinder and piston are to be made of considerable thickness and strength as the steam is supplied to it at a very high pressure. On the other hand low-pressure cylinders are of large size as they do not have to work under high-pressure steam and need not to have thick walls. This results in significant saving in material and the weight.
4. Economy in steam consumption due to use of high pressure steam.
5. In a single cylinder steam engine the condensation of steam takes place when it enters the engine cylinder and comes in contact with cool cylinder walls. The condensed steam re-evaporates partly as the cylinder walls pick up heat during the later part of the stroke.
6. Steam leakage across the piston reduces due to the small pressure difference across the piston. The small pressure difference across the piston also reduces the extent of stresses in the engine parts.
7. Due to small temperature differences, the engine parts are subjected to low thermal stresses.
8. The failure of one cylinder does not stop the engine working as it can continue working on other cylinder at a reduced load.

9. The engine can start in any position of the crank because of multi cylinder construction employing crank at different angular positions. This feature of compound steam engine is of great value in marine, locomotive and mining work.
10. For the same power developed, compound steam engine costs less as compared to the single cylinder steam engine.
11. The compound steam engines are designed to run at high speed and to produce high power because of uniform turning moment in compound steam engine and low weight of the reciprocating parts of the engine with reduced vibration and noise level.

Disadvantages of compounding

1. Proper and timely care is required due to complicated arrangement.
2. High wear and tear.
3. Increased problems related with lubricants.
4. Need more floor area.
5. Increased heat losses due to large cylindrical surface area.

9.14 CLASSIFICATION OF COMPOUND STEAM ENGINE

Compound steam engines can be classified in three categories:

(*a*) Tandem compound steam engine

(*b*) Woolfe compound steam engine

(*c*) Receiver compound steam engine.

9.14.1 Tandem Compound Steam Engine

The schematic diagram of a tandem compound steam engine is shown in Fig. (9.12). It consists of two cylinders namely high and low-pressure cylinders whose axes lie in the same straight line. In this engine, two pistons are mounted on a common piston rod. The angular difference between the crank is zero degree. The steam is supplied to one side of the high-pressure cylinder piston. After the partial expansion in high-pressure cylinder, steam passes to the low-pressure cylinder where it further expands. The valves should be operated in such a way that continuous supply of steam should be ensured in high-pressure as well as low-pressure cylinders of the steam engine simultaneously.

The advantage of this arrangement allows the direct flow of the steam from high-pressure cylinder to the low-pressure cylinder. As the engine has common crank and piston rod, the operation in both the cylinders is in the same phase therefore the maximum and minimum torque would occur at the same crank angle which causes lot of variation in the turning moment. This variation in turning moment will require a large and heavy flywheel so as to absorb the jerks, vibration and fluctuations in the crankshaft.

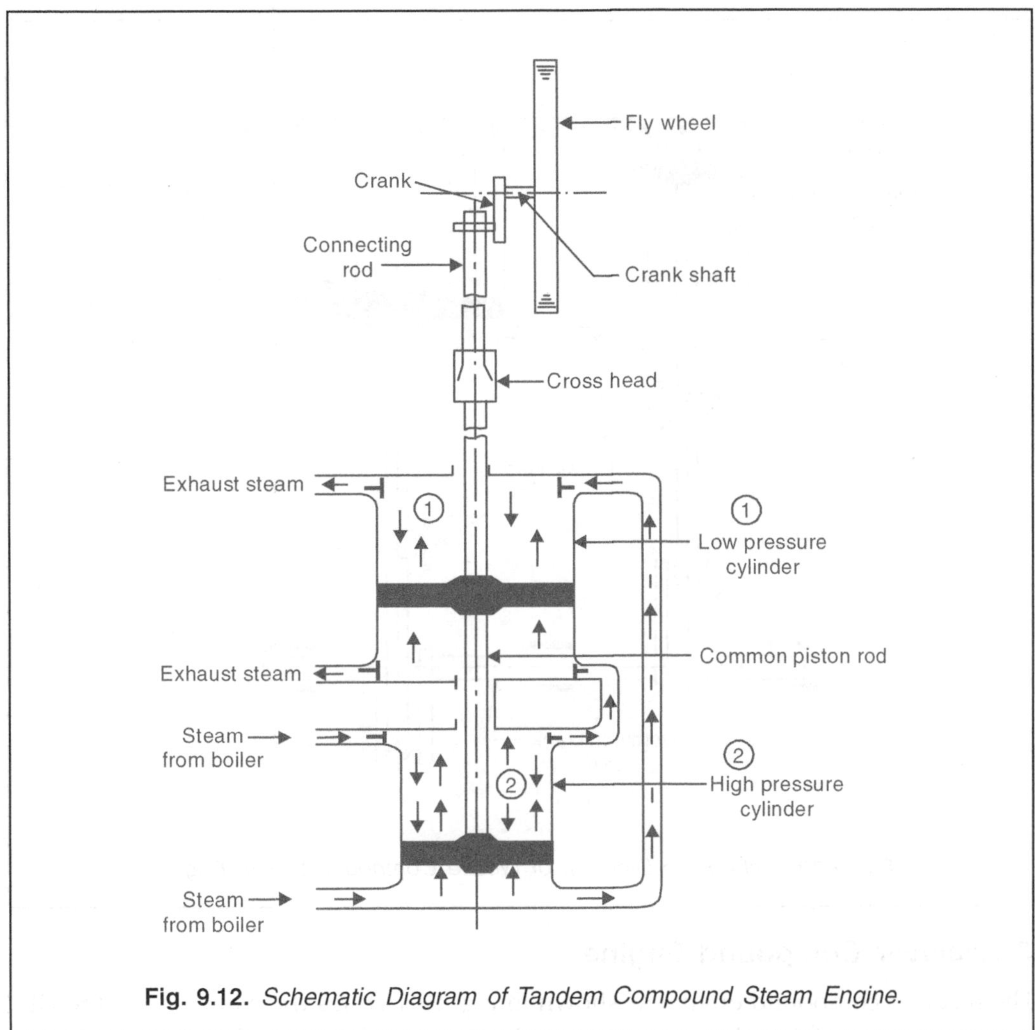

Fig. 9.12. *Schematic Diagram of Tandem Compound Steam Engine.*

9.14.2 Woolfe Compound Steam Engine

The schematic diagram of a Woolfe compound steam engine is shown in Fig. (9.13). In this type of engine, the axes of HP and LP cylinders are parallel to each other. The cranks of the cylinders are at 180° to each other *i.e.,* when the piston of high-pressure cylinder is at the inner dead centre position, the piston of low-pressure cylinder will be at the outer dead centre position. Both the HP and LP cylinders will have their individual connecting rods, cross heads and cranks attached to the same crankshaft as shown in Fig. (9.13). Since the engine has cranks at 180° apart therefore the two cycles are in phase and cause large variation in the turning moment on the crankshaft. Hence a large flywheel will be required to absorb fluctuations in the crankshaft. In Woolfe compound steam engine, the steam from the boiler is supplied to one side of HP cylinder whereas the exhaust steam from the HP cylinder enters directly in the LP cylinder where it expands further.

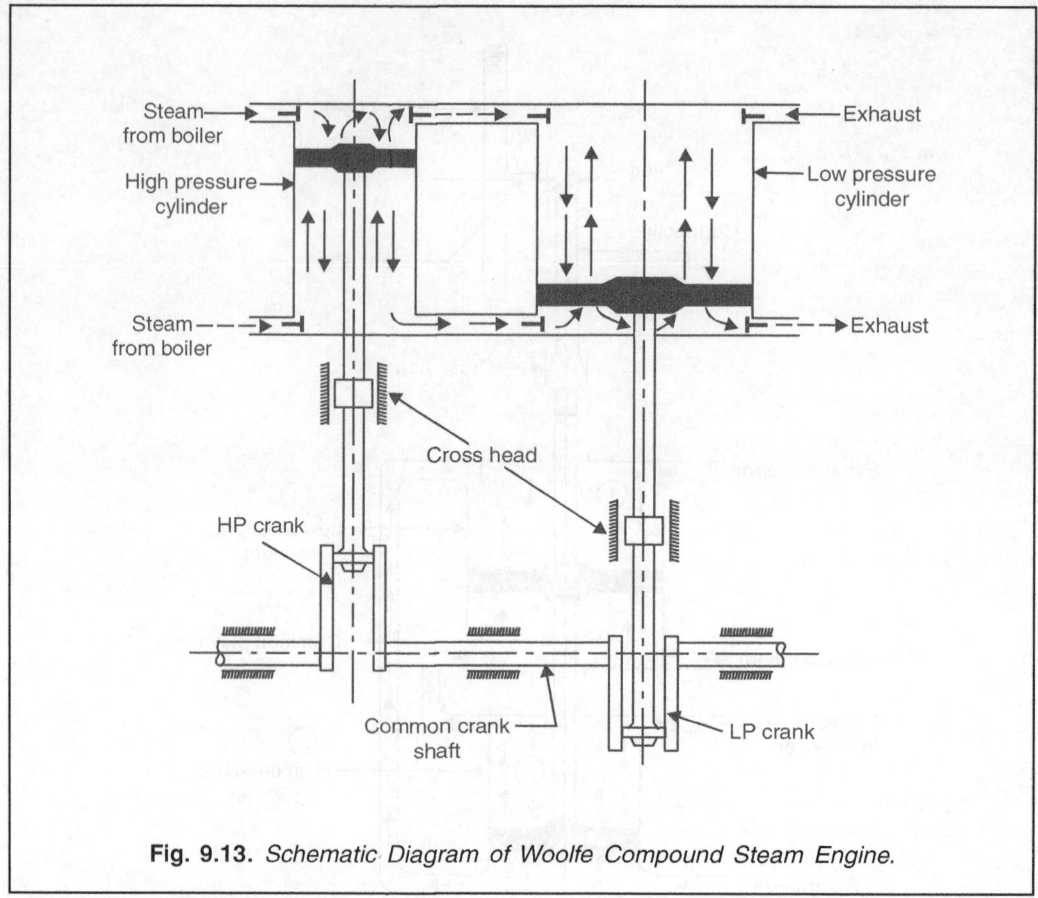

Fig. 9.13. *Schematic Diagram of Woolfe Compound Steam Engine.*

9.14.3 Receiver Compound Engine

The receiver compound engine as shown in Fig. (9.14) has the axes of both the HP (high-pressure) cylinder and LP (low-pressure) cylinders parallel to each other. In this engine, pistons are out of phase, therefore strokes in the two cylinders do not begin simultaneously. Because of this, steam from HP cylinder is not directly passed into LP cylinder but exhaust steam from HP cylinder is first discharged into a chamber called receiver as shown in Fig. (9.14). The receiver acts as a reservoir of steam from where the steam is passed to low-pressure cylinder during its admission (suction) stroke. In this engine, the cranks are at 90° and each cylinder has its own individual piston rod, connecting rod, cross head and crank but attached to a common crankshaft. A large capacity reservoir is useful in minimizing the fluctuations in steam pressure. The 90° phase difference between the cranks reduces the variation in the torque. This necessitates a small flywheel and low balancing weight to absorb the fluctuations on the crankshaft. This is the main advantage of this engine as compared to other two types of compound engines.

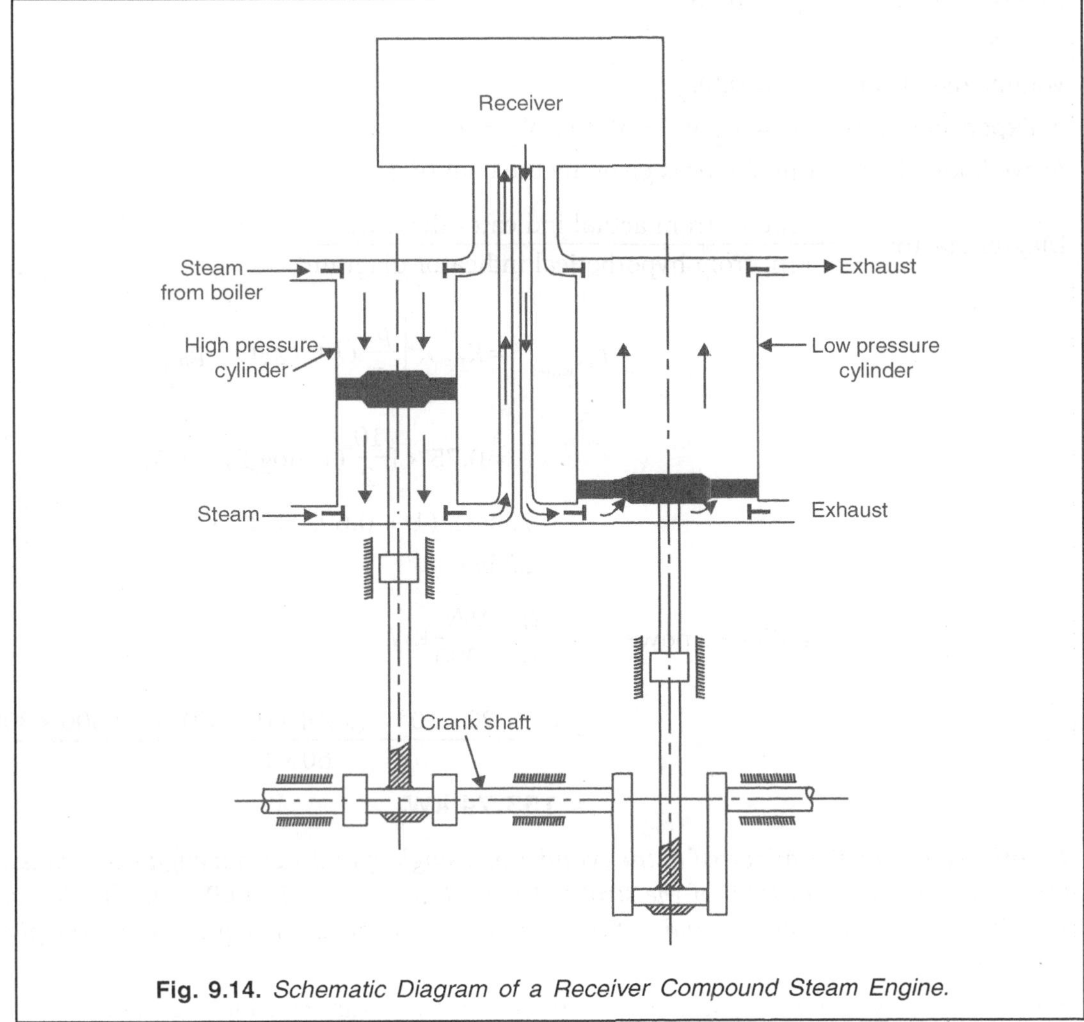

Fig. 9.14. *Schematic Diagram of a Receiver Compound Steam Engine.*

SOLVED PROBLEMS

Problem 1. *A single cylinder double acting steam engine is supplied steam at 10 bar. The cylinder has 250 mm diameter and 600 mm stroke length. Cut-off of the steam supply takes place at half the stroke length. The back pressure is 1.5 bar. Find out the indicated power of the engine when it runs at 300 RPM. Take diagram factor as 0.75.*

Solution: Given,

Steam pressure, p = 10 bar

Cylinder diameter, D = 250 mm = 250/1000 = 0.250 m

Length of stroke L = 600 mm = 600/1000 = 0.600 m = 0.6 m

Back pressure p_b = 1.5 bar

Engine speed N = 300 RPM

Diagram factor $K_{DF} = 0.75$

As given

Volume of cut-off $V_1 = 0.5V_2$

∴ Expansion ratio $r = V_2/V_1 = V_2/0.5\,V_2 = 2$

As we know the diagram factor is given by the equation

$$\text{Diagram factor} = \frac{\text{m.e.p. from actual indicator diagram}}{\text{m.e.p. from hypothetical indicator diagram}}$$

$$\therefore \qquad p_{m_{\text{actual}}} = K_{DF} \times \left[\frac{p_1}{r}(1+\log_2 r) - p_b\right]$$

$$= 0.75 \times [\frac{10}{2}(1+\log_e 2) - 1.5]$$

$$= 0.75 \times [5\,(1 + 0.6931) - 1.5]$$

$$= 5.22 \text{ bar}$$

$$\therefore \qquad \text{Indicated power} = \frac{2p_m ALN}{60 \times 1000} \text{ kW}$$

$$= \frac{2 \times 5.22 \times 10^5 \times \{\pi/4 \times (0.250)^2\} \times 0.600 \times 300}{60 \times 1000}$$

$$= \mathbf{153.74\ kW.}$$

Problem 2. *Steam is admitted to the cylinder of a single cylinder steam engine at a pressure of 1500 kPa and it cut-off at 0.25 of the stroke. The back pressure is 120 kPa and the clearance volume is 9% of the swept volume. Find out the hypothetical mean effective pressure on the piston during the stroke.*

Solution: As we know hypothetical mean effective pressure considering the clearance is given by equation 9.8.

$$\text{Hypothetical}_{\text{m.e.p.}} = P_1\left\{\frac{1}{r} + \left(\frac{1}{r} + C\right)\log_e\left(\frac{1+C}{\frac{1}{r}+C}\right)\right\} - P_b \qquad \text{... (9.8)}$$

Given:

Cut-off $\frac{1}{r} = 0.25$

Clearance volume = 9% of the swept volume C = 0.09.

Admission pressure of steam P_1 = 1500 kPa

Back pressure of steam P_b = 120 kPa

∴ $\text{Hypothetical}_{\text{m.e.p.}}$ = 1500 [0.25 + (0.25 + 0.09) {$\log_e$ (1 + 0.09)/(0.25 + 0.09)}] − 120

$$= 1500\ [0.25 + (0.34) \log_e (3.206)] - 120$$

$$= 1500\ [0.25 + 0.34 \times 1.165] - 120$$

$$= \mathbf{849.15\ kPa.}$$

Problem 3. *During a test on a double acting single cylinder steam engine running at 150 RPM, the indicated power was 80 kW, cylinder diameter 30 cm, stroke 45 cm, steam pressure at admission 1200 kPa, cut-off at 1/3rd stroke and back pressure 35 kPa. Find out the diagram factor neglecting the effect of clearance volume.*

Solution: Given that

Cut-off = $1/3^{rd}$ stroke, Indicated power = 80 W,

$\therefore$ Expansion ratio $= r = \dfrac{1}{\text{cut-off}} = 3$

Steam pressure at admission P_1 = 1200 kPa

Back pressure P_b = 35 kPa

Hypothetical mean effective pressure can be calculated by using equation (9.3).

$$\text{Hypothetical}_{\text{mep}} = \frac{P_1}{r}\{1 + \log (r)\} - P_b$$

$$= \frac{1200}{3}\ [1 + \log_e(3)] - 35$$

$$= 839.44 - 35 = 804.44 \text{ kPa}$$

As we know the indicated power of double acting steam $= \dfrac{2p_m ALN}{60 \times 1000} kW$

Where p_m = actual m.e.p. in N/m^2

A = area of the piston in m^2 = $\dfrac{\pi d^2}{4}$ and d = 30 cm = 0.30 m

L = length of stroke in metre = 45 cm = 0.45 m

N = engine speed in RPM = 150 RPM

$$\therefore \quad 80 = \frac{2p_m \times \frac{\pi}{4}(0.30)^2 \times 0.45 \times 150}{60 \times 1000}$$

$$p_m = 503008.22 \text{ N/m}^2$$

$$= 503008.22 \text{ Pa}$$

$$= 503 \text{ kPa}$$

$$\text{Diagram factor } K_{DF} = \text{m.e.p.}_{\text{actual}} / \text{m.e.p.}_{\text{hypothetical}}$$

$$= \mathbf{503/804.44 = 0.625.}$$

Problem 4. *The following data were obtained during a trial of a single cylinder, double acting steam engine:*

Speed = 104.5 rpm, cylinder dia. = 220 mm, piston stroke = 300 mm, mean effective pressure (MEP) (each end) = 120 kPa, effective brake diameter = 1.2 m, dead load on the brake = 726 N, spring balancing reading = 176 N, pressure of steam supplied = 700 kPa, steam used/hr. = 57 kg. Condensate temperature = 50°C. Steam supplied was dry and saturated. Calculate the mechanical efficiency, indicated and brake thermal efficiencies of the engine.

(RGPV Jan./Feb. 2006)

Solution: Given:

$$\text{Speed of the engine } N = 104.5 \text{ rpm}$$

$$\text{Cylinder diameter } D = 220 \text{ mm} = \frac{220}{1000}\text{m} = 0.220 \text{ m}$$

$$\text{Piston stroke } L = 300 \text{ mm} = \frac{300}{1000}\text{m} = 0.300 \text{ m}$$

$$\text{Mean effective pressure at each end } m_p = 120 \text{ kPa}$$

$$\text{Effective brake diameter } D_{\text{brake}} = 1.2 \text{ m}$$

$$\text{Dead load on the brake } W = 726 \text{ N}$$

$$\text{Spring balance reading } S = 176 \text{ N}$$

$$\text{Steam pressure } p = 700 \text{ kPa}$$

$$\text{Steam consumption} = 57 \text{ kg/hr}$$

$$\text{Condensate temperature} = 50°\text{C}$$

As we know that IP for a double acting steam engine is given by

$$IP = \frac{2\times p_m ALN}{60\times 1000}$$

Where unit of p_m is in *pascal*

$$IP = \frac{2\times 120\times 1000\times \frac{\pi}{4}(0.02)^2\times 0.3\times 104.5}{60\times 1000} = 4.76 \text{ kW}$$

Brake power is given by

$$BP = \frac{(W-S)\pi DN}{60\times 1000}$$

$$BP = \frac{(726-176)\pi\times 1.2\times 104.5}{60000}$$

$$BP = 3.61 \text{ kW}$$

Mechanical efficiency is given by

$$\eta_{\text{mech}} = \frac{BP}{IP} = \frac{3.61}{4.76} = 0.7584 = 75.84\%$$

From steam tables

Since pressure of steam is given in kPa while pressure unit followed in the steam table is bar so pressure of steam first should be changed in bar.

Pressure of dry saturated steam is 7 bar

Specific enthalpy of dry saturated steam at 700 kPa (=7bar) is 2763.5 kJ/kg

Specific heat of feed water at 50°C is = 4.187 × (50–0) = 209.35 kJ/kg

Heat supplied for producing steam in the boiler

$$Q_s = m_s(h_1 - h_w) = \frac{57}{3600}(2763.5 - 209.35) = 40.44 \text{ kW}$$

$$\therefore \quad \text{Indicated thermal efficiency} = \eta_{ith} = \frac{IP}{Q_s} = \frac{4.76}{40.44} = 0.1177 = 11.77\%$$

and

$$\therefore \quad \text{Brake thermal efficiency} = \eta_{bth} = \frac{BP}{Q_s} = \frac{3.61}{40.44} = 0.0892 = 8.92\% \textbf{ Ans.}$$

Problem 5. *The following readings were taken during a trial of a single cylinder double acting non-considering steam engine running at 240 RPM.*

Cylinder dia. = 250 mm, piston rod dia. = 50mm, stroke of the engine = 350 mm, cut-off = 30% of stroke, length of indicator diagram = 53 mm, area of indicator diagram for cover end = 1570 mm^2, area of indicator diagram for crank end = 1440 mm^2, spring number 0.12 bar/mm, circumference of brake wheel = 5.25 m, circumference of brake rope = 62.5 mm, dead load on the brake = 1900 N, reading of the spring balance = 190 N, Pressure of the steam supplied = 10.5 bar, Dryness fraction of steam supplied = 0.92

Find out:

(i) Indicated power (IP) *(ii) Brake power (BP)*

(iii) Mechanical (η) *(iv) Brake thermal (η)* *(RGPV Dec. 2006)*

Solution:

Indicated power for a single cylinder double acting steam engine is

$$IP = \frac{p_{m1} A_1 LN}{60 \times 1000} + \frac{p_{m2} A_2 LN}{60 \times 1000} \quad ...(i)$$

$$p_m = \frac{\text{Area of indicator diagram} \times \text{Spring number}}{\text{Length of the indicator diagram}}$$

$$p_{m1} = \frac{A_1 \times S}{L} = \frac{1570 \times 0.12}{53} = 3.55 \text{ bar}$$

$$p_{m2} = \frac{A_2 \times S}{L} = \frac{1440 \times 0.12}{53} = 3.26 \text{ bar}$$

$$\text{Area of piston on cover end side } A_1 = \frac{\pi}{4}(D)^2 = \frac{\pi}{4}(0.25)^2 = 0.049 \text{ m}^2$$

$$\text{Area of piston on crank end } A_2 = \frac{\pi}{4}\left(D^2 - d^2\right) = \frac{\pi}{4}\left((0.25)^2 - (0.05)^2\right)$$

$$= 0.047 \text{ m}^2$$

Now substituting these values in equation (*i*)

$$IP = \frac{3.55\times10^5 \times 0.049 \times 0.35 \times 240}{60\times1000} + \frac{3.26\times10^5 \times 0.047 \times 0.35 \times 240}{60\times1000}$$

$$IP = 24.35 + 21.45$$

$$IP = 45.80 \text{ kW} \quad \textbf{Ans.}$$

(*b*) Brake power

$$BP = \frac{(W-S)(D+d)\pi N}{60\times1000}$$

$$BP = \frac{(1900-190)(5.25+0.0625)\times\pi\times240}{60000}$$

$$BP = 36.33 \text{ kW}$$

(*iii*) Mechanical Efficiency is given by

$$\eta_{mech} = \frac{BP}{IP} = \frac{36.33}{45.80} = 0.7932 = 79.32\%$$

(*iv*) Specific steam consumption:

Specific steam consumption per hour,

$$m_s = \frac{\text{Volume of steam consumed per revolution}}{\text{specific volume}} \times RPM \times 60$$

for cover end

$$m_{s_1} = \left(\frac{\pi}{4}D^2 L\rho\right)\times\frac{60\times N}{xV_g}$$

Where V_g is the specific volume of steam whose value can be taken from steam table and ρ is the cut-off ratio = 0.30 (given)

The pressure of steam supplied is 10.5 bar

∴ from steam table $V_g = 0.1855 \text{ m}^3/\text{kg}$

$$\therefore \quad m_{s1} = \left(\frac{\pi}{4}D^2 L\rho\right)\times\frac{60\times N}{xV_g}$$

$$m_{s1} = \left(\frac{\pi}{4}\times0.25^2\times0.35\times0.30\right)\times\frac{60\times240}{0.9\times0.1855}$$

$$m_{s1} = 444.56 \text{ kg/hr}$$

For crank end:

$$m_{s2} = \left[\frac{\pi}{4}\left(D^2 - d^2\right)L\rho\right] \times \frac{60 \times N}{xV_g}$$

$$m_{s2} = \left[\frac{\pi}{4}\left(0.25^2 - 0.05^2\right) \times 0.35 \times 0.30\right] \times \frac{60 \times 240}{0.9 \times 0.1855}$$

$$m_{s2} = 426.78 \text{ kg/hr}$$

Total steam consumption m_s = 444.56 + 426.78 = 871.34 kg/hr

As we know that specific steam consumption is defined as the steam consumed per hour per kW power produced.

$\therefore$ Specific steam consumption on IP basis $= \frac{871.34}{45.8} = 19.02$ kg/hr

$\therefore$ Specific steam consumption on BP basis $= \frac{871.34}{36.33} = 23.98$ kg/hr

(*d*) From steam tables at 10.5 bar

$$h_f = 772.3 \text{ kJ/kg}$$
$$h_{fg} = 2007.2 \text{ kJ/kg}$$
$$v_g = 0.1855 \text{ m}^3\text{/kg}$$
$$t_s = 182.02°\text{C}$$

Heat supplied to boiler for generating steam is given by

$$h_1 = h_f + xh_{fg}$$
$$= 772.3 + 0.9 \times 2007.2$$
$$= 2578.78 \text{ kJ/kg}$$

The enthalpy of feed water h_w can be ignored as the engine is non-condensing.

Brake thermal efficiency

$$\eta_{brake(thermal)} = \frac{BP}{m_s(h_1 - h_w)} = \frac{BP}{m_s \times h_1}$$

$$= \frac{36.33}{\frac{871.34}{3600} \times 2578.78} = 0.058$$

= 5.8% **Ans.**

Unsolved Problems

1. A double acting steam engine has a cylinder 200 mm bore by 300 mm stroke and cut-off occurs at 0.4 stroke. The admission and exhaust pressures of steam are 7 bar and 0.38 bar respectivley. If the diagram factor is 0.8, calculate the indicated power of the engine at 200 RPM, neglecting the clearance and assuming hyperbolic expansion.

   (*RGPV-Dec. 2003*) [**Ans.** 17.8 kW]

2. Following reading were taken during the test at full load on a single cylinder, double acting condensing type throttled governed steam engine:

Diameter of the cylinder	: 400 mm
Stroke of the engine	: 600 mm
cut-off	: 50% of stroke
Pressure of steam supplied	: 11 bar
Back pressure	: 0.8 bar
Brake wheel diameter	: 4.5 m
Net load on the brake	: 4900 N
Speed of the engine	: 150 RPM
Diagram factor	: 0.82

   Find the indicator power, brake power and mechanical efficiency of the engine.

   (*RGPV-June 2004*) [**Ans.** 263.1 kW, 173.2 kW, 65.8%]

3. A single cylinder double-acting steam engine is supplied with steam at 1MPa. The cylinder has 275 mm diameter and 650 mm stroke length. Cut-off of the steam supply takes place at half the stroke length. The back pressure of steam is 150 kPa. Assuming a diagram factor of 0.75, find out the indicated power of the engine when it runs at 350 RPM.

   [**Ans.** 235.13 kW]

4. A single cylinder double acting steam engine develops 75 kW when steam is admitted at 10.5 bar. The cut-off is at 1/3rd of the stroke and the back pressure is 0.275. The diagram factor is 0.72 and the mechanical efficiency is 84%. If the mean piston speed is 250m/min and the stroke to bore ratio is 1.2, determine the bore and stroke of the engine.

5. A single cylinder double acting steam engine develops 22.5 kW when running at 240 RPM. It is supplied with dry saturated steam at 9 bar. The exhaust pressure is 1.4 bar. Steam supply is cut-off at 0.4 of the stroke. Taking diagram factor to be 0.8, and stroke-bore ratio as 12.5, find out the bore and stroke of the engine. Assuming expansion to be hyperbolic, determine the steam consumption per hour.

   [**Ans.** 187 mm, 234 mm, 33.97 kg]

6. The bore of a single cylinder double acting steam engine is 300 mm and its stroke length is 440 mm. Steam is supplied at a pressure of 600 kPa. The steam is cut-off at 0.4 of stroke and compression starts from 0.8 return stroke, the clearance is 10% of the volume swept by the piston in one stroke. The engine speed is 150 RPM. Assuming hyperbolic expansion and compression find out (*a*) the indicated power of the engine, (*b*) steam consumption per kWh. take diagram factor 0.9. [**Ans.** 47.72 kW, 16.5 kg/kWh]

Review Questions

1. Define a steam engine?
2. Give a detailed classification of steam engines.
3. Why is steam engine known as prime mover?
4. Distinguish between the external and internal combustion engine.
5. Discuss briefly the functions of the following parts of a steam engine:

   (*i*) Eccentric (*ii*) Cross-head

   (*iii*) Piston (*iv*) Crank

   (*v*) D-slide valve
6. Explain the following terms in case of a steam engine

   (*i*) Indicated power (*ii*) Brake power

   (*iii*) Mechanical efficiency
7. Derive an expression for mean effective pressure (m.e.p.) from a hypothetical indicator diagram for a simple steam engine assuming compression and clearance. (*RGPV-Dec. 2001*)
8. Describe with sketches the working of a steam engine.
9. How the governing of steam engine is done? Explain. (*RGPV-June 2001*)
10. List down the advantages of compound steam engine over single stage steam engine. (*RGPV-Dec. 2003*)
11. Name the different methods used for governing compound steam engine. Discuss their relative advantages and disadvantages.
12. Discuss in brief the following compound steam engines:

    (*i*) Woolfe type (*ii*) Receiver type (*RGPV-June 2004*)
13. Draw a hypothetical indicator diagram and actual indicator diagram of a steam engine. Explain the difference between these two diagrams.
14. What do you mean by the term 'Diagram Factor'? Mention its average value.
15. Derive an expression for the hypothetical mean effective pressure in a steam engine cylinder neglecting the clearance volume, state clearly the assumptions you make?
16. State the methods adopted to improve diagram factor.
17. Explain what do you mean by 'Willan's line'?

    What is the function of a stuffing box? Where stuffing boxes are used in a steam engine? Describe a stuffing box with a sketch.
18. Write expression for different efficiencies associated with a steam engine. (*RGPV-June 2002*)
19. Give name and function of important parts of a simple steam engine. (*RGPV-Dec. 2002*)
20. Explain briefly the following methods of governing of steam engines:

    (*i*) Throttle governing (*ii*) Cut-off governing. (*RGPV-Dec. 2003*)

OBJECTIVE TYPE QUESTIONS

Select the most appropriate answer:

1. Steam engine is known as
(A) external combustion engine
(B) internal combustion engine
(C) external and internal combustion engine
(D) none of the above.

2. Diagram factor is defined as the ratio of
(A) actual m.e.p. to steam pressure
(B) actual m.e.p. to theoretical m.e.p
(C) theoretical m.e.p. to actual m.e.p
(D) none of the above.

3. Cut-off ratio in steam engine is the ratio of volume at cut-off to the
(A) clearance volume (B) cylinder volume
(C) swept volume (D) clearance +swept volume.

4. The diagram factor is always
(A) zero (B) more than one
(C) equal to one (D) less than one.

5. Which one does not belong to steam engine?
(A) D-slide valve (B) stuffing box
(C) cross head (D) carburettor.

6. Steam acts on both sides of the piston in
(A) single acting steam engine (B) double acting steam engine
(C) triple steam engine (D) none of the above.

7. The function of .... is to guide the motion of the piston rod and prevents it from bending.
(A) cross head (B) piston
(C) eccentric (D) flywheel.

8. In a compound steam engine the diameter of high pressure cylinder as compared to low pressure cylinder is
(A) more (B) less
(C) equal (D) none of the above.

9. Willan's law holds good for
(A) throttle governed engine (B) cut-off governed engine
(C) combination of A and B (D) none of the above.

10. The cross head has
(A) rotary motion (B) reciprocating motion
(C) oscillating motion (D) none of the above.

11. The .... is used to exhaust steam from the cylinder at the proper moment.

(A) flywheel (B) eccentric

(C) D-slide valve (D) piston.

12. Diagram factor's value range from

(A) 0.25 to 0.50 (B) 0.5 to 0.6

(C) 0.7 to 0.9 (D) 0.9 to 1.5.

13. The speed of crankshaft for the low speed steam engine is ....

(A) over 200 RPM (B) over 500 RPM

(C) over 100 RPM (D) less than 100 RPM.

14. A smaller flywheel is required in

(A) tandem type compound steam engine

(B) receiver type compound steam engine

(C) woolfe type compound steam engine

(D) none of the above.

15. The cylinder of steam engine is usually ....

(A) water cooled (B) air cooled

(C) gas cooled (D) steam jacketed.

ANSWERS

1. (A)	**2.** (B)	**3.** (C)	**4.** (D)	**5.** (D)	**6.** (B)
7. (A)	**8.** (B)	**9.** (A)	**10.** (B)	**11.** (C)	**12.** (C)
13. (D)	**14.** (B)	**15.** (D).			

Refrigeration

10.1 INTRODUCTION

The term refrigeration may be defined as the process of cooling a substance (*i.e.*, process of removing heat from a substance). A refrigerator is a cyclic device for producing a low temperature atmosphere. The working substance used in the refrigeration cycles are called as refrigerants.

Applications

- For making and storing ice
- Cooling of storage chamber for preserving food
- For cooling water
- In the manufacturing and preservation of medicines and chemicals
- Transport refrigeration
- In the field of air-conditioning
- In the field of cryogenics.

Unit of Refrigeration

The amount of heat extracted (or removed) in a given time is termed as the refrigerating effect. The unit used in the field of refrigeration is known as "Tonne of Refrigeration". A tonne of refrigeration is defined as the quantity of heat removed to produce 1 ton of ice in 24 hours from and at 0°C. Usually one tonne of refrigeration is taken as 210 kJ/min or 3.5 kJ/s.

Coefficient of Performance

The performance of any refrigeration system is judged by a factor known as coefficient of performance (*COP*). The *COP* of a refrigeration system is defined as the ratio of refrigerating effect produced to the work done on the refrigerant.

$$COP = \frac{\text{refrigerating effect produced}}{\text{Work done on refrigerant}}$$

Since the heat extracted in the refrigerator is the refrigerating effect, the equation for *COP* can be written as

$$COP = \frac{\text{Heat extracted from a cold body}}{\text{Work done on refrigerant}} = \frac{N}{W}$$

10.2 AIR REFRIGERATION SYSTEM

Refrigerators using air as working substance operate either on a reversed Carnot cycle or reversed joule cycle. The reversed Joule cycle is also known as 'Bell Coleman cycle'. Air as a refrigerant has two major advantages:

1. Air is available free of cost.
2. Air leakage does not cause any problem.

10.2.1 Reversed Carnot Cycle

The reversed Carnot cycle is used for producing refrigeration. It is shown in Fig. 10.1.

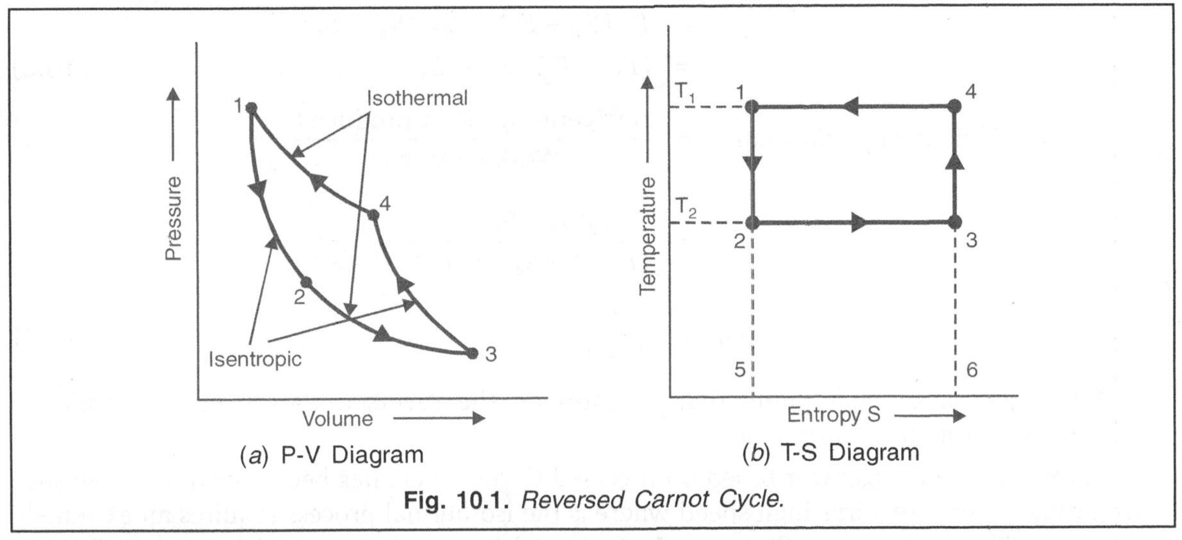

Fig. 10.1. *Reversed Carnot Cycle.*

The sequence of operation followed by the reversed Carnot cycle is mentioned below: starting the cycle at state 1, the clearance space of the cylinder is full of air. The air is expanded isentropically up to state 2.

Process 1–2

Isentropic expansion → Here the air is expanded isothermally from state 1 to state 2. This results in temperature falling from T_1 to T_2.

Process 2–3

Isothermal expansion → The air is now further expanded isothermally from state 2 to state 3 at tempeature T_2. During this process, heat is absorbed from the cold body.

Process 3–4

Isentropic compression → The air is next compressed isentropically from state 3 to state 4 by using external power which increases the temperature from T_2 to T_1. During this process no heat is absorbed or rejected by the air.

Process 4–1

Isothermal compression → The last operation of this cycle is the isothermal compression 4–1. Here the air is compressed isothermally from state 4 to state 1. During this process heat is rejected by air to the hot body.

From the T-S diagram

Heat rejected to hot body = area 1–5–6–4–1 = $T_1\ (S_4 - S_1)$ ...(10.1)

Heat abstracted from cold body = area 2–3–6–5–2 = $T_2\ (S_3 - S_2) = T_2\ (S_4 - S_1)$ as we know that ($S_3 = S_4$ and $S_1 = S_2$)

$$\therefore \quad \text{Work required per cycle} = \text{heat rejected} - \text{heat abstracted}$$

$$= T_1\ (S_4 - S_1) - T_2\ (S_4 - S_1)$$

$$= (T_1 - T_2)\ (S_4 - S_1) \qquad \text{...(10.2)}$$

$$\therefore \quad \text{Coefficient of performance} = \frac{\text{refrigerating effect produced}}{\text{Work required}}$$

$$= \frac{T_2(S_4 - S_1)}{(T_1 - T_2)(S_4 - S_1)}$$

$$\text{COP} = \frac{T_2}{T_1 - T_2} \qquad \text{...(10.3)}$$

A refrigerator or heat pump that operates on the reversed Carnot cycle is called a Carnot refrigerator or a Carnot heat pump.

In real life, no refrigerator based on reversed Carnot cycle has been constructed because as isentropic process requires high speed whereas the isothermal process requires an extremely slow speed. This variation in flow speed of air and hence piston speed is not possible.

10.2.2 Bell Coleman Cycle

Bell Coleman cycle is a modification of ideal reversed Carnot cycle so as to make it practicable. In this cycle, the isothermal operations of Carnot cycle are replaced by constant pressure operations. This cycle is also known as reversed Joule cycle. Figure 10.2(*a*) shows an open air refrigeration system. In this system, air is discharged into cold chamber at atmospheric pressure and is allowed to come in contact with the cold body. The compressor sucks the air from the cold chamber, compresses it and delivers it to the air cooler. The high pressure cooled air is then expanded in the expansion cylinder (air motor). The isentropic expansion cools the air to a temperature below that of the cold storage chamber. The low temperature air leaving the expansion cylinder then enters the cold chamber and abstracts heat from the substances kept there in the cold chamber. PV diagram of this cycle is shown in Fig. 10.2(*b*).

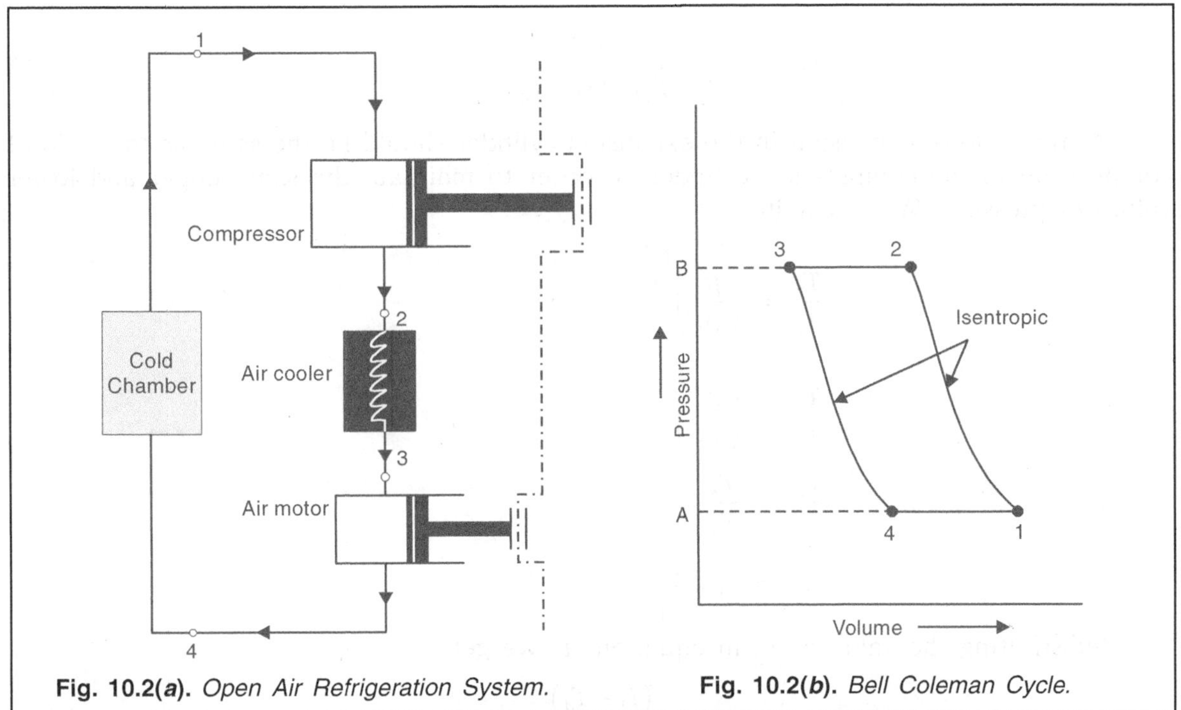

Fig. 10.2(*a*). *Open Air Refrigeration System.* **Fig. 10.2(*b*).** *Bell Coleman Cycle.*

The process A–1 represents the suction of air into the compressor. The process 1–2 shows the isentropic compression of air by the compressor. The process 2–B represents the discharge of high pressure air from the compressor into the air cooler. Due to the cooling of air in the air cooler, volume reduction is from 2–3. Next on the PV diagram, process 3–4 represents the isentropic expansion of air in the expansion cylinder (air motor). The process 4–1 represents the absorption of heat at constant pressure. The air enters the cooling chamber corresponding to state 4.

As shown in the P–V diagram of the cycle, area A–1–2–B–A is the work done on the air in the air compressor and area A–4–3–B–A is the work done by air in the motor and area 1–2–3–4–1 represents work input to the refrigerator.

Let T_1, T_2, T_3 and T_4 be the absolute temperatures at points 1, 2, 3 and 4 respectively.

The heat absorbed from the cold chamber

$$N = C_p\ (T_1 - T_4) \text{ per kg of air}$$

Heat rejected to the cooler

$$= C_p\ (T_2 - T_3) \text{ per kg of air}$$

As we know work required is the difference of heat rejected and heat absorbed

Work required $\quad W = C_p\ (T_2 - T_3) - C_p\ (T_1 - T_4)$ per kg of air

Therefore *COP* of the air refrigeration system is given by

$$COP = \frac{N}{W} = \frac{C_p(T_1 - T_4)}{C_p(T_2 - T_3) - C_p(T_1 - T_4)}$$

$$COP = \frac{(T_1 - T_4)}{(T_2 - T_3) - (T_1 - T_4)} \qquad \text{...}(i)$$

As the ratio of expansion in the expansion cylinder should be the same as the ratio of compression in the compression cylinder, in order to maintain the same upper and lower values of pressure. We can write

$$\frac{T_2}{T_1} = \left(\frac{p_2}{p_1}\right)^{\frac{\gamma-1}{\gamma}}$$

$$\frac{T_3}{T_4} = \left(\frac{p_2}{p_1}\right)^{\frac{\gamma-1}{\gamma}}$$

$$\therefore \quad \frac{T_2}{T_1} = \frac{T_3}{T_4}$$

$$\therefore \quad T_2 = T_1 \times \frac{T_3}{T_4}$$

Substituting the value of T_2 in equation (*i*) we get

$$COP = \frac{(T_1 - T_4)}{\left(T_1 \frac{T_3}{T_4} - T_3\right) - (T_1 - T_4)}$$

$$= \frac{(T_1 - T_4)}{\frac{T_3}{T_4}(T_1 - T_4) - (T_1 - T_4)}$$

$$= \frac{(T_1 - T_4)}{(T_1 - T_4)\left(\frac{T_3}{T_4} - 1\right)}$$

$$= \frac{1}{\left(\frac{T_3}{T_4} - 1\right)}$$

$$COP = \frac{T_4}{(T_3 - T_4)} \qquad \text{...}(ii)$$

The *COP* as expressed by equation (*ii*) is applicable under the following conditions:

(*a*) The compression and expansion takes place isentropically.

(*b*) Pressure loss is not considered during the cycle.

But in real-life practice, the compression and expansion takes place polytropically and there is a pressure loss in the air cooler and cooling chamber.

The *COP* of Bell Coleman cycle (open air refrigeration cycle) in comparison with reversed Carnot cycle is very low and therefore the power consumption is more for a given refrigerating effect.

Disadvantages

(*a*) Main disadvantage of open air refrigeration cycle is the freezing of the moisture in the air. This moisture carried away with air may frost at the end of the expansion process. The frost formation may clog the line through which the working substance flows.

(*b*) As the air is supplied to the cooling chamber (refrigerated space) at atmospheric pressures, volume of air handled by the compressor, cooler and expander is very large which in turn increases the size of these componenets.

10.3 CLOSED AIR REFRIGERATION CYCLE

Before this we have discussed the open air refrigeration cycle. Here we are discussing the closed air refrigeration cycle which is also known as 'dense air refrigeration cycle'. Under this cycle, the circulated working substance *i.e.,* air does not come in contact with the substances kept in cold chamber (refrigerated space). This option of air not making contact with refrigerated space enables us to overcome the limitations of the open air refrigeration system. In this cycle, air is always passed through pipes of various components of the refrigeration system at all times. Here the conditioned air is used to absorb heat from other fluid. Thus the cooled fluid is made to circulate in to the space to be cooled.

10.4 VAPOUR COMPRESSION REFRIGERATION SYSTEM

Vapour compression refrigeration system is the most widely used cycle for refrigerators and air-conditioning system. This cycle uses vapour as the working substance which easily evaporates and condenses. The vapour used does not leave the refrigerated system but keep on circulating again and again through the system alternatively condensing and re-evaporating.

In evaporating, it absorbs heat from the cold body which is used as source of latent heat and gets converted from liquid to vapour. In condensing, it rejects heat to external hot body which is used for cooling it.

Basic Components

The basic components of the vapour compression refrigeration system are:

(*i*) Compressor

(*ii*) Condenser

(*iii*) Expansion valve

(*iv*) Evaporator.

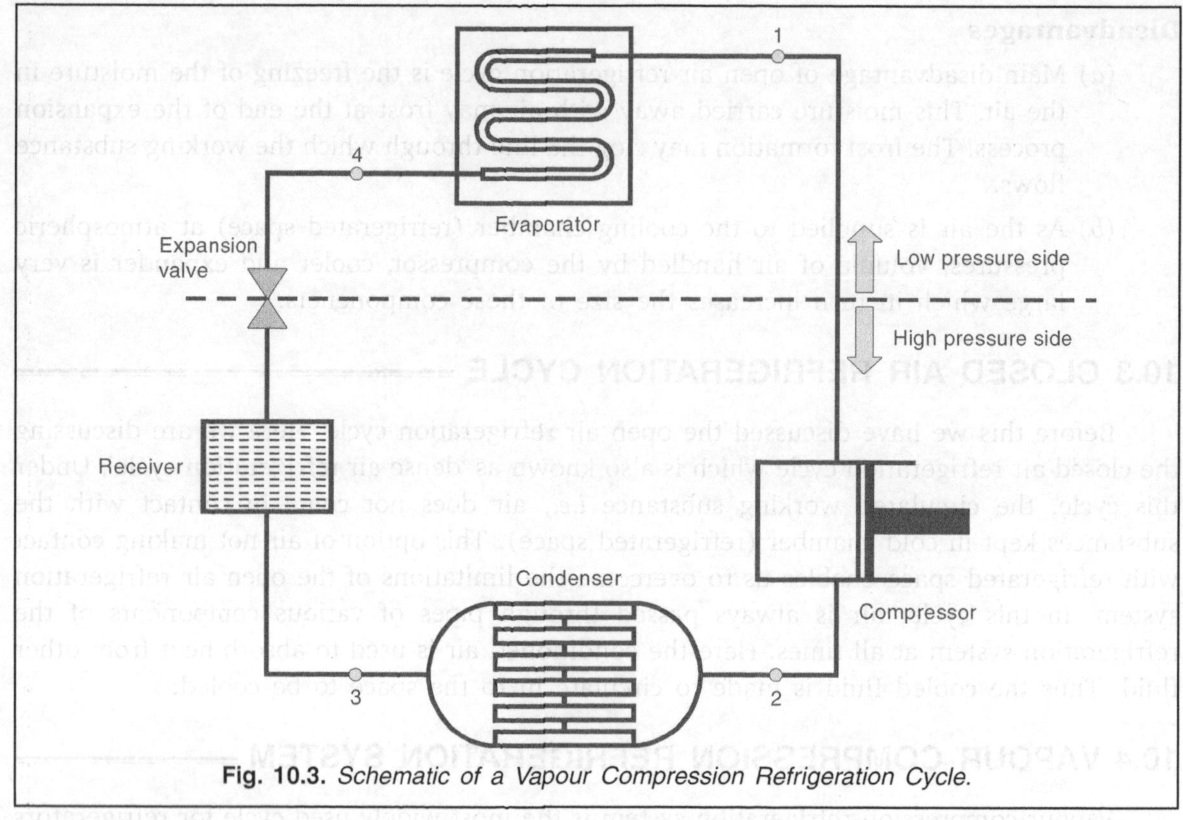

Fig. 10.3. *Schematic of a Vapour Compression Refrigeration Cycle.*

Compressor

The function of a compressor is to receive refrigerant at a particular temperature and pressure and to deliver it after compression at higher temperature and pressure. The compressors used may be of reciprocating type or rotary type depending on the requirements of handling the amount of vapours.

Condenser

The high pressure and high temperature compressed vapour is discharged into the condenser where heat is transferred to the cooling fluid. Water or air is generally preferred as cooling fluid. The vapour cools and then condenses at saturation temperature which corresponds to the pressure in the condenser. For example, the coils behind the domestic refrigerator from where heat is dissipated to kitchen air serve as the condenser. The vapour after condensing is sent to the liquid storage vessel. The high pressure liquid then enters the expansion valve.

Expansion valve

It is a valve with narrow passage through which high pressure liquid expands from high pressure to a low pressure at constant enthalpy. The liquid after expansion becomes low temperature wet vapour and is transferred to the receiver from where it enters the evaporator through a control value.

Evaporator

It is the main component where refrigeration effect takes place. In the evaporator, liquid vapour absorbs heat from the space to be cooled. The evaporator is in the form of a plain, finned or ribbed tube as the case may be through which liquid vapour flows. The evaporator accomplishes the actual purpose of cooling the refrigerated space through the absorption of latent heat. For example, in domestic refrigerators, the coils in the freezer compartment where heat is absorbed by the refrigerant, act as the evaporator. The evaporated vapour is sucked by the compressor from the evaporator and delivered to the condenser. Thus the cycle is completed.

Figure 10.3 shows a schematic of the vapour compression refrigeration system. The vapour compression refrigeration cycle consists of four processes.

- **1–2 Isentropic compression** (takes place in compressor).
- **2–3 Heat rejection at constant pressure** (this takes place in condenser).
- **3–4 Expansion process** (this takes place in an expansion device).
- **4–1 Heat absorption at constant pressure** (this takes place in an evaporator).

Figure 10.4 shows the T–S diagram for ideal vapour compression refrigeration cycle.

At point 1, the fairly dry vapour enters the compressor after leaving the evaporator. This vapour is compressed isentropically from point 1 to point 2. During compression, the pressure and temperature of the vapour increases. The refrigerant vapour then enters the condenser as super heated vapour at point 2 and leaves as saturated liquid at point 3 as a result of heat rejection to the surrounding cooling medium which normally is water. The temperature of vapour at this point is still above the temperature of the surroundings. The high pressure saturated liquid leaves the condenser at point 3 and enters the expansion value at point 3. The expansion of the high pressure saturated liquid takes place through the narrow opening of the expansion valve. During this process which takes place point 3 to point 4 lowers the pressure and temperature of the working substance and at the same time causes it to partly evaporate. The temperature of the vapour drops below the temperature of the refrigerated space during the expansion process 3–4. The low temperature vapour

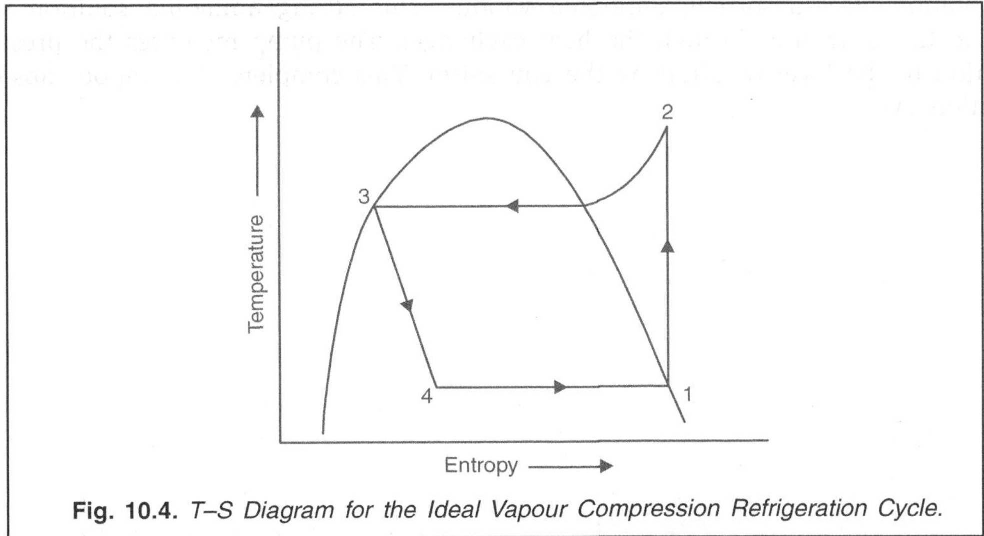

Fig. 10.4. *T–S Diagram for the Ideal Vapour Compression Refrigeration Cycle.*

enters the evaporator at point 4 and it completely evaporates by absorbing heat from the space where refrigeration effect is required. Here the wet vapour is changed in dry vapour and will next passes to compressor at point 1 thus completing the cycle.

10.5 VAPOUR ABSORPTION REFRIGERATION SYSTEM

In 1824, the well known scientist Michael Faraday has discovered the principle of vapour absorption. However, this concept was applied to refrigeration in 1860 by a French scientist Ferninand Carve. The commonly used refrigerant for vapour absorption system is ammonia NH_3. Heat energy is utilized in vapour absorption system whereas mechanical energy is utilized in vapour compression system so as to change the conditions and phases of working substance *i.e.,* refrigerant.

An important limitations of the vapour compression refrigeration system is that it requires a compressor to compress large volume of vapours which in turn requires large power for its operation. In vapour absorption system, a compressor is replaced by an absorber, a pump and a generator. Otherwise the two systems are same as both the systems need a condenser, an expansion valve, and an evaporator. Vapour absorption refrigeration system can be economical for large-scale applications and it does not employ ozone depleting refrigerants.

Figure 10.5 shows a schematic diagram of a vapour absorption refrigeration system. Ammonia vapour is produced in the generator at high pressure from the strong solution of ammonia (NH_3) by using an external heat source. The water vapour carried with ammonia is removed in the rectifier and only dehydrated ammonia gas vapour enters into the condenser. High pressure ammonia vapour is condensed in the condenser. The cooled ammonia solution is passed through an expansion valve and both the pressure and temperature of the refrigerant are reduced below the temperature maintained in the evaporator. The low temperature refrigerant enters the evaporator and absorbs the required heat from the evaporator and leaves the evaporator as saturated vapour. At low pressure, fairly dry vapour leaving the evaporator passes to the absorber where it is dissolved in the weak ammonia solution contained in the absorber. On entering the absorber, this weak ammonia solution becomes strong solution after absorbing ammonia vapour. This strong ammonia solution is then pumped to the generator through the heat exchanger. The pump increases the pressure of the solution to the level required by the condenser. This completes the vapour absorption refrigeration cycle.

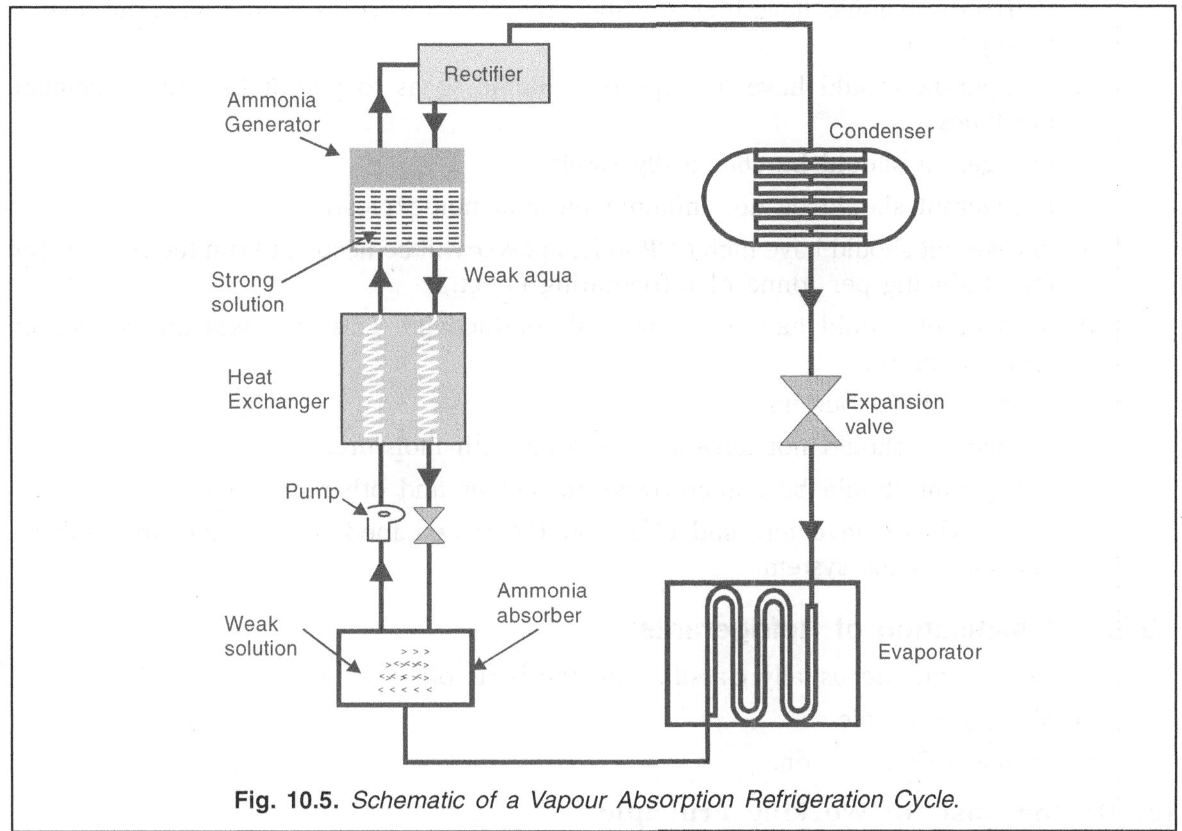

Fig. 10.5. *Schematic of a Vapour Absorption Refrigeration Cycle.*

10.6 REFRIGERANTS

A refrigerant is a working substance used in refrigerating machines that absorbs heat through expansion and releases it through condensation. In last few years the selection and use of refrigerant has become a matter of great concern due to global environmental issues related to the depletion of ozone layer and global warming or greenhouse effect.

10.6.1 Properties of Ideal Refrigerant

An ideal refrigerant should possess the following desirable properties:

1. A refrigerant should have suitable condensing and evaporating pressures. Higher pressure in the refrigerating system requires heavy construction (*i.e.,* pipe and fittings etc.)
2. The refrigerant should have low boiling point at atmospheric pressure to give the desired temperature in the evaporator.
3. Refrigerant should have high latent heat because higher the latent heat, smaller will be the requirement of the refrigerant.
4. Refrigerant should have low specific heat of liquid so as to minimize the amount of vapour formed during throttling in the expansion valve. This enhances the refrigerating effect.

5. Refrigerant should have high specific heat of vapour so as to minimize the degree of superheat.
6. Refrigerant should have low specific volume so as to permit the use of smaller machines.
7. Refrigerant should be chemically stable.
8. Refrigerant should be non-inflammable and non-explosive.
9. Refrigerant should have high *COP* so less power will be needed to run the compressor for producing per tonne of refrigerating effect.
10. Refrigerant should have high thermal conductivity. This increases the condenser and evaporator.
11. It should be odourless.
12. Refrigerant should not have any reaction with moisture.
13. Refrigerant should be non-corrosive to ferrous and other materials.
14. It should not have any bad effect on the stored food items when any leakage develops in the system.

10.6.2 Classification of Refrigerants

The refrigerants are usually classified on the basis of:

(*a*) Working Principle

(*b*) Chemical Composition.

(*a*) On the Basis of Working Principle

On the basis of working principle, refrigerants can be classified in two categories:

(*i*) Primary refrigerants

(*ii*) Secondary refrigerants.

Primary Refrigerants:

The primary refrigerants are those refrigerants which are directly involved in refrigeration system. The primary refrigerant undergoes a cyclic process involving compression, condensation, expansion, and evaporation with change in phase.

Secondary Refrigerants:

The secondary refrigerants are first cooled by primary refrigerants and then are subsequently used as a medium for producing cooling effect. The secondary refrigerants do not undergo any cyclic process. No phase change takes place in these refrigerants and these produce the cooling effect by absorption of sensible heat.

(*b*) On the Basis of Chemical Composition

The refrigerants are classified in four categories. These are:

- Halocarbon Refrigerants
- Azetropes
- hydrocarbon refrigerants
- Inorganic refrigerants

Halocarbon Refrigerants:

These refrigerants contain one or more of three halogens, chlorine, fluorine and bromine. These refrigerants are traded in the market under the brand names of Freon, Genetron, Isotron and Aretron. Halocarbons particularly chlorine based halocarbons (CFCs: Chloro Fluoro Carbon) were in extensive use before 1990s. Their depleting effect on Ozone layer has forced the world community to phase out these refrigerants. The efforts are now being made to replace CFCs by the class of the refrigerants having hydrogen in place of chlorine. Such types of eco-friendly refrigerants are known as hydrofluoro carbons (HFCs) for example, R–134a can replace R–12.

The halocarbon refrigerants have extensive application in domestic, commercial and industrial installations. Some of the important refrigerants belonging to this group are given in Table 10.1.

Table 10.1: Halocarbon Refrigerants

Name of Refrigerants	*Chemical Name*	*Chemical Formula*
R–11	Trichloro mono fluoro methane	CCl_3F
R–12	Dichloro–difluoro methane	CCl_2F_2
R–22	Mono chloro difluoro methane	$CHClF_2$
R–30	Methylene chloride	CH_2Cl_2
R–40	Methyle chloride	CH_3Cl
R–100	Ethyl chloride	C_2H_5Cl
R–113	Trichloro trifluoroethane	$C_2F_3Cl_3$
R–114	Tetrafluoro dichloroethane	$Cl_2F_4Cl_2$
R–152	Difluoro-ethane	$C_2H_6F_2$

Azetropes:

The refrigerants from this group consist of mixtures of different substances. These refrigerants have fixed thermodynamic properties and these do not separate into their constituents by distillation or with the change in temperature and pressure. For example, R–500 is azetrope which contains 73.8% R–12 and 26.2% of R–152.

Hydrocarbon Refrigerants:

The refrigerants of this group are essentially the compounds of carbon and hydrogen. These refrigerants possess satisfactory thermodynamic properties but are highly flammable. The few important refrigerant of this group are shown in Table 10.2.

Table 10.2: Hydrocarbon Refrigerants

Name of Refrigerants	*Chemical Name*	*Chemical Formula*
R–50	Methane	CH_4
R–170	Ethane	C_2H_6
R–290	Propane	C_2H_8
R–600	Butane	C_4H_{10}

Inorganic Refrigerants:

These are non–hydrocarbon refrigerants and are available in nature. Before the advent of hydrocarbon group of refrigerants these refrigerants are used for different applications such as ice plants, aero-refrigeration etc., the important refrigerant of this group are given in Table 10.3.

Table 3: Inorganic Refrigerants

Name of Refrigerants	*Chemical Name*	*Chemical Formula*
R–717	Ammonia	NH_4
R–718	Water	H_2O
R–729	Air	—
R–744	Carbon dioxide	CO_2
R–764	Sulphur dioxide	SO_2

10.6.3 Commonly Used Refrigerants

(A) Halocarbon Compounds

R–11

It is composed of one carbon, three chlorine and one fluorine atoms (or parts by weight).

Properties

- It is non-corrosive, non-toxic and non-flammable.
- It mixes completely with mineral lubricating oil under all conditions.
- Low pressure and high boiling point liquid
- Leakage is detected by halide torch.

Applications

- Commercial air-conditioning systems such as cinema theatres, malls, office buildings.

R–12

Properties

- It is non-toxic, non-flammable, and non-explosive.
- Chemical stable up to 550°C.

- It does not break even under adverse operating conditions.
- It has small specific volume, low working pressure and high critical temperature.
- It is odourless and colourless.
- Well suited for reciprocating and rotary compressors.

Applications

- It is suitable for low, medium and high temperature applications.
- It finds extensive applications in domestic refrigerators, freezers, water coolers etc.
- Due to its excellent electric insulating behaviour it is universally used in sealed type compressors.

R–22

Properties

- Non-toxic, non-flammable, non-corrosive and non-irritating.
- Significant reduction (60% less) in compressor displacement per tonne of refrigeration as compared with R–12.
- Fully miscible with oil at temperatures usually prevailing on the high side of the plant, whereas it is only partly miscible at usual evaporating temperature.
- The biggest problem as compared to R–12 is the high discharge temperature which requires water cooling of the compressor head and cylinder.

Applications

- The range of applications includes small commercial units to medium size industrial plants.

10.7 ECO-FRIENDLY REFRIGERANTS

In last few years there has been a radical change as regards the choice of refrigerants both for the domestic and commercial purposes. The reason for this was the depleting ozone layer in our atmosphere. The emission of chlorofluorocarbons (CFCs), halons and bromide is one of the major factors in the depletion of ozone layer. The Montreal Protocol in 1987 agreed that the production of these chemicals would be phased out by 1995 and alternative fuels will be developed. The signatories' countries of that protocol have ceased the production of these chemicals.

Table 10.4: Application of Refrigerants before 1990s

Typical Application	*Refrigerant Used*
Domestic refrigerators and freezers	R12
Malls, Cinema Theatres	R12, R22, R502
Air-conditioning	R11, R114, R12, R22
Industrial	R717, R22, R502, R13B1

The increasing emphasis being put on global warming issues have forced the users to develop the technology and explore the new eco-friendly refrigerants. The few eco-friendly refrigerants are:

- Ammonia (R717)
- Carbon dioxide (R744)
- Water (R718)
- Air
- Hydrocarbons

The important properties of these eco-friendly refrigerants and their applications are given below.

Ammonia (R717)

Properties

- It has the highest refrigerating effect per kg of refrigerant.
- It has high latent heat and high heat transfer coefficient. Both these properties are needed in a good refrigerant.
- It is highly toxic and flammable.
- Less expensive refrigerant and low running cost.
- Low volumetric displacement due to its low specific volume.
- Low mass needs to be circulated per tonne of refrigeration.
- Easy to detect leakage due to its pungent odour. The leakage can be detected by bringing a sulphur torch near the leakage; dense white smoke comes out if ammonia vapour is present.
- It attacks many non-ferrous metals (copper, zinc, brass etc.) in the presence of little moisture but it has no effect on iron and steel.

Applications

- It is widely used refrigerant in large ice making and cold storage plants.
- It finds extensive use with vapour absorption refrigeration system.

Hydrocarbons

Properties

- These are highly inflammable.
- Negligible impact on ozone layer
- Negligible impact on other global warming causes.
- High *COP*

Applications

These are used in all the fields of refrigeration and air-conditioning viz:

- Domestic refrigeration
- Industrial refrigeration and air-conditioning
- Transport refrigeration

Sulphur Dioxide (R–764)

Properties

- It is non-flammable and non-explosive
- It has very pungent and suffocating odour.
- It works at low pressures and possesses small latent heat of vaporization.
- It has low refrigerating effect and high specific volume. Due to this, large sized compressors at high speeds need to be used for producing the same refrigerant effect as with other refrigerants.
- It is very corrosive when in contact with moisture as it forms sulphuric acid when comes in contact with moisture.

Applications

- It was used in household refrigerators and small commercial units.

Carbon Dioxide (R–744)

Properties

- It has a small specific volume and so it needs small sized compressor for producing relatively large refrigerating effect.
- The high working pressure, low critical temperature and low *COP* and high power consumption has restricted its use.
- It is colourless and odourless. These properties make its leakage difficult to detect. However its leak detection is done by soap solution.
- It is chemically stable as it does not react with oils and metals in the presence of moisture.
- It is nontoxic but will suffocate in high concentration.
- It is non-explosive and non-flammable.

Applications

- It has limited use due to high power consumptions. In old days it was used in marine applications.
- Presently it is used to manufacture dry ice (solid CO_2).

Water R–118

Properties

- It has excellent thermodynamic properties with single exception being its freezing point which is 0°C Celsius.
- It is non-toxic and non-flammable.
- It is available in abundance at no cost.
- It has high heat transfer coefficient which permits smaller dimensions for condensers and evaporators.
- It has low operating pressure which allows the use of light components.
- The specific volume of water at atmospheric temperature is very high and this prevents the use of reciprocating compressors but centrifugal compressors can be used.

Applications

- Water is used as ice to produce cooling effect.
- It is used in steam refrigeration.
- It is used as solvent in vapour refrigeration system.

Air

Properties

- It is available free of cost and is completely safe.
- It has low *COP*.
- It is non-toxic and non-flammable.

Applications

- It finds extensive application in aviation industry.
- It is used in few applications in trucks and buses.

SOLVED PROBLEMS

Problem 1. *A refrigerating system operates on the reversed Carnot cycle. The higher temperature of the refrigerant in the system is 37°C and the lower is –10°C. The capacity of refrigerating machine is 12 tonnes. Neglecting all losses find out (i) the coefficient of performance, (ii) the heat rejected from the system per hour, (iii) power required.*

Solution.

Given:

$$T_1 = 37 + 273 = 310 \text{ K}$$

$$T_2 = -10 + 273 = 263 \text{ K}$$

Refrigerating capacity = 12 tonnes

(*i*) *COP* for a reversed Carnot cycle is given by

$$COP = \frac{T_2}{T_1 - T_2} = \frac{263}{310 - 263}$$

$$= 5.596 \quad \textbf{Ans.}$$

(*ii*) As we know

$$1 \text{ tonne of refrigeration} = 210 \text{ kJ/min}$$

$$\therefore \quad \text{Refrigerating effect} = 12 \times 210 \times 60 = 151200 \text{ kJ/hr}$$

$$COP = \frac{\text{desired cooling effect produced}}{\text{work done in producing the cooling effect}}$$

$$\therefore \quad \text{Work done} = \frac{\text{Refrigerating effect}}{\text{COP}} = \frac{151200}{5.596}$$

$$= 27019.3 \text{ kJ/hr}$$

Heat rejected from the system per hour

$$= \text{Refrigerating effect/hr} + \text{Work done/hr}$$

$$= 151200 + 27019.3$$

$$= 178219.3 \text{ kJ/hr } \textbf{Ans.}$$

(*iii*) Power required $= \dfrac{\text{Work done / hr}}{3600} = \dfrac{27019.3}{3600}$

$$= 7.50 \text{ kJ/s} = 7.5 \text{ kW } \textbf{Ans.}$$

Problem 2. *As shown in Fig. 10.6. E is a heat engine with efficiency of 0.4 and R is a refrigerator. Given that $Q_2 + Q_4 = 3Q_1$. Find out the COP of the refrigerator.*

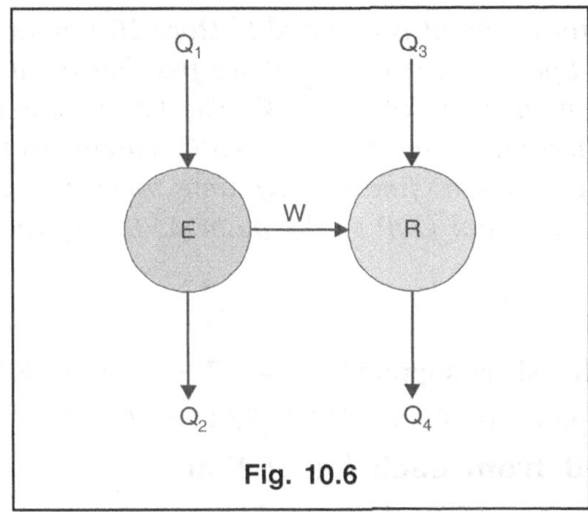

Fig. 10.6

Solution.

As we know that efficiency (or *COP*) of the heat engine is given by

$$\eta = \frac{Q_1 - Q_2}{Q_1} \qquad ...(i)$$

Given $\eta = 0.4$ and $Q_2 + Q_4 = 3Q_1$

$$\therefore \quad 0.4 = \frac{Q_1 - Q_2}{Q_1}$$

$$\therefore \quad 0.4\ Q_1 = Q_1 - Q_2$$

$$\therefore \quad 0.6\ Q_1 = Q_2$$

From Fig. 10.6 we can write for a refrigerator

$$W + Q_3 = Q_4 \qquad ...(ii)$$

and from Fig 10.6, it is clear that $W = Q_1 - Q_2$ therefore substituting this value of W in equation (*ii*) we get

$$Q_1 - Q_2 + Q_3 = Q_4$$

$$\therefore \quad Q_1 + Q_3 = Q_2 + Q_4$$

$\therefore \qquad Q_1 + Q_3 = 3Q_1$

$\therefore \qquad Q_3 = 2Q_1$

Now *COP* of a refrigerator is given by

$$COP_{Ref} = \frac{Q_3}{W} = \frac{Q_3}{Q_1 - Q_2}$$

$$= \frac{2Q_1}{Q_1 - 0.6Q_1} = \frac{2Q_1}{0.4Q_1}$$

$$= 5$$

$\therefore \qquad COP_{Ref} = 5$ **Ans.**

Problem 3. *A cold storage plant is required to store 10 tonnes of fish. The fish is supplied at a temperature of 27 °C. Specific heat of fish above freezing point is 2.95 kJ/kg/°C. Specific heat of fish below freezing point is 1.30 kJ/kg/°C. The fish is stored in cold storage which is maintained at –10°C. The freezing point of fish is –5°C. Latent heat of fish is 235.1 kJ/kg. If the plant requires 75 kW to drive it, find out (i) time taken to achieve the cooling, (ii) the capacity of the plant. Assume actual COP of the plant as 0.4 of the Carnot COP.*

Solution.

Given:

Temperature at which fish is supplied $T_1 = 27 + 273 = 300$ K

Fish is stored at temperature $T_2 = -10 + 273 = 263$ K

(*i*) **Heat removed from each kg of fish**

$$= 2.95\,[27 - (-5)] + 235.1 + 1.30\,[-5 - (-10)]$$

$$= 336 \text{ kJ/kg}$$

$\therefore$ Heat removed from 10 tonnes of fish

$$= 10 \times 1000 \times 336$$

$$= 3360000 \text{ kJ}$$

Theoretical *COP* is given by

$$= \frac{T_2}{T_1 - T_2} = \frac{263}{300 - 263}$$

$$= 7.108$$

$$\text{Actual } COP = 0.4 \times 7.108 = 2.843$$

$$\text{Work done per minute} = 75 \times 60 = 4500 \text{ kJ/min}$$

$\therefore$ Heat removed by the plant $= 4500 \times 2.843$

$$= 12793.5 \text{ kJ/min}$$

$$\text{Time required to achieve cooling} = \frac{3360000}{12793.5}$$

$$= 262.63 \text{ min} \quad \textbf{Ans.}$$

(*ii*) **Capacity of plant**

As we know that one tonne of refrigeration = 210 kJ/min

$$\therefore \quad \text{Capacity of the plant} = \frac{12793.5}{210}$$

$$= 60.92 \text{ tonnes of refrigeration} \quad \textbf{Ans.}$$

Problem 4. *A refrigeration system has working temperature of –36°C and 47°C. What is the maximum COP possible? Determine the actual refrigerating effect produced per kW hour if the actual COP is 75% of the maximum.*

Solution.

As mentioned in the problem

Higher or the source temperature is $T_1 = 47 + 273 = 320$ K

Lower or the sink temperature is $T_2 = -36 + 273 = 237$ K

Theoretical or maximum *COP* is given by

$$= \frac{T_2}{T_1 - T_2} = \frac{237}{320 - 237}$$

$$= 2.855$$

and $\quad$ Actual $COP = 0.75 \times 2.855 = 2.141$

As we know that one tonne of refrigeration = 210 kJ/min

Actual refrigerating effect produced per kW hour = $COP_{\text{actual}} \times W_{\text{actual}}$

$$= 60 \times 2.141$$

$$= 128.5 \text{ kJ/min}$$

$$= \frac{128.5}{210}$$

$$= 0.612 \text{ tonne of refrigeration} \quad \textbf{Ans.}$$

Problem 5. *A refrigerator operating between –6°C and 30°C has the capacity of 900 tonnes. Find out the mass of ice produced per day at 0°C from water at 27°C. Also find the power required to drive the unit. Assume that the cycle operates on reversed Carnot cycle and latent heat for ice formation from water may be taken as 335 kJ/kg. Take specific heat of water as 4.187 kJ/kg K.*

Solution

As we know that

$$\text{One tonne of refrigeration} = 3.517 \text{ kJ/s}$$

$$= 210 \text{ kJ/min}$$

∴ Refrigeration capacity or Capacity of the plant to extract heat is equal to

$$= 900 \times 210 \text{ kJ/min}$$

$$= 189000 \text{ kJ/min}$$

Heat removed from water to produce one kg of ice at 0°C from water at 27°C

$$= (27-0) \times 4.187 + 335$$

$$= 448.05 \text{ kJ/kg}$$

Let m be the rate of ice formation in kg/min

$$\text{Rate of ice formation} = \frac{\text{Refrigeration capacity}}{\text{heat removed in kJ per kg of ice}}$$

$$= \frac{189000}{448.05}$$

$$= 421.83 \text{ kg/min}$$

$$\text{Mass of ice produced per day} = 421.83 \times 60 \times 24$$

$$= 607435.2 \text{ kg}$$

$$= 607.43 \text{ tonnes} \quad \textbf{Ans.}$$

As we know *COP* is expressed by

$$COP_{REF} = \frac{T_2}{T_1 - T_2}$$

In this problem

Higher or the source temperature is $T_1 = 30 + 273 = 303$ K

Lower or the sink temperature is $T_2 = -6 + 273 = 267$ K

$$COP_{Ref} = \frac{267}{303 - 267}$$

$$= 7.416$$

As $$COP_{Ref} = \frac{\text{Refrigeration capacity}}{\text{Work done}}$$

$$\therefore \quad \text{Work done} = \frac{\text{Refrigeration capacity}}{COP_{Ref}}$$

$$\therefore \quad \text{Work done} = \frac{189000}{7.416}$$

$$= 25485.44 \text{ kJ/min}$$

$$\therefore \quad \text{Power required to drive the unit} = \frac{25485.44}{60}$$

$$= 424.76 \text{ kW} \quad \textbf{Ans.}$$

Problem 6. *In a winter season, a heat pump is used to maintain the inside temperature of a home at 25°C. COP of the heat pump is 3. On a particular day when the ambient temperature is 5°C, the heat losses from the home is 21600 kJ/hr. find out the (i) power consumed by the heat pump and (ii) rate of heat absorption from outside.*

Solution.

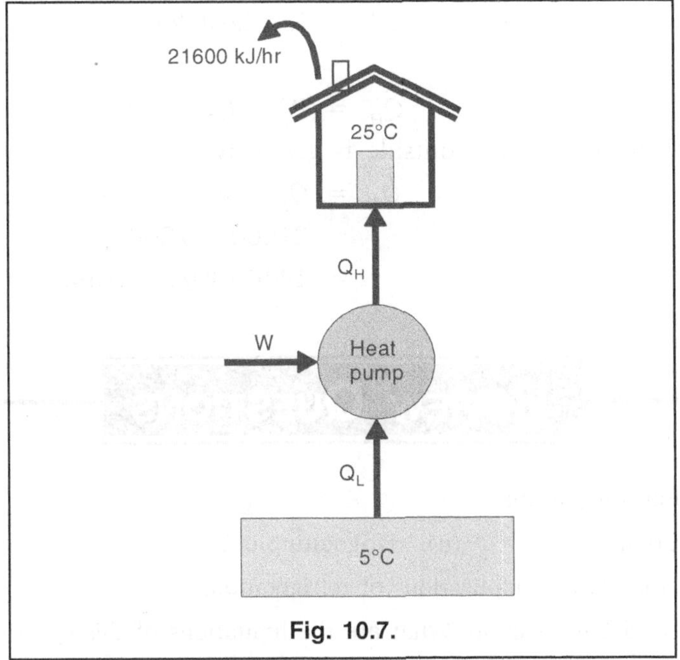

Fig. 10.7.

As we know that *COP* of a heat pump is given by

$$COP_{Pump} = \frac{\text{heating effect produced}}{\text{work input}}$$

Since the pump is being used to maintain the home at temperature of 25°C. It means that rate of heat gained by the home is equal to the heat loss by the home.

Therefore,

$$Q_H = 21600 \text{ kJ/hr}$$

∴ Work input or power consumed by the heat pump is

$$\text{Work input} = \frac{\text{heating effect produced}}{\text{COP}_{Pump}}$$

$$= \frac{Q_H}{COP_{Pump}}$$

$$= \frac{21600}{3}$$

$$= 7200 \text{ kJ/hr}$$

Since the power consumed is usually expressed in Watt, so for converting the above value in kJ/s or watt, divide it by 3600.

$$\text{Power consumed} = \frac{7200}{3600} = 2\text{kJ/s}$$

$$= 2 \text{ kW}$$ **Ans.**

(*ii*) Now using the principle of energy conservation we can write

$$\text{Net heat gained} = \text{Net heat lost}$$

In this case

$$Q_H = W + Q_1$$

∴ Rate of heat absorbed from outside is given by

$$Q_1 = Q_H - W$$
$$= 21600 - 7200$$
$$= 14400 \text{ kJ/hr} \quad \textbf{Ans.}$$

Review Questions

1. Define the following terms:
   (*i*) refrigeration (*ii*) refrigerating effect
2. Write down the various applications of refrigeration.
3. Explain reversed Carnot cycle. What are the limitations of this cycle?
4. Distinguish between the open air and closed or dense air refrigeration cycles.
5. Define unit of refrigeration.
6. Explain the vapour compression cycle with the help of a neat sketch.
7. Classify the refrigerants and write down the desirable properties of a refrigerant.
8. Write down a comprehensive note on eco-friendly refrigerants.
9. A refrigeration system has working temperature of 250 K and 310 K. What is the maximum *COP* possible? Determine the actual refrigerating effect produced per kW hour if the actual *COP* is 75% of the maximum. **[4.166, 0.893 tonne of refrigeration]**
10. A refrigerator operating on reversed Carnot cycle extracts 600 kJ/min heat from a refrigerated space being maintained at –15°C and rejects heat to atmosphere at 27°C. Determine the work input required to run the refrigerator. **[97.67 kJ/min]**
11. In a Bell-Coleman refrigerating machine, the air is drawn in from the cold chamber at 267 K and atmospheric pressure of 102 kPa, and after isentropic compression to 510 kPa the air at this constant pressure is cooled to 288 K. The air is then expanded isentropically to atmospheric pressure and discharged into the cold chamber. Find out (*i*) the coefficient of performance, (*ii*) the heat abstracted from the cold chamber per kg of air. Take C_p = 1.005 kJ/kg K and C_v = 0.712 kJ/kg K. **[1.7, 84.8 kJ/kg]**
12. An air refrigeration system is working between 2.7 MPa and 270 kPa for producing a cooling effect of 5000 kJ/min, the temperature of air leaving the cold chamber is –5°C and at leaving the cooler is 27°C. Neglecting losses and clearance in the compressor and expander, find out (*i*) *COP*, (*ii*) mass of air circulated per hour (*iii*) power required to drive the machine.

Objective Type Questions

Select the most appropriate answer:

1. The *COP* of a refrigerator on a reversed Carnot cycle is 5. The ratio of higher absolute temperature to the lower temperature *i.e.*, $\frac{T_2}{T_1}$ is

(A) 1.25 (B) 1.3

(C) 1.4 (D) 1.2

2. The difference between vapour compression and vapour absorption cycle is that

(A) more work input is required in absorption cycle

(B) less work input is required in absorption cycle

(C) a pump is required in vapour compression cycle

(D) there is no difference between these two cycles

3. Air refrigeration cycle is used in

(A) Commercial refrigerators (B) Domestic refrigerators

(C) Gas liquefaction (D) Air conditioning

4. The refrigerant which is used for refrigeration purpose in aeroplanes is ....

(A) CO_2 (B) Freon–12

(C) Ammonia (D) Air

5. The *COP* of a domestic refrigerator

(A) is greater than one (B) is less than one

(C) is equal to one (D) any value

(E) depends upon the size

6. The capacity of a refrigerator is one tonne. It means that the heat removing capacity of the refrigerator is:

(A) 1000 kcal/hr (B) 2000 kcal/hr

(C) 3000 kcal/hr (D) 10000 kcal/hr

7. One tonne of refrigeration corresponds to

(A) 100 kcal/min (B) 50 kcal/min

(C) 80 kcal/min (D) 50 kcal/hr

8. Match List A with List B and select the correct answer using the codes given below the lists.

List–A	List–B
A. Bell Coleman refrigeration	1. Compressor
B. Vapour compression refrigeration	2. Generator
C. Absorption Refrigeration	3. Flash chamber
D. jet Refrigeration	4. Expansion Cylinder

(A) A1–B4–C3–D2 (B) A4–B1–C3–D2

(C) A1–B4–C2–D3 (D) A4–B1–C2–D3

9. The desirable combination of properties for a refrigerant include
(A) High specific heat and low specific volume
(B) High heat transfer coefficient and low latent heat
(C) High thermal conductivity and low freezing point
(D) High specific heat and low boiling point

10. The refrigerating effect means
(A) the heat rejected by the system per hour
(B) the work done on the system
(C) heat absorbed from the space or the body to be cooled
(D) heating the particular space

11. The refrigerant for a refrigerator should have
(A) high total heat (B) high sensible heat
(C) low sensible heat (D) high latent heat
(E) low latent heat

12. Refrigerating system used for most of the domestic refrigerators is:
(A) Vapour compression refrigeration system
(B) Vapour absorption refrigeration system
(C) Electrolux refrigerator
(D) Air refrigeration system

13. One ton of the refrigeration is
(A) The refrigerating effect to freeze one ton of water at 0°C into the ice at 0°C in 24 hours
(B) The refrigerating effect to produce one ton of ice at NTP conditions.
(C) The cooling effect produced by melting one ton of ice
(D) None of the above

14. The function of a compressor in a vapour absorption system is performed by the
(A) absorber (B) absorber and pump
(C) pump and generator (D) absorber, pump and generator

15. The significant advantage of using ammonia as a refrigerant is its
(A) characteristic odour (B) high latent heat
(C) solubility in water (D) inflammability

16. Choose the wrong statement
(A) a refrigerant should have high latent heat
(B) a refrigerant should have positive evaporating pressure
(C) a refrigerant should be cheap
(D) a refrigerant should have low thermal conductivity

17. Air refrigeration system operates on
(A) reversed Carnot cycle (B) reversed Brayton cycle
(C) reversed Otto cycle (D) none of the above

18. In a refrigerator, the frost on cooling coils
(A) reduces heat transfer (B) increases heat transfer
(C) reduces power consumption (D) protects the coils

19. In SI units, one tonne of refrigeration is equal to

(A) 21 kJ/min (B) 420 kJ/min

(C) 210 kJ/min (D) 300 kJ/min

20. In a refrigeration cycle, the flow of refrigerant is controlled by

(A) compressor (B) condenser

(C) evaporator (D) expansion valve

(E) all of the above

ANSWERS

1. (D)	**2.** (B)	**3.** (C)	**4.** (D)	**5.** (A)	**6.** (C)
7. (B)	**8.** (D)	**9.** (D)	**10.** (C)	**11.** (D)	**12.** (B)
13. (A)	**14.** (D)	**15.** (B)	**16.** (B)	**17.** (B)	**18.** (A)
19. (C)	**20.** (D).				

Air Standard Cycles

11.1 INTRODUCTION

An internal combustion engine does not work on a thermodynamic cycle because the working fluid does not go through a complete cycle in the engine even though the engine operates in a mechanical cycle. An IC engines takes a mixture of air and fuel or air and fuel separately but it does not return to its original state after completion of cycle.

The cycles encountered in actual devices such as IC engines are difficult to analyze because of the presence of many complicating factors such as friction, and the lack of time for achieving equilibrium conditions during the cycle. In order to analyze the internal combustion engines feasibly, idealized cycle known as air standard cycle is used. The conclusions derived from the analysis of ideal cycles are also applicable to actual cycles.

The theoretical thermal efficiency of the ideal cycle is known as the air standard efficiency, since it is worked out on the basis of the working fluid being air throughout the cycle. The use of air eliminates the effect of the calorific value of fuel.

The analysis of all air standard cycles is based upon the following assumptions:

Assumptions

1. The working fluid is air throughout the cycle and follows the perfect gas laws $pV = mRT$.
2. The mass of air in the cycle remains fixed *i.e.,* the engine operates as a closed cycle so that working fluid is restored to its initial state at the end of each cycle. Thus there is no inlet process or exhaust process.
3. The combustion process is replaced by a heat transfer process from an external source.
4. The cycle is completed by heat transfer to the surroundings.
5. All processes are internally reversible, and compression and expansion processes are reversible adiabatic processes.
6. The specific heats of air C_p, C_v and their ratio have constant values.

11.2 TERMINOLOGY USED IN THERMODYNAMIC CYCLES

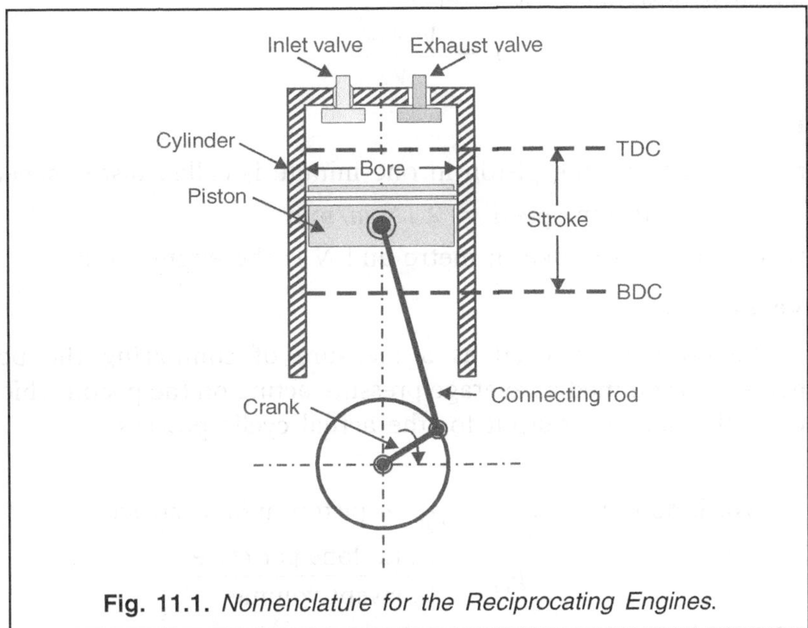

Fig. 11.1. *Nomenclature for the Reciprocating Engines.*

Cylinder Base or Bore (*d*): The nominal inside diameter of the cylinder is called cylinder bore or simply bore. It can be designated by letter '*d*'.

Stroke (L): The nominal distance through which a working piston moves between two successive reversals of its direction of motion is called the stroke. It is equal to twice the radius of crank. It can be designated by letter 'L'.

Piston area (A): The area of a circle of diameter equal to the cylinder bore is called the piston area. It can be designated by letter 'A'.

Top Dead Centre (TDC): The extreme position of the piston at the top side of the cylinder head is known as Top Dead Centre (TDC). In this case, the piston is farthest to the crankshaft. In case of horizontal engine this position of the piston is called Inner Dead Centre (IDC).

Bottom Dead Centre (BDC): The extreme position of the piston at the bottom side of the cylinder head is known as Bottom Dead Centre (BDC). In this case, the piston is nearest to the crankshaft. In case of horizontal engine this position of the piston is called Outer Dead Centre (ODC).

Displacement or Swept Volume (V_s): The volume swept by the piston during one stroke *i.e.,* while moving from top dead centre to bottom dead centre is called swept volume.

Clearance Volume (V_c): The nominal volume of the combustion chamber above the piston when it is at the top dead centre is known as clearance volume. It can be designated by letter V_c.

Volume of the Cylinder: It is the sum of the clearance volume and swept volume.

$$\therefore \qquad \text{Cylinder volume } V = V_c + V_s$$

Compression Ratio (*r*): It is the ratio of the total cylinder volume to that of the clearance volume. It is designated by letter '*r*'.

$$r = \frac{V_c + V_s}{V_c} \quad ...(11.1)$$

Piston Speed

The distance travelled by the piston in one minute is called piston speed.

$$\text{Piston speed} = 2\,LN\ m/\text{min} \quad ...(11.2)$$

Where L is the length of stroke in metre and N is the engine rpm.

Mean Effective Pressure

Mean effective pressure is used as a measure of comparing the performance of reciprocating engines. It is defined as average pressure acting on the piston which will produce the same output as the network output for the actual cyclic process.

Thus

$$\text{Work done per cycle} = p_m \times \text{piston area} \times stroke$$

and

$$p_m = \frac{\text{work done per cycle}}{\text{swept volume}}$$

$$\text{Work done per minute} = P_m \times A \times L \times n$$

Where n is the number of cycles per minute.

An engine having large p_m will produce more power at the same rated speed.

Mean effective pressure is the mean height of the pressure-volume diagram as shown in Fig. 11.2.

Mean effective pressure is defined as

$$m.e.p. = \frac{\text{area of the diagram}}{\text{length of the diagram}} \quad ...(11.3)$$

Mean effective pressure is usually expressed in bar or kPa.

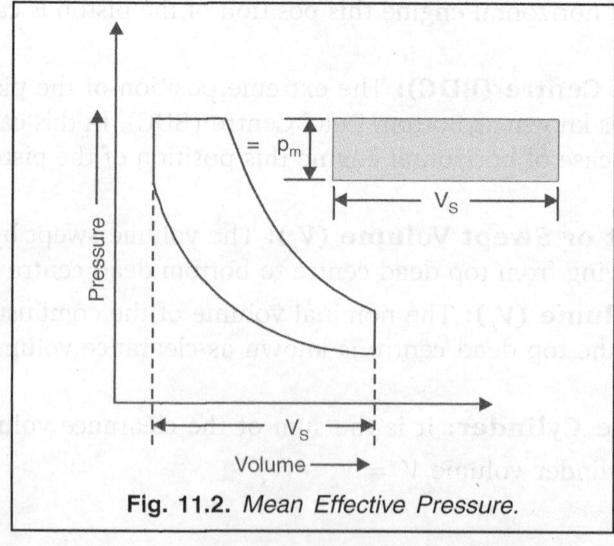

Fig. 11.2. *Mean Effective Pressure.*

Work Ratio

In any cycle, both positive and negative work transfers take place and the network transfer is the difference of these two work transfers. The ratio of the network transfer to the positive work transfer is called the work ratio.

$$\text{Work ratio} = \frac{\text{Network transfer}}{\text{positive work transfer}} \qquad ...(11.4)$$

11.3 CARNOT CYCLE

In 1824, a French engineer Sadi Carnot first proposed an engine using perfect gas undergoing a cyclic process and consisting of reversible processes. An engine based on the working of a Carnot cycle is termed as Carnot engine. A Carnot engine is a hypothetical device and is not practical to construct as it consists of reversible processes which proceeds at a very slow piston speed.

Carnot engine is used as a yardstick to measure the performance of any other engine. If the direction of Carnot cycle is reversed than it can be used for refrigeration/heat pump.

The Carnot cycle consists of four reversible processes which include two isothermal and two adiabatic processes. The P–V and T–S diagram of the Carnot cycle is shown in Fig. 11.3.

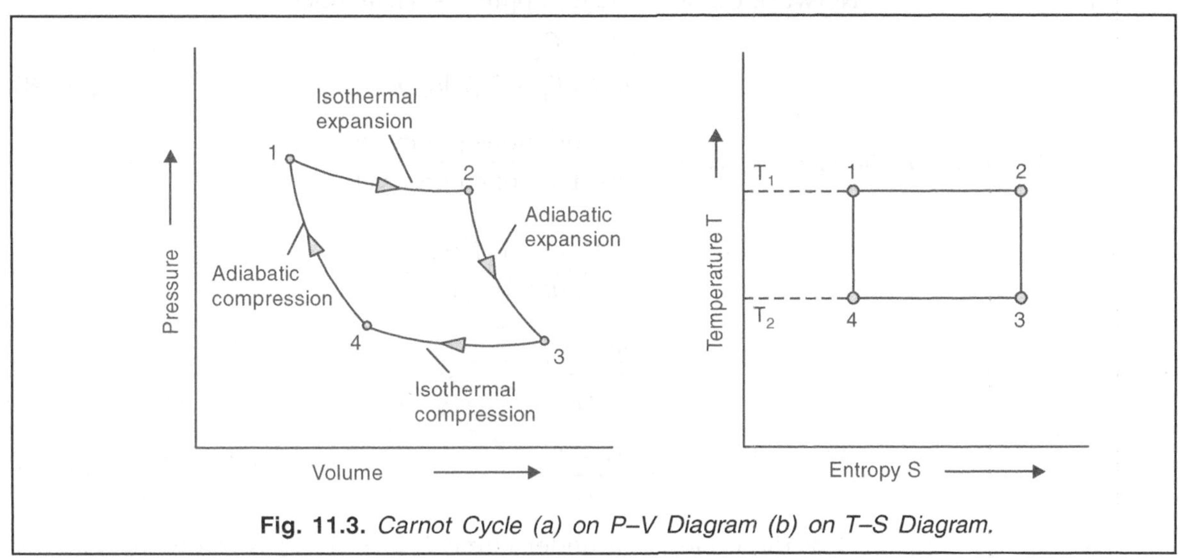

Fig. 11.3. *Carnot Cycle (a) on P–V Diagram (b) on T–S Diagram.*

The four reversible processes are as follows:

Process 1–2 represents isothermal expansion. Heat is supplied at constant temperature T_1 during this process and it is equal to work done during the process.

$$\text{Heat energy supplied } Q_1 = p_1V_1 \log_e \frac{V_2}{V_1}$$

$$= mRT_1 \log_e r$$

Where, $$r = \frac{V_2}{V_1} \quad \text{...(11.5)}$$

Process 2–3 is an adiabatic expansion process. The working fluid *i.e.,* air expands adiabatically until its temperature drops from T_1 to T_2.

Work done during this process is $\frac{p_2V_2 - p_3V_3}{\gamma - 1}$ ...(11.6)

Process 3–4: During the process 3–4 reversible isothermal heat rejection at constant temperature T_2. The working fluid is slowly compressed up to point 4. The heat is rejected by the air to the sink.

$$\text{Heat rejected } Q_2 = p_3V_3 \log_e \frac{V_3}{V_4} \text{ as we know } r = \frac{V_3}{V_4} = \frac{V_2}{V_1}$$

$$\therefore \quad Q_2 = mRT_2 \log_e r \quad \text{...(11.7)}$$

Process 4–1: The system returns back to the initial state by adiabatic compression. The temperature of air rises from T_2 to T_1 and work is done on the air.

$$\therefore \text{ Work done during compression} = \frac{p_4V_4 - p_1V_1}{\gamma - 1}$$

$$\begin{aligned} \therefore \quad \text{Network done} &= \text{Heat supplied} - \text{Heat rejected} \\ &= Q_1 - Q_2 \\ &= mR\,(T_1 - T_2) \log_e r \quad \text{...(11.8)} \end{aligned}$$

$$\therefore \quad \text{Efficiency of the Carnot cycle} = \frac{\text{Work done per cycle}}{\text{heat supplied per cycle}}$$

$$\eta = \frac{mR\,(T_1 - T_2) \log_e r}{mRT_1 \log_e r}$$

$$= \frac{T_1 - T_2}{T_1}$$

$$\eta = \frac{T_1 - T_2}{T_1} \quad \text{...(11.9)}$$

As shown in Fig. 11.3 of the cycle, area under curve 1–2–3 (P–V) is the work done by the gas during the expansion part of the cycle and the area under curve 3–4–1 is the work done on the gas during the compression part of the cycle. The area 1–2–3–4–1 represents the network done during the Carnot cycle.

11.4 OTTO CYCLE (CONSTANT VOLUME CYCLE)

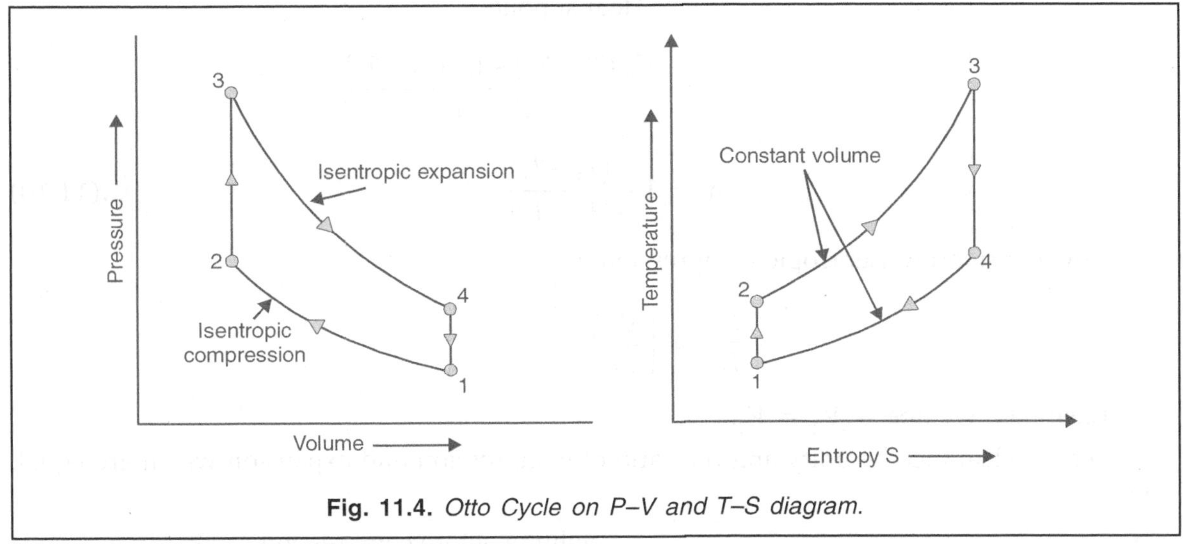

Fig. 11.4. *Otto Cycle on P–V and T–S diagram.*

This cycle was first proposed by Beau De Rochas in 1862. Dr. A.N. Otto in 1876 has successfully applied this cycle to an internal combustion engine. The Otto cycle consists of two constant volume and two adiabatic processes. The ideal P–V and T–S diagrams of this cycle are shown in Fig. 11.4.

Consider one kg of air in the engine cylinder at point 1. The sequence of operation followed by the cycle is given below:

Process 1–2: At point 1, the piston is at the beginning of in stroke. The air is compressed isentropically to point 2 during which the piston moves from bottom dead centre to top dead centre.

Process 2–3: During this process the piston is at rest at TDC. Heat is now added at constant volume from an external hot source up to point 3. At point 3, the hot source is then withdrawn. In real engine, the heat supplied is achieved by the combustion of the air-fuel mixture in the cylinder by spark.

Process 3–4: During this process, the air expands isentropically to point 4 and work is done on the piston.

Process 4–1: During this process, heat is now rejected at constant volume to an external cold source. This cold source is then removed after the air is brought to its initial conditions and thus the cycle is completed.

The heat is supplied during constant volume process (2–3) and rejected during constant volume process 4–1. There is no exchange of heat during the two isentropic processes (1–2) and (3–4).

$$\text{Heat supplied during the process 2–3} = C_v\,(T_3 - T_2)$$

$$\text{Heat rejected during the process 4–1} = C_v\,(T_4 - T_1)$$

$$\text{Net work done} = \text{heat supplied} - \text{heat rejected}$$

$$= C_v\,(T_3 - T_2) - Cv\,(T_4 - T_1)$$

$$\text{Efficiency, } \eta = \frac{\text{Net work done}}{\text{Heat supplied}}$$

$$\eta = \frac{C_v (T_3 - T_2) - C_v (T_4 - T_1)}{C_v (T_3 - T_2)}$$

$$\eta = 1 - \frac{(T_4 - T_1)}{(T_3 - T_2)} \quad \text{...(11.10)}$$

Now considering isentropic compression 1–2,

$$\frac{T_2}{T_1} = \left(\frac{V_1}{V_2}\right)^{\gamma - 1}$$

Clearance volume $= V_2 = V_3$.

Let the clearance be unity and the ratio of compression and expansion which are equal, be r.

$$\therefore \quad \text{Compression ratio} = r = \frac{\text{volume before compression}}{\text{volume after compression}} = \frac{V_1}{V_2}$$

$$\therefore \quad \frac{T_2}{T_1} = r^{\gamma-1}$$

$$T_2 = T_1 \times r^{\gamma-1} \quad \text{...(11.11)}$$

Again considering isentropic expansion (3–4)

$$\frac{T_3}{T_4} = \left(\frac{V_4}{V_3}\right)^{\gamma-1} = \left(\frac{V_1}{V_2}\right)^{\gamma-1} \quad \text{as } V_4 = V_1 \text{ and } V_3 = V_2$$

$$\therefore \quad \frac{T_3}{T_4} = r^{\gamma-1}$$

$$T_3 = T_4 \times r^{\gamma-1} \quad \text{...(11.12)}$$

Substituting the value of T_2 and T_3 in equation (11.10), we get

$$\text{Air standard efficiency} = \eta = 1 - \frac{(T_4 - T_1)}{(T_3 - T_2)}$$

$$\eta = 1 - \frac{T_4 - T_1}{T_4 \times r^{\gamma-1} - T_1 \times r^{\gamma-1}}$$

$$\eta = 1 - \frac{1}{r^{\gamma-1}} \left\{\frac{(T_4 - T_1)}{(T_4 - T_1)}\right\}$$

$$\eta = 1 - \frac{1}{r^{\gamma-1}} \quad \text{...(11.13)}$$

From equation (11.13) it is cleat that the air standard efficiency of IC engines working on Otto cycle is a function of the compression ratio r only and thus efficiency increases as the compression ratio increases.

In real engines, the compression ratio varies from 5 to 8 and depends on the quality of fuel. Figure 11.5 shows the plot between the efficiency of the cycle and compression ratio. In the beginning the efficiency increases with increasing the compression ratio and after that it starts slowing down.

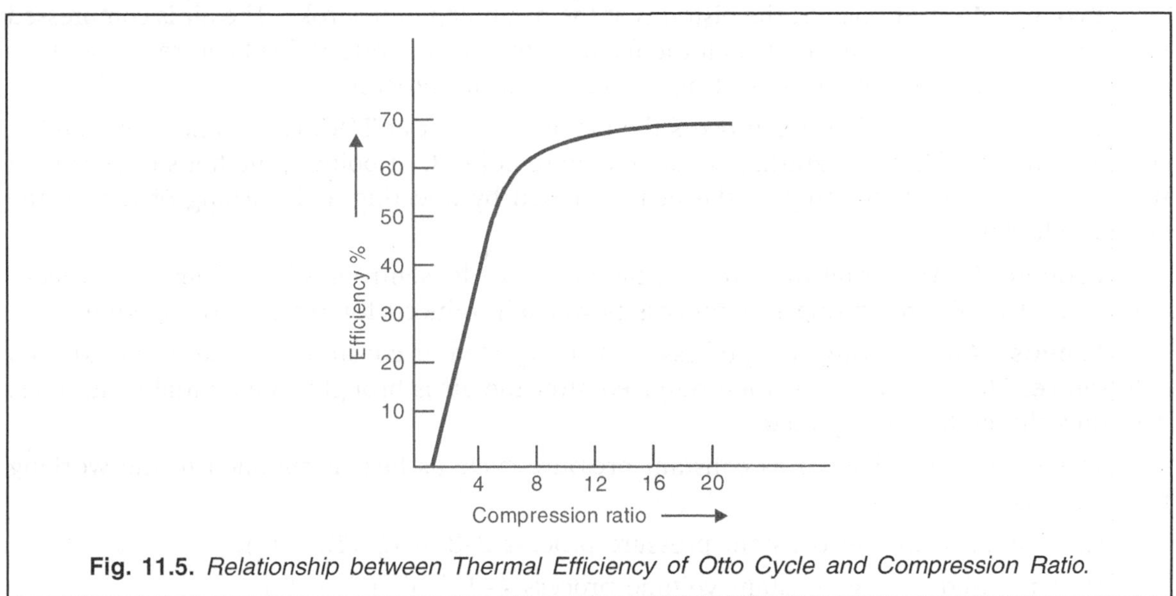

Fig. 11.5. *Relationship between Thermal Efficiency of Otto Cycle and Compression Ratio.*

11.5 DIESEL CYCLE (CONSTANT PRESSURE CYCLE)

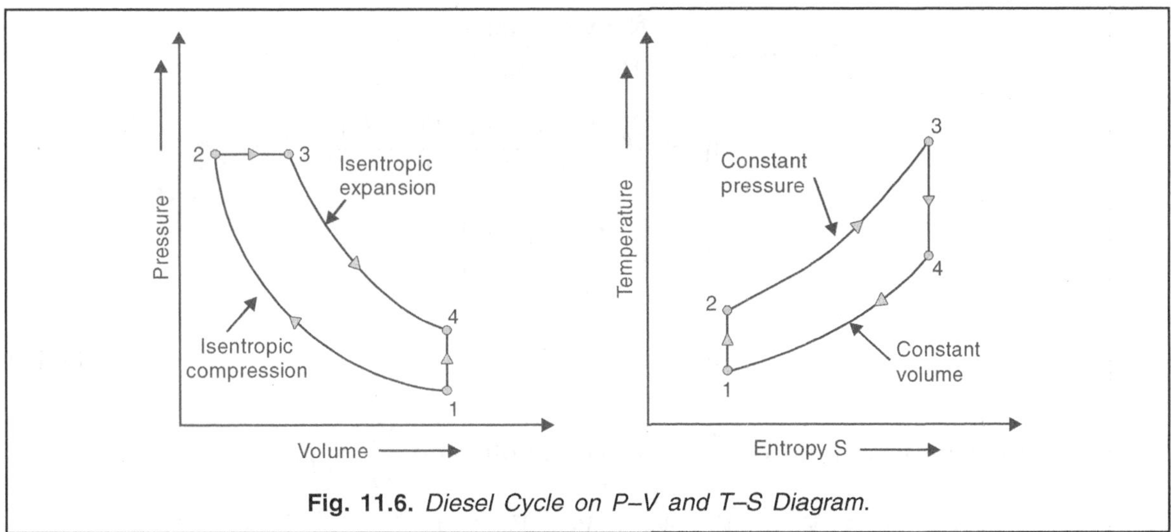

Fig. 11.6. *Diesel Cycle on P–V and T–S Diagram.*

This cycle was introduced by a German engineer Rudolph Diesel in 1893 and presently used in all Diesel (compression ignition) engines. The diesel cycle consists of a constant pressure, a constant volume and two isentropic processes. The ideal P–V and T–S diagrams of Diesel cycle are shown in Fig. 11.6.

In this cycle heat is supplied during constant pressure process 2–3, heat is rejected during the constant volume process 4–1, Isentropic compression takes place along process 1–2 and isentropic expansion by process 3–4.

Consider one kg of air in the engine cylinder at point 1. The sequence of operation followed by the cycle is given below:

Process 1–2: At point 1, the piston is at the beginning of in stroke. The air is compressed isentropically to point 2 during which the piston moves from bottom dead centre to top dead centre. Pressure and temperature of air increases during process 1–2.

Process 2–3: During this process the piston is at rest at TDC. Heat is now supplied at constant pressure from an external hot source up to point 3. At point 3, the hot source is then withdrawn. In real Diesel engine, the heat is added by injecting and burning of fuel in the engine cylinder.

Process 3–4: During this process, the air expands isentropically to point 4 and work is done on the piston. During this process power is produced for running the engine.

Process 4–1: During this process, heat is rejected at constant volume to an external cold source. This cold source is then removed after the air is brought to its initial conditions. And thus the cycle is completed.

This cycle is also known as constant pressure cycle as heat is supplied to the working fluid at constant pressure.

Heat supplied during constant pressure process 2–3 $= C_p\ (T_3 - T_2)$

Heat rejected during constant volume process 4–1 $= C_v\ (T_4 - T_1)$

Net work done by the working fluid (*i.e.*, air) = heat supplied – heat rejected

$$\text{Network done} = C_p\ (T_3 - T_2) - C_v\ (T_4 - T_1)$$

$$\text{Efficiency, } \eta = \frac{\text{Net work done}}{\text{Heat supplied}}$$

$$\eta = \frac{C_p\,(T_3 - T_2) - C_v\,(T_4 - T_1)}{C_p\,(T_3 - T_2)}$$

$$\eta = 1 - \frac{C_v}{C_p}\frac{(T_4 - T_1)}{(T_3 - T_2)}$$

$$\eta = 1 - \frac{1}{\gamma}\left(\frac{T_4 - T_1}{T_3 - T_2}\right) \quad \text{...(11.14)}$$

Let the compression ratio be $\dfrac{V_1}{V_2} = r$ and cut-off ratio $\dfrac{V_3}{V_2} = \rho$.

Then expansion ratio $\dfrac{V_4}{V_3} = \dfrac{V_4}{V_2} \times \dfrac{V_2}{V_3} = \dfrac{V_1}{V_2} \times \dfrac{V_2}{V_3} = \dfrac{r}{\rho}$ as $V_1 = V_4$

Considering constant pressure heat addition process 2–3, we can write

$$\frac{p_2V_2}{T_2} = \frac{p_3V_3}{T_3}$$

As $p_2 = p_3$ hence $\qquad \frac{T_3}{T_2} = \frac{V_3}{V_2} = \rho$

$$\therefore \qquad T_2 = \frac{T_3}{\rho} \qquad ...(11.15)$$

Process 1–2 is isentropic compression, so we can write

$$\frac{T_2}{T_1} = \left(\frac{V_1}{V_2}\right)^{\gamma-1} = r^{\gamma-1}$$

$$\therefore \qquad T_2 = T_1 \times r^{\gamma-1} \qquad ...(11.16)$$

Substituting the value of T_2 in equation (11.15), we get

$$T_3 = T_2 \times \rho = T_1 \times r^{\gamma-1} \times \rho$$

$$T_3 = T_1 \times r^{\gamma-1} \times \rho \qquad ...(11.17)$$

Process 3–4 is isentropic expansion, so we can write

$$\frac{T_3}{T_4} = \left(\frac{V_4}{V_3}\right)^{\gamma-1} = \left(\frac{r}{\rho}\right)^{\gamma-1}$$

$$\therefore \qquad T_4 = \frac{T_3(\rho)^{\gamma-1}}{(r)^{\gamma-1}}$$

$$= \frac{T_1 \times r^{\gamma-1} \times \rho(\rho)^{\gamma-1}}{(r)^{\gamma-1}} \qquad ...(11.18)$$

$$\therefore \qquad T_4 = T_1 \times \rho^{\gamma}$$

Now substituting the values of T_2, T_3 and T_4 from equations (11.16), (11.17) and (11.18) in equation (11.14), we get

$$\eta = 1 - \frac{1}{\gamma}\left(\frac{T_4 - T_1}{T_3 - T_2}\right)$$

$$\eta = 1 - \frac{1}{\gamma}\left(\frac{T_1 \times \rho^{\gamma} - T_1}{T_1 \times r^{\gamma-1} \times \rho - T_1 \times r^{\gamma-1}}\right)$$

$$\eta = 1 - \frac{1}{\gamma}\frac{T_1(\rho^{\gamma} - 1)}{T_1 r^{\gamma-1}(\rho - 1)}$$

$$\eta = 1 - \frac{1}{\gamma}\frac{(\rho^{\gamma} - 1)}{r^{\gamma-1}(\rho - 1)}$$

$$\eta = 1 - \frac{1}{r^{\gamma-1}} \times \left[\frac{\rho^{\gamma}-1}{\gamma(\rho-1)}\right] \quad ...(11.19)$$

Equation (11.19) gives the air standard efficiency of the diesel cycle.

11.6 DUAL COMBUSTION CYCLE

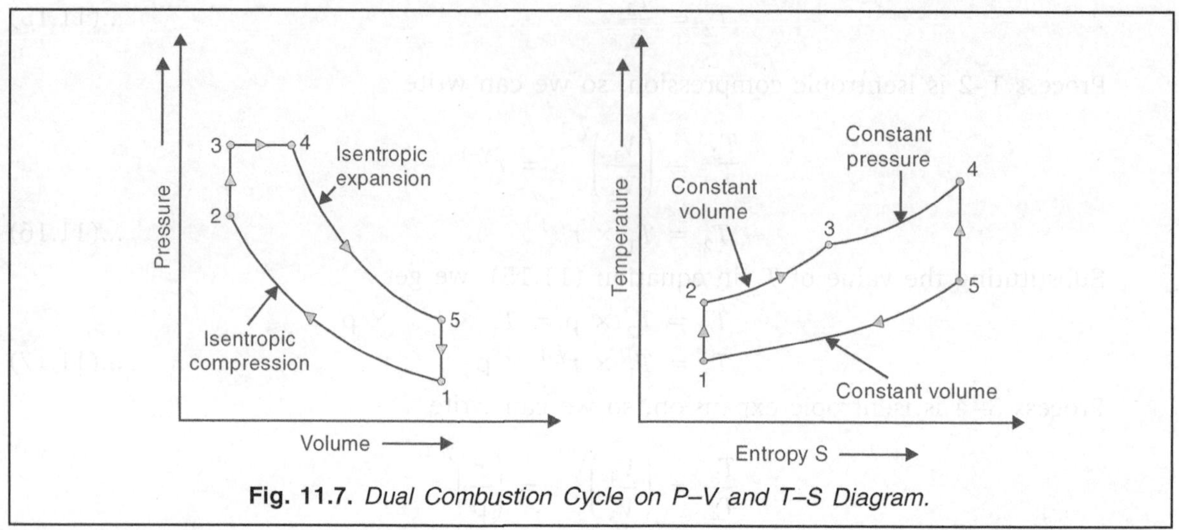

Fig. 11.7. *Dual Combustion Cycle on P–V and T–S Diagram.*

This cycle is a combination of Otto and diesel cycles. It is known as a dual combustion or composite cycle as the combustion of fuel takes place partly at constant volume and partly at constant pressure. This cycle is used in modern high speed oil engines and semi-Diesel engines. The ideal P–V and T–S diagrams of this cycle are shown in Fig. 11.7.

The process 1–2 is isentropic compression, 2–3 heat addition volume, 3–4 heat addition at constant pressure, 4–5 isentropic expansion and 5–1 heat rejection at constant volume.

Consider one kg of air in the engine cylinder at point 1. The sequence of operation followed by the cycle is given below:

Process 1–2: At point 1, the piston is at the beginning of in stroke. The air is compressed isentropically to point 2 during which the piston moves from bottom dead centre to top dead centre. Pressure and temperature of air increases during process 1–2.

Process 2–3: During this process the piston is at rest at TDC. Heat is now supplied at constant volume from an external hot source up to point 3. The hot source is still maintained for the next portion of the cycle.

Process 3–4: During this part of the cycle, heat is supplied at constant pressure. The hot source is withdrawn at point 4.

Process 4–5: Air now expands isentropically to point 5 and work is done on the piston. During this process power is produced for running the engine.

Process 5–1: During this process, heat is rejected at constant volume to an external cold source. This cold source is then removed after the air is brought to its initial conditions. And thus the cycle is completed.

Heat supplied = heat added during constant volume process + heat added during constant pressure process

$$\text{Heat supplied} = C_v (T_3 - T_2) + C_p (T_4 - T_3)$$

and

$$\text{Heat rejected} = C_v (T_5 - T_1)$$

$$\text{Network done} = \text{Heat supplied} - \text{Heat rejected}$$

$$= C_v (T_3 - T_2) + C_p (T_4 - T_3) - C_v (T_5 - T_1)$$

The thermal efficiency of the cycle is:

$$\text{Efficiency, } \eta = \frac{\text{Network done}}{\text{Heat supplied}}$$

$$\eta = \frac{C_v (T_3 - T_2) + C_p (T_4 - T_3) - C_v (T_5 - T_1)}{C_v (T_3 - T_2) + C_p (T_4 - T_3)}$$

$$\eta = 1 - \frac{C_v (T_5 - T_1)}{C_v (T_3 - T_2) + C_p (T_4 - T_3)}$$

$$\eta = 1 - \frac{(T_5 - T_1)}{(T_3 - T_2) + \gamma (T_4 - T_3)} \quad \text{...(11.20)}$$

Let compression ratio, $r = \frac{V_1}{V_2}$ and cut-off ratio $\rho = \frac{V_4}{V_3}$

Then expansion ratio $\frac{V_5}{V_4} = \frac{V_1}{V_2} \times \frac{V_3}{V_4} = \frac{r}{\rho}$ (as $V_5 = V_1$ and $V_2 = V_3$)

Let pressure ratio or explosion ratio $\beta = \frac{p_3}{p_2}$.

Considering process 1–2 (isentropic compression), we can write

$$\frac{T_2}{T_1} = \left(\frac{V_1}{V_2}\right)^{\gamma-1} = r^{\gamma-1}$$

$$\therefore \quad T_2 = T_1 \times r^{\gamma-1} \quad \text{...(11.21)}$$

Considering process 2–3 (constant volume) we can write

$$V_3 = V_2 \text{ and } V_2 = \frac{V_1}{r} \quad \text{...(11.22)}$$

$$\frac{p_2 V_2}{T_2} = \frac{p_3 V_3}{T_3}$$

$$\frac{p_2}{T_2} = \frac{p_3}{T_3}$$

$$\therefore \quad T_3 = T_2 \times \frac{p_3}{p_2}$$

$$\therefore \quad T_3 = T_1 \times r^{\gamma-1} \times \beta \quad \text{...(11.23)}$$

Considering process 3–4 (constant pressure process)

$$\frac{p_3V_3}{T_3} = \frac{p_4V_4}{T_4}$$

$$\frac{V_3}{T_3} = \frac{V_4}{T_4} \text{ as } p_3 = p_4$$

$$\therefore \quad T_4 = T_3 \times \frac{V_4}{V_3} = T_3 \times \rho$$

$$\therefore \quad T_4 = T_1\, \rho\, \beta\, r^{\gamma-1} \qquad \text{...(11.24)}$$

Considering process 4–5 (isentropic expansion)

$$\frac{T_4}{T_5} = \left(\frac{V_5}{V_4}\right)^{\gamma-1} = \left(\frac{r}{\rho}\right)^{\gamma-1}$$

$$T_5 = \frac{T_4}{\left(\frac{r}{\rho}\right)^{\gamma-1}}$$

Substituting value of T_4 from equation (11.24) we get

$$\therefore \quad T_5 = \frac{T_4(\rho)^{\gamma-1}}{(r)^{\gamma-1}}$$

$$= \frac{T_1\, \rho\, \beta\, r^{\gamma-1}\, \rho^{\gamma-1}}{r^{\gamma-1}} = T_1\, \beta\, \rho^{\gamma}$$

$$\therefore \quad T_5 = T_1\, \beta\, \rho^{\gamma} \qquad \text{...(11.25)}$$

Now substituting the values of T_2, T_3, T_4 and T_5 in equation (11.20) we get

$$\eta = 1 - \frac{(T_5 - T_1)}{(T_3 - T_2) + \gamma(T_4 - T_3)}$$

$$\eta = 1 - \frac{T_1\, \beta\, \rho^{\gamma} - T_1}{\left(T_1\, \beta\, r^{\gamma-1} - T_1\, r^{\gamma-1}\right) + \gamma\left(T_1\, \rho\, \beta\, r^{\gamma-1} - T_1\, \beta\, r^{\gamma-1}\right)}$$

$$\eta = 1 - \frac{1}{r^{\gamma-1}}\left[\frac{T_1\,(\beta\, \rho^{\gamma} - 1)}{T_1(\beta - 1) + \gamma\, T_1\,(\rho\, \beta - \beta)}\right]$$

$$\eta = 1 - \frac{1}{r^{\gamma-1}}\left[\frac{(\beta\, \rho^{\gamma} - 1)}{(\beta - 1) + \gamma\, \beta(\rho - 1)}\right] \qquad \text{...(11.26)}$$

The equation (11.26) gives the air standard efficiency of the dual-combustion cycle.

SOLVED PROBLEMS

Problem 1. *The bore and stroke of an engine working on ideal Otto cycle are 180 mm and 350 mm respectively. If the clearance volume is 0.003 m^3, then find out the air standard efficiency.*

Solution.

Air standard efficiency of Otto cycle is given by

$$\eta_{otto} = 1 - \frac{1}{r^{\gamma-1}}$$

Where r is the compression ratio

$$r = \frac{\text{stroke volume} + \text{clearance volume}}{\text{clearance volume}}$$

$$\text{Stroke volume} = V_s = \frac{\pi}{4} \times (D)^2 \times L_s$$

$$V_s = \frac{\pi}{4} \times (0.180)^2 \times 0.350$$

$$V_s = 8.906 \times 10^{-3} \text{ m}^3$$

$$\text{Clearance volume} = V_c = 0.003 \text{ m}^3$$

$$\therefore \quad \text{compression ratio } r = \frac{8.906 \times 10^{-3} + 0.003}{0.003} = 3.97$$

$$\eta_{otto} = 1 - \frac{1}{r^{\gamma-1}}$$

$$\eta_{otto} = 1 - \frac{1}{3.97^{0.4}}$$

$$\eta_{otto} = 1 - 0.576$$

$$\eta_{otto} = 0.424 = 42.4\% \textbf{ Ans.}$$

Problem 2. *A gas engine working on the Otto cycle has a cylinder of 150 mm diameter and stroke 200 mm, the clearance volume is 1560 cc. Find out the air standard efficiency. Assume C_p = 1.004 kJ/kg K and C_v = 0.717 kJ/kg K for air.*

Solution. $$\text{Stroke volume} = V_s = \frac{\pi}{4} \times (D)^2 \times L_s$$

$$V_s = \frac{\pi}{4} \times (15)^2 \times 20$$

$$= 3534.29 \text{ cm}^3 = 3534.29 \text{ cc}$$

Clearance volume is given in the problem

$$\therefore \quad V_c = 1560 \text{ cc}$$

$$r = \frac{\text{stroke volume} + \text{clearance volume}}{\text{clearance volume}}$$

$$r = \frac{3534.29 + 1560}{1560} = \frac{5094.29}{1560} = 3.26$$

$$\eta_{otto} = 1 - \frac{1}{r^{\gamma-1}}$$

$$\eta_{otto} = 1 - \frac{1}{3.26^{0.4}}$$

$$\eta_{otto} = 1 - 0.623$$

$$\eta_{otto} = 0.376 = 37.6\% \textbf{ Ans.}$$

Problem 3. *An I.C. engine is working on an Otto cycle. In this cycle, the temperature and pressure at the beginning of isentropic compression is 27°C and 180 kPa respectively. What should be the compression ratio for $\gamma = 1.4$ and thermal efficiency of 46%? Also find out the temperature and pressure at the end of compression.*

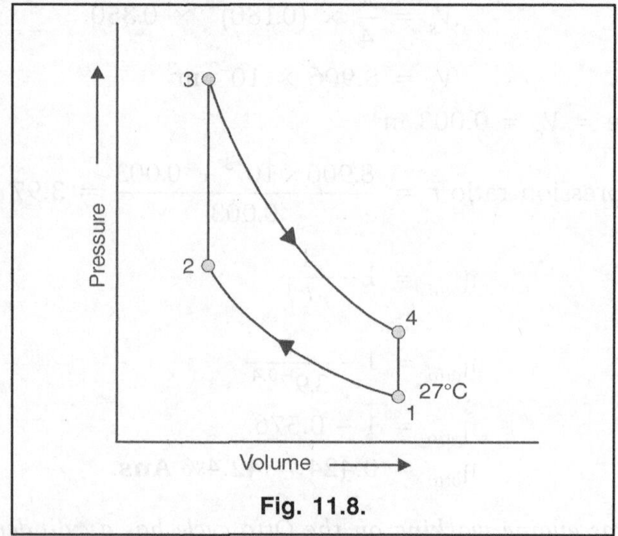

Fig. 11.8.

Solution.

Given: With reference to above Fig. 11.8

$$\text{Temperature } T_1 = 27 + 273 = 300 \text{ K}$$

$$\text{Pressure } p_1 = 180 \text{ kPa}$$

$$\text{Efficiency } \eta_{otto} = 46\%$$

$$T_2 = ?$$

$$p_2 = ?$$

$$\eta_{otto} = 1 - \frac{1}{r^{\gamma-1}}$$

$$0.46 = 1 - \frac{1}{r^{0.4}}$$

$$\frac{1}{r^{0.4}} = 1 - 0.46 = 0.54$$

$$r^{0.4} = \frac{1}{0.54}$$

$$r = \left(\frac{1}{0.54}\right)^{1/0.4} = 4.66$$

As we know that $p_1 V_1^{\gamma} = p_2 V_2^{\gamma}$

$$\therefore \quad p_2 = p_1 \left(\frac{V_2}{V_1}\right)^{\gamma} = 180\ (4.66)^{1.4} = 1552.43 \text{ kPa}$$

and

$$\frac{T_2}{T_1} = \left(\frac{p_2}{p_1}\right)^{\frac{\gamma-1}{\gamma}} = \left(\frac{1552.43}{180}\right)^{\frac{1.4-1}{1.4}}$$

$$T_2 = T_1 \times (8.62)^{0.286}$$

$$T_2 = 300 \times 1.85 = 555.5 \text{ K}$$

$$T_2 = 282.5°\text{C} \ \textbf{Ans.}$$

Problem 4. *Pressure and temperature of air at the entry to an Otto cycle are 120 kPa and 298 K respectively. The compression ratio is 7. If the temperature of air after expansion is 540 K, then find out (a) thermal efficiency, (b) the maximum pressure and temperature and (c) the mean effective pressure of the cycle.*

Solution.

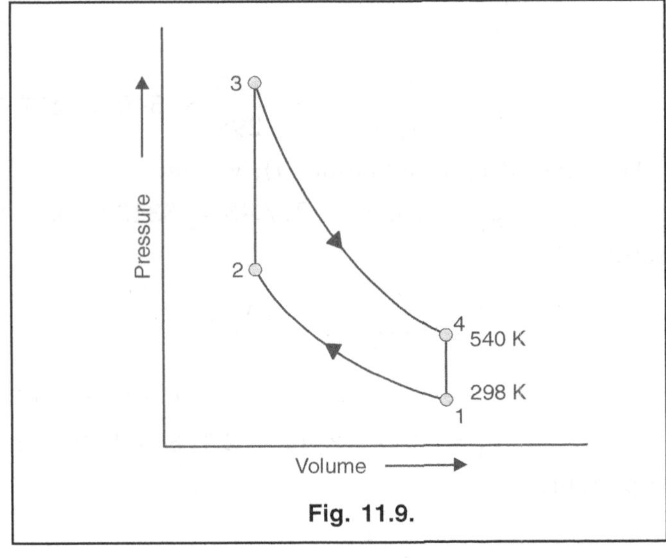

Fig. 11.9.

Given: With ref. to Fig. 11.9

$$P_1 = 120 \text{ kPa}$$
$$T_1 = 298 \text{ K}$$
$$T_4 = 540 \text{ K}$$
$$r = 7$$

to find out max. temp. $= T_3 = ?$

$$\eta_{otto} = ? \; p_m = ?$$

(*a*) Thermal efficiency

$$\eta_{otto} = 1 - \frac{1}{r^{\gamma-1}}$$

$$\eta_{otto} = 1 - \frac{1}{7^{0.4}}$$

$$\eta_{otto} = 1 - 0.4591$$

$$\eta_{otto} = 0.5408 = 54.08\%$$

(*b*) The maximum pressure

$$\frac{p_3}{p_4} = \left(\frac{V_4}{V_3}\right)^{\gamma} = \left(\frac{V_1}{V_2}\right)^{\gamma} = (7)^{1.4} = 15.24$$

$$p_3 = 15.24 \times p_4 \qquad \text{...}(i)$$

Now applying perfect gas equation across process 4–1, we get

$$\frac{p_4 V_4}{T_4} = \frac{p_1 V_1}{T_1}$$

$\because$ $$V_4 = V_1$$

$\therefore$ $$\frac{p_4}{T_4} = \frac{p_1}{T_1}$$

$\therefore$ $$p_4 = \frac{p_1}{T_1} \times T_4 = \frac{120}{298} \times 540 = 217.45 \text{ kPa}$$

Now substituting the value of p_4 in equation (*i*), we get

$$p_3 = 15.24 \times 217.45 = 3313.93 \text{ kPa } \textbf{Ans.}$$

Maximum temperature (T_3)

$$\left(\frac{T_3}{T_4}\right) = \left(\frac{V_4}{V_3}\right)^{\gamma-1} = \left(\frac{V_1}{V_2}\right)^{\gamma-1}$$

$$= (r)^{\gamma-1} = (7)^{1.4-1} = (7)^{0.4} = 2.178$$

$\therefore$ $$T_3 = 2.178 \times T_4 = 2.178 \times 540 = 1176.07 \text{ K } \textbf{Ans.}$$

(*c*) Mean effective pressure

We know

$$p_m = \frac{\text{Network done}}{\text{Stroke volume}}$$

$$\text{Stroke volume} = V_1 - V_2$$

Using perfect gas equation

$$p_1 V_1 = RT_1$$

$$120 \times 1000 \times V_1 = 287 \times 298$$

$$\therefore \quad V_1 = \frac{287 \times 298}{120 \times 1000} = 0.731 \text{ m}^3/\text{kg}$$

As we know $\quad r = \dfrac{V_1}{V_2}$

$$\therefore \quad V_2 = \frac{V_1}{r} = \frac{0.713}{7} = 0.102 \text{ m}^3/\text{kg}$$

$$\therefore \quad \text{Stroke volume} = V_1 - V_2 = 0.713 - 0.102 = 0.611 \text{ m}^3/\text{kg}$$

$$\text{Work done} = \text{heat added} - \text{heat rejected}$$

$$\frac{T_2}{T_1} = \left(\frac{V_1}{V_2}\right)^{\gamma-1}$$

$$\therefore \quad T_2 = T_1 \times (7)^{1.4-1} = 298 \times 7^{0.4} = 649 \text{ K}$$

$$\text{Work done per kg} = C_v \, [(T_3 - T_2) - (T_4 - T_1)]$$

$$= 0.718 \, [(1176.07 - 649) - (540 - 298)]$$

$$= 204.7 \text{ kJ/kg}$$

$$p_m = \frac{\text{Network done}}{\text{Stroke volume}}$$

$$\therefore \quad p_m = \frac{204.7}{0.611} = 335 \text{ kPa}$$

$p_m = 335$ kPa **Ans.**

Problem 5. *Calculate the air standard efficiency of a Diesel cycle having a compression ratio of 18 and cut-off ratio of 2.5. Assume $\gamma = 1.4$.*

Solution. Given:

$$\text{Compression ratio } r = 18$$

$$\text{Cut-off ratio } \rho = 2.5$$

and efficiency of diesel cycle is given by

$$\eta_{Diesel} = 1 - \frac{1}{r^{\gamma-1}}\left[\frac{(\rho)^{\gamma} - 1}{\gamma(\rho - 1)}\right]$$

$$\eta_{Diesel} = 1 - \frac{1}{18^{1.4-1}}\left[\frac{(2.5)^{1.4} - 1}{1.4(2.5 - 1)}\right]$$

$$\eta_{Diesel} = 1 - 0.315\left[\frac{3.606 - 1}{1.4 \times 1.5}\right] = 1 - 0.315 \times 1.24$$

$$\eta_{Diesel} = 1 - 0.3909 = 0.6091$$

$$\eta_{Diesel} = 60.91\% \text{ Ans.}$$

Problem 6. *A Diesel engine is working with a compression ratio of 16 and expansion ratio of 9. Calculate the air standard efficiency of the cycle.*

Solution.

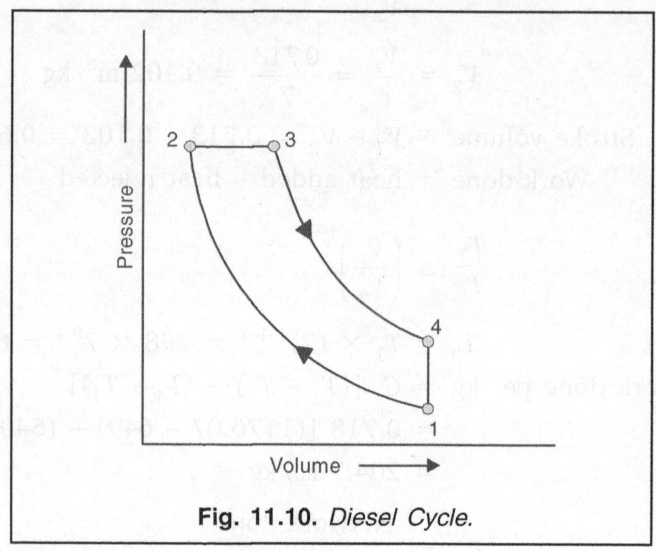

Fig. 11.10. *Diesel Cycle.*

Given:

$$\text{Compression ratio } r = \frac{V_1}{V_2} = 16$$

$$\text{Expansion ratio} = 9 = \frac{r}{\rho} = \frac{V_4}{V_3} = \frac{16}{\rho}$$

$$\therefore \quad \text{Expansion ratio} = \frac{16}{\rho}$$

$$9 = \frac{16}{\rho}$$

$$\therefore \quad \rho = \frac{16}{9} = 1.77$$

and efficiency of diesel cycle is given by

$$\eta_{Diesel} = 1 - \frac{1}{r^{\gamma-1}}\left[\frac{(\rho)^{\gamma} - 1}{\gamma(\rho - 1)}\right]$$

$$\eta_{Diesel} = 1 - \frac{1}{16^{1.4-1}}\left[\frac{(1.77)^{1.4} - 1}{1.4\,(1.77 - 1)}\right]$$

$$\eta_{Diesel} = 1 - \frac{1}{16^{0.4}}\left[\frac{2.238 - 1}{1.4 \times 0.77}\right] = 1 - 0.329 \times \frac{1.238}{1.078}$$

$$\eta_{Diesel} = 1 - 0.378 = 0.622$$

$$\eta_{Diesel} = 62.2\% \textbf{ Ans.}$$

Problem 7. *An air standard Otto cycle is designed to operate with the following data:*
Maximum cycle pressure and temperature: 5 Mpa and 2250 K
Minimum cycle pressure and temperature: 0.1 Mpa and 300 K
Determine the net work output per unit mass of working fluid and the thermal efficiency.

(RGPV Jan./Feb. 2008)

Solution.

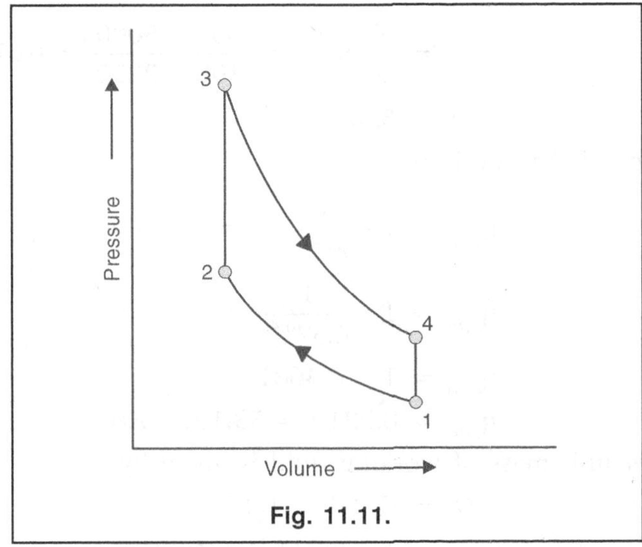

Fig. 11.11.

Given with reference to Fig. 11.11

Maximum cycle pressure p_3 = 5 MPa = 5000 kPa

Maximum cycle temperature T_3 = 2250 K

Minimum cycle pressure p_1 = 0.1 MPa = 100 kPa

Minimum cycle temperature T_1 = 300 K

To find out the efficiency of the cycle, first the value of compression ratio is to be determined.

Using perfect gas equation

$$\frac{p_1 V_1}{T_1} = \frac{p_2 V_2}{T_2}$$

$$\therefore \quad \frac{V_1}{V_2} = \frac{p_2 T_1}{p_1 T_2}$$

$$r = \frac{V_1}{V_2}$$

$$r = \frac{p_2 T_1}{p_1 T_2} \quad \text{...(ii)}$$

In this equation (*ii*) the values of p_1 and T_1 are given and p_2 and T_2 are not known. So we need to find out the values of p_2 and T_2 from the given data.

Therefore let us consider process 2–3 which is a constant volume heat addition process, so we can write

$$\frac{p_2}{T_2} = \frac{p_3}{T_3}$$

Substituting this value in equation (*ii*), we get

$$r = \frac{p_2 T_1}{p_1 T_2} = \frac{T_1}{p_1} \times \frac{p_2}{T_2} = \frac{T_1}{p_1} \times \frac{p_3}{T_3}$$

$$r = \frac{T_1}{p_1} \times \frac{p_3}{T_3} = \frac{300}{100} \times \frac{5000}{2250} = 6.67$$

$$r = 6.67$$

Thermal efficiency of Otto cycle is:

$$\eta_{otto} = 1 - \frac{1}{r^{\gamma-1}}$$

$$\eta_{otto} = 1 - \frac{1}{6.67^{0.4}}$$

$$\eta_{otto} = 1 - 0.4681$$

$$\eta_{otto} = 0.5319 = 53.19\% \quad \textbf{Ans.}$$

Heat supplied per unit mass of working fluid is given by

$$Q_s = C_v (T_3 - T_2)$$

For this we need to find out T_2 whose value is not given.

Therefore considering process 1–2 (isentropic compression), we can write

$$\frac{T_2}{T_1} = \left(\frac{V_1}{V_2}\right)^{\gamma-1}$$

$$\therefore \quad T_2 = T_1 \times (6.67)^{1.4-1}$$

$$= 300 \times 6.67^{0.4} = 640.87 \text{ K}$$

$$\therefore \quad Q_s = 0.718\ (2250 - 640.87) = 1155.35 \text{ kJ/kg}$$

Now as we know

$$\text{Efficiency, } \eta = \frac{\text{Net work done}}{\text{Heat supplied}}$$

$$\text{Net work done} = \text{Efficiency} \times \text{Heat supplied} = \eta \times Q_s$$

$$= 0.5319 \times 1155.35 = 614.53 \text{ kJ/kg} \quad \textbf{Ans.}$$

Problem 8. *In an air standard diesel cycle with compression ratio 14, the conditions of air at the start of compression stroke are 1 bar and 300 K. After addition of heat at constant pressure, the temperature rises to 2775 K. Determine the thermal efficiency of the cycle, net work done per kg of air and mean effective pressure.* **(RGPV Jan./Feb. 2008)**

Solution.

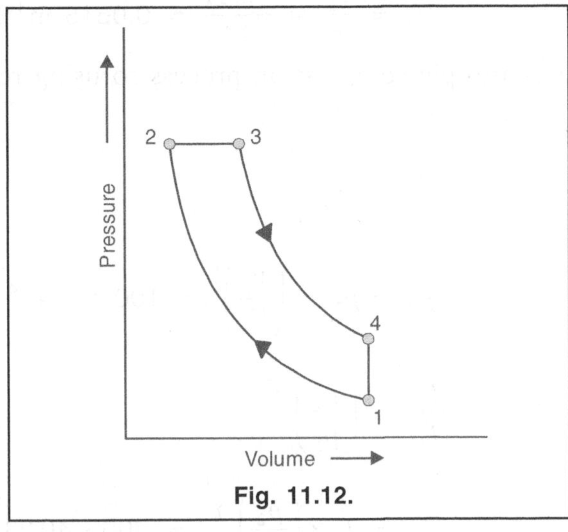

Fig. 11.12.

Given:

$$\text{Compression ratio } r = 14$$

$$p_1 = 1 \text{ bar} = 100 \text{ kPa}$$

$$T_1 = 300 \text{ K}$$

$$T_3 = 2775 \text{ K}$$

As we know

$$\eta_{Diesel} = 1 - \frac{1}{r^{\gamma-1}} \times \left[\frac{\rho^\gamma - 1}{\gamma(\rho - 1)}\right]$$

Where Compression ratio $r = \dfrac{V_1}{V_2}$ and

$$\text{Cut off ratio } \rho = \frac{V_3}{V_2}$$

So in order to find out the efficiency of the cycle we need to find out the values of V_1, V_2 and V_3.

So considering process 1–2 and using perfect gas equation we can write

$$p_1V_1 = mRT_1$$

Since air is the working fluid

$$\therefore \quad R = 0.287 \text{ kJ/kg K}$$

$$\therefore \quad V_1 = \frac{RT_1}{p_1} = \frac{0.287 \times 300}{100} = 0.861 \text{ m}^3/\text{kg}$$

As compression ratio is given so using that expression V_2 can be determined.

$$\therefore \quad r = \frac{V_1}{V_2}$$

$$V_2 = \frac{V_1}{r} = \frac{0.861}{14} = 0.0615 \text{ m}^3/\text{kg}$$

Since process 1–2 is isentropic compression process so using relation $pV^\gamma = C$ we can write

$$\frac{p_1}{p_2} = \left(\frac{V_2}{V_1}\right)^\gamma$$

$$p_2 = p_1 \times \left(\frac{V_1}{V_2}\right)^\gamma = 100 \times 14^{1.4} = 4023.3 \text{ kPa}$$

Now using
$$\frac{T_2}{T_1} = \left(\frac{p_2}{p_1}\right)^{\frac{\gamma-1}{\gamma}}$$

$$\therefore \quad T_2 = T_1 \times \left(\frac{p_2}{p_1}\right)^{\frac{\gamma-1}{\gamma}} = 300 \times 40.23^{\frac{0.4}{1.4}} = 862.13 \text{ K}$$

From constant pressure process 2–3, we get

$$\frac{V_3}{V_2} = \frac{T_3}{T_2}$$

$$\therefore \quad V_3 = V_2 \times \frac{T_3}{T_2} = 0.0615 \times \frac{2775}{862.13}$$

$$= 0.198 \text{ m}^3/\text{kg}$$

$$\therefore \quad \text{cut-off ratio} = \rho = \frac{V_3}{V_2} = \frac{0.198}{0.0615} = 3.22$$

$$\therefore \quad \eta_{Diesel} = 1 - \frac{1}{r^{\gamma-1}} \times \left[\frac{\rho^\gamma - 1}{\gamma(\rho - 1)}\right]$$

$$\eta_{Diesel} = 1 - \frac{1}{14^{0.4}} \times \left[\frac{3.22^{1.4} - 1}{1.4(3.22 - 1)}\right]$$

$$\eta_{Diesel} = 1 - 0.4636$$

$$\eta_{Diesel} = 0.5364$$

η_{Diesel} = 53.64% **Ans.**

Heat supplied per unit mass of working fluid is given by

$$Q_s = C_p (T_3 - T_2)$$

$$Q_s = 1.005 (2775 - 862.13) = 1922.43 \text{ kJ/kg}$$

$$\text{Efficiency, } \eta = \frac{\text{Network done}}{\text{Heat supplied}}$$

$$\text{Network done} = \text{Efficiency} \times \text{Heat supplied} = 0.5364 \times 1922.43$$

$$= 1031.2 \text{ kJ/kg} \quad \textbf{Ans.}$$

Problem 9. *Compare the power output for Diesel and Otto engine from the following available data:*

Maximum temperature = 1227°C

Exhaust temperature = 477°C

Ambient conditions – atm. Pressure = 1 bar

Temperature = 27°C

Air consumption = 2 kg/min

Assume: $C_p = 1.005$ *kJ/kg K,* $C_v = 0.718$ *kJ/kg K,* $\gamma = 1.4$ **(RGPV Jan./Feb. 2007)**

Solution.

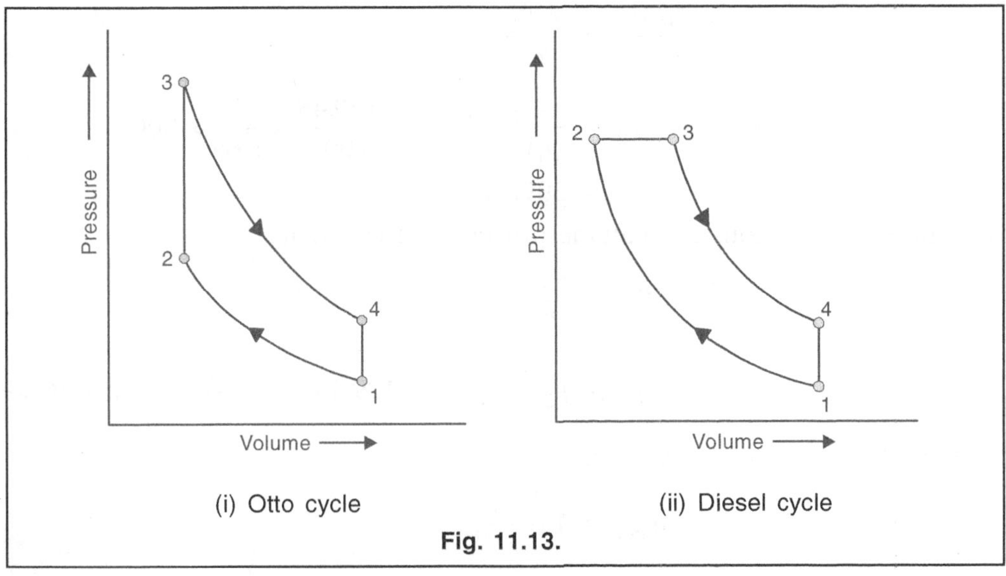

Fig. 11.13.

Given:

$$\text{Maximum temperature } T_3 = 1227°\text{C} = 1500 \text{ K}$$

$$\text{Exhaust temperature } T_4 = 477°\text{C} = 750 \text{ K}$$

Ambient condition:

$$T_1 = 27°\text{C} = 300 \text{ K}$$

$$p_1 = 1 \text{ bar} = 100 \text{ kPa}$$

(A) For Otto cycle

$$\frac{T_3}{T_4} = \left(\frac{V_4}{V_3}\right)^{\gamma-1} = \left(\frac{V_1}{V_2}\right)^{\gamma-1} = (r)^{\gamma-1}$$

$$\frac{T_3}{T_4} = (r)^{\gamma-1}$$

$$r^{1.4-1} = \frac{1500}{750} = 2$$

$$r^{0.4} = 2$$

$$\therefore \quad r = (2)^{1/0.4} = (2)^{2.5} = 5.65$$

Compression ratio $r = 5.65$

Considering process 1–2 (isentropic compression), we get

$$p_1 V_1^\gamma = p_2 V_2^\gamma$$

$$\therefore \quad p_2 = p_1 \times \left(\frac{V_1}{V_2}\right)^\gamma = 100 \times (5.65)^{1.4}$$

$$\therefore \quad p_2 = 1129.45 \text{ kPa}$$

Now using perfect gas equation

$$\frac{p_1 V_1}{T_1} = \frac{p_2 V_2}{T_2}$$

$$T_2 = \frac{p_2 V_2}{p_1 V_1} \times T_1 = \frac{1129.45}{100} \times \frac{1}{5.65} \times 300$$

$$= 599.71 \text{ K}$$

Using process 2–3 (constant volume process) we can write

$$\frac{p_2}{T_2} = \frac{p_3}{T_3}$$

$$p_3 = p_2 \times \frac{T_2}{T_3} = 1129.45 \times \frac{599.71}{1500} = 451.56 \text{ kPa}$$

Thermal efficiency of Otto cycle is:

$$\eta_{otto} = 1 - \frac{1}{r^{\gamma-1}}$$

$$\eta_{otto} = 1 - \frac{1}{5.65^{0.4}}$$

$$\eta_{otto} = 1 - 0.5002$$

$$\eta_{otto} = 0.4998$$

$$\eta_{otto} = 49.98\%$$

$$\text{Heat supplied } Q_s = C_v \,(T_3 - T_2)$$

$$= 0.718 \,(1500 - 599.71) = 646.41 \text{ kJ/kg}$$

$$\text{Heat rejected } Q_R = C_v \,(T_4 - T_1)$$

$$= 0.718 \,(750 - 300) = 323.1 \text{ kJ/kg}$$

$$\text{Power output per kg} = Q_s - Q_R$$
$$= 646.41 - 323.1 = 323.31 \text{ kJ/kg}$$

$$\text{Total power} = \frac{2}{60} \times 323.31 = 10.77 \text{ kJ/sec.} = 10.77 \text{ kW}$$ **Ans.**

(B) For Diesel cycle

Considering process 1–4 (constant volume heat rejection process) we can write

$$\frac{p_4 V_4}{T_4} = \frac{p_1 V_1}{T_1} \quad \text{as } V_4 = V_1$$

$$\therefore \quad \frac{p_4}{T_4} = \frac{p_1}{T_1}$$

$$\therefore \quad p_4 = p_1 \times \frac{T_4}{T_1} = 100 \times \frac{750}{300} = 250 \text{ kPa}$$

For the process 3–4 (isentropic expansion) we can write

$$\frac{T_3}{T_4} = \left(\frac{p_3}{p_4}\right)^{\frac{\gamma-1}{\gamma}}$$

$$\therefore \quad p_3 = p_4 \times \left(\frac{T_3}{T_4}\right)^{\frac{\gamma}{\gamma-1}} = 250 \times \left(\frac{1500}{750}\right)^{\frac{1.4}{0.4}}$$
$$= 2828.43 \text{ kPa}$$

For a constant pressure process 2–3

$$\therefore \quad p_2 = p_3 = 2828.43 \text{ kPa}$$

From process 1–2, we get

$$\frac{T_2}{T_1} = \left(\frac{p_2}{p_1}\right)^{\frac{\gamma-1}{\gamma}}$$

$$\therefore \quad T_2 = T_1 \times \left(\frac{p_2}{p_1}\right)^{\frac{\gamma-1}{\gamma}} = 300 \times \left(\frac{2828.43}{100}\right)^{\frac{0.4}{1.4}}$$
$$= 779.56 \text{ K}$$

$$\frac{T_2}{T_1} = (r)^{\gamma-1}$$

$$\therefore \quad \frac{779.56}{300} = r^{0.4}$$

$$r = 2.6^{\frac{1}{0.4}} = 2.6^{2.5} = 10.9$$

Efficiency of diesel cycle is given by

$$\eta = 1 - \frac{1}{\gamma}\left(\frac{T_4 - T_1}{T_3 - T_2}\right)$$

$$\eta = 1 - \frac{1}{1.4}\left(\frac{750 - 300}{1500 - 779.56}\right)$$

$$\eta_{Diesel} = 1 - \frac{1}{1.4} \times \frac{450}{720.44}$$

$$\eta_{Diesel} = 1 - 0.4461$$

$$\eta_{Diesel} = 0.5539$$

$$\eta_{Diesel} = 55.39\%$$

$$\text{Heat supplied } Q_s = C_p\,(T_3 - T_2)$$

$$= 1.005\,(1500 - 779.56) = 724.04 \text{ kJ/kg}$$

$$\text{Heat rejected } Q_R = C_v\,(T_4 - T_1)$$

$$= 0.718\,(750 - 300) = 323.1 \text{ kJ/kg}$$

$$\text{Power output per kg} = Q_s - Q_R$$

$$= 724.04 - 323.1 = 400.94 \text{ kJ/kg}$$

$$\text{Total power} = \frac{2}{60} \times 400.94 = 13.36 \text{ kJ/sec.} = 1336 \text{ kW}$$ **Ans.**

Problem 10. *An Otto cycle takes in air at 300°K. The ratio of maximum to minimum temperature is 6. Find out the optimum compression ratio for the maximum work output of the cycle.* **(RGPV Jan./Feb. 2007)**

Solution.

As we know that for Otto cycle the optimum compression ratio r for maximum work done is given by

$$r = \left(\frac{T_3}{T_1}\right)^{1.25}$$

Where T_3 is the maximum cycle temperature

T_1 is the minimum cycle temperature

r is the compression ratio

Given data is as follows:

Air intake temperature is nothing but it is T_1 $\therefore$ $T_1 = 300$ K

and

$$\frac{T_3}{T_1} = 6$$

$\therefore$ optimum compression ratio $r = (6)^{1.25} = 9.4$ **Ans.**

Problem 11. *An engine working on a dual combustion cycle has a compression ratio of 14. The maximum temperature in the cycle is 750°C. If the working fluid receives equal amounts of heat during both the constant volume and constant pressure processes, then find out the air standard efficiency of the cycle. Assume temperature at the start of cycle is 37°C,* C_p *= 1.005 kJ/kg K and* C_v *= 0.718 kJ/kg K.*

Solution.

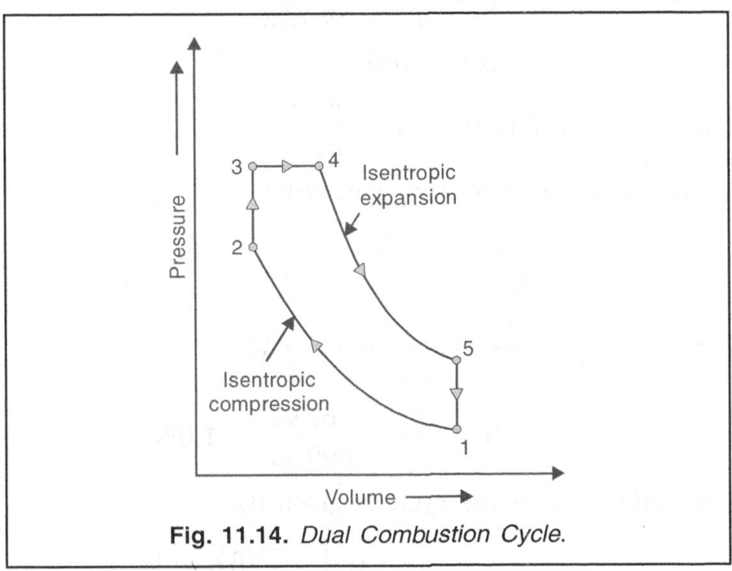

Fig. 11.14. *Dual Combustion Cycle.*

Given:

Maximum temperature in cycle = T_4 = 750 + 273 = 1023 K

Temperature at the start of cycle is T_1 = 37 + 273 = 310 K

$$\text{Compression ratio } r = \frac{V_1}{V_2} = 14$$

Considering process 1–2 (isentropic compression), we can write

$$\frac{T_2}{T_1} = \left(\frac{V_1}{V_2}\right)^{\gamma-1} = r^{\gamma-1}$$

$$\therefore \quad T_2 = T_1 \times r^{\gamma-1}$$

$$\therefore \quad T_2 = 310 \times (14)^{1.4-1} = 890.86 \text{ K}$$

As mentioned in the problem

Heat added at constant pressure = heat added at constant volume

$$\therefore \quad C_p(T_4 - T_3) = C_v(T_3 - T_2)$$

$$1.005\,(1023 - T_3) = 0.718\,(T_3 - 890.86)$$

$$1028.11 - 1.005\,T_3 = 0.718\,T_3 - 639.64$$

$$T_3\,(1.005 + 0.718) = 1028.11 + 639.64$$

$$T_3 = \frac{1667.75}{1.723} = 967.93 \text{ K}$$

From constant pressure process 3–4 we can write

$$\frac{T_4}{T_3} = \frac{V_4}{V_3} = \rho$$

Where ρ is the cut-off ratio.

$$\rho = \frac{T_4}{T_3} = \frac{1023}{967.93}$$

$$\rho = 1.057$$

Let pressure ratio or explosion ratio $\beta = \frac{p_3}{p_2}$

From constant volume process 2–3 we can write

$$\frac{p_2V_2}{T_2} = \frac{p_3V_3}{T_3}$$

$$\frac{T_3}{T_2} = \frac{p_3}{p_2} \text{ as } V_2 = V_3$$

$$\therefore \quad \beta = \frac{T_3}{T_2} = \frac{967.93}{890.86} = 1.086$$

The air standard efficiency of the cycle is given by

$$\eta = 1 - \frac{1}{r^{\gamma-1}}\left[\frac{(\beta\,\rho^{\gamma} - 1)}{(\beta - 1) + \gamma\,\beta\,(\rho - 1)}\right]$$

$$\eta = 1 - \frac{1}{14^{1.4-1}}\left[\frac{\left(1.086 \times (1.057)^{1.4} - 1\right)}{(1.086 - 1) + 1.4 \times 1.086\,(1.057 - 1)}\right]$$

$$\eta = 1 - 0.348\left[\frac{0.1736}{0.1726}\right]$$

$$\eta = 1 - 0.3498$$

$$\eta = 0.6502$$

$\eta = 65.02\%$ **Ans.**

REVIEW QUESTIONS

1. Find out the loss in air standard efficiency of diesel engine of compression ratio of 15 and the cut-off changes from 5 to 15% of the stroke. [**Ans.** 6.55]
2. An air engine working on the Otto cycle has the compression ratio increased from 5 to 6. Calculate the change in efficiency of the engine due to this rise. Assume $\gamma = 1.4$. [**Ans.** 8.12%]

3. In an Otto cycle air at 27°C and 1.5 bar is compressed adiabatically until the pressure is 18 bar. Heat is added at constant volume until the pressure rises to 45 bar. Calculate the air standard efficiency, the compression ratio, and the mean effective pressure for the cycle. Assume C_v = 0.717 kJ/kg and R = 8.314 kJ/kg mol K.

4. A four cylinder petrol engine working on the air standard cycle has a temperature and pressure of 297 K and 102 kPa respectively at the beginning of compression. The maximum cycle temperature is 1673 K. The engine has a total swept volume of 200 m^3 and the clearance volume in each cylinder is 60 cm^3. Determine (*a*) the air standard efficiency, (*ii*) the mean effective pressure. **[Ans.** 59.07%, 538.6 kPa]

5. An air engine is working on an ideal cycle in which heat is received and rejected at constant volume. The pressure and temperature at the beginning of compression are 1 bar and 315 K respectively. The pressure at the end of adiabatic compression is 15 times that at the start. If the temperature reached at the end of constant volume heat addition is 2220 K, then find out (*a*) the heat supplied per kg of air, (*b*) the air standard efficiency, (*c*) the temperature and pressure at the end of compression.

6. For a diesel engine operating on air standard cycle with γ = 1.4, determine the thermal efficiency for cut off ratios of 1.25, 1.5 and 2. The pressure and temperature at the beginning of compression is 0.98 bar and 300 K and the compression ratio is 14. Comment the impact of increase in cut-off ratios on thermal efficiency of the diesel cycle.

[Ans. 63.61%, 62%, 59.26%]

[**Comment:** The increase in cut-off ratio leads to decrease in cycle efficiency]

7. An engine working on a dual cycle has a cylinder bore of 200 mm and stroke 400 mm, the compression ratio is 14 and the pressure ratio of the constant volume process is 2. The constant pressure process is cut off at 5% of the stroke. Find out (*a*) the temperatures at the end of heat addition and at the end of expansion (*b*) air standard efficiency of the cycle. The initial conditions are 1 atm and 300 K.

■■■

12 Internal Combustion Engines

12.1 INTRODUCTION

A heat engine is a device that transforms the chemical energy of fuel into thermal energy and utilizes this energy to produce mechanical work.

Heat engines can be classified in two categories:

(1) Internal combustion engine (I.C. Engine).

(2) External combustion engine.

In an external combustion engine, the combustion of fuel takes place outside the engine cylinder whereas in case of an internal combustion engine, the combustion of fuel takes place in the engine cylinder.

In internal combustion engine the working substance (fluid) is the products of combustion of air-fuel mixture. Internal combustion engines find wide applications in automobile sector, power boats, ships, locomotives and small power plants etc.

The common examples of external combustion engines are: *steam engine* and *steam turbine* where the working fluid is steam. In case of an external combustion engine, the combustion of fuel takes place in a boiler where steam is produced by using the heat energy of combustion. Steam thus produced is utilized to produce power when steam acts on the piston of a steam engine cylinder or on the blades of a steam turbine. Another example of an external combustion engine is a *closed cycle gas turbine* in which heat of combustion from an outside furnace is transferred to air (or gas), which is used in a gas turbine. In external combustion engine the working fluid is entirely different from air-fuel mixture.

The **advantages** of external combustion engines are:

- Use of low cost fuel (solid, liquid or gas)
- High starting torque
- Used for electric power generation.

In case of an internal combustion engine, the thermal energy resulting from the combustion of fuel is converted directly to produce mechanical power. Since combustion takes place inside the engine cylinder, very high temp. (2000°C) is produced inside the cylinder. It is therefore necessary to remove some portion of the heat from the cylinder to prevent damage to the metal of the cylinder either by circulating the water through jacket around the cylinder or by cooling it through atmospheric air. The thermal efficiency of an internal combustion engine is high as compared to a steam engine plant as internal combustion engines do not have losses that arise due to use of boilers, condensers etc., in steam engine plant.

Thermal efficiency of I.C. engine varies from 30 to 40%.
Thermal efficiency of steam plant varies from 20 to 30%.

Advantages of internal combustion engine

- High Thermal Efficiency
- Greater mechanical simplicity
- Lower weight to power ratio
- Low initial cost
- Low water requirement for dissipation of energy through cooling system
- Free from auxiliary devices such as boiler and condenser etc.
- Easy and instantaneous starting from cold conditions
- Require less space.

12.2 BASIC ENGINE PARTS AND NOMENCLATURE

In order to understand the working of an internal combustion engine it is very important to have first hand idea about important engine components and nomenclature associated with the engine. A large number of components are required to perform their function satisfactorily to produce the power. Figure (12.1) shows the cross-section of a single cylinder spark ignition internal combustion engine.

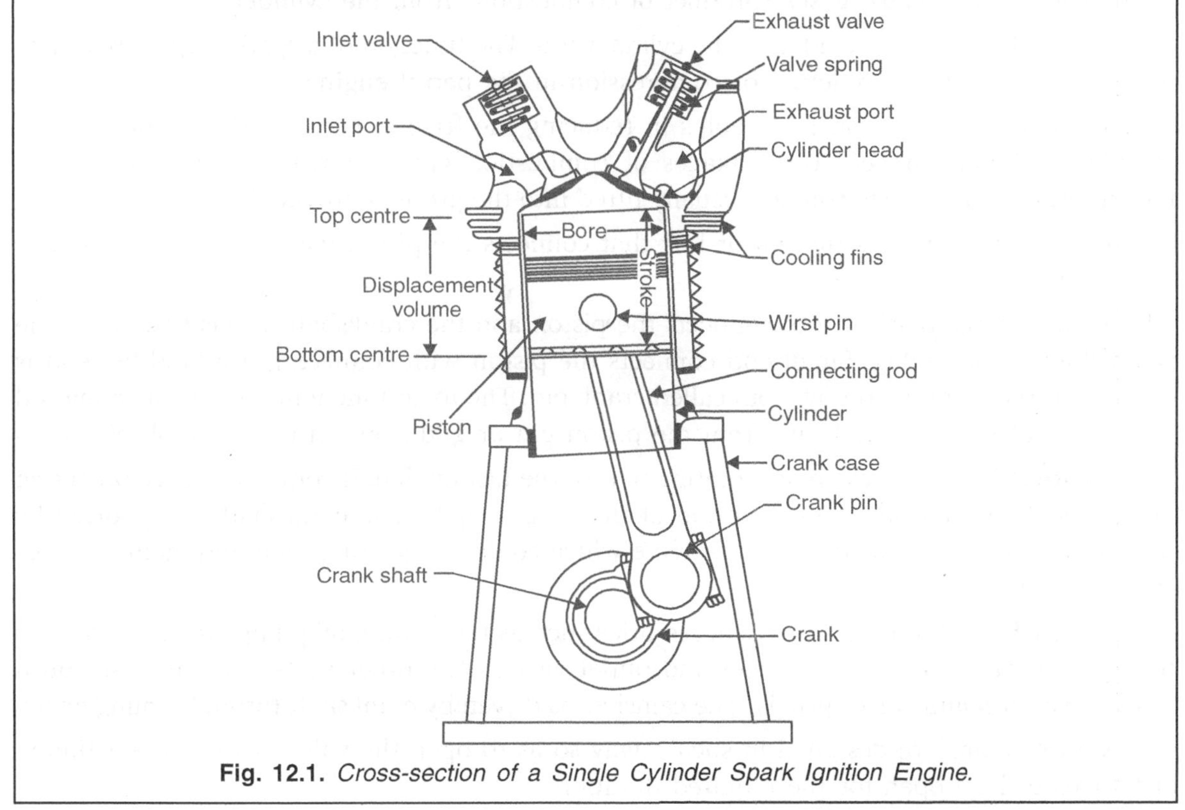

Fig. 12.1. *Cross-section of a Single Cylinder Spark Ignition Engine.*

12.2.1 Main Engine Parts

Cylinder. I.C. engines have one or more cylinders in which combustion of fuel takes place. The cylinder is supported in position by the cylinder block. It is a cylindrical vessel in which the piston makes a reciprocating motion.

Combustion Chamber. The space enclosed between the upper portion of cylinder and the piston top during the combustion process is called combustion chamber.

Piston. It is the link which transmits the energy of the expanding gases to the output shaft. Piston is a gas tight movable cylindrical part which moves up and down in the cylinder against which the combustion pressure acts.

Ports. The openings in the cylinder head for inlet and exhaust valves are called ports.

Inlet Manifold. The pipe in the single cylinder engine or system of pipes in multi-cylinder engine which connects the intake system to the inlet valve and through which air-fuel mixture is drawn inside the cylinder is called inlet manifold.

Exhaust Manifold. The pipe which connects the exhaust system to the exhaust valve of the engine and through which the products of combustion (exhaust gases) are discharged into the atmosphere is called exhaust manifold.

Inlet Valve. It is provided either on the cylinder head or on the cylinder side for regulating the charge (fuel) coming into the cylinder.

Exhaust Valve. It is also provided either on the cylinder head or on the cylinder side for removing the exhaust gases (product of combustion) from the cylinder.

Spark Plug. It is provided at the cylinder top. The function of a spark plug is to initiate the combustion after completion of compression in the petrol engine.

Piston Rings. Piston rings helps in reducing the friction between the piston and the cylinder walls and prevent the leakages of combustion gases from the combustion space. Piston rings, made of cast iron, are usually fitted into the grooves provided around the piston.

Gudgeon Pin. It is the pin or link that connects the piston with the small end of the connecting rod.

Connecting Rod. It interconnects the piston and the crankshaft. It has two ends, one is small and other is big. Small end connects the piston with connecting rod and big end is connected to the crank arm by a pin called crank pin. The main function of the connecting rod is to transmit the push and pull from the piston pin or gudgeon pin to the crank pin.

Crank Shaft. It is the main rotating part of the I.C. Engine. It converts the reciprocating motion of the piston into rotary motion of the output shaft. The crankshaft is supported by main bearings. The main body of the engine which contains the crankshaft and main bearings is called crank case.

Cam Shaft. Cam shaft drives accessories such as fuel pump, oil pump etc., and controls the opening and closing of the inlet and outlet valves. The other parts associated are push rods, rocker arm and valve springs. The camshaft is driven by crankshaft through timing gears.

Cams. Cams are designed in such a way so as to open the valve at the correct timing and to keep them open for the required duration.

Flywheel. It is a wheel mounted on the crank shaft which stores excess energy during the power stroke and returns that energy during the other strokes when needed and thus enables to keep the output torque constant.

12.2.2 I.C. Engines Terminology

Cylinder Base or Bore (d): The nominal inside diameter of the cylinder is called cylinder bore or simply bore. It is designated by letter 'd'.

Stroke (L): The nominal distance through which a working piston moves between two successive reversal of its direction of motion is called the stroke. It is equal to twice the radius of crank. It is designated by letter 'L'.

Piston Area (A): The area of a circle of diameter equal to the cylinder bore is called the piston area. It is designated by letter 'A'.

Top Dead Center: The extreme position of the piston at the topside of the cylinder head is known as Top Dead Center (TDC). In this case, the piston is farthest to the crankshaft. In case of horizontal engine this position of the piston is called Inner Dead Center (IDC).

Bottom Dead Center: The extreme position of the piston at the bottom of the cylinder is known as Bottom Dead Center (BDC). In case of horizontal engine this is known as the outer dead center. In this case the piston is nearest to the crankshaft.

Displacement or Swept Volume (V_S): The volume swept by the piston during one stroke *i.e.,* while moving from top dead center to bottom dead center is called swept volume.

Clearance Volume (V_C): The nominal volume of the combustion chamber above the piston when it is at the top dead center is known as clearance volume. It can be designated by letter V_C.

Volume of the Cylinder: It is the sum of the clearance volume and swept volume.

$\therefore$ Cylinder volume $V = V_C + V_S$

Compression Ratio (r): It is ratio of the total cylinder volume to that of the clearance volume. It is designated by letter *r.*

$$r = \frac{(V_c + V_s)}{V_c}.$$

Higher the compression ratio, better will be the ideal efficiency of an engine.

12.3 CLASSIFICATION OF INTERNAL COMBUSTION ENGINES

The internal combustion engines can be classified on the basis of:

- The number of strokes.
- Types of fuel used.
- Types of ignition.
- Types of cooling.
- Types of operating cycles.
- Types of cylinder arrangement.
- Speed.
- Action of products of combustion on the piston.
- Methods of charging.

Number of Strokes. On the basis of the number of piston strokes, the I.C. Engines can be classified in two categories:

- Four-stroke engine.
- Two-stroke engine.

Types of Fuel Used. On the basis of the types of fuel used, I.C. Engine can be classified as:

- Gasoline or Petrol Engine.
- Diesel Engine.
- CNG Engine.
- High-Speed Diesel Oil Engine.
- LPG Engine.
- Light Diesel Oil Engine.
- Multi Fuel Engine.

Working Cycle or Cycle of Operation. On the basis of cycle of operation, I.C. Engines are of two types:

- Otto Cycle Engine (also known as Spark Ignition (SI) engine or constant volume heat addition cycle engine).
- Diesel Cycle Engine (also called Compression Ignition (CI) engine or constant pressure heat addition cycle engine).

Types of Cooling. On the basis of cooling system, the I.C. Engines are of two types:

- Air-cooled I.C. Engine.
- Water cooled I.C. Engine.

Types of Ignition. Spark ignition engines need an external source of energy for the initiation of spark for starting the combustion process. Two types of ignition system are used in S.I. Engine.

- *Battery Ignition System:* Here battery is the primary source of energy for producing the spark.
- *Magneto Ignition System:* A magneto (generator) is the main source of energy for producing sparks.

Method of Charging. On the basis of the types of charging, the I.C. Engines are two types:

- Natural aspirated engines.
- Supercharged engines.

In case of naturally aspirated I.C. Engine, the air or air-fuel mixture is admitted at atmospheric pressure, whereas in case of supercharged I.C. Engine, air or air fuel mixture is admitted above atmospheric pressure.

Speed. On the basis of speed, the I.C. Engines can be classified as:

- High speed engine (speed is more than 1000 rpm).
- Low speed engine (speed is less than 400 rpm).

On the basis of the action of products of combustion upon the piston. This classification is according to the combustion of fuel taking place on one side or on both side of piston. I.C. Engine with combustion taking place on one side of the piston is called *Single Acting I.C. Engine* and if combustion takes place on both side of piston it is called *Double Acting I.C. Engine.*

On the basis of applications. I.C. engines can be classified as:

- Automotive engine for surface transport.
- Marine engine for water transport.
- Aircraft engine for aircraft propulsion.
- Industrial engine/Prime Mover for electric power generation.

Arrangement of cylinders. (Only applicable to multi-cylinder engines)

- *In-line Engine.* All cylinders are arranged linearly and transmits power to a single crankshaft.
- *V-Engine.* Two in-line engines inclined at an angle to each other and with one crankshaft.
- *Radial Engine* is one where more than two cylinders in each row are equally spaced around the crankshaft.
- *Opposed Cylinder Engine*
- *Opposed Piston Engine.*

12.4 SEQUENCE OF OPERATIONS IN I.C. ENGINE

A certain sequence of operations takes place in an I.C. Engine if it has to work successfully. The sequence is quite rigid and cannot be changed. The sequence of operations in an I.C. Engine is as follows:

- ☞ The charge (air-fuel mixture in case of petrol engine and only air in case of diesel engine) in correct proportion must be supplied to the engine cylinder.
- ☞ The air-fuel mixture in case of petrol engine and pure air in case of diesel engine is compressed in the engine cylinder during compression stroke. In case of petrol engine (SI Engine) the compression ratio is from 6 to 10 and in case of diesel engine (CI Engine), the compression ratio is from 14 to 20. In case of diesel engine the charge consists of pure air only therefore fuel (diesel oil) is injected at the end of the compression by means of a nozzle called injector.
- ☞ After compression, the ignition of charge takes place. The compressed mixture of fuel and air in case of petrol engine is ignited by means of a spark generated by a spark plug fitted at top of the cylinder whereas in case of diesel engine the ignition of fuel takes place automatically as soon as it is injected in the cylinder due to the sharp rise in temperature and pressure of the charge due to higher compression ratio. However if the temperature and pressure of the charge is not sufficiently high to cause self ignition (as it happens in winter) some external means have to be provided to produce ignition.

- ☞ After the combustion of charge (*i.e.*, fuel), the energy released causes the piston to move away from the top dead center (expansion of products of combustion takes place) on its power stroke and rotates the crankshaft. The crankshaft in turn drives the machine connected to it.
- ☞ When the expansion of products of combustion is over, the burnt out gases are exhausted (discharged) from the engine cylinder so as to allow the fresh charge (air and fuel in case of petrol engine and only air in case of diesel engine) to enter the cylinder.
- ☞ Thus the cycle is completed.

12.5 CYCLES OF OPERATION

Every I.C. engine has to go through these operations as mentioned above in order to complete one cycle of operation. A cycle of operation can be completed either in two-strokes of the piston or the four-strokes of the piston. If in an engine, a cycle of operations is completed in two revolution of crankshaft or four-strokes of the piston, the engine is called four-stroke cycle engine. If on the other hand, the cycle of operation is completed in two-stroke of the piston or in one revolution of the crank shaft, then engine is called two-stroke cycle engine. In four-stroke cycle engine there is one power stroke and three idle strokes. Whereas in two-stroke cycle engine there is one power stroke and one idle stroke. Most of the spark ignition engines are four-stroke type.

12.6 FOUR-STROKE PETROL ENGINE

Figure (12.2) depicts the operations of a four-stroke petrol engine.

Description. The credit of inventing the spark ignition engine goes to Nicolas A. Otto in 1876. In four-stroke petrol engine, the cycle of operations is completed in four-strokes of the piston or two revolutions of the crankshaft. It works on the constant volume cycle or Otto cycle. The five operations to be completed in four-strokes of the piston are suction, compression, combustion, expansion and exhaust. Crankshaft rotates by 180 degree for each of the four-strokes to be executed by the piston. This type of engine finds use in passenger cars and small aeroplanes. The working of a four-stroke petrol engine is based on these 4-strokes. These are:

(1) Suction stroke
(2) Compression stroke
(3) Expansion stroke
(4) Exhaust stroke.

Suction stroke. The suction stroke (0–1) (See Fig. 12.3) starts when the piston is at the top dead center and about to move downwards. At this time, the inlet valve is open and the exhaust valve is closed. As the piston moves downwards, suction is created in the cylinder and fresh charge of air-fuel mixture enters the cylinder via inlet valve. When the piston reaches the bottom dead center the suction stroke ends and the inlet valve closes.

Compression stroke. During compression stroke, both inlet and exhaust valves are closed, the piston moves up in the cylinder and compresses the charge of air-fuel mixture into the clearance volume (process 1–2). The compression ratio used in this type of engine is from 6 to 10 which is insufficient to ignite the air-fuel mixture therefore just before

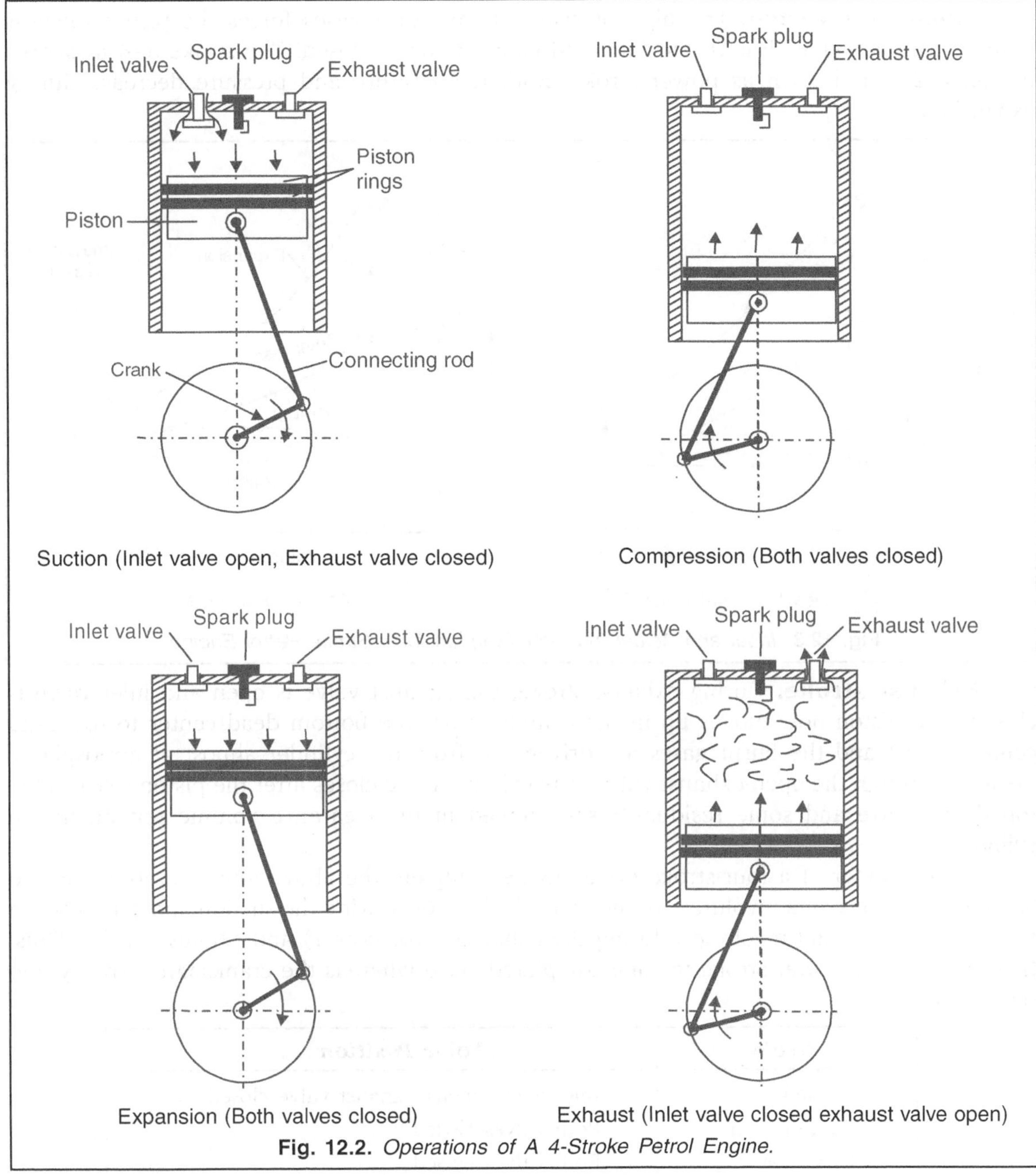

Fig. 12.2. *Operations of A 4-Stroke Petrol Engine.*

the end of compression stroke, the mixture is ignited with the help of an electric spark between the electrodes of the spark plug fitted on the cylinder head. Ignition (burning) takes place rapidly which results in rapid rise in pressure of gases and combustion takes place at constant volume as shown by line $2 \rightarrow 3$ on the indicator diagram. During the burning process, the chemical energy of the fuel is converted into heat energy producing a temperature rise of about 2000°C (process 2–3). The pressure at the end of the combustion process is significantly increased due to the heat released.

Expansion stroke. The high pressure of the burnt gases forces the piston towards bottom dead center with both the inlet and exhaust valves closed. The stroke 3–4 as shown in Fig. (12.3) is known as power stroke. Both temperature and pressure decrease during expansion.

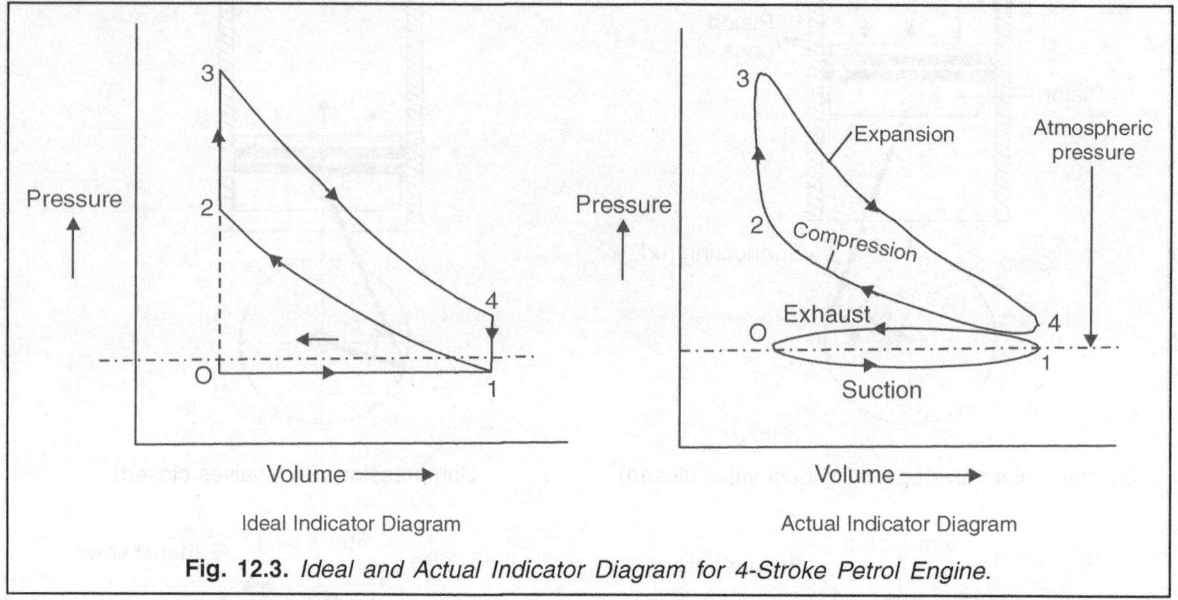

Fig. 12.3. *Ideal and Actual Indicator Diagram for 4-Stroke Petrol Engine.*

Exhaust stroke. During exhaust stroke, the exhaust valve is open and inlet valve is closed. The piston now moves up in the cylinder from the bottom dead center to top dead center (4→0) and the burnt gases are driven out from the cylinder almost at atmospheric pressure through the open exhaust valve. The exhaust valve closes after the piston reaches the top dead centre and some residual gases trapped in the clearance volume remain in the cylinder.

Each cylinder of a four-stroke cycle engine complete the above four operations in two engine revolutions, one revolution of the crankshaft occurs during the suction and compression stroke and the second revolution during the expansion (or power) and exhaust stroke. Thus, there is only one power stroke for one completed cycle whereas the crankshaft turns by two revolutions.

Stroke	***Valve Position***
Suction	Inlet valve opens/exhaust valve closed
Compression	Both valves closed
Expansion	Both valves closed
Exhaust	Inlet valve closed/exhaust valve open

The ideal as well as actual indication diagrams of four-stroke petrol engine are shown in Fig. (12.3). The suction line 0–1 is below the atmospheric pressure line. The reason being the restricted area of inlet valve through which the charge can not flow into the cylinder in sufficient quantity to keep up the atmospheric pressure with the fast moving piston.

The exhaust line 4–0 is slightly above the atmospheric pressure line. This is due to the restricted area of the exhaust passage which do not allow the exhaust gases to leave the engine cylinder quickly as a result of which the exhaust pressure remains somewhat higher than the atmospheric pressure.

The loop, which has area 4-0-1 is called negative loop. This negative loop represents the pump loss due to admission of fresh charge and removal of exhaust gases. The loop 1-2-3-4 represents the positive area. It is the gross work done by the piston during the cycle.

0-1 Suction
1-2 Compression
2-3 Combustion
3-4 Expansion
4-0 Exhaust

12.7 FOUR-STROKE DIESEL ENGINE

German engineer Rudolf Diesel had invented the diesel engine (also known as compression ignition engine) in 1892. Four-stroke Compression Ignition (CI) engine is similar to four-stroke SI engine except that the high compression ratio is used in CI engine. The compression ratio of a SI engine varies from 6 to 10 while it is 14 to 20 for a CI engine. Only air is drawn in cylinder during suction stroke in the C.I engine instead of an air-fuel mixture as in case of a petrol engine. Due to the use of high compression ratio, the temperature at the end of compression stroke is very high to self ignite the fuel which is injected into the combustion chamber. The major difference between the SI engine and CI engine is that in the former, ignition takes place at constant volume while in the case of CI engine ignition takes place at constant pressure.

The CI engine has three valves namely air suction valve, fuel valve and the exhaust valve. This type of engine is provided with a high-pressure fuel pump and an injector for injecting the fuel into the combustion chamber.

☞ The sequence of operations for the four-strokes diesel engine is shown in Fig. 12.4.

Suction Stroke (0 → 1)

Suction stroke starts with the piston at the Top Dead Center (TDC). With the inlet valve is open and exhaust valve is closed, only air (charge) is drawn in the cylinder through the inlet valve during the suction stroke when the piston moves downwards or outwards.

Compression Stroke (1 → 2)

Compression stroke starts when both the inlet as well as exhaust valves are closed, the piston moves up and compresses the air inducted during the suction stroke into the clearance space. It is compressed to a high pressure and temperature (about 3500 kPa and 600°C).

Expansion Stroke (3 → 4)

During expansion stroke both the valves (*i.e.* inlet and exhaust) remain closed. Fuel is injected through the fuel valve at the end of the compression stroke. The fuel is injected at such a rate that the pressure is maintained constant as long as fuel valve remains open. The heat produced by the high compression of air raises the temperature of air very high to ignite the fuel as soon as it is injected into the cylinder. After the injection of fuel is completed, the products of combustion expand. The products of combustion drive the piston down on its power stroke to the Bottom Dead Center (BDC).

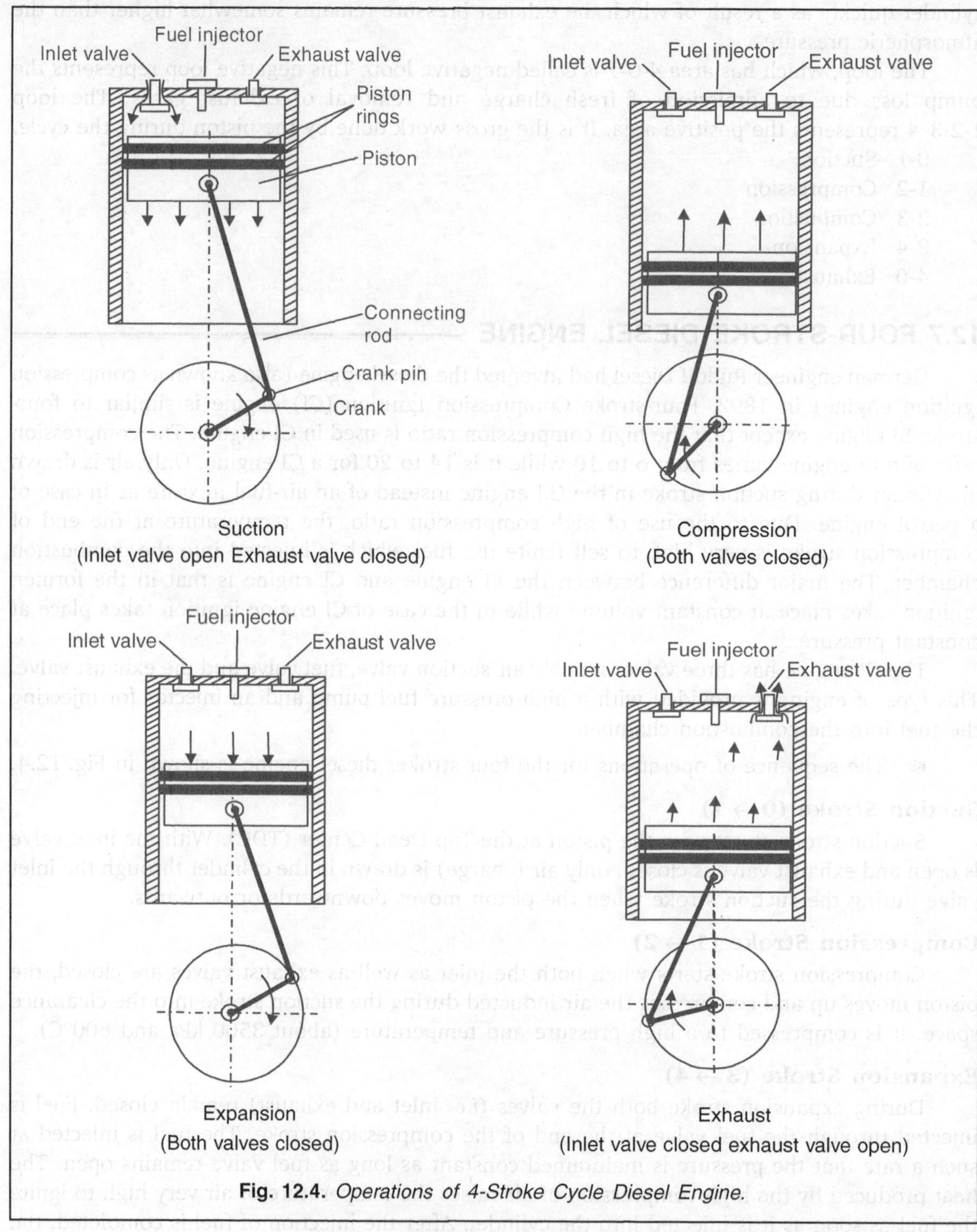

Fig. 12.4. *Operations of 4-Stroke Cycle Diesel Engine.*

Exhaust Stroke (1 → 0)

During this stroke intake valve is closed and exhaust valve is open. The piston while moving from BDC to TDC, drives the burnt gases (products of combustion) into the atmosphere through the open exhaust valve so as to make room for fresh charge (only air) to enter in the cylinder. The exhaust valve closes after the piston reaches the TDC. This completes the cycle.

☞ Due to the use of high pressure, the diesel engine is heavier than petrol engines. Diesel engine has high thermal efficiency because of high compression ratio and greater expansion. The ideal as well as actual indicator diagrams of four-stroke diesel engine are shown in Fig. (12.5).

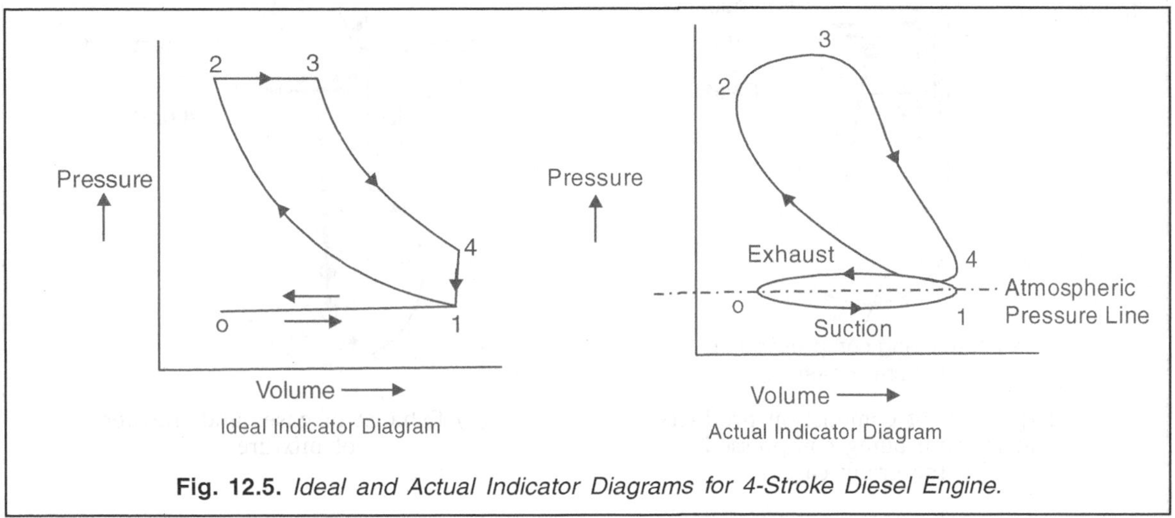

Fig. 12.5. *Ideal and Actual Indicator Diagrams for 4-Stroke Diesel Engine.*

12.8 TWO-STROKE PETROL ENGINE

In two-stroke engine, the cycle of operations is completed in two-strokes *i.e.,* in one revolution of the crankshaft. The main difference between the two-stroke and four-stroke engines is in the way of taking the fresh charge and removing the burnt gases from the cylinder.

The two-stroke engine cycle was invented by a British engineer Sir Dugald Clark (1878), which could be completed in two-strokes of the piston rather than in four-strokes. His work was based on the concept that if the suction and exhaust could be carried out by an alternative arrangement specially without the movement of the piston then there will be a power stroke for each revolution of the crankshaft. In such type of arrangement theoretically the power output of the engine can be doubled for the same speed as compared to a four-stroke engine. *Two-stroke petrol engines are used in scooters, motor-cycles etc.*

In two-stroke engine, suction is accomplished by air compressed in crank case. The induction of compressed charge moves out the product of combustion through exhaust ports. Therefore, no piston strokes are required for carrying out suction and exhaust operations. Two-strokes are sufficient to complete the cycle, one for compressing the fresh charge and the other for expansion or power stroke.

Figure (12.6*a*) shows the two-stroke petrol engine using crank case compression. In this engine the charge of air-fuel mixture is compressed in the crankcase by the underside of the piston during the expansion or power stroke. Figure (12.6*d*) shows the charge (air and fuel mixture) is sucked through the inlet valve when the pressure in the crankcase falls due to upward motion of the piston during compression stroke.

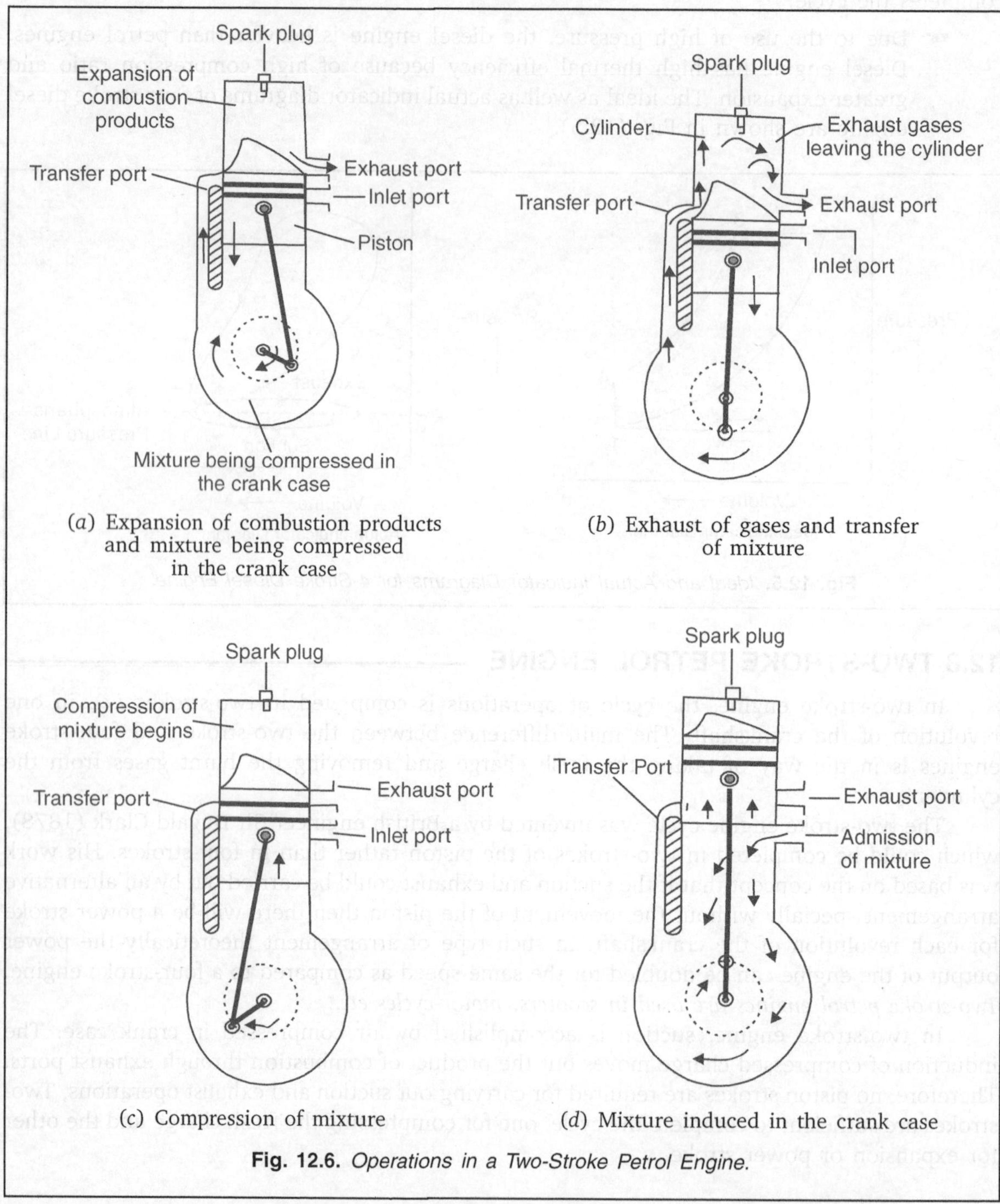

(*a*) Expansion of combustion products and mixture being compressed in the crank case

(*b*) Exhaust of gases and transfer of mixture

(*c*) Compression of mixture

(*d*) Mixture induced in the crank case

Fig. 12.6. *Operations in a Two-Stroke Petrol Engine.*

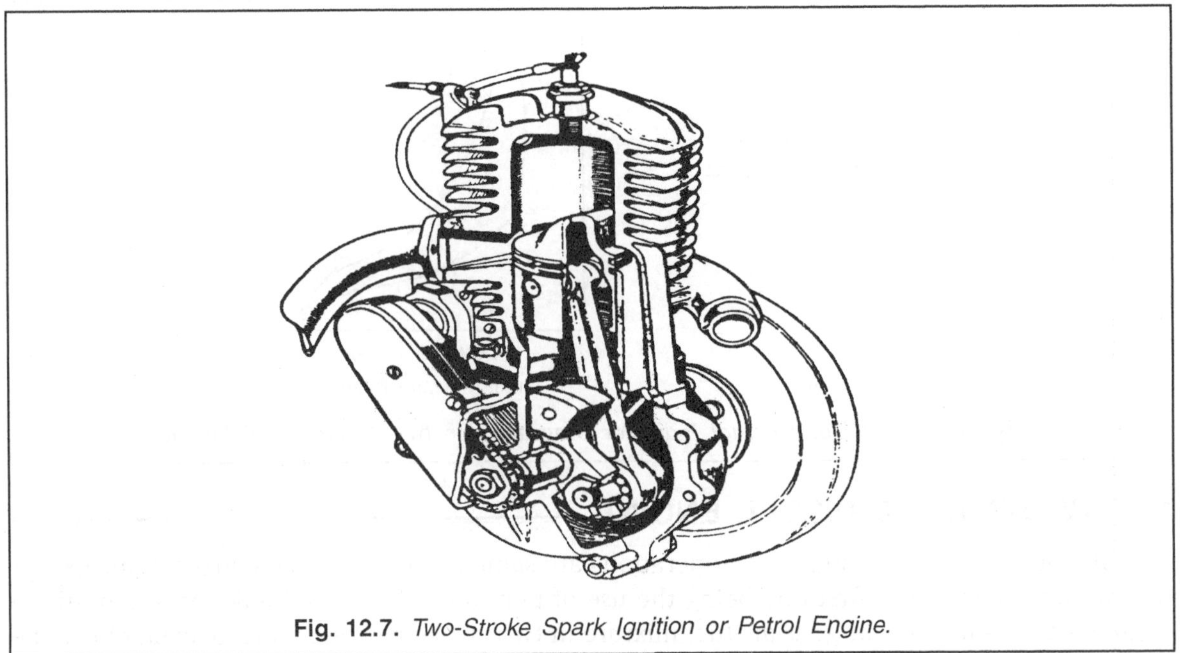

Fig. 12.7. *Two-Stroke Spark Ignition or Petrol Engine.*

After the compression of charge, ignition of charge takes place with the help of a spark plug located at the top of the cylinder head, which is followed by expansion of the products of combustion. During the expansion stroke, charge in the crankcase is compressed. When expansion is about to complete, the piston opens (uncovers) the exhaust port Fig. (12.6*a*). The expanded gases leave the cylinder through the exhaust port after doing work on the piston. The cylinder pressure drops to atmospheric pressure as the combustion products leave the cylinder. A little later the transfer port is uncovered by the piston. This allows the transfer of compressed charge in crankcase to enter the engine cylinder Fig. (12.6*b*). The piston is so designed that fresh charge of fuel and air (compressed in crankcase) will sweep up to the top of the cylinder and push out the remaining exhaust gases through the exhaust port. The top of the piston is designed with a deflector and the ports are inclined at 45°. Both these measures help in scavenging. Further the deflector and specially shaped ports prevent the fresh charge from flowing directly to the exhaust port and being lost. The purpose of scavenging is to clear the upper portion of the cylinder from exhaust gases so as to prevent the loss of fresh charge going out through exhaust port. There is however a loss of unburnt fuel through the exhaust port while scavenging and charging the cylinder with fresh charge. Figure (12.7) shows a close up of the two-stroke spark ignition or Petrol engine.

☞ During compression stroke, the pressure in the crank case lowers and the fresh charge is sucked in the crank case through inlet valve provided on the body of the crankcase. The compression of this charge takes place during the power (expansion) stroke of the piston. Thus, in crankcase scavenging engine, both sides of piston are working sides. On one side the positive power is obtained and the other side, (*i.e.*, crankcase side) acts as a compressor absorbing power. Net power produced by the engine is the difference of the two. Figure (12.8) shows the ideal and actual indicator diagrams for two-stroke petrol engine.

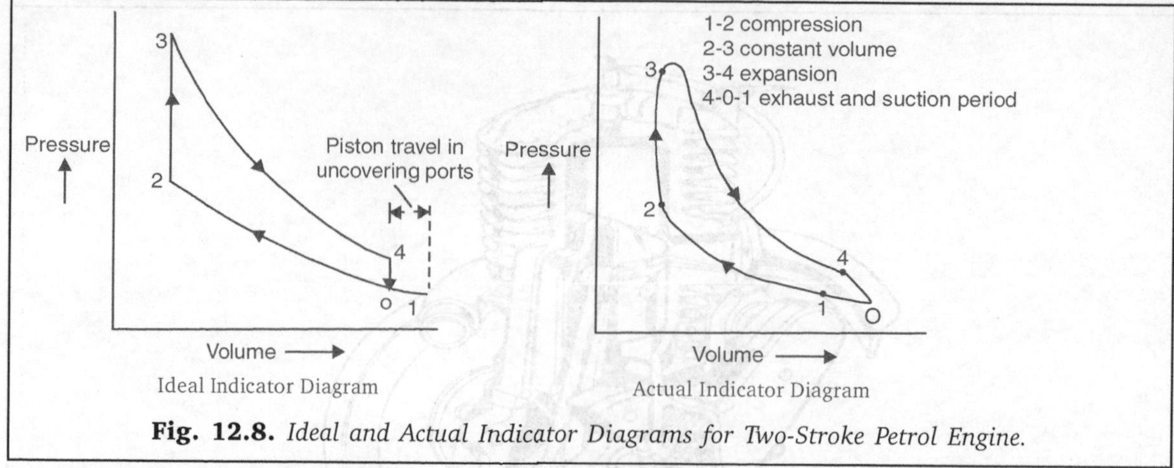

Fig. 12.8. *Ideal and Actual Indicator Diagrams for Two-Stroke Petrol Engine.*

12.9 TWO-STROKE DIESEL ENGINE

In two-stroke diesel engines all operations are same as in two-stroke petrol engine except the two differences. The first one being the use of pure air only as the intake charge in diesel engine whereas intake charge is air-fuel mixture in two-stroke petrol engine. Second one is the use of spark plug in two-stroke petrol engine whereas fuel injector is fitted to supply fuel in two-stroke diesel engine for combustion purposes. All other things remain the same. Figure (12.9) depicts the different operations being carried out in a two-stroke diesel engine. The ideal and actual indicator diagrams for the two-stroke diesel engine are plotted in Fig. (12.10).

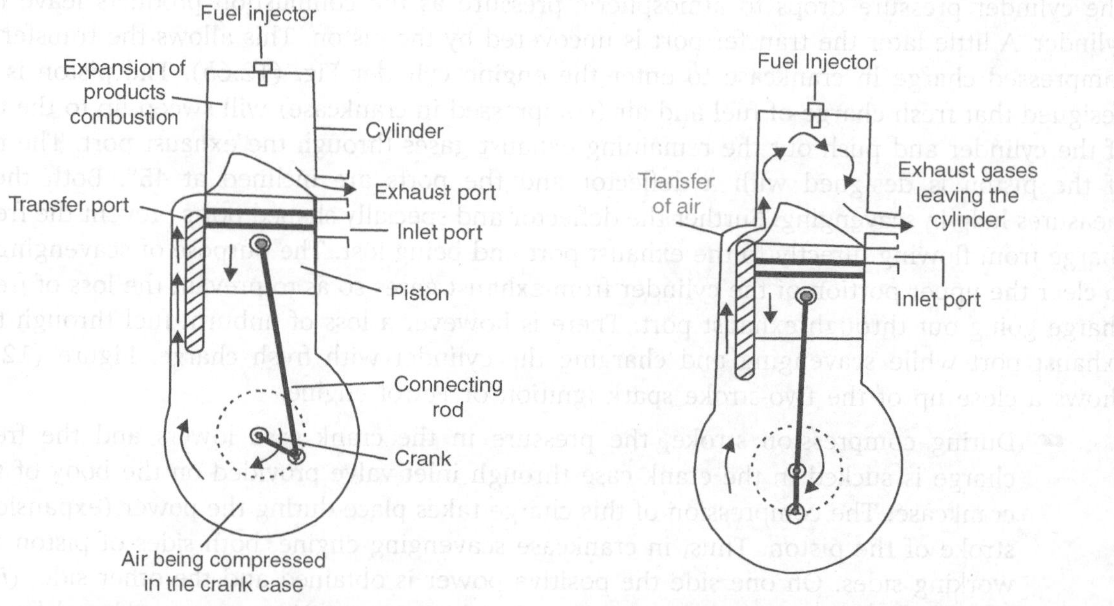

(*a*) Expansion of products in the cylinder and compression of air in the crank case

(*b*) Exhaust of burnt-out gases and transfer of charge (air only)

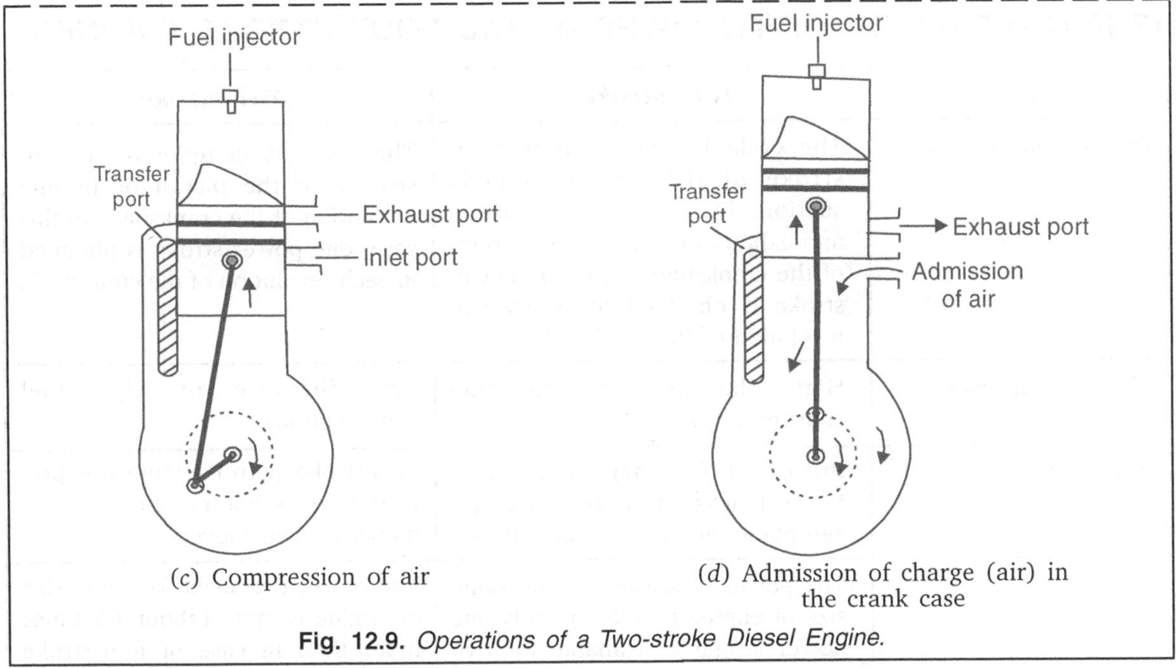

Fig. 12.9. *Operations of a Two-stroke Diesel Engine.*

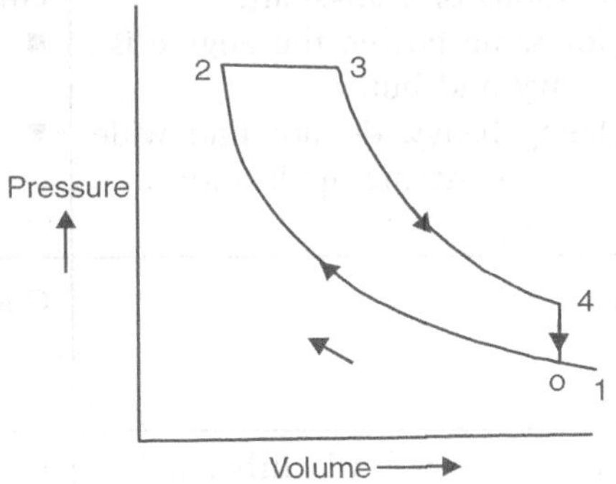

Ideal Indicator Diagram

Fig. 12.10(*a*). *Ideal Indicator Diagram for a Two-Stroke Cycle Diesel Engine.*

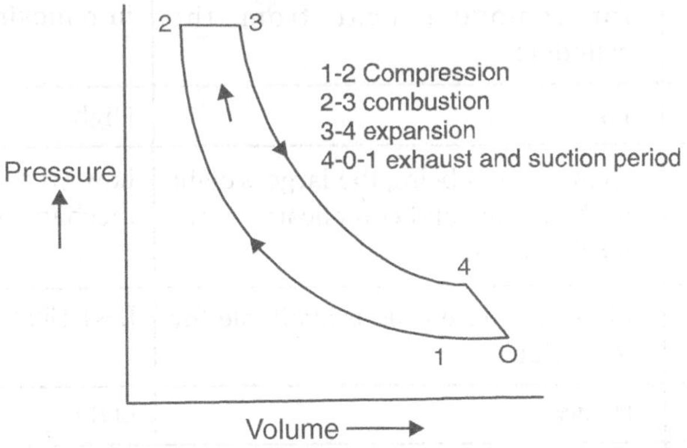

1-2 Compression
2-3 combustion
3-4 expansion
4-0-1 exhaust and suction period

Actual Indicator Diagram

Fig. 12.10(*b*). *Actual Indicator Diagram for a Two-Stroke Cycle Diesel Engine.*

12.10 COMPARISON OF TWO-STROKE AND FOUR-STROKE ENGINES

Basis	***Four-stroke***	***Two-stroke***
Completion of cycle	The cycle is completed in four-stroke of the piston (namely suction, compression, expansion and exhaust) or in two revolutions of the crankshaft. Thus one power stroke is obtained for every two revolution of the crankshaft.	The cycle is completed in two-strokes of the piston or in one revolution of the crankshaft. In this case, one power stroke is obtained in each revolution of the crankshaft.
Thermal efficiency	High (because of low fuel consumption).	Low (because of high fuel consumption).
Applications	Preferred for large size cars, buses, trucks, industrial engines, aeroplanes and power generators.	Two-stroke petrol engine are preferred for scooters, motorcycles, mopeds, lawn movers.
Power produced	The power produced for the same size of engine is small as only one power stroke is available for two revolutions of crankshaft. ■ for same power, the engine is heavy and bulky. ■ being heavy, do not find wide use for marine applications.	The power produced for same size of engine is more (about 1.3 times more than in case of four-stroke engine). ■ for the same power the engine is light and compact. ■ being light and compact find, extensive use for marine applications.
Scavenging (removal of exhaust gases)	Better	Good
Cooling and lubrications requirements	Low cooling and lubrication requirements (because of one power stroke in two revolutions of crankshaft as more time is available for removing heat from the cylinder).	High cooling and lubrication requirement because of one power stroke in one revolution of crankshaft as less time is available in removing heat from the cylinder.
Wear and tear	Low	High
Initial cost	High (reason being the large weight of the engine and complicated valve mechanism).	Low (reason being absence of valve mechanism).
Volumetric efficiency	More (as more time is available for induction).	Less (less time for induction).
Flywheel	Heavy	Light
Valve and valve mechanism	Contains valves and complicated valve mechanism to open and close the valves.	No valves but only ports.

12.11 COMPARISON OF SPARK IGNITION (SI) AND COMPRESSION IGNITION (CI) ENGINE

	Description	*SI Engine*	*CI Engine*
1.	Thermodynamic cycle used	Otto cycle (constant volume combustion cycle).	Diesel cycle (constant pressure combustion cycle).
2.	Fuel used	Petrol	Diesel.
3.	Air-fuel ratio	10: 1 to 20: 1	18: 1 to 100: 1.
4.	Compression ratio	6 to 10	14 to 20.
5.	Fuel Induction	Fuel and air are inducted as gaseous mixture during suction stroke. A carburettor is necessary to supply the air-fuel mixture.	Fuel is directly injected into combustion chamber at high pressure at the end of the compression stroke. A fuel pump and injector are necessary.
6.	Ignition system	Requires an ignition system with spark plug in the combustion chamber. Primary voltage is provided with a battery.	Self ignition occurs due to high temperature of air because of the high compression ratio. Spark plug is not required.
7.	Speed	High (3000 to 6000 rpm).	Low–500 rpm. Medium (500 to 1500 rpm). High (1500 to 3600 rpm).
8.	Running cost	High	Low
9.	Thermal efficiency	Low (Because of low compression ratio, thermal efficiency is low (30%)).	High (Because of high compression ratio, thermal efficiency is high (40%)).
10.	Weight	Light due to low peak pressure.	High due to high peak pressure.
11.	Space required	Less	More.

12.12 FUEL SUPPLY IN PETROL ENGINES

In SI engine, air-fuel mixture for combustion is prepared outside the engine cylinder. The process of formation of a combustible air-fuel mixture by mixing the proper amount of fuel with air before admitting it to cylinder for combustion is called carburetion. This process is a part of the induction system as shown in Fig. (12.11). **Carburettor** is a device for mixing the air and fuel (petrol) in correct proportion. The carburettor is the most important part of the fuel induction system.

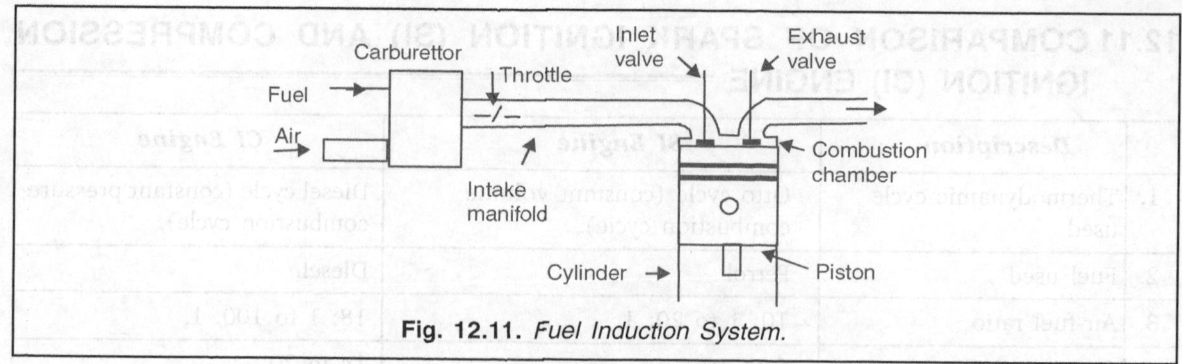

Fig. 12.11. *Fuel Induction System.*

Proper air-fuel mixture should be supplied to engine cylinder as engine needs to run at different speeds as per the requirements. Three types of air-fuel mixture in common use are:

- *Chemically correct mixture* (also known as "Stoichiometric mixture").

  Chemically correct mixture is one, which possesses just enough air for the complete combustion of the fuel. The value of chemically correct mixture is 15 : 1 (15 parts of air and 1 part of petrol).

- *Rich mixture*

  A mixture that contains less air than the chemically correct mixture is called a rich mixture. For example 12 : 1, 10 : 1 are called rich mixtures.

- *Lean mixture*

  A mixture that contains more air than the chemically correct mixture is called a lean mixture. For example 18 : 1, 21 : 1 come under lean mixture's category.

12.12.1 A Simple Carburettor

The main functions of a carburettor are:

- To maintain a small reserve of petrol at constant level in the float chamber.
- To vaporize the petrol by means of the engine suction and to produce homogeneous air-fuel mixture as per the operating requirements of the engine.
- The air-fuel mixture produced by it must be within combustible range.
- To maintain a correct mixture at all speeds.

The Fig. (12.12) shows a simple carburettor. It consists of a float chamber, nozzle with metering orifice, a venturi and a throttle valve. Petrol enters the float chamber through a filter and a needle valve system. A small quantity of petrol is maintained at constant level in the float chamber. When the level of petrol in the float chamber goes down, valve spindle moves up and thus fresh fuel *i.e.*, petrol enters the chamber. When the correct level is reached the float closes the fuel supply valve. When engine runs, the air is sucked through the choke (or venturi tube) by the engine suction. During suction stroke air is drawn through venturi. Venturi is a tube of decreasing cross-section with minimum area at the throat. The choke tube surrounding the top of the fuel jet is reduced in diameter so as to increase the velocity of air at this point and reduce its pressure. Atmospheric pressure exists on top of the float chamber being produced by air vent (hole). Because of the pressure difference between the float chamber and the choke tube at the throat, fuel is discharged into air stream and gets vaporized. The tip of the jet is placed slightly higher by about 1.5 mm than the normal level of petrol in the float chamber so as to avoid leakage of petrol in the float chamber when there is no air flow or when the engine is at rest. The amount of petrol coming out from the main jet depends on the velocity of air through the choke tube. In motor cars the throttle valve is operated by the accelerator. The air-petrol ratio for chemically correct combustion is 15 : 1.

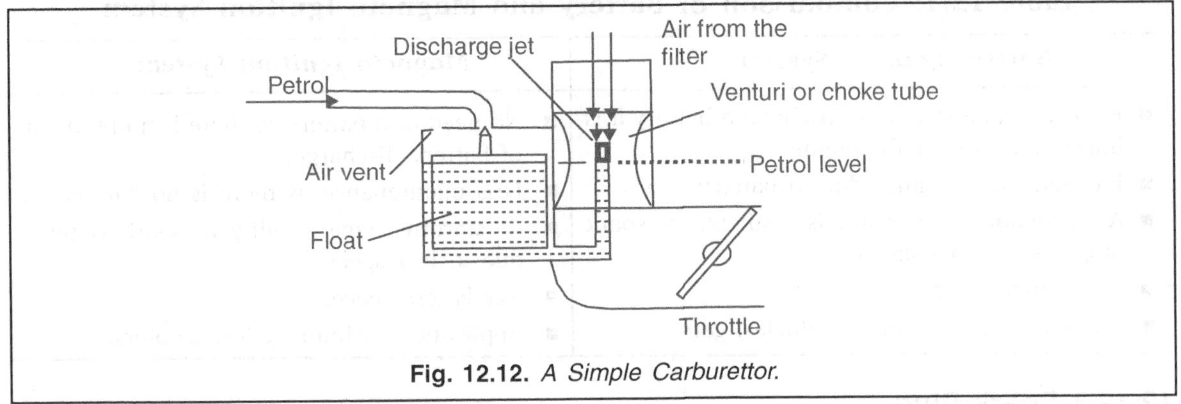

Fig. 12.12. *A Simple Carburettor.*

As the speed increases the flow of petrol and air will not increase in the same proportion. Petrol is more responsive to suction than air. The laws regulating the flow of liquid for a jet and air through the venturi are not the same as one is liquid and other is a gas. Consequently as the engine speed increases the flow of petrol into the engine increases faster than the flow of air and this makes the mixture too rich at high speed.

To overcome this shortcoming of the carburettor two jets are used for this purpose. One is the main jet and the other is compensating jet.

12.12.2 Ignition System of Petrol Engine

In petrol engine ignition takes place by means of an electric spark at the end of the compression stroke. There are three systems of ignition for a petrol engine. These are:

- Battery or Coil ignition system
- Magneto ignition system
- Electronic ignition system.

Battery or coil ignition system consists of the following components:

(*i*) Battery (6 or 12 volts)

(*ii*) Ammeter

(*iii*) Switch

(*iv*) Induction coil

(*v*) Contact breaker

(*vi*) Condenser

(*vii*) Distributor

(*viii*) Spark plug.

The magneto ignition system is similar in principle to the coil ignition system except the difference in the source of electrical energy. Battery ignition system uses a battery whereas the magneto ignition system uses a magneto (generator) to supply low voltage, all other system components being similar. Primary coils of a few turns and secondary coils of large number of turns are wound across the magneto or generator.

Table (12.1) gives the comparison of battery and magneto ignition system.

Table 12.1: Comparison of Battery and Magneto Ignition System

Battery Ignition System	*Magneto Ignition System*
■ Battery is a must as with discharged battery it is impossible to start the engine.	■ No need of a battery so there is no problem of battery discharge.
■ Frequent maintenance due to battery.	■ Less maintenance as there is no battery.
■ A reasonably good spark is available at spark plug even at low speed.	■ At starting stage, quality of spark is poor due to low speed.
■ Needs more space.	■ Needs less space.
■ Applications: light motor vehicles, cars.	■ Applications: Motor cycles, scooters.

12.12.3 Spark Plug

Spark plug is the means of transforming the requisite voltage generated by the ignition system into a spark within the combustion chamber of the cylinder in which the piston has reached Top Dead Centre (TDC) at the end of its compression stroke and the combustible mixture is ready for ignition.

A spark plug as shown in Fig. (12.13) consists of a steel shell, an insulator and two electrodes. The central electrode is enclosed in a ceramic insulator. The other electrode is welded to the steel shell of the plug and thereby automatically grounded when the

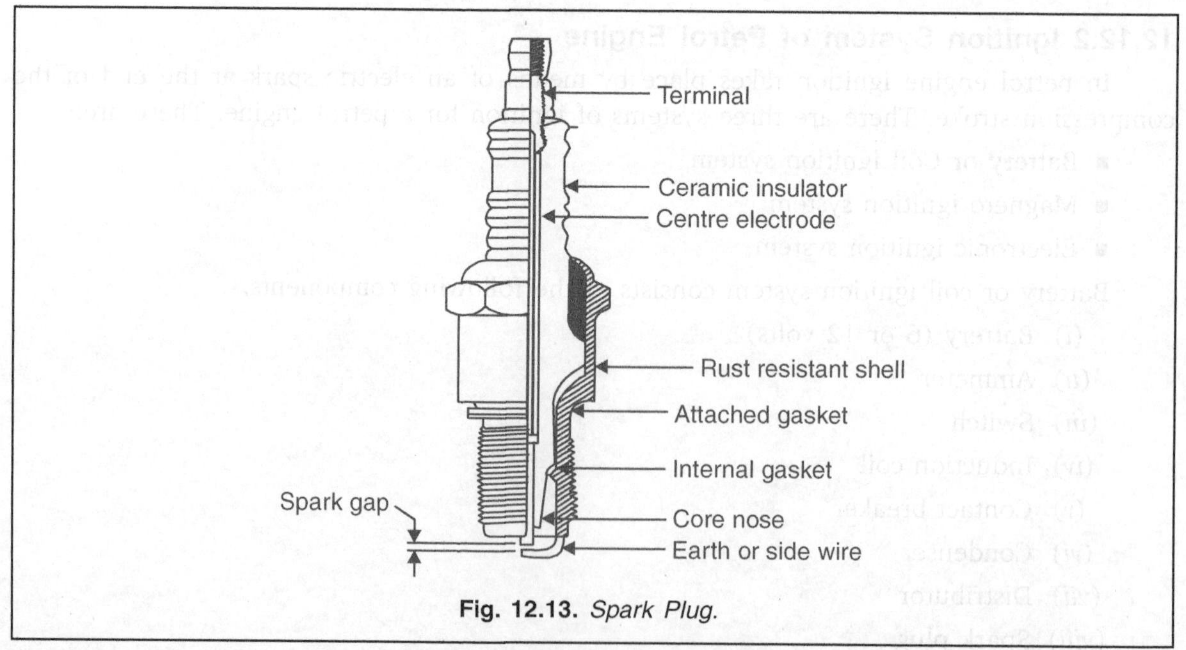

Fig. 12.13. *Spark Plug.*

plug is fitted on the cylinder head of the engine. The electrodes are usually made of high nickel alloy to withstand the severe erosion and corrosion during their use. A spark plug consists of a threaded outcasing which can be screwed in the combustion chamber. Within the casing is an insulated electrode sealed to prevent leakage to the spark plug gap. The spark jumps the air gap between the ends of the two electrodes. The gap between the electrodes varies from 0.45 mm to 0.65 mm.

12.13 FUEL SUPPLY IN DIESEL ENGINE

In diesel engine, pure air is compressed in the cylinder and then fuel oil is injected in the engine cylinder against the pressure of compression that is about 20–35 bar. Diesel engines must have some external source of forcing the fuel oil into the cylinder. For diesel engines the fuel oil may be injected by two methods.

- Air injection method.
- Solid injection or airless injection method.

The complete system requires the following parts:

(*i*) A fuel pump to deliver fuel oil under pressure to the injection nozzle.

(*ii*) An injector to inject the fuel oil into the cylinder in a finely atomized state.

(*iii*) A governor to regulate the fuel supply according to load.

(*iv*) An air compressor in case of air injection engine.

In air injection system the fuel is injected into the cylinder by means of compressed air. This system needs a multi stage air compressor and because of this it is no more preferred.

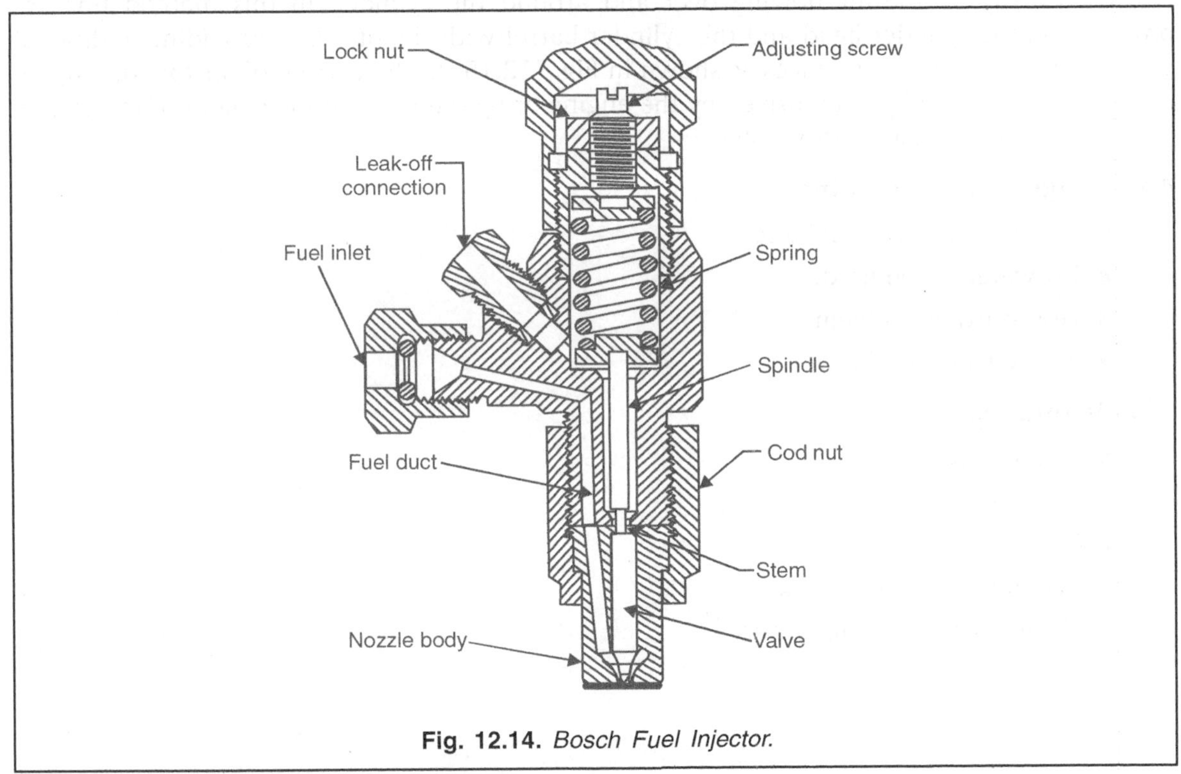

Fig. 12.14. *Bosch Fuel Injector.*

In solid injection system the fuel is injected directly into the combustion chamber without using the compressed air. Hence, it is known as airless injection. This method employs a mechanically operated fuel pump which raises the pressure of the fuel and inject it in atomized form through a fuel injection nozzle (injector) under high pressure. The injection pressure normally varies from 10000 kPa or more than this. The schematic diagram of a fuel injector is shown in Fig. (12.14).

12.14 COOLING OF I.C. ENGINE

In an I.C. engine as the combustion takes place, very high temperature is developed in the cylinder which raises the temperature of the cylinder. Heat is also developed due to friction between the piston rings and cylinder. If the heat is not taken away it will cause excessive rise in the temperature. This will cause:

1. Seizure of piston in cylinder.
2. Burning of lubricating oil.
3. Cracking of certain parts such as cylinder and piston due to thermal stresses.

Therefore, the temperature is not allowed to exceed above 150°C by cooling the engine. Too much cooling on the other hand will reduce the thermal efficiency of the engine. The amount of heat to be removed is about 30% of the heat of the combustion. The cooling may be done by air or water.

12.14.1 Air Cooling

It is used in case of two wheelers (bikes and scooters) and aircrafts. In this method heat is carried away by the air flowing over and around the cylinder. In this method fins are provided on the cylinder head and the cylinder barrel with the object of providing additional conductive and radiating surfaces as shown in Fig. (12.15*a*). The current of air for cooling the fins may be provided by a fan driven by the engine or by the forward movement of the engine itself as in case of bikes and scooters.

Advantages of air cooling

- Cheap and easy installation.
- No water is required.
- Less starting problem.
- No freezing troubles.

Disadvantages

- Limited size.
- Non-uniform cooling.
- Noisy.
- Maintenance is not easy.
- Higher working temperature.

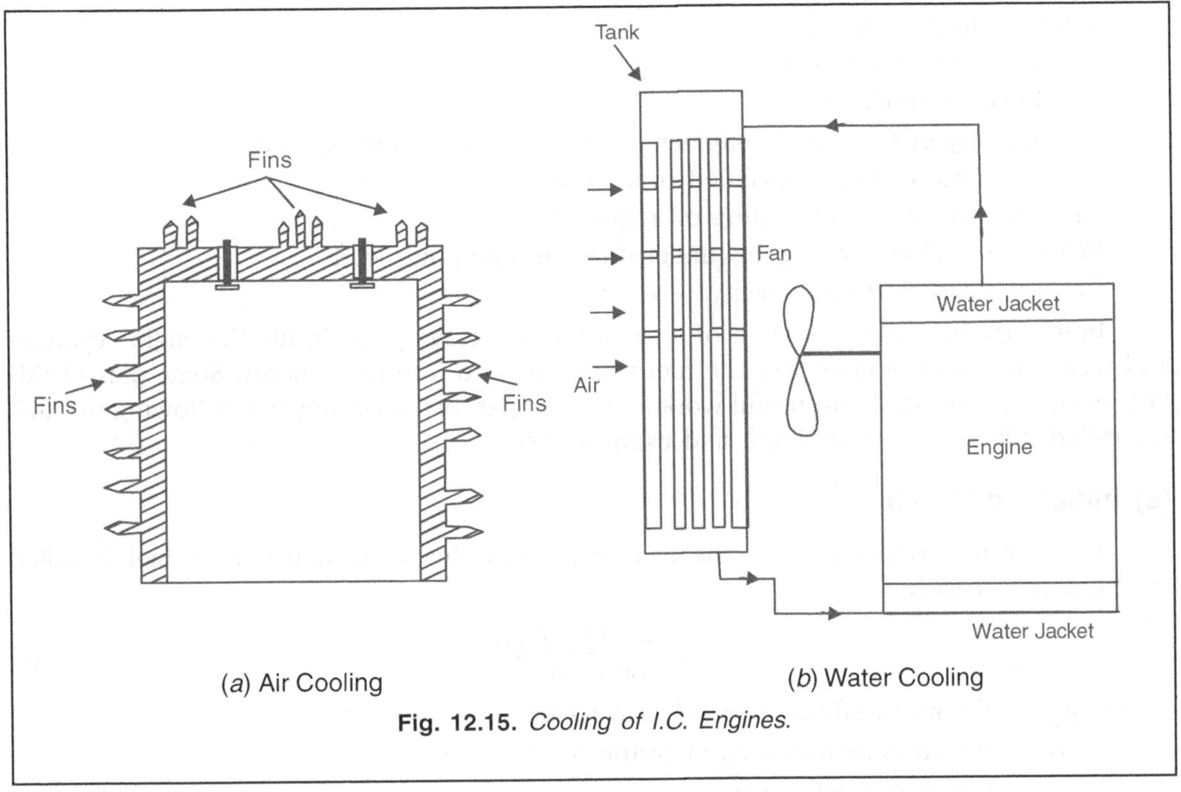

Fig. 12.15. *Cooling of I.C. Engines.*

12.14.2 Water Cooling

In water cooled IC engines, water jackets are provided around the cylinder in which cooling water flows. This method gives better cooling as compared to air cooling on account of the superior convective and conductive properties of water. Figure (12.15*b*) shows the water circulation system of an automobile. Heat is transferred from cylinder walls to the water. The water becomes heated in its passage through the jackets and is cooled itself by means of an air cooled radiator system. Water is only a media which takes away heat and is cooled in radiator tubes by flowing air.

12.15 PERFORMANCE OF I.C. ENGINES

Engine performance is an indication of the degree of success with which the engine performs its assigned task *i.e.*, the conversion of chemical energy contained in the fuel into useful mechanical work.

Engine performance can be judged by selecting certain basic parameters. The effect of operating conditions, design concepts, and modifications on these basic parameters are studied so as to know the performance of the engine. The important basic parameters which affect the engine performance are:

- Power output
- Mechanical efficiency
- Air-fuel ratio
- Thermal efficiency

- Mean effective pressure
- Specific fuel consumption
- Volumetric efficiency

The majority of tests on IC engines are carried out to find out the:

(*i*) rated power under specified conditions
(*ii*) the quantity of lubricating oil required
(*iii*) the steadiness of engines under different operating loads
(*iv*) Overload carrying capacity.

In IC engines, the combustion of air-fuel mixture takes place inside the engine cylinder and there the chemical energy of the fuel gets converted into heat energy. Some part of this heat energy is converted into useful work and some gets lost to cooling water flowing through the cylinder jackets, surroundings, and exhaust gases.

(*a*) Indicated Power

The power developed inside the engine cylinder due to combustion of fuel is called indicated power (IP).

$$IP = \frac{p_m A\, L\, k\, n}{60 \times 100} kW \qquad \ldots(i)$$

where p_m = the mean effective pressure in N/m^2

A = the cross section area of piston in m^2

L = length of stroke in m

N = engine speed in rpm,

k = number of cylinders

n = the number of working strokes per minute

for a 2-stroke engine $n = N$

for a 4-stroke engine $n = \frac{N}{2}$

(*b*) Brake Power

The power developed by an engine at the output shaft is called the brake power (BP). It is usually measured by means of a brake mechanism *i.e.*, prony brake or rope brake dynamometer.

Let

W = brake load in N

l = length of arm in m

N = engine speed in rpm

T = Torque in Nm

ω = angular speed in rad/s

$$BP = \frac{T \times \omega}{60 \times 1000} kW$$

$$= \frac{2\pi NT}{60 \times 1000} kW \qquad \ldots(ii)$$

(*i*) When torque is to be determined by using prony brake dynamometer

Let

$$\text{Torque} = \text{load} \times \text{length of arm}$$

$$T = W \times l$$

$$\therefore \quad BP = \frac{2\pi NWl}{60 \times 1000} \; kW \qquad ...(iii)$$

(*ii*) When torque is to be determined by using prony brake dynanometer

Let

W = Weight at the end of the rope in N

S = spring balance reading in N

N = engine speed in rpm

D = diameter of the brake drum in m

d = diameter of the rope in m

(D + d) = effective diameter of the brake wheel

$$\therefore \quad BP = \frac{(W - S)\,\pi\,(D + d)\,N}{60 \times 1000} kW \qquad ...(iv)$$

when diameter of the rope is neglected then

$$BP = \frac{(W - S)\,\pi\,DN}{60 \times 1000} kW$$

(*c*) Friction Power

The difference between indicated power and brake power is called the friction power (FP).

$$FP = IP - BP \qquad ...(v)$$

(*d*) Thermal Efficiency

Thermal efficiency of an engine is the indicator of the conversion of heat supplied into work energy. It is the ratio of power developed by the engine to the heat supplied by the fuel.

Mathematically

$$\text{Thermal efficiency} = \frac{\text{Power developed}}{\text{Heat supplied by the fuel}} \qquad ...(vi)$$

It is based either on Indicated Power (IP) or on Brake Power (BP).

(*e*) Indicated Thermal Efficiency

The indicated thermal efficiency is the ratio of the indicated power to the heat supplied by the fuel.

$$\text{Indicated thermal efficiency} = \frac{\text{Indicated Power}}{\text{Heat supplied by the fuel}}$$

$$\eta_i = \frac{IP}{\dot{m}_f \times CV} \quad ...(vii)$$

Where

$\dot{m}_f$ = mass of fuel consumed in kg/s

= fuel consumed in litres per second × specific gravity of fuel

CV = calorific value of fuel in kJ/kg

IP = indicated power in kJ/s

(*f*) Brake Thermal Efficiency

It is the ratio of brake power to heat supplied by the fuel. In IC engines this is also known as overall efficiency.

$$\eta_b = \frac{BP}{\dot{m}_f \times CV} \quad ...(viii)$$

$\dot{m}_f$ = mass of fuel consumed in kg/s

CV = calorific value of fuel in kJ/kg

BP = brake power in kJ/s

(*g*) Mechanical Efficiency

It is the ratio of brake power to the indicated power.

$$\eta_{mech} = \frac{BP}{IP} \quad ...(ix)$$

The mechanical efficiency depends on the design of the engine, its condition and load. For different types of engines, its value usually varies from 75 to 90%.

Mechanical efficiency is defined as ratio of brake thermal efficiency to indicated thermal efficiency.

$$\eta_{mech} = \frac{\eta_b}{\eta_i}$$

$$\eta_b = \eta_i \times \eta_{mech} \quad ...(x)$$

(*h*) Efficiency Ratio or Relative Efficiency

This is the ratio of indicated thermal efficiency to the corresponding air standard cycle efficiency. It shows how close the actual engine comes to the theoretically possible performance.

$$\text{Efficiency ratio} = \frac{\text{indicated thermal efficiency}}{\text{air standard efficiency}} \quad ...(xi)$$

(*i*) Specific Fuel Consumption

It is the mass of the fuel consumed per kW power developed per hour.

Brake specific fuel consumption

$$bsfc = \frac{\dot{m}_f}{BP} \text{ kg/kWh} \quad ...(xii)$$

SOLVED PROBLEMS

Problem 1. *An inventor claims that his petrol engine operating between temperatures of the sink and source are 727°C and 2100°C will produce 0.75 kW consuming 0.12 kg/hr of petrol having 46000 kJ/kg calorific value. Check the validity of the claim.*

Solution.

Given:

$$T_3 = 727 + 273 = 1000 \text{ K}$$
$$T_1 = 2127 + 273 = 2400 \text{ K}$$
$$\text{work done} = 0.75 \text{ kW}$$
$$\text{fuel consumption rate} = 0.12 \text{ kg/hr}$$
$$\text{CV} = 46000 \text{ kJ/kg}$$

As we know that the maximum efficiency between two specific temperatures is given by the Carnot cycle efficiency

$$\eta_{Carnot} = \frac{T_1 - T_3}{T_1} = \frac{2400 - 1000}{2400}$$
$$\eta_{Carnot} = 0.5833$$
$$\eta_{Carnot} = 58.33\%$$

We also know that the heat supplied by fuel to engine is given by

$$Q_s = \text{fuel consumed} \times \text{CV}$$
$$Q_s = 0.12 \times 46100$$
$$Q_s = 0.5532 \text{ kJ/hr}$$
$$Q_s = 1.536 \text{ kJ/s}$$

$$\text{work done} = 0.75 \text{ kW} = 0.75 \text{ kJ/s}$$

$$\text{Efficiency claimed} = \frac{\text{Work done}}{\text{Heat supplied}}$$

$$\text{Efficiency claimed} = \frac{0.75}{1.536} = 0.4883$$

$$\text{Efficiency claimed} = 48.83\%$$

The efficiency claimed by the inventor is 48.83% and the maximum possible efficiency as given by Carnot cycle is 58.33%. Since the eficiency claimed is less than the Carnot cycle efficiency therefore the inventor's claim is valid.

Problem 2. *The following particulars refer to a 4-stroke engine:*

Bore = 300 mm

Stroke = 450 mm

Crank speed = 250 RPM

Net load on the brake drum = 1200 N

Diameter of brake drum = 2000 mm

Indicated mean effective pressure = 6 bar

Fuel consumption = 0.0013 kg/s

Specific gravity of fuel = 0.78

Calorific value of fuel = 44000 kJ/kg

Find out (i) Brake power, (ii) Indicated power, (iii) frictional power, (iv) mechanical efficiency (v) Indicated and brake thermal efficiency.

Solution.

***(i)* Indicated power** is given by using equation

$$\text{IP} = \frac{p_m ALkn}{60 \times 1000} \ kW \quad \text{...(i)}$$

where

p_m = the mean effective pressure in bar = 6 bar = 6 × 10^5 N/m²

A = the cross section area of piston in m² = $\frac{\pi}{4} \times \left(\frac{300}{1000}\right)^2$ = 0.0707 m²

L = length of stroke in m = 450 mm = $\frac{450}{1000}$ m = 0.45 m

N = engine speed in rpm = 250 rpm

k = number of cylinders = 1

n = the number of working strokes per minute

for a 4–stroke engine $n = \frac{N}{2} = \frac{250}{2}$ = 125 rpm

$$\therefore \quad \text{IP} = \frac{p_m ALkn}{60 \times 1000} \ kW$$

$$\text{IP} = \frac{6 \times 10^5 \times 0.0707 \times 0.45 \times 1 \times \frac{250}{2}}{60 \times 1000} = 39.76 \text{ kW}$$

***(ii)* Brake Power**

Given

Load on brake drum = 1000 N

Diameter of brake drum = 2000 mm = 2 m

∴ Radius of brake drum = 1m

$$\text{BP} = \frac{2\pi NT}{60 \times 1000} \ kW \quad \text{...(ii)}$$

Torque on the brake drum T = load × radius

∴ T = 1200 × 1 = 1200 Nm

$$\text{BP} = \frac{2 \times \pi \times 250 \times 1200}{60000} = 31.41 \text{ kW}$$

(*iii*) Frictional Power

$$FP = IP - BP$$
$$= 39.76 - 31.41 = 8.35 \text{ kW}$$

(*iv*) Mechanical Efficiency

$$\eta_{mech} = \frac{BP}{IP}$$

$$\eta_{mech} = \frac{31.41}{39.76}$$

$$\eta_{mech} = 0.7899 = 79\%$$

(*v*) Indicated Thermal Efficiency

$$\eta_i = \frac{IP}{\dot{m}_f \times CV}$$

$$\eta_i = \frac{39.76}{0.0013 \times 44000} = 0.6951$$

$$\eta_i = 69.51\%$$

(*vi*) Brake Thermal Efficiency

$$\eta_b = \frac{BP}{\dot{m}_f \times CV}$$

$$\eta_b = \frac{31.41}{0.0013 \times 44000} = 0.5491$$

$$\eta_b = 54.91\%$$

Problem 3. *A six cylinder, four-stroke spark ignition engine of 10 cm × 12 cm (bore/stroke) with a compression ratio of 6 is tested at 4800 rpm on a dynanometer of arm 55 cm. During a 10-minute test, dynamometer reads 45 kg and the engine consumed 5 kg of petrol of calorific value 45 MJ/kg. The carburettor receives the air at 29°C and 1 bar at the rate of 10 kg/min. Find out (i) the brake power, (ii) the brake mean effective pressure, (iii) the brake specific fuel consumption, (iv) the brake specific air consumption, (v) the brake thermal efficiency (vi) the air-fuel ratio.*

Solution.

(*i*) the Brake Power (BP)

$$BP = \frac{2\pi NT}{60 \times 1000} \text{ kW}$$

Let us find out torque T

$$T = 45 \times 9.81 \times \frac{55}{100} = 242.79 \text{ Nm}$$

$$BP = \frac{2 \times \pi \times 4800 \times 242.79}{60 \times 1000} \text{ kW}$$

$$BP = 122.04 \text{ kW}$$

(*ii*) the brake mean effective pressure

$$BP = \frac{p_{bm} \times ALNnk}{60 \times 1000}$$

$$122.04 = \frac{p_{bm} \times \frac{\pi}{4} \times \left(\frac{10}{100}\right)^2 \times \frac{12}{100} \times \frac{4800}{2} \times 6}{60 \times 1000}$$

$$p_{bm} = \frac{122.04 \times 60000}{13.57}$$

$$p_{bm} = 539535.26 \text{ N/m}^2$$

$$p_{bm} = 5.39 \text{ bar}$$

(*iii*) brake specific fuel consumption

$$bsfc = \frac{\text{fuel consumption in kg / hr}}{\text{BP in kW}}$$

Since 5 kg of fuel is consumed in 10 minutes

$$\therefore \quad mf = \frac{5}{10} \text{ kg/min} = 0.5 \times 60 = 30 \text{ kg/hr}$$

$$bsfc = \frac{30}{122.04} = 0.245 \text{ kg/kW hr}$$

(*iv*) Brake specific air consumption

$$m_a = 10 \text{ kg/min} = 10 \times 60 = 600 \text{ kg/hr}$$

$$bsac = \frac{\text{air consumption in kg / hr}}{\text{BP in kW}} = \frac{600}{122.04}$$

$$bsac = 4.916 \text{ kg/kW – hr}$$

(*v*) Brake thermal efficiency

$$\eta_b = \frac{BP}{\dot{m}_f \times CV}$$

$$\eta_b = \frac{122.04 \times 1000}{\frac{30}{3600} \times 45 \times 10^6} = 0.3254$$

$$\eta_b = 32.54\%$$

(*vi*) air fuel ratio

$$\text{air fuel ratio} = \frac{m_a}{m_f} = \frac{600}{30} = 20:1$$

Problem 4. *A–4 cylinder, two-stroke cycle, petrol engine develops 30 kW at 2500 rpm. The mean effective pressure on each piston is 8 bar and mechanical efficiency is 80%. Calculate the diameter and stroke of each cylinder, stroke to bore ratio is 1.5. Also calculate the fuel consumption of the engine. If the brake thermal efficiency is 28%, the calorific value of the fuel is 43900 kJ/kg.* ***(RGPV Jan./Feb. 2006)***

Solution.

Given:

$$BP = 30 \text{ kW}$$
$$N = 2500 \text{ rpm}$$
$$\eta_{mech} = 80\% = 0.80$$
$$p_m = 8 \text{ bar} = 8 \times 10^5 \text{ N/m}^2$$
$$L = 1.5 \text{ D}$$
$$\eta_b = 28\% = 0.28$$
$$CV = 43900 \text{ kJ/kg}$$

As we know

$$\eta_{mech} = \frac{BP}{IP}$$

$$0.80 = \frac{30}{IP}$$

$$\therefore \quad IP = \frac{30}{0.80}$$

$$\therefore \quad IP = 37.5 \text{ kW}$$

and indicated power is given by equation

$$\therefore \quad \text{IP} = \frac{p_m ALkn}{60 \times 1000}$$

$$\therefore \quad 37.5 = \frac{8 \times 10^5 \times \frac{\pi}{4}(D)^2 \times 1.5D \times 4 \times 2500}{60 \times 1000}$$

$$\therefore \quad L = 1.5\ D = 1.5 \times 62 = 93 \text{ mm}$$

$$\text{Cylinder diameter } D = 62 \text{ mm}$$

$$\text{Length of stroke} = 93 \text{ mm} \quad \textbf{Ans.}$$

As we know brake thermal efficiency is expressed by the equation

$$\eta_b = \frac{BP}{\dot{m}_f \times CV}$$

$$0.28 = \frac{30}{\dot{m}_f \times 43900}$$

$$\text{specific brake fuel consumption} = \frac{\dot{m}_f}{BP} = \frac{1}{0.28 \times 43900}$$

$$= 8.135 \times 10^{-5} \text{ kg/kWs}$$

$$= 8.135 \times 10^{-5} \times 3600 \text{ kg/kWh}$$

$$= 0.293 \text{ kg/kWh} \quad \textbf{Ans.}$$

REVIEW QUESTIONS

1. Discuss the field of applications of Internal Combustion Engines.
2. Describe the operating principle of a four-stroke diesel engine. Label various parts with their functions. *(RGPV–June 2001)*
3. What are the differences between the SI and CI engines? *(RGPV–June 2002, Dec. 2002)*
4. Draw a neat sketch of a 4-stroke petrol engine and label various parts with their function. *(RGPV–June 2002)*
5. Compare two-stroke and four-stroke I.C. Engines. *(RGPV–Dec. 2001)*
6. Describe with the help of a neat sketch a Bajaj Scooter engine. *(RGPV–Dec. 2001, RGPV-June 2004)*

   ☞ *(Note: Bajaj Scooter engine is a two-stroke petrol engine.)*
7. What are the relative advantages and disadvantages of I.C. engine working on four-stroke and two-stroke cycles? *(RGPV–Dec. 2003)*
8. Explain why Otto cycle engines need some method of ignition but Diesel cycle engines generally do not need ignition devices?
9. Tick the correct choice in the below mentioned situations:

Particular Situation	*Petrol Engine*	*Diesel Engine*
(*a*) Spark plug is used	☐	☐
(*b*) Combustion takes place at constant volume	☐	☐
(*c*) Compression ratio is 10	☐	☐
(*d*) LML Scooter	☐	☐
(*e*) Bajaj Tempo	☐	☐
(*f*) Carburettor is used	☐	☐
(*g*) Fuel Injector is used	☐	☐
(*h*) Maruti 800	☐	☐
(*i*) Pressure in the cylinder at the end of compression is 35 bar.	☐	☐
(*j*) Mixture of air and fuel is taken During the suction stroke.	☐	☐

10. Describe the functions of a carburettor. Explain briefly, giving suitable sketches the working of a simple carburettor.

11. In what respect four-stroke diesel cycle engine differs from four-stroke cycle spark engines? Discuss the working of a four-stroke diesel engine. *(RGPV–June 2004)*

12. A four cylinder 4-stroke petrol engine develops 25 kW at 1440 rpm. The mean effective pressure is 7.5 bar. Determine the bore and stroke of the engine, if the length of the stroke is 1.5 times the bore. [**Ans.** 83.85 mm, 125.77 mm]

13. A single cylinder four-stroke diesel engine running at 1800 rpm has a bore of 85 mm and a stroke of 110 mm. It takes 0.56 kg of ait per minute and develops a brake power output of 6 kW while the air-fuel ratio is 20:1, the calorific value of the fuel used is 42550 kJ/kg, and the ambient air density is 1.18 kg/m^3. Find out (*a*) the volumetric efficiency, (*b*) Brake specific fuel consumption. [**Ans.** 84.5%, 0.28 kg/kWh]

14. A test on a petrol engine produces following data:

IP = 30 kW

BP = 26 kW

Speed = 1000 RPM

Fuel consumption per BP hour = 0.35 kg

Calorific value of fuel used = 44000 kJ/kg

Find out the (*a*) indicated thermal efficiency, (*b*) brake thermal efficiency, (*c*) mechanical efficiency. [**Ans.** 26.97, 23.37, 86.66]

15. The following particulars refer to a two-stroke diesel engine: bore = 15 cm, stroke = 20 cm, piston speed = 400 m/min, torque developed = 60 Nm, mechanical efficiency = 80%, indicated thermal efficiency = 40%, calorific value of fuel used = 44000 kJ/kg. Determine: (*a*) indicated power, (*ii*) indicated mean effective pressure, and (*iii*) fuel consumption per kW hour on brake power basis.

16. A six cylinder, four-stroke cycle, petrol engine is to be designed to produce brake power of 320 kW at 2400 rpm. The stroke to bore ratio is to be 1.25 to 1. Assuming a mechanical efficiency of 80% and indicated mean effective pressure of 950 kPa, determine the required cylinder bore and stroke. [**Ans.** 15.3 cm, 19.125 cm]

17. Following data refer to a four-stroke double acting diesel engine having cylinder diameter 200 mm and piston stroke 300 mm.

Speed = 450 rpm

Diameter of piston rod = 20 mm

Mean effective pressure on cover side = 6.5 bar

Mean effective pressure on crank end = 7 bar

Dead load on the brake = 1370 N

Spring balance reading = 150 N

Brake wheel diameter = 1.2 m

Brake rope diameter = 20 mm

Find out the mechanical efficiency of the engine. [**Ans.** 73.89]

18. A four stoke single cylinder diesel engine runs with 900 rpm has the following specifications:

Area of the indicator diagram = 500 sq mm

Length of the indicator diagram = 70 mm

Spring scale rading = 50 kPa/mm

Bore drum diameter = 120 mm

Length to diameter ratio = 1

Brake drum diameter = 420 mm

Rope diameter = 7 mm

Dead weight on the weight hanger = 15 kg

Spring balance reading = 20N

Determine the indicated mean effective pressure, indicated power, brake power and mechanical efficiency. **[Ans.** 357 kPa, 3.635 kW, 2.558 kW, 70.4%]

19. A four cylinder four-stroke petrol engine has a bore of 70 mm and stroke of 100 mm. during test, the engine produces a torque of 65 Nm while running at 3000 rpm. If the clearance volume in each cylinder is 63 cm^3, the brake efficiency ratio based on air standard efficiency is 0.5 and calorific value of petrol is 43900 kJ/kg. Find out the fuel consumption in kg/hr and brake mean effective pressure in bar. *[6.16 kg/hr, 5.30 bar]*

20. The output of an IC engine is measured by a rope brake dynamometer. The diameter of brake pulley is 750 mm, and rope diameter is 50 mm, the dead load on the tight side of the rope is 450 N and the spring balance reading is 100 N. The engine consumes 4.5 kg/h of fuel at 1000 rpm. The calorific value of fuel is 44000 kJ/kg. Determine

(*a*) Brake specific fuel consumption

(*b*) Brake thermal efficiency **[Ans.** 0.307 kg/kWh, 26.6%]

21. An eight cylinder automobile engine of 80 mm diameter and 90 mm stroke with a compression ratio of 7, is tested at 4000 RPM on a dynamometer of 600 mm arm length. During a ten minutes test period at a dynamometer scale reading of 450 N, 4.8 kg of gasoline having a calorific value of 45000 kJ/kg was burnt and air at 27°C and 1.0 bar was supplied to the carburettor at the rate of 6.6 kg/min. Find (*i*) the brake power delivered, (*ii*) the brake mean effective pressure, (*iii*) the brake specific fuel consumption, (*iv*) brake thermal efficiency, (*v*) the volumetric efficiency. **[Ans.** 113 kW, 8.32 bar, 0.233 kg/kW hr, 31.4%, 75%]

22. A large diesel engine runs on a 4-stroke cycle at 2000 rpm. The engine has a displacement of 25 litres and a brake mean effective pressure of 0.6 MPa. It consumes 0.018 kg/of fuel having CV of 42000 kJ/kg. Determine the brake power and brake thermal efficiency. **[Ans.** 250 kW, 33.06%]

OBJECTIVE TYPE QUESTIONS

Select the most appropriate answer:

1. In your opinion the meaning of 100 cc parameter in a 100 cc engine is

(A) fuel tank capacity (B) weight of the vehicle

(C) swept volume (D) cylinder volume.

2. The lubrication in I.C. engine is used to

(A) cool the cylinder (B) dissipate heat due to friction

(C) reduce friction (D) none of the above.

3. Cooling system in I.C. engine removes

(A) 85% of heat (B) 50% of heat

(C) 30% of heat (D) 100% of heat.

4. The type of cooling used in I.C. engine is

(A) Air and liquid cooling (B) air cooling

(C) liquid cooling (D) air, cooling and gas cooling.

5. Air cooling in I.C. engine is used in case of

(A) buses (B) trucks

(C) cars (D) motor cycles.

6. The efficiency of I.C. engine is ....

(A) 15–20% (B) 20–25%

(C) 25–30% (D) 30–35%.

7. The spark plug gap is generally kept from

(A) 0.1 to 0.2 mm (B) 0.2 to 0.4 mm

(C) 0.4 to 0.6 mm (D) 0.8 to 1.0 mm.

8. Fins are provided over the engine cylinder in scooters for

(A) increasing the strength of cylinder (B) effective cooling

(C) good appearance (D) higher efficiency.

9. What will happen if by mistake diesel is filled in a scooter fuel tank.

(A) The scooter will not run (B) The engine will explode

(C) Black smoke will be produced (D) The scooter will give more mileage.

10. Theoretically correct air-fuel ratio for petrol engine is

(A) 10: 1 (B) 15: 1

(C) 20: 1 (D) 5: 1.

11. Which does not belong to C.I. engine?

(A) fuel pump (B) fuel injector

(C) carburettor (D) flywheel.

12. When choke is applied in petrol engine, the engine gets

(A) extra air supply (B) weak mixture

(C) rich mixture (D) lean mixture.

13. Compression ratio for petrol engine is

(A) 6 to 10 (B) 10 to 15

(C) 15 to 25 (D) 25 to 30.

14. Compression ratio for diesel engine is

(A) 6 to 10 (B) 10 to 15

(C) 15 to 25 (D) 25 to 30.

15. What will happen if by mistake petrol is taken in place of diesel in a truck fuel tank?

(A) the engine will produce black smoke

(B) the engine will burst into flames

(C) the truck will not run at all

(D) the truck will run with high knocking.

16. The self ignition temperature of petrol is of the order of ....

(A) Less than 100°C
(B) Less than 200°C
(C) More than 200°C but less than 300°C
(D) More than 500°C.

17. The thermal efficiency of diesel engine is:

(A) 50%
(B) 35%
(C) 75%
(D) 90%.

18. The thermal efficiency of petrol engine is

(A) 50%
(B) 30%
(C) 60%
(D) 80%.

ANSWERS

1. (C)	**2.** (B)	**3.** (C)	**4.** (A)	**5.** (D)	**6.** (D)
7. (C)	**8.** (B)	**9.** (A)	**10.** (B)	**11.** (C)	**12.** (C)
13. (A)	**14.** (C)	**15.** (D)	**16.** (D)	**17.** (B)	**18.** (B).

■■■

Index

D

E

L

M

N

O

P

STEAM TABLES

and

Mollier Diagram

(*S.I. Units*)

CONTENTS

SYMBOLS AND UNITS USED IN THE TABLES

t = Temperature, °C

t_s = Saturation temperature, °C

p = Pressure, bar

h_f = Specific enthalpy of saturated liquid, kJ/kg

h_{fg} = Specific enthalpy of evaporation (latent heat), kJ/kg

h_g = Specific enthalpy of saturated vapour, kJ/kg

s_f = Specific entropy of saturated liquid, kJ/kg K

s_{fg} = Specific entropy of evaporation, kJ/kg K

s_g = Specific entropy of saturated vapour, kJ/kg K

v_f = Specific volume of saturated liquid, m^3/kg

v_g = Specific volume of saturated steam, m^3/kg

TABLE I
Saturated Water and Steam (Temperature) Tables

Temp. (°C)	Absolute pressure (bar)	Specific enthalpy (kJ/kg)			Specific entropy (kJ/kg K)			Specific volume (m^3/kg)	
t	p	h_f	h_{fg}	h_g	s_f	s_{fg}	s_g	v_f	v_g
0	0.0061	– 0.02	2501.4	2501.3	– 0.0001	9.1566	9.1565	0.0010002	206.3
0.01	0.0061	0.01	2501.3	2501.4	0.000	9.156	9.156	0.0010002	206.2
1	0.0065	4.2	2499.0	2503.2	0.015	9.115	9.130	0.0010002	192.6
2	0.0070	8.4	2496.7	2505.0	0.031	9.073	9.104	0.0010001	179.9
3	0.0076	12.6	2494.3	2506.9	0.046	9.032	9.077	0.0010001	168.1
4	0.0081	16.8	2491.9	2508.7	0.061	8.990	9.051	0.0010001	157.2
5	0.0087	21.0	2489.6	2510.6	0.076	8.950	9.026	0.0010001	147.1
6	0.0093	25.2	2487.2	2512.4	0.091	8.909	9.000	0.0010001	137.7
7	0.0100	29.4	2484.8	2514.2	0.106	8.869	8.975	0.0010002	129.0
8	0.0107	33.6	2482.5	2516.1	0.121	8.829	8.950	0.0010002	120.9
9	0.0115	37.8	2480.1	2517.9	0.136	8.789	8.925	0.0010003	113.4
10	0.0123	42.0	2477.7	2519.7	0.151	8.750	8.901	0.0010004	106.4
11	0.0131	46.2	2475.4	2521.6	0.166	8.711	8.877	0.0010004	99.86
12	0.0140	50.4	2473.0	2523.4	0.181	8.672	8.852	0.0010005	93.78
13	0.0150	54.6	2470.7	2525.3	0.195	8.632	8.828	0.0010007	88.12
14	0.0160	58.8	2468.3	2527.1	0.210	8.595	8.805	0.0010008	82.85
15	0.0170	63.0	2465.9	2528.9	0.224	8.557	8.781	0.0010009	77.93
16	0.0182	67.2	2463.6	2530.8	0.239	8.519	8.758	0.001001	73.33
17	0.0194	71.4	2461.2	2532.6	0.253	8.482	8.735	0.001001	69.04
18	0.0206	75.6	2458.8	2534.4	0.268	8.444	8.712	0.001001	65.04
19	0.0220	79.8	2456.5	2536.3	0.282	8.407	8.690	0.001002	61.29
20	0.0234	84.0	2454.1	2538.1	0.297	8.371	8.667	0.001002	57.79
21	0.0249	88.1	2451.8	2539.9	0.311	8.334	8.645	0.001002	54.51
22	0.0264	92.3	2449.4	2541.7	0.325	8.298	8.623	0.001002	51.45
23	0.0281	96.5	2447.0	2543.5	0.339	8.262	8.601	0.001002	48.57
24	0.0298	100.7	2444.7	2545.4	0.353	8.226	8.579	0.001003	45.88
25	0.0317	104.9	2442.3	2547.2	0.367	8.191	8.558	0.001003	43.36
26	0.0336	109.1	2439.9	2549.0	0.382	8.155	8.537	0.001003	40.99
27	0.0357	113.2	2437.6	2550.8	0.396	8.120	8.516	0.001004	38.77
28	0.0378	117.4	2435.2	2552.6	0.409	8.086	8.495	0.001004	36.69
29	0.0401	121.6	2432.8	2554.5	0.423	8.051	8.474	0.001004	34.73
30	0.0425	125.8	2430.5	2556.3	0.437	8.016	8.453	0.001004	32.89

Temp. (°C)	Absolute pressure (bar)	Specific enthalpy (kJ/kg)			Specific entropy (kJ/kg K)			Specific volume (m^3/kg)	
t	p	h_f	h_{fg}	h_g	s_f	s_{fg}	s_g	v_f	v_g
31	0.0450	130.0	2428.1	2558.1	0.451	7.982	8.433	0.001005	31.17
32	0.0476	134.2	2425.7	2559.9	0.464	7.948	8.413	0.001005	29.54
33	0.0503	138.3	2423.4	2561.7	0.478	7.915	8.393	0.001005	28.01
34	0.0532	142.5	2421.0	2563.5	0.492	7.881	8.373	0.001006	26.57
35	0.0563	146.7	2418.6	2565.3	0.505	7.848	8.353	0.001006	25.22
36	0.0595	150.9	2416.2	2567.1	0.519	7.815	8.334	0.001006	23.94
37	0.0628	155.0	2413.9	2568.9	0.532	7.782	8.314	0.001007	22.74
38	0.0663	159.2	2411.5	2570.7	0.546	7.749	8.295	0.001007	21.60
39	0.0700	163.4	2409.1	2572.5	0.559	7.717	8.276	0.001007	20.53
40	0.0738	167.6	2406.7	2574.3	0.573	7.685	8.257	0.001008	19.52
41	0.0779	171.7	2404.3	2576.0	0.586	7.652	8.238	0.001008	18.57
42	0.0821	175.9	2401.9	2577.8	0.599	7.621	8.220	0.001009	17.67
43	0.0865	180.1	2399.5	2579.6	0.612	7.589	8.201	0.001009	16.82
44	0.0911	184.3	2397.2	2581.5	0.626	7.557	8.183	0.001010	16.02
45	0.0959	188.4	2394.8	2583.2	0.639	7.526	8.165	0.001010	15.26
46	0.1010	192.6	2392.4	2585.0	0.652	7.495	8.147	0.001010	14.54
47	0.1062	196.8	2390.0	2586.8	0.665	7.464	8.129	0.001011	13.86
48	0.1118	201.0	2387.6	2588.6	0.678	7.433	8.111	0.001011	13.22
49	0.1175	205.1	2385.2	2590.3	0.691	7.403	8.094	0.001012	12.61
50	0.1235	209.3	2382.7	2592.1	0.704	7.372	8.076	0.001012	12.03
52	0.1363	217.7	2377.9	2595.6	0.730	7.312	8.042	0.001013	10.97
54	0.1502	226.0	2373.1	2599.1	0.755	7.253	8.008	0.001014	10.01
56	0.1653	234.4	2368.2	2602.6	0.781	7.194	7.975	0.001015	9.149
58	0.1817	242.8	2363.4	2606.2	0.806	7.136	7.942	0.001016	8.372
60	0.1994	251.1	2358.5	2609.6	0.831	7.078	7.909	0.001017	7.671
62	0.2186	259.5	2353.6	2613.1	0.856	7.022	7.878	0.001018	7.037
64	0.2393	267.9	2348.7	2616.5	0.881	6.965	7.846	0.001019	6.463
66	0.2617	276.2	2343.7	2619.9	0.906	6.910	7.816	0.001020	5.943
68	0.2859	284.6	2338.8	2623.4	0.930	6.855	7.785	0.001022	5.471
70	0.3119	293.0	2333.8	2626.8	0.955	6.800	7.755	0.001023	5.042
75	0.3858	313.9	2321.4	2635.3	1.015	6.667	7.682	0.001026	4.131
80	0.4739	334.9	2308.8	2643.7	1.075	6.537	7.612	0.001029	3.407
85	0.5783	355.9	2296.0	2651.9	1.134	6.410	7.544	0.001033	2.828
90	0.7014	376.9	2283.2	2660.1	1.192	6.287	7.479	0.001036	2.361
95	0.8455	397.9	2270.2	2668.1	1.250	6.166	7.416	0.001040	1.982
100	1.0135	419.0	2257.0	2676.0	1.307	6.048	7.355	0.001044	1.673

TABLE II
Saturated Water and Steam (Pressure) Tables

Absolute pressure (bar)	Temp. (°C)	Specific enthalpy (kJ/kg)			Specific entropy (kJ/kg K)			Specific volume (m^3/kg)	
p	t_s	h_f	h_{fg}	h_g	s_f	s_{fg}	s_g	v_f	v_g
0.006113	0.01	0.01	2 501.3	2 501.4	0.000	9.156	9.156	0.0010002	206.14
0.010	7.0	29.3	2 484.9	2 514.2	0.106	8.870	8.976	0.0010000	129.21
0.015	13.0	54.7	2 470.6	2 525.3	0.196	8.632	8.828	0.0010007	87.98
0.020	17.0	73.5	2 460.0	2 533.5	0.261	8.463	8.724	0.001001	67.00
0.025	21.1	88.5	2 451.6	2 540.1	0.312	8.331	8.643	0.001002	54.25
0.030	24.1	101.0	2 444.5	2 545.5	0.355	8.223	8.578	0.001003	45.67
0.035	26.7	111.9	2 438.4	2 550.3	0.391	8.132	8.523	0.001003	39.50
0.040	29.0	121.5	2 432.9	2 554.4	0.423	8.052	8.475	0.001004	34.80
0.045	31.0	130.0	2 428.2	2 558.2	0.451	7.982	8.433	0.001005	31.13
0.050	32.9	137.8	2 423.7	2 561.5	0.476	7.919	8.395	0.001005	28.19
0.055	34.6	144.9	2 419.6	2 565.5	0.500	7.861	8.361	0.001006	25.77
0.060	36.2	151.5	2 415.9	2 567.4	0.521	7.809	8.330	0.001006	23.74
0.065	37.6	157.7	2 412.4	2 570.1	0.541	7.761	8.302	0.001007	22.01
0.070	39.0	163.4	2 409.1	2 572.5	0.559	7.717	8.276	0.001007	20.53
0.075	40.3	168.8	2 406.0	2 574.8	0.576	7.675	8.251	0.001008	19.24
0.080	41.5	173.9	2 403.1	2 577.0	0.593	7.636	8.229	0.001008	18.10
0.085	42.7	178.7	2 400.3	2 579.0	0.608	7.599	8.207	0.001009	17.10
0.090	43.8	183.3	2 397.7	2 581.0	0.622	7.565	8.187	0.001009	16.20
0.095	44.8	187.7	2 395.2	2 582.9	0.636	7.532	8.168	0.001010	15.40
0.10	45.8	191.8	2 392.8	2 584.7	0.649	7.501	8.150	0.001010	14.67
0.11	47.7	199.7	2 388.3	2 588.0	0.674	7.453	8.117	0.001011	13.42
0.12	49.4	206.9	2 384.2	2 591.1	0.696	7.390	8.086	0.001012	12.36
0.13	51.0	213.7	2 380.2	2 593.9	0.717	7.341	8.058	0.001013	11.47
0.14	52.6	220.0	2 376.6	2 596.6	0.737	7.296	8.033	0.001013	10.69
0.15	54.0	226.0	2 373.2	2 599.2	0.754 9	7.254 4	8.009 3	0.001014	10.022
0.16	55.3	231.6	2 370.0	2 601.6	0.772 1	7.214 8	7.986 9	0.001015	9.433
0.17	56.6	236.9	2 366.9	2 603.8	0.788 3	7.177 5	7.965 8	0.001015	8.911
0.18	57.8	242.0	2 363.9	2 605.9	0.803 6	7.142 4	7.945 9	0.001016	8.445
0.19	59.0	246.8	2 361.1	2 607.9	0.818 2	7.109 0	7.927 2	0.001017	8.027
0.20	60.1	251.5	2 358.4	2 609.9	0.832 1	7.077 3	7.909 4	0.001017	7.650
0.21	61.1	255.9	2 355.8	2 611.7	0.845 3	7.047 2	7.892 5	0.001018	7.307
0.22	62.2	260.1	2 353.3	2 613.5	0.858 1	7.018 4	7.876 4	0.001018	6.995
0.23	63.1	264.2	2 350.9	2 615.2	0.870 2	6.990 8	7.861 1	0.001019	6.709
0.24	64.1	268.2	2 348.6	2 616.8	0.882 0	6.964 4	7.846 4	0.001019	6.447

Absolute pressure (bar)	Temp. (°C)	Specific enthalpy (kJ/kg)			Specific entropy (kJ/kg K)			Specific volume (m^3/kg)	
p	t_s	h_f	h_{fg}	h_g	s_f	s_{fg}	s_g	v_f	v_g
0.25	65.0	272.0	2 346.4	2 618.3	0.893 2	6.939 1	7.832 3	0.001020	6.205
0.26	65.9	275.7	2 344.2	2 619.9	0.904 1	6.914 7	7.818 8	0.001020	5.980
0.27	66.7	279.2	2 342.1	2 621.3	0.914 6	6.891 2	7.805 8	0.001021	5.772
0.28	67.5	282.7	2 340.0	2 622.7	0.924 8	6.868 5	7.793 3	0.001021	5.579
0.29	68.3	286.0	2 338.1	2 624.1	0.934 6	6.846 6	7.781 2	0.001022	5.398
0.30	69.1	289.3	2 336.1	2 625.4	0.944 1	6.825 4	7.769 5	0.001022	5.229
0.32	70.6	295.5	2 332.4	2 628.0	0.962 3	6.785 0	7.747 4	0.001023	4.922
0.34	72.0	301.5	2 328.9	2 630.4	0.979 5	6.747 0	7.726 5	0.001024	4.650
0.36	73.4	307.1	2 325.5	2 632.6	0.995 8	6.711 1	7.707 0	0.001025	4.408
0.38	74.7	312.5	2 322.3	2 634.8	1.011 3	6.677 1	7.688 4	0.001026	4.190
0.40	75.9	317.7	2 319.2	2 636.9	1.026 1	6.644 8	7.670 9	0.001026	3.993
0.42	77.1	322.6	2 316.3	2 638.9	1.040 2	6.614 0	7.654 2	0.001027	3.815
0.44	78.2	327.3	2 313.4	2 640.7	1.053 7	6.584 6	7.638 3	0.001028	3.652
0.46	79.3	331.9	2 310.7	2 642.6	1.066 7	6.556 4	7.623 1	0.001029	3.503
0.48	80.3	336.3	2 308.0	2 644.3	1.079 2	6.529 4	7.608 6	0.001029	3.367
0.50	81.3	340.6	2 305.4	2 646.0	1.091 2	6.503 5	7.594 7	0.001030	3.240
0.55	83.7	350.6	2 299.3	2 649.9	1.119 4	6.442 8	7.562 3	0.001032	2.964
0.60	86.0	359.9	2 293.6	2 653.6	1.145 4	6.387 3	7.532 7	0.001033	2.732
0.65	88.0	368.6	2 288.3	2 656.9	1.169 6	6.336 0	7.505 5	0.001035	2.535
0.70	90.0	376.8	2 283.3	2 660.1	1.192 1	6.288 3	7.480 4	0.001036	2.369
0.75	92.0	384.5	2 278.6	2 663.0	1.213 1	6.243 9	7.457 0	0.001037	2.217
0.80	93.5	391.7	2 274.0	2 665.8	1.233 0	6.202 2	7.435 2	0.001039	2.087
0.85	95.1	398.6	2 269.8	2 668.4	1.251 8	6.162 9	7.414 7	0.001040	1.972
0.90	96.7	405.2	2 265.6	2 670.9	1.269 6	6.125 8	7.395 4	0.001041	1.869
0.95	98.2	411.5	2 261.7	2 673.2	1.286 5	6.090 6	7.377 1	0.001042	1.777
1.0	99.6	417.5	2 257.9	2 675.4	1.302 7	6.057 1	7.359 8	0.001043	1.694
1.1	102.3	428.8	2 250.8	2 679.6	1.333 0	5.994 7	7.327 7	0.001046	1.549
1.2	104.8	439.4	2 244.1	2 683.4	1.360 9	5.937 5	7.298 4	0.001048	1.428
1.3	107.1	449.2	2 237.8	2 687.0	1.386 8	5.884 7	7.271 5	0.001050	1.325
1.4	109.3	458.4	2 231.9	2 690.3	1.410 9	5.835 6	7.246 5	0.001051	1.236
1.5	111.3	467.1	2 226.2	2 693.4	1.433 6	5.789 8	7.233 4	0.001053	1.159
1.6	113.3	475.4	2 220.9	2 696.2	1.455 0	5.746 7	7.201 7	0.001055	1.091
1.7	115.2	483.2	2 215.7	2 699.0	1.475 2	5.706 1	7.181 3	0.001056	1.031
1.8	116.9	490.7	2 210.8	2 701.5	1.494 4	5.667 8	7.162 2	0.001058	0.977
1.9	118.6	497.8	2 206.1	2 704.0	1.512 7	5.631 4	7.144 0	0.001060	0.929

Absolute pressure (bar)	Temp. (°C)	Specific enthalpy (kJ/kg)			Specific entropy (kJ/kg K)			Specific volume (m^3/kg)	
p	t_s	h_f	h_{fg}	h_g	s_f	s_{fg}	s_g	v_f	v_g
2.0	120.2	504.7	2 201.6	2 706.3	1.530 1	5.596 7	7.126 8	0.001061	0.885
2.1	121.8	511.3	2 197.2	2 708.5	1.546 8	5.563 7	7.110 5	0.001062	0.846
2.2	123.3	517.6	2 193.0	2 710.6	1.562 7	5.532 1	7.094 9	0.001064	0.810
2.3	124.7	523.7	2 188.9	2 712.6	1.578 1	5.501 9	7.080 0	0.001065	0.777
2.4	126.1	529.6	2 184.9	2 714.5	1.592 9	5.472 8	7.065 7	0.001066	0.746
2.5	127.4	535.3	2 181.0	2 716.4	1.607 1	5.444 9	7.052 0	0.001068	0.718
2.6	128.7	540.9	2 177.3	2 718.2	1.620 9	5.418 0	7.038 9	0.001069	0.693
2.7	129.9	546.2	2 173.6	2 719.9	1.634 2	5.392 0	7.026 2	0.001070	0.668
2.8	131.2	551.4	2 170.1	2 721.5	1.647 1	5.367 0	7.014 0	0.001071	0.646
2.9	132.4	556.5	2 166.6	2 723.1	1.659 5	5.342 7	7.002 3	0.001072	0.625
3.0	133.5	561.4	2 163.2	2 724.7	1.671 6	5.319 3	6.990 9	0.001074	0.606
3.1	134.6	566.2	2 159.9	2 726.1	1.683 4	5.296 5	6.979 9	0.001075	0.587
3.2	135.7	570.9	2 156.7	2 727.6	1.694 8	5.274 4	6.969 2	0.001076	0.570
3.3	136.8	575.5	2 153.5	2 729.0	1.705 9	5.253 0	6.958 9	0.001077	0.554
3.4	137.8	579.9	2 150.4	2 730.3	1.716 8	5.232 2	6.948 9	0.001078	0.538
3.5	138.8	584.3	2 147.4	2 731.6	1.727 3	5.211 9	6.939 2	0.001079	0.524
3.6	139.8	588.5	2 144.4	2 732.9	1.737 6	5.192 1	6.929 7	0.001080	0.510
3.7	140.8	592.7	2 141.4	2 734.1	1.747 6	5.172 9	6.920 5	0.001081	0.497
3.8	141.8	596.8	2 138.6	2 735.3	1.757 4	5.154 1	6.911 6	0.001082	0.486
3.9	142.7	600.8	2 135.7	2 736.5	1.767 0	5.135 8	6.902 8	0.001083	0.473
4.0	143.6	604.7	2 133.0	2 737.6	1.776 4	5.117 9	6.894 3	0.001084	0.462
4.2	145.4	612.3	2 127.5	2 739.8	1.794 5	5.083 4	6.877 9	0.001086	0.441
4.4	147.1	619.6	2 122.3	2 741.9	1.812 0	5.050 3	6.862 3	0.001088	0.423
4.6	148.7	626.7	2 117.2	2 743.9	1.828 7	5.018 6	6.847 3	0.001089	0.405
4.8	150.3	633.5	2 112.2	2 745.7	1.844 8	4.988 1	6.832 9	0.001091	0.390
5.0	151.8	640.1	2 107.4	2 747.5	1.860 4	4.958 8	6.819 2	0.001093	0.375
5.2	153.3	646.5	2 102.7	2 749.3	1.875 4	4.930 6	6.805 9	0.001094	0.361
5.4	154.7	652.8	2 098.1	2 750.9	1.889 9	4.903 3	6.793 2	0.001096	0.348
5.6	156.2	658.8	2 093.7	2 752.5	1.904 0	4.876 9	6.780 9	0.001098	0.337
5.8	157.5	664.7	2 089.3	2 754.0	1.917 6	4.851 4	6.769 0	0.001099	0.326

Absolute pressure (bar)	Temp. (°C)	Specific enthalpy (kJ/kg)			Specific entropy (kJ/kg K)			Specific volume (m^3/kg)	
p	t_s	h_f	h_{fg}	h_g	s_f	s_{fg}	s_g	v_f	v_g
6.0	158.8	670.4	2 085.0	2 755.5	1.930 8	4.826 7	6.757 5	0.001101	0.315
6.2	160.1	676.0	2 080.9	2 756.9	1.943 7	4.802 7	6.746 4	0.001102	0.306
6.4	161.4	681.5	2 076.8	2 758.2	1.956 2	4.779 4	6.735 6	0.001104	0.297
6.6	162.6	686.8	2 072.7	2 759.5	1.968 4	4.756 8	6.725 2	0.001105	0.288
6.8	163.8	692.0	2 068.8	2 760.8	1.980 2	4.734 8	6.715 0	0.001107	0.280
7.0	165.0	697.1	2 064.9	2 762.0	1.991 8	4.713 4	6.705 2	0.001108	0.273
7.2	166.1	702.0	2 061.1	2 763.2	2.003 1	4.692 5	6.695 6	0.001110	0.265
7.4	167.2	706.9	2 057.4	2 764.3	2.014 1	4.672 1	6.686 2	0.001111	0.258
7.6	168.3	711.7	2 053.7	2 765.4	2.024 9	4.652 2	6.677 1	0.001112	0.252
7.8	169.4	716.3	2 050.1	2 766.4	2.035 4	4.632 8	6.668 3	0.001114	0.246
8.0	170.4	720.9	2 046.5	2 767.5	2.045 7	4.613 9	6.659 6	0.001115	0.240
8.2	171.4	725.4	2 043.0	2 768.5	2.055 8	4.595 3	6.651 1	0.001116	0.235
8.4	172.4	729.9	2 039.6	2 769.4	2.065 7	4.577 2	6.642 9	0.001118	0.229
8.6	173.4	734.2	2 036.2	2 770.4	2.075 3	4.559 4	6.634 8	0.001119	0.224
8.8	174.4	738.5	2 032.8	2 771.3	2.084 8	4.542 1	6.626 9	0.001120	0.219
9.0	175.4	742.6	2 029.5	2 772.1	2.094 1	4.525 0	6.619 2	0.001121	0.215
9.2	176.3	746.8	2 026.2	2 773.0	2.103 3	4.508 3	6.611 6	0.001123	0.210
9.4	177.2	750.8	2 023.0	2 773.8	2.112 2	4.492 0	6.604 2	0.001124	0.206
9.6	178.1	754.8	2 019.8	2 774.6	2.121 0	4.475 9	6.596 9	0.001125	0.202
9.8	179.0	758.7	2 016.7	2 775.4	2.129 7	4.460 1	6.589 8	0.001126	0.198
10.0	179.9	762.6	2 013.6	2 776.2	2.138 2	4.444 6	6.582 8	0.001127	0.194
10.5	182.0	772.0	2 005.9	2 778.0	2.158 8	4.407 1	6.565 9	0.001130	0.185
11.0	184.1	781.1	1 998.5	2 779.7	2.178 6	4.371 1	6.549 7	0.001133	0.177
11.5	186.0	789.9	1 991.3	2 781.3	2.197 7	4.336 6	6.534 2	0.001136	0.170
12.0	188.0	798.4	1 984.3	2 782.7	2.216 1	4.303 3	6.519 4	0.001139	0.163
12.5	189.8	806.7	1 977.4	2 784.1	2.233 8	4.271 2	6.505 0	0.001141	0.157
13.0	191.6	814.7	1 970.7	2 785.4	2.251 0	4.240 3	6.491 3	0.001144	0.151
13.5	193.3	822.5	1 964.2	2 786.6	2.267 6	4.210 4	6.477 9	0.001146	0.146
14.0	195.0	830.1	1 957.7	2 787.8	2.283 7	4.181 4	6.465 1	0.001149	0.141
14.5	196.7	837.5	1 951.4	2 788.9	2.299 3	4.153 3	6.452 6	0.001151	0.136
15.0	198.3	844.7	1 945.2	2 789.9	2.314 5	4.126 1	6.440 6	0.001154	0.132
15.5	199.8	851.7	1 939.2	2 790.8	2.329 2	4.099 6	6.428 9	0.001156	0.128
16.0	201.4	858.6	1 933.2	2 791.7	2.343 6	4.073 9	6.417 5	0.001159	0.124
16.5	202.8	865.3	1 927.3	2 792.6	2.357 6	4.048 9	6.406 5	0.001161	0.120
17.0	204.3	871.8	1 921.5	2 793.4	2.371 3	4.024 5	6.395 7	0.001163	0.117

Absolute pressure (bar)	Temp. (°C)	Specific enthalpy (kJ/kg)			Specific entropy (kJ/kg K)			Specific volume (m^3/kg)	
p	t_s	h_f	h_{fg}	h_g	s_f	s_{fg}	s_g	v_f	v_g
17.5	205.7	878.3	1 915.9	2 794.1	2.384 6	4.000 7	6.385 3	0.001166	0.113
18.0	207.1	884.6	1 910.3	2 794.8	2.397 6	3.977 5	6.375 1	0.001168	0.110
18.5	208.4	890.7	1 904.7	2 795.5	2.410 3	3.954 8	6.365 1	0.001170	0.107
19.0	209.8	896.8	1 899.3	2 796.1	2.422 8	3.932 6	6.355 4	0.001172	0.105
19.5	211.1	902.8	1 893.9	2 796.7	2.434 9	3.911 0	6.345 9	0.001174	0.102
20.0	212.4	908.6	1 888.6	2 797.2	2.446 9	3.889 8	6.336 6	0.001177	0.0995
20.5	213.6	914.3	1 883.4	2 797.7	2.458 5	3.869 0	6.327 6	0.001179	0.0971
21.0	214.8	920.0	1 878.2	2 798.2	2.470 0	3.848 7	6.318 7	0.001181	0.0949
21.5	216.1	925.5	1 873.1	2 798.6	2.481 2	3.828 8	6.310 0	0.001183	0.0927
22.0	217.2	931.0	1 868.1	2 799.1	2.492 2	3.809 3	6.301 5	0.001185	0.0907
22.5	218.4	936.3	1 863.1	2 799.4	2.503 0	3.790 1	6.293 1	0.001187	0.0887
23.0	219.5	941.6	1 858.2	2 799.8	2.513 6	3.771 3	6.284 9	0.001189	0.0868
23.5	220.7	946.8	1 853.3	2 800.1	2.524 1	3.752 8	6.276 9	0.001191	0.0849
24.0	221.8	951.9	1 848.5	2 800.4	2.534 3	3.734 7	6.269 0	0.001193	0.0832
24.5	222.9	957.0	1 843.7	2 800.7	2.544 4	3.716 8	6.261 2	0.001195	0.0815
25.0	223.9	962.0	1 839.0	2 800.9	2.554 3	3.699 3	6.253 6	0.001197	0.0799
25.5	225.0	966.9	1 834.3	2 801.2	2.564 0	3.682 1	6.246 1	0.001199	0.0783
26.0	226.0	971.7	1 829.6	2 801.4	2.573 6	3.665 1	6.238 7	0.001201	0.0769
26.5	227.1	976.5	1 825.1	2 801.6	2.583 1	3.648 4	6.231 5	0.001203	0.0754
27.0	228.1	981.2	1 820.5	2 801.7	2.592 4	3.632 0	6.224 4	0.001205	0.0740
27.5	229.1	985.9	1 816.0	2 801.9	2.601 6	3.615 8	6.217 3	0.001207	0.0727
28.0	230.0	990.5	1 811.5	2 802.0	2.610 6	3.599 8	6.210 4	0.001209	0.0714
28.5	231.0	995.0	1 807.1	2 802.1	2.619 5	3.584 1	6.203 6	0.001211	0.0701
29.0	232.0	999.5	1 802.6	2 802.2	2.628 3	3.568 6	6.196 9	0.001213	0.0689
29.5	233.0	1 004.0	1 798.3	2 802.2	2.637 0	3.553 3	6.190 2	0.001214	0.0677
30.0	233.8	1 008.4	1 793.9	2 802.3	2.645 5	3.538 2	6.183 7	0.001216	0.0666
30.5	234.7	1 012.7	1 789.6	2 802.3	2.653 9	3.523 3	6.177 2	0.001218	0.0655
31.0	235.6	1 017.0	1 785.4	2 802.3	2.662 3	3.508 7	6.170 9	0.001220	0.0645
31.5	236.5	1 021.2	1 781.1	2 802.3	2.670 5	3.494 2	6.164 7	0.001222	0.0634
32.0	237.4	1 025.4	1 776.9	2 802.3	2.678 6	3.479 9	6.158 5	0.001224	0.0624
32.5	238.3	1 029.6	1 772.7	2 802.3	2.686 6	3.465 7	6.152 3	0.001225	0.0615
33.0	239.2	1 033.7	1 768.6	2 802.3	2.694 5	3.451 8	6.146 3	0.001227	0.0605
33.5	240.0	1 037.8	1 764.4	2 802.2	2.702 3	3.438 0	6.140 3	0.001229	0.0596
34.0	240.9	1 041.8	1 760.3	2 802.1	2.710 1	3.424 4	6.134 4	0.001231	0.0587
34.5	241.7	1 045.8	1 756.3	2 802.1	2.717 7	3.410 9	6.128 6	0.001233	0.0579

Absolute pressure (bar)	Temp. (°C)	Specific enthalpy (kJ/kg)			Specific entropy (kJ/kg K)			Specific volume (m³/kg)	
p	t_s	h_f	h_{fg}	h_g	s_f	s_{fg}	s_g	v_f	v_g
35.0	242.5	1 049.8	1 752.2	2 802.0	2.725 3	3.397 6	6.122 8	0.001234	0.0570
35.5	243.3	1 053.7	1 748.2	2 801.8	2.732 7	3.384 4	6.117 1	0.001236	0.0562
36.0	244.2	1 057.6	1 744.2	2 801.7	2.740 1	3.371 4	6.111 5	0.001238	0.0554
36.5	245.0	1 061.4	1 740.2	2 801.6	2.747 4	3.358 5	6.105 9	0.001239	0.0546
37.0	245.7	1 065.2	1 736.2	2 801.4	2.754 7	3.345 8	6.100 4	0.001242	0.0539
37.5	246.5	1 069.0	1 732.3	2 801.3	2.761 8	3.333 2	6.095 0	0.001243	0.0531
38.0	247.3	1 072.7	1 728.4	2 801.1	2.768 9	3.320 7	6.089 6	0.001245	0.0524
38.5	248.1	1 076.4	1 724.5	2 800.9	2.775 9	3.308 3	6.084 2	0.001247	0.0517
39.0	248.8	1 080.1	1 720.6	2 800.8	2.782 9	3.296 1	6.078 9	0.001249	0.0511
39.5	249.6	1 083.8	1 716.8	2 800.5	2.789 7	3.284 0	6.073 7	0.001250	0.0504
40.0	250.3	1 087.4	1 712.9	2 800.3	2.796 5	3.272 0	6.068 5	0.001252	0.0497
41.0	251.8	1 094.6	1 705.3	2 799.9	2.809 9	3.248 3	6.058 2	0.001255	0.0485
42.0	253.2	1 101.6	1 697.8	2 799.4	2.823 1	3.225 1	6.048 2	0.001259	0.0473
43.0	254.6	1 108.5	1 690.3	2 798.8	2.836 0	3.202 3	6.038 3	0.001262	0.0461
44.0	256.0	1 115.4	1 682.9	2 798.3	2.848 7	3.179 9	6.028 6	0.001266	0.0451
45.0	257.4	1 122.1	1 675.6	2 797.7	2.861 2	3.157 9	6.019 1	0.001269	0.0440
46.0	258.7	1 128.8	1 668.3	2 797.0	2.873 5	3.136 2	6.009 7	0.001272	0.0430
47.0	260.1	1 135.3	1 661.1	2 796.4	2.885 5	3.114 9	6.000 4	0.001276	0.0421
48.0	261.4	1 141.8	1 653.9	2 795.7	2.897 4	3.093 9	5.991 3	0.001279	0.0412
49.0	262.6	1 148.2	1 646.8	2 794.9	2.909 1	3.073 3	5.982 3	0.001282	0.0403
50.0	263.9	1 154.5	1 639.7	2 794.2	2.920 6	3.052 9	5.973 5	0.001286	0.0394
51.0	265.1	1 160.7	1 632.7	2 793.4	2.931 9	3.032 8	5.964 8	0.001289	0.0386
52.0	266.4	1 166.8	1 625.7	2 792.6	2.943 1	3.013 0	5.956 1	0.001292	0.0378
53.0	267.6	1 172.9	1 618.8	2 791.7	2.954 1	2.993 5	5.947 6	0.001296	0.0371
54.0	268.7	1 178.9	1 611.9	2 790.8	2.965 0	2.974 2	5.939 2	0.001299	0.0363
55.0	269.9	1 184.9	1 605.0	2 789.9	2.975 7	2.955 2	5.930 9	0.001302	0.0356
56.0	271.1	1 190.8	1 598.2	2 789.0	2.986 3	2.936 4	5.922 7	0.001306	0.0349
57.0	272.2	1 196.6	1 591.4	2 788.0	2.996 7	2.917 9	5.914 6	0.001309	0.0343
58.0	273.3	1 202.3	1 584.7	2 787.0	3.007 1	2.899 5	5.906 6	0.001312	0.0336
59.0	274.4	1 208.0	1 578.0	2 786.0	3.017 2	2.881 4	5.898 6	0.001315	0.0330
60.0	275.5	1 213.7	1 571.3	2 785.0	3.027 3	2.863 5	5.890 8	0.001318	0.0324
61.0	276.6	1 219.3	1 564.7	2 784.0	3.037 2	2.845 8	5.883 0	0.001322	0.0319
62.0	277.7	1 224.8	1 558.0	2 782.9	3.047 1	2.828 3	5.875 3	0.001325	0.0313
63.0	278.7	1 230.3	1 551.5	2 781.8	3.056 8	2.810 9	5.867 7	0.001328	0.0308
64.0	279.8	1 235.7	1 544.9	2 780.6	3.066 4	2.793 8	5.860 1	0.001332	0.0302

Absolute pressure (bar)	Temp. (°C)	Specific enthalpy (kJ/kg)			Specific entropy (kJ/kg K)			Specific volume (m^3/kg)	
p	t_s	h_f	h_{fg}	h_g	s_f	s_{fg}	s_g	v_f	v_g
65.0	280.8	1 241.1	1 538.4	2 779.5	3.075 9	2.776 8	5.852 7	0.001335	0.0297
66.0	281.8	1 246.5	1 531.9	2 778.3	3.085 3	2.760 0	5.845 2	0.001338	0.0292
67.0	282.8	1 251.8	1 525.4	2 777.1	3.094 6	2.743 3	5.837 9	0.001341	0.0287
68.0	283.8	1 257.0	1 518.9	2 775.9	3.103 8	2.726 8	5.830 6	0.001345	0.0283
69.0	284.8	1 262.2	1 512.5	2 774.7	3.112 9	2.710 5	5.823 3	0.001348	0.0278
70.0	285.8	1 267.4	1 506.0	2 773.5	3.121 9	2.694 3	5.816 2	0.001351	0.0274
71.0	286.7	1 272.5	1 499.6	2 772.2	3.130 8	2.678 2	5.809 0	0.001355	0.0269
72.0	287.7	1 277.6	1 493.3	2 770.9	3.139 7	2.662 3	5.802 0	0.001358	0.0265
73.0	288.6	1 282.7	1 486.9	2 769.6	3.148 4	2.646 5	5.794 9	0.001361	0.0261
74.0	289.6	1 287.7	1 480.5	2 768.3	3.157 1	2.630 9	5.788 0	0.001364	0.0257
75.0	290.5	1 292.7	1 474.2	2 766.9	3.165 7	2.615 3	5.781 0	0.001368	0.0253
76.0	291.4	1 297.6	1 467.9	2 765.5	3.174 2	2.599 9	5.774 2	0.001371	0.0249
77.0	292.3	1 302.5	1 461.6	2 764.2	3.182 7	2.584 6	5.767 3	0.001374	0.0246
78.0	293.2	1 307.4	1 455.3	2 762.8	3.191 1	2.569 5	5.760 5	0.001378	0.0242
79.0	294.1	1 312.3	1 449.1	2 761.3	3.199 4	2.554 4	5.753 8	0.001381	0.0239
80.0	294.9	1 317.1	1 442.8	2 759.9	3.207 6	2.539 5	5.747 1	0.001384	0.0235
81.0	295.8	1 321.9	1 436.6	2 758.4	3.215 8	2.524 6	5.740 4	0.001387	0.0232
82.0	296.7	1 326.6	1 430.3	2 757.0	3.223 9	2.509 9	5.733 8	0.001391	0.0229
83.0	297.5	1 331.4	1 424.1	2 755.5	3.232 0	2.495 2	5.727 2	0.001394	0.0225
84.0	298.4	1 336.1	1 417.9	2 754.0	3.239 9	2.480 7	5.720 6	0.001397	0.0222
85.0	299.2	1 340.7	1 411.7	2 752.5	3.247 9	2.466 3	5.714 1	0.001401	0.0219
86.0	300.1	1 345.4	1 405.5	2 750.9	3.255 7	2.451 9	5.707 6	0.001404	0.0216
87.0	300.9	1 350.0	1 399.3	2 749.4	3.263 6	2.437 6	5.701 2	0.001408	0.0213
88.0	301.7	1 354.6	1 393.2	2 747.8	3.271 3	2.423 5	5.694 8	0.001411	0.0211
89.0	302.5	1 359.2	1 387.0	2 746.2	3.279 0	2.409 4	5.688 4	0.001414	0.0208
90.0	303.3	1 363.7	1 380.9	2 744.6	3.286 7	2.395 3	5.682 0	0.001418	0.0205
91.0	304.1	1 368.3	1 374.7	2 743.0	3.294 3	2.381 4	5.675 7	0.001421	0.0202
92.0	304.9	1 372.8	1 368.6	2 741.4	3.301 8	2.367 6	5.669 4	0.001425	0.0199
93.0	305.7	1 377.2	1 362.5	2 739.7	3.309 3	2.353 8	5.663 1	0.001428	0.0197
94.0	306.4	1 381.7	1 356.3	2 738.0	3.316 8	2.340 1	5.656 8	0.001432	0.0194
95.0	307.2	1 386.1	1 350.2	2 736.4	3.324 2	2.326 4	5.650 6	0.001435	0.0192
96.0	308.0	1 390.6	1 344.1	2 734.7	3.331 5	2.312 9	5.644 4	0.001438	0.0189
97.0	308.7	1 395.0	1 338.0	2 733.0	3.338 8	2.299 4	5.638 2	0.001442	0.0187
98.0	309.4	1 399.3	1 331.9	2 731.2	3.346 1	2.285 9	5.632 1	0.001445	0.0185
99.0	310.2	1 403.7	1 325.8	2 729.5	3.353 4	2.272 6	5.625 9	0.001449	0.0183

Absolute pressure (bar)	Temp. (°C)	Specific enthalpy (kJ/kg)			Specific entropy (kJ/kg K)			Specific volume (m^3/kg)	
p	t_s	h_f	h_{fg}	h_g	s_f	s_{fg}	s_g	v_f	v_g
100.0	311.1	1 408.0	1 319.7	2 727.7	3.360 5	2.259 3	5.619 8	0.001452	0.0181
102.0	312.4	1 416.7	1 307.5	2 724.2	3.374 8	2.232 8	5.607 6	0.001459	0.0176
104.0	313.8	1 425.2	1 295.3	2 720.5	3.388 9	2.206 6	5.595 5	0.001467	0.0172
106.0	315.3	1 433.7	1 283.1	2 716.8	3.402 9	2.180 6	5.583 5	0.001474	0.0168
108.0	316.6	1 442.2	1 270.9	2 713.1	3.416 7	2.154 8	5.571 5	0.001481	0.0164
110.0	318.0	1 450.6	1 258.7	2 709.3	3.430 4	2.129 1	5.559 5	0.001488	0.0160
112.0	319.4	1 458.9	1 246.5	2 705.4	3.444 0	2.103 6	5.547 6	0.001496	0.0157
114.0	320.7	1 467.2	1 234.3	2 701.5	3.457 4	2.078 3	5.535 7	0.001504	0.0153
116.0	322.1	1 475.4	1 222.0	2 697.4	3.470 8	2.053 1	5.523 9	0.001511	0.0149
118.0	323.4	1 483.6	1 209.7	2 693.3	3.484 0	2.028 0	5.512 1	0.001519	0.0146
120.0	324.6	1 491.8	1 197.4	2 689.2	3.497 2	2.003 0	5.500 2	0.001527	0.0143
122.0	325.9	1 499.9	1 185.0	2 684.9	3.510 2	1.978 2	5.488 4	0.001535	0.0139
124.0	327.1	1 508.0	1 172.6	2 680.6	3.523 2	1.953 3	5.476 5	0.001543	0.0137
126.0	328.4	1 516.0	1 160.1	2 676.1	3.536 0	1.928 6	5.464 6	0.001551	0.0134
128.0	329.6	1 524.0	1 147.6	2 671.6	3.548 8	1.903 9	5.452 7	0.001559	0.0131
130.0	330.8	1 532.0	1 135.0	2 667.0	3.561 6	1.879 2	5.440 8	0.001567	0.0128
132.0	332.0	1 540.0	1 122.3	2 662.3	3.574 2	1.854 6	5.428 8	0.001576	0.0125
134.0	333.2	1 547.9	1 109.5	2 657.4	3.586 8	1.830 0	5.416 8	0.001584	0.0123
136.0	334.3	1 555.8	1 096.7	2 652.5	3.599 3	1.805 3	5.404 7	0.001593	0.0120
138.0	335.5	1 563.7	1 083.8	2 647.5	3.611 8	1.780 7	5.392 5	0.001602	0.0117
140.0	336.6	1 571.6	1 070.7	2 642.4	3.624 2	1.756 0	5.380 3	0.001611	0.0115
142.0	337.7	1 579.5	1 057.6	2 637.1	3.636 6	1.731 3	5.367 9	0.001619	0.0112
144.0	338.8	1 587.4	1 044.4	2 631.8	3.649 0	1.706 6	5.355 5	0.001629	0.0110
146.0	339.9	1 595.3	1 031.0	2 626.3	3.661 3	1.681 8	5.343 1	0.001638	0.0108
148.0	341.1	1 603.1	1 017.6	2 620.7	3.673 6	1.656 9	5.330 5	0.001648	0.0106
150.0	342.1	1 611.0	1 004.0	2 615.0	3.685 9	1.632 0	5.317 9	0.001658	0.0103
152.0	343.2	1 618.9	990.3	2 609.2	3.698 1	1.607 0	5.305 1	0.001668	0.0101
154.0	344.2	1 626.8	976.5	2 603.3	3.710 3	1.581 9	5.292 2	0.001678	0.00991
156.0	345.3	1 634.7	962.6	2 597.3	3.722 6	1.556 7	5.279 3	0.001689	0.00971
158.0	346.3	1 642.6	948.5	2 591.1	3.734 8	1.531 4	5.266 3	0.001699	0.00951

Absolute pressure (bar)	Temp. (°C)	Specific enthalpy (kJ/kg)			Specific entropy (kJ/kg K)			Specific volume (m^3/kg)	
p	t_s	h_f	h_{fg}	h_g	s_f	s_{fg}	s_g	v_f	v_g
160.0	347.3	1 650.5	934.3	2 584.9	3.747 1	1.506 0	5.253 1	0.001710	0.00931
162.0	348.3	1 658.5	920.0	2 578.5	3.759 4	1.480 6	5.239 9	0.001721	0.00911
164.0	349.3	1 666.5	905.6	2 572.1	3.771 7	1.455 0	5.226 7	0.001733	0.00893
166.0	350.3	1 674.5	891.0	2 565.5	3.784 2	1.429 0	5.213 2	0.001745	0.00874
168.0	351.3	1 683.0	875.6	2 558.6	3.797 4	1.402 1	5.199 4	0.001757	0.00855
170.0	352.3	1 691.7	859.9	2 551.6	3.810 7	1.374 8	5.185 5	0.001769	0.00837
172.0	353.2	1 700.4	844.0	2 544.4	3.824 0	1.347 3	5.171 3	0.001783	0.00819
174.0	354.2	1 709.0	828.1	2 537.1	3.837 2	1.319 8	5.157 0	0.001796	0.00801
176.0	355.1	1 717.6	811.9	2 529.5	3.850 4	1.292 2	5.142 5	0.001810	0.00784
178.0	356.0	1 726.2	795.6	2 521.8	3.863 5	1.264 3	5.127 8	0.001825	0.00767
180.0	356.9	1 734.8	779.1	2 513.9	3.876 5	1.236 2	5.112 8	0.001840	0.00750
182.0	357.8	1 743.4	762.3	2 505.8	3.889 6	1.207 9	5.097 5	0.001856	0.00733
184.0	358.7	1 752.1	745.3	2 497.4	3.902 8	1.179 2	5.082 0	0.001872	0.00717
186.0	359.6	1 760.9	727.9	2 488.8	3.916 0	1.150 1	5.066 1	0.001889	0.00701
188.0	360.5	1 769.7	710.1	2 479.8	3.929 4	1.120 5	5.049 8	0.001907	0.00684
190.0	361.4	1 778.7	692.0	2 470.6	3.942 9	1.090 3	5.033 2	0.001926	0.00668
192.0	362.3	1 787.8	673.3	2 461.1	3.956 6	1.059 4	5.016 0	0.001946	0.00652
194.0	363.2	1 797.0	654.1	2 451.1	3.970 6	1.027 8	4.998 3	0.001967	0.00636
196.0	364.0	1 806.6	634.2	2 440.7	3.984 9	0.995 1	4.980 0	0.001989	0.00620
198.0	364.8	1 816.3	613.5	2 429.8	3.999 6	0.961 4	4.961 1	0.002012	0.00604
200.0	365.7	1 826.5	591.9	2 418.4	4.014 9	0.926 3	4.941 2	0.002037	0.00588
202.0	366.5	1 837.0	569.2	2 406.2	4.030 8	0.889 7	4.920 4	0.002064	0.00571
204.0	367.3	1 848.1	545.1	2 393.3	4.047 4	0.851 0	4.898 4	0.002093	0.00555
206.0	368.2	1 859.9	519.5	2 379.4	4.065 1	0.809 9	4.875 0	0.002125	0.00538
208.0	368.9	1 872.5	491.7	2 364.2	4.084 1	0.765 7	4.849 8	0.002161	0.00521
210.0	369.8	1 886.3	461.3	2 347.6	4.104 8	0.717 5	4.822 3	0.002201	0.00502
212.0	370.6	1 901.5	427.4	2 328.9	4.127 9	0.663 9	4.791 7	0.002249	0.00483
214.0	371.3	1 919.0	388.4	2 307.4	4.154 3	0.602 6	4.756 9	0.002306	0.00462
216.0	372.1	1 939.9	341.6	2 281.6	4.186 1	0.529 3	4.715 4	0.002379	0.00439
218.0	372.9	1 967.2	280.8	2 248.0	4.227 6	0.434 6	4.662 2	0.002483	0.00412
220.0	373.7	2 011.1	184.5	2 195.6	4.294 7	0.285 2	4.579 9	0.002671	0.00373
221.2	374.1	2 107.4	0.0	2 107.4	4.442 9	0.0	4.442 9	0.003170	0.00317

TABLE III
Superheated Steam at Various Pressures and Temperatures

↓p (bar) (t_s)	t (°C) →	50	100	150	200	250	300	400	500
0.01	v	149.1	172.2	195.3	218.4	241.5	264.5	310.7	356.8
(7.0)	u	2445.4	2516.4	2588.4	2661.6	2736.9	2812.2	2969.0	3132.4
	h	2594.5	2688.6	2783.6	2880.0	2978.4	3076.8	3279.7	3489.2
	s	9.242	9.513	9.752	9.967	10.163	10.344	10.671	10.960
0.05	v	29.78	34.42	39.04	48.66	48.28	52.9	62.13	71.36
(32.9)	u	2444.8	2516.2	2588.4	2661.9	2736.6	2812.6	2969.6	3133.0
	h	2593.7	2688.1	2783.4	2879.9	2977.6	3076.7	3279.7	3489.2
	s	8.498	8.770	9.009	9.225	9.421	9.602	9.928	10.218
0.1	v	14.57	17.19	19.51	21.82	24.14	26.44	31.06	35.68
(45.8)	u	2443.9	2515.5	2587.9	2661.3	2736.0	2812.1	2968.9	3132.3
	h	2592.6	2687.5	2783.0	2879.5	2977.3	3076.5	3279.6	3489.1
	s	8.175	8.448	8.688	8.904	9.100	9.281	9.608	9.898
0.5	v		34.18	3.889	43.56	4.821	5.284	6.209	7.134
(81.3)	u		2511.6	2585.6	2659.9	2735.0	2811.3	2968.5	3132.0
	h		2682.5	2780.1	2877.7	2976.0	3075.5	3278.9	3488.7
	s		7.695	7.940	8.158	8.356	8.537	8.864	9.155
0.75	v		2.27	2.587	2.900	3.211	3.520	4.138	4.755
(92.0)	u		2509.2	2584.2	2659.0	2734.4	2810.9	2968.2	3131.8
	h		2679.4	2778.2	2876.5	2975.2	3074.9	3278.5	3488.4
	s		7.501	7.749	7.969	8.167	8.349	8.677	8.967
1.0	v		1.696	1.936	2.172	2.406	2.639	3.103	3.565
(99.6)	u		2506.2	2582.8	2658.1	2733.7	2810.4	2967.9	3131.6
	h		2676.2	2776.4	2875.3	2974.3	3074.3	3278.2	3488.1
	s		7.361	7.613	7.834	8.033	8.216	8.544	8.834
1.01325	v			1.912	2.146	2.375	2.603	3.062	3.519
(100)	u			2582.6	2658.0	2733.6	2810.3	2967.8	3131.5
	h			2776.3	2875.2	2974.2	3074.2	3278.1	3488.0
	s			7.828	7.827	8.027	8.209	8.538	8.828
1.5	v			1.285	1.143	1.601	1.757	2.067	2.376
(111.4)	u			2579.8	2656.2	2732.5	2809.5	2967.3	3131.2
	h			2772.6	2872.9	2972.7	3073.1	3277.4	3487.6
	s			7.419	7.643	7.844	8.027	8.356	8.647

↓p (bar) (t_s)	t (°C) →	50	100	150	200	250	300	400	500
	v			0.960	1.080	1.199	1.316	1.549	1.781
2.0	u			2576.9	2654.4	2731.2	2808.6	2966.7	3130.8
(120.2)	h			2768.8	2870.5	2971.0	3071.8	3276.6	3487.1
	s			7.279	7.507	7.709	7.893	8.222	8.513
	v			0.764	0.862	0.957	1.052	1.238	1.424
2.5	u			2574.7	2655.7	2734.9	2813.8	2973.9	3139.6
(127.4)	h			2764.5	2868.0	2969.6	3070.9	3275.9	3486.5
	s			7.169	7.401	7.604	7.789	8.119	8.410
	v			0.634	0.716	0.796	0.875	1.031	1.187
3.0	u			2570.8	2650.7	2728.7	2806.7	2965.6	3130.0
(133.5)	h			2761.0	2865.6	2967.6	3069.3	3275.0	3486.1
	s			7.078	7.311	7.517	7.702	8.033	8.325
	v			0.471	0.534	0.595	0.655	0.773	0.889
4.0	u			2564.5	2646.8	2726.1	2804.8	2964.4	3129.2
(143.6)	h			2752.8	2860.5	2964.2	3066.8	3273.4	3484.9
	s			6.930	7.171	7.379	7.566	7.899	8.191

↓p (bar) (t_s)	t (°C) →	200	250	300	350	400	450	500	600
	v	0.425	0.474	0.523	0.570	0.617	0.664	0.711	0.804
5.0	u	2642.9	2723.5	2802.9	2882.6	2963.2	3045.3	3128.4	3299.6
(151.8)	h	2855.4	2960.7	3064.2	3167.7	3271.9	3377.2	3483.9	3701.7
	s	7.059	7.271	7.460	7.633	7.794	7.945	8.087	8.353
	v	0.352	0.394	0.434	0.474	0.514	0.553	0.592	0.670
6.0	u	2638.9	2720.9	2801.0	2881.2	2962.1	3044.2	3127.6	3299.1
(158.8)	h	2850.1	2957.2	3061.6	3165.7	3270.3	3376.0	3482.8	3700.9
	s	6.967	7.182	7.372	7.546	7.708	7.859	8.002	8.267
	v	0.300	0.336	0.371	0.406	0.440	0.473	0.507	0.574
7.0	u	2634.8	2718.2	2799.1	2879.7	2960.9	3043.2	3126.8	3298.5
(165.0)	h	2844.8	2953.6	3059.1	3163.7	3268.7	3374.7	3481.7	3700.2
	s	6.886	7.105	7.298	7.473	7.635	7.787	7.930	8.196

↓p (bar) (t_s)	t (°C) →	200	250	300	350	400	450	500	600
8.0 (170.4)	v	0.261	0.293	0.324	0.354	0.384	0.414	0.443	0.502
	u	2630.6	2715.5	2797.2	2878.2	2959.7	3042.3	3126.0	3297.8
	h	2839.3	2950.1	3056.5	3161.7	3267.1	3373.4	3480.6	3699.4
	s	6.816	7.038	7.233	7.409	7.572	7.724	7.867	8.133
9.0 (175.4)	v	0.230	0.260	0.287	0.314	0.341	0.367	0.394	0.446
	u	2626.3	2712.7	2795.2	2876.7	2958.5	3041.3	3125.2	3297.3
	h	2833.6	2946.3	3053.8	3159.7	3265.5	3372.1	3479.6	3698.6
	s	6.752	6.979	7.175	7.352	7.516	7.668	7.812	8.078
10.0 (179.9)	v	0.206	0.233	0.258	0.282	0.307	0.330	0.354	0.401
	u	2621.9	2709.9	2793.2	2875.2	2957.3	3040.3	3124.4	3296.8
	h	2827.9	2942.6	3051.2	3157.8	3263.9	3370.7	3478.5	3697.9
	s	6.694	6.925	7.123	7.301	7.465	7.618	7.762	8.029
15.0 (198.3)	v	0.132	0.152	0.169	0.187	0.203	0.219	0.235	0.267
	u	2598.8	2695.3	2783.1	2867.6	2951.3	3035.3	3120.3	3293.9
	h	2796.8	2923.3	3037.6	3147.5	3255.8	3364.2	3473.1	3694.0
	s	6.455	6.709	6.918	7.102	7.269	7.424	7.570	7.839
20.0 (212.4)	v		0.111	0.125	0.139	0.151	0.163	0.176	0.200
	u		2679.6	2772.6	2859.8	2945.2	3030.5	3116.2	3290.9
	h		2902.5	3023.5	3137.0	3247.6	3357.5	3467.6	3690.1
	s		6.545	6.766	6.956	7.127	7.285	7.432	7.702
25 (223.9)	v		0.0870	0.0989	0.109	0.120	0.130	0.140	0.159
	u		2662.6	2761.6	2851.9	2939.1	3025.5	3112.1	3288.0
	h		2880.1	3008.8	3126.3	3239.3	3350.8	3462.1	3686.3
	s		6.408	6.644	6.840	7.015	7.175	7.323	7.596
30 (233.8)	v		0.0706	0.0811	0.0905	0.0994	0.108	0.116	0.132
	u		2644.0	2750.1	2843.7	2932.8	3020.4	3108.0	3285.0
	h		2855.8	2993.5	3115.3	3230.9	3344.0	3456.5	3682.3
	s		6.287	6.539	6.743	6.921	7.083	7.234	7.509
40 (250.4)	v			0.0588	0.0664	0.0734	0.080	0.0864	0.0989
	u			2725.3	2826.7	2919.9	3010.2	3099.5	3279.1
	h			2960.7	3092.5	3213.6	3330.3	3445.3	3674.4
	s			6.362	6.582	6.769	6.936	7.090	7.369

↓p (bar) (t_s)	t (°C) →	200	250	300	350	400	450	500	600
50 (263.9)	v			0.0453	0.0519	0.0578	0.0633	0.0686	0.0787
	u			2698.0	2808.7	2906.6	2999.7	3091.0	3273.0
	h			2924.5	3068.4	3195.7	3316.2	3433.8	3666.5
	s			6.208	6.449	6.646	6.819	6.976	7.259
60 (275.5)	v			0.0362	0.0422	0.0474	0.0521	0.0567	0.0653
	u			2667.2	2789.6	2892.9	2988.9	3082.2	3266.9
	h			2884.2	3043.0	3177.2	3301.8	3422.2	3658.4
	s			6.067	6.333	6.541	6.719	6.880	7.168
70 (285.8)	v			0.0295	0.0352	0.0399	0.0442	0.0481	0.0557
	u			2632.2	2769.4	2878.6	2978.0	3073.4	3260.7
	h			2838.4	3016.0	3158.1	3287.1	3410.3	3650.3
	s			5.931	6.228	6.448	6.633	6.798	7.089

↓p (bar) (t_s)	t (°C) →	350	375	400	450	500	550	600	700
80 (294.9)	v	0.02995	0.03222	0.03432	0.03817	0.04175	0.04516	0.04845	0.05481
	h	2987.3	3066.1	3138.3	3272.0	3398.3	3521.0	3642.0	3882.4
	s	6.130	6.254	6.363	6.555	6.724	6.878	7.021	7.281
90 (303.3)	v	0.0258	0.02796	0.02993	0.03350	0.03677	0.03987	0.04285	0.04857
	h	2956.6	3041.3	3117.8	3256.6	3386.1	3511.0	3633.7	3876.5
	s	6.036	6.169	6.285	6.484	6.658	6.814	6.959	7.222
100 (311.0)	v	0.02242	0.02453	0.02641	0.02975	0.03279	0.03564	0.03837	0.04358
	h	2923.4	3015.4	3096.5	3240.9	3373.7	3500.9	3625.3	3870.5
	s	5.944	6.089	6.212	6.419	6.597	6.756	6.903	7.169
110 (318.0)	v	0.01961	0.02169	0.02351	0.02668	0.02952	0.03217	0.03470	0.03950
	h	2887.3	2988.2	3074.3	3224.7	3361.0	3490.7	3616.9	3864.5
	s	5.853	6.011	6.142	6.358	6.540	6.703	6.851	7.120
120 (324.6)	v	0.01721	0.01931	0.02108	0.02412	0.02680	0.02929	0.03164	0.03610
	h	2847.7	2958.9	3051.3	3208.2	3348.2	3480.4	3608.3	3858.4
	s	5.760	5.935	6.075	6.300	6.487	6.653	6.804	7.075
130 (330.8)	v	0.01511	0.01725	0.01900	0.02194	0.0245	0.02684	0.02905	0.03322
	h	2803.3	2927.9	3027.2	3191.3	3335.2	3469.9	3599.7	3852.3
	s	5.663	5.859	6.009	6.245	6.437	6.606	6.759	7.033

↓p (bar) (t_s)	t (°C) →	350	375	400	450	500	550	600	700
140	v	0.01322	0.01546	0.01722	0.02007	0.02252	0.02474	0.02683	0.03075
(336.6)	h	2752.6	2894.5	3001.9	3174.0	3322.0	3459.3	3591.1	3846.2
	s	5.559	5.782	5.945	6.192	6.390	6.562	6.712	6.994
150	v	0.01145	0.01388	0.01565	0.01845	0.02080	0.02293	0.02491	0.02861
(342.1)	h	2692.4	2858.4	2975.5	3156.2	3308.6	3448.6	3582.3	3840.1
	s	5.442	5.703	5.881	6.140	6.344	6.520	6.679	6.957
160	v	0.00975	0.01245	0.01426	0.01701	0.01930	0.02134	0.02323	0.02674
(347.3)	h	2615.7	2818.9	2947.6	3138.0	3294.9	3437.8	3573.5	3833.9
	s	5.302	5.622	5.188	6.091	6.301	6.480	6.640	6.922
170	v		0.01117	0.01302	0.01575	0.01797	0.01993	0.02174	0.02509
(352.3)	h		2776.8	2918.2	3119.3	3281.1	3426.9	3564.6	3827.7
	s		5.539	5.754	6.042	6.259	6.442	6.604	6.889
180	v		0.00996	0.01190	0.01462	0.01678	0.01868	0.02042	0.02362
(356.9)	h		2727.9	2887.0	3100.1	3267.0	3415.9	3555.6	3821.5
	s		5.448	5.689	5.995	6.218	6.405	6.570	6.858
190	v		0.00881	0.01088	0.01361	0.01572	0.01756	0.01924	0.02231
(361.4)	h		2671.3	2853.8	3080.4	3252.7	3404.7	3546.6	3815.3
	s		5.346	5.622	5.948	6.179	6.369	6.537	6.828
200	v		0.00767	0.00994	0.01269	0.9477	0.01655	0.01818	0.02113
(365.7)	h		2602.5	2818.1	3060.1	3238.2	3393.5	3537.6	3809.0
	s		5.227	5.554	5.902	6.140	6.335	6.505	6.799
210	v		0.00645	0.00907	0.01186	0.01390	0.01564	0.01722	0.02006
(369.8)	h		2511.0	2779.6	3039.3	3223.5	3382.1	3528.4	3802.8
	s		5.075	5.483	5.856	6.103	6.301	6.474	6.772
220	v		0.00482	0.00825	0.01110	0.01312	0.01481	0.01634	0.01909
(373.7)	h		2345.1	2737.6	3017.9	3208.6	3370.6	3519.2	3796.5
	s		4.810	5.407	5.811	6.066	6.269	6.444	6.745

TABLE IV
Supercritical Steam

p (bar)	**t (°C) →**	**350**	**375**	**400**	**425**	**450**	**500**	**600**	**700**	**800**
230	*v*	0.00162	0.00221	0.00748	0.00915	0.01040	0.01239	0.01554	0.01821	0.02063
	h	1632.8	1912.2	2691.2	2869.2	2995.8	3193.4	3510.0	3790.2	4056.2
	s	3.137	4.137	5.327	5.587	5.765	6.030	6.415	6.719	6.980
250	*v*	0.00160	0.00197	0.00600	0.00788	0.00916	0.01112	0.01414	0.01665	0.01891
	h	1623.5	1848.0	2580.2	2806.3	2949.7	3162.4	3491.4	3775.5	4047.1
	s	3.680	4.032	5.142	5.472	5.674	5.959	6.360	6.671	6.934
300	*v*	0.00155	0.00179	0.00279	0.00530	0.00673	0.00868	0.01145	0.01366	0.01562
	h	1608.5	1791.5	2151.1	2614.2	2821.4	3081.1	3443.9	3745.6	4024.2
	s	3.643	3.930	4.473	5.150	5.442	5.790	6.233	6.561	6.833
350	*v*	0.00152	0.00110	0.00210	0.00343	0.00496	0.00693	0.00953	0.01153	0.01328
	h	1597.1	1762.4	1987.6	2373.4	2672.4	2994.4	3395.5	3713.5	4001.5
	s	3.612	3.872	4.213	4.775	5.196	5.628	6.118	6.463	6.745
400	*v*	0.00149	0.00164	0.00191	0.00253	0.00369	0.00562	0.00809	0.00994	0.01152
	h	1588.3	1742.8	1930.9	2198.1	2512.8	2903.3	3346.4	3681.2	3978.7
	s	3.586	3.829	4.113	4.503	4.946	5.470	6.011	6.375	6.666
500	*v*	0.00144	0.00156	0.00173	0.00201	0.00249	0.00389	0.00611	0.00773	0.00908
	h	1575.3	1716.6	1874.6	2060.0	2284.0	2720.1	3247.6	3616.8	3933.6
	s	3.542	3.764	4.003	4.273	4.588	5.173	5.818	6.219	6.529
600	*v*	0.00140	0.00150	0.00163	0.00182	0.00209	0.00296	0.00483	0.00627	0.00746
	h	1566.4	1699.5	1843.4	2001.7	2179.0	2567.9	3151.2	3553.5	3889.1
	s	3.505	3.764	3.932	4.163	4.412	4.932	5.645	6.082	6.411
700	*v*	0.00137	0.00146	0.00157	0.00171	0.00189	0.00247	0.00398	0.00526	0.00632
	h	1560.4	1687.7	1822.8	1967.2	2122.7	2463.2	3061.7	3492.4	3845.7
	s	3.473	3.673	3.877	4.088	4.307	4.762	5.492	5.961	6.307
800	*v*	0.00135	0.00142	0.00152	0.00163	0.00177	0.00219	0.00339	0.00452	0.00548
	h	1556.4	1679.4	1808.3	1943.9	2086.9	2394.0	2982.7	3434.6	3803.8
	s	3.444	3.638	3.833	4.031	4.232	4.642	5.360	5.851	6.213
900	*v*	0.00133	0.00139	0.00147	0.00157	0.00169	0.00201	0.00297	0.00397	0.00484
	h	1553.9	1673.4	1797.7	1927.2	2062.0	2346.7	2915.6	3381.1	3763.8
	s	3.419	3.607	3.795	3.984	4.174	4.554	5.247	5.753	6.128
1000	*v*	0.01308	0.00137	0.00144	0.00152	0.00163	0.00189	0.00267	0.00355	0.00434
	h	1552.7	1669.4	1790.0	1914.8	2043.8	2312.8	2859.8	3332.3	3726.1
	s	3.396	3.579	3.762	3.944	4.126	4.485	5.151	5.664	6.050

TABLE V
Conversion Factors

Force

1 newton	= 1 kg-m/sec^2
	= 0.012 kgf
1 kgf	= 9.81 N

Pressure

1 bar	= 750.06 mm Hg
	= 0.9869 atm
	= 10^5 N/m^2
	= 10^3 kg/m-sec^2
1 N/m^2	= 1 pascal
	= 10^{-5} bar
	= 10^{-2} kg/m-sec^2
1 atm	= 760 mm Hg
	= 1.03 kgf/cm^2 = 1.01325 bar
	= 1.01325×10^5 N/m^2

Work, Energy or Heat

1 joule	= 1 newton metre
	= 1 watt-sec
	= 2.7778×10^{-7} kWh
	= 0.239 cal
	= 0.239×10^{-3} kcal
1 cal	= 4.184 joule
	= 1.1622×10^{-6} kWh
1 kcal	= 4.184×10^3 joule
	= 427 kgfm
	= 1.1622×10^{-3} kWh
1 kWh	= 8.6×10^5 cal
	= 860 kcal
	= 3.6×10^6 joule
1 kgfm	= $\left(\frac{1}{427}\right)$ kcal = 9.81 joules

Power

1 watt	= 1 joule/sec = 0.86 kcal/h
1 h.p.	= 75 mkgf/sec = 0.1757 kcal/sec
	= 735.3 watt
1 kW	= 1000 watts
	= 860 kcal/h

Specific heat

1 kcal/kg °K	= 4.18 kJ/kg K

Thermal conductivity

1 watt/m-K = 0.8598 kcal/h-m °C

1 kcal/h-m °C = 1.16123 watt/m-K

= 1.16123 joules/s-m-K

Heat transfer co-efficient

1 watt/m²-K = 0.86 kcal/m²-h °C

1 kcal/m²-h °C = 1.163 watt/m²-K

IMPORTANT ENGINEERING CONSTANTS AND EXPRESSIONS IN S.I. UNITS

S.No.	Engineering constants and expressions	M.K.S. system	S.I. units
1.	Value of g_0	9.81 kg-m/kgf-sec²	1 kg-m/N-sec²
2.	Universal gas constant	848 kgf-m/kg mole °K	848 × 9.81 = 8314 J/kg-mole °K (∵ 1 kgf-m = 9.81 joules)
3.	Gas constant (R)	29.27 kgf m/kg °K for air	$\frac{8314}{29}$ = 287 joules/kg K for air
4.	Specific heats (for air)	c_v = 0.17 kcal/kg °K c_p = 0.24 kcal/kg °K	c_v = 0.17 × 4.184 = 0.71128 kJ/kg K c_p = 0.24 × 4.184 = 1 kJ/kg K
5.	Flow through nozzle-exit velocity (C_2)	91.5 √U where U is in kcal	44.7 √U where U is in kJ
6.	Refrigeration 1 ton	= 50 kcal/min	= 210 kJ/min
7.	**Heat transfer** The Stefan Boltzman Law is given by :	Q = σT⁴ kcal/m²-h when σ = 4.9 × 10^{-8} kcal/h-m² °K⁴	Q = σT⁴ watts/m²-h when σ = 5.67 × 10^{-8} W/m²K⁴

MOLLIER DIAGRAM
(Enthalpy vs Entropy)
in
SI UNITS

upto 1000 bar (10^5 N/m^2) and 800 °C

Entropy , s kJ/kgK →

Constant Pressure (bar)

Constant Specific volume (m^3/kg)

Constant Temperature (°C)

Saturation Line

Constant Dryness Fraction

Quality of wet Steam (x)

x = 0.95

x = 0.9

x = 0.85

x = 0.80

x = 0.75

x = 0.70

Entropy , s kJ/kgK →

STEAM TABLES